TODAY'S TECHNICIAN

Classroom Manual for
Medium/Heavy Duty Truck Diesel Engines

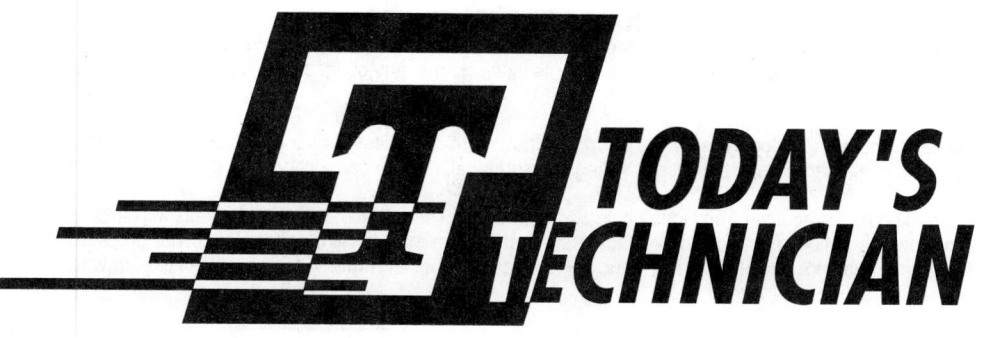

TODAY'S TECHNICIAN

Classroom Manual for
Medium/Heavy Duty Truck Diesel Engines

John F. Kershaw, Ed. D.
C-Tec, Inc.

Sean Bennett
Centennial College
Toronto, CANADA

Jack Erjavec
Series Advisor
Professor Emeritus, Columbus State Community College
Columbus, Ohio

DELMAR
THOMSON LEARNING

Australia   Canada   Mexico   Singapore   Spain   United Kingdom   United States

NOTICE TO THE READER

Delmar Staff

Director: Alar Elken
Executive Editor: Sandy Clark
Developmental Editor: Allyson Powell
Editorial Assistant: Matthew Seeley
Executive Marketing Manager: Maura Theriault
Channel Manager: Mona Caron

Executive Production Manager: Mary Ellen Black
Senior Production Coordinator: Karen Smith
Project Editor: Christopher Chien
Art/Design Coordinator: Cheri Plasse

**Library of Congress
Cataloging-in-Publication Data**

Kershaw, John F.
  Shop manual for medium/heavy duty truck diesel engines / John F. Kershaw, Sean Bennett.
    p. cm. — (Today's technician)
  Includes index.
  ISBN 0-8273-7221-3
  1. Trucks—Motors (Diesel)—Maintenance and repair. I. Title: Medium/heavy duty truck diesel engines. II. Bennett, Sean. III. Title. IV. Series.

  TL230.2 .K47 2000
  629.25′06′0288—dc21   00-060212

Asia (including India):
Thomson Learning
60 Albert Street, #15-01
Albert Complex
Singapore 189969
Tel 65 336-6411
Fax 65 336 7411

Australia/New Zealand:
Nelson
102 Dodds Street
South Melbourne, Victoria 3205
Australia
Tel 61 (0)3 9685-4111
Fax 61 (0)3 9685-4199

Latin America:
Thomson Learning
Seneca 53
Colonia Polanco
11560 Mexico D. F. Mexico
Tel (525) 281-2906
Fax (525) 281-2656

Canada:
Nelson
1120 Birchmount Road
Toronto, Ontario
Canada M1K 5G4
Tel (416) 752-9100
Fax (416) 752-8102

UK/Europe/Middle East:
Thomson Learning
Berkshire House
168-173 High Holborn
London WC1V 7AA
United Kingdom
Tel 44 (0)171 497-1422
Fax 44 (0)171 497-1426

Business Press
Berkshire House
168-173 High Holborn
London WC1V 7AA
United Kingdom
Tel 44 (0)171 497-1422
Fax 44 (0)171 497-1426

Spain:
Parainfo
Calle Magallanes 25
28015 Madrid
España
Tel 34 (0)91 446-3350
Fax 34 (0)91 445-6218

Distribution Services:
ITPS
Cheriton House
North Way
Andover,
Hampshire SP10 5BE
United Kingdom
Tel 44 (0)1264 34-2960
Fax 44 (0)1264 34-2759

International Headquarters:
Thomson Learning
International Division
290 Harbor Drive, 2nd Floor
Stamford, CT 06902-7477
USA
Tel (203) 969-8700
Fax (203) 969-8751

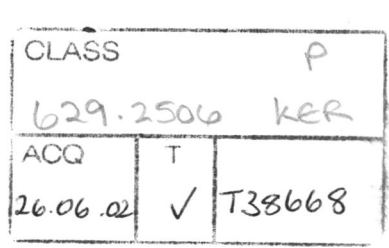

CONTENTS

PREFACE

Unlike yesterday's mechanic, the technician of today and for the future must know the underlying theory of all systems and be able to service and maintain those systems. Today's technician must also know how these individual systems interact with each other. Standards and expectations have been set for today's technician, and these must be met in order to keep the world's medium and heavy duty trucks running efficiently and safely.

The *Today's Technician* series, by Delmar Thomson Learning, features textbooks that cover all mechanical and electrical systems of medium and heavy duty trucks. Principal titles correspond with the eight major areas of ASE (National Institute for Automotive Service Excellence) certification.

Each title is divided into two manuals: a Classroom Manual and a Shop Manual. Dividing the material into two manuals provides the reader with the information needed to begin a successful career as a medium and heavy duty truck technician without interrupting the learning process by mixing cognitive and performance-based learning objectives.

Each Classroom Manual contains the principles of operation for each system and subsystem. It also discusses the design variations used by different manufacturers. The Classroom Manual is organized to build upon basic facts and theories. The primary objective of this manual is to allow the reader to gain an understanding of how each system and subsystem operates. This understanding is necessary to diagnose the complex truck systems.

The understanding acquired by using the Classroom Manual is required for competence in the skill areas covered in the Shop Manual. All of the high priority skills, as identified by ASE, are explained in the Shop Manual. The Shop Manual also includes step-by-step instructions for diagnostic and repair procedures. Photo Sequences are used to illustrate many of the common service procedures. Other common procedures are listed and are accompanied with fine-line drawings and photographs that allow the reader to visualize and conceptualize the finest details of the procedure. The Shop Manual also contains the reasons for performing the procedures, as well as when that particular service is appropriate.

The two manuals are designed to be used together and are arranged in corresponding chapters. Not only are the chapters in the manuals linked together, the contents of the chapters are also linked. Both manuals contain clear and thoughtfully selected illustrations. Many of the illustrations are original drawings or photos prepared for inclusion in this series. This means that the art is a vital part of each manual.

The page layout is designed to include information that would otherwise break up the flow of information presented to the reader. The main body of the text includes all of the "need-to-know" information and illustrations. In the side margins are many of the special features of the series. Items such as definitions of new terms, common trade jargon, tools lists, and cross-references are placed in the margin, out of the normal flow of information so as not to interrupt the thought process of the reader.

Jack Erjavec, Series Advisor

Classroom Manual

To stress the importance of safe work habits, the Classroom Manual dedicates one full chapter to safety. Included in this chapter are common safety practices, safety equipment, and safe handling of hazardous materials and wastes. This includes information on MSDS sheets and OSHA regulations. Other features of this manual include:

Cognitive Objectives

These objectives define the contents of the chapter and define what the student should have learned upon completion of the chapter.

Each topic is divided into small units to promote easier understanding and learning.

Marginal Notes

New terms are pulled out and defined. Common trade jargon also appears in the margin and gives some of the common terms used for components. This allows the reader to speak and understand the language of the trade, especially when conversing with an experienced technician.

Cautions and Warnings

Throughout the text, cautions are given to alert the reader to potentially hazardous materials or unsafe conditions. Warnings are also given to advise the student of things that can go wrong if instructions are not followed or if a nonacceptable part or tool is used.

References to the Shop Manual

Reference to the appropriate topic in the Shop Manual is given whenever necessary. Although the chapters of the two manuals are synchronized, material covered in other chapters of the Shop Manual may be fundamental to the topic discussed in the Classroom Manual.

CHAPTER 7

Air Induction and Exhaust System and Engine Brakes

Upon completion and review of this chapter, you should be able to:

❑ Identify the intake and exhaust system components that comprise the engine breathing circuit.

❑ Describe how intake air is routed to the engine cylinders and exhaust gases are routed to the tailpipe.

❑ Outline the principle of operation of a Roots blower and its function on a two-stroke cycle engine.

❑ Identify the main subcomponents on a truck diesel engine turbocharger and the role they perform; explain the operation of the turbocharger.

❑ Define the role of a charge air cooler and the relative efficiencies of each type.

❑ Relate the parallel port and crossflow valve configurations and valve seat angle to breathing efficiency and cylinder gas dynamics.

❑ Define the term "*tuned*" in relation to intake and exhaust manifold design.

❑ Define thermocouple pyrometer and outline its use on a diesel engine.

❑ Explain the role and operating principles of the exhaust muffler (silencer), and identify the two ways mufflers reduce combustion noise in an engine.

❑ Describe the operating principles of a truck catalytic converter.

❑ Identify the different types of cold engine starting aids used in a diesel engine.

❑ Identify and differentiate between the three types of engine brakes used on a diesel engine, and explain their operating principles and the relative advantages and disadvantages of each.

Introduction

Turbochargers are used on almost all current commercial vehicle diesel engines. Information that could be handled separately under the headings "intake systems" and "exhaust systems" will be dealt with in this section on engine breathing to facilitate the study of the component common to intake and exhaust systems. The turbocharger charges the intake manifold at pressures above atmospheric. Turbocharged engines are commonly described as having manifold boost. The manifold-boosted compression ignition (CI) engine cylinder is *charged* with air rather than induced. In a naturally aspirated engine, air or an air/fuel mixture is drawn into the engine cylinder by the atmospheric pressure created in the engine cylinder on the intake stroke of the piston. This air or air/fuel mixture is therefore induced into the cylinder. The correct term to describe the components responsible for delivering breathing air in naturally aspirated engines is *induction circuit*. The preferred term for describing the breathing air delivery components in an engine with manifold boost is *air intake circuit*.

The function of the air intake system (Figure 7-1) in a diesel engine is to supply a charge of clean, cool air to the engine cylinders for combustion, cooling, and scavenging. Diesel engines are designed for lean burn operation. Essentially, this means that the air charged to the engine cylinder always substantially exceeds that required to oxidize the fuel. Expectations of engine longevity have lengthened dramatically during the past decade, and this has altered the design of air intake system components.

The function of the exhaust system is to minimize both engine noise and noxious emissions while restricting the exhaust gas flow to a minimum degree. The turbocharger is a component common to both intake and exhaust systems. Turbochargers allow the exhaust system to recapture

In a **naturally aspirated** engine, air or an air/fuel mixture is drawn into the engine cylinder by the atmospheric pressure created in the engine cylinder on the intake stroke of the piston.

The function of the exhaust system is to minimize both engine noise and noxious emissions while restricting the exhaust gas flow to a minimum degree.

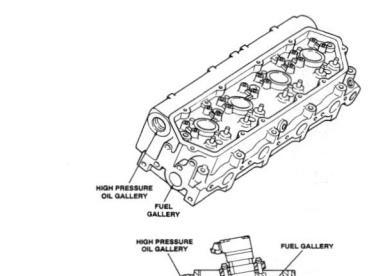

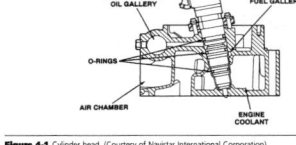

Figure 4-1 Cylinder head. (Courtesy of Navistar International Corporation)

a cylinder head means observing the original equipment manufacturer's (OEM) torque increments and sequencing.

⚠ **WARNING:** Torquing an engine cylinder head to the correct specification, but without observing the incremental steps, can damage the gaskets by deforming them.

◼ **CAUTION:** If you remove a diesel engine cylinder head, make sure that the cylinder head lifting fixtures used are secure and the correct ones. Diesel cylinder heads are very heavy; if they fall, personal injury could result.

The objective of torquing cylinder head gaskets is to ensure that the required amount of clamping force is obtained and the gasket conforms to its engineered yield shape. Most head gaskets require no applied sealant and, in fact, may fail if sealants are used. Head gaskets are designed to seal engine components in a region where both the temperatures and pressure are at their highest; they must do this and at that same time accommodate a large amount of component creep. **Creep** is the relative movement of clamped engine components due to different

Shop Manual Chapter 4, page 129.

Shop Manual Chapter 4, page 130.

99

Oil Pump

Engine oil pumps (Figure 5-5) are usually of the positive displacement type and have pumping capacities that greatly exceed the requirements of the engine. They are gear driven and usually located in the crankcase close to the oil they pump, although in some Cummins and Caterpillar applications they are external. Most oil pumps located in the crankcase are driven by a vertical shaft and pinion engaged with a drive gear on the camshaft. A pick-up is located close to, but not contacting, the base of the oil pan (see Figure 5-5). Two basic types of oil pump are used, external gear and Gerotor.

A BIT OF HISTORY

The original cylinder lubrication on Rudolf Diesel's engine occurred through grooved pieces as part of the piston skirt. These piston pieces dipped into the oil reservoir and splash lubricated the engine components including the bearings. Supercharging quickly proved this design unreliable because of the oil spray being transported with the air via the receiver into the combustion chamber. This design was replaced by a piston oil pump.

External Gear External gear pumps (Figure 5-6) consist of two meshed gears, one driving the other within a housing machined with an inlet (suction) port and outlet (charge) port. As the gears rotate, the teeth entrap inlet oil and force it outside between the gear teeth and the gear housing to the outlet port. Where the teeth mesh in the center, a seal is formed, which prevents any back flow of oil to the inlet. This is by far the most common design of oil pump on current engines. Gear-type engine oil pumps seldom malfunction; when they show evidence of wear, the underlying reason is usually contaminated engine lube.

Figure 5-5 Oil pump assembly.

124

A Bit of History

This feature gives the student a sense of the evolution of truck systems. This feature not only contains nice-to-know information, but also should spark some interest in the subject matter.

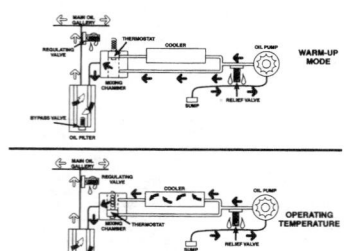

Figure 5-15 Navistar temperature control circuit. (Courtesy of Navistar International Corporation)

Summary

❑ A diesel engine lubricant performs a number of roles, including minimizing friction, supporting hydrodynamic suspension, cooling, and cleaning. Good heavy duty diesel engine oil should have good oxidation stability, a high natural viscosity index, and proven field performance.

❑ Viscosity describes a fluid's resistance to sheer. Lubricity denotes the flow characteristics of a fluid. Lubricity in engine oils is affected by temperature, with hot oils tending to flow more readily than cold oils.

❑ Most diesel engine OEMs recommend the use of multigrade oils over straight grades and approve the use of synthetic engine oils, especially for operation in conditions of severe cold.

❑ The pour point of an engine oil is the temperature at which the oil starts to gel. Oils formulated for winter use have pour point depressant additives.

❑ Sludged oil is usually a result of oil degeneration caused by prolonged low-load, cold weather operation. Ash in used engine oils is mineral residue caused by oxide and sulfate incineration. High ash levels indicate high temperature operation. High oil sump levels can cause oil aeration.

❑ When interpreting API oil classifications, the S prefix denotes oil formulated for spark-ignited engines, and the C denotes oil formulated for compression-ignition engines. Most of the field research indicates that synthetic oils substantially outperform traditional oils. However, because they cost much more than traditional oils, they will probably not be used extensively by operators until engine OEMs endorse extended oil change intervals.

❑ It is important to maintain the correct engine oil level. The technician should be aware that the consequences of an excessively high oil level could be as severe as those for low oil level.

❑ Positive displacement pumps of the external gear and Gerotor types are used on current engines. Most are of the external gear pump design. Oil pumps are designed to

Terms To Know
API
Boundary lubrication
Bundle
Bypass filter
Bypass valve
Centrifugal filter
Dry sump
Fluid friction
Full flow filter
Gerotor
Hydrodynamic suspension
Lubrication
Lubricity

Terms to Know

A list of new terms appears next to the Summary. Definitions for these terms can be found in the Glossary at the end of the manual.

Summaries

Each chapter concludes with summary statements that contain the important topics of the chapter. These are designed to help the reader review the contents.

greater volumes of oil than that required to lubricate the engine. An adjustable, pressure-regulating valve defines the peak system pressure.

❑ The filters used on a heavy duty lubrication system may be classified as full-flow or bypass, depending on how they are plumbed into the circuit. Full-flow filters are located in series between the oil pump and the lubrication circuit. **Bypass filters** are arranged in parallel and receive oil from an oil gallery and returning it directly to the oil sump. Filters are usually rated by their mechanical straining specification in microns, but they also filter by absorption and adsorption.

❑ Oil coolers are heat exchangers used on most heavy duty, highway diesel engines. The cooling medium used is engine coolant. The common type of oil cooler is the bundle, which consists of a housing within which coolant is pumped through a cylindrical "bundle" of copper tubes in one direction while engine oil is pumped around the tubes in the opposite direction.

❑ Engine oil pressure measurement in most modern engines is performed by a variable capacitance type sensor, which signals the management ECM, the dash gauge, or display. It is, therefore, an ECM output. Engine oil pressures usually have to fall to very low levels before programmed electronic failure strategies are triggered.

Review Questions

Short Answer Essays

1. Outline the function of the main components in a typical diesel engine lubrication circuit.

2. List the required properties of heavy duty engine oil.

3. Define *hydrodynamic suspension* and describe how this principle is used in a typical diesel engine.

4. List API classifications and SAE viscosity grades.

5. Describe the two types of oil pump commonly used on diesel engines, and outline the operating principles of each.

6. Describe the principle of operation of the pressure-regulating valve and its role in the lubrication circuit.

7. Define positive filtration.

8. Outline the differences between full-flow and bypass filters.

9. Describe the operating principles of the various types of heavy duty oil filters and the relative operating efficiencies of each.

10. Outline the objectives of the oil cooler in the lubrication circuit and its operating principles.

Fill-in-the-Blanks

1. Heavy duty diesel engine oil should have good _____ _____, a high natural viscosity index, and proven _____ _____.

2. Viscosity describes a fluid's resistance to _____.

3. Lubricity denotes the _____ characteristics of a fluid. Lubricity in engine oils is affected by _____.

4. The _____ _____ of engine oil is the temperature at which the oil starts to gel.

5. When interpreting API oil classifications, the __ prefix denotes oil formulated for spark-ignited engines, and the __ denotes oil formulated for compression-ignition engines.

134

Review Questions

Short answer essay, fill-in-the-blank, and multiple-choice type questions follow each chapter. These questions are designed to accurately assess the student's competence in the stated objectives at the beginning of the chapter.

Shop Manual

To stress the importance of safe work habits, the Shop Manual also dedicates one full chapter to safety. Other important features of this manual include:

Performance Objectives

These objectives define the contents of the chapter and define what the student should have learned upon completion of the chapter.

Although this textbook is not designed to simply prepare someone for the certification exams, it is organized around the ASE task list. These tasks are defined generically when the procedure is commonly followed and specifically when the procedure is unique for specific vehicle models. Imported and domestic model trucks are included in the procedures.

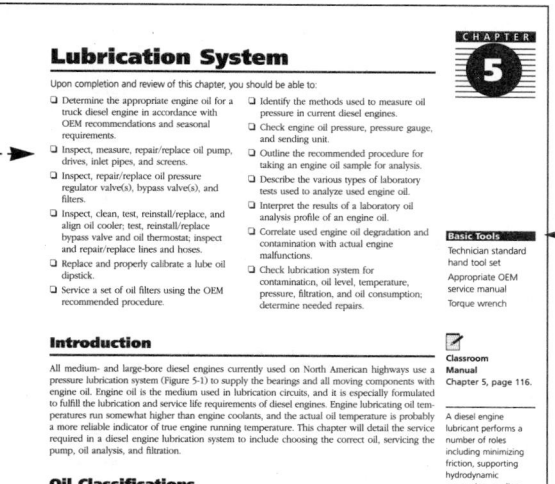

CHAPTER 5

Lubrication System

Upon completion and review of this chapter, you should be able to:

❏ Determine the appropriate engine oil for a truck diesel engine in accordance with OEM recommendations and seasonal requirements.
❏ Inspect, measure, repair/replace oil pump, drives, inlet pipes, and screens.
❏ Inspect, repair/replace oil pressure regulator valve(s), bypass valve(s), and filters.
❏ Inspect, clean, test, reinstall/replace, and align oil cooler; test, reinstall/replace bypass valve and oil thermostat; inspect and repair/replace lines and hoses.
❏ Replace and properly calibrate a lube oil dipstick.
❏ Service a set of oil filters using the OEM recommended procedure.

❏ Identify the methods used to measure oil pressure in current diesel engines.
❏ Check engine oil pressure, pressure gauge, and sending unit.
❏ Outline the recommended procedure for taking an engine oil sample for analysis.
❏ Describe the various types of laboratory tests used to analyze used engine oil.
❏ Interpret the results of a laboratory oil analysis profile of an engine oil.
❏ Correlate used engine oil degradation and contamination with actual engine malfunctions.
❏ Check lubrication system for contamination, oil level, temperature, pressure, filtration, and oil consumption; determine needed repairs.

Basic Tools
Technician standard hand tool set
Appropriate OEM service manual
Torque wrench

Classroom Manual
Chapter 5, page 116.

A diesel engine lubricant performs a number of roles including minimizing friction, supporting hydrodynamic suspension, cooling, and cleaning.

Classroom Manual
Chapter 5, page 119.

Viscosity is a fluid's resistance to sheer.

Introduction

All medium- and large-bore diesel engines currently used on North American highways use a pressure lubrication system (Figure 5-1) to supply the bearings and all moving components with engine oil. Engine oil is the medium used in lubrication circuits, and it is especially formulated to fulfill the lubrication and service life requirements of diesel engines. Engine lubricating oil temperatures run somewhat higher than engine coolants, and the actual oil temperature is probably a more reliable indicator of true engine running temperature. This chapter will detail the service required in a diesel engine lubrication system to include choosing the correct oil, servicing the pump, oil analysis, and filtration.

Oil Classifications

API Classifications

The **American Petroleum Institute (API)** classifies all engine oil sold in North America. There are two main classifications, designated by the prefix letters S and C. The S classes of engine oil designate those oils suitable for passenger cars and light trucks. The S represents spark-ignited, or SI. The C classes of engine oils designate oils suitable for heavy duty trucks, buses, and industrial and agricultural equipment. The C represents compression-ignited (CI) engines or diesel engines. Both S and C category oil classifications are listed and described here because many fleets use engine oils that claim to be suitable for both C and S classifications.

⚠️ **WARNING:** In most cases, OEMs have specific requirements for engine lubricants, which should be observed as failure to do so could result in higher HC emissions.

161

Tools Lists

Each chapter begins with a list of the Basic Tools needed to perform the tasks included in the chapter. Whenever a Special Tool is required to complete a task, it is listed in the margin next to the procedure.

Marginal Notes

Page numbers for cross-referencing appear in the margin. Some of the common terms used for components, and other bits of information, also appear in the margin. This provides an understanding of the language of the trade and helps when conversing with an experienced technician.

Photo Sequences

Many procedures are illustrated in detailed Photo Sequences. These detailed photographs show the students what to expect when they perform particular procedures. They also can provide a student a familiarity with a system or type of equipment, which the school may not have.

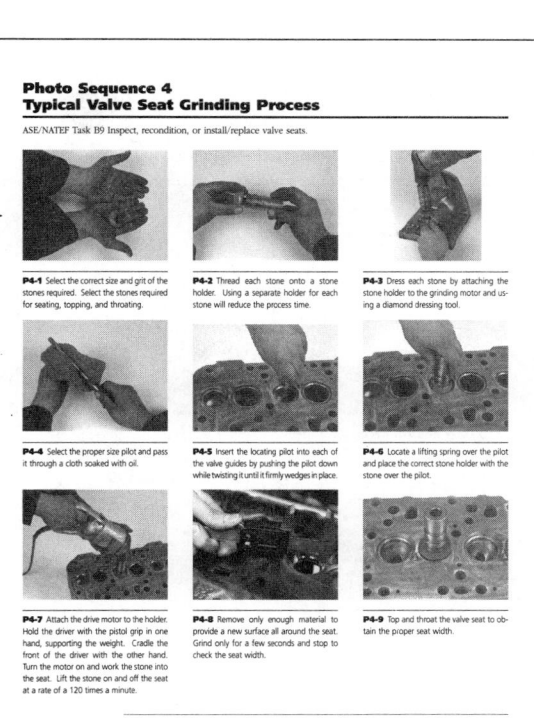

Photo Sequence 4
Typical Valve Seat Grinding Process

ASE/NATEF Task B9 Inspect, recondition, or install/replace valve seats.

P4-1 Select the correct size and grit of the stones required. Select the stones required for seating, topping, and throating.

P4-2 Thread each stone onto a stone holder. Using a separate holder for each stone will reduce the process time.

P4-3 Dress each stone by attaching the stone holder to the grinding motor and using a diamond dressing tool.

P4-4 Select the proper size pilot and pass it through a cloth soaked with oil.

P4-5 Insert the locating pilot into each of the valve guides by pushing the pilot down while twisting it until it firmly wedges in place.

P4-6 Locate a lifting spring over the pilot and place the correct stone holder with the stone over the pilot.

P4-7 Attach the drive motor to the holder. Hold the driver with the pistol grip in one hand, supporting the weight. Cradle the front of the driver with the other hand. Turn the motor on and work the stone into the seat. Lift the stone on and off the seat at a rate of a 120 times a minute.

P4-8 Remove only enough material to provide a new surface all around the seat. Grind only for a few seconds and stop to check the seat width.

P4-9 Top and throat the valve seat to obtain the proper seat width.

138

Service Tips

Whenever a short-cut or special procedure is appropriate, it is described in the text. These tips are generally those things commonly done by experienced technicians.

Customer Care

This feature highlights those little things a technician can do or say to enhance customer relations.

References to the Classroom Manual

Reference to the appropriate topic in the Classroom Manual is given whenever necessary. Although the chapters of the two manuals are synchronized, material covered in other chapters of the Classroom Manual may be fundamental to the topic discussed in the Shop Manual.

Cautions and Warnings

Throughout the text, cautions are given to alert the reader to potentially hazardous materials or unsafe conditions. Warnings are also given to advise the student of things that can go wrong if instructions are not followed or if a nonacceptable part or tool is used.

Job Sheets

Located at the end of each chapter, the Job Sheets provide a format for students to perform procedures covered in the chapter. A reference to the ASE and/or NATEF tasks addressed by the procedure is referenced on the Job Sheet.

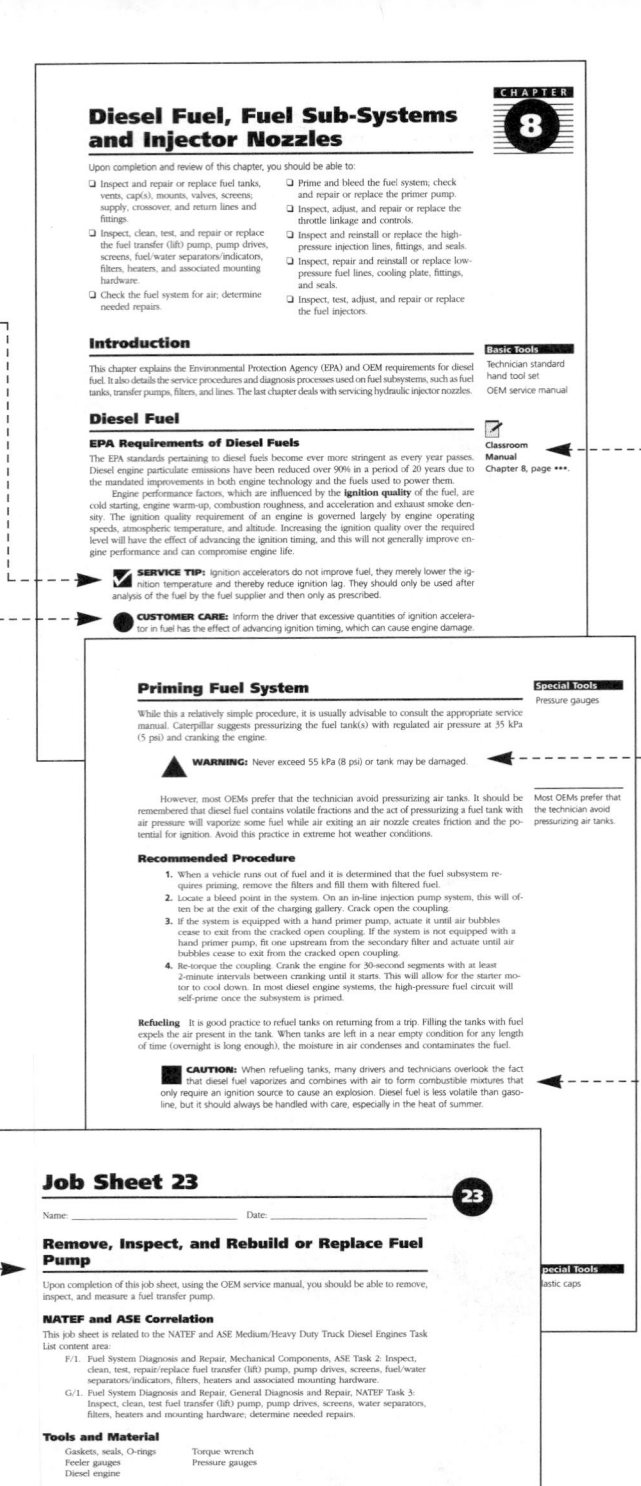

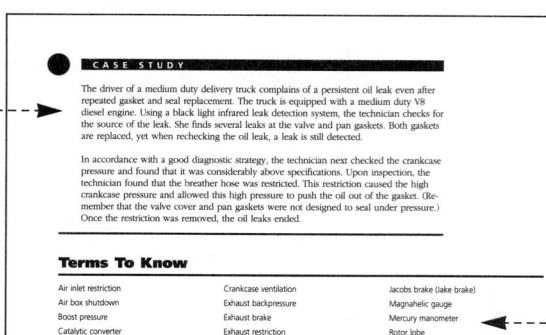

Case Studies

Case Studies concentrate on the ability to properly diagnose the systems. Each chapter ends with a case study in which a vehicle has a problem, and the logic used by a technician to solve the problem is explained.

ASE-Style Review Questions

Each chapter contains ASE-style review questions that reflect the performance objectives listed at the beginning of the chapter. These questions can be used to review the chapter as well as to prepare for the ASE certification exam.

ASE Practice Examination

A 50-question ASE practice exam, located in the appendix, is included to test students on the content of the complete Shop Manual.

CASE STUDY

The driver of a medium duty delivery truck complains of a persistent oil leak even after repeated gasket and seal replacement. The truck is equipped with a medium duty V8 diesel engine. Using a black light infrared leak detection system, the technician checks for the source of the leak. She finds several leaks at the valve and pan gaskets. Both gaskets are replaced, yet when rechecking the oil leak, a leak is still detected.

In accordance with a good diagnostic strategy, the technician next checked the crankcase pressure and found that it was considerably above specifications. Upon inspection, the technician found that the breather hose was restricted. This restriction caused the high crankcase pressure and allowed this high pressure to push the oil out of the gasket. (Remember that the valve cover and pan gaskets were not designed to seal under pressure.) Once the restriction was removed, the oil leaks ended.

Terms To Know

Air inlet restriction	Crankcase ventilation	Jacobs brake (Jake brake)
Air box shutdown	Exhaust backpressure	Magnahelic gauge
Boost pressure	Exhaust brake	Mercury manometer
Catalytic converter	Exhaust restriction	Rotor lobe
Center rotating housing assembly	Impeller wheel	Wait-to-start
Charge air-cooling pressure test	Inlet restriction gauge	Water manometer

ASE-Style Review Questions

1. A diesel engine develops too much boost at high engine rpm. *Technician A* says the oil bath air filter is dry. *Technician B* says the wastegate is malfunctioning. Who is right?
   A. A only C. Both A and B
   B. B only D. Neither A nor B

2. *Technician A* says that if the turbocharger lines and hoses become frayed or nicked, they must be replaced not repaired. *Technician B* says turbocharger oil tubes must be checked for obstructions. Who is right?
   A. A only C. Both A and B
   B. B only D. Neither A nor B

3. *Technician A* says that gray or black smoke could be

this condition is caused by an intake breather tube restriction or clogged filter. *Technician B* says you can use a pressure gauge graduated in inches of mercury to measure this pressure. Who is right?
   A. A only C. Both A and B
   B. B only D. Neither A nor B

5. A recently overhauled engine develops cylinder liquid lock. After the liquid was cleared from the cylinder, the heads were inspected, and the head gaskets were replaced, but the problem reoccured. *Technician A* says the intercooler is defective. *Technician B* says the head has a hairline crack. Who is right?
   A. A only C. Both A and B
   B. B only D. Neither A nor B

Terms to Know

Terms in this list can be found in the Glossary at the end of the manual.

Table 4-12 NATEF and ASE Task

Inspect, clean, and/or replace precombustion chambers where specified by manufacturer; determine needed repairs.

Problem Area	Symptoms	Possible Causes	Classroom Manual	Shop Manual
CYLINDER HEAD PRECHAMBER	White smoke Engine misfire	1. Missing prechamber 2. Cracked prechamber 3. Excessive prechamber protrusion	98	127

Table 4-13 NATEF and ASE Task

Inspect, reinstall, and time the drive gear train. (Includes checking timing sensors, gear wear, and backlash of crankshaft, camshaft, auxiliary, drive, and idler gears; service shafts, bushings, and bearings.)

Problem Area	Symptoms	Possible Causes	Classroom Manual	Shop Manual
TIMING GEAR POSITION	Low power	1. Incorrectly timed gear train	112	143

Table 4-14 NATEF and ASE Task

Inspect and/or replace valve bridges (crossheads) and guides; adjust bridges (crossheads).

Problem Area		
SEAT FAILURE		

Diagnostic Chart

Chapters include detailed diagnostic charts linked with the appropriate ASE and/or NATEF task. These charts list common problems and most probable causes. They also list a page reference in the Classroom Manual for better understanding of the system's operation and a page reference in the Shop Manual for details on the procedure necessary for correcting the problem.

Appendix A

ASE Practice Examination

1. A recently overhauled engine has evidence of liquid in one cylinder. After clearing the liquid from the cylinder, inspecting the heads, and replacing the head gaskets, the problem reoccurs. *Technician A* says the intercooler may be defective. *Technician B* says the head may have a hairline crack. Who is correct?
   A. A only C. Both A and B
   B. B only D. Neither A nor B

2. When pressure testing an engine block, *Technician A* says the block need only be submerged in water for about 20–30 minutes and the water checked for bubbles. *Technician B* says cracked cylinder blocks must be replaced. Who is correct?
   A. A only C. Both A and B
   B. B only D. Neither A nor B

3. After honing a cylinder liner, the cylinder liner surface should have:
   A. 60-degree crossover angle pattern with a 15–29 microinch crosshatch.
   B. 40-degree crossover angle pattern with a 15–20 microinch crosshatch.
   C. 45-degree crossover angle pattern with a 20–30 microinch crosshatch.
   D. 60-degree crossover angle pattern with a 15–20 microinch crosshatch.

4. When installing the cylinder insert and liner, all of these statements are true EXCEPT:
   A. An aluminum block should be heated by submerging it in a tank of hot water for 20 minutes.
   B. The block bore and counterbore must be clean before insert and liner installation.
   C. A hydraulic press must be used to install the liner in the block.
   D. The liner must be pushed downward until the liner flange contacts the insert.

5. While removing, replacing, and servicing viscous-type vibration dampers:
   A. the vibration damper may be disassembled and repaired.
   B. most engine manufacturers recommend damper replacement during a major engine overhaul.
   C. if a puller is not available, a hammer and punch may be used to remove the damper.
   D. dents in the damper housing have no effect on damper operation.

6. When assembling pistons and connecting rods, *Technician A* says the markings on the connecting rod and bearing cap may face either side of the engine. *Technician B* says on some two-stroke cycle diesel engines the piston and connecting rod assembly should be installed in the liner before installation in the engine. Who is correct?
   A. A only C. Both A and B
   B. B only D. Neither A nor B

7. *Technician A* says the cylinder head warpage should be measured transversely across the cylinder head at each end and between each cylinder. *Technician B* says a blown head gasket between two cylinders may indicate a transversely warped cylinder head. Who is correct?
   A. A only C. Both A and B
   B. B only D. Neither A nor B

8. When testing a cylinder head for potential coolant leakage, *Technician A* says the coolant needs to be heated to operating temperature. *Technician B* says the coolant does not need to be under pressure. Who is correct?
   A. A only C. Both A and B
   B. B only D. Neither A nor B

9. *Technician A* says a lifter bottom should be smooth and slightly concave. *Technician B* says if a lifter bottom is pitted, the camshaft mating lobe may be scored. Who is correct?
   A. A only C. Both A and B
   B. B only D. Neither A nor B

10. When checking valve head height, the required tool combination is:
    A. a straight edge and feeler gauge.
    B. a dial indicator, step block, steel ruler, and outside caliper.
    C. a vernier micrometer and point gauge.
    D. a steel ruler and T-square.

11. A misfiring cylinder is diagnosed on a diesel engine with unit injectors. The technician replaces the injector in the misfiring cylinder, but the cylinder continues to misfire. The cause of this problem could be:
    A. a worn cam follower roller.
    B. a restricted fuel filter.
    C. high fuel-pump pressure.
    D. a leaking injector tube.

457

Reviewers

We would like to extend a special thank you to those who saw things we overlooked and for their contributions:

David Biegel
Madison Area Technical College

Douglas Bradley
Utah Valley State College

Dennis Chapin
Rogue Community College

Steve Johnson
Illinois Central College

Kenneth W. Kephart
Central Texas College

Dave Lieber
Northeastern Junior College

Randolph B. Turnage
Wilson Technical Community College

Dennis Vickerman
Southeast Technical Institute

The publisher and author also would like to thank Don Knowles for his contributions to this text.

Contributing Companies

We would also like to thank these companies who provided technical information and art for this edition:

American Isuzu Motors Inc.
Butterworth Publishers - Oxford
Caterpillar Inc.
Chevron Products Company
CRC Industries, Inc.
Cummins Engine Company, Inc.
Deere & Company
Detroit Diesel Corporation
DuPont Company
Federal-Mogul Corporation
Freightliner Corporation
General Fire Extinguisher Corporation
Hennessy Industries, Inc.
Jacobs Vehicle Systems™
Kleer-Flo Company
Mac Tools
Mack Trucks, Inc.
MPSI

Navistar International Engine Group
Parker Hannefin
Robert Bosch Corporation
Schwitzer Engine Components, Schwitzer Group
The Sherwin-Williams Company
Siebe North, Inc.
Snap-on Tools Company
The Maintenance Council (TMC) of the American Trucking Assoc.
Utah Valley State College
Vibratech® Inc.
Volvo Trucks North America, Inc.
Williams Controls Inc.

Portions of materials contained herein have been reprinted with permission of General Motors Corporation, Service Operations.

Safety Practices

Upon completion and review of this chapter, you should be able to:

❏ Recognize shop hazards and take the necessary steps to avoid personal injury or property damage.

❏ Explain the purposes of the Occupational Safety and Health Act.

❏ Identify the necessary steps for personal safety in the truck shop.

❏ State the reasons for prohibiting drug and alcohol use in the truck shop.

❏ Describe how to properly lift a heavy object.

❏ Explain the steps required to provide electrical safety in the truck shop.

❏ Identify the steps required to provide safe handling and storage of gasoline and diesel fuel.

❏ Describe the necessary housekeeping safety steps.

❏ Explain the essential general truck shop safety practices.

❏ Identify the steps required to provide fire safety in the truck shop.

❏ Describe typical fire extinguisher operating procedure.

❏ Describe four different types of fires, and identify the type of fire extinguisher required for each kind of fire.

❏ Describe three pieces of truck shop safety equipment other than fire extinguishers.

❏ Identify hazardous waste materials as defined by state and federal regulations.

❏ Follow proper safety precautions when handling hazardous waste materials.

Introduction

Safety is extremely important in the truck shop. Knowledge and practice of safety precautions prevent serious personal injury and expensive property damage. Medium/Heavy Duty Truck students and technicians must be familiar with shop hazards and all types of safety including personal, gasoline and diesel fuel handling, housekeeping, general shop, fire, and **hazardous material** handling and disposal. The first step in providing a safe shop is learning about all types of safety precautions. The second, and most important, step in this process is applying our knowledge of safety precautions when working in the shop. In other words, we must actually develop safe working habits in the shop from our understanding of various safety precautions. When shop employees have a careless attitude toward safety, accidents are more likely to occur. All shop personnel must develop a serious attitude toward safety in the shop. The result of this serious attitude is that shop personnel will learn and adopt all shop safety rules.

Shop personnel must be familiar with their rights regarding hazardous waste disposal. **Right-to-know laws** explain these rights. Shop personnel must also be familiar with hazardous materials in the truck shop and the proper disposal methods for these materials according to state and federal regulations.

Shop Manual Chapter 1, page 11.

A BIT OF HISTORY

In 1975, 102,508 diesel-powered trucks were sold in the United States, and 84,878 of these trucks had a gross vehicle weight rating (GVWR) over 33,000 pounds (Class 6 truck). In 1998, total sales of diesel-powered trucks in the United States were 629,730, and 220,456 of these trucks had a GVWR over 33,000 pounds.

Occupational Safety and Health Act

The United States Congress passed the **Occupational Safety and Health Act (OSHA)** in 1970. The purposes of this legislation are: (1) To assist and encourage the citizens of the United States in their efforts to ensure safe and healthful working conditions by providing research, information, education, and training in the field of occupational safety and health. (2) To ensure safe and healthful working conditions for working men and women by authorizing enforcement of the standards developed under the Act. Since approximately 25% of workers are exposed to health and safety hazards on the job, the OSHA standards are necessary to monitor, control, and educate workers regarding health and safety in the workplace. Employers and employees in Canada should be familiar with Workplace Hazardous Materials Information Systems (WHMIS).

Shop Hazards

Service technicians and students encounter many hazards, especially when working with large and heavy diesel engines and fuel in a truck shop. With known hazards, technicians must follow basic shop safety rules and procedures to avoid personal injury. Some of the hazards in a truck shop are as follows:

1. Flammable liquids such as gasoline and paint represent a fire hazard and must be handled and stored properly.
2. Flammable materials such as oily rags must be stored properly to avoid a fire hazard.
3. Batteries contain a **corrosive** sulfuric acid solution and produce explosive hydrogen gas while charging.
4. Loose sewer and drain covers may cause foot or toe injuries.
5. Caustic liquids, such as those in hot or cold cleaning tanks, are harmful to skin and eyes.
6. High-pressure air in the shop compressed-air system can be very dangerous if it penetrates the skin and enters the bloodstream. Blood poisoning may result.
7. Frayed cords on electric equipment and lights may result in severe electrical shock.
8. Hazardous waste material such as batteries must be handled properly to avoid harmful effects.
9. Carbon monoxide and particulates from truck exhaust are harmful to human beings.
10. Loose clothing or long hair may become entangled in rotating parts on equipment or vehicles, resulting in serious injury.
11. Dust and vapors generated during some repair jobs are harmful.
12. Asbestos dust generated during brake lining service and clutch service is a contributor to lung cancer.
13. High noise levels from shop equipment such as an air chisel may cause hearing loss.
14. Oil, grease, water, or parts cleaning solutions on shop floors may cause someone to slip and fall, resulting in serious injury.

Safety in the Truck Shop

Each person in a truck shop must follow certain basic shop safety rules to remove the danger from shop hazards. Personal injury and vehicle and property damage are prevented when all personnel in the shop follow basic shop safety rules.

Personal Safety

Personal safety is the responsibility of each technician in the shop. Always follow these safety practices:

1. Always use the correct tool for the job. If the wrong tool is used, it may slip and cause hand injury.
2. Follow the truck manufacturer's recommended service procedures.
3. Always wear eye protection such as safety glasses or a face shield (Figure 1-1).
4. Wear protective gloves when cleaning parts in hot or cold tanks and when handling hot parts such as exhaust manifolds.
5. Do not smoke when working on a truck. A spark from a cigarette or lighter may ignite flammable materials in the work area.
6. When working on a running engine, keep hands and tools away from rotating parts.
7. Do not wear loose clothing, and keep long hair tied behind your head. Loose clothing or long hair can easily become entangled in rotating parts.
8. Wear safety shoes or boots.
9. Do not wear watches, jewelry, or rings when working on a truck. Severe burns occur when jewelry makes contact between an electrical terminal and ground.
10. Always place a shop exhaust hose on the truck exhaust pipe if the engine is running, and be sure the exhaust fan is running. Carbon monoxide in the vehicle exhaust is harmful or fatal to the human body.
11. Be sure that the shop has adequate ventilation.
12. Make sure the work area has adequate lighting.
13. Use trouble lights with steel or plastic cages around the bulb. If an unprotected bulb breaks, it may ignite flammable materials in the area.
14. When servicing trucks, always apply the parking brake.
15. Avoid working on a truck parked on an incline.
16. Never work under a truck unless the vehicle chassis is supported securely on safety stands.
17. When you raise one end of a truck, place wheel chocks on both sides of the wheels remaining on the floor.
18. Know the location of shop first-aid kits and eyewash fountains.
19. Familiarize yourself with the location of all shop fire extinguishers.
20. Do not use any type of open flame heater to heat the work area.
21. Collect oil, fuel, brake fluid, and other liquids in the proper safety containers.
22. Use only approved cleaning fluids and equipment. Do not use diesel fuel or gasoline to clean parts.

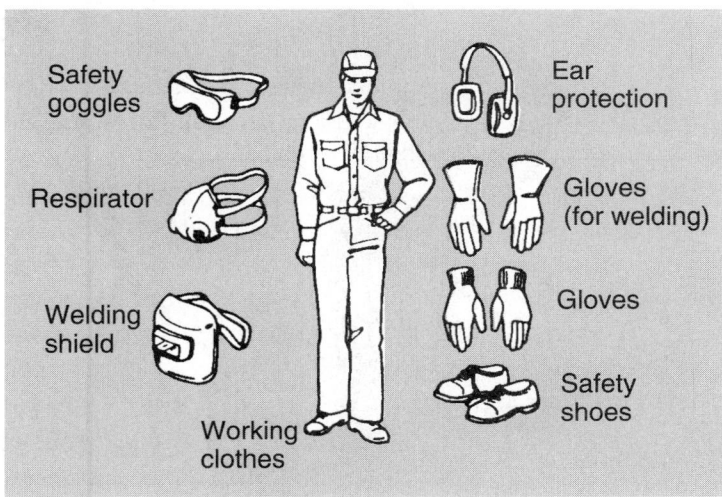

Figure 1-1 Shop safety equipment. (Courtesy of General Motors Corporation, Service Operations)

23. Obey all state and federal safety, fire, and hazardous material regulations.

24. Always operate equipment according to the equipment manufacturer's recommended procedure.

25. Do not operate equipment unless you are familiar with the correct operating procedure.

26. Do not leave running equipment unattended.

27. Do not use electrical equipment, including trouble lights, with frayed cords.

28. Be sure the safety shields are in place on rotating equipment.

29. Before operating electrical equipment, be sure the power cord has a ground connection.

30. When working in an area where extreme noise levels are encountered, wear ear plugs or covers.

31. Always wear boots or shoes that provide adequate foot protection. Heavy-duty work boots or shoes with steel toecaps are best for working in the truck shop. Footwear must protect against heavy falling objects, flying sparks, and corrosive liquids. Soles on footwear must protect against punctures by sharp objects.

32. Wear a respirator to protect your lungs when working in dusty conditions.

33. When working on, around, or under a truck or tractor, disconnect the batteries and tag the driver's door, steering wheel, or start button so no one starts the vehicle. The tag should read, "Do not start."

34. If a truck or tractor has a brake failure or the parking brake chambers have been caged for towing the vehicle, the unit should be tagged on the door, steering wheel, and start button, "Do not start or move. No Brakes."

35. Do not raise or lower a tilt cab with the engine running (Figure 1-2). This action may cause components to contact rotating belts, pulleys, and other engine parts.

Figure 1-2 Tilt cab. (Courtesy of General Motors Corporation, Service Operations)

Smoking, Alcohol, and Drugs in the Shop

Do not smoke when working in the shop. If the shop has designated smoking areas, smoke only in these areas. Do not smoke in a customer's truck. A nonsmoker may not appreciate cigarette odor in his or her truck. A spark from a cigarette or lighter may ignite flammable materials in the workplace.

Never use drugs or alcohol while working in the shop. Even a small amount of drugs or alcohol will affect your reaction time. In an emergency a slow reaction time may cause personal injury. If a heavy object falls off the workbench and drugs or alcohol slows your reaction time, you may not get your foot out of the way in time to avoid an injury. If a fire starts in the workplace and you are a few seconds slower getting a fire extinguisher into operation because of alcohol or drug use, it could make the difference between extinguishing the fire and having expensive fire damage.

> The improper or excessive use of alcoholic beverages and/or drugs is termed *substance abuse*.

Lifting and Carrying

Diesel engine service jobs often require heavy lifting. You should know your maximum weight-lifting ability; do not attempt to lift more than this amount of weight. Truck technicians have a comprehensive array of lifting equipment available to them that many do not use enough. Lifting a 75-pound cylinder head would be easy if the head were shaped like a barbell where the mass is concentrated in the middle and there are handgrips. However, performing this task in the cramped conditions of an engine compartment is an invitation to back problems. If a heavy part exceeds your weight-lifting ability, use the appropriate lifting apparatus or have a co-worker help with the lifting job. Follow these steps when lifting or carrying an object:

1. If an object or engine component is beyond your lifting ability, use a lifting tool, such as a chain hoist or lifting crane (cherry picker).
2. If the object is to be carried, be sure your path is free of obstructions or loose parts or tools.
3. Position your feet close to the object and position your back reasonably straight for proper balance (Figure 1-3).
4. Keep your back and elbows as straight as possible. Continue to bend your knees until your hands reach the best lifting location on the object you are lifting.
5. Be certain any container you are lifting is in good condition. If it falls apart during the lifting operation, parts may drop out resulting in foot injury or part damage.

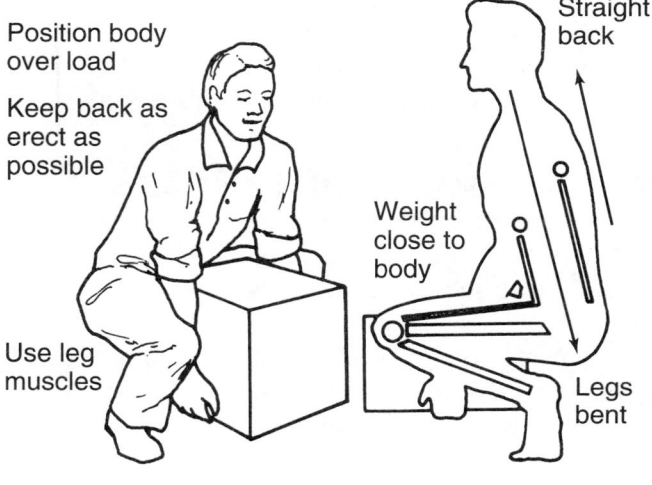

Position body over load

Keep back as erect as possible

Use leg muscles

Straight back

Weight close to body

Legs bent

Figure 1-3 Proper lifting procedure.

6. Maintain a firm grip on the object, and do not attempt to change your grip while lifting is in progress.

7. Straighten your legs to lift the object, and keep the object close to your body. Use leg muscles rather than back muscles.

8. If you have to change direction of travel, turn your whole body instead of twisting it.

9. Do not bend forward to place an object on a workbench or table. Position the object on the front surface of the workbench and slide it back. Do not pinch your fingers under the object while setting it on the front of the bench.

10. When placing an object on the floor or a low surface, bend your legs to lower the object. Do not bend your back forward, because this movement strains back muscles.

11. When you place a heavy object on the floor, put suitable blocks under the object to prevent jamming your fingers under the object.

Hand Tool Safety

Shop Manual Chapter 1, page 15.

Improper use and care of hand tools causes many shop accidents. Follow these tool safety steps:

1. Maintain tools in good condition and keep them clean. Worn tools may slip and cause hand injury. If you use a hammer with a loose head, the head may fly off and cause personal injury or vehicle damage. If your hand slips off a greasy tool, some part of your body may hit the vehicle. For example, your head may hit the vehicle hood.

2. Use the right tool for the job. Using the wrong tool may cause damage to the tool, fastener, or your hand if the tool slips. For example, if you use a screwdriver for a chisel or pry bar, the blade may shatter, causing serious personal injury.

3. Use sharp-pointed tools with caution. Always check your pockets before sitting on the vehicle seat. A screwdriver, punch, or chisel in your back pocket may put an expensive tear in the upholstery. Do not lean over fenders with sharp tools in your pockets.

4. Keep tool tips that are intended to be sharp in a sharp condition. Sharp tools such as chisels will do the job faster with less effort.

Electrical Equipment Safety

1. Replace or repair frayed cords on electrical equipment such as shop lights, drills, grinders, wheel aligners, wheel balancers, overhead electric hoists, and cleaning equipment.

2. All cords from lights and electrical equipment must have a ground connection. The ground connector is the round terminal in a three-prong electrical plug. Do not use a two-prong adapter to plug in a three-prong electrical cord. Three-prong electrical outlets should be mandatory in all shops.

3. Do not leave electrical equipment running and unattended.

Gasoline and Diesel Fuel Safety

CAUTION: Gasoline is a very explosive liquid! One exploding gallon of gasoline has a force equal to fourteen sticks of dynamite.

The expanding vapors that come from gasoline are extremely dangerous. These vapors are present even in cold temperatures. Vapors formed in gasoline tanks on many trucks are controlled, but vapors from gasoline storage may escape from the can, resulting in a hazardous situation. Therefore, always place gasoline storage containers in a well-ventilated space. Although diesel fuel is not as volatile as gasoline, the same basic rules apply to both diesel fuel and gasoline storage. Follow these safety precautions regarding gasoline or diesel fuel containers:

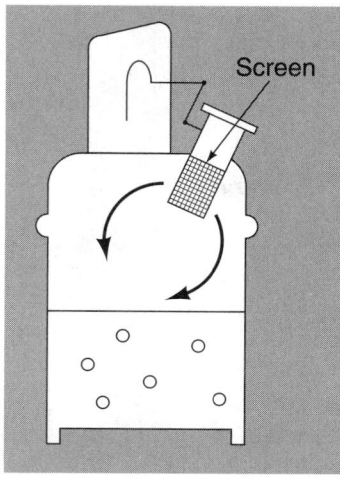

Figure 1-4 Gasoline storage can.

1. Always use a red, approved gasoline container. These storage cans have a flash-arresting screen at the outlet (Figure 1-4) that prevents external ignition sources from igniting the gasoline within the can when someone pours the gasoline or diesel fuel.

2. Always use a red, approved gasoline container to allow for proper identification of a hazardous substance.

3. Do not fill gasoline containers completely full. Always leave the level of gasoline at least one inch from the top of the container to allow expansion of the gasoline at higher temperatures. If a gasoline container is completely full, and the gasoline expands, the expansion will force gasoline from the can and create a dangerous spill.

4. If gasoline or diesel fuel containers must be stored, place them in a designated storage locker or facility.

5. When a gasoline container must be transported, be sure it is secured against upsets.

6. Do not store a partially filled gasoline container for long periods, because it may give off vapors and produce a potential danger. Never leave gasoline containers open except while filling or pouring gasoline from the container.

7. Do not prime an engine with gasoline while cranking the engine.

8. Never use gasoline as a cleaning agent.

9. Always connect a ground strap to containers when filling or transferring fuel or other flammable products from one container to another to prevent static electricity that could result in explosion and fire.

Housekeeping Safety

1. Keep shop floors clean! Always clean shop floors immediately after a spill.

2. Store paint and other flammable liquids in a closed, steel cabinet (Figure 1-5).

3. Store oily rags in approved, covered garbage containers (Figure 1-6). A slow generation of heat occurs from oxidation of oil on rags. Heat may continue to be generated until the ignition temperature is reached. The oil and the rags then begin to burn causing a fire. This action is called spontaneous combustion. However, if the oily rags are in an airtight, approved container, the fire cannot get enough oxygen to continue burning.

4. Keep the shop neat and clean. Always pick up tools and parts, and do not leave creepers (a wheeled device on which to lie while working underneath a truck) lying on the floor.

5. Keep the workbenches clean. Do not leave heavy objects, such as parts, on the bench after you are finished with them.

Figure 1-5 Flammable liquids storage cabinet. (Courtesy of The Sherwin Williams Company)

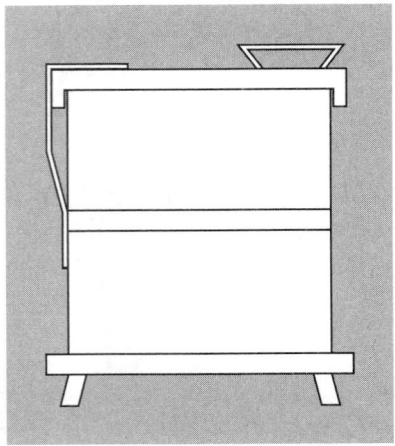

Figure 1-6 Oily rag storage container.

General Truck Shop Safety

Shop safety equipment must be easily accessible and in good working condition.

Safety in the truck shop is extremely important. When shop safety rules are observed, shop productivity is increased because employees are not absent from work due to work-related injuries. Observing shop safety rules involves teamwork. Everyone in the shop must obey shop safety rules to provide a safe working area. If one employee does not obey safety rules, another employee may be injured. For example, if an employee leaves a hydraulic jack handle sticking out from under a vehicle, another employee may trip over this handle and injure him- or herself. All technicians must observe and follow these general shop safety rules in the truck shop:

1. Be sure that all sewer covers fit properly and are securely in place.
2. Always wear a face shield, protective gloves, and protective clothing when necessary. Wear gloves when working with solvents and caustic solutions, handling hot metal, or grinding metal. Shop coats and coveralls are the most common types of protective clothing.
3. Never direct high-pressure air from an air gun against human flesh. Air may penetrate the skin and enter the bloodstream, which can cause serious health problems and even death. Always keep air hoses in good condition. If an end blows off an air hose, the hose may whip around and cause an injury. Use only OSHA approved air gun nozzles. Be especially careful with diesel injection nozzle pop-testers because these testers spray test fluid under very high pressure. When observing opening pressure and spray pattern, keep hands and fingers out of the way of the pressure spray. Test fluid may penetrate the skin and enter the bloodstream, causing serious health problems or death.
4. Handle all hazardous waste materials according to state and federal regulations. (A more detailed explanation regarding these regulations is found later in this chapter.)
5. Always place a shop exhaust hose on the exhaust pipe of a truck if the engine is running in the shop, and be sure the shop exhaust fan is on.
6. Keep hands, long hair, and tools away from rotating parts on running engines such as fan blades and belts. Remember that an electric-drive fan may start turning at any time.
7. Always use the correct tool for the job. For example, never strike a hardened steel component, such a piston pin, with a steel hammer. This type of component may shatter, and fragments can penetrate eyes or skin.
8. Follow the diesel engine or truck manufacturer's recommended service procedures.

9. Collect oil, fuel, brake fluid, and other liquids in the proper safety containers.
10. Use only approved cleaning fluids and equipment. Do not use diesel fuel or gasoline to clean parts.
11. Obey all state and federal safety, fire, and hazardous material regulations.
12. Always operate equipment according to the manufacturer's recommended procedure.
13. Do not operate equipment unless you are familiar with the correct operating procedure.
14. Do not leave running equipment unattended.
15. Be sure the safety shields are in place on rotating equipment.
16. Follow a regular schedule for maintenance and adjustment of all shop equipment.
17. If the shop has safety lines around equipment, always work within these lines when operating equipment.
18. Be sure that heating equipment is well ventilated.
19. Do not run in the shop or engage in horseplay.
20. Post emergency telephone numbers near the telephone. These numbers should include a doctor, ambulance, fire department, hospital, and police. Keep the following safety hotline numbers near the telephone in case of emergency:

Post emergency telephone numbers near the telephone.

❏ Chemical Emergency Preparedness Hotline (CERCLA SARA Title III) 1-800-535-0202
❏ Chemical Transportation Emergency Center (CHEMTREC) 1-800-424-9300
❏ Chemical Referral Center (CMA) 1-800-CMA-8200
❏ EPA RCRA, Superfund Hazardous Waste Hotline 1-800-424-9346
❏ EPA Small Business Hotline 1-800-368-5888
❏ National Response Center 1-800-424-8802
❏ National Safety Council 1-312-527-4800
❏ NIOSH The National Institute of Occupational Safety & Health 1-800-356-4674
❏ OSHA, Health Standards 1-202-523-7075
❏ Safe Drinking Water Hotline 1-800-426-4791
❏ Substance Identification 1-800-848-6538
❏ US Department of Transportation (DOT) Hotline 1-202-366-4488

21. Do not place hydraulic jack handles where someone could trip over them.
22. Keep aisles clear of debris.

Fire Safety

Observing fire safety rules may prevent a fire in the truck shop. Following the proper fire safety rules and procedures may also make the difference between extinguishing a fire with minimum damage and having a fire get out of control causing very expensive damage. Follow these fire safety rules and procedures in the truck shop:

1. Familiarize yourself with the location and operation of all shop fire extinguishers.
2. If a fire extinguisher is used, report it to management so the extinguisher can be recharged.
3. Do not use any type of open flame heater in the work area.
4. Do not turn on the ignition switch or crank the engine with a fuel line disconnected.
5. Store all combustible materials such as gasoline, paint, and oily rags in approved safety containers.
6. Clean up gasoline, diesel fuel, oil, or grease spills immediately.
7. Always wear clean shop clothes. Do not wear oil-soaked clothes.
8. Do not allow sparks and flames near batteries.
9. Securely fasten welding tanks in an upright position.
10. Do not block doors, stairways, or exits.
11. Do not smoke when working on vehicles.

12. Do not smoke or create sparks near flammable materials or liquids.

13. Store combustible shop supplies, such as paint, in a closed, steel cabinet.

14. Keep gasoline and diesel fuel in approved safety containers.

15. If a fuel tank is removed from a vehicle, do not drag the tank on the shop floor.

16. Know the approved fire escape route from your classroom or shop to the outside of the building.

17. If a fire occurs, do not open doors or windows. This action creates extra draft that makes the fire worse.

18. Do not put water on a gasoline fire, because water will make the fire worse by spreading the gasoline over a wider area.

19. Call the fire department as soon as a fire begins, and then attempt to extinguish the fire.

20. If possible stand 6–10 feet from the fire and aim the fire extinguisher nozzle at the base of the fire with a sweeping action.

21. If a fire produces a lot of smoke in the room, remain close to the floor to obtain oxygen and avoid breathing smoke.

22. If the fire is too hot or the smoke makes breathing difficult, get out of the building.

23. Do not re-enter a burning building.

24. Keep solvent containers covered except when pouring from one container to another. When flammable liquids are transferred from bulk storage, the bulk container should be grounded to a permanent shop fixture such as a metal pipe. During this transfer process, ground the bulk container to the portable container (Figure 1-7). These ground wires prevent the buildup of a static electric charge, which could result in a spark and disastrous explosion. Always discard or clean empty solvent containers, because fumes in these containers are a fire hazard.

25. Familiarize yourself with different types of fires and fire extinguishers, and know the type of extinguisher to use on each fire.

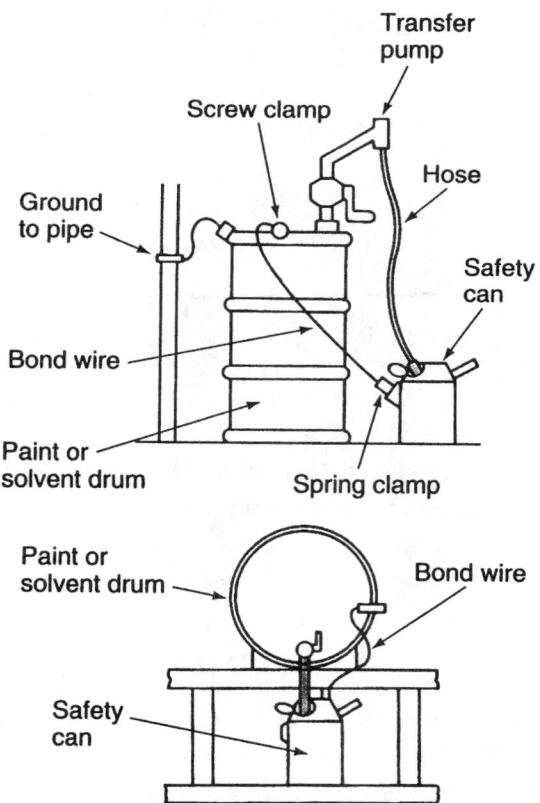

Figure 1-7 Grounding the bulk container. (Courtesy of DuPont Automotive Finishes)

Truck Shop Safety Equipment

Fire Extinguishers

Fire extinguishers are one of the most important pieces of safety equipment. All shop personnel must know the location of the fire extinguishers in the shop. If you have to waste time looking for an extinguisher after a fire starts, the fire could get out of control before you get the extinguisher into operation. Fire extinguishers should be easily accessible at all times. Everyone working in the shop must know how to operate the fire extinguishers. There are several different types of fire extinguishers, but their operation usually involves these steps:

1. Get as close as possible to the fire without jeopardizing your safety.
2. Grasp the extinguisher firmly and aim the extinguisher at the fire.
3. Pull the pin from the extinguisher handle.
4. Squeeze the handle to dispense the contents of the extinguisher.
5. Direct the fire extinguisher nozzle at the base of the fire, and dispense the contents of the extinguisher with a sweeping action back and forth across the fire. Most extinguishers discharge their contents in 8–25 seconds.
6. Always be sure the fire is extinguished.
7. Always keep an escape route open behind you so a quick exit is possible if the fire gets out of control.

A decal on each fire extinguisher identifies the type of chemical in the extinguisher and provides operating information (Figure 1-8). Shop personnel should be familiar with the following types of fires and fire extinguishers:

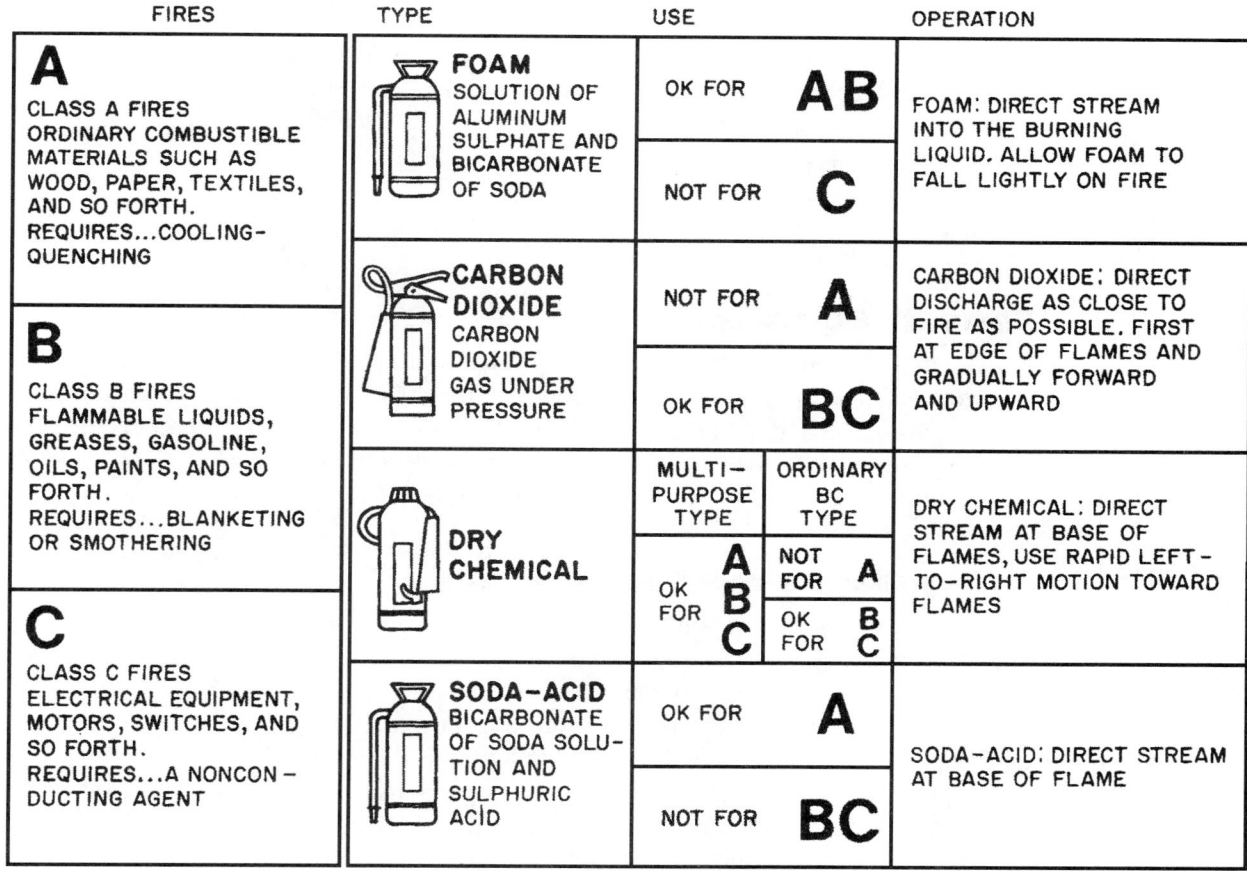

Figure 1-8 Fire extinguisher label. (Courtesy of General Fire Extinguisher Corporation)

Class A fires involve ordinary combustible materials such as paper, wood, clothing, and textiles. Multipurpose **dry chemical extinguishers** are used on these fires.

Class B fires involve the burning of flammable liquids such as gasoline, diesel fuel, oil, paint, solvents, and greases. These fires may be extinguished with multipurpose dry chemical extinguishers. Fire extinguishers containing **halogen or halon** are effective on class B fires. The chemicals in this type of extinguisher attach to the hydrogen, hydroxide, and oxygen molecules to stop the combustion process almost instantly. However, the resultant gases from the use of halogen-type extinguishers are very toxic and harmful to the operator of the extinguisher. Halon fire extinguishers are now illegal because of chlorofluorocarbon (CFC) regulations.

Class C fires involve the burning of electrical equipment such as wires, motors, and switches. These fires may be extinguished with multipurpose dry chemical extinguishers.

Class D fires involve the combustion of metal chips, turnings, and shavings. A dry chemical extinguisher is the only type of extinguisher recommended for these fires.

Additional information regarding types of extinguishers for various types of fires is provided in Figure 1-8.

Causes of Eye Injuries

Eye injuries may occur in various ways in an automotive shop. Some of the common eye accidents are these:

1. Thermal burns from excessive heat
2. Irradiation burns from excessive light, such as from an arc welder
3. Chemical burns from strong liquids, such as battery electrolyte
4. Foreign material in the eye
5. Penetration of the eye by a sharp object
6. A blow from a blunt object

Wearing safety glasses and observing shop safety rules will prevent most eye accidents.

Eyewash Fountains

If a chemical gets in your eyes, it must be washed out immediately to prevent a chemical burn. An eyewash fountain is the most effective way to wash the eyes. This fountain is similar to a drinking water fountain, but the eyewash fountain has water jets placed throughout the fountain top. Every shop should be equipped with some type of eyewash facility (Figure 1-9). Be sure you know the location of the eyewash fountain in the shop.

Safety Glasses and Face Shields

The mandatory use of eye protection with safety glasses or a face shield is one of the most important safety rules in a truck shop. Face shields protect the face; safety glasses protect the eyes. When grinding, you must wear safety glasses or a face shield. Many shop insurance policies require the use of eye protection in the shop. Some medium/heavy duty truck technicians have been blinded in one or both eyes because they did not bother to wear safety glasses. All safety glasses must be equipped with safety glass, and they should provide some type of side protection (Figure 1-10). Safety glasses should feel comfortable on your face. If they are uncomfortable, you may tend to take them off, leaving the eyes unprotected. You should wear a face shield when handling hazardous chemicals or when using an electric grinder or buffer (Figure 1-11).

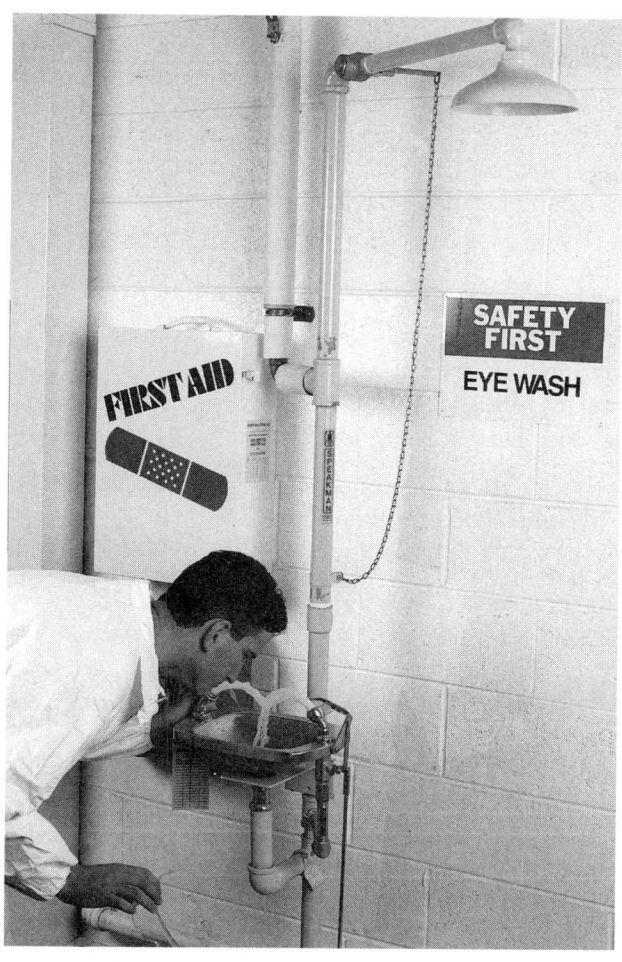

Figure 1-9 Eye wash.

Figure 1-10 Safety glasses. (Courtesy of Siebe North)

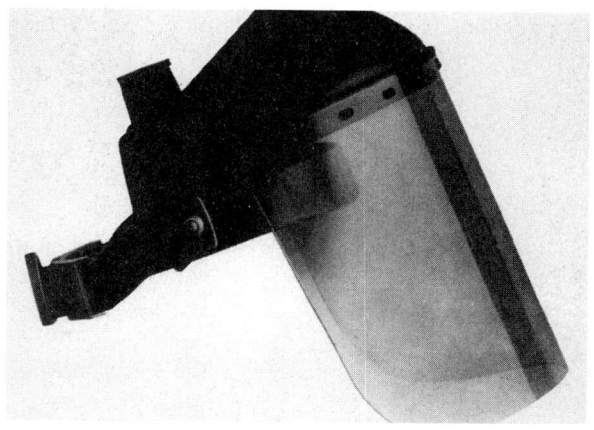

Figure 1-11 Face shield. (Courtesy of Siebe North)

First-Aid Kits

First-aid kits should be clearly identified and conveniently located (Figure 1-12). These kits contain such items as bandages and ointment required for minor cuts. All shop personnel must be familiar with the location of first-aid kits. At least one of the shop personnel should have basic first-aid training, and this person should be in charge of administering first aid and keeping first-aid kits filled.

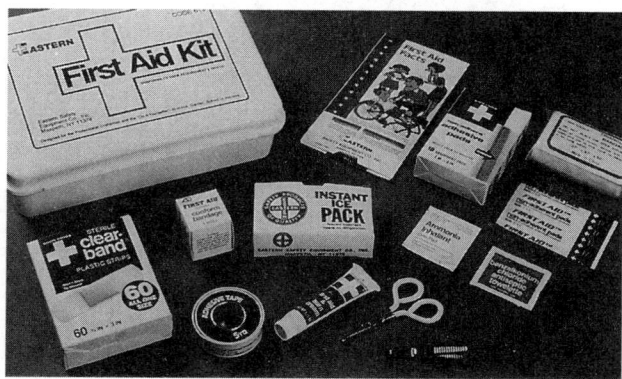

Figure 1-12 First-aid kit.

Hazardous Waste Disposal

■ **CAUTION:** When handling hazardous waste material always wear the proper protective clothing and equipment detailed in the right-to-know laws. This includes respirator equipment (Figure 1-13). All recommended procedures must be followed accurately. Personal injury may result from improper clothing, equipment, or procedures when handling hazardous materials.

Hazardous waste materials in truck shops are chemicals or components that the shop no longer needs. These materials pose a danger to the environment and people if they are disposed of in ordinary garbage cans or sewers. However, no material is considered hazardous waste until the shop has finished using it and is ready to dispose of it. The Environmental Protection Agency (EPA) publishes a list of hazardous materials that is included in the Code of Federal Regulations. The EPA considers waste hazardous if it is included on the EPA list of hazardous materials, or it has one or more of these characteristics:

1. **Reactive:** Any material that reacts violently with water or other chemical is considered hazardous. If a material releases cyanide gas, hydrogen sulfide gas, or similar gases when exposed to low pH acid solutions, it is considered hazardous.
2. **Corrosive:** A material is hazardous if it burns the skin or dissolves metals and other materials.
3. **Toxic:** Materials are hazardous if they leach one or more of eight heavy metals in concentrations greater than 100 times primary drinking water standard.
4. **Ignitable:** A liquid is hazardous if it has a flash point below 140° F (60° C); a solid is hazardous if it ignites spontaneously.
5. **Radioactive:** Any substance that emits measurable levels of radiation is considered hazardous. When individuals bring containers of highly radioactive substances into the truck shop environment, qualified personnel with the appropriate equipment must test them.

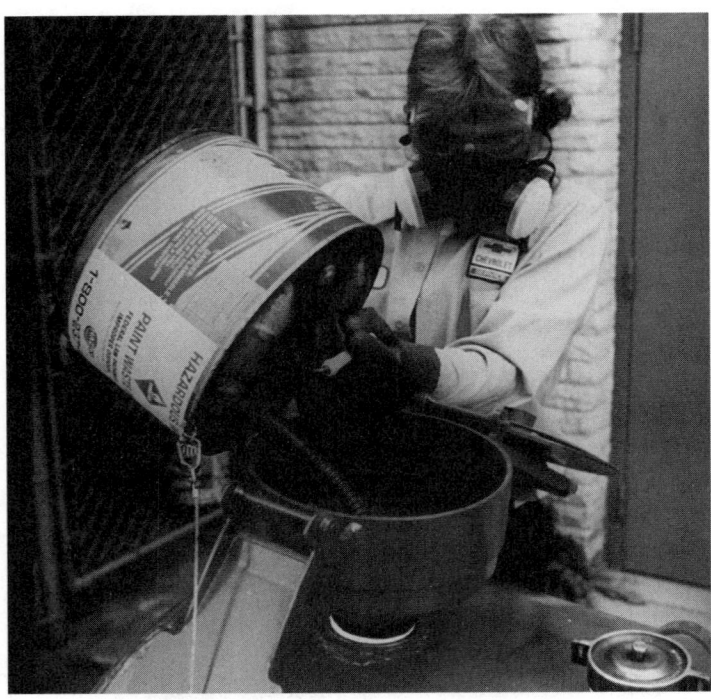

Figure 1-13 Respirator. (Courtesy of DuPont Automotive Finishes)

CAUTION: Hazardous waste disposal laws include serious penalties for anyone caught breaking these laws.

Federal and state laws control the disposal of hazardous waste materials. Every shop employee must be familiar with these laws. Hazardous waste disposal laws include the Resource Conservation and Recovery Act (RCRA), which states that hazardous material users are responsible for hazardous materials from the time they become a waste until proper disposal is completed. Many truck shops hire an independent hazardous waste hauler to dispose of hazardous waste material (Figure 1-14). The shop owner or manager should have a written contract with the hazardous waste hauler.

Rather than have hazardous waste material hauled to an approved hazardous waste disposal site, a shop may choose to recycle the material in the shop. In this case the user must store hazardous waste material properly and safely and be responsible for the transportation of this material until it arrives at an approved hazardous waste disposal site and is processed according to the law. The RCRA controls these types of automotive wastes:

1. Paint and body repair products waste
2. Solvents for parts and equipment cleaning
3. Batteries and battery acid
4. Mild acids used for metal cleaning and preparation
5. Waste oil and engine coolants or antifreeze
6. Air conditioning refrigerants
7. Engine oil filters

Observe the following rules when disposing of hazardous waste material:

1. Never pour hazardous wastes on weeds to kill them
2. Never pour hazardous wastes on gravel streets to prevent dust
3. Never throw hazardous wastes in a dumpster
4. Never dispose of hazardous wastes anywhere but at an approved disposal site
5. Never pour hazardous wastes down sewers, toilets, sinks, or floor drains

The **right-to-know laws** state that employees have a right to know when the materials they use at work are hazardous. The right-to-know laws started with the **Hazard Communication Standard** published by the OSHA in 1983. Originally, this document was intended for

The Resource Conservation and Recovery Act (RCRA) states that hazardous material users are responsible for hazardous materials from the cradle to the grave.

Figure 1-14 Waste hauler. (Courtesy of DuPont Automotive Finishes)

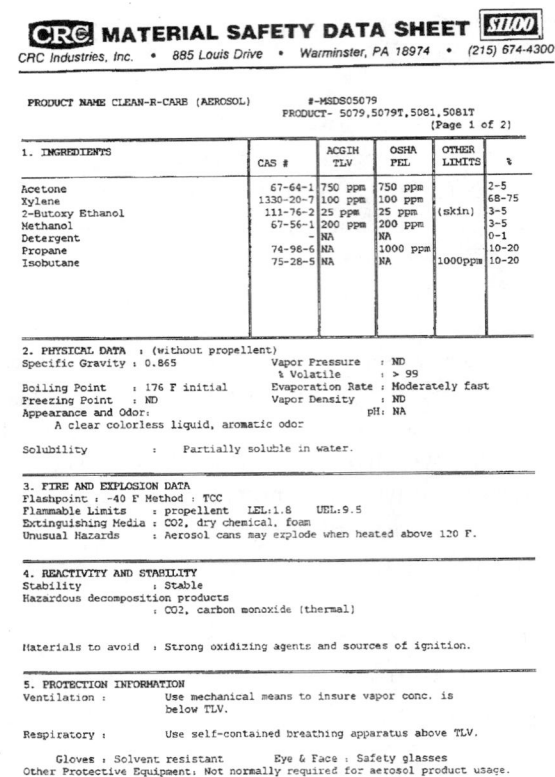

Figure 1-15 MSDS sheet. (Courtesy of Storm Vulcan Mattoni)

The right-to-know laws state that employees have a right to know when the materials they use at work are hazardous. The right-to-know laws started with the Hazard Communication Standard published by the Occupational Safety and Health Administration (OSHA) in 1983.

chemical companies and manufacturers that required employees to handle hazardous materials in their work situation. Now most states have established their own right-to-know laws. Meanwhile, the federal courts have decided to apply these laws to all companies, including truck service shops. Under the right-to-know laws, employers have three responsibilities regarding the handling of hazardous materials by their employees. First, all employees must be trained about the types of hazardous materials they will encounter in the workplace. Employees must be informed about their rights under legislation regarding the handling of hazardous materials. All hazardous materials must be properly labeled, and information about each hazardous material must be posted on **material safety data sheets (MSDS)** available from the manufacturer (Figure 1-15). In Canada, MSDS sheets are called Workplace Hazardous Materials Information Systems (WHMIS).

The employer has a responsibility to place MSDS sheets where they are easily accessible by all employees. The MSDS sheets provide extensive information about the hazardous material, such as:

1. Chemical name
2. Physical characteristics
3. Protective equipment required for handling
4. Explosion and fire hazards
5. Other incompatible materials
6. Health hazards, such as signs and symptoms of exposure
7. Medical conditions aggravated by exposure, and emergency and first-aid procedures
8. Safe handling procedures
9. Spill and leak procedures

Second, the employer must make sure that all hazardous materials are properly labeled. The label information must include health, fire, and reactivity hazards posed by the material and the protective equipment necessary to handle the material. The manufacturer must supply all warning and precautionary information about hazardous materials, and this information must be read and understood by the employee before handling the material. Third, employers are responsible for maintaining permanent files regarding hazardous materials. These files must include information on hazardous materials in the shop, proof of employee training programs, and information about accidents such as spills or leaks of hazardous materials. The employer's files must also include proof that employees' requests for hazardous material information, such as MSDS sheets, have been met. The employer must maintain a general right-to-know compliance procedure manual.

Antifreeze and Solvent Handling

Used antifreeze and spent solvent are hazardous waste materials. These materials may be stored on site in durable, labeled containers and then transported to an approved waste disposal site. A technician may use an antifreeze recycler to recycle antifreeze (Figure 1-16). After the antifreeze is recycled, a small amount of a rust inhibitor additive is added to the proper additive balance.

It is the responsibility of the repair shop to determine if their spent solvent is hazardous waste. Waste solvents that are considered hazardous waste have a flash point below 140° F (60° C). Hot water or aqueous parts cleaners may be used to avoid disposing of spent solvent as hazardous waste. Solvent-type parts cleaners with filters are available to greatly extend solvent life and reduce spent solvent disposal costs (Figure 1-17). Solvent reclaimers are available that clean and restore the solvent so it lasts indefinitely (Figure 1-18).

Figure 1-16 Antifreeze recycler. (Courtesy of Kleer-Flo Company, Eden Prairie, MN)

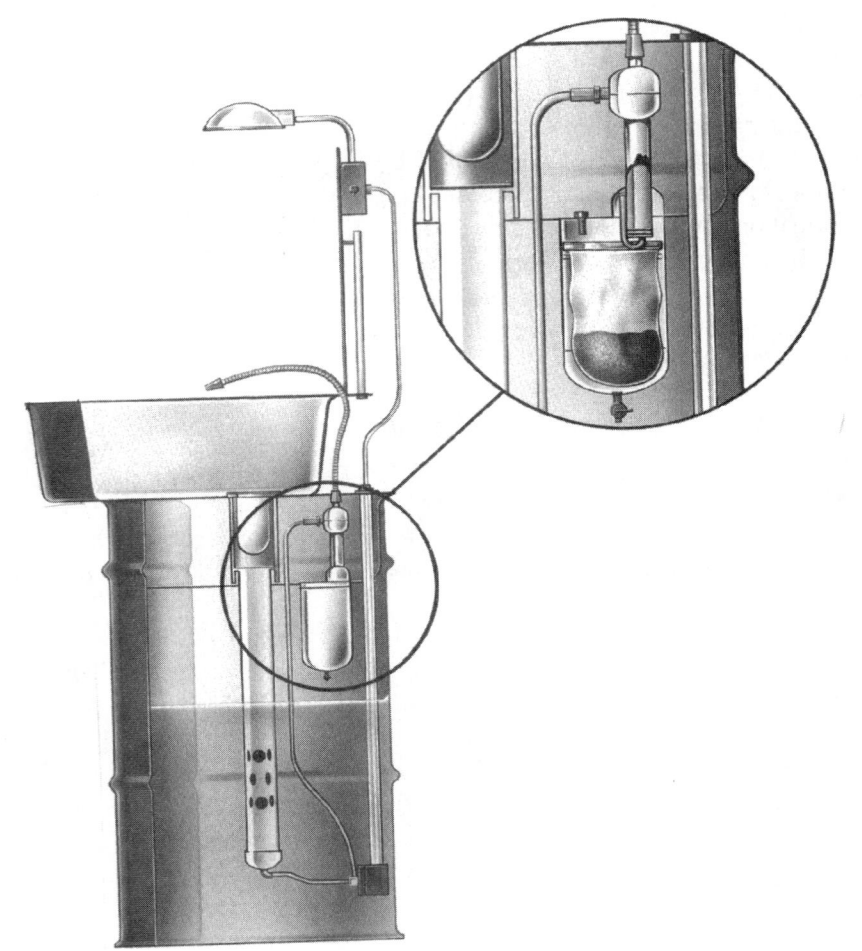

Figure 1-17 Solvent parts cleaner. (Courtesy of Kleer-Flo Company, Eden Prairie, MN)

Figure 1-18 Solvent reclaimer. (Courtesy of Kleer-Flo Company, Eden Prairie, MN)

Summary

❏ The United States Occupational Safety and Health Act of 1970 ensured safe and healthful working conditions and authorized enforcement of safety standards.

❏ Many hazardous materials and conditions can exist in a truck shop, including flammable liquids and materials, corrosive acid solutions, loose sewer covers, caustic liquids, high-pressure air, frayed electric cords, hazardous waste materials, carbon monoxide, improper clothing, harmful vapors, high noise levels, and spills on shop floors.

❏ Material safety data sheets (MSDS) and workplace hazardous material information systems (WHMIS) provide information regarding hazardous materials, labeling, and handling.

❏ The danger regarding hazardous conditions and materials can be avoided by applying the necessary safety precautions. These precautions include all areas of safety such as personal safety, gasoline and diesel fuel handling safety, housekeeping safety, general shop safety, fire safety, and hazardous waste handling safety.

❏ The truck shop must supply required shop safety equipment, and all shop personnel must be familiar with the location and operation of this equipment.

❏ Shop safety equipment includes gasoline and diesel fuel safety cans, steel storage cabinets, combustible material containers, fire extinguishers, eyewash fountains, safety glasses and face shields, first-aid kits, and hazardous waste disposal containers.

Terms to Know

Corrosive

Halogen and halon fire extinguishers

Hazard Communication Standard

Hazardous material

Ignitable

Material Safety Data Sheets (MSDS)

Multipurpose dry chemical fire extinguisher

Occupational Safety and Health Administration (OSHA)

Radioactive

Reactive

Right-to-know laws

Safety glasses

Toxic

Review Questions

Short Answer Essays

1. Explain the purposes of the Occupational Safety and Health Act.

2. Identify twelve shop hazards and explain why each hazard is dangerous.

3. Describe five steps that are necessary for personal protection in the truck shop.

4. Explain why smoking is dangerous in the shop.

5. Describe the danger in drug or alcohol use in the shop.

6. Explain three safety precautions related to electrical safety in the shop.

7. Define six essential safety precautions regarding gasoline handling.

8. Describe five steps required to provide housekeeping safety in the shop.

9. List twenty rules related to general shop safety.

10. Describe how to put out a fire with a fire extinguisher.

Fill-in-the-Blanks

1. A poisonous gas in vehicle exhaust is _____.

2. Heavy-duty boots with _____ steel toecaps are best for working in the shop.

3. One gallon of gasoline has a force equal to_____ sticks of dynamite.

4. Breathing asbestos dust may cause _____.

5. Class C fires involve the burning of _____equipment.

6. Irradiation eye burns may be caused by excessive light from an _____.

7. Hazardous wastes in a truck shop include _____.

8. The right-to-know laws state that employees have a right to know when the materials they handle at work are _____.

9. Material safety data sheets (MSDS) supply specific information regarding _____.

10. Hazardous materials must never be dumped in or _____.

ASE-Style Review Questions

1. *Technician A* says breathing asbestos dust may cause high blood pressure. *Technician B* says oily rags should be stored in an OSHA approved container to avoid a fire hazard. Who is correct?
   A. A only
   B. B only
   C. Both A and B
   D. Neither A nor B

2. Two technicians are discussing shop hazards. *Technician A* says high-pressure air from an air gun may penetrate the skin. *Technician B* says air in the bloodstream may be fatal. Who is correct?
   A. A only
   B. B only
   C. Both A and B
   D. Neither A nor B

3. A technician is lifting a heavy object. *Technician A* says to bend your back while grasping the object. *Technician B* says to place your feet as close as possible to the object. Who is correct?
   A. A only
   B. B only
   C. Both A and B
   D. Neither A nor B

4. *Technician A* says that Class B fires are extinguishable using water. *Technician B* says that Class D fires are extinguishable using a dry chemical extinguisher. Who is correct?
   A. A only
   B. B only
   C. Both A and B
   D. Neither A nor B

5. Two technicians are discussing hazardous waste disposal. *Technician A* says the right-to-know laws require employers to train employees regarding hazardous waste materials. *Technician B* says the right-to-know laws do not require employers to keep permanent records regarding hazardous waste materials. Who is correct?
   A. A only
   B. B only
   C. Both A and B
   D. Neither A nor B

6. Two technicians are discussing hazardous waste disposal. *Technician A* says the right-to-know laws require employers to train employees regarding hazardous waste materials. *Technician B* says the right-to-know laws do not require employers to inform employees of their rights regarding hazardous waste materials. Who is correct?
   A. A only
   B. B only
   C. Both A and B
   D. Neither A nor B

7. *Technician A* says a solid that ignites spontaneously is considered a hazardous material. *Technician B* says a liquid with a flash point above 140° F (60° C) is considered a hazardous material. Who is correct?
   A. A only
   B. B only
   C. Both A and B
   D. Neither A nor B

8. *Technician A* says certain types of hazardous waste react violently with water. *Technician B* says hazardous waste users are responsible for hazardous waste materials from the time they become waste until the proper waste disposal is completed. Who is correct?
   A. A only
   B. B only
   C. Both A and B
   D. Neither A nor B

9. When discussing electrical, gasoline, and diesel fuel safety, which of these statements is correct?
   A. A two-prong adapter may be used to plug in a three-prong electrical cord.
   B. Electric cords from lights and electrical equipment do not require a ground connection.
   C. When stored in the shop, gasoline or diesel fuel containers should be completely full.
   D. Gasoline containers must be secured when they are transported.

10. Two technicians are discussing personal safety. *Technician A* says that if the wrong tool is used, it may slip and cause hand injury. *Technician B* says it is permissible to smoke while working on a truck. Who is correct?
   A. A only
   B. B only
   C. Both A and B
   D. Neither A nor B

Principles of Operation

Upon completion and review of this chapter, you should be able to:

❑ Define basic engine components.

❑ Define *bore* and *stroke,* and calculate displacement.

❑ Define and calculate *compression ratio.*

❑ Explain the following performance factors: speed, work, torque, and horsepower.

❑ Differentiate among the different types of horsepower.

❑ Calculate torque and horsepower for a specific diesel engine.

❑ Explain the difference between torque and torque rise.

❑ Differentiate between volumetric and mass efficiency.

❑ Define the different engine classifications.

❑ Differentiate between open combustion chambers and prechambers.

❑ Explain the two- and four-stroke cycles of operation.

❑ Identify and define the three periods of diesel combustion.

❑ Explain power and speed control in a diesel engine.

Introduction

Diesel engines are typically more economical than gasoline engines and normally get more miles per gallon of fuel than a gasoline engine of similar size. However, they require heavier components because of the higher compression ratio. Historically, diesel engines accelerated more slowly, had a narrower **speed** range, and were more expensive to manufacture due to their heavy components and fuel injection requirements. They are somewhat noisier than gasoline engines, especially during idle, and they require more precise machining of some components.

Engineering improvements have reduced the disadvantages of the diesel in recent years; these improvements and the need for greater fuel economy are helping to increase the demand for diesel engines. The development of special alloys has reduced the difference in weight between diesel and gas engines. Engineering advances have improved acceleration and widened the speed range. High fuel savings are now offsetting more expensive engine construction. The noise level, although still higher than in gas engines, has been reduced.

The diesel engine operates on a different principle than the gas engine, because fuel is not mixed with air entering the cylinder during the intake **stroke.** Air alone is compressed during the compression stroke and the diesel fuel is injected or sprayed into the cylinder at the end of the compression stroke. The compression ratio is much higher and provides compressed air temperatures as high as 1,000° F. The fuel temperature at this instant is high enough to ignite spontaneously, when the injector sprays or injects it into the cylinder.

Combustion of the fuel is controlled by the speed at which the fuel is injected into the cylinder. Thus, in the diesel engine, the combustion is not a rapid burning of the fuel already present in the cylinder, as in a gas engine, but a slower burning that produces an even increase in pressure. The compression-ignited diesel engine has a superior way of controlling load. Because the diesel operates with a stratified fuel–air mixture in the cylinder, combustion occurs as the fuel mixes with the air. An inlet throttle is unnecessary.

Diesel engines reduce load at a given speed by injecting less fuel into an essentially constant mass of cylinder air. The average overall fuel–air ratio in the cylinder becomes very lean at light load and idle. This occurs well beyond the flammability limit of the fuel. This is possible because a major share of the diesel combustion process takes place in the smaller areas, as atomized fuel mixes with compressed cylinder air. The fuel injection pump varies engine timing according to speed. Control of load through variation of the fuel–air ratio benefits efficiency and frees the engine from needing a throttle.

Diesels provide a thermodynamic advantage. The average specific heat of the cylinder gas is lowered at part load during combustion and expansion, because of both the leaner fuel–air ratio and the resulting lower average temperature. More of the heat value of the fuel in the burning process goes to heat the air, with less lost to the cooling system and exhaust. This reduction in heat loss serves to increase the work realized from a unit of fuel.

Differences Between Diesel and Gasoline Engines

A gasoline (spark ignited) engine possesses greater explosive power than does a diesel engine. The gasoline engine can produce high output power and revolutions per minute (rpm) despite its smaller size. It also operates with less noise and has a broader speed range.

Under full-load running conditions, the diesel engine has effective mixture strength of 85% of the chemically correct mixture with enough air for complete combustion. The gasoline engine possesses a mixture equivalent of 120% of the chemically correct value and, therefore, cannot produce complete combustion, because the mixture is 20% too rich. The resulting power is higher due to the complete combustion of the hydrogen part of the hydrocarbon fuel, which has a higher calorific value than the carbon part. Table 2-1 shows the difference between the calorific values of hydrogen, carbon to carbon dioxide and carbon to carbon monoxide.

Under part-load conditions, the efficiency relationship is more favorable to diesel fuel since the fuel quantity is merely reduced, producing weaker mixtures. The mixture strength of the gasoline engine remains fairly constant at a stoichiometric (chemically correct) rate of 14.7 parts air to 1 part fuel. The term *stoichiometric* defines the relationship between the relative quantities of substances involved in a reaction. In motive power technology, managing a stoichiometric burn ratio means controlling fuelling, so that the air in the engine cylinder is precisely managed to completely oxidize the fuel. A more comprehensive definition of stoichiometric is the actual ratio of the reactants in any reaction (not necessarily just a combustion reaction) to the exact ratios required to complete the reaction. Although the heat released during the two combustion periods is similar, the temperature reached in the less dense gasoline engine charge is much higher. Subsequent heat losses in a gasoline engine during expansion are proportionately much greater than the diesel engine. The diesel operates over a mixture strength (fuel–air ratio) range of 0 to 85%, where the gasoline-fueled, spark-ignition engine mixtures vary from 85% to 120%.

Efficiency

Why is the compression-ignition engine more efficient than the spark-ignition gasoline fueled engine? The heat balance indicates that a major part of the efficiency credit in the diesel engine comes from lower exhaust and coolant losses. These two factors result from the fuller use of heat released during combustion and expansion, or power stroke. The higher expansion ratio of the diesel engine ensures that at the end of the power stroke, the gases will have cooled to a lower temperature and will have released more heat. The slower burning rate of the diesel fuel resulting from controlled fuel injection also ensures that the cylinder pressure distribution occurs at a more favorable crank angle. Table 2-2 summarizes the various diesel engine qualities.

Table 2-3 shows a detailed summary comparison between the gasoline and diesel engines.

Table 2-1 Calorific table

Hydrogen = 61, 500 Btu per Pound
Carbon (to CO_2) = 14,600 Btu per Pound
Carbon (to CO) = 4,400 Btu per Pound

Thermodynamics is the study of procedures in which some material goes through a cyclic process and one form of energy, such as heat, is in part converted to another form of energy such as the mechanical energy of a shaft.

Calorific refers to the quantity of heat released on the complete combustion of a unit weight or volume of fuel.

British thermal unit (Btu) is a common heat measure.

Stoichiometric is the actual ratio of the reactants in any reaction (not necessarily just a combustion reaction) to the exact ratios required to complete the reaction.

Table 2-2 Diesel efficiencies

Higher thermal efficiency
Lower fuel consumption
Superior way of controlling load
Lower specific heat at part throttle
Lower average charge temperature
Increased work (torque) from a unit of fuel
Diesel fuel has 11% more energy per unit of
  volume than gasoline
	No throttle to restrict air intake.
	No negative pumping due to a closed throttle valve
	Compress only air
	Possess higher compression ratios.
	No spark ignition system.
	Diesels inject fuel directly into combustion chamber.
	Diesels control power and speed by amount of fuel
	  injected.

Table 2-3 Comparison of gas and diesel engines

Phase	Gasoline Engine	Diesel Engine
Intake Compression	Draws in air–fuel mixture Because they compress an air–fuel mixture, gas engines are limited to about 8:1 to 10.5:1 compression ratio, resulting in about 130 psi (895 kPa) compression pressure and a temperature of about 280° C (540° F).	Draws in only air Because they compress only air, diesels are limited by metallurgical, air–fuel mixing, and engine-life considerations. Therefore, ratios of 16:1 to 23:1 are common, resulting in compression pressures of 400 to 600 psi (3,080 to 4,135 kPa) and temperatures of about 1,000° F (540° C).
Ignition	They use an electrical spark to ignite the fuel–air mixture.	Diesels use the heat of compression to auto-ignite the mixture. This provides a faster and more complete combustion, and a much greater heat rise
Power	Maximum pressure during combustion is limited by lower compression pressures and lower heat value of the fuel to about 450 psi (3,100 kPa)	Mixing air & fuel and the metallurgical limits of the engine limit diesels. The maximum pressure is about 1,200 psi (8,275 kPa)
Exhaust	Exhaust temperature ranges from 1,300° to 1800° F (700° C − 980° C), because burning is still taking place in the exhaust manifold. This also results in higher levels of carbon monoxide (CO).	Diesel exhaust temperatures are much lower because the combustion is more complete.
Thermal efficiency	21% to 28%	32% to 38% higher than a gas engine due to the combination of high compression and combustion temperatures. The atomized injection of a correctly metered amount of fuel results in much more complete burning and greater thermal efficiency.

Process and Fuel Control Differences

1. The first difference involves the intake stroke. The fuel–air mixture is not homogeneous. During the intake stroke the engine takes in only air. No throttle exists, so the cylinder is completely filled with air at the inlet manifold pressures. The air mixes with any residual gas In the cylinder. Fuel is added near the end of the compression stroke by an injector (nozzle) mounted in the **prechamber.** The fuel injector sprays and atomizes the fuel into the chamber in a short burst (about 20°–30° of rotation), forming a fuel-rich core surrounded by air zones.

2. The second difference is that when you change the amount of fuel injected you control power and speed.

3. The third difference is that ignition in a diesel engine occurs without a spark. Once fuel is sprayed into the chamber, it starts to vaporize and mix with the hot compressed air. After 0.001 second, any zones that are hot enough and have the correct fuel–air mixture ratio will ignite. Since ignition occurs shortly after the fuel injection process starts (often before it's finished) fuel cannot be injected until near the end of the compression stroke or the engine will experience heavy knock and poor performance.

4. The fourth difference between gas and diesel engines is that the flame does not spread through the combustion chamber as in a gasoline engine, because the core of fuel produced by the injector is too rich to burn and only air exists in other parts of the chamber. Thus combustion occurs only at the interface where fuel and air come together (called a *diffusion flame*). The time required for ignition to occur is about 1/4 to 1/10 of the time required to mix the fuel with the air. Therefore, for diesel engines, the rate of combustion is controlled by how rapidly the fuel is mixed with the air. Once the engine burns the first fuel, a rich core of fuel remains along with zones of air. The remaining fuel is burned as it mixes with the air.

BIT OF HISTORY

The term *diesel* originates from Rudolf Diesel, shown in Figure 2-1, who is widely acknowledged as the inventor of the engine that bears his name. Dr. Diesel produced the first diesel, or compression-ignition, engine in 1893. Many have credited other contemporary engineers with the diesel invention, and certain aspects have been the subjects of much acrimonious controversy. However, it is certain that Dr. Diesel was the first engineer to realize the full thermodynamic potential of the compression ignition principle and of the cycle now known as the *diesel cycle*. Rudolf Diesel published his theories in 1892 in a work titled, "The Theory and Construction of an Economical Thermal Motor."

In 1892, the spark ignition engine running on the *4-stroke (Otto) cycle* was limited to compression ratios of around 4:1 because of low-grade fuels and the preignition, or more correctly,

Figure 2-1 Rudolf Diesel.

detonation problems associated with low ignition temperatures. Since compression ratio limits the **thermal efficiency** of the engine, a substantial improvement was only possible if air alone entered during induction. The inventor therefore proposed that this air be compressed to a temperature at which subsequent fuel introduction would mean spontaneous ignition. The rate of fuel injection would then control the rate of combustion in the cylinder. The inventor hoped to achieve a constant combustion pressure during the main part of the power stroke, thus reproducing the smooth torque characteristic of the steam engine. The superior efficiency of the internal combustion engine would then ensure wide appeal in applications served by steam power.

Diesels never achieve a true constant pressure cycle, but the efficiency of the diesel engine was higher than any engine of the time and is still easily the most economical heat motor known to man. Alkroyd Stuart, a prominent English engineer, did produce an invention known as the "hot bulb engine" that was very similar to the diesel version and, in fact, existed before the first true diesel engine. Stuart sought to overcome pre-ignition and his practical experiments provided for an initial external ignition source. It was subsequently found however that with the compression temperatures involved, the ignition was superfluous, and that once the engine was running, combustion became spontaneous.

Basic Diesel Engine Components

Cylinder Block The cylinder block (Figure 2-2A) is a complex structure that contains the following: sleeves (liners), which house the pistons and rods; cooling passages; lubrication passages; in-block camshaft and bearings; and crankshaft bearing supports in the bottom end. It can have one-piece cast-in type cylinders, (sometimes called a parent-bore block) or inserted liners or sleeves.

Crankshaft and Main Bearings The crankshaft (Figure 2-2A) of the engine takes the pressure of combusted fuel–air generated on top of the piston and transmits it through the connecting rod to the crankshaft throw and into output. It converts power into rotary motion and conveys this power to the outside. The main bearings support the tremendous pressure caused by these actions to the crankshaft.

Cylinder Liners (Sleeves) Most large-bore diesel engine use replaceable liners (sleeves) (Figure 2-2A), so that if a cylinder bore becomes worn, it can be replaced without having to rebore or replace the main cylinder block. The liners (sleeves) come in three types:

- ❏ Dry-type: A sleeve fitted into a bored hole in the block with or without a flange at the top, which contains no O-rings or other sealing devices.
- ❏ Wet-type: A cylinder liner that fits into the block and comes in contact with the coolant. Because coolant circulates around it, the line uses seals at the top and bottom. Sealing is done on the top end by fitting the liner into a counterbore cut into the block. O-Rings made of oil- and water-resistant neoprene are fitted to the bottom of the sleeves and prevent the coolant water from leaking into the crankcase or oil pan.
- ❏ Semi wet/dry: The upper half of the liner is like a dry-type and the lower half is like a wet-type liner.

Piston, Rings, and Connecting Rod The function of the piston and the rings that are fitted in grooves on the piston is to transmit pressure from the burning fuel and air to the connecting rod

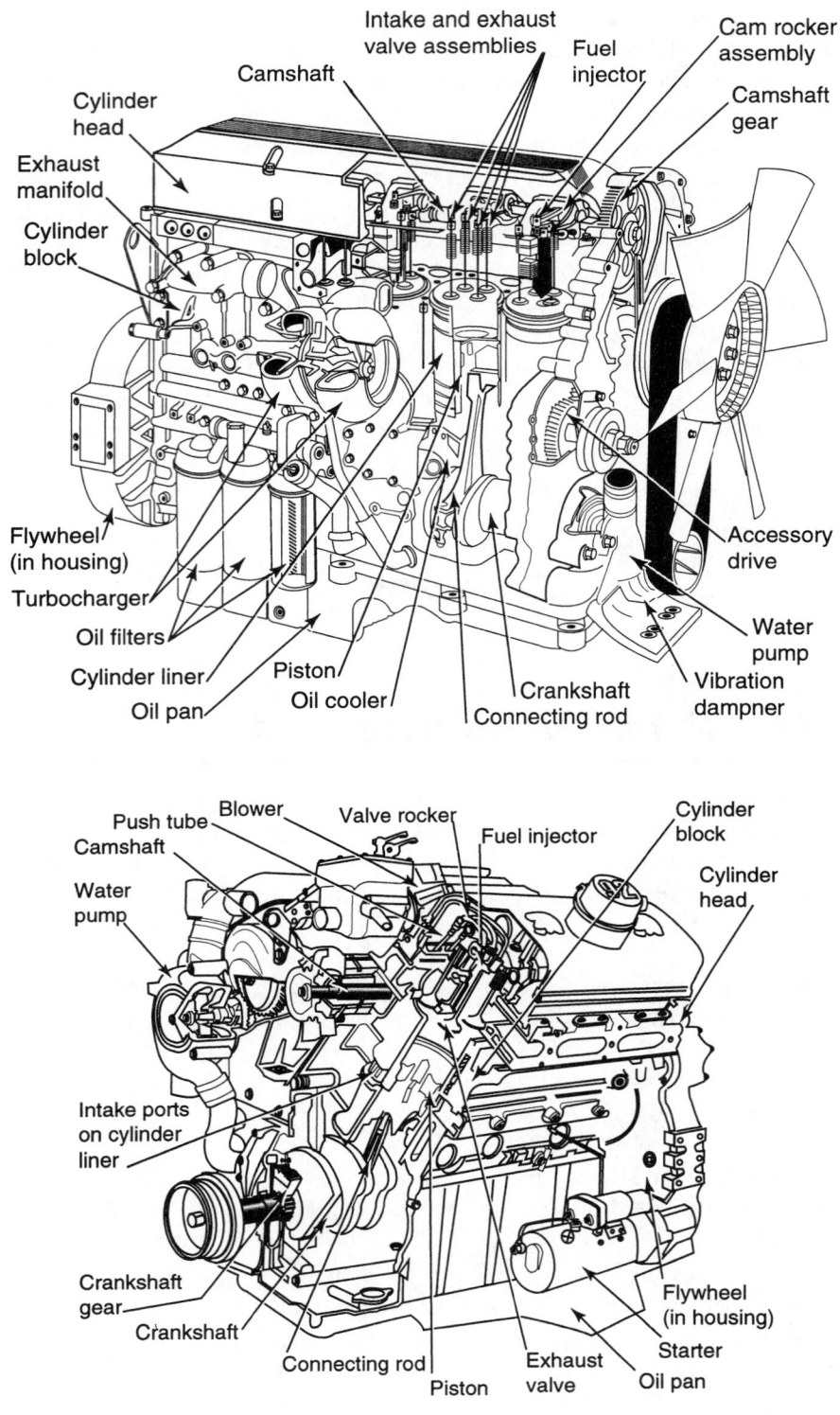

Figure 2-2 Diesel engine, exploded view. (Reprinted by permission of copyright owner Detroit Diesel Corporation, all rights reserved.)

that is connected to the crankshaft (Figure 2-2A). The connecting rod's function, as the name implies, is to connect the piston to the crankshaft. Holding the piston and connecting rod together is the piston pin, usually a full-floating type, which means that the pin floats in both the piston and the rod.

Camshaft and Timing Gears The camshaft (Figure 2-2A) in a diesel engine operates the intake and exhaust valve, and in some engines may drive the oil pump and/or injection pump. The camshaft is timed to the crankshaft by a timing gear or camshaft gear that is meshed into a gear on the front of the crankshaft. This drives the camshaft and ensures that the engine valves will stay in time with the crankshaft and pistons.

Cam Followers The cam followers (Figure 2-2A/B) (sometimes called *lifters* or *tappets*) are mounted in drilled holes in the block and ride on the cam lobes. Inserted into the cam followers are long rods or hollow tubes called *push rods,* which operate the valves. (These are not used on overhead camshaft engines).

Cylinder Head and Valves The cylinder head's (Figure 2-2A) main function is to provide a cap for the cylinder. In addition, it provides a passageway that allows air into the cylinder and allows exhaust gases to pass out. The ports are opened and closed by poppet-type valves that fit into guides in the cylinder head.

Rocker Arms and Push Rods Rocker arms (Figure 2-2A/B) are mounted on a shaft with one end on the valves and the other end on a push rod. Movement of the push rod causes the arm to rock on its pivot shaft, hence the name *rocker arm*. Push rods are solid or hollow rods that fit into the cam followers and transmit the cam action of the camshaft.

Oil Pan This is a pan-shaped cover that bolts onto the bottom of the block and acts as a reservoir for the engine oil (Figure 2-2).

Lubricating Oil Pump This is generally a positive displacement-type gear (a pump that delivers a given quantity of oil every revolution) (Figure 2-2A). It supplies oil under pressure to the engine.

Water Pump This is a non-positive displacement centrifugal pump (a pump that does not deliver a given amount of water every revolution) (Figure 2-2A). It aids the flow of coolant water through the engine block and radiator. The pump is generally mounted on the front of the engine block, driven by a V-belt or multi-groove belt from the crankshaft, and connected in the cooling system with rubber hoses.

Radiator This is a device designed to allow water to flow through it, thereby cooling the water by radiation.

Oil Cooler This is a device used to cool the engine oil during engine operation. The construction of the cooler allows coolant water and engine oil to circulate through the cooler simultaneously without being mixed. Heat is removed by convection between the hotter oil and the engine coolant.

Flywheel (housing) This is a circular housing bolted to the back of the engine that serves as an engine and transmission mount (Figure 2-2A). Enclosed in the flywheel housing is the flywheel. A single cylinder of an automotive engine provides only one power impulse for every two revolutions of the crankshaft, and thereby delivers power only one-fourth of the time. To provide a more continuous flow of power, modern engines use four, six, or eight cylinders. The cranks on the crankshaft of a six-cylinder automotive engine are set 120 degrees apart, and the firing order of the cylinders is 1–5–3–6–2–4. Thus three power strokes occur for every revolution of the crankshaft, and there is a fairly continuous and uniform delivery of power to the crankshaft. However, this power delivery can be improved further to produce smoother engine performance. For this purpose, a flywheel is attached to the output end of the engine to make the engine run with minimal variation in speed. The principal characteristic of the flywheel is its inertia—its ability to resist and moderate any fluctuation in speed (in short, to steady the speed). The speed of the flywheel increases during the power stroke of the engine and decreases as it delivers power to the engine during the exhaust, intake, and compression stroke. The flywheel, because of its opposition to a change in speed, prevents the engine from speeding up during the power stroke and similarly from slowing down during the powerless phase.

Torsional Vibration Damper (Harmonic Balancer) The main function of the vibration damper is to dampen out torsional twisting vibration that occurs when the engine runs. The damper is mounted on the front of the crankshaft.

Intake Manifold This component is bolted to the cylinder head or intake port. The intake manifold provides passageways for the passage of clean air from the air cleaner to the engine (Figure 2-2A).

Exhaust Manifold Connected to the engine exhaust outlets, the exhaust manifold provides a means for collecting the exhaust gases and routing them to the muffler (Figure 2-2A).

Fuel System The fuel system delivers the correct amount of fuel to the engine cylinders at the correct time, depending on the engine load and speed.

Starter This is an auxiliary motor used to crank the engine for starting. It may be electric, air, or hydraulically powered.

Turbocharger An exhaust-driven air pump that supplies air to the engine under pressure (Figure 2-2A). This device increases engine efficiency.

Roots Blower (Air Box) The roots blower (Figure 2-2B) is a gear-driven, positive displacement air pump used to scavenge the engine cylinders in Detroit Diesel Corporation (DDC) two-stroke cycle engines. It produces relatively low peak pressures compared to those produced by turbochargers. As a scavenging component, the primary function of the Roots blower is to displace end gases from the engine cylinders. It may also produce a small amount of manifold boost. All DDC 2-stroke cycle engines used in automotive applications (that is, highway trucks and buses) use Roots blowers for cylinder scavenging. These engines may also use a turbocharger to produce manifold boost. Under conditions of high engine loading and therefore peak turbocharger efficiency, the turbocharger may be capable of performing cylinder scavenging.

Engine Fundamentals

Bore

Bore (Figure 2-3) is the diameter of the cylinder. If the bore is larger than the stroke, the engine is said to be oversquare, and most of its power is dependent upon revolutions per minute (rpm) and is generated at higher rpm values. When the stroke is larger than the bore, the engine is undersquare, with most of its power dependent upon torque.

Stroke

Stroke is the movement of the piston from top to bottom, or the distance the piston travels. In Figure 2-4, when the piston is at the top of its stroke, it is said to be at top dead center, or TDC. At TDC the crank is at a point nearest the cylinder. When the piston is at the bottom of its stroke,

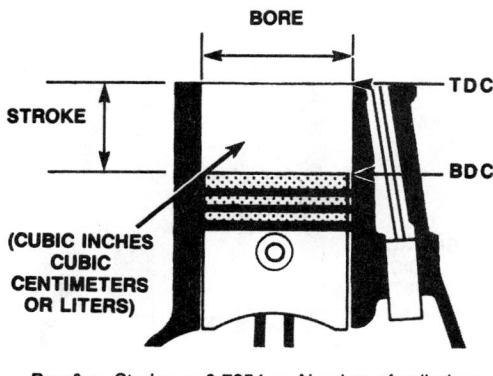

Bore² × Stroke × 0.7854 × Number of cylinders

Figure 2-3 Piston bore.

it is said to be at bottom dead center, or BDC. At BDC the crank is at a point farthest from the cylinder. The space between the piston and the cylinder head when the piston is at TDC is the *clearance space*. It is sometimes called the *compression* or *combustion space*. The volume of this space is known as the *clearance volume*.

❏ **TDC: Top Dead Center** The uppermost point of the piston travel in the engine cylinder.

❏ **BDC: Bottom Dead Center** The lowest point of piston travel in the engine cylinder.

❏ **BTDC: Before Top Dead Center** A point of piston travel through its upstroke.

❏ **ATDC: After Top Dead Center** A point of piston travel through its downstroke.

Square Engine Term used to describe an engine in which the cylinder bore diameter is exactly equal to the piston stroke dimension. When bore and stroke values are expressed, bore always appears before stroke.

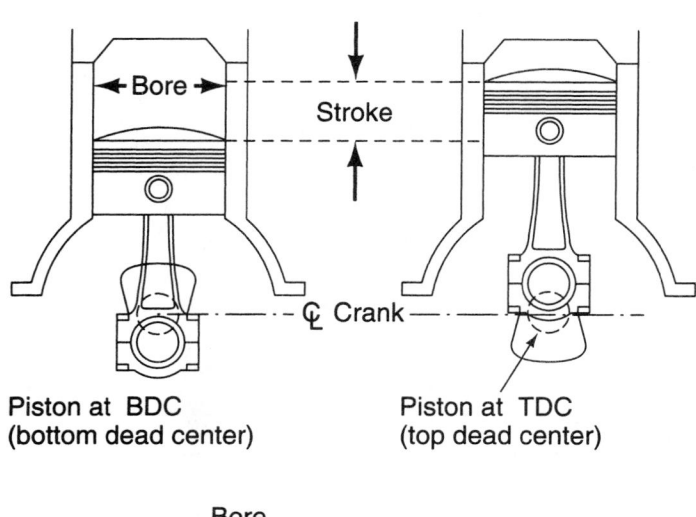

Piston at BDC
(bottom dead center)

Piston at TDC
(top dead center)

$$\frac{\text{Bore}}{\text{Stroke}} = \text{Bore stroke ratio}$$

Figure 2-4 Piston stroke.

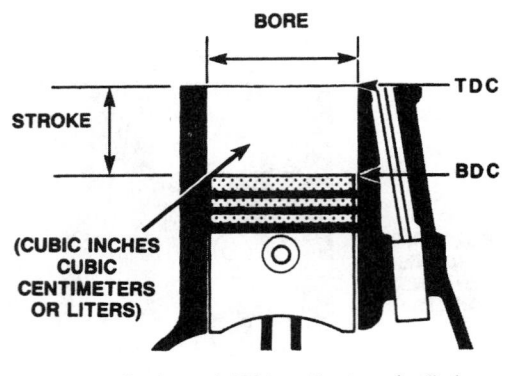

Bore² × Stroke × 0.7854 × Number of cylinders

Figure 2-5 Engine displacement.

Oversquare Engine Term used to describe an engine in which the cylinder bore diameter is larger than the stroke dimension.

Undersquare Engine Term used to describe an engine in which the cylinder bore diameter is smaller than the stroke dimension. Most truck and bus diesel engines are undersquare.

Engine (Piston) Displacement

One states the total (piston) **engine displacement** (Figure 2-5) as the volume in cubic inches or liters (cubic centimeters) of the space swept through by all the pistons of the engine during one complete revolution of the crankshaft. The total piston displacement is easily found by multiplying the displacement of one piston by the number of pistons. The formula is expressed as follows:

$$V_d = .7854d^2 \times s \text{ (height)}$$

d = Bore or diameter of the piston   s = the stroke or piston travel **length**
V = Volume of the cylinder

Sample Problem: A six-cylinder diesel engine has a bore of 5.4 inches and a stroke of 6.5 inches. What is the total piston displacement?
Formula:

$$V_d = .7854 \times (5.4)^2 \times 6.5 = 149 \text{ cubic inches} \times 6 = 894 \text{ in}^3.$$

One can convert these values into metric values as shown in Table 2-4.

Table 2-4 System conversion table

Engine Cubic Inches	Conversion Factor	Answer
149 cubic inches (in³)	Multiply by .01639	2.44 liters
894 cubic inches (in³)	Multiply by .01639	14.7 liters
149 cubic inches (in³)	Multiply by 16.387	2441 cubic centimeters (cm³)
894 cubic inches (in³)	Multiply by 16.387	14,650 cubic centimeters (cm³)

16.387 cubic centimeters (cm³) = 1 cubic inch (in³)
61.024 cubic inches (in³) = 1 liter

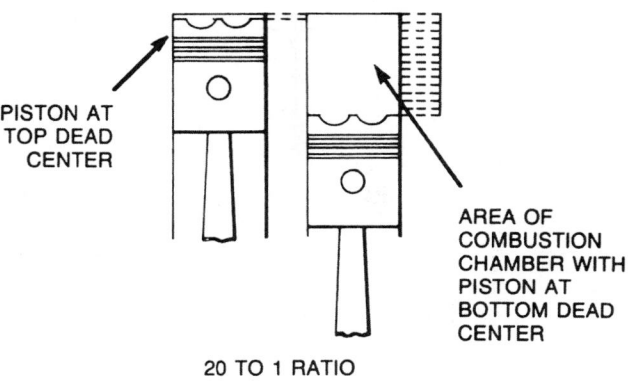

PISTON AT TOP DEAD CENTER

AREA OF COMBUSTION CHAMBER WITH PISTON AT BOTTOM DEAD CENTER

20 TO 1 RATIO

Figure 2-6 Compression ratio.

Compression Ratio

The term **compression ratio** is a misnomer, because it is actually a volume ratio. In an internal combustion engine, compression ratio (Figure 2-6) is the ratio of the total cylinder volume to the clearance volume. It is the volume of the cylinder at the beginning of the compression stroke divided by the volume of the cylinder at the end of the compression stroke. This ratio can be expressed by the following formula:

$$r = \frac{V_d + V_c}{V_c}$$

Performance Factors

The function of the internal combustion engine is to convert chemical energy released from the combustion of fuel into useful mechanical work. Certain performance factors are used to describe how well the engine performs. These are *speed, work, torque,* and *horsepower.*

Speed

In engine testing, **speed** is really angular velocity (rate of change of position at a particular angle) and is defined in crankshaft revolutions per minute (rpm) of time. It is shown in formulas as the letter "N." Speed measurement accuracy depends on the combined accuracy of the revolution count and the time interval of that count. Technicians typically perform speed measurements using a photo tachometer on a mechanical fuel-injected engine, and a scan tool or a laptop computer on an electronically injected engine.

Work

Work (Figure 2-7) is defined as the product of a force and the distance through which the force acts. Simply expressed, work is force (in pounds or Newtons) times distance (the foot-meter), hence the foot-pound or Newton-meter. For an engine and dynamometer, if the speed of the engine is known, the below formula can be expressed as

$$Wk = 2\pi FRN$$

Wk = work, in foot-pounds per revolution $\quad \pi = 3.1416$, a constant
  F = force, in pounds or Newtons $\quad\quad$ R = radius of shaft, in feet or meters

The distance through which the restraining force acts in one shaft revolution is $2\pi R$, in which R is the radius of the shaft and π is 3.1416. The engine turning the dynamometer and producing a

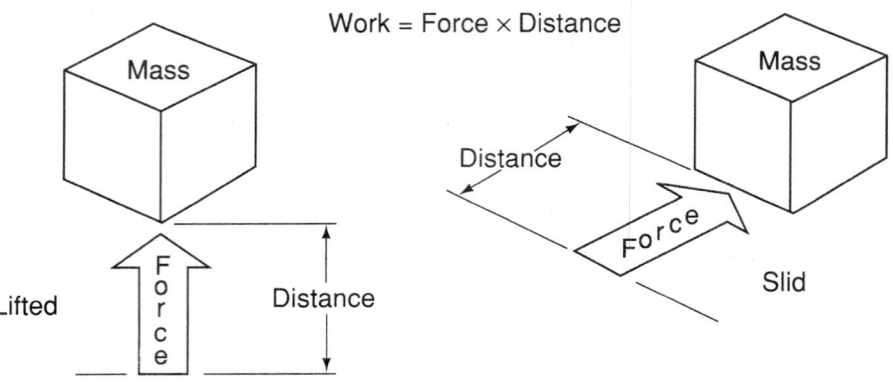

Figure 2-7 Work.

force is comparable to the engine crankshaft standing still and the dynamometer being turned around it with a force, F, acting though a radius arm, R. When the speed is not known, work expressed as a formula and produced per revolution is as follows:

$$Wk = 2\pi FR$$

Wk = work, in foot-pounds per revolution   F = force, in pounds or Newton's
  N = engine speed in RPM   R = radius of shaft, in feet or meters
  π = 3.1416, a constant

Torque

Torque (Figure 2-8) is a force that tries to turn or twist something around, and it is defined as the product of a force and the perpendicular distance between the line of action of the force and the axis of rotation. Torque is force (F) multiplied by length. The length of the lever is the length of the throw of the crankshaft journal. It is a twisting effort expressed as the capacity for the engine to do work, whereas horsepower is defined as the rate at which the engine can do work. Diesel engines produce torque by combustion force pushing down on top of the piston moving a lever that is the throw of the crankshaft. Torque is generally expressed in foot-pounds or Newton-meters (1 ft-lb = 1.355 Newton-meters and .737 ft-lb = 1 Newton-meter). Torque is expressed mathematically in the following formula:

$$T = FR$$

T = torque, in foot-pounds or   F = force, in pounds or Newtons
    Newton-meters   R = radius, or torque-arm distance, in
                                           feet or meters

$$Torque = \frac{hp \times 5{,}252}{rpm}$$

The mathematical constant 5,252 is derived from the basic horsepower formula, and is calculated as follows:

$$6.28 = 2\pi$$
1 shaft revolution = $2\pi R$ (the circumference of circle = $2\pi R$)
33,000 divided by 6.28 = 5,252.

Engine torque can be measured using either a cradle absorption dynamometer or a transmission dynamometer. The cradle absorption dynamometer is capable of absorbing the power output of the engine. The transmission dynamometer measures power transmitted from an engine to some other device, but it does not absorb any of the power. Chassis rolls or inertia wheel dy-

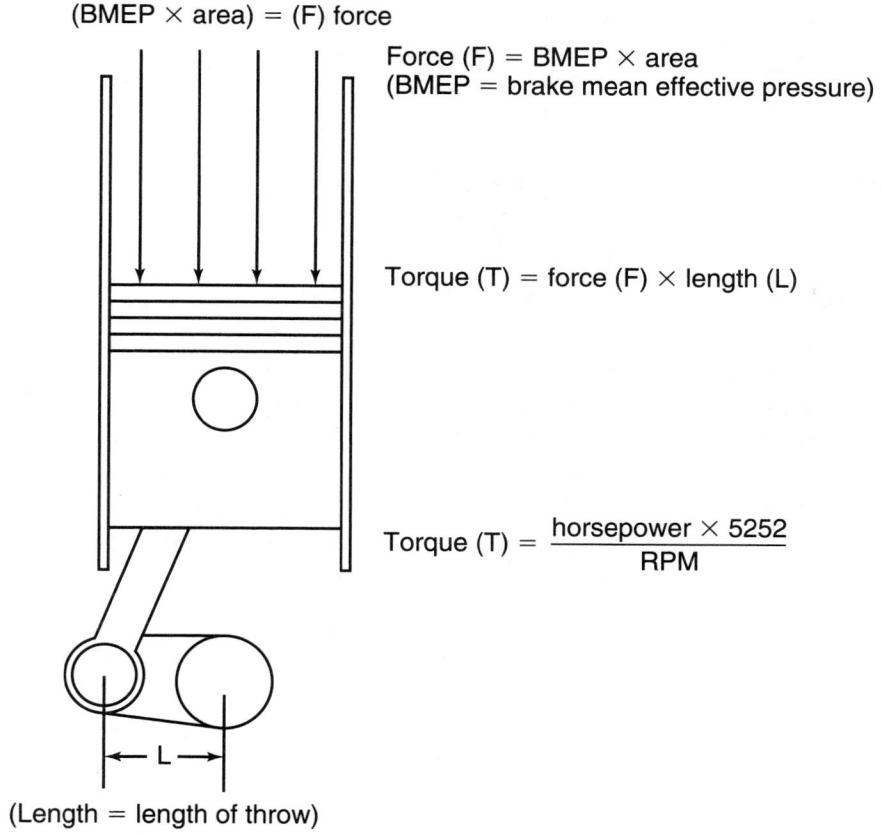

$$(BMEP \times area) = (F) \text{ force}$$

Force (F) = BMEP × area
(BMEP = brake mean effective pressure)

Torque (T) = force (F) × length (L)

$$\text{Torque (T)} = \frac{\text{horsepower} \times 5252}{\text{RPM}}$$

A (Length = length of throw)

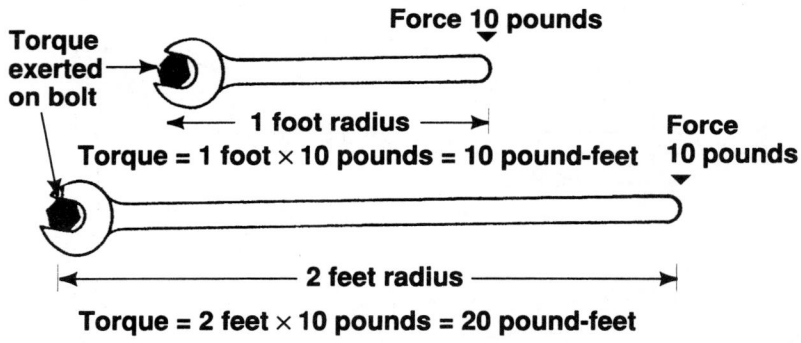

Force 10 pounds

Torque exerted on bolt

← **1 foot radius** →

Torque = 1 foot × 10 pounds = 10 pound-feet

Force 10 pounds

← **2 feet radius** →

B **Torque = 2 feet × 10 pounds = 20 pound-feet**

Figure 2-8 (A) Formula for torque; (B) Force applied to a wrench produces torque; if the bolt turns work happens.

namometers are examples of transmission dynamometers, such as those used for IM240 emission testing.

Torque Rise

Torque rise expresses in a percentage value the increase in engine torque as the engine speed is reduced from its maximum no-load rpm, or rated speed. For example, an engine develops 1000 ft-lb (1356 Nm) of torque at its rated speed of 2100 rpm and this torque increases to 1500 ft-lb (2034 Nm) peak torque at 1200 rpm. When the rpm is reduced to 1200, (known as the *peak torque speed*), the rate of torque rise is equal to 50%. If the 900-rpm drop from the rated peak torque speed divides this 50% torque rise to the peak torque rpm, this engine has developed 5.55% torque rise from every 100-rpm decrease. Torque rise is expressed in the following formula:

$$\text{Torque rise} = \frac{\text{Peak torque} - \text{Rated speed torque}}{\text{Rated speed torque}}$$

$$\text{Torque rise} = \frac{1{,}500 \text{ ft-lb} - 1{,}000 \text{ ft-lb}}{1{,}000 \text{ ft-lb}} = 50\%$$

Horsepower

Horsepower (Figure 2-9) is defined as work per unit of time. It is the rate of doing work or torque. The horsepower is the unit of measurement that expresses the power output of an internal combustion engine. One horsepower is equivalent to 33,000 ft-lb (foot-pounds) of work done in 1 minute. For example, if a weight of 550 lb is lifted vertically a distance of 60 ft, the work done is 550 × 60 = 33,000 ft-lb; and this is done in 1 minute. One horsepower is expended, since 33,000 ft-lb per minute is equal to 1 hp. In the metric system, horsepower is expressed in kilowatts: 0.746 kw = 1 hp, or 1.341 hp = 1 kw. The horsepower of an engine as measured by a dynamometer is expressed by the formula:

$$\text{hp} = \frac{2\pi \text{FRN}}{33{,}000}$$

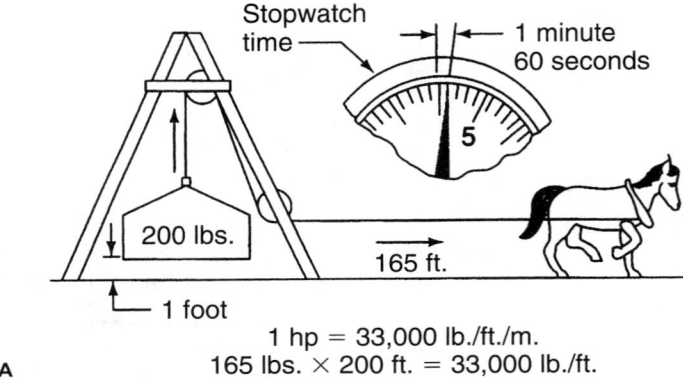

1 hp = 33,000 lb./ft./m.
165 lbs. × 200 ft. = 33,000 lb./ft.

A

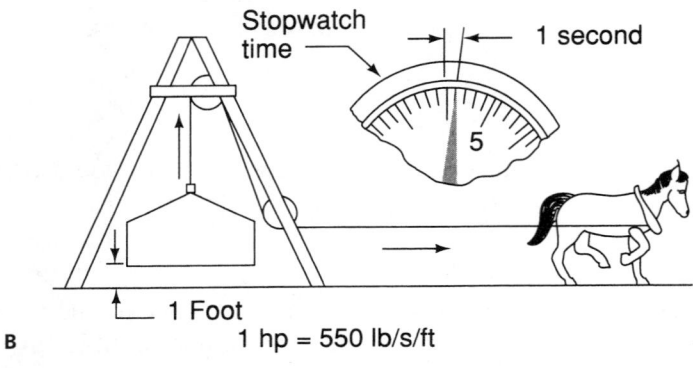

1 hp = 550 lb/s/ft

B

Figure 2-9 (A) One horsepower can do 33,000 foot-pounds of work in 1 minute; (B) One horsepower is produced when 550 pounds are moved 1 foot in 1 second.

hp = horsepower

  F = dynamometer load, in pounds

  R = radius arm of dynamometer, in feet

33,000 = conversion factor, foot-pounds per minute, to produce one hp

N = engine speed, in revolutions per minute (rpm)

π = 3.1416, a constant

In general, work is done whenever a force overcomes a resistance. The horsepower required overcoming resistance, or drag is found by the following formula:

$$hp = \frac{DV}{33,000}$$

D = velocity, in feet per minute   V = drag, or resistance, in pounds

Sample Problem: The speed of a diesel-powered truck is 60 mph, and the total resistance of the wind plus rolling resistance is 125 lb. Determine the horsepower.

Solution: Since there are 5,280 ft in 1 mile and 60 minutes in an hour, the truck is traveling 1 mile per minute, or 5,280 fpm (feet per minute). In 1 minute the diesel engine has to supply 5,280 $\times$ 125 = 660,000 ft-lb of work to overcome the resistance and maintain the 60 mph speed. Since 33,000 ft-ib per minute is equivalent to 1 hp, the horsepower required at the truck rear wheels is obtained by applying the first horsepower formula:

$$hp = \frac{DV}{33,000} = \frac{5,289 \text{ ft} \times 125 \text{ lbs}}{33,000} = \frac{660,000 \text{ ft-lbs}}{33,000} = 20 \text{ hp}$$

Brake Horsepower (BHP)

Brake or shaft horsepower is the power delivered at the shaft of the engine. The term *brake horsepower* comes from the method of testing the early engines. This consisted of putting a mechanical brake on the engine and measuring the force required to bold the brake from turning. The energy produced was dissipated as heat. Water or air was used to cool the friction surfaces of the brake. Brakes of this type, called *Prony brakes,* were an early type of the absorption dynamometer. The term *horsepower* originated as follows: 1 hp is produced when a horse walks 165 feet in 1 minute pulling a 200-lb weight: 165 ft. $\times$ 200 lb. = 33,000 ft-lb.

Brake horsepower can be measured with either a transmission or an absorption dynamometer and is determined by the same formula as the one listed under *horsepower.* Stated for brake horsepower:

$$bhp = \frac{2\pi FRN}{33,000}$$

bhp = Brake horsepower

  F = Dynamometer load, in pounds

  R = Radius arm of dynamometer, in feet

33,000 = Conversion factor, foot-pounds per minute, to produce 1 hp

N = Engine speed, in revolutions per minute (rpm)

π = 3.1416, a constant

t = torque

When applied to a Prony Brake Dynamometer the formula becomes:

$$bhp = \frac{F \times 6.28 \text{ RN}}{33,000} = \frac{FRN}{5,252} = \frac{t \times rpm}{5,252}$$

$$bhp = \frac{t \times rpm}{5,252}$$

t = torque
F = dynamometer load (in pounds)
N = engine speed in rpm

r = effective length of the lever (in feet)
radius arm of the dynamometer

The constant 5,252 is calculated as follows:

$$6.28 = 2\pi$$
$$1 \text{ shaft revolution} = 2\pi R \text{ (the circumference of circle } = 2\pi R)$$
$$33,000 \text{ divided by } 6.28 = 5,252 \text{ (the constant)}$$

Sample Problem: A small diesel engine being tested is coupled to a cradled dynamometer that has a radius arm of 1.75 feet. The test data show a speed of 3,000 RPM and a load of 80 pounds. Calculate the torque and brake horsepower.
Solution:

$$t = FR \ (t = 80 \text{ pounds} \times 1.75 \text{ feet} = 140 \text{ ft-lb.})$$

$$bhp = \frac{2\pi FRN}{33,000} = \frac{2 \times 3.1416 \times 140 \times 3000}{33,000} = 80 \text{ bhp}$$

$$bhp = \frac{t \times rpm}{5,252} = \frac{140 \times 3000}{5,252} = 80 \text{ bhp}$$

The Intermittent Continuous Horsepower rating provides for engines that are primarily continuous yet have some variations in load and or speed.

Society of Automotive Engineers (SAE) Standard Conditions for bhp Rating:

❏ Net horsepower available from the engine with specified injector size and engine speed.
❏ Rating guaranteed within 57%
❏ Corrected to SAE standard ambient conditions.
❏ Air temperature 85° F (29.4° C).
❏ Elevation 500 ft (159.4 m).
❏ Dry air density .0705 lb/cu ft (11.29 grams per cubic meter [g/m³]).
❏ The engine may be without the accessories required for the application (compressor, fan, generator, etc.).

Mean Effective Pressure

Mean effective pressure (MEP) is the hypothetical constant pressure that, if acting on the piston, would produce the same work that the actual varying pressures produce. There are two mean pressure values normally used in engine testing: brake mean effective pressure (BMEP) and indicated mean effective pressure (IMEP).

Indicated Horsepower

Indicated horsepower (IHP) is the power developed inside the engine at the face of the piston. The power delivered by an engine to the generator with which it is connected is always less than the actual power developed in the cylinders. This is because some of the indicated horsepower is used to overcome the fiction of the moving parts of the engine. Indicated horsepower, therefore, is equal to the sum of the brake horsepower plus the friction horsepower (FHP). By means of a dynamic pressure transducer and an oscilloscope, it is possible to obtain a pressure-time (pt) diagram for an engine. From this diagram, with the stroke and connecting rod length known, it is possible to construct a pressure-volume (pV) diagram. The pV diagram represents the pressures related to piston position that exist in the cylinder throughout a complete crankshaft revolution.

The area of the pV diagram is proportional to the energy developed in the cylinder. The area of the diagram can be measured in several ways, and is divided by the length of the diagram to obtain the mean or average height of the diagram. As height on the diagram represents

Shaft Horsepower is the net horsepower available at the output shaft of an application, e.g., horsepower measured at the output flange of a marine gearbox.

pressure in the cylinder, multiplying the mean height by the scale of the spring used in the indicator mechanism gives the average, or mean, effective pressure acting in the cylinder during the revolution in which the diagram was made. Once the IMEP is known, you can determine the indicated horsepower (IHP), which is the power developed at the piston head. Other factors that must be known are as follows:

$$IHP = \frac{PiLANK}{33,000}$$

Pi = indicated mean effective pressure in PSI

L = length of stroke in feet

A = area of piston in square inches.

N = number of power strokes per minute.

K = number of cylinders

Friction Horsepower

Friction horsepower represents the power needed to overcome the friction of the crankshaft and camshaft in their bearings, the friction of the pistons in the cylinders, the friction of the gearing, and the power used to drive the pumps. Friction horsepower is the difference between the indicated horsepower and the brake horsepower, and it can be approximated by using a pV diagram with no load on the engine. If there is no load on the engine, the only resistance is that caused by the friction of the working part of the engine. In other words, friction horsepower is the indicated horsepower when the engine runs under no load. The friction horsepower can also be found by motoring the engine with the fuel turned off and finding the power required to overcome friction. IHP = FHP + BHP, and BHP = IHP − FHP

Brake Mean Effective Pressure (BMEP)

For modern high-speed diesel engines, the indicated mean effective pressure (IMEP) generally cannot be measured directly, so BMEP is used. BMEP is theoretical constant pressure acting on the piston during each power stroke. BMEP produces a power equal to the brake or shaft horsepower of the engine. This term indicates how well an engine uses its displacement to produce work, and it is a better parameter for comparing engines than torque or horsepower. The BMEP force on the top of the piston is the average pressure exerted on each square inch of the top of the piston during the power stroke, multiplied by the area on top of the piston (Force [f] = area × BMEP). BMEP is determined by the following formula:

$$BMEP = P_b = \frac{bhp \times 33,000}{LANK}$$

L = length of stroke (in feet)

A = area of piston (in square inches)

N = number of power strokes per minute

K = number of cylinders

Efficiency

Thermal Efficiency

The thermal efficiency (Figure 2-10) of an engine is the percentage of the fuel-supplied heat energy that appears at the crankshaft as useful work. If two engines produce the same horsepower at the crankshaft and one burns less fuel than the other, the thermal efficiency is greater for the more economical engine.

The heat energy supplied by the fuel is not all converted into power for useful work. Some of it is lost in the hot exhaust gases. The cooling water in the water jackets carries some away, and some is used in overcoming friction. The remainder is available for useful work.

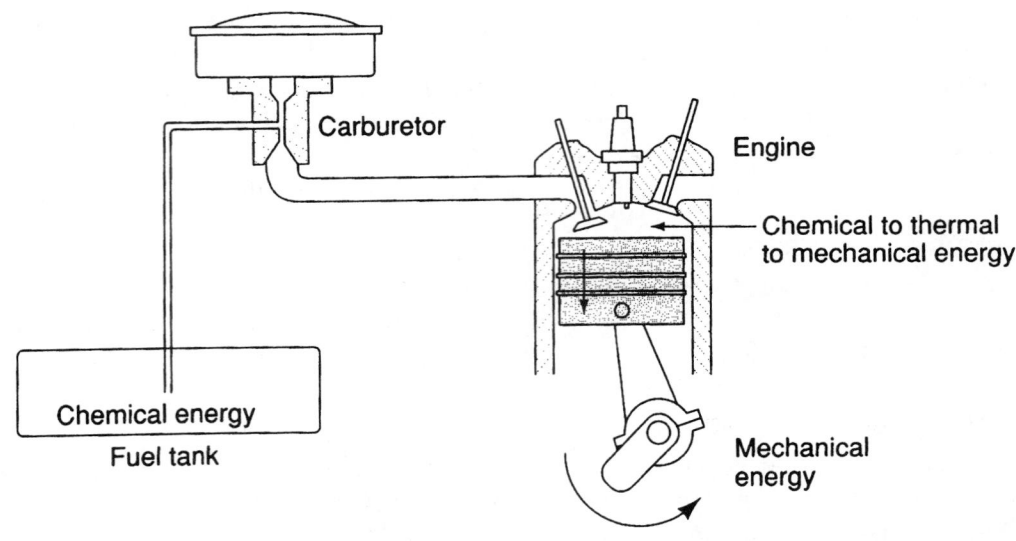

Figure 2-10 Thermal efficiency in a spark-ignited internal combustion engine.

Mechanical Efficiency

The fraction of the indicated horsepower delivered at the crankshaft is called the mechanical efficiency of the engine. Since the brake horsepower is always less than the indicated horsepower by the amount of the friction horsepower, the mechanical efficiency must be less than 1, or less than 100%. Dividing the brake horsepower by the indicated horsepower and multiplying the result by 100 yields the mechanical efficiency. Expressed in mathematical form:

$$n_m = \frac{bhp}{ihp} \times 100$$

N_m = mechanical efficiency in percent   ihp = indicated horsepower of engine
bhp = brake horsepower of engine

Sample Problem: A diesel engine developing a 540 ihp has a friction horsepower (fhp) of 57. What is its mechanical efficiency?

Solution: The brake horsepower is 540 − 57 = 483 bhp and yields a mechanical efficiency of n_m = bhp/ihp × 100 483 ÷ 540 × 100 = 89.4

Volumetric Efficiency

Volumetric efficiency is actually calculated on a weight rather than volume basis. *Ambient conditions* are the pressure and temperature of the air entering during the intake stroke.

Power extracted from an internal combustion piston engine is related to the amount of air that is consumed and retained by the cylinders. The higher the percentage of retained air, the larger the quantity of fuel that can be injected and completely burned. Volumetric efficiency (VE) is defined as the ratio of the weight of air actually entering the cylinder to the weight of air that could enter the cylinder, measured at ambient conditions, if the piston-displaced volume were completely filled.

The actual air consumption may not be at ambient conditions but at a slightly different pressure and temperature. If it is at a slightly different pressure and temperature, this should be taken into account when calculating the specific weight of the metered air. This discussion and example establishes that when the actual air consumption of an engine is metered at ambient conditions, the volumetric efficiency may be taken as the ratio of the actual volume flow of air to the theoretical volume flow of air. Volumetric efficiency is a measure of how well an engine "breathes," or takes in air for combustion. Factors affecting volumetric efficiency are the air cleaner, valve timing, manifolds, porting, exhaust pipe, oxidation catalyst, muffler, particulate trap, and turbocharger.

In naturally aspirated (NA) non-turbocharged or blower-equipped engines that rely on atmospheric air pressure to force its way into the cylinder, the resistance to airflow caused by the intake ducting (such as the diameter, number of bends, length, and air-cleaner restriction) and intake manifold design lowers volumetric efficiency. The volumetric efficiency of an NA engine is always less than atmospheric pressure (14.7 psi or 101.35 kPa) at sea level.

When a turbocharger or gear-driven blower is added to a two- or four-stroke cycle engine, the volumetric efficiency can exceed atmospheric pressure. The critical factor in determining the cylinder air pressure before the start of the compression stroke is the timing of the intake valve closing in a four-stroke cycle engine, or the liner port and exhaust valve closing on a two-stroke cycle Detroit Diesel engine.

Technicians often talk about an engine being "supercharged" and believe that if an engine is fitted with a turbocharger or gear-driven blower that it automatically becomes supercharged. Remember that in technical terms, the intake valve timing on a four-stroke cycle engine and the port and exhaust valve timing on a two-stroke cycle model determine if the engine is actually supercharged. If the cylinder air pressure at the start of the compression stroke is higher than the atmospheric pressure, then the engine is considered supercharged. The degree of supercharging, however, is directly related to the actual cylinder air pressure charge.

Atmospheric Pressure Atmospheric pressure decreases above sea level. At 5,000 ft (1,524 m) above sea level, a 1 square-inch column of air (Figure 2-11) from the earth's surface to the outer edge of the atmosphere is 5,000 ft (1,524 m) shorter than the same column at sea level. Therefore the weight of this column of air is less at 5,000 ft (1,524 m) elevation than at sea level. As altitude continues to increase, atmospheric pressure continues to decrease. At an altitude of several hundred miles above sea level the earth's atmosphere ends, and only vacuum exists beyond that point.

Vacuum When a space contains a vacuum, it contains a smaller quantity of air compared with the quantity of air the space is capable of containing with atmospheric pressure in the space. Vacuum can be measured in pounds per square inch, but inches of mercury (in. Hg) are most commonly used for this measurement. Let us assume that a plastic "U" tube is partially filled with mercury, and atmospheric pressure is allowed to enter one end of the tube. If vacuum is supplied to

> Vacuum may be described as the absence of atmospheric pressure.

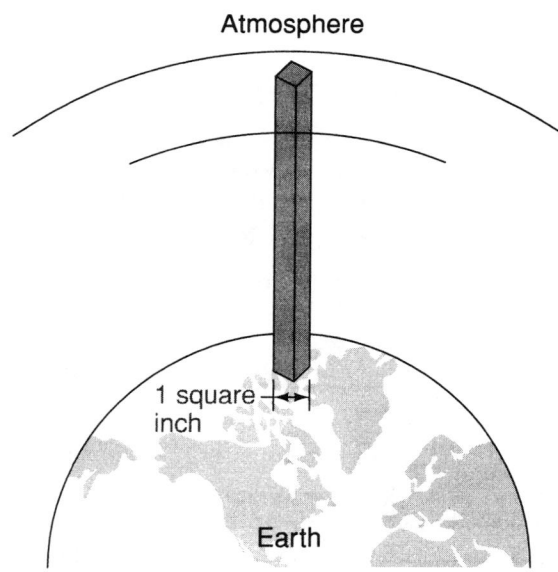

Figure 2-11 A column of air that represents atmospheric pressure.

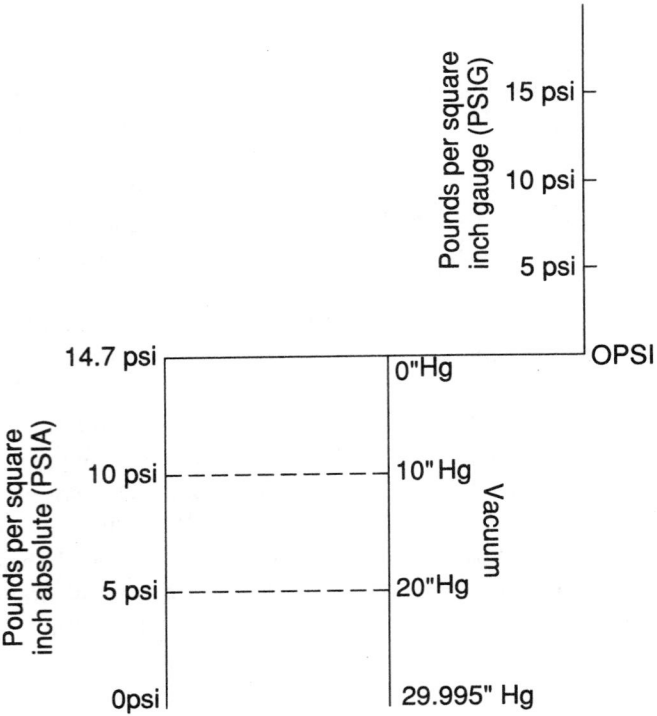

Figure 2-12 Atmospheric pressure and vacuum scales.

the other end of the "U" tube, the mercury is forced downward by the atmospheric pressure. When this movement occurs, the mercury also moves upward on the side where the vacuum is supplied. If the mercury moves downward 10 in. (25.4 cm) where the atmospheric pressure is supplied and upward 10 in. (25.4 cm) where the vacuum is supplied, 20 in. Hg (67.6 kPa) is supplied to the "U" tube. If a complete vacuum exists in a space, the pounds per square inch absolute (PS1A) is zero, and this is equal to 29.9 in. Hg (101.06 kPa) (Figure 2-12).

Vacuum and atmospheric pressure are used in several truck systems. For example, atmospheric pressure is available outside the engine air intake. When a piston moves downward with an intake valve open, a vacuum is created in the cylinder above the piston. The air moves rapidly from the high pressure outside the air intake to the lower pressure in the cylinder.

Brake-Specific Fuel Consumption

Brake-specific fuel consumption (BSFC) is the rate of fuel consumption per unit of brake horsepower output. This parameter demonstrates how efficiently the engine converts fuel energy into work. Most original equipment manufacturers (OEMs) prefer BSFC to thermal efficiency because it is determined using the practical units of weight, horsepower, and time. The following formula determines BSFC:

$$BSFC = \frac{W}{BHP}$$

W = weight of fuel component in pounds   BHP = brake horsepower
  per hour

This fuel has an API (American Petroleum Institute) gravity rating of approximately 36. The API rating of the fuel determines its heat value and therefore the Btu (British thermal unit) of heat content available from a pound or a gallon.

Engine manufacturers' sales data literature lists BSFC, shown in either lb/bhp-hr or g/kW-h (grams/kilowatt-hour). One lb/hp-hr is equal to 608.277 g/kW-h. The BSFC for a Caterpillar 3406E engine rated at 475 hp (354 kW) at 1,800 rpm and peak torque of 1,750 lb/ft (2373 Nm) at 1200 rpm is approximately 0.316 lb/hp-hr (192 g/kWh) for a fuel rate of 21.3 U.S. gallons (80.8 L/h) when running at 1,800 rpm. At the peak torque rating of 1,200 rpm, the fuel rate is 0.304 lb/hp-hr (185 g/kW-h) for a fuel consumption rate of 16.9 U.S. gallons (64 L/h).

The gallons per hour or the fuel rate can be determined as follows. Consider the example listed with an engine speed of 1,800 rpm and 475 hp (354 kW) where the BSFC is shown as 0.316 lb/hp-hr (192 g/kW-hr). Multiplying 475 by 0.316 yields 150.1 lb/hr of fuel consumed, which is equivalent to 21.3 gallons/hr (80.8 L/hr). If we divide 150.1 lb by 21.3, the weight of the fuel per U.S. gallon is 7.046 lb (3.196 kg). Based on the foregoing information, we can use the following formula to determine BSFC:

$$\text{BSFC} = \frac{W}{bhp} = \frac{150.11 \text{ lb/hr}}{475 \text{ hp}} = 0.316 \text{ lb/hr (192 g/kW)}$$

Air Standard Cycles

Diesel Pressure-Volume (pV) Curves

The diesel engine does not use a throttle to control the speed of the engine as the gasoline engine does. Except in electronically controlled systems that have a drive-by-wire control, the diesel engine throttle is connected to a fuel-control mechanism to vary the amount of fuel that is injected into the cylinder. No throttling of air takes place, which means that the diesel operates with a stratified charge, or with fuel and air in the cylinder, under all operating conditions.

The net result of the unthrottled air in the diesel engine is that at idle rpm and light loads, the air to fuel ratio in the cylinder is very lean. Because of the excess amount of air (oxygen) contained in the cylinder of a diesel engine during idle and light-load situations, the average specific heat of the cylinder gases is lowered, which increases the indicated work obtained from a given amount of fuel. This condition, along with that of unthrottled airflow, means that the diesel engine's efficiency increases as load is reduced in contrast to gasoline engines. Figure 2-13 is a pV diagram for a turbocharged and direct-injected high-speed heavy-duty four-stroke cycle diesel

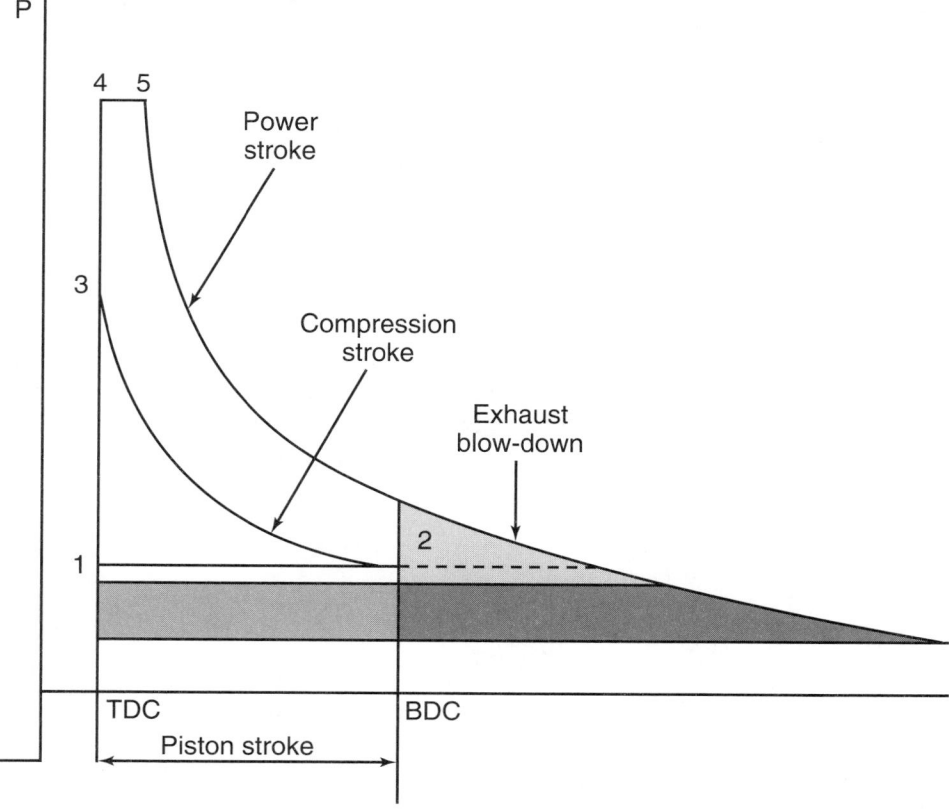

Figure 2-13 Diesel pressure volume chart showing the 4-stroke cycle.

engine. Notice that from position 1 to 2 the piston moves down the cylinder on the intake stroke as it is charged with turbo-boost air at higher than atmospheric pressure, as indicated in line PI. Depending on the valve timing, actual inlet valve closure will control the degree of trapped cylinder air pressure. In this example, compression starts at position 2 as the piston moves up the cylinder. Fuel is injected at a number of degrees BTDC, and both pressure and temperature rise as the fuel starts to burn from positions 3 to 4. As the piston moves away from TDC on its power stroke, positions 4 to 5, the continuous injection of fuel provides a condition of somewhat "constant counterpart," a condition that occurs much of the time in city driving modes.

Engine Classification

Figure 2-14 contains the typical engine cylinder arrangements. Engines in which cylinders are arranged vertically with the head end on top are called *upright* or *in-line* engines. If they are vertical with the head end down, the engine is called an *inverted* engine. A *flat* or *horizontal* engine has the cylinders arranged in a horizontal position. Most diesel engine manufacturers build in-line six-cylinder engine configurations as the standard design in large-bore diesel engines. However, V-type configurations are still in production as built by Detroit Diesel, General Motors Powertrain Division, Mack, and Navistar.

Engine Types

Designers classify internal combustion engines as two-stroke cycle, four-stroke cycle, spark-ignition, or compression-ignition engines. A two-stroke cycle engine completes one cycle of operation with two strokes of the piston and one complete revolution of the crankshaft. A four-stroke cycle engine completes one cycle of operation with four strokes of the piston and two revolutions of the crankshaft. In a spark-ignition engine, an electric spark ignites the fuel–air mixture, whereas in a compression-ignition engine, high compression produces a temperature sufficient to auto-ignite the mixture. Diesel engines are also classified by combustion chamber design. A **direct injected (DI)** diesel injects fuel directly into the combustion chamber. An **indirect injected (IDI)** diesel injects fuel indirectly by going into a prechamber before going into the main combustion chamber.

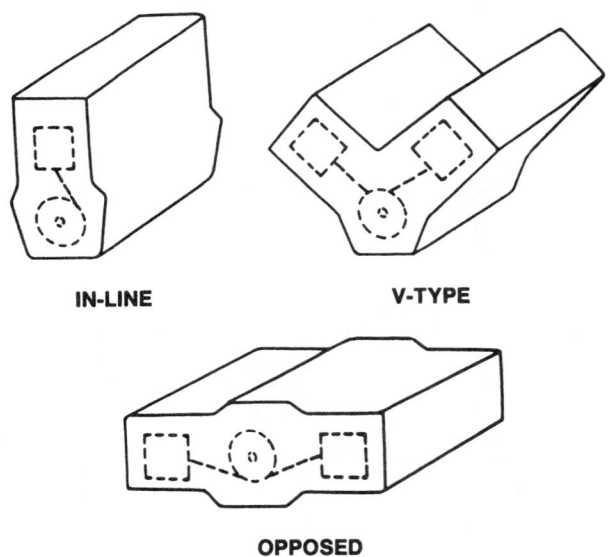

IN-LINE   **V-TYPE**

OPPOSED

Figure 2-14 Typical cylinder arrangements. (Courtesy of Deere and Company)

Diesel Combustion Chambers

Diesel combustion chambers are designed so that the compressed air will seek out and mix with the fuel. The design must make maximum use of the available oxygen and induce total combustion from the auto-ignition mixture nucleus generated from the air–fuel mixture. High oxygen utilization depends at present on the use of a high level of air motion **(turbulence).** Figure 2-15 shows turbulence in a DI Mann-type combustion chamber. Efficient conversion of combustion energy into work requires:

- ❏ Complete combustion of as much fuel as possible by the air in the chamber.
- ❏ Combustion completion early in the power stroke and timing the combustion peak to be near top dead center (TDC).

The increased heat transfer between the fuel droplet and the air can be compared to that between a unit heater with the fan running and a person standing in front of it. Without the fan running, heat transfer between the heater and the individual is slight; with the fan running (turbulence) the individual is immediately subjected to a blast of hot air.

Combustion chamber designs come in two distinct groups: an open combustion chamber with a direct injection (DI) system (Figure 2-16A); and a pre-combustion chamber with an indirect injection (IDI) system (Figure 2-16B). In the DI system, fuel injection and subsequent combustion takes place within the actual working chamber or cylinder of the engine. The IDI system employs a separate combustion chamber or cell that is remote from the working cylinder but connected to it by a channel or passage. It is generally referred to as a prechamber or antechamber may be designed in several different ways.

Open Combustion Chamber—Direct Injection (DI)

The **direct injection (DI)** diesel has no prechamber, and the combustion chamber (Figure 2-16A) is open. Fuel is injected directly into the space between the cylinder head and the top of the piston. This is also referred to as an *open chamber* design. The piston often contains a bowl or has a specially shaped crown to aid in the mixing process for good combustion. One such design, the Toroidal Piston, was a development of the plain menispherical piston cavity introduced by the Saurer Company of Switzerland in 1934. The combustion chamber section is heart shaped

Air Turbulence is the extreme disturbance of the compressed air in the combustion chamber. It causes the air molecules to move in all directions, colliding with each other and causing friction and heat.

The Detroit Diesel two-stroke uses a swirl design called the *Hesselman Helical Loop.*

The term *toroidal* denotes a tornado-like swirl of air in the cup or bowl inside the piston crown.

Swirl is the orderly rotation of gases (more or less as a whole) in the cylinder to improve mixing and heat transfer.

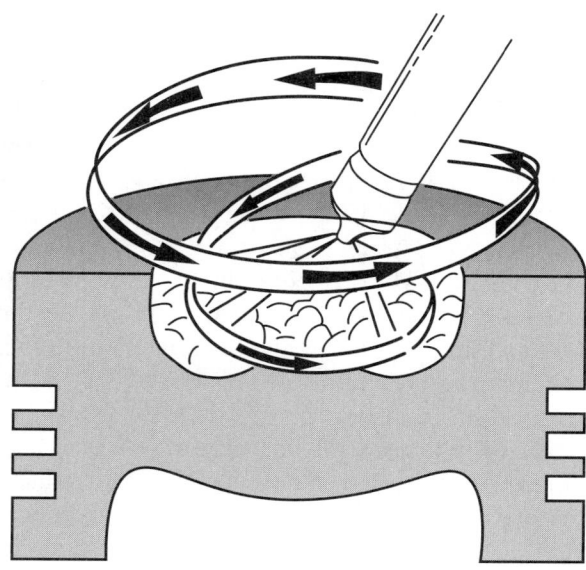

Figure 2-15 Turbulence. (Courtesy of the Society of Automotive Engineers)

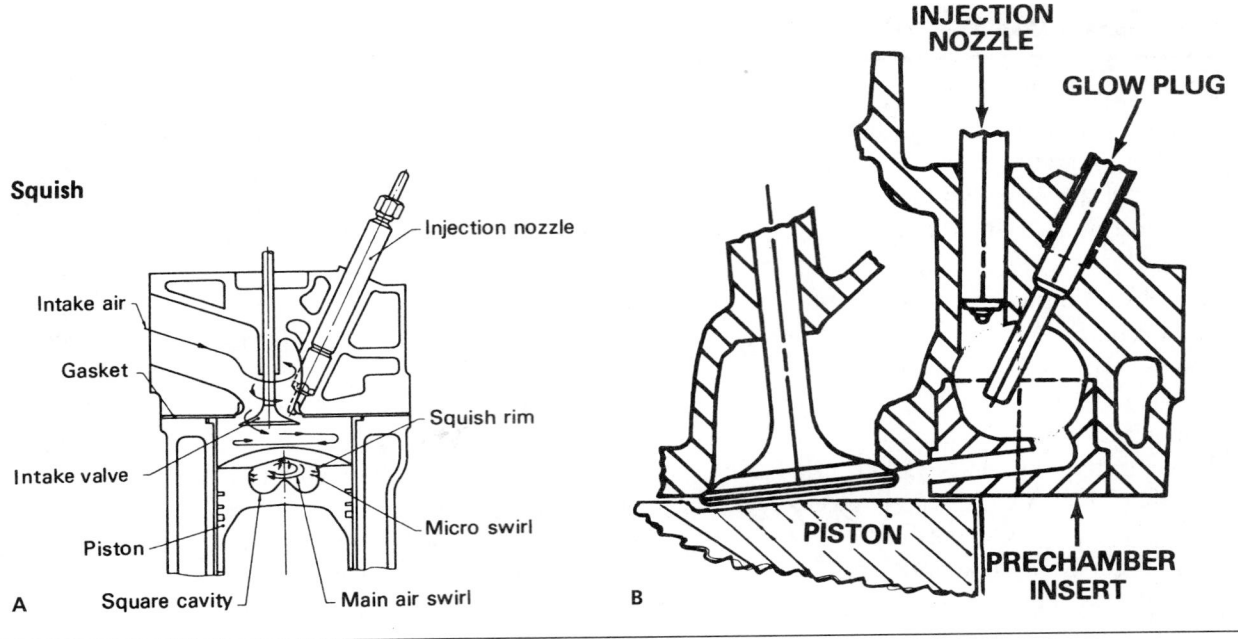

Figure 2-16 (A) Direct injection (DI) combustion chamber showing squish; (B) Indirect injection (IDI) precombustion chamber.

Squish is air motion caused by a small clearance space over part of the piston. Squish usually induces good air and fuel mixing.

Quench (squish) *Area* is the thin space between the piston crown and cylinder head where the conductive cooling of a portion of the cylinder head gases occurs during combustion.

Glow plugs are electrically heated metal pieces that are screwed into the combustion chamber. They are turned on only during a cold start for a limited time, and they aid in evaporation of fuel droplets.

and is designed to superimpose a piston displacement (squish) rotary swirl at right angles to the induction-produced swirl around the piston axis. The resulting dual turbulence is of a spiraling form around the piston axis and resembles a toroid or tornado. Other piston crown designs include the Mexican-hat–shaped piston used by Detroit Diesel, Caterpillar, Cummins, and Mack; and the M-Type or MAN Piston that employs an elliptical bowl and is used by Perkins, Cummins, Caterpillar, and Navistar.

In addition to piston cavities, other designs to produce air turbulence include masked intake valves, spiral-shaped intake ports, turbocharging, aftercooling, and angular-shaped liner air inlet ports on Detroit Diesel two-stroke engines Figure 2-17. A spiral intake port acts as a forcing cone and twists the incoming air into turbulence. It accelerates the inlet air speed and then transmits a twisting turbulence (swirl) that is further enhanced by the toroidal shape of the piston cavity (bowl). As the piston compresses this air mass, it develops hundreds of miniature tornadoes circling in a violent vortex. DI engines usually do not require a glow plug system.

The nozzle is centrally placed about the combustion space and is usually of the multi-hole type. (Four holes are most common). The symmetrically placed spray ensures even distribution and utilization of all available air. Hole dimensions are arranged to give the required penetration and aid in the mixing of air and fuel. In general, the more intense the swirl, the fewer holes required. DI systems require high pressures in the area of 18,000 to 30,000 psi for this penetration. The hole orifices are located so that the spray pattern fits the combustion chamber without impinging on the cylinder walls or piston. Despite the inferior power at high speed, the direct injection system has the more desirable torque curve shape for road vehicle applications and has an overall 5% to 10% efficiency advantage over the indirect (IDI) chamber design. This is mainly due to the lower direct heat losses that result from lower combustion chamber surface area to volume ratios. The advantages and disadvantages of the direct injection engine are presented in Table 2-5.

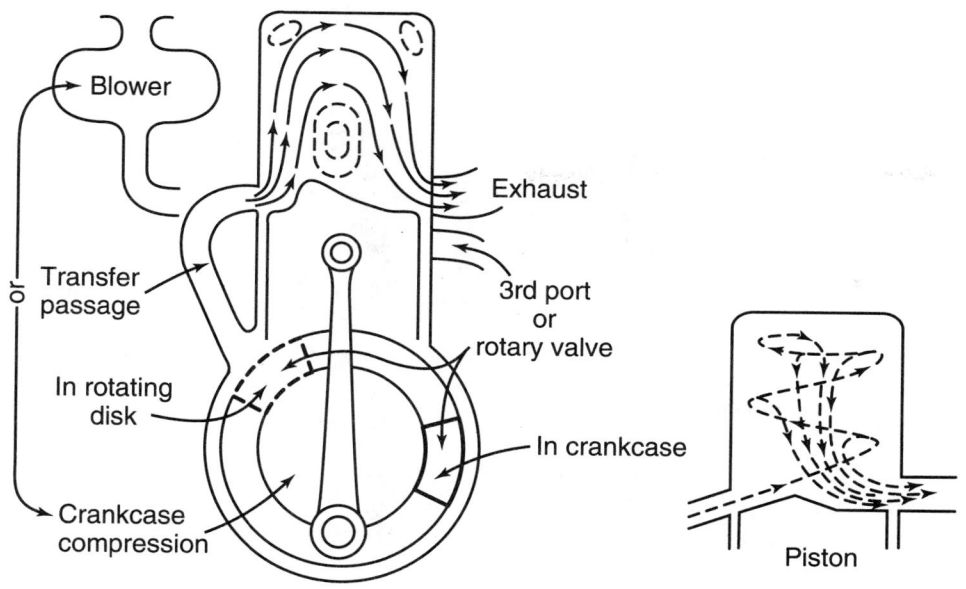

Figure 2-17 Two-stroke diesel showing the Hesselman helical loop method of air turbulence for mixing air and fuel.

Table 2-5 DI diesel advantages and disadvantages

DI Advantages	DI Disadvantages
10%-20% better fuel economy than IDI Higher thermal efficiency than IDI Lower exhaust temperature than IDI Low surface-to-volume ratio Glow plugs not always necessary Lower friction and heat losses due to lower compression ratio (16/18 : 1) Combustion energy release more advanced than IDI	Higher combustion pressure and noise Critical nozzle performance Higher exhaust emissions High sensitivity to fuel properties Limited speed range More expensive fuel system

Precombustion Chambers/Indirect Injection (IDI)

Automotive-type diesel engines use a precombustion chamber in an indirect injection engine (IDI). The IDI diesel incorporates a prechamber that is joined to the main chamber above the piston by a connecting flow passage. Fuel is injected into this prechamber, which may also contain a glow plug to help in cold starting. Combustion begins in the prechamber as shown in Figure 2-18, but spills into the main chamber as fuel, air, and burned and burning gases emit from the prechamber while the piston descends on the power stroke. Most IDI prechamber engines require a glow plug system to aid in start-up.

IDI diesel engines use either swirl-type (high turbulence) or air-cell non-swirl design prechambers. Swirl prechambers have a spherical shape that mixes the air and fuel by air swirl. This assists in promoting high turbulence by creating a swirling mass of air in the prechamber.

A precombustion chamber system (prechamber) (Figure 2-18) has the major chamber in the cylinder head and only a small space between the piston and the cylinder head. Close piston clearance produces high turbulence in the prechamber because most of the air in the cylinder is forced through a small opening into the prechamber in a very short amount of time.

PRE CHAMBER DESIGN

Figure 2-18 IDI prechamber design showing combustion starting in the prechamber and moving into the main combustion chamber. (Courtesy of General Motors Corporation, Service Operations)

Prechambers promote rapid combustion. The charge is forced out of the small opening, agitating the entire mixture and resulting in a more complete combustion. The advantage of this design over the open chamber direct injection system that is used in a constant load engine is that it has a broader operating range. It provides less noise with effective emission control. It is also insensitive to fuel characteristics and works well with a less expensive fuel injection system.

A BIT OF HISTORY

Sir Harry Ricardo, a British engineer, developed the swirl prechamber system called the Ricardo Comet Head (Figure 2-19) in 1931. His system also employed cavities within the piston besides the *retort*-shaped combustion prechamber in the cylinder head that were designed to improve the distribution of air swirl when compared with earlier chambers of the same type. By partly forming the combustion cell within a heat-insulated member, and by utilizing compression-induced swirl only, an improved maximum power rating was obtained. These essential differences

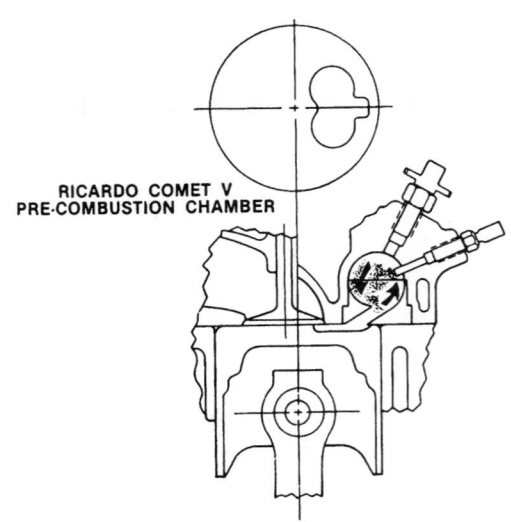

RICARDO COMET V
PRE-COMBUSTION CHAMBER

Figure 2-19 Ricardo Comet V prechamber. (Courtesy of General Motors Corporation, Service Operations)

Table 2-6   Prechamber advantages and disadvantages

Turbulent Swirl Prechamber Advantages	Turbulent Swirl Prechamber Disadvantages
Wide range of fuels can be used. Good utilization of available oxygen in the air charge.	The extra work required to compress the air charge and produce the swirl. The high heat loss from the scrubbing action of the gases over the chamber walls. A higher fuel rate than an open-chamber engine (DI engine).

with the direct combustion type ensured a reduced delay period and minimum interference with breathing, both important factors in the preservation of high-speed performance. As with most other indirect systems, a pintle-type injection nozzle is typically used.

In direct injection engines (no prechamber), mixing limits engine speeds to 3,000–4,000 rpm. Very recent engine designs have pushed that up to around 5,000 rpm, but with high hydrocarbon emissions. For an efficient engine the combustion duration must be less than about 80 degrees of crankshaft rotation. At 3,000 rpm the combustion duration is 0.0044 second. The slow mixing of direct injection engines has resulted in the development of prechamber engines. In prechamber engines, fuel is injected into a prechamber where combustion starts. To improve mixing, the reacting mixture expands through an orifice in the pre-cup into the air-filled main chamber. Intense mixing occurs in the main chamber, resulting in fast, complete burning of the remaining fuel. Not only does this increase the speed of reaction, it also results in greatly reduced hydrocarbon and soot emissions. Engine speeds of 5,000 rpm are common for prechamber engines. At 5,000 rpm with 80-degree crank angle combustion duration, the combustion time is 0.0026 second. The advantages and disadvantages of prechamber engines are presented in Table 2-6.

Diesel Engine Operation

1. Diesel engines operate on the principle that high compression heat is obtained through rapid compression of air in the cylinder, and that fuel is injected into this air. After the fuel mixes with the air, it auto-ignites. This is called *spontaneous combustion*.

2. The diesel engine is a thermal (heat) engine. It converts heat energy by fuel combustion (Figure 2-20) into mechanical energy through the pistons and crankshaft.

3. The diesel engine is an internal combustion engine: the fuel and air are burned inside the cylinders.

4. It is a reciprocating type, meaning it converts the back and forth motion of the pistons into rotary motion by a crank mechanism.

5. There are three necessary ingredients for combustion. In order for an internal combustion engine to operate it must have:
   - ❏ *Air:* a supporter of combustion, a source of oxygen to burn the fuel;
   - ❏ *Fuel:* to supply the force, pressure, or energy as it burns and expands;
   - ❏ *Ignition:* in some form, to ignite fuel, a source of heat to cause a fire.

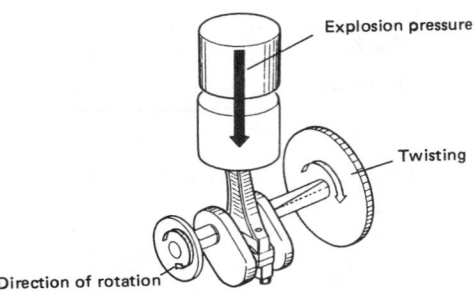

Figure 2-20 Combustion. (Reprinted with permission by American Isuzu Motors, Inc.)

There are two types of operating cycles commonly used:

1. Four-stroke cycle
2. Two-stroke cycle

4-Stroke Cycle

This cycle is virtually the same as the Otto cycle or gasoline engine 4-stroke cycle. There are, however, differences in combustion, power control, and compression ratio. The cycle consists of intake, compression, power, and exhaust. When all four are done, the engine has completed one full cycle.

Intake Stroke The intake stroke (Figure 2-21) on a diesel engine starts with the piston at top dead center (TDC). A lobe on the camshaft opens the intake valve either directly or through a follower, pushrod, and rocker arm assembly.

The piston moves down in the bore due to the rotation of the crankshaft. As the piston moves down, it pulls outside air through the air cleaner and into the air crossover manifold past the open intake valve and into the cylinder. The downward movement of the piston creates a low-pressure area above the piston (as the volume increases the pressure decreases). Air rushes

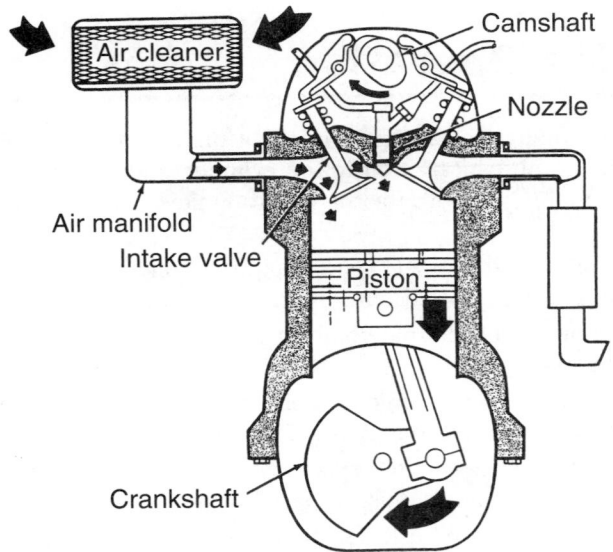

Figure 2-21 Intake stroke. (Courtesy of General Motors Corporation, Service Operations)

in to fill the space left by the downward movement of the piston, because atmospheric pressure is greater than the pressure in the cylinder. Actually, the piston tries to "inhale" a volume equal to its own displacement.

The fuel-air mixture is not homogeneous. During the intake stroke, only air is inducted. No throttle exists, so the cylinder is completely filled with air at the inlet manifold pressure. The air mixes with any residual gasses in the cylinder. The energy needed to move the piston from TDC downward comes from either the flywheel or the overlapping power strokes on a multiple cylinder engine. As the piston nears BDC it slows down nearly to a stop. When the piston reaches BDC, the intake valve closes, sealing the cylinder filled with air, and the compression stroke begins.

Compression Stroke The turning crankshaft now forces the piston upward (Figure 2-22). Both valves are closed; there is no way (except past the rings) for the air to get out. The volume is decreasing as the piston rises, so air is compressed.

In the compression of a gas, volume decreases; pressure and temperature rise as external work is done on the gas. This causes collisions of the air molecules within the cylinder. In an IDI prechamber engine, air compression forces air into the swirl prechamber in a tornado-like fashion, creating a hot swirl of air. For example, if a volume of air is compressed to 1/22 of its original volume, as it is in a diesel, the open space between the molecules is greatly reduced, increasing the number of collisions and the pressure between them. These collisions cause heat due to the kinetic energy of the molecules. Compression ratio is the ratio of the volume at BDC to the volume at TDC (clearance volume). A higher compression ratio means higher thermal efficiency or that portion of the heat supplied to the engine that is turned into work. As the compression ratio increases, the expansion ratio also increases, thus thermal efficiency increases. Diesel engines usually have compression ratios in the area of 16 to 23.5: 1.

$$\text{Compression Ratio} = \frac{\text{Volume BDC}}{\text{Volume TDC}}$$

The internal energy of the gas is increased as heat is added to the gas. High heat generated by this greater compression will cause the fuel when injected to atomize or break up into finely divided particles, allowing it to mix easily with the air. In the IDI engine, mixing is further enhanced by the addition of more heat through the spinning action of the spherical shaped prechamber. Ignition will occur as the fuel mixes with the air. The temperature of the compressed

Boyle's Law states that the pressure of a perfect gas varies inversely to its volume at a constant temperature, or the product of the pressure and volume is a constant.

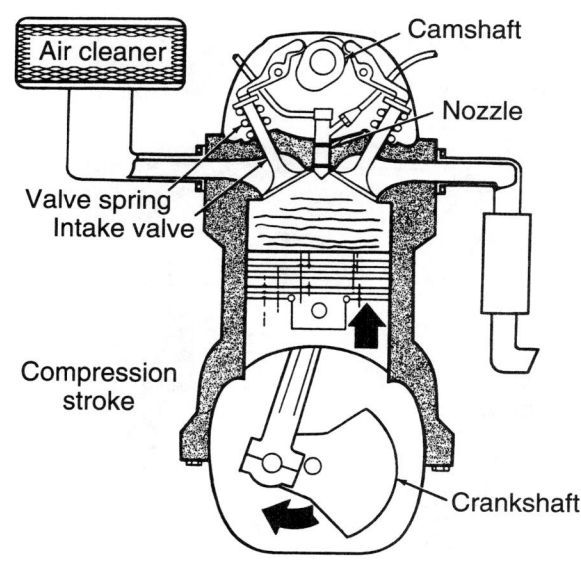

Figure 2-22 Compression stroke. (Courtesy of General Motors Corporation, Service Operations)

air is approximately 1,000° F. The temperature is generally higher than the spontaneous ignition point of the fuel; which is approximately 558° F (292° C).

Near the end of the compression stroke, fuel will be sprayed into either the prechamber in an indirect injection (IDI) engine or the combustion chamber in a direct injection (DI) engine. In a DI engine, the compressed air will swirl in a Toroidal piston cavity, increasing the friction between air molecules, which will promote mixing with the fuel. The IDI diesel prechamber is joined to the main combustion chamber above the piston by a connecting flow passage. A swirl-precombustion chamber has a spherical shape that mixes the air and fuel by air swirl. This assists in promoting high turbulence by creating a swirling mass of air in the prechamber. Turbulence is the extreme disturbance of the compressed air in the combustion chamber. It causes the air molecules to move in all directions. They collide with each other and cause friction and heat. This will increase the transfer of heat between the cool liquid-fuel droplets and this hotter air.

Power Stroke The power stroke (Figure 2-23) begins shortly after the fuel is sprayed into the main combustion chamber or prechamber by the fuel injector. The fuel is sprayed into the chamber by a nozzle or injector mounted in the main chamber or prechamber. The fuel nozzle sprays and atomizes the fuel into the chamber in a short burst (at approximately 20 to 30 degrees of rotation), forming a fuel-rich core surrounded by air zones.

During this period, fuel has entered the combustion chamber, but has not begun to burn. The temperature of the air is much higher than the fuel, so some of it has evaporated, but some has formed into small droplets. It starts to vaporize and mix with the hot compressed air. After about 0.001 second, any zones that are hot enough and have the correct fuel–air mixture ratio will auto-ignite. It is important to note that ignition will take place only where air meets fuel. The power stroke for an IDI engine begins in the prechamber and spills into the main chamber as fuel, air, and burned and burning gases issue from the prechamber while the piston descends on the power stroke.

The first fuel burns very rapidly. This rapid burning causes a sudden rise in pressure, which in turn results in a highly localized pressure that causes an audible noise known as diesel knock. The noise level of the knock depends on the speed of the pressure rise. The high pressures in the cylinder push down on the piston. This pressure forces the piston down into the bore, which

Charles' Law states that the increase in temperature in a perfect gas produces the same increase in volume if the pressure remains constant.

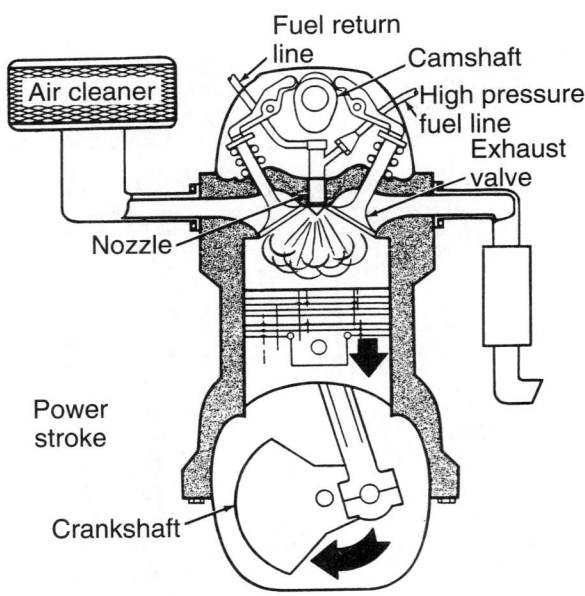

Figure 2-23 Power stroke. (Courtesy of General Motors Corporation, Service Operations)

causes the crankshaft to rotate (translation to rotation). The pressure falls as the volume increases. The temperature falls, as the gas does external work.

The oxygen and fuel burn, and the nitrogen expands, pushing the piston down under power. As the piston continues downward, these gases in the cylinder expand and cool as they give up their energy. The power stroke is the only stroke in which energy is used from the fuel, and cylinder pressure is the highest in this stroke. The prechamber may also contain a glow plug to help in cold starting.

Exhaust Stroke As the piston nears the bottom of its travel (Figure 2-24), the exhaust valve, operated by a lobe on the camshaft, begins to open. The piston then begins to rise in the cylinder, beginning the exhaust stroke. The upward movement of the piston forces the spent gases past the exhaust valve and out of the cylinder. As the piston nears the top of its movement, the camshaft lobe again opens the intake valve and the cycle repeats itself. The exhaust valve is allowed to close, by spring pressure, shortly after the piston begins its downward movement. This is a stroke that produces no work but expends a quantity of energy to push exhaust gases from the cylinder.

2-Stroke Cycle

The 2-stroke cycle (Figure 2-25) completes two strokes or actions in one turn of the engine. One full cycle consists of compression and power. The 4-stroke cycle requires two turns of the engine to complete the cycle. Two-stroke cycle operation is as follows:

1. The intake valves have been eliminated and replaced by a row of ports around the cylinder liner.
2. A Roots type blower is placed on the engine to blow air into the cylinder whenever the piston uncovers the ports. Each port has an angular inlet called the *Hesselman Helical Loop* to swirl the air, creating turbulence with a resultant improvement in the fuel–air mixing.
3. The unit injector injects fuel into the cylinder at the correct time.
4. The exhaust valve or valves open to allow the blower to push the exhaust gases out of the cylinder.

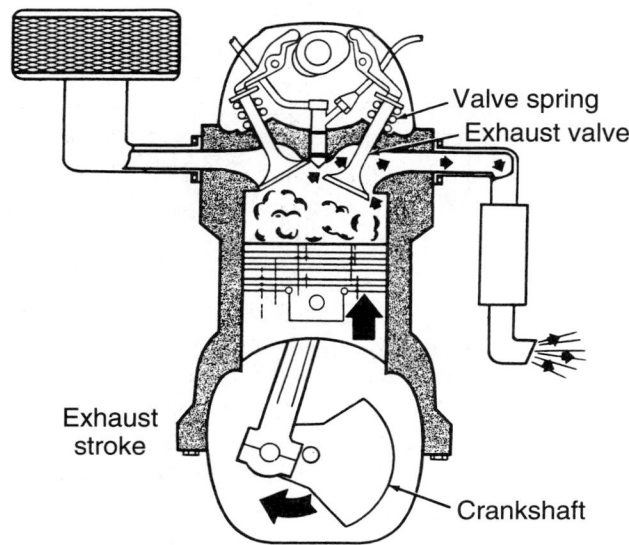

Figure 2-24 Exhaust stroke. (Courtesy of General Motors Corporation, Service Operations)

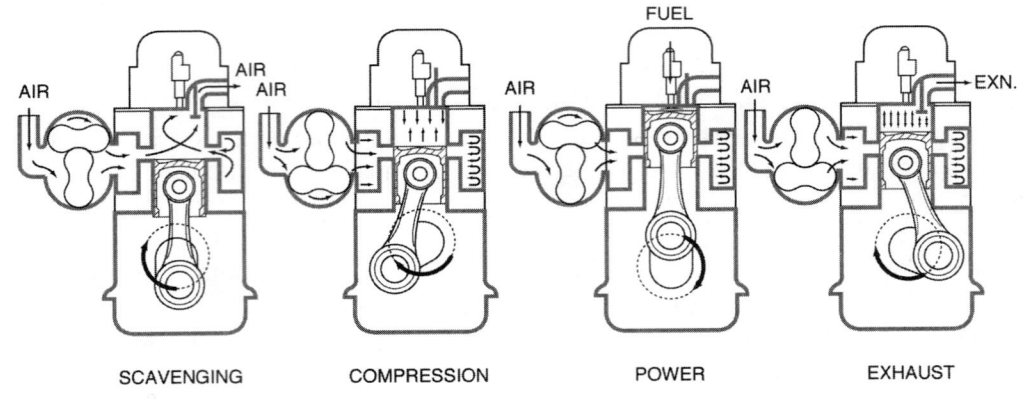

AIR → AIR AIR → AIR → AIR FUEL AIR → EXN.

SCAVENGING   COMPRESSION   POWER   EXHAUST

Figure 2-25 Two-stroke cycle of diesel operation.

Two-Stroke Cycle Operation

1. The blower blows intake air alone into the cylinder when the ports are uncovered. The piston and the exhaust valve(s) then close, covering the ports.
2. Compression of the air occurs as the piston approaches TDC, beating the air.
3. The injector is timed to inject fuel, finely atomized, into the hot compressed air in the cylinder. The rapid expansion of the burning gases forces the piston down toward BDC.
4. The exhaust valve(s) open. The piston uncovers the inlet ports, and the cylinders are swept clean with scavenging air.

4-stroke and 2-stroke cycles are compared in Table 2-7.

Valve and Port Timing

As in 4-stroke cycle gasoline engines, intake and exhaust valves are used with 4-stroke cycle diesel engines. The purpose of the intake valve is to permit the intake of clean air for combustion during the intake, or suction stroke. The exhaust valve permits release of the burned gases to the atmosphere during the exhaust stroke. The size and operating mechanisms of the valve train are designed so these operations will occur efficiently at the right time in the cycle.

Table 2-7 Advantages of four-stroke & two-stroke cycles

4-Stroke Advantages	2-Stroke Advantages
Stable operation over a wide range of loads and speeds	50%-80% greater power output per unit of piston displacement at the same speed, depending on scavenging.
Flatter torque curve	Twice as many power pulses per cylinder per revolution
Higher piston speeds	
Cooler pistons	Produces greater power from the same displacement
Common crank in multicylinder engines	
No fuel loss during exhaust	Fewer number of parts
Higher mechanical efficiency	Lower crank pin pressures
Lower brake–specific fuel consumption	
Lower pumping losses	
Easier to start	
Lower exhaust emissions	

The term *valve timing* is used to express the time the valve functions in an engine operating cycle. Valve action is stated in degrees of crank rotation about top or bottom dead center positions of the piston in the cylinder.

In the four-stroke cycle engine, the intake stroke occurs only during a down movement of the piston, and the exhaust stroke occurs only during an up stroke of the piston. Moving air has inertia and allows opening the valves longer than the piston stroke in a working engine.

The piston moves at varying speeds during its stroke. It must stop at each end of the stroke, gains speed toward midstroke, and slows down to a stop at the opposite end. The inertia of the air is used to keep the exhaust gas or intake air moving even though the piston may be stopped. If the valve is held open, the intake stroke now lasts longer than the down travel of the piston. The exhaust stroke also lasts longer than the up stroke of the piston.

Exhaust Valve Timing The exhaust valve is timed to open before BDC (Figure 2-26) in the power stroke, and to close after TDC in the exhaust stroke. The reason for this timing is to achieve the best possible exhaust conditions. The valve is opened before BDC in the power stroke to release the low pressure remaining from combustion and to ensure full opening of the valve when the exhaust stroke begins. This avoids excessive backpressure against the piston during the start of the exhaust stroke and reduces withdrawing of the gases through the valve and seat. There is a small power loss by doing this, but it is canceled out by a gain through release of the exhaust gases. The exhaust valve is open past the end of the exhaust stroke to provide an overlap of exhaust valve and intake valve opening. This permits thorough clearing of exhaust gases from the clearance space and also creates a partial vacuum. The inertia of this flow of exhaust gases creates this suction or partial vacuum in the clearance volume. All the burned gases are therefore expelled, and the charge of intake air is fresh air.

Intake Valve Timing In a 4-stroke cycle diesel engine, the intake valve is timed to open BTDC (Figure 2-26) and close after BDC of the intake stroke. This is to achieve the highest volumetric or mass efficiency. The opening of the intake valve close to the end of the exhaust stroke, and before the closing of the exhaust valve, allows clearing of burned gases before the beginning of the intake stroke.

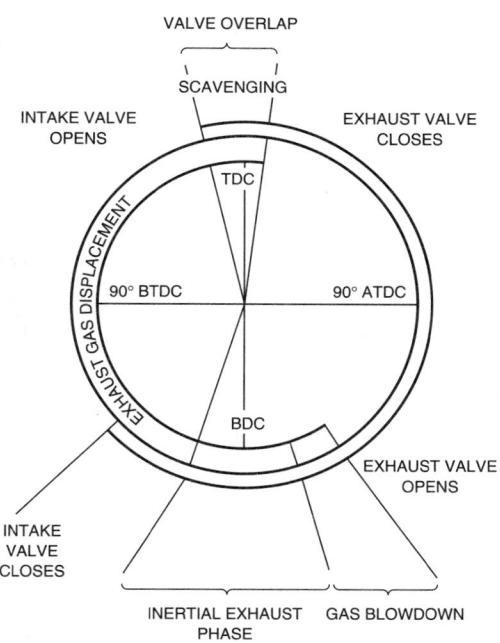

Figure 2-26 Valve timing in a typical diesel engine.

The incoming air partially displaces the small amount of exhaust that remains and reduces the amount of contamination of the intake air. This is sometimes referred to as the scavenging period in the 4-stroke cycle. During the period that both valves are open, near the end of the exhaust stroke and the start of the intake stroke, the piston movement is slight, but the exhaust gas that has been forced out into the exhaust path by the piston continues outward through the exhaust passage.

For similar reasons the intake valve is held open after the piston has reached the end of the intake stroke. Air tends to continue flowing into the cylinder due to the partial vacuum created by the piston during suction and the inertia imparted to the column of incoming air.

Diesel Combustion Process

Combustion Chemistry in Heat Engines

The operation of an internal combustion engine depends on the quick and complete shift of the chemical energy of fuel into heat energy in the cylinder. This requires that the liquid fuel be introduced into the cylinder, vaporized, and mixed with air before combustion. The carbon and hydrogen of the fuel, in combination with the oxygen in the air, results in combustion. The process differs for spark-ignition and compression-ignition engines. At the start of compression in all internal combustion engines, there is more air than fuel in the cylinder. Combustion products from the prior cycle fill the clearance volume in the cylinder and dilute the incoming fuel and air. For example, in a gasoline engine when a fuel injector supplies fuel, the incoming air carries liquid fuel droplets along with vaporized fuel to be compressed before the combustion process. In diesel engines, fuel is introduced during the compression process.

Composition of the Atmosphere Dry atmospheric air consists of oxygen, nitrogen, and small amounts of other gases such as carbon dioxide, argon, and hydrogen. We consider nitrogen as the other gases for most calculations. Table 2-8 shows the percentages, by weight and volume, of nitrogen and oxygen in the atmosphere. The average molecular weight of dry air is taken as 29.

Unfortunately, atmospheric air contains varying percentages of water vapor that affect specific heats and heat loss. The text will not attempt to modify calculations to account for this error. Its presence in most cases simply means an additional amount of inert material. By volume, there is 3.76 times more nitrogen than oxygen in the atmosphere, and by weight there is 3.32 times more nitrogen.

By mass, nitrogen represents the largest ingredient in the engine cylinder during the combustion process in any engine aspirated with air. At ambient temperatures and pressures, nitrogen is relatively *inert,* or unlikely to become involved in chemical reactions. However, subjected to engine cylinder heat and pressure, nitrogen may become involved in the oxidation process, producing compounds of nitrogen known collectively as NO_x:

1. *Nitric oxide (NO)* is a colorless, odorless gas that in the presence of oxygen will convert to NO_2.
2. *Nitrogen dioxide (NO_2)* is a reddish-orange gas that has corrosive and toxic properties.
3. *Nitrous oxide (N_2O)* is a colorless, odorless gas better known as *laughing gas.*

Table 2-8 Atmospheric Composition

Element	% By Weight	% By Volume
Nitrogen	77	79
Oxygen	23	21

NO_x emission is generally the result of a number of fuel/engine parameters; the most important of which are actual reaction temperatures and the excess air factor. Generally but not consistently, high combustion temperatures will tend to produce higher NO_x emission. NO_x is a major contributor to photochemical smog.

Requirements for Combustion A substance is said to be combustible if it is composed of elements that will burn when mixed with oxygen and ignited. The elements of a compound that will burn are called *combustibles*. The most common combustibles are carbon and hydrogen. Combustion continues due to the supply of oxygen with which the carbon and hydrogen can unite, and the air furnishes this oxygen; thus air is a supporter of combustion.

The compounds that are created in the process of combustion are called the *products of combustion*, and are usually gases or vapors. So, if a hydrocarbon burns in air, the products of combustion are carbon dioxide, water vapor, and nitrogen. The nitrogen is inactive, and takes no part in the actual burning. The water vapor condenses to water when it cools.

Three conditions are necessary to produce combustion. First, there must be a combustible; (the fuel or hydrocarbon). Second, there must be a supporter of combustion (the oxygen in the air). Third, this mixture of fuel and air must be brought to the proper temperature for fuel vaporization so combustion will start. To ignite the fuel, a source of heat is necessary to cause a fire. This ignition source may be a flame, electric spark, or the heat of compression in a diesel engine that causes auto-ignition, thus leading to spontaneous combustion.

Diesel Combustion Periods

James Ricardo established that the diesel combustion (compression-ignition engine) process takes place in three stages or periods. They are:

   1. Delay period (ignition delay)
   2. Period of rapid combustion
   3. Period of controlled combustion

Delay Period (Ignition Delay-ID) Figure 2-27 shows graphically the **Ignition Delay (ID) period,** also known as the delay period. This is the time between the *start of injection* and the *start of ignition,* which starts the pressure rise due to combustion. It may also be called *ignition lag,* or ignition delay (ID). It occurs at the end of compression, after the start of injection, and the fuel does not ignite immediately. After the start of injection, the cylinder pressure follows the compression curve. At the end of this period the pressure rises at an increasing rate until the maximum pressure slope is reached. ID is much longer than the active combustion period, particularly during the part-load operation. Fuel injection also continues during the ID period. The longer the delay period (ID), the greater will be the amount of fuel accumulated in the combustion chamber before the start of combustion. This results in higher rates of pressure rise and maximum gas pressures and temperatures. The length of the Delay Period depends mainly on the pressures and temperatures that exist in the cylinder gases during ID and the combustion chamber surface temperature against which the fuel hits. There is much emphasis connected to ID in diesel combustion because ID affects:

   1. The rate of pressure rise and maximum gas pressure. These affect noise, vibrations, and stress.
   2. The maximum gas temperature. This affects the NO_x emission, fuel consumption, power, cooling system thermal loads, and the temperature and thermal stressing of the combustion chamber walls.
   3. The time for evaporation and mixing before ignition. (This depends only on the ID length.)

The ID period contains a series of preliminary reactions that occur before the appearance of flame. These reactions start in the vapor that surrounds the fuel droplet surfaces as each drop enters into the cylinder. Ignition starts in this vapor layer surrounding the drop, and the combustion

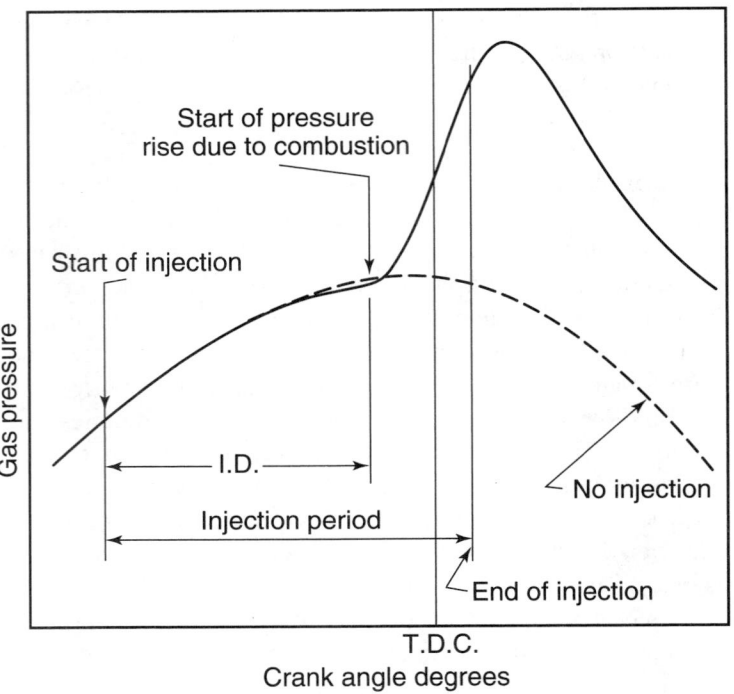

Figure 2-27 Ignition delay period.

rate of fuel drops is limited by their evaporation rate. The burning rate decreases as the fraction of oxygen in the surrounding air decreases. The single biggest factor that changes ID is the average temperature of the cylinder contents (fuel–air ratio) during the ID period.

Diesel injection occurs in layers (stratified like the atmospheric stratosphere). In premixed charges the fuel–air ratio can change the time between the flame appearance and the completion of the compression process. With diesel fuel injection, as long as the fuel is not completely evaporated, the complete range of fuel–air ratios from 0 (no fuel) to infinite (no air within the fuel droplets) must be present, and ignition will occur where the local fuel–air ratio is most ignitable. The ID period is affected by four factors:

1. Nozzle droplet size diameter formation and spray disintegration (fuel properties).
2. Air temperature and combustion chamber pressure prior to fuel injection, leading to liquid fuel heating and evaporation. High pressure and temperature can shorten the delay period.
3. Turbulence of combustion chamber air and diffusion of fuel vapor into the air to form a combustible mixture.
4. Chemical fuel composition, which includes the cetane number and the decomposition of heavier fuel components into lighter components.

In a diesel engine the ID is independent of speed and varies according to the above four reasons. ID does not vary in direct relation to engine speed. In a spark-ignited (gasoline) engine *flame speed* is nearly proportional, meaning that it increases when engine speed increases. Therefore, the number of crank angles occupied by the combustion process is nearly independent of rpm. However, in a diesel engine, *flame speed* as such does not exist. Instead, the combustion rate is determined by how rapidly the fuel and air are mixed. As engine speed increases, mixing increases, but not enough to compensate for the reduced amount of time available for combustion. Thus, the upper speed of the engine is limited.

Injection Timing Effect: Injection advance results in longer IDs because the fuel is injected into air at a lower temperature and pressure. If the increase in ID in crank degrees is less than the injection advance, autoignition occurs earlier in the cycle. Longer ID causes more fuel evaporation and mixing in the lean fuel region of the combustion chamber along with higher NO_x. Retarding injection timing is one of the effective ways of reducing NO_x emissions, but results in a loss in brake mean effective pressure (BMEP) and fuel economy. Typically, retarded timing can cause white or black smoke in a diesel engine and advanced timing causes black smoke.

Period of Rapid Combustion In the **period of rapid combustion,** injection occurs at or just before top dead center (TDC) to take advantage of the highest compression temperature (Figure 2-28). The ID is always long enough (approximately .001 to .003 seconds) that, when ignition happens, there is an appreciable amount of evaporated and finely divided fuel well mixed with air. Figure 2-28 shows that ID delay lasts about 7 crank angle degrees, as shown in area "B." Once ignited, this fuel tends to burn very rapidly because of multiplicity of ignition points and the high temperature already existing in the combustion chamber. After about 0.001 second, any zones that are hot enough and have the correct fuel-air mixture ratio will auto-ignite. Flame does not propagate through the combustion chamber as in a gasoline engine. This is because the core of fuel produced by the injector is too rich to burn, and only air exists in other parts of the chamber. Thus combustion occurs only at the interface where fuel and air come together (commonly called a diffusion flame).

The first fuel is burned at this point. This rapid burning causes a sudden rise in pressure, creating a highly localized pressure that produces an audible noise known as diesel Knock. It is interesting to note this knock happens at the beginning of the combustion cycle as opposed to gasoline engine spark-knock (detonation) that occurs at the end of the combustion cycle. The amplitude of the knock depends on the speed of the pressure rise. Both the rate and the extent of pressure during this period increases as the ID increases, since both mixing time and the fraction of fuel taking part in rapid combustion increase. When the ID is longer than the injection time, the amount of fuel involved is unaffected by ID length. The rate of combustion is controlled by how rapidly the fuel is mixed with the air. Once the first fuel is burned, a rich core of fuel

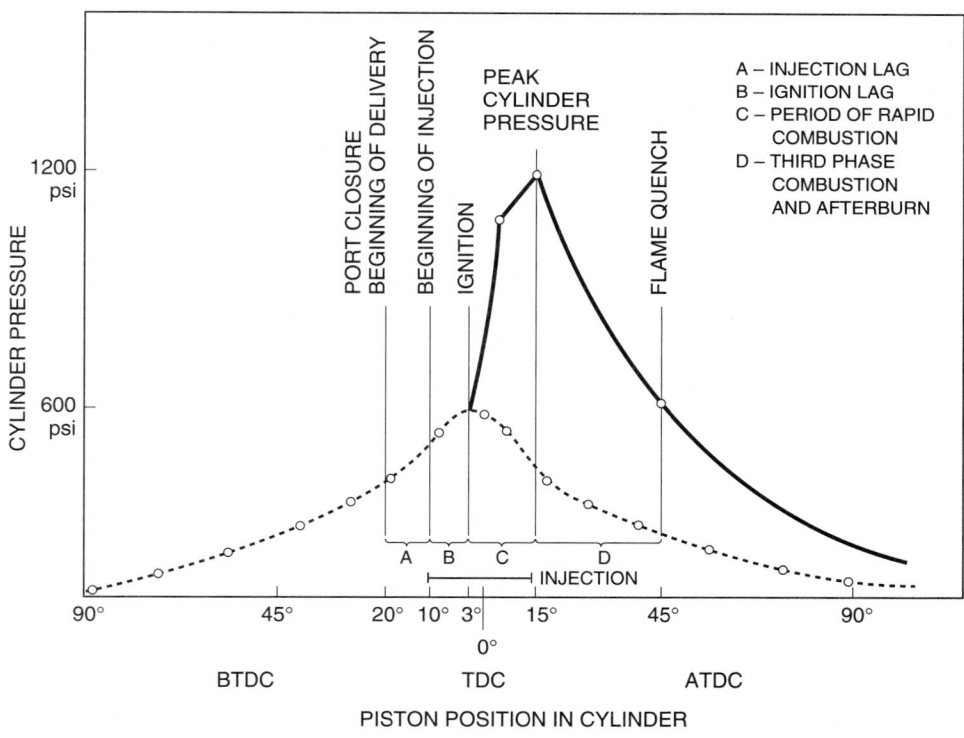

Figure 2-28 Diesel pV curve showing the crank angle duration of combustion.

remains, along with zones of air. The remaining fuel is burned as it mixes with the air. The mixing time is the reason some diesel engines have limited speeds and long combustion periods.

A fast rise in pressure can be caused by a long ignition delay time. Since most of the fuel is injected during this ID, a large amount of fuel causes a fast pressure rise. Also during a long ID, more time is available for mixing air and fuel. The sudden rise in pressure can be enhanced by good air-to-fuel mixing. High speeds and good nozzle atomization can increase mixing. If the fuel and air mix well, the burning takes place very quickly and a loud knock occurs. If the air and fuel mix poorly, the time required for the air zones to interface with the fuel core increases, the burning takes place more slowly, and the knock is less severe.

This period of rapid combustion or diesel knock is increased by long delay periods resulting in more increased mixing. In a diesel engine, the fuel should ignite very rapidly and burn quickly to keep the period of rapid combustion to a minimum. To change this delay period, a fuel additive known as cetane is used. The ignition quality of diesel fuels is expressed in terms of a cetane number or rating. A high cetane number indicates a short delay period and a low cetane number indicates a long delay period. If the engine is equipped with glow plugs as a cold starting aid, it is not very sensitive to cetane ratings.

Period of Controlled Combustion **The period of controlled combustion** is the period of time from maximum pressure to the point when combustion is measurably complete. Once the first fuel is injected, it vaporizes and mixes with the hot compressed air. It is burned and injection continues, and a rich core of fuel remains along with zones of air. The remaining fuel is burned as it mixes with the air. During this period, when the ID is longer than the injection time, the amount of fuel involved is affected only if it has not mixed with the necessary oxygen during the period of rapid combustion.

Fuel injection continues through the delay period and period of rapid combustion. Fuel injected after the period of rapid combustion (first fuel) is burned as it is mixed with the air zones. Because the fuel only burns as it is injected into the cylinder, the rate of pressure rise is moderate. This is the period of controlled combustion.

Flame speed does not exist in diesel engines. Instead, combustion rate is determined by how rapidly the fuel and air mix together. As engine speed increases, mixing increases, but not enough to compensate for the reduced amount of time available for combustion. Thus the upper speed of the engine is limited. One can go too far with mixing. If all of the fuel could be instantaneously mixed with the air, then the mixture would all burn immediately upon ignition. The result would be very, very loud knock sound and engine damage shortly thereafter. The trick in diesel engine design is to get the fuel mixed with the air in a way that provides quiet, fast combustion.

Because a diesel engine has a high compression ratio, it also has a high expansion ratio. This high expansion makes the diesel engine more efficient than a gasoline engine. Diesel engine efficiency is also demonstrated by cooler exhaust. Due to the long combustion period, the fuel is more completely consumed. In a gasoline engine the fuel is not as completely consumed and is still burning as it enters the exhaust. The diesel engine converts more of the fuel's energy into useful work than any other internal combustion engine. Gas engine thermal efficiency is about 21% to 28%; diesel engine thermal efficiency ranges from 32% to 38% efficiency. Diesel combustion also results in much lower carbon monoxide emissions.

Combustion Summary For optimum efficiency, the fuel injected must be evenly distributed in the compressed air so it forms a homogeneous charge when burning occurs. To ensure that the combustion or oxidation process occurs quickly, the fuel must be atomized or broken up into finely divided particles in order to present the largest possible surface area to the surrounding air. This atomization ensures a rapid transfer of heat from the compressed air to the fuel particles and minimizes the inevitable delay period before spontaneous combustion begins. Once started, atomization will accelerate the burning off or vapor burning of the fuel surface and bring about early combustion completion.

All compression ignition engines have a delay period between first injection and commencement of burning, which is followed by a rapid pressure rise due to the spontaneous ignition of a portion of the total injected fuel. This is termed the period of rapid combustion, and is followed by progressive combustion with the introduction of the remainder of the fuel during the period of controlled combustion burning phase. Some delay period is essential to ensure that the outer zones of the combustion chamber receive fuel, and that distribution is not hampered by premature combustion.

The rate of fuel injection for optimum distribution is dependent upon the air turbulence or the air velocities that flow past the nozzle. Good air turbulence maintains the optimum balance required between uncontrolled and controlled burning phases. Burning must be complete before the end of the power stroke and overall combustion pressure distribution should be at a favorable crank angle. In general, the higher the working speed range for a given combustion chamber design, the higher the rate of injection overall. This is particularly so with DI systems.

Air temperature in the combustion chamber during the compression stroke generally averages between 900 and 1,200 psi (6,205 to 8,274 kPa) on IDI engines. On DI Diesel engines this pressure can run between 1,800 and 2,300 psi (12,411 to 13,790 kPa). These high pressures reached in diesel engines result in peak combustion chamber temperatures as high as 3,500° to 4,000° F (1,926° to 2,482° C).

Late combustion of previously unburned fuel or intermediary products of combustion happens before the after burning period is complete. Because of ignition delay and combustion time, high-speed engines cannot attain constant-pressure combustion. It is therefore necessary to start injection earlier, with the result that much of the combustion occurs at a constant volume, similar to a gasoline engine, while the remainder of combustion occurs at constant or near constant pressure. More efficient burns occur at a constant volume rather than at a constant pressure.

Because of incomplete mixing of the injected fuel with the air in the combustion chamber, some fuel droplets do not burn until late in the cycle. This means that the combustion chamber temperature is lower and there is also less oxygen to sustain the remaining burn; as a result, incomplete combustion occurs. This results in smoke coming out the exhaust. White smoke results from incomplete combustion in overlean combustion chamber areas, fuel spray impingement on metal surfaces, and also from low temperatures in the cylinder, such as when starting an engine (especially on cold days). Fuel with too high a cetane rating can also cause white smoke when used in high-ambient-temperature conditions. White smoke occurs more often in an IDI engine from retarded timing. Very late timing on a DI engine causes white smoke.

Gray or black smoke is the result of incomplete combustion in rich combustion chamber areas, caused by such conditions as engine overload, insufficient fuel injector spray penetration, late ignition due to retarded injection timing, or poor fuel evaporation and mixing due to advanced timing. Air starvation is the major cause of black smoke. Owing to the extremely short time available for mixing; as the fuel–air rate increases beyond a certain value, an appreciable fraction of the fuel fails to find the necessary oxygen for combustion and passes through the cylinder unburned or partially burned.

Power and Speed Control

A throttle valve that controls the fuel–air mixture mass flow or density controls the power and speed in a gasoline engine. In a diesel engine, the air throttle valve does not restrict intake. Power and speed in a diesel engine are controlled by the amount of fuel injected, and when it is injected.

The same amount of air enters the cylinder during each cycle whether at idle or full power. Some air is lost past the rings at low speed, when leakage has time to occur. Varying the amount of fuel (injection duration) injected into the cylinder controls power. For low power, such as at idle, a short injection period and a small quantity of fuel are used. When high power is required, the injection period is longer. A larger quantity is forced into the cylinder and most of the air is used in the combustion process.

An advantage of the diesel engine is that it has good efficiency at low power because the fuel–air ratio is so lean at reduced power. The fuel–air ratio in a diesel engine varies from 100:1 at idle to 20:1 at full load. At high power, so much fuel may be injected that some of the fuel does not combine with the air in the cylinder. This results in unburned fuel being exhausted as smoke. Fuel heated without air is smoke. The fuel–air ratio at this point is called the smoke limit. The smoke limit is reached whenever maximum power is needed. The formation of black smoke begins when the oxygen percentage entering the engine (EGR) is equal to the percentage consumed by the combustion process.

Fuel Injection Timing

Ignition delay in a diesel engine is related to such factors as the fuel's distillation range (vaporization/temperature), the pressure and temperature of the compressed air within the combustion chamber, air turbulence, and engine load and speed. Engine manufacturers determine the best fuel injection timing point by empirical tests in a test cell with the engine on a dynamometer. Actual fuel injection timing is then determined after consideration of the following factors:

❑ Federally mandated exhaust emissions limits that apply to NO_x (oxides of nitrogen), HC (hydrocarbons), CO_1 (carbon monoxide), CO_2 (carbon dioxide), and PM (particulate matter)

❑ Horsepower output

❑ Fuel consumption

❑ Engine noise

❑ Exhaust gas density due to incomplete combustion (black soot)

❑ Exhaust gas temperatures

The actual start of fuel injection varies among makes and models of engines due to design differences; at an idle speed the variance can be anywhere between 5 and 15 degrees BTDC. As the engine speed is increased and a greater volume of fuel is injected, timing must be advanced to allow the fuel to burn to completion because of the now-higher piston speed, since there will be less time available. Consider that in an engine having a piston stroke of 6 in. (152.4 mm), at an idle speed of 600 rpm the speed of the piston will be 2 x 6 in., since the piston will move up the cylinder once and down the cylinder once for every 360 degrees, or each complete turn of the crankshaft. Therefore, piston speed can be determined by the following formula:

$$\text{Piston Speed} = \frac{2 \times \text{stroke length} \times \text{rpm}}{12}$$

At a 600-rpm engine idle speed, a certain piston will travel 600 ft/min. In 1 hour the piston travels 36,000 ft (60 × 600). Dividing 36,000 by 5,280 ft, yields its speed in miles per hour, which in this case is 6.81 mph (11 km/h). At a maximum engine speed of 2,100-rpm, the piston will travel at 2,100 ft/min, or 23.86 mph (38.39 km/h). If the start of fuel pressurization began at the same number of degrees before TDC at the high-speed setting as at the low-speed setting, the piston would be closer to the top of its stroke before fuel injection actually began, while running at the higher speed. The start of fuel injection has therefore been retarded (so that it begins later in the compression stroke of the upward moving piston) at the higher speed. It becomes necessary to advance the start of fuel injection (inject fuel earlier) in the cylinder with an increase in engine speed. Fuel pressurization (compression) begins within the pumping plunger and barrel bore. In an in-line multiple-plunger injection pump that uses long fuel lines to transfer the fuel from the pump to the injector and nozzle, the line length must also be considered.

Speed Advance A speed timing advance device in the injection pump is used to achieve better engine performance through the operating speed range of the engine. Timing advance is needed to compensate for two delay periods:

1. The injection pressure wave traveling the length of the injection line,
2. The ignition delay period.

High-pressure injection lines connect the injection pump to the nozzles. Pump pressure travels to the nozzle as a pressure wave, resulting in a time lag. Compensating for inherent injection lag improves high-speed performance. Starting delivery of fuel to the nozzle earlier when the engine is operating at higher speeds ensures that combustion takes place when the piston is in its most effective position to produce optimum power with minimum fuel consumption and minimum smoke.

Four-Trace Oscilloscope Pattern Figure 2-29 shows a four-trace oscilloscope pattern for a medium duty diesel engine equipped with a computer-controlled electronic solenoid distributor-type injection pump. The engine is set at 2,700 rpm at a full fuel load, with a residual line pressure of 800 psi. The oscilloscope traces in Figure 2-29 illustrate why it is necessary to provide sufficient fuel injection advance to compensate for ignition delay and line length:

❏ Trace 2 (from the top) shows the start of pumping at 16 degrees before top dead center (BTDC). The peak line pressure rises to 4400 psi.

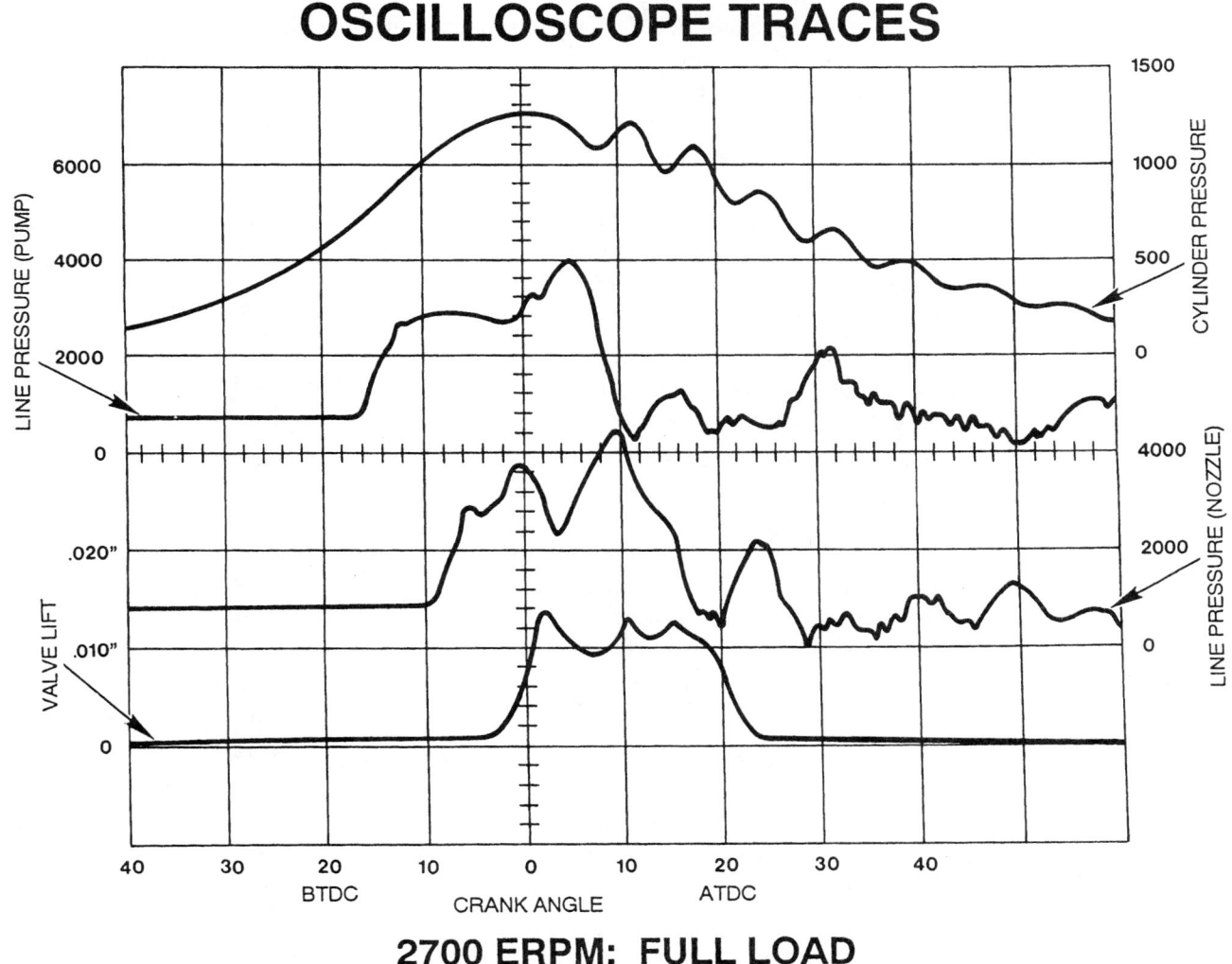

Figure 2-29 Oscilloscope pattern. (Courtesy of General Motors Corporation, Service Operations)

- ❑ Trace 4 (from the top) shows the injector nozzle needle valve lifting at 3 degrees before top dead center (BTDC). The pressure rises to over 4,000 psi with needle lift duration of 29 degrees.
- ❑ Fuel ignition occurs at 8 degrees after top dead center (ATDC)
- ❑ The ignition delay is 11 degrees from the time that the fuel injection begins to ignition, or the 3 degrees BTDC plus the 8 degrees ATDC when ignition occurs.

Summary

- ❑ The cylinder block contains the following: sleeves (liners), which house the pistons and rods; cooling passages; lubrication passages; in-block camshaft and bearings; and crankshaft bearing supports in the bottom end. It can have one-piece cast in type cylinders, called a parent-bore block, or inserted liners or sleeves. The crankshaft transmits the power through the connecting rod to the crankshaft throw and into output. It converts power into rotary motion and conveys this power to the outside. The main bearings support the tremendous pressure caused by these actions to the crankshaft. The camshaft operates the intake and exhaust valve. The camshaft is timed to the crankshaft by a timing gear or camshaft gear that is meshed into a gear on the front of the crankshaft. The cam followers are mounted in drilled holes in the block and ride on the cam lobes. Inserted into the cam followers are pushrods, which operate the valves (except on overhead camshafts). The rocker arms are mounted on a shaft with one end on the valves and the other end on a push rod. Movement of the push rod causes the arm to rock on its pivot shaft, hence the name *rocker arm*. Push rods transmit the cam action of the camshaft. The main function of the vibration damper is to dampen out torsional twisting vibration.
- ❑ *Bore* is the diameter of the cylinder. When the bore is larger than the stroke, the engine is oversquare. When the stroke is larger than the bore, the engine is undersquare, with most of its power dependent upon torque. Stroke is the distance the piston travels. Total piston displacement is stated as the volume in cubic inches or liters (cubic centimeters) of the space swept through by all the pistons of the engine during one complete revolution of the crankshaft.
- ❑ *Compression ratio* is the ratio of the total cylinder volume to the clearance volume. It is the volume of the cylinder at the beginning of the compression stroke divided by the volume of the cylinder at the end of the compression stroke.
- ❑ *Speed* is angular velocity or the rate of change of position at a particular angle. It is defined in crankshaft revolutions per minute (RPM) of time and shown in formulas as the letter *N*. *Work* is the product of a force and the distance through which the force acts. *Torque* is a force that tries to turn or twist something around, and is defined as the product of a force and the perpendicular distance between the line of action of the force and the axis of rotation. Torque is force multiplied by length. Torque rise expresses in a percentage value the increase in engine torque as the engine speed reduces from its maximum no-load rpm or rated speed.
- ❑ Horsepower is as work per unit of time. It is the rate of doing work or torque. There are six types of horsepower: brake, continuous, shaft, road, indicated, and friction.
- ❑ *Volumetric efficiency (VE)* is the ratio of the weight of air actually entering the cylinder to the weight of air that could enter the cylinder, measured at ambient conditions, if the piston-displaced volume were completely filled. This efficiency, although called a volumetric efficiency, is actually calculated on a weight basis and is more appropriately considered mass efficiency. Volumetric efficiency is a measure of how well an engine breathes or takes in air for combustion.
- ❑ Diesel engines are classified as in-line, V-type, or horizontal design. They use either direct (fuel) injection (DI) in an open combustion chamber with fuel directly injected into the combustion chamber, or indirect (fuel) injection (IDI), with fuel injected into a precombustion chamber.

❑ Diesel engines operate on a 2-stroke cycle or a 4-stroke cycle. The four-stroke cycle consists of intake, compression, power, and exhaust. When the engine completes all four, the engine has completed one full cycle. The two-stoke cycle completes two strokes or actions in one turn of the engine; i.e., compression and power. The four-stroke cycle requires two turns of the engine to complete the cycle where the two-stroke cycle only requires one turn.

❑ Optimum efficiency requires that the fuel injected be evenly distributed in the compressed air, forming a homogeneous charge when burning occurs. In order for the combustion to be as rapid as possible, the fuel must be atomized or broken up during injection into finely divided particles so that it presents the largest possible surface area to surrounding air. In diesel engines there occurs a delay period between first injection and commencement of burning, which is then followed by a rapid pressure rise due to the spontaneous ignition of a portion of the total injected fuel. This is termed the *period of rapid combustion,* and is followed by progressive combustion with the introduction of the remainder of the fuel during the period of controlled combustion burning phase. These conditions fall within the general requirements that burning must be complete before the end of the power stroke.

❑ Varying the amount of fuel that is injected into the cylinder and when it is injected controls the power in a diesel engine. For low power, such as at idle, only a short injection period and a small quantity of fuel is used. When high power is required, the injection period is longer. An advantage of the diesel engine is that it has good efficiency at low power because the fuel-air ratio is so lean at reduced power. The fuel-air ratio in a diesel engine varies from 100:1 at idle to 20:1 at full load.

Review Questions

Short Answer Essays

1. Define the basic components of a diesel engine.

2. Define the terms *bore, stroke,* and *engine displacement.*

3. Explain the relationship between torque and horsepower.

4. Define two types of horsepower.

5. Calculate the torque and horsepower for a known diesel engine.

6. Define two classifications of diesel engines.

7. Explain the difference between the 2-stroke and 4-stroke cycle of engine operation.

8. Differentiate between an open combustion chamber and a prechamber engine.

9. Explain the three stages of diesel combustion.

10. Explain how a diesel engine controls power and speed.

Fill-in-the-Blanks

1. The cylinder block is a complex structure that contains the following: sleeves (liners), which house the _____ and _____; cooling passages; lubrication passages; in-block camshaft and bearings; and crankshaft bearing supports in the bottom end.

2. The crankshaft transmits the power through the connecting rod to the _____ _____ and into output.

3. The main function of the vibration damper is to dampen out _____ _____.

4. If the bore is larger than the stroke, the engine is said to be _____ _____.

5. Compression ratio is the ratio of the total _____ _____ to the clearance volume.

6. _____ _____ expresses in a percentage value the increase in engine torque as the engine speed reduces from its maximum no-load rpm or rated speed.

7. Volumetric efficiency (VE) is the ratio of the _____ of air actually entering the cylinder to the weight of air that could enter the cylinder, measured at ambient conditions, if the piston-displaced volume were _____ filled.

8. Diesel engines are classified as in-line, _____, or horizontal design.

9. In a diesel engine there occurs a _____ _____ between first injection and commencement of burning, which is then followed by a _____ _____ rise due to the spontaneous ignition of a proportion of the total injected fuel.

10. The power in a diesel engine is controlled by varying the _____ of fuel that is injected into the cylinder and when it is _____.

ASE-Style Review Questions

1. *Technician A* says when energy is released to do work it is called chemical energy. *Technician B* says stored energy is referred to as potential energy. Who is correct?
A. A only   **C.** Both A and B
B. B only   **D.** Neither A nor B

2. Which of the following combustion chamber designs provide the best fuel economy when used in medium and heavy diesel truck applications?
A. Prechamber design
B. Swirl prechamber
C. Direct injection
D. Indirect diesel injection

3. Excessive black smoke is noticed coming from a diesel engine. What action should be taken first?
A. Perform a compression test
B. Set the injection timing
C. Check for air restrictions
D. Replace the fuel injectors

4. Air used in a diesel engine for combustion is composed of oxygen and nitrogen. *Technician A* says that by volume and weight there is more oxygen than nitrogen in a given amount of air. *Technician B* says that there has to be more oxygen to sustain combustion. Who is right?
A. A only   **C.** Both A and B
B. B only   **D.** Neither A nor B

5. *Technician A* says that a glow plug system is not required for the initial starting of an IDI precombustion diesel engine. *Technician B* says that a DI diesel does not always require the use of a glow system for start-up. Who is right?
A. A only   **C.** Both A and B
B. B only   **D.** Neither A nor B

6. *Technician A* says volumetric efficiency is related to the amount of air that enters the cylinder. *Technician B* says the lower this amount retained air, the larger the quantity of fuel that can be injected into the cylinder. Who is right?
A. A only   **C.** Both A and B
B. B only   **D.** Neither A nor B

7. All of the following are disadvantages of a turbulent swirl precombustion chamber EXCEPT:
A. A wide range of fuel can be used.
B. More work is required to compress the air.
C. A high heat loss.
D. A higher fuel rate.

8. Which of these is one of the three ingredients for combustion?
A. Oxygen   **C.** Helium
B. Water   **D.** Argon

9. Which one of these strokes does a two-stroke diesel engine contain?
A. Intake   **C.** Power
B. Exhaust   **D.** Squish

10. Which of these is the smallest part of a compound?
A. Atom   **C.** Proton
B. Molecule   **D.** Neutron

Cylinder Block Assembly

Upon completion and review of this chapter, you should be able to:

❏ Identify the types of cylinder liners used in diesel engines and list the relative merits of each type.

❏ Explain how cavitation erosion occurs in wet liners.

❏ Define the functions of the piston assembly.

❏ Identify trunk and articulating pistons and list their advantages and disadvantages.

❏ Outline the advantages of the Mexican hat open combustion chamber design in a DI diesel engine.

❏ Explain how rings lubricate the cylinder walls and seal the cylinder.

❏ Identify commonly used diesel engine piston rings.

❏ Outline the conditions required to enable rings to seal the cylinder.

❏ Describe the role of connecting rods and outline their stresses.

❏ Outline the normal forces applied to diesel engine crankshafts.

❏ Define the term *hydrodynamic suspension*.

❏ Explain the function of the harmonic balancer (tortional [vibration] damper) and flywheel assemblies.

Introduction

The engine cylinder block contains the components that make up the bottom end and driving end of the diesel engine. The **cylinder block** acts as the central frame of the engine: all the other engine components are in some way attached to it, including the remaining engine housing components. The pistons, connecting rods and crankshaft are responsible for transmitting the gas pressures developed in engine cylinders to the engine power take-off mechanism, usually a flywheel. This group of components will be defined as the *engine power train*. This chapter will explain the design and operating characteristics of the following components or systems:

1. The engine cylinder block, including the cylinder liners

2. Pistons and connecting rods

3. Piston rings and pins

4. Crankshafts

5. Crankshaft and rod bearings

6. Oil pans

7. Torsional damper (harmonic balancer)

8. Flywheel

Engine Cylinder Block

The engine **cylinder block,** as shown in Figure 3-1, is the frame of the engine around which all the other components are assembled, in much the same way that sub-components are assembled around a truck frame. The cylinder block houses the engine cylinders and the engine crankcase. North American highway medium- and large-bore diesel engines all use in-line or V-cylinder configurations requiring a single crankshaft and therefore use cylinder blocks cast as a single unit. In all these engines, the cylinder block is manufactured from essentially a cast iron alloy. Aluminum and plastics may be used as a material for oil pans. For example, the Caterpillar 3176 engine uses an aluminum spacer deck. Cast iron alloy, however, remains the material of choice. The pursuit of lighter, stronger engines that reduce exhaust emissions and last longer determines the way a cylinder block is alloyed.

The cylinder block houses the engine cylinders and the engine crankcase.

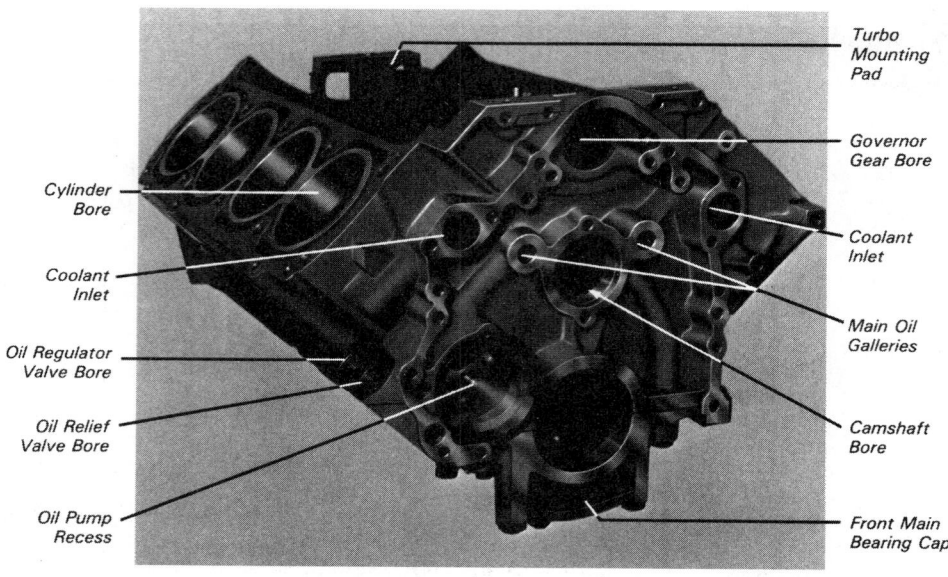

Figure 3-1 Cylinder block terms.

In most large-bore highway diesel engines, the cylinder block is bored and fitted with cylinder **liners** or **sleeves,** which can be replaced when the engine is overhauled. The crankshaft is supported in a cradle of main bearings referred to as the **crankcase.**

The in-line, 6-cylinder engine configuration as shown in Figure 3-2 has become the configuration of choice in medium- and large-bore truck diesel engines. Some engine manufacturers build V configurations, usually V-8s with V angles ranging from 90 degrees to a relatively tight 50 degrees. Only one larger-bore, V-8 engine has prevailed over the years in truck applications: Mack Trucks 90 degree V E9 engine. The story is somewhat different in bus and coach applications, where the relatively lower height of the V configuration makes it a good fit in the engine compartment located under the rear passenger compartment. In these applications, Detroit Diesel Corporation's (DDC) range of 60 degree V engines have been dominant until fairly recently, especially in low-profile city buses.

Cylinder blocks are bored to contain the camshaft cast with coolant passages and a water jacket. Even in a liquid-cooled engine the cylinder block plays a role in dissipating rejected heat to the atmosphere.

Cylinder Block Materials

Most truck and bus cylinder blocks are manufactured from Grey cast irons usually alloyed with at least some silicon to improve the metallurgical characteristics (reduce the brittleness) of the block material. Almost all current medium- and large-bore highway diesel engines use cylinder liners, meaning that the block base materials are not directly subject to the engine cylinder pressures and temperatures. This permits the use of relatively uncomplicated cast iron alloys as cylinder block materials. The cylinder block incorporates the following (Figure 3-3):

1. Bores for the piston assemblies.
2. Main bearing bores to mount the crankshaft.
3. Bores/supports to mount the camshaft (in some cases).
4. Coolant passages/water jacket.
5. Lubricant passages/drillings.
6. Mounting locations for other engine components.

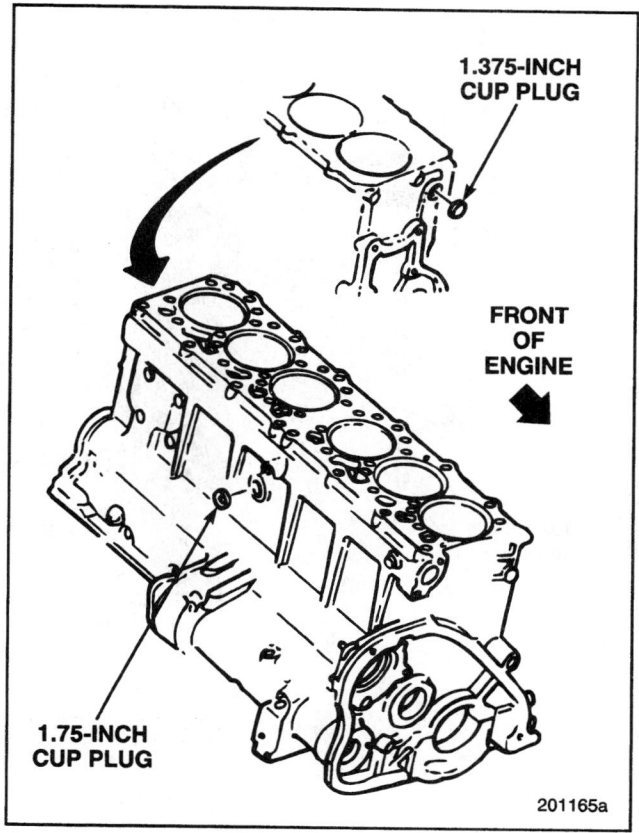

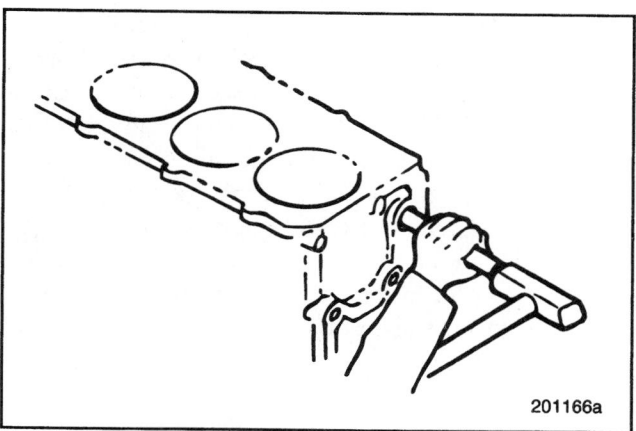

Figure 3-2 In-line 6-cylinder engine.

Cylinder blocks used in truck and bus CI engines can be categorized by type:

1. Integral cylinder bore.
2. Wet sleeve/liner.
3. Dry sleeve/liner.
4. Combination wet/dry liners.

Other factors such as 2-stroke cycle and air-cooled will also be reflected in the block design.

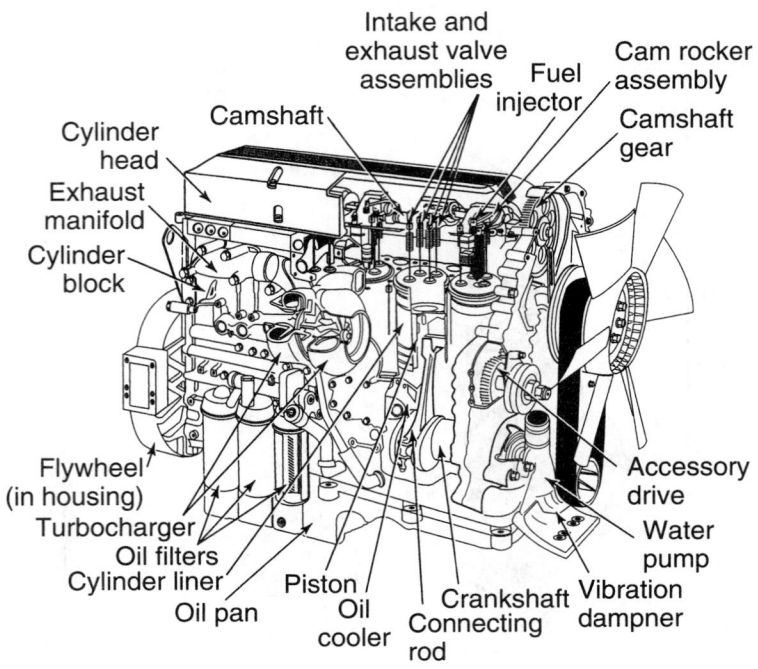

Figure 3-3 Diesel engine exploded view. (Reprinted by permission of copyright owner Detroit Diesel Corporation, all rights reserved.)

Cylinder Liners and Sleeves

Integral Cylinder Bore

Most automobile and small-bore engines use integral or **parent-bore** (cylinder) **blocks,** as shown in Figure 3-4. Typically, large-bore dimension diesel engines do not use a parent bore. The Caterpillar 3208, GM 6.5L, and Navistar 7.3L (T444E) are examples of parent-bore assemblies. The initial cost saving in producing a block of this design is compromised at engine overhaul, when the block will require either replacement or a boring/sleeving machining operation. OEMs seldom use this design, because large-bore diesel engines can easily go 500,000 miles before overhaul, so they are designed with the premise that they will probably undergo reconditioning at least once in their operational life. In the case of parent-bore blocks, where this use is predominantly in gasoline, spark ignited engines, the cost is less and they typically are used for that long a time period.

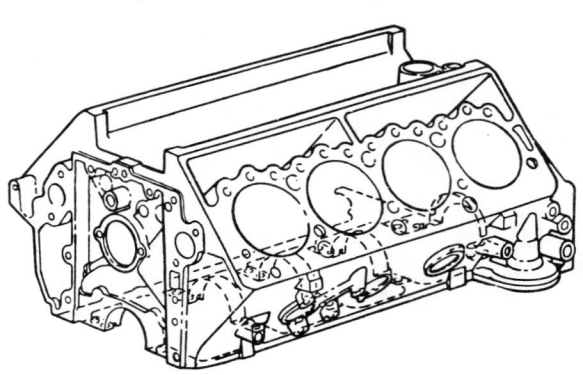

Figure 3-4 Parent bore cylinder block. (Courtesy of General Motors Corporation, Service Operations)

■ **CAUTION:** Never wear jewelry when working around diesel engine blocks. Gold, silver, and copper are excellent conductors of electricity. If jewelry comes in contact with a current-carrying wire running along the cylinder block, severe injury may result.

Wet Liner (Sleeve)

Liners (sleeves) house the piston and rod, thus they compose the cylinder. Wet liners are designed so that the water jacket is in direct contact with the liner in the block bore. The **liner (sleeve)** must therefore have a wall thickness sufficient to sustain the engine's peak combustion pressures (see Figure 3-5). **Wet liners** transfer heat efficiently into to the coolant and are easily replaced at overhaul. Their main disadvantages are that a seal must be maintained for the life of the liner, and that an O-ring failure results in contaminated engine lube.

Shop Manual Chapter 3, page 74.

▲ **WARNING:** To avoid liner pitting from cavitation erosion you must maintain the cooling system.

Coolant **cavitation** erosion caused by vapor bubble implosion can shorten the life of wet liners, but this is seldom a problem if the coolant is properly monitored when ethylene glycol and propylene glycol coolants are used. Extended life coolant (ELC) coolant requires no monitoring. Despite the fact that a wet liner is constructed from a sizeable mass of cast iron or steel, it is relatively easily flexed. Try installing a telescoping gauge into the bore of a wet liner on the bench and locking it into position: gentle hand pressure is sufficient to flex the liner enough for

Shop Manual Chapter 3, page 73.

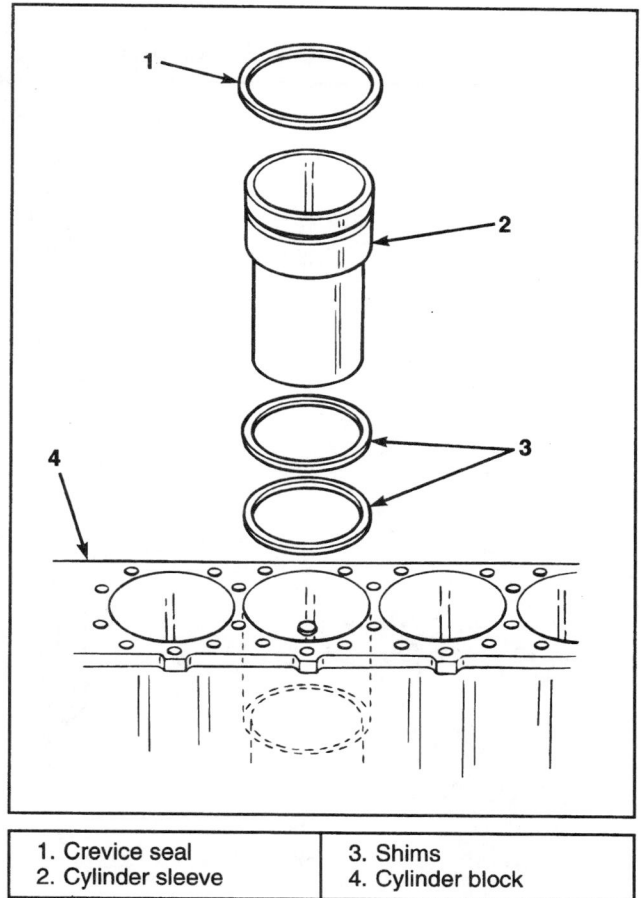

| 1. Crevice seal | 3. Shims |
| 2. Cylinder sleeve | 4. Cylinder block |

Figure 3-5 Wet cylinder liner. (Courtesy of Mack Trucks, Inc.)

the gauge to unseat and drop to the bottom of the liner. When subject to cylinder pressures which may be as high as 2000 psi (18.78 MPa), wet liners expand outward into the wall of coolant that surrounds them. They then contract or shrink, creating a low-pressure void or bubble, which is made of "boiled-off" coolant vapor. This *bubble collapse* occurs almost immediately, causing the wall of coolant to impact on the liner exterior wall. This condition repeats itself at high frequency (17 times per second at 2,000 rpm) and has been tested to produce pressures of up to 60,000 psi. Bubble collapse results in cavitation unless the coolant provides protection in the form of a coating on the exterior of the liner wall. Cavitation can be identified by pitting/erosion that usually appears outside the thrust faces of the piston.

The O rings used to seal wet liners are made from a variety of rubber-type compounds, and it is important that the OEM installation recommendations be observed; they may be installed dry or coated in coolant, soap, engine oil, and various other substances. Wet liners are often alloyed so they are metallurgically superior to the block casting, which increases longevity.

Dry Liner (Sleeve)

The **dry liner** as shown in Figure 3-6 uses thinner walled sleeves than the wet liner. Dry liners (sleeves) are installed into the block bore, usually with a marginally loose fit, and are retained by the cylinder head. The dry sleeves do not transfer heat as efficiently as the wet liners, but they are easily replaced and do not present coolant-sealing problems.

In older applications, dry liners were made of cast iron material almost identical to that of the engine block; these were installed with a fractional interference for optimum heat transfer. Current dry liners tend to be alloyed for superior toughness. The liner's ability to transfer heat is dependant on maximizing its surface contact area in the block, so they are manufactured with a slightly greater coefficient of heat expansion than cast iron. Engine OEMs design dry liners to be installed loose so that they expand into the block bore when heated.

▲ **WARNING:** If these dry liners are installed with the interference fit mandated in older liners they will buckle in use, greatly reducing their service life.

Precise measuring of the block inner diameter (ID), the liner outer diameter (OD), and selective fitting ease installation and increase engine longevity. Loose fits of around .035 mm (.0015 in.) are common.

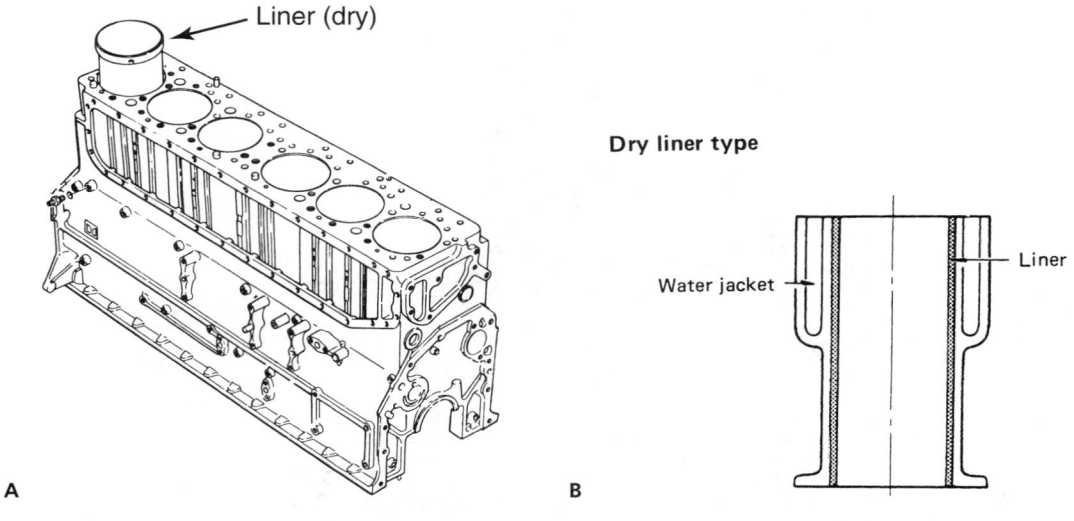

Figure 3-6 (A) Showing a dry cylinder liner installed in a cylinder block; (B) Cross-section of a dry liner wall. (Reprinted with permission by American Isuzu Motors, Inc.)

Combination Wet/Dry Sleeve (Liner)

The wet/dry sleeve is designed so that the hottest part of the liner at the top is in direct contact with the coolant in the water jacket, and the lower portion fits directly to the cylinder block with a fractional loose or interference fit. Consequently, the upper portion in direct contact with the water jacket must have considerably more mass, because it has to contain the cylinder combustion pressures. The wet/dry liner must also seal the water jacket and with the wet liners, O-rings are usually the means.

Piston Assemblies

Shop Manual Chapter 3, page 81.

By definition, a **piston** as shown in Figure 3-7, includes a circular plug that seals an engine cylinder bore and reciprocates. Pistons are subject to the gas within the cylinder on which it will either impart or receive force. The piston assembly consists of the piston, the piston rings (which are used both to seal the cylinder and lubricate the cylinder walls), and a wrist pin that connects the piston to the connecting rod. The upper face of the piston is called the *crown,* and is exposed directly to the cylinder chamber and therefore the effects of combustion. A connecting rod links the piston assembly to a throw on the crankshaft. The crankshaft throw is offset from the centerline of the crankshaft. As the crankshaft is rotated, the piston reciprocates in the cylinder bore, thus the linear reciprocating movement of the piston in the cylinder bore is translated into rotary movement at the crankshaft. In such an arrangement, the throw dimension defines piston stroke or piston travel distance. Stroke is the distance from the centerline of the crankshaft main bearing to the centerline of the crank throw.

Diesel engine pistons absorb up to 20% of the heat of the combustion gases. A piston's ability to rapidly dissipate heat is essential to engine life. For example, high-output, turbocharged engines often have piston cooling jets that spray lubricating oil on the underside of the piston to help take the heat away from the piston. Piston crown temperatures are always high, with the rings assisting in the cooling process by conducting some of this heat to the cylinder walls. The

Stroke is the distance from the centerline of the crankshaft main bearing to the centerline of the crank throw.

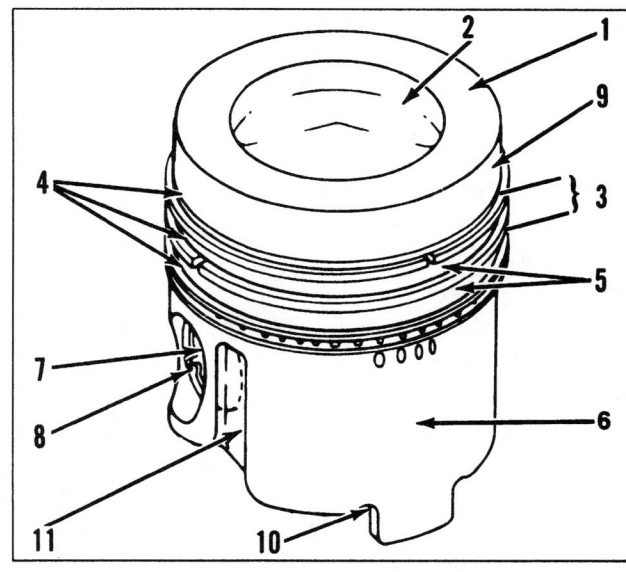

Illustration 67. 3400 Series Engine piston nomenclature. 1. Crown   2. Crater   3. Ring band   4. Ring grooves   5. Ring lands   6. Skirt   7. Pin bore   8. Snap ring groove   9. Top land   10. Cooling jet relief   11. Skirt relief.

Figure 3-7 Trunk type piston assembly. (Courtesy of Caterpillar Inc.)

crown geometry (the shape of the piston leading edge) has much to do with the gas dynamics (swirl and squish) produced during the compression stroke, which determines the fuel-air mixing characteristics and ignition point. The Mexican hat piston crown design is fairly common in direct injected (DI) diesel engines that tend to have low clearance volumes: at TDC. The piston may rise so high in the cylinder bore that recesses are required to accommodate the protruding cylinder valves. These recesses are known as **valve pockets**.

> ⚠️ **WARNING:** It cannot be emphasized enough how critical it is for a piston to efficiently transfer the heat it is exposed to, as combustion temperatures may exceed the melting point of the material from which it is made. Combustion temperatures rise to transient spikes of 2000° C (3630° F). Aluminum melts at 660° C (1220° F), and cast iron melts at 1540° C (2800° F).

Two basic piston designs are used in current truck and bus diesel engines. Trunk-type pistons (Figure 3-7) were until the 1990s the piston of choice in most diesels in the 200 BHP (150 kW) to 500 BHP (375 kW) power range. These have given way more recently to articulating two piece pistons. Engine manufacturers have used articulating pistons for many years, and some technicians familiar with Detroit Diesel engines know them as crosshead pistons. There are advantages and disadvantages to both trunk and articulating designs.

Trunk-Type Pistons

Single piece **trunk-type pistons** (Figure 3-7) are usually manufactured from aluminum alloys. The rationale for using a single piece piston in a truck diesel engine is that it minimizes piston mass, so because of weight, cast-iron trunk pistons are seldom used. The aluminum alloying substance is most often a small percentage of silicone, which considerably toughens the aluminum. The low melting temperature of aluminum combined with a lack of toughness when compared with cast iron will require most aluminum trunk pistons to use a ring groove insert for at least the top compression ring groove. The insert is usually a Ni-Resist™ insert, a nickel bearing, iron alloy with great resistance to high temperatures and wear, but also with a coefficient of heat expansion nearly identical to that of aluminum. The Ni-Resist™ insert has a high bonding ability to the aluminum piston. Figure 3-8 details trunk piston terminology and typical tolerance specifications.

Heat treatment also serves to improve the metallurgical characteristics of aluminum. In some cases, hypereutectic (an alloying process) aluminum pistons are used to provide increased fatigue resistance. Another method of increasing the toughness and high-temperature performance of aluminum without compromising its primary advantage of light weight is the CFA (ceramic fiber reinforced) process used on some current smaller-bore engines. The CFA process eliminates the need for a Ni-Resist™ insert and reinforces the piston at the top of the ring belt extending up into the crown periphery. This design permits both a higher top ring location (an emission advantage) and lowers piston-operating temperatures because the coefficient of heat transfer is higher than the Ni-Resist™. To toughen the crown area of the piston, some manufacturers use a fiber reinforcement manufacturing practice known as squeeze cast fiber reinforced (SCFR); this is an aluminum fiber manufacturing process.

Aluminum pistons are light in weight and have a high coefficient of heat transfer, which means they dissipate heat rapidly. Aluminum also has a high coefficient of heat expansion, which dictates that many diesel engine trunk pistons be *cam ground.* A cam-ground piston is slightly elliptical when cold. As the piston heats, the greater mass of material around the pin boss will expand more than the thinner skirt area between the pin bosses. The idea is that at running temperatures, the piston should expand to a circular shape. Aluminum trunk pistons are cast to strengthen the piston where it is most vulnerable, so apart from the reinforcement at the pin boss, they have increased mass at the crown. Aluminum trunk-type pistons may also be surface treated to improve their wear characteristics. Some common methods are as follows:

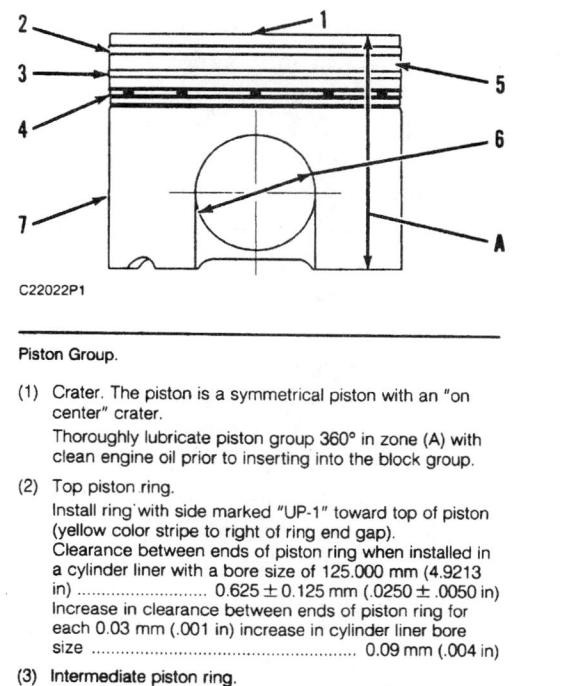

C22022P1

Piston Group.

(1) Crater. The piston is a symmetrical piston with an "on center" crater.

Thoroughly lubricate piston group 360° in zone (A) with clean engine oil prior to inserting into the block group.

(2) Top piston ring.

Install ring with side marked "UP-1" toward top of piston (yellow color stripe to right of ring end gap).
Clearance between ends of piston ring when installed in a cylinder liner with a bore size of 125.000 mm (4.9213 in) .... 0.625 ± 0.125 mm (.0250 ± .0050 in)
Increase in clearance between ends of piston ring for each 0.03 mm (.001 in) increase in cylinder liner bore size .... 0.09 mm (.004 in)

(3) Intermediate piston ring.

Install ring with side marked "UP-2" toward top of piston (green color stripe to right of ring end gap).
Width of groove in piston for intermediate ring (new) .... 3.061 ± 0.013 mm (.1032 ± .0005 in)
Depth of groove in piston for intermediate ring (new) .... 3.727 mm (.1467 in)

Thickness of intermediate ring
(new) .... 2.990 ± 0.010 mm (.1177 ± .0004 in)
Clearance between groove and intermediate ring
(new) .... 0.048 to 0.094 mm (.0002 to .0037 in)
Clearance between ends of piston ring when installed in a cylinder liner with a bore size of 125.000 mm (4.9213 in) .... 0.625 ± 0.125 mm (.0250 ± .0050 in)
Increase in clearance between ends of piston ring for each 0.03 mm (.001 in) increase in cylinder liner bore size .... 0.09 mm (.004 in)

(4) Oil regulating piston ring.

Oil ring spring ends to be assembled 180° from ring end gap (white colored portion of spring must be visible at ring end gap).
Width of groove in piston for oil ring
(new) .... 4.033 ± 0.013 mm (.1588 ± .0005 in)
Depth of groove in piston for oil ring (new) ..... 3.727 mm (.1467 in)
Thickness of oil ring (new) .... 3.987 ± 0.013 mm (.1570 ± .0005 in)
Clearance between groove and oil ring
(new) .... 0.020 to 0.072 mm (.0008 to .0028 in)
Clearance between ends of piston ring when installed in a cylinder liner with a bore size of 125.000 mm (4.9213 in) .... 0.55 ± 0.15 mm (.022 ± .006 in)
Increase in clearance between ends of piston ring for each 0.03 mm (.001 in) increase in cylinder liner bore size .... 0.09 mm (.004 in)

After piston rings have been installed, rotate rings so the end gaps are 120° apart.

(5) Crown assembly.

(6) Piston pin bore diameter in piston skirt .... 51.15 ± 0.15 mm (2.014 ± .006 in)

Thoroughly lubricate piston pin with clean engine oil prior to inserting into piston group and rod assembly.

(7) Piston skirt.

Figure 3-8 Trunk piston terminology. (Courtesy of Caterpillar Inc.)

1. Tin plating (tinning can also be used to repair scores on pistons)
2. Anodizing
3. Chrome plating (usually of crown)
4. Squeeze cast fiber reinforced (more than just surface treatment; the structural integrity of the piston is increased)

Piston pins, when not full floating, are usually press fit to the piston boss and float on the rod eye. Heating the piston to 95° C (200° F) in boiling water facilitates pin assembly. Trunk pistons are prone to piston slap. This is the tilting action on the piston when the piston is thrust loaded by cylinder gas pressure. Tapering the piston so that the outside diameter at the lower skirt slightly exceeds the outside diameter over the ring belt region can minimize piston slap. The ring belt region is exposed to more heat and will expand more as the piston heats to operating temperatures. Advantages of aluminum alloy, trunk type pistons are as follows:

1. Lightweight. This reduces the piston mass and therefore the inertia forces that the connecting rod and crankshaft must sustain during tensile stressing of the connecting rod and crankshaft are significantly lower. This permits the use of lighter weight components throughout the engine powertrain.

2. Cooler piston crown temperatures. The ability of aluminum alloy pistons to rapidly dissipate the combustion heat they are subjected to permits lower crown temperatures.

3. Quieter. Engines using aluminum alloy, trunk-type pistons generally produce less non–combustion-related noise than comparatively configured and sized engines with articulating piston assemblies.

Articulating Pistons

Most OEMs are now manufacturing at least some engines equipped with articulating pistons, usually in their high-output models. The **articulating piston** shown in Figure 3-9 consists of a crown, usually manufactured out of a cast iron alloy, and a separate skirt, usually (but not always) manufactured out of an aluminum alloy. A wrist pin links the two components to the connecting rod eye. The Detroit Diesel Corporation (DDC) has used *crosshead* pistons for many years. The crosshead piston is a two-piece piston assembly consisting of a crown and a skirt united by a semi-floating **wrist pin**; because both the crown and skirt assemblies have some degree of independent movement (i.e. they pivot) the assembly *articulates*. DDC wrist pins are bolted to the rod assembly, so they are described as semi-floating. The true articulating piston assembly has a free-floating piston pin and therefore bearing surface with both the pin boss and the connecting rod eye. Otherwise, it is similar to the crosshead piston.

Most piston crowns are manufactured from a cast-iron alloy suitable for the high pressures and temperatures it is directly exposed to, while the skirt can be made out of a lighter material, usually an aluminum alloy. Two-piece, articulating pistons offer greater longevity and reduced tendency to piston slap. However, the reason they have become the piston design of choice in recent engines has more to do with the higher combustion pressures, temperatures, and emissions standards required of today's optimum-efficiency low-emission diesel engines. Engine designers are fueling diesel engines with ever higher injection pressures to maximize the complete combustion of the fuel injected into the cylinder, a consequence of which is the extension of the time duration of peak combustion temperature. Also, the volume in the cylinder above the top compression ring and below the crown leading edge tends to be unaffected by cylinder turbulence and the effects of cylinder scavenging, so the cylinder contains dead gas. Placing the top compression ring as close as possible to the crown leading edge can minimize this volume, called *headland volume*. Because the tensile strength of cast iron is more than twice that of the aluminum alloy trunk piston and requires no groove insert, the upper ring groove can be located optimally close to the crown leading edge. This is known as *headland piston design* and it not only diminishes smoking but also improves fuel economy.

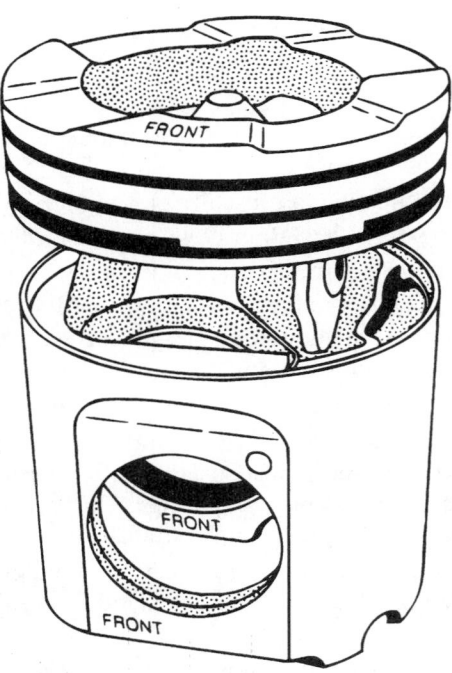

Figure 3-9 Articulating piston. (Courtesy of Mack Trucks, Inc.)

A major disadvantage of articulating piston assemblies over the aluminum trunk design is significantly increased weight, which increases tensional loading on the power train, requiring the use of beefed-up engine cylinder block and powertrain components, especially the connecting rods and crankshaft.

Advantages of Articulating Pistons

1. **Reduced piston slap.** When the piston crown is subjected to cylinder gas pressure both on the compression and power strokes, the thrust loading tends to cock (pivot off vertical centerline) the piston in its bore. This action is minimized with the articulating piston design, as the skirt assembly is separate and not subject to the vertical load forces of the crown.

2. **Reduced thermal distortion.** The cast iron used as the crown material is less subject to temperature distortion within engine combustion temperature parameters. The skirt, whether manufactured from aluminum alloy or ferrous metal, is to some extent isolated from the crown, permitting cooler and more consistent running temperatures and therefore closer fit tolerances.

3. **Greater life.** The superior toughness of the cast-iron crown allows it to withstand considerably more rigorous operating expectations and routine abuse.

4. **Improved emissions.** Locating the top ring close to the piston leading edge is only possible with the superior toughness of cast iron, which permits the headland crown design.

5. **Improved fuel economy.** Despite the substantial increase in piston mass of articulating pistons over equivalent aluminum alloy trunk pistons, the greater toughness and temperature tolerance of cast iron alloy piston crowns permit higher cylinder pressures and temperatures to maximize thermal efficiency of the engine.

Piston Thrust Faces   As cylinder gas pressure acts on a piston, there is a tendency for it to cock (pivot off a vertical centerline) in the cylinder bore because it pivots on the wrist pin. This action creates thrust surfaces on either side of piston. The **major thrust face** is on the inboard side of the piston as its throw rotates through the power stroke. The **minor thrust face** is on the outboard side of the piston as its throw rotates through its power stroke as shown in Figure 3-10. Major thrust face is sometimes simply called the thrust face while the minor *thrust face* is called the *anti-thrust face*. It is important for the diesel technician to identify the thrust faces of a piston for purposes of failure analysis.

Combustion Chamber Designs

Most contemporary truck and bus diesel engines are direct injected (DI). The fuel charge is therefore injected directly into the engine cylinder. In an indirect injected (IDI) diesel engine, the fuel charge is injected into a cell connected to but not integral with the cylinder cavity and in most cases ignition takes place within this cell. The cell is known by several names, according to type and location. Among these names are *precombustion chamber, energy cell,* and *turbulence chamber.* The truck and bus diesel technician is only likely to have encountered IDI technology in the now obsolete Caterpillar precombustion chambers used with their poppet nozzles. Today, while many small-bore diesel engines still use IDI engines, all of the medium and heavy duty engines manufactured in North America are direct injected.

Direct injected engines use an open-combustion chamber principle. In an open-combustion chamber, the injector is usually located in the cylinder head and positioned over the piston crown. The shape of the piston crown will therefore define the type of combustion chamber. Since the mixing of the fuel charge with air takes place within the engine cylinder, the gas dynamics (swirl and turbulence) are critical in determining the mixing efficiency and the actual point at which ignition occurs. Because the piston is the only moving component in the cylinder after the intake valves close, the geometry of the piston crown is critical in defining the gas

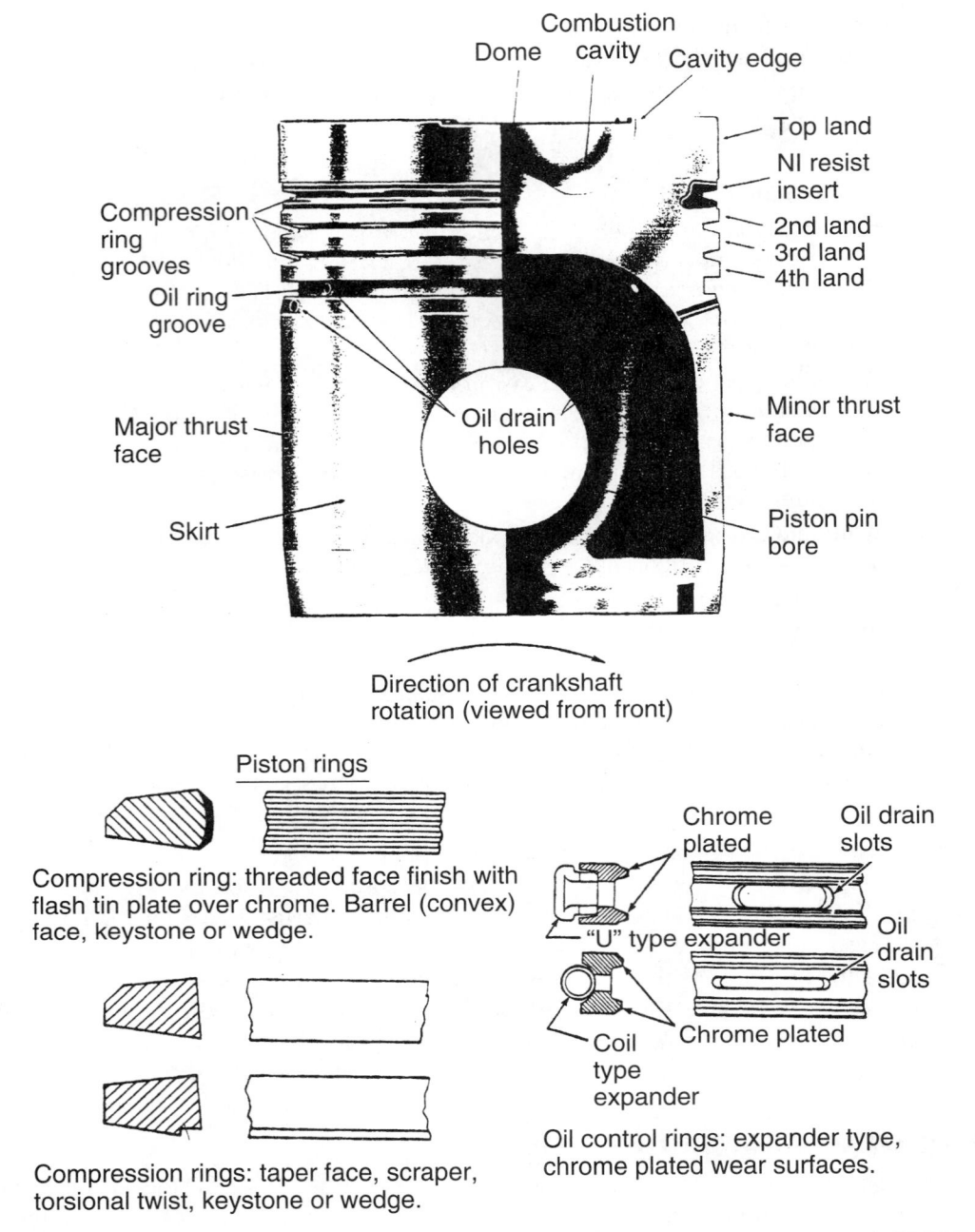

Figure 3-10 Piston and ring terminology. (Courtesy of Mack Trucks, Inc.)

dynamics. High-turbulence gas dynamics are those that are designed to produce more vigorous gas movement than quiescent gas dynamics, which use a lesser amount of swirl and turbulence. Most DI engines use one of three basic piston crown designs.

Mexican Hat The **Mexican hat piston** crown design is by far the most common and the title perfectly describes its shape. The central area of piston crown is recessed below the piston leading edge, which forms a kind of wall in the center of the recess in a cone shaped protrusion (see Figure 3-11). This shape produces desirable gas dynamics (turbulence) through the compression stroke, and the fuel injector is located so that it directs atomized fuel around the conical protrusion within the recessed cavity of the piston crown where the swirl effect is greatest. The bowl depth of the Mexican hat piston crown usually determines how much gas movement is generated.

Deep bowl designs produce greater turbulence and are often used with fuel systems with lower peak injection pressures. Some engine OEMs describe their engines as having quiescent gas dynamics, meaning that relatively low turbulence is generated. A shallower Mexican hat bowl is used with quiescent systems, which in most cases require the of use higher fuel injection pressures. Engine OEMs frequently use the Mexican hat piston crown design because it provides desirable gas dynamics, low risk of fuel burn out on the piston below the injector, and long service life.

Mann Type (or "M" Type) The **Mann type piston** (Figure 3-12) crown is named after the German Company that first designed them. It is usually used with trunk-type pistons and consists of a spherical recess or bowl located under the injector, not necessarily in the center of the piston crown. Depending on the depth of the recess, the Mann type combustion chamber generally produces high turbulence but is more vulnerable to localized burnout in the bowl. Navistar uses this design.

Dished A slightly concave to almost flat piston crown design that produces low turbulence when compared with the previous types. It is more likely to be encountered in an IDI application.

Ricardo Turoidal Piston Figure 3-13 shows a Ricardo Turoidal piston used in IDI engines with the Ricardo Comet five precombustion chambers. The top of the piston has what appears to be a half–four-leaf clover recess in the head. As the piston rises on the compression stroke, air swirl begins in these recesses. When the piston reaches TDC and the prechamber opening, the air swirl

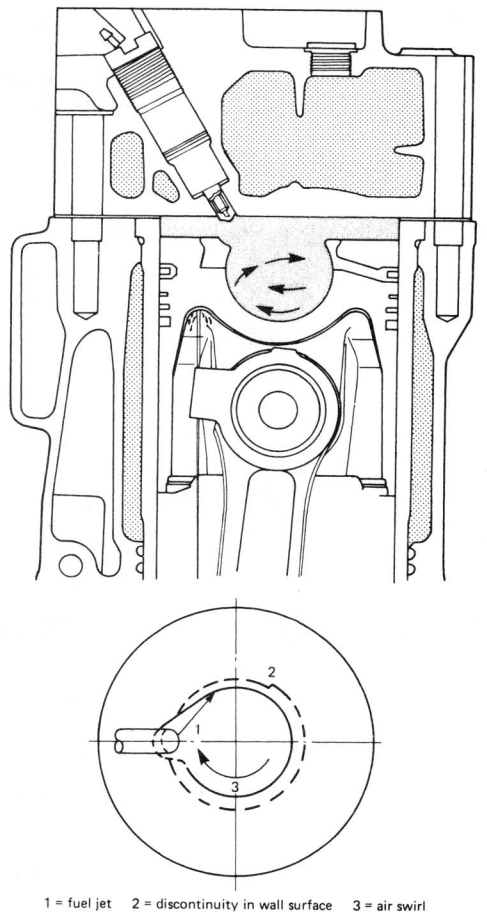

1 = fuel jet   2 = discontinuity in wall surface   3 = air swirl

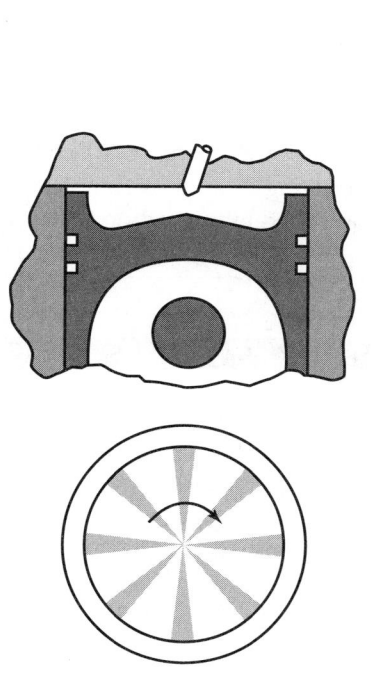

Figure 3-11 Mexican hat piston. (Courtesy of SAE)

Figure 3-12 "M" Mann-type piston. (Courtesy of Butterworths Diesel Technology)

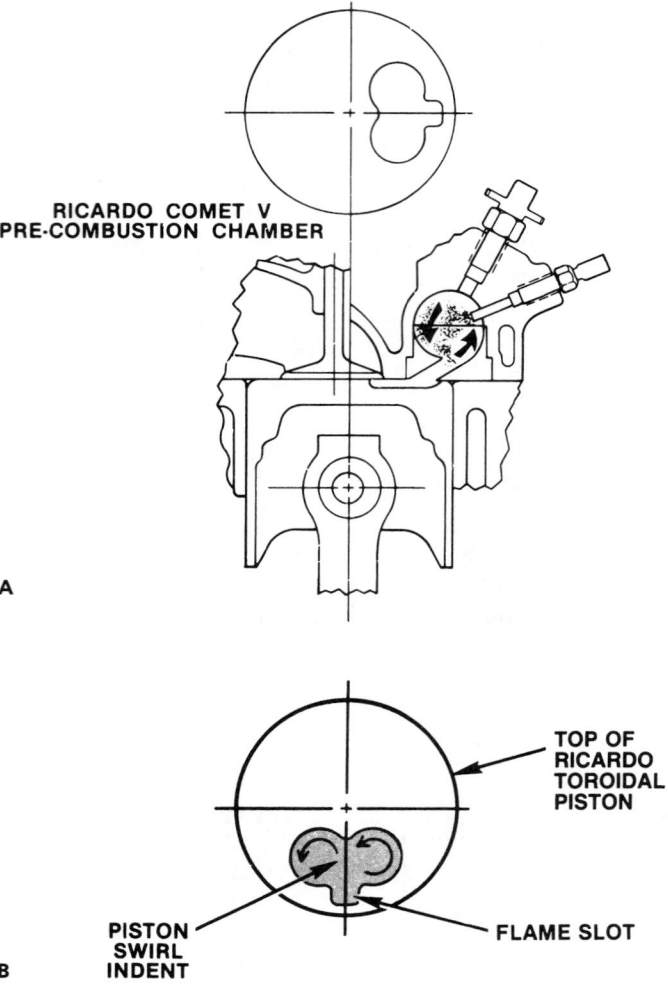

RICARDO COMET V
PRE-COMBUSTION CHAMBER

A

TOP OF
RICARDO
TOROIDAL
PISTON

PISTON
SWIRL
INDENT

FLAME SLOT

B

Figure 3-13 (A) Ricardo turoidal piston cross-section; (B) Top of the Ricardo turoidal piston. (Courtesy of General Motors Corporation, Service Operations)

mass enters the flame slot at the entrance to the prechamber. The charge is forced out of the prechamber throat area agitating the entire mixture and resulting in a more complete mixing of air and fuel and better combustion. This design has a broad speed range and produces low noise and effective emission control.

A BIT OF HISTORY

Sir Harry Ricardo, a British engineer, developed the swirl prechamber system called The Ricardo Comet Head in 1931. His system also employed cavities within the piston besides the retort-shaped combustion prechamber in the cylinder head that are designed to improve the distribution of air swirl when compared with earlier chambers of the same type. By partly forming the combustion cell within a heat insulated member, and by utilizing compression-induced swirl only, an improved maximum power rating was obtained.

Piston Cooling

The size of the piston, BMEP, and whether the engine is boosted will determine the need for *piston cooling* and the type used (Figure 3-14). Because combustion temperatures may peak at values that exceed the melting temperature of the cylinder materials, it is essential that the heat not converted to useable energy (rejected heat) is either exhausted directly or dissipated through the cylinder materials. Therefore, some of this heat must be transferred through the piston assembly. For obvious reasons, the cooling of aluminum, trunk-type pistons is more critical than with articulating pistons. However, it should be remembered that a slightly mis-aimed piston-cooling jet can cause a rapid failure. Three methods of cooling pistons are generally used:

Shaker: Oil is delivered through the connecting rod to a cell in the underside of the crown; the motion of the piston shakes the oil like a cocktail shaker, which cools the piston crown.

Circulation: Oil is delivered through the connecting rod and the wrist pin, and is subsequently circulated through a series of grooves machined into the underside of the piston crown. It then drains back into the crankcase.

Spray: A stationary jet, block-mounted below the cylinder liner and fed by a lubricating oil gallery, is directed at the underside of the piston. This oil cools the piston crown and may also lubricate the wrist pin. The jet must be precisely aimed on installation to be effective. OEM design accomplishes aiming by using a clear Perspex template that fits over the fire ring groove on the cylinder block deck, and an aim rod inserted in the jet orifice. A direction or target window

BMEP is theoretical constant pressure acting on the piston during each power stroke. BMEP produces a power equal to the brake or shaft horsepower of the engine. This term indicates how well an engine uses its displacement to produce work and it is a better parameter for comparing engines than torque or horsepower.

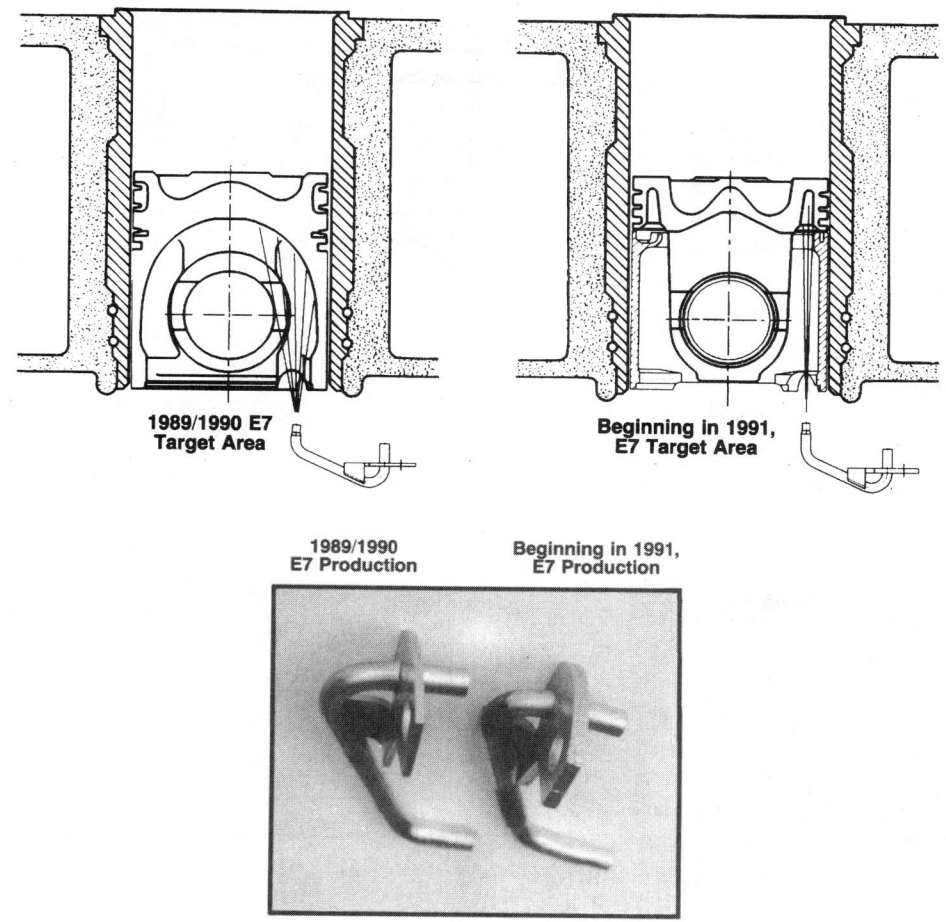

1989/1990 E7 Target Area

Beginning in 1991, E7 Target Area

1989/1990 E7 Production

Beginning in 1991, E7 Production

Figure 3-14 Piston cooling jets. (Courtesy of Mack Trucks, Inc.)

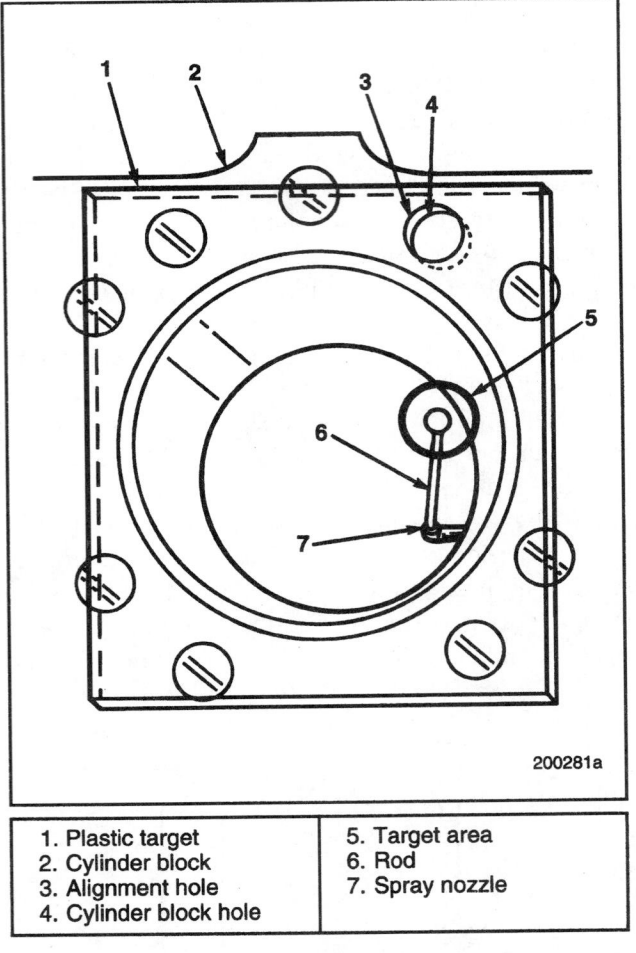

1. Plastic target	5. Target area
2. Cylinder block	6. Rod
3. Alignment hole	7. Spray nozzle
4. Cylinder block hole	

Figure 3-15 Spray-nozzle targeting. (Courtesy of Mack Trucks, Inc.)

is scribed in the Perspex template and the aim rod must be positioned within the window (Figure 7-6 demonstrates this procedure). The spray cooling method is used in most current turbocharged diesel engines (Figure 3-15).

Engines may use one of these piston cooling methods or a combination of them.

Some Facts

1. Heat conductivity, or the rate of heat flow, is approximately three times greater in aluminum than cast iron. Therefore, aluminum will dissipate the heat it is exposed to much more quickly than cast irons and steels.

2. The weight of aluminum is .097 lb per cubic inch.

3. The weight of cast iron is .284 lb per cubic inch.

Piston Ring Design

The function of **piston rings**, as shown in Figure 3-16, is to seal the piston in the cylinder bore. Most pistons require rings to effectively seal, and those that do not are usually found in automobile racing applications using special piston materials and, more importantly, run at high RPM, which permit little time for cylinder leakage. Rings have three important functions:

Shop Manual Chapter 3, page 81.

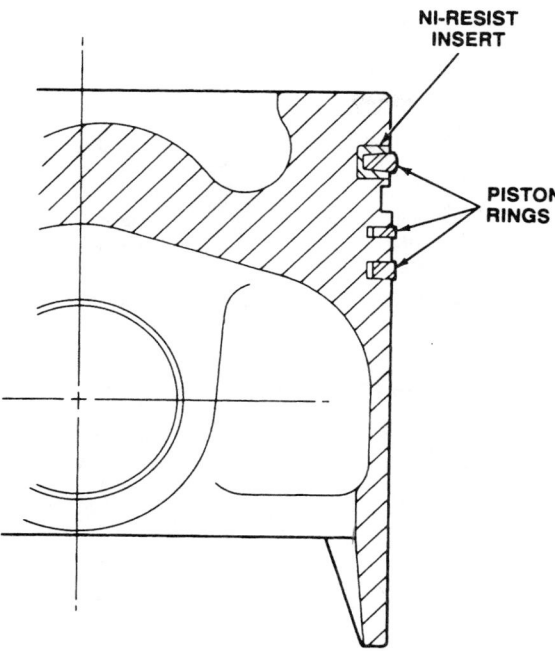

Figure 3-16 Piston with rings. (Courtesy of Navistar International Engine Group)

1. Sealing: They are designed to seal compression and combustion gases within the engine cylinder.
2. Lubrication: They are designed to apply and regulate a film of lubricant to the cylinder walls.
3. Cooling: They provide a path for heat to be transferred from the piston to the cylinder walls.

Piston rings are located in the circular recesses in the piston known as ring grooves. Ring grooves are located between the ring lands. Piston rings may be broadly categorized as *compression rings* and *oil control rings.* **Compression rings** are responsible for sealing the engine cylinder and play a role in helping to dissipate piston heat to the cylinder walls. **Oil control rings** are responsible for lubricating the cylinder walls and also provide a path to dissipate piston heat to the cylinder walls. Piston rings are designed with an uninstalled diameter larger than the cylinder bore, so that when they are installed, radial pressure is applied to the cylinder wall. Until recently, most piston compression rings were manufactured from cast-iron alloys and tended to be more brittle than today's versions. Now, most piston rings are metallurgically more similar to stainless steel than cast iron, and as a consequence, their susceptibility to fracture has diminished. In fact, many are highly flexible. The open graphite structure of the cast-iron rings gave them desirable self-lubricating properties but a not-so-desirable shorter service life. In most current diesels, these have given way to much tougher and more flexible rings which often show little evidence of wear at engine overhaul. The wall section of the piston in which the set of rings is located is known as the ring belt.

Ring Action

The major sealing force of piston rings is high-pressure gas. Piston rings have a small side clearance. The result of this minimal side clearance is that when cylinder pressure acts on the upper sectional area of the ring, it is first forced down into the land. This enables cylinder pressure to get behind the ring and drive it into the cylinder wall (Figure 3-17). Sealing efficiencies increase with cylinder pressure values.

Compression rings are responsible for sealing the engine cylinder and play a role in helping to dissipate piston heat to the cylinder walls.

Oil control rings are responsible for lubricating the cylinder walls and also provide a path to dissipate piston heat to the cylinder walls.

COMPRESSION RING TYPES

KEYSTONE/TRADEZONAL RING BARREL-FACED RING RECTANGULAR RING INSIDE BEVEL RING TAPER-FACED RING

RINGS MAY USE COMBINATIONS OF THE ABOVE GEOMETRY

TYPES OF RING JOINTS

STRAIGHT JOINT ANGLE JOINT STEP JOINT

RING GEOMETRY TERMINOLOGY

PISTON LEADING EDGE
NI-RESIST INSERT
RING SIDE CLEARANCE
BARREL FACED KEYSTONE RING
RING CLEARANCE

COMPRESSION RING ACTION

CYLINDER GAS PRESSURE

CYLINDER PRESSURE FORCES RING INTO THE LAND, THEN GETS BEHIND IT TO FORCE IT INTO THE CYLINDER WALL

PISTON

CYLINDER WALL

Figure 3-17 Piston ring geometry and action. (Courtesy of Diesel Engineering Handbook)

The engine manufacturer determines the number of rings used. Factors considered are bore size, engine speed, and engine configuration. Time is probably the major factor in determining the number of compression rings; the slower the maximum running speed of the engine, the greater the total number of rings because there is more time for gas blowby to occur. Most current truck and bus medium- and large-bore diesel engines have rated speeds that are in the 2,000 rpm range, and engine OEMs commonly use a three-ring configuration of two compression rings and a single oil control ring. However, four- and five-ring configurations can still be seen.

The top compression ring will get the greatest sealing assist from cylinder pressures. Gas blowby from the top compression ring is that used to seal the second compression ring. Gas that blows by all the rings enters the crankcase, so crankcase pressure values are often used an indication of the overall health of an engine. As a rule of thumb, gas pressure diminishes by around 50% beyond the top compression ring. Because rings seal the cylinder, some cylinder leakage past the ring belt is inevitable, but the limiting factor is time. In an engine operated at 2,000 RPM, one full stroke of the piston takes place in 15 milliseconds (0.015 second, so there is insufficient time for significant cylinder leakage to take place. Gas blowby will be more pronounced at low speed or during a lugging condition. *Lugging* is a running condition in which cylinder pressures are peaking because there is more time.

Lugging is a running condition in which cylinder pressures are peaking because there is more time.

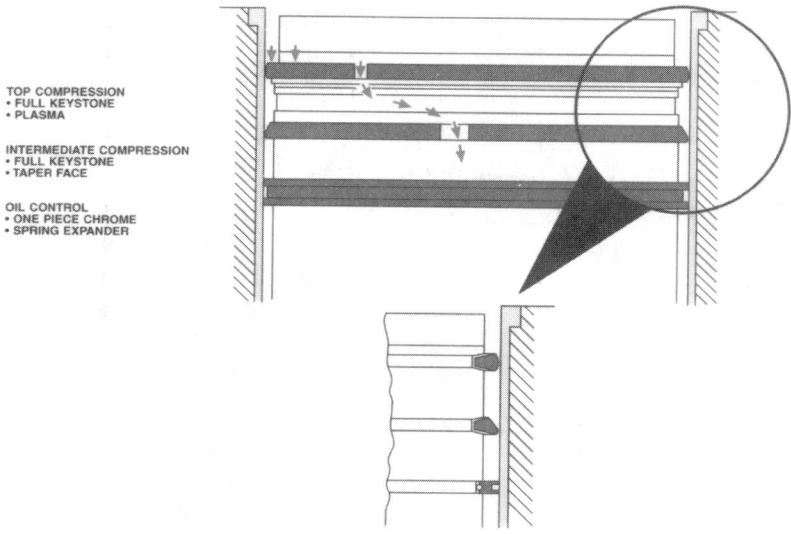

Figure 3-18 Ring stagger and pressure balance. (Courtesy of Caterpillar Inc.)

It should be noted that all the piston rings play a role in controlling the oil film on the cylinder wall, including the top ring. It may be categorized as a compression ring, and that role becomes increasingly more important as emissions standards become tougher and more rigorously enforced (Figure 3-18).

Piston Ring Types

There are many different types of piston ring categorized by function and geometry. When classified by function there are three types.

Compression Rings   These rings are designed with the primary objective of sealing cylinder compression and combustion pressures, though they also play a role in controlling the oil film that is applied to the cylinder wall. There are many different designs, and a variety of materials are used. Malleable and cast irons have generally given way to highly alloyed steels containing molybdenum, silicon, chromium, vanadium and other alloys. Most lack the brittleness of the older cast iron rings; on the contrary, they can be substantially deformed without fracturing. *Compression rings* are usually coated by tinning, plasma and chrome cladding to diminish friction (see Figure 3-19). Some ring coatings are break-in coatings designed to facilitate run-in. These temporary coatings wear and end up in the crankcase lube, and this should be taken into account when examining oil sample analyses. Sometimes the upper compression ring is known as the **fire ring**, but more often this term is used to describe the cylinder seal at the top of the liner that is often integral with the cylinder head gasket (Figure 3-20).

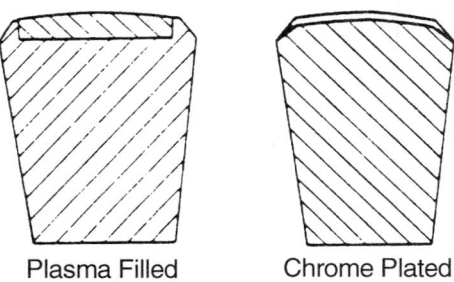

Plasma Filled   Chrome Plated

Figure 3-19 Comparison of plasma-filled and chrome-plated ring face. (Courtesy of Caterpillar Inc.)

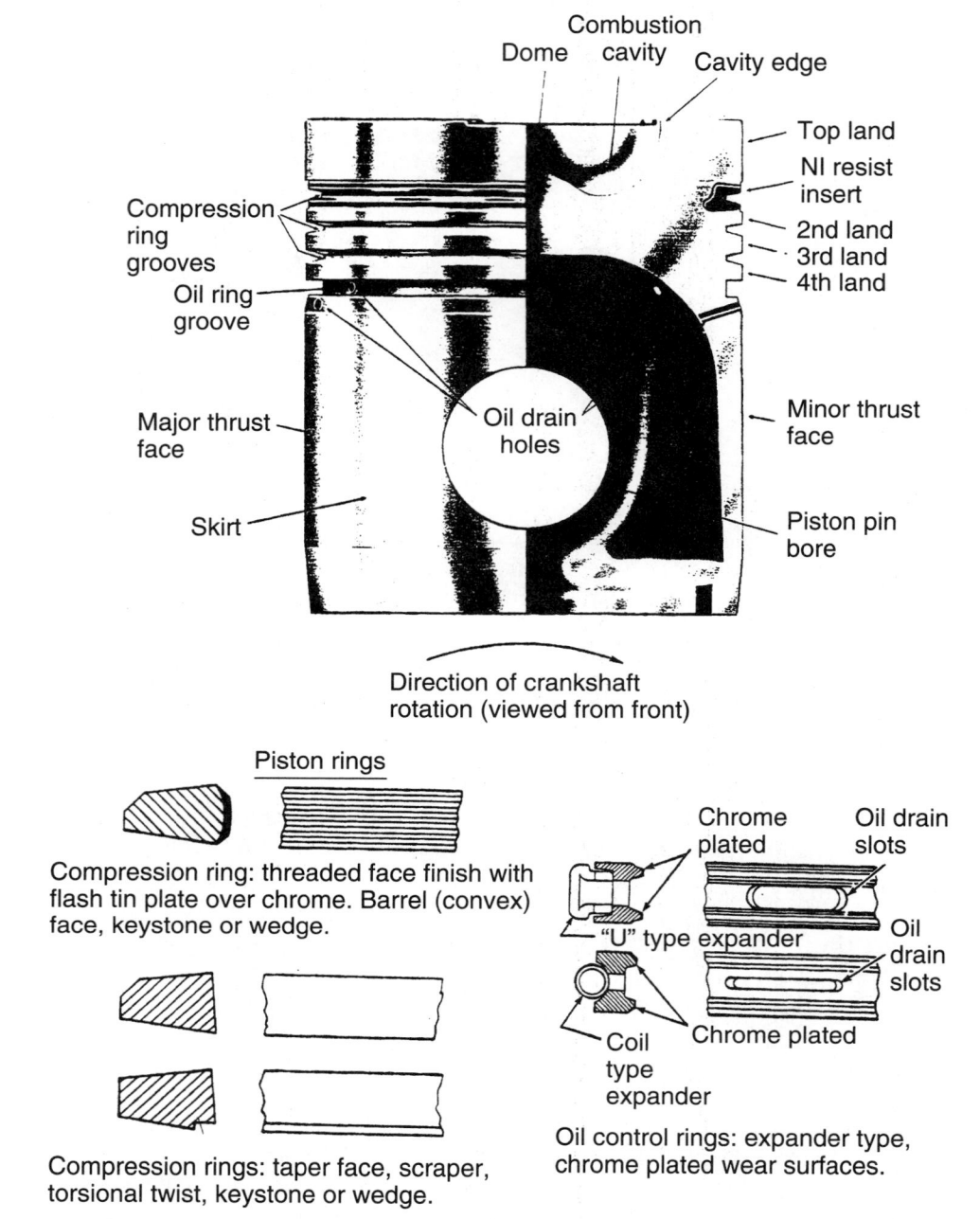

Figure 3-20 Piston and ring terminology. (Courtesy of Mack Trucks, Inc.)

Combination Compression and Scraper Rings Designed both to assist in sealing combustion gases that have blown by the ring above it and to assist in controlling the oil film on the cylinder wall. Manufacturers who use this term are referring to a ring or rings located in the intermediate area of the ring belt, under the top compression ring and above the oil control ring(s).

Oil Control Rings Oil control rings are designed to control the oil film on the cylinder wall. This has to be achieved with some precision; if there is too much oil on the cylinder wall, some will end up in the combustion chamber. Too little will result in scoring and scuffing of the cylinder wall. The action of first applying and then wiping lubricant from the cylinder wall is also responsible for removing heat from the cylinder, transferring it to the engine lube. Most *oil control rings* use circumferal scraper rails forced into the cylinder wall by a circumferal expander, usually a coiled spring. They are sometimes known as *conformable rings* because they will flex to conform to a moderately distorted liner bore.

Piston Ring Geometry

Ring geometry describes the physical shape of the piston rings. Most piston rings use combinations of the characteristics outlined here. Some examples follow:

Keystone or Trapezoidal Rings   Keystone rings are wedge shaped and fitted to a wedge-shaped ring groove. The design is commonly used for the top compression ring because its shape allows the gas pressure exerted on its upper sectional area to get behind the ring and act on its inner circumference to load it into the cylinder wall. Keystone rings also inhibit the build-up of carbon deposits due to the angled scraping action caused by the relative motion of the ring within its groove, so they are less susceptible to sticking. They will also function effectively as they wear. These rings are used in most headland-designed pistons. They have become the ring design of choice among the middle- and large-bore diesel OEMs.

Rectangular Rings   The sectional shape of the piston is rectangular. Commonly used in the past when the ring material of choice was cast iron, this ring design is loaded evenly (that is, with relatively little twist) into the cylinder wall when subjected to gas pressure, resulting in lower unit sealing pressures but greater longevity.

Barrel Faced   The outer face is "barrelled," with a radius (rounded), usually with the objective of increasing its service life: there is no sharp edge to bite into the cylinder wall when the ring subjected to gas pressure twists within the groove. Keystone rings are often barrel faced.

Inside Bevel   A peripheral recess is machined into the inner circumference of the ring to facilitate cylinder gas getting behind the ring and causing it to twist within the groove. When a ring twists within its groove the result is usually high unit sealing pressures—that is, the ring is allowed to bite into the cylinder wall to maximize the seal.

Taper Faced   The design is much the same as the rectangular ring with the exception that the outer face is angled, giving it a sharp lower edge. Once again, this enables the ring to achieve high unit-sealing pressures, that is, to bite into the cylinder wall when loaded with cylinder gas pressure.

Channel Section   This design is used exclusively for oil control rings. They usually consist of a channel sectioned or grooved ring with a number of slots to permit oil to be both applied and removed from the cylinder walls. Often an expander ring is used in conjunction with this design to improve its conformability: this is a coiled or trussed spring installed to a groove behind the face rails.

Ring Joint Geometry

Piston rings must be designed so that when heated to operating temperatures, they do not expand sufficiently to permit the joint edges to contact each other. They must also seal with some efficiency when cold and when cylinder pressures are low. Three types of joint design are shown in Figure 3-17:

1. Straight. The split edges of the ring abut. This design has the disadvantage of affording the most potential for gas blow-by at the ring joint. It is, however, the most commonly used.
2. Stepped. This design uses an L-shaped step at the joint and affords the least potential for cylinder leakage at the ring joint.
3. Angled. The ring is faced with complimentary angles at the abutting joint and seals fairly efficiently at the ring joint.

Ring Materials and Coatings

Molybdenum is harder than chrome, has a lower coefficient of friction and due to its high melting point, 2,620° C (4,750° F) vs. 1,766° C (3,210° F) for chrome and 1,538° C (2,800° F) for cast iron, is highly scuff (localized welding) resistant.

Base metals used are Grey cast-iron, die-formed stainless steels and nodular cast irons. Nodular cast irons have tensile strengths exceeding 200,000 psi and are ductile to the extent that they can be completely deformed without fracturing. Some cast-iron rings with a stainless steel (molybdenum) inlay have been used in diesel applications. Stainless steels are commonly used in diesel oil control rings and sometimes those are clad or faced with chromium.

Rings coated with tin or electroplated with cadmium are less common in today's engines; plasma-sprayed molybdenum is used to coat piston rings in engines designed for longevity. Some heavy-duty diesel engines use ceramic ring coatings composed of a mixture of aluminum and titanium oxides.

Piston and Cylinder Wall Lubrication

Oil-control rings are designed to maintain a precisely managed oil film on the engine cylinder wall. They further assist with the sealing of cylinder gasses and coating the critical surface areas that are in direct contact with the piston assembly and loaded with extremely high unit pressures. On the piston downstroke, when not loaded by cylinder pressure, lubricating oil is forced into the lower part of the ring groove while the ring is contacting the upper ledge of the land. When the piston changes direction to travel upwards, the ring is forced into the lower land of the ring groove, allowing the lubricating oil to pass around the ring to be applied to the liner wall. While the action of simultaneously applying and scraping oil from the cylinder walls ensures that the applied film thickness is minimal, all engines will burn some oil. In current low emissions engines, this burned oil may be minute, but it does exist.

Wrist Pins

In the United Kingdom, wrist pins are referred to as gudgeon pins.

Wrist or piston pins, shown in Figure 3-21, are used to connect the piston assembly with the connecting rod eye or small end. In the two-piece articulating piston assembly they also serve to link the crown with the skirt: both crown and skirt are permitted independent movement on the wrist pin. The wrist pin transmits loading from the piston assembly and the connecting rod while acting as the axis for the angular movement of the connecting rod as the crankshaft rotates. The extent of the loads to be transmitted will determine whether the pin is solid or bored through: the weight of the wrist pin adds to the piston assembly mass, so it is engineered to be as light as possible while sustaining the loads it is subjected to. Maximum engine speed and peak cylinder pressures will determine the actual wrist pin design and material. In most cases, wrist pins are manufactured from mild steel and surface hardened, but middle-alloy steels are used in some heavy-duty applications. Their bearing surfaces are lubricated by engine oil directed upwards through a rifle bore in the connecting rod. *Full-floating* piston pins are fitted to both the rod eye and the piston with minimal clearance, usually between .0025 mm and .0250 mm (.0001 in.–.001 in.).

Piston Pin Retention

All full-floating piston pins require a means of preventing the pin from exiting the pin boss and contacting the cylinder walls. Snap rings and plugs are used. When installing the internal snap rings used by most engine OEMs, it is important to observe the installation instructions. These usually require that the split joint be installed perpendicular to the piston (on a vertical plane to piston travel), usually up but sometimes down, depending on the manufacturer.

▲ **WARNING:** Snap rings are subject to inertia, which increases proportionally with piston speed. When snap rings are installed with the split joint at right angles to the direction of piston travel, they have been known to separate from their retention groove in the pin boss, causing an engine failure.

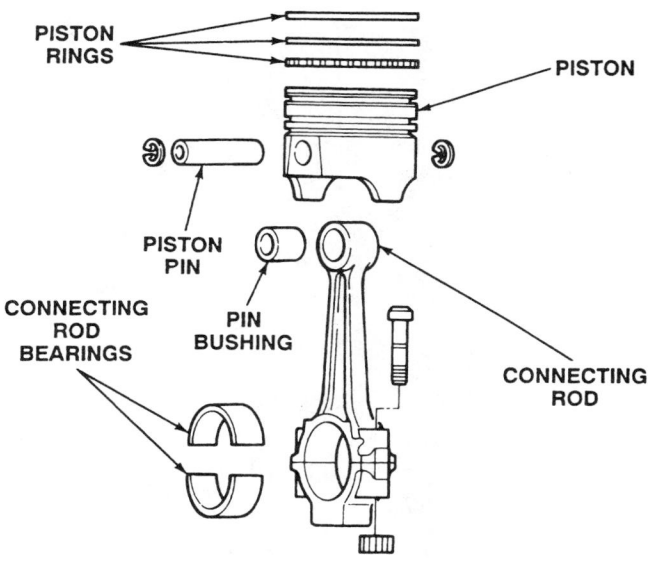

Figure 3-21 Piston and rod assembly. (Courtesy of Navistar International Engine Group)

Semi-floating wrist pins, such as those used in crosshead piston assemblies, are bolted directly to the connecting rod. In DDC two-stroke cycle engines, a press fit, sealing cap is used. This component should be vacuum leak-tested when assembled. Failure of this cap to seal will result in wrist pin lubricant bleeding to the cylinder walls, and air box pressures will charge the crankcase.

Connecting Rods

Connecting rods (also called *con-rods*) transmit the force developed in the cylinder and acting on the piston to the throw on the crankshaft (Figure 3-22). The end of the connecting rod that links to the piston wrist pin is known as either the *rod eye* or the *small end*, while the end that links it to the crankshaft throw is known as the *big end*. Both the rod eye and big end have bearing surfaces. In this way the linear force that acts on the piston and drives it through its stroke can be converted to rotary force or torque by the crank throw, which rotates around the centerline of the crankshaft. Connecting rods used with crosshead-type pistons have no rod eye bearing and instead have a saddle that bolts directly to the wrist pin. The piston pin boss bearing therefore provides upper bearing action. Most truck and bus diesel engines use two-piece conning rods. The rod is usually forged in one piece, the big end cap being subsequently cut off, faced and fastened (bolted) for machining. Most rods use an I-beam section design but round section has also been used. Many rods are rifle drilled from big to small ends to carry lubricant from the crank throw up to the wrist pin for purposes of both lubrication and cooling. Keystone eye rods have become commonplace as they reduce bending stresses on the wrist pin by increasing the loaded sectional area. Connecting rods are subjected to two types of loading discussed in the following text.

Compression Loading

During the compression and power strokes of the cycle, the connecting rod is subjected to compression loading. The extent of compression loading on a rod can be calculated if the cylinder pressure and the piston sectional area values are known. Truck and bus diesel engine

Shop Manual
Chapter 3, page 91.

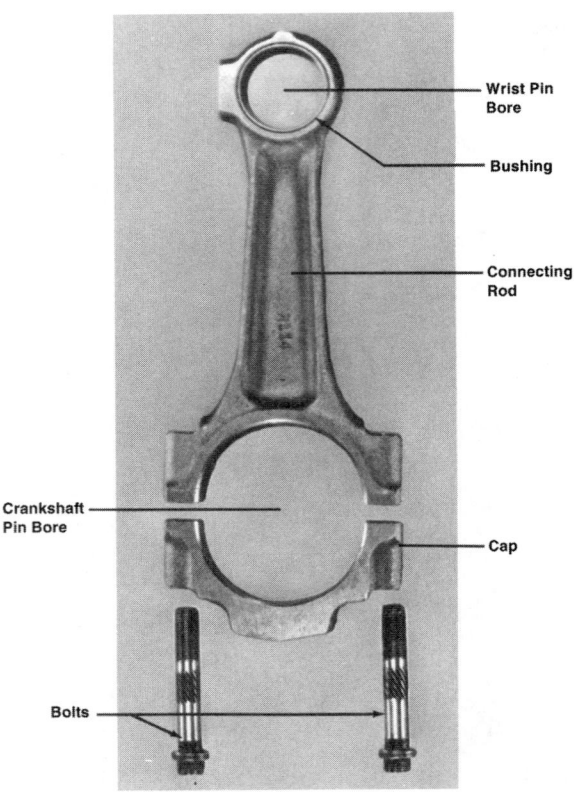

Figure labels: Wrist Pin Bore, Bushing, Connecting Rod, Crankshaft Pin Bore, Cap, Bolts

Figure 3-22 Connecting rod. (Reprinted by permission of copyright owner Detroit Diesel Corporation, all rights reserved.)

connecting rods seldom fail due to compression overloading of con-rods, and when they do, it is usually coincidental with another failure, such as hydraulic lock. Hydraulic lock is usually the result of a cylinder head gasket failure that has permitted coolant to leak into the cylinder.

Tensional Loading

At the completion of each stroke, the piston actually stops in the cylinder at either TDC or BDC. This reversal of motion occurs nearly 70 times per second when an engine is run at 2000 rpm. The greater the mass of the piston assembly, the greater the inertial forces, and therefore the greater the tensile stress to which the rod and crank throw are subject. This can be extreme in modern engines using heavy articulating piston assemblies. Tensile loading on connecting rods increases with engine rpm and the resulting piston speed increase.

 WARNING: When an engine overspeeds, the increased tensile loading on the connecting rod can lead to failure.

Connecting Rod Materials

Connecting rods are metallurgically sophisticated because of the punishment they must endure under normal operation. Desirable characteristics are a degree of elasticity, lightness and the ability to absorb the piston compression loads. Rods may be heat-treated, and some are shot peened.

Connecting Rod Bearings   Most engine OEMs use a single-piece bushing, press-fit to the rod eye, and two-piece friction-bearing shells at the big end. Rod eye bearings should be removed using an appropriately sized mandrel or driver and an arbor or hydraulic press.

WARNING: Using a hammer and any type of driver that is not sized to the rod eye bore is not recommended, as the chances of damage are high.

New one-piece bushings should be installed using a press and mandrel, ensuring that the oil hole is properly aligned, and then sized using a broach or hone. Split big end bearings should also be installed respecting the oil hole location and the throw journal to bearing clearance should be measured. Split bearing shells and rod eye bushings should both be installed to a clean dry bore: remove any packing protective coating from the bearings by washing them in solvent, followed by compressed air drying.

Crankshafts and Bearings

A **crankshaft,** as shown in Figure 3-23, is a shaft with offset throws or journals to which piston assemblies are connected by means of connecting rods. When rotated, the offset **crankshaft throws** convert the linear reciprocating movement of the pistons in the cylinder bores into rotary motion at the crankshaft. Figure 3-24, view A, shows some typical crank throw arrangements, and view B shows crankshaft terminology. Crankshafts are supported by friction bearings at the main journals and require pressurized lubrication at all times. Engine oil pressure from an operating engine enables *hydrodynamic suspension* within the bearing bores. **Hydrodynamic suspension** is the supporting of a rotating shaft on a fluid wedge of constantly changing, pressurized engine oil. This oil suspension prevents direct main journal-to-bearing contact. An oil film protects the friction-bearing shell when the engine is stationary or cold cranked. Friction bearings are also used at the crank throw/connecting rod big end. The lubrication required for these bearings is supplied from oil passages routed through the crankshaft and oil holes located at each journal. Offsetting the oil hole angle from the crankshaft axis generally provides a thicker wedge of oil for the crankshaft to ride and for this reason, most OEMs do this. Crankshafts are designed for dynamic balance and use counterweights to oppose the unbalancing forces generated by the pistons. These forces tend to diminish as the number of cylinders in an engine increases and companion throws contribute a counter balancing effect. Crankshafts are subjected to the following two types of force:

Bending Forces

Bending stress occurs between the supporting main journals between each power stroke. Crankshafts are engineered to withstand the considerable bending forces that result from subjecting a crank throw to the compression and combustion pressures developed in the cylinder. This *normal* bending stress takes place between the main journals any time the crank throw between them is loaded by cylinder pressure and it therefore peaks when engine cylinder pressures peak.

Shop Manual Chapter 3, page 92.

Hydrodynamic suspension is the supporting of a rotating shaft on fluid wedge of constantly changing, pressurized engine oil.

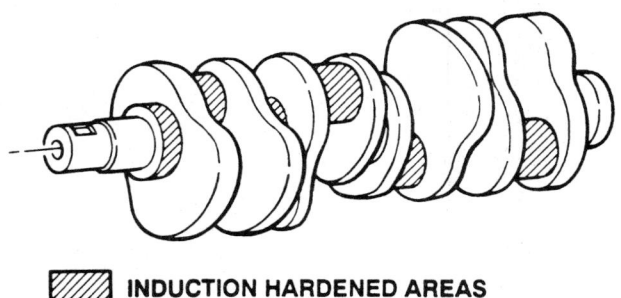

INDUCTION HARDENED AREAS

Figure 3-23 Crankshaft showing induction-hardened bearing surfaces. (Courtesy of Navistar International Engine Group)

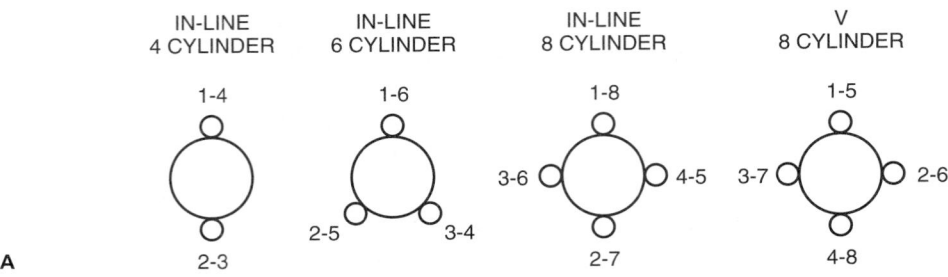

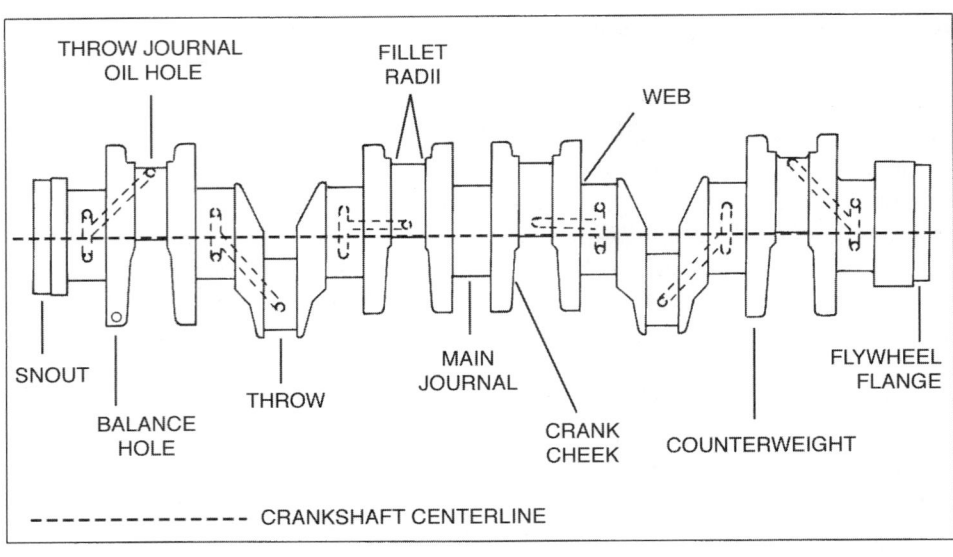

Figure 3-24 (A) Crankshaft throw configuration; (B) Crankshaft terminology. (Courtesy of Caterpillar Inc.)

Torsional Forces

Torsional stresses are the twisting vibrations that a crank is subjected to at high speed. Crankshaft torsional vibration occurs because while under compression (that is, driving the piston assembly attached to it upward on the compression stroke) a given crank throw will slow to a speed marginally less than *average* crank speed. Subsequently, when the throw receives a power stroke, it will accelerate to a speed marginally greater than *average* crank speed. These twisting vibrations or oscillations take place at high frequencies, and crankshaft design, materials, and hardening methods must take them into account. Torsional stresses tend to peak at crank journal oil holes at the flywheel end of the shaft. Torsional oscillations are amplified when an engine is run at slower speeds with high cylinder pressures, because the real time duration between cylinder firing pulses is extended. This type of running (lower speed/high load) in known as *lugging,* but with many new generation, high torque-rise engines, engine torsional oscillations can be projected through the drivetrain.

Crankshaft Construction

Most crankshafts are made of steel forging, but special cast-iron alloys have been used. All crankshafts are tempered (heat treated) to provide a tough core with the required flexing characteristics to withstand the bending and torsional punishment it will be subjected to. An understanding of journal hardening procedures is important, as the reconditionability of the crankshaft is often dependent on this.

Journal Surface Hardening Methods

1. **Flame hardening.** Used on plain carbon steels, and therefore seldom used on current truck/bus applications. Consists of the application of heat followed by quenching with oil or water. The result is relatively shallow surface hardening. The actual hardness depends on the carbon and other alloy content of the steel.

2. **Nitriding.** This is used on SAE 4130/4140 steels and involves higher temperatures than flame hardening; the surface hardens to a greater depth—around .65 mm (.025 in.).

3. **Induction hardening.** The area to be hardened is enclosed by an applicator coil through which A/C (alternating current) is pulsed heating the surface. Tempering is achieved by blast air or liquid quenching. This process results in hardening to depths of up to 1.75 mm (.085 in.), providing a much wider wear and machinability margin than the preceding methods. Most current crankshafts are surface hardened by this method.

Rod and Main Bearings

It makes sense for the truck technician to have a rudimentary appreciation of bearing construction and to be able to diagnose characteristic failures. Most truck and bus engine manufacturers make available excellent bearing failure analysis charts and booklets. These use high-definition color photography and examples of actual failures that make diagnosing failures easy.

Shop Manual Chapter 3, page 95.

Construction and Design   Two basic designs are used in current applications:

1. Concentric wall—uniform wall thickness.
2. Eccentric wall—wall thickness is greater at crown than parting faces; also known as *deltawall bearings*.

Materials   Rod and main friction bearings consist of a steel base or backing plate, onto which is layered copper, lead, tin, aluminum and sintered combinations of metals, often with a zinc or tin outer protective coating (Figure 3-25). Friction bearings are designed to have a degree of embedability—that is, they permit small abrasives to penetrate the outer shell (known as the overlay), to a depth where they will cause a minimum amount of scoring to the crank journals.

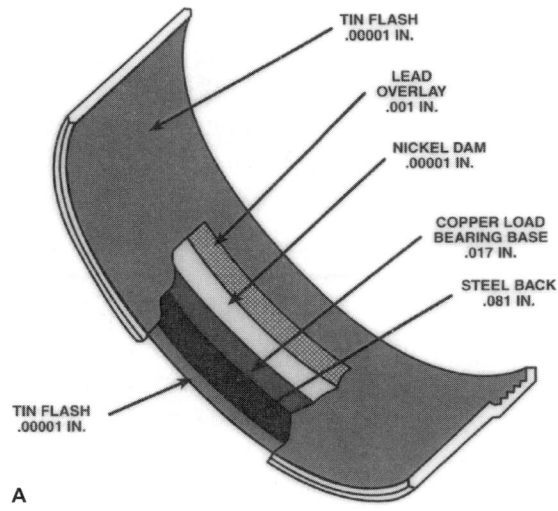

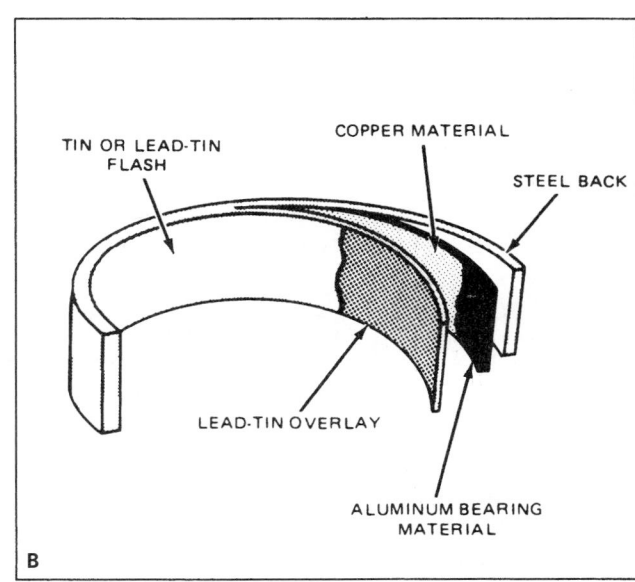

Figure 3-25 (A) Typical bearing construction (Courtesy of Navistar International Engine Group); (B) Copper-bonded bearing (Courtesy of Caterpillar Inc.).

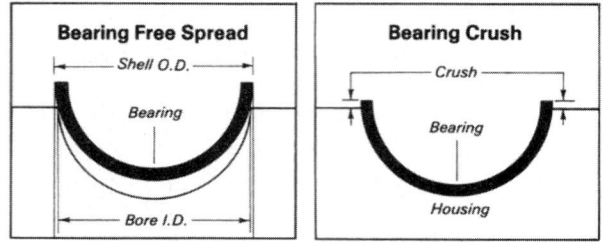

Figure 3-26 Bearing spread and crush. (Reprinted by permission of copyright owner Detroit Diesel Corporation, all rights reserved.)

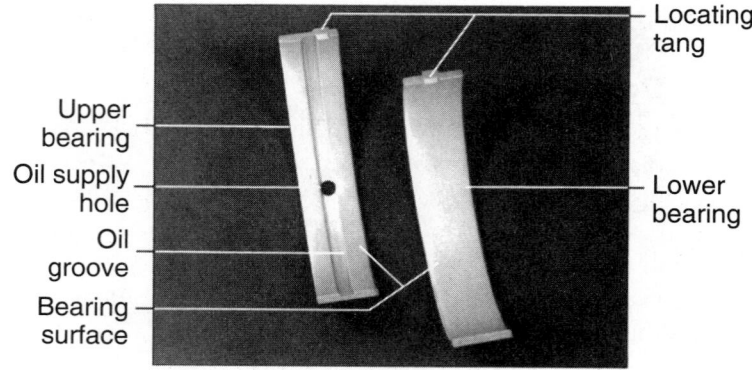

Figure 3-27 Bearing tangs. (Reprinted by permission of copyright owner Detroit Diesel Corporation, all rights reserved.)

Crankshaft Endplay One of the main bearings is usually flanged to define crank endplay; this is known as the **thrust bearing**. Thrust bearings are available in several sizes to accommodate some thrust surface wear and some thrust face dressing in the crankshaft. Alternatively, split rings known as *thrust washers* may be used to control crank endplay. Endplay specifications would typically be in the .02–.03 mm (.008–.012 in.) range.

Bearing Retention Bearings are retained primarily by crush. The outside diameter of a pair of uninstalled bearing shells slightly exceeds the bore to which it is installed (see Figure 3-26). This creates radial pressure that acts against the bearing halves and provides good heat transfer. The bearing halves are also slightly elliptical to allow the bearing to be held in place during installation; this is known as *bearing spread*. Tangs in bearings (Figure 3-27) are inserted into notches in the bearing bore to minimize longitudinal movement, prevent bearing rotation, and align oil holes.

Vibration Damper (Balancer)

A **torsional (vibration) damper** (*harmonic balancer*) (Figure 3-28) is mounted on the free end of the crankshaft, usually at the front of the engine. Both terms describe the same component. Its function is to reduce the level of vibration and add to the flywheel's mass in establishing rotary inertia. In other words, its primary function is to reduce crankshaft torsional vibration.

The vibration damper consists of a damper drive or housing and inertia ring. The housing is coupled to the crankshaft and using springs, rubber or viscous medium, drives the inertia ring: the objective is to drive the inertia ring at *average* crankshaft speed. Viscous-type dampers (Figure 3-29) have become almost universal in truck and bus diesels. The annular housing is hollow and bolted to the crankshaft. Within the hollow housing, the inertia ring is suspended in and

Inertia is a resistance to the state of motion. An object at rest tends to stay at rest and an object in motion tends to stay in motion.

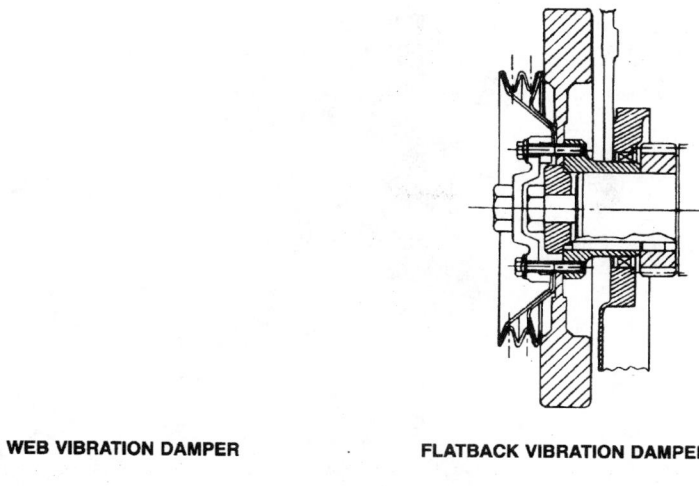

CENTER WEB VIBRATION DAMPER   **FLATBACK VIBRATION DAMPER**

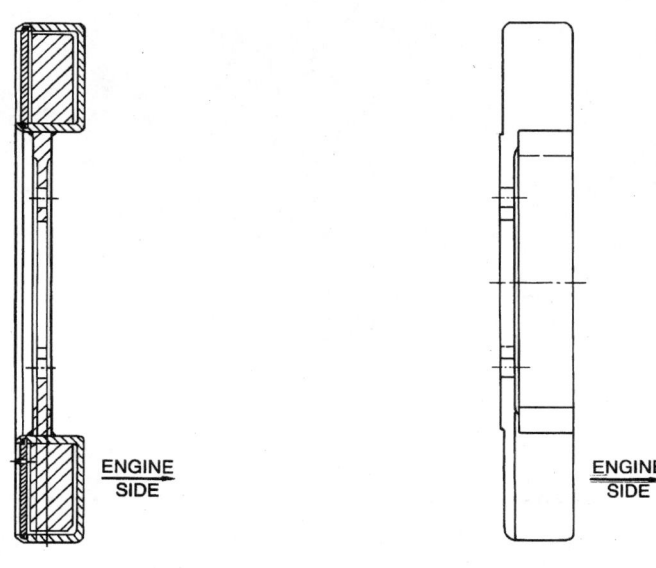

ENGINE
SIDE

ENGINE
SIDE

Figure 3-28 Vibration damper. (Courtesy of Mack Trucks, Inc.)

driven by silicone gel. The shearing of the viscous fluid film between the drive ring and the inertia ring creates the damping action.

Most OEMs recommend replacement of the harmonic balancer at each major overhaul, but this is seldom done due to the expense. Practice has shown that these components frequently exceed OEM-projected expectations. The consequences of not replacing the damper when scheduled are economic, as it may result in a failed crankshaft. The shearing action of the silicone gel produces friction, which is released as heat. This leads to eventual breakdown of the silicone gel, a result of prolonged service life or old age. Drive housing damage is another common reason for viscous damper failure, and this is probably caused by careless service facility practice in most cases.

Rubber-type vibration dampers are not often used on today's heavy truck diesel engines because the heavier torque requires something more effective. This type consists of a drive hub bolted to the crankshaft: a rubber ring is bonded both to the drive hub and the inertia ring. The rubber ring therefore acts both as the drive and the damping medium. The inherent elasticity of rubber enables it to function as a damping medium, but the internal friction generates heat, which eventually hardens the rubber and renders it less effective and vulnerable to shear failures. These dampers are found mostly on light duty diesel- and gasoline-fueled engines.

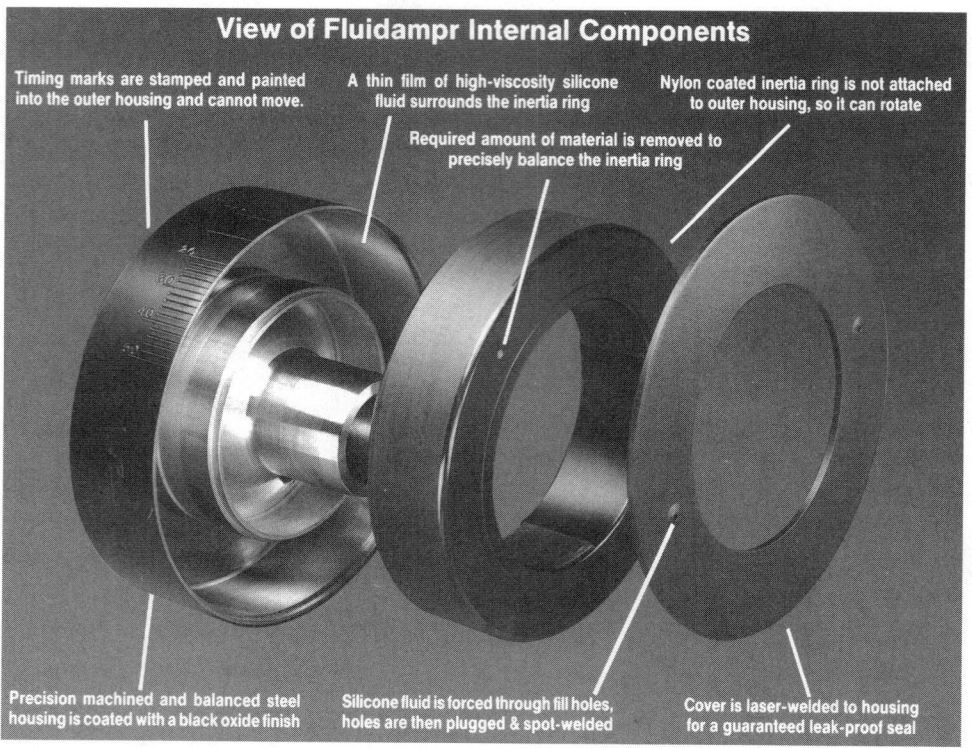

Figure 3-29 Viscous vibration damper. (Courtesy of Vibratech® Inc.)

**Shop
Manual**
Chapter 3, page 101.

Flywheel

The engine flywheel in the typical truck/bus diesel engine is normally mounted at the rear of the engine. **Flywheels** have the following three basic functions:

1. They store kinetic energy (the energy of motion), in the form of inertia, and both help smooth out the power pulses in the engine and establish an even crankshaft rotational speed.

2. They provide a mounting for engine output: it is the power take-off device to which a clutch or torque converter is bolted.

3. They provide a means of driving the engine by cranking motor during start-up.

As a energy storage device, the flywheel plays a major role in dampening the torsional vibrations that act on the crankshaft during engine operation, and its mass helps rotate the engine between firing pulses. Flywheel mass depends on a number of factors, such as whether the engine is two- or four-stroke cycle, the number of engine cylinders, and the engine operating rpm range. Because the number of crank angle degrees between power strokes on a two-stroke cycle is half that on a four-stroke cycle of the equivalent number of cylinders, generally less flywheel mass is required.

❏ Six-cylinder two-stroke cycle frequency of power strokes: 60 crank angle degrees
❏ Six-cylinder four-stroke cycle frequency of power strokes: 180 crank angle degrees

The flywheel therefore stores kinetic energy in the form of inertia. Engines designed to run at consistently high rotational speeds require less flywheel mass, and it should be noted that a heavy flywheel will inhibit rapid response to acceleration demand. While a four-stroke cycle, single-cylinder diesel engine is unlikely to power any modern truck, it is worth taking a look at such an engine and noting its flywheel mass, a dominant characteristic of its appearance.

Types of Flywheels

The flywheels used on all North American built trucks are categorized by the SAE according to size, shape, and bolt configuration. Two basic geometric shapes are used: the *flat face* design and the *pot* design. Most trucks equipped with current medium- and large-bore highway diesel engines use one of two SAE flywheel sizes:

- ❏ SAE # 4 accommodates a 15 1/2″ clutch assembly
- ❏ SAE # 5 accommodates a 14″ clutch assembly

Flywheel Construction

Flywheels used on most truck and bus diesel engines are machined from cast iron or cast steel. Rim velocities are factored by the rotational speed (engine rpm) and the clutch diameter. However, rim stresses seldom generate metallurgical failures in truck and bus diesel engine flywheels due to their relatively low rotational speeds. Stresses tend to peak at the juncture of the hub with the rim, an area that is subject to torsional as well as centrifugal loads, but failures are rare.

Summary

❏ The engine cylinder block is the main frame of an engine, the component to which all others are attached. Most truck diesel engines use either wet or dry liners to facilitate engine overhaul and extend engine longevity. Dry liners are fitted to the cylinder block fractionally loose or with a fractional interference fit. Dry liners do not transfer heat from the engine cylinder to the coolant in the water jacket as efficiently as wet liners. Wet liners have much greater mass than dry liners, as they must fully support the combustion pressures and transfer heat to the water jacket efficiently as the coolant surrounds them. Wet liners must seal the water jacket using O-rings.

❏ Trunk-type pistons manufactured out of aluminum alloys were widely used by truck and bus engine OEMs until the late 1980s because of their light weight and ability to rapidly transfer heat. Two-piece articulating-piston assemblies are more common in today's low emission, extended operational life medium- and large-bore highway diesel engines. Most aluminum-alloy, trunk-type pistons used in truck diesel engines support the top compression ring with a Ni-resist insert and are both cam ground and tapered.

❏ The Mexican hat piston crown, open combustion chamber is the more commonly used in current direct injected diesel engines. Engine oil is used to help cool pistons in three ways: the shaker, circulation, and spray jet methods.

❏ Piston rings seal when cylinder pressure acts on the exposed sectional area of the ring which first forces it down into the land and then gets behind it to load the ring face into the cylinder wall. The efficiency with which piston rings seal a cylinder increases with increased cylinder pressure. Gases that manage to pass by the piston rings enter the crankcase and are known as blow-by gasses. The keystone or trapezoidal ring design is the most commonly used top compression ring in today's highway diesel engines.

❏ Oil control rings are designed to apply a film of oil to the cylinder wall on the up stroke of the piston and "scrape" it on the downstroke. All the piston rings play a role in controlling the oil film on the cylinder.

❏ Full-floating wrist pins have a bearing surface with both the piston boss and the connecting rod eye. Crosshead pistons articulate but have a semi-floating wrist pin that bolts directly to the rod small end. Full-floating wrist pins are retained in the piston boss by snap rings. DDC two-stroke cycle engines use press-fit caps to seal the pin boss to isolate the crankcase from the air box.

❏ Connecting rods are subjected to compression and tensional loads in normal service operation.

Terms to Know

Articulating piston
Cam ground piston
Cavitation
Compression ring
Connecting rod
Crankcase
Crankshaft
Cylinder block
Dry liners
Fire rings
Flywheel
Hydrodynamic suspension
Main bearings
Mexican hat piston
Major thrust area
Mann-type piston
Minor thrust face
Liners
Oil control ring
Parent bore block
Piston
Piston rings
Sleeves
Thrust bearing
Torsional damper
Trunk-type piston
Valve pockets
Wet liners
Wrist pin

- ❏ Crankshafts must be designed to withstand considerable bending and torsional stress. Most medium- and large-bore highway diesel engines use induction-hardened crankshafts. The friction bearings used in crankshaft throw and main journals are retained by crush.
- ❏ Harmonic balancer or vibration dampers consist of a drive plate, drive medium, and inertia ring. The viscous-type damper is the most commonly used on today's truck and bus diesels: the hollow drive ring is bolted directly to the crankshaft and suspended in gelled silicone within a solid inertia ring. The shearing action of the silicone drive medium between the drive ring and the inertia ring effect the damping.
- ❏ The flywheel stores kinetic energy in the form of inertia to help smooth out the power pulses delivered to the engine power train. Size and shape categorize flywheels by the SAE.

Review Questions

Short Answer Essays

1. Define the function of a piston assembly.

2. Identify the types of cylinder liners used in diesel engines and list the relative merits of each type.

3. Explain how cavitation erosion occurs in wet liners.

4. Identify trunk and articulating pistons and list their advantages and disadvantages.

5. Outline the advantages of the Mexican hat, open-combustion chamber design in a DI diesel engine.

6. Outline the conditions required to enable rings to seal the cylinder.

7. Describe the role of connecting rods and outline their stresses.

8. Identify common crankshaft arrangements and match to the appropriate cylinder block configuration.

9. Define the term hydrodynamic suspension.

10. Outline the roles played by the harmonic balancer (tortional [vibration] damper) and flywheel assemblies.

Fill-In-The-Blanks

1. The engine _____ _____ can be considered to be the main frame of an engine, the component to which all others are attached.

2. Wet liners must seal the water jacket using _____.

3. Most aluminum alloy, trunk-type pistons used in truck diesel engines support the top compression ring with a _____ insert and are both ____ ground and tapered.

4. Engine oil is used to help cool pistons in three ways, the _____, _____ and _____ ____ methods.

5. Gases that manage to pass by the piston rings enter the crankcase and are known as _____ _____.

6. Crosshead or articulating pistons articulate, but have a _____ _____ wrist pin that bolts directly to the rod small end.

7. Full-floating wrist pins are retained in the piston boss by _____ _____.

8. Connecting rods are subjected to _____ and _____ loads in normal service operation.

9. The friction bearings used in crankshaft throw and main journals are retained by _____.

10. The flywheel stores kinetic energy in the form of inertia to help _____ ____ the power pulses delivered to the engine power train.

ASE-Style Review Questions

1. Which of the following cylinder block designs would be the most common in medium and heavy duty truck diesel engines?
   A. 8-cylinder, 90-degree V configuration
   B. In-line 6-cylinder
   C. 6-cylinder, 60-degree V configuration
   D. In-line 8-cylinder

2. The reason for bench pressure testing a diesel engine cylinder head is to:
   A. test cylinder gas leakage.
   B. check for air leaks.
   C. check for exhaust leaks.
   D. check for coolant leaks.

3. When *selective fitting* a set of dry liners to a cylinder block, which of the following statements should be true?
   A. The liner with the largest OD is fitted to the bore with the largest id.
   B. The liners are installed with an interference fit.
   C. The liner with the smallest OD is fitted to the bore with the largest id.
   D. The liners are installed with a fractionally loose fit.

4. Most diesel engine cylinder blocks are manufactured from:
   A. cast aluminum alloy.
   B. cast iron alloys.
   C. composite fibers.
   D. stainless steel.

5. Two technicians are discussing the advantages and disadvantages of trunk and articulating pistons. *Technician A* says reduced piston slap occurs on an articulating piston because the thrust loading tends to cock (pivot off vertical centerline) the piston in its bore. This action is minimized, as the skirt assembly is separate and not subject to the vertical load forces of the crown. *Technician B* says trunk style pistons are generally noisier than articulating pistons because they produce more non–combustion-related noise than comparatively configured and sized engines with articulating piston assemblies. Who is right?
   A. A only   C. Both A and B
   B. B only   D. Neither A nor B

6. Which of these conditions is the most likely reason that enables rings to seal the cylinder:
   A. Slower engine speed
   B. High cylinder pressure
   C. Excess side clearance
   D. Low cylinder pressure

7. *Technician A* says crankshaft torsional stresses are the twisting vibrations that a crank is subject to at high speed. *Technician B* says crankshaft torsional vibration occurs because a given crank throw under compression will slow to a speed marginally less than average crank speed. Who is right?
   A. A only   C. Both A and B
   B. B only   D. Neither A nor B

8. All of the following processes are used to strengthen crankshaft journals *except*:
   A. flame hardening.
   B. nitriding.
   C. induction hardening.
   D. annealing.

9. Which of these is the *most* likely reason to use a vibration or torsional damper:
   A. Smooth ride
   B. Prevent crankshaft breakage
   C. Reduce engine vibration
   D. Mounting drive accessories

10. Which of these is the *least* likely use of an engine flywheel:
   A. To mount the ring gear
   B. Smooth out the power pulses
   C. Engine output device
   D. Stop the engine

Cylinder Head and Valve Train

Upon completion and review of this chapter, you should be able to:

❏ Explain the role of the cylinder head in the managing of engine functions.

❏ Define component *creep* and *gasket yield*.

❏ Define the role of the camshaft in a typical diesel engine.

❏ Interpret the terminology used to describe camshaft geometry.

❏ Identify the role the valve train components play in the running of an engine.

❏ List the types of tappet/cam follower used in diesel engines.

❏ Describe the role of the rocker assembly in the engine feedback assembly.

❏ Define the term *rocker ratio*.

❏ Define the role of the cylinder head valves and interpret the terminology used to describe them.

❏ Describe how valve rotators operate.

❏ Define the role of valve seat inserts.

❏ Describe the role of the engine timing gear train in managing engine functions.

Introduction

The cylinder head and camshaft make up the feedback assembly of an engine. As the engine's mechanical management apparatus, it includes timing and accessory drive gearing, the camshaft, tappets, valve and unit injector trains, and fuel pumping mechanisms. Its components are driven by, and often timed to, the engine crankshaft. The drive mechanism for medium and heavy duty highway diesel engine feedback assemblies is almost always a gear set. In lighter duty engines, pulleys and belts or chains and sprockets may be used as drives.

Cylinder Head Construction

Cylinder heads function to seal the engine cylinders and manage cylinder breathing.

Cylinder heads (Figure 4-1) function to seal the engine cylinders and manage cylinder breathing. Diesel applications are usually made of cast iron machined with breathing tracts and ports, cooling and lubrication circuit manifolds, fuel manifolds, and injector bores. They may also support rocker assemblies (Figure 4-2A) and camshafts when the overhead camshaft design (Figure 4-2B) is used. There are several configurations used in diesel engines:

1. Multicylinder, single casting.

2. Single cylinder.

3. Integral with block (requires use of valve cage assembly).

Only the first two are found in current North American truck and bus engines. Cylinder heads usually contain the valve assemblies, breathing ports, injector bores, coolant, and lubricant passages. On IDI applications, precombustion chambers are either integral or installed as units in the head.

Shop Manual
Chapter 4, page 119.

Cylinder Head Gaskets

Yield point means that a malleable gasket is crushed to conform to the profile designed to produce optimum sealing.

Most current cylinder head gaskets (Figure 4-3) are *integral*, meaning the fire rings and grommets are manufactured integrally with the gasket plate. If not, great care should be taken to ensure that the *fire rings* (sealing rings at the liner flange) and grommets are properly positioned while the cylinder head is being installed to the block. The head gasket must be properly torqued to ensure that its **yield point** is attained for proper sealing. Yield point means that a malleable gasket is crushed to conform to the profile designed to produce optimum sealing. Properly torquing

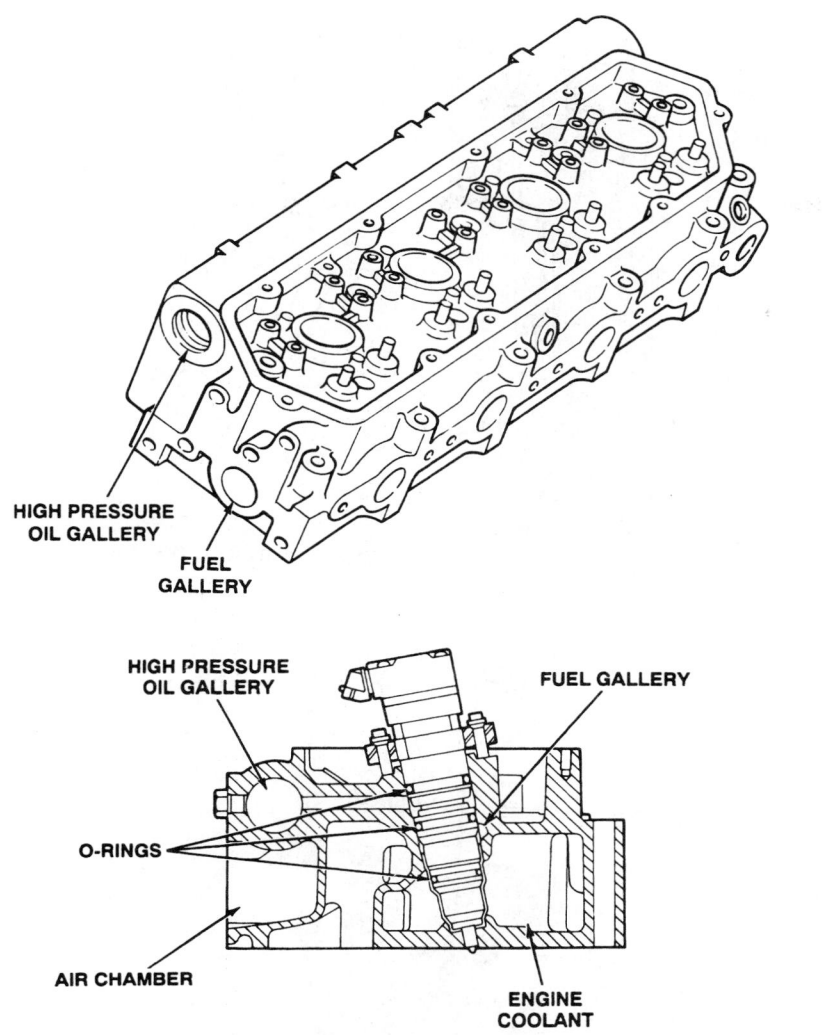

HIGH PRESSURE OIL GALLERY

FUEL GALLERY

HIGH PRESSURE OIL GALLERY

FUEL GALLERY

O-RINGS

AIR CHAMBER

ENGINE COOLANT

Figure 4-1 Cylinder head. (Courtesy of Navistar International Engine Group)

a cylinder head means observing the original equipment manufacturer's (OEM) torque increments and sequencing.

⚠ **WARNING:** Torquing an engine cylinder head to the correct specification, but without observing the incremental steps, can damage the gaskets by deforming them.

■ **CAUTION:** If you remove a diesel engine cylinder head, make sure that the cylinder head lifting fixtures used are secure and the correct ones. Diesel cylinder heads are very heavy; if they fall, personal injury could result.

The objective of torquing cylinder head gaskets is to ensure that the required amount of clamping force is obtained and the gasket conforms to its engineered yield shape. Most head gaskets require no applied sealant and, in fact, may fail if sealants are used. Head gaskets are designed to seal engine components in a region where both the temperatures and pressure are at their highest: they must do this and at that same time accommodate a large amount of component creep. **Creep** is the relative movement of clamped engine components due to different

Shop Manual Chapter 4, page 129.

Shop Manual Chapter 4, page 130.

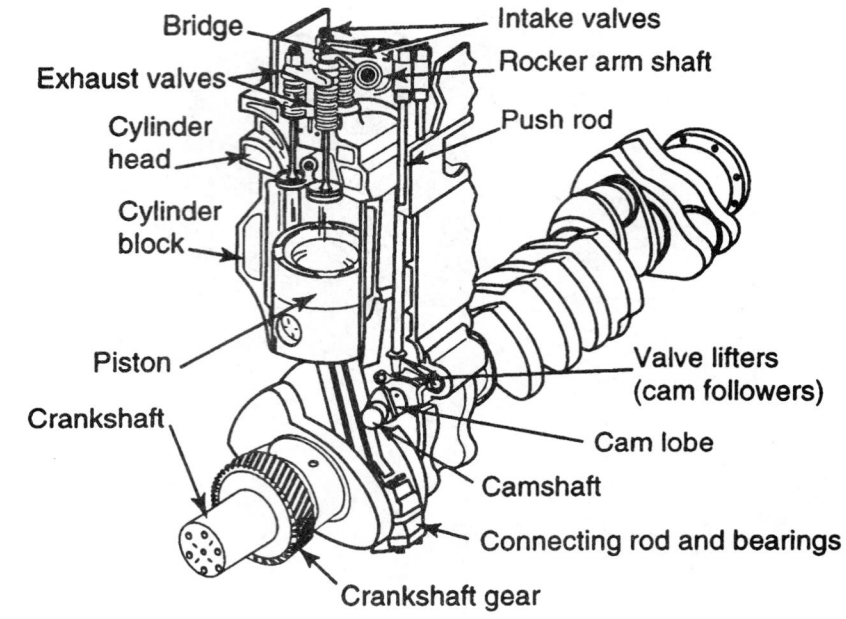

Cam-in-block

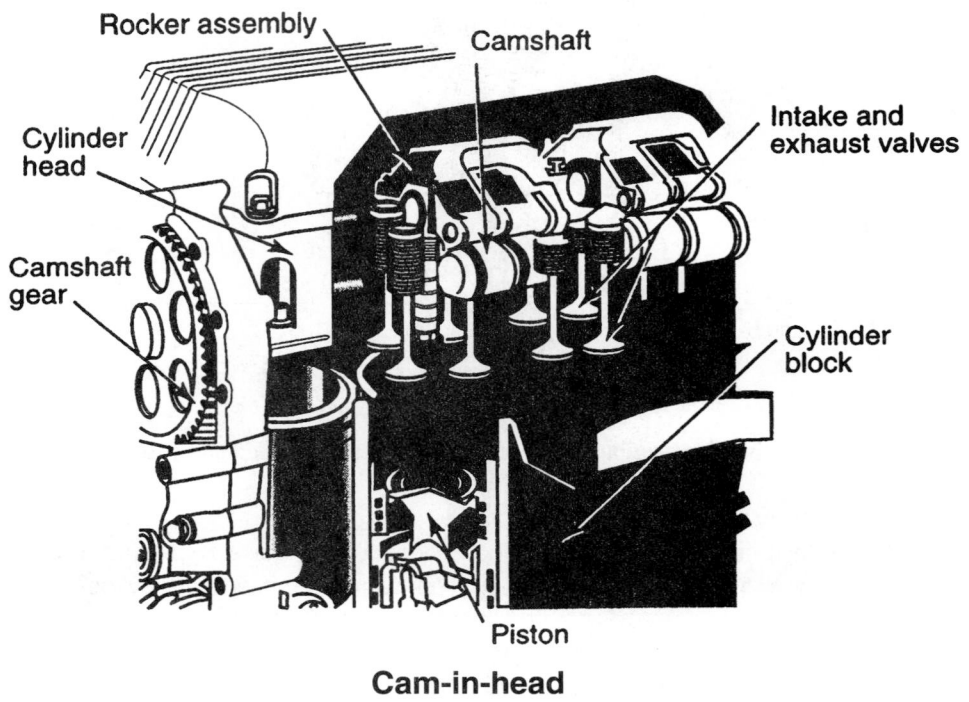

Cam-in-head

Figure 4-2 Cylinder head cutaway. (Courtesy of Cummins Engine Company and reprinted by permission of copyright owner Detroit Diesel Corporation, all rights reserved.)

Creep is the relative movement of clamped engine components due to different coefficients of heat expansion or different mass.

coefficients of heat expansion or different mass. For instance, multiple cast iron cylinder heads clamped to a cast iron cylinder blocks are going to expand and arrive at their operating temperature well before the cylinder block does; the opposite will occur on cooling. The head gasket is engineered to seal under all the operating conditions of the engine. Cylinder head fasteners should be lightly lubed with engine oil prior to installation.

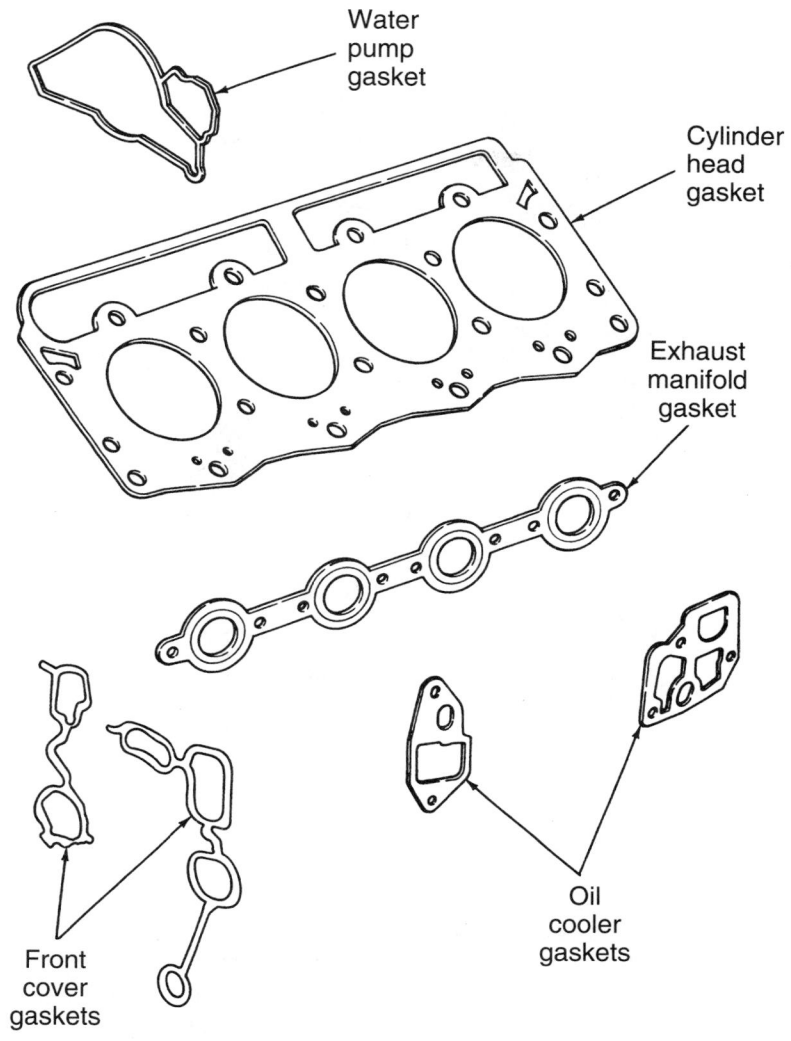

Water pump gasket

Cylinder head gasket

Exhaust manifold gasket

Front cover gaskets

Oil cooler gaskets

Figure 4-3 Cylinder head gasket. (Courtesy of Navistar International Engine Group)

 WARNING: Oil should never be poured into the block threads, because a hydraulic lock may result.

In-Block Camshaft Design

The **camshaft** (Figure 4-4) in most diesel engines is gear driven by the crankshaft through one revolution per complete cycle of the engine. In a 4-stroke cycle engine, to complete the full cycle the engine must be turned through two revolutions, or 720 degrees, during which the crankshaft would turn one revolution. Camshaft speed would therefore be one half engine speed. In a 2-stroke cycle engine, the full cycle is completed in a single rotation, or 360 degrees. Camshaft speed and crankshaft or engine speed would therefore be geared so they were equal. The **camshaft** in a diesel engine actuates the **valve trains** (the term *train* can describe any components that ride the cam profile and are actuated by it) and ever more commonly. The injection pumping apparatus, such as unit pumps, mechanically or electronically control unit injectors. The camshaft is supported at its journals by bushings or bearings, and in most cases is pressure lubricated.

Shop Manual Chapter 4, page 131.

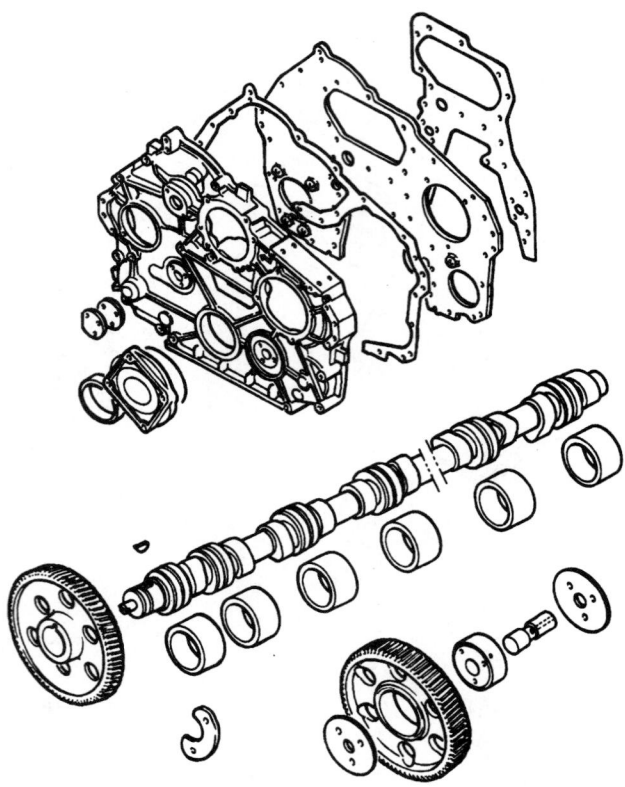

Figure 4-4 In-block camshaft. (Courtesy of Mack Trucks, Inc.)

Overhead Camshaft Design

Overhead camshafts (Figure 4-5) are not as common in current truck diesel engines but there are signs this trend may be changing. Caterpillar 3406E, Detroit Diesel Corporation (DDC) series 60, and Cummins Signature 600 engines are examples of overhead camshaft design. Cam geometry and the camshaft gear timing to the engine crankshaft dictate valve train timing and unit injection pump or unit injector stroke. In some applications, such as Cummins PT engines, this may be adjustable within a small window by the use of offset camshaft-to-camshaft gear locating keys or the shimming of cam follower housings.

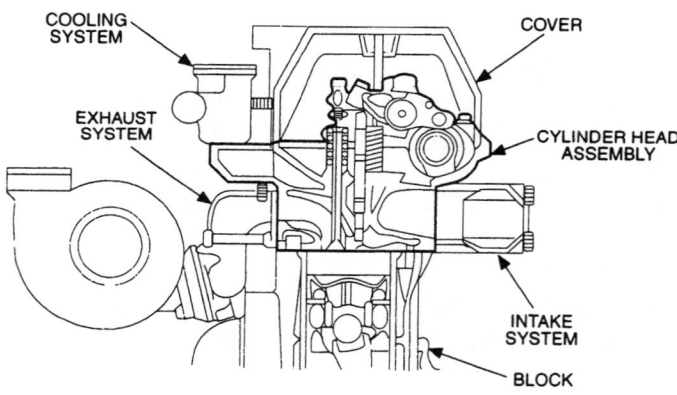

Figure 4-5 Overhead camshaft. (Reprinted by permission of copyright owner Detroit Diesel Corporation, all rights reserved.)

Camshaft Geometry

Cam geometry (Figure 4-6), or the profile outside of **base circle**, will actuate the trains riding the profile and convert the rotary movement of the camshaft into reciprocating motion. The camshaft is subjected to loading whenever a cam is actuating the train that rides its profile. This loading can be considerable, especially when the camshaft actuates engine compression brake hydraulics and the fuel injection pumping apparatus. Cam bushings are routinely replaced during engine overhaul, but if they are to be reused, they must be measured with a dial bore gauge or telescoping gauge and micrometer to ensure that they are within the OEM reuse parameters. Interference fit, cylinder block located, camshaft bushings are removed sequentially, usually starting from the front to the back of the cylinder block using a correctly sized bushing driver (mandrel) and slide hammer.

Hardening is usually by Nitriding or other hard-facing process, and finish grinding follows this hardening. Resurfacing of the journals is possible, but most camshaft failures are related to the cam lobe. Camshafts are supported by bearing journals within a longitudinal bore in the cylinder block, or, in the case of an overhead cam, on a pedestal arrangement in the cylinder head.

Cams are eccentrically machined to convert rotary movement to linear movement. The smallest radius of a cam that is concentric with the camshaft center line is known as the **base circle** or **inner base circle (IBC)**. The largest radial dimension from the camshaft centerline is known as **outer base circle (OBC)**. The shaping of the **cam profile** that connects cam base circle with its outer base circle is described as ramping. A cam may be designed so that the larger percentage of its periphery (circumference) is base circle; in this case, the train that it is responsible for actuating will be unloaded for the larger percentage of the cycle. Cams used to actuate engine cylinder valves are of this design. Alternatively, cams may be designed so that most of their periphery is outer base circle, in which case, the train that it actuates will be loaded for most of the cycle. Cummins PT injectors and Detroit Diesel electronic unit injectors (EUIs) and mechanical unit injectors (MUIs) are actuated by cam profiles of this design (Figure 4-7). Either cam cap crush on overhead camshafts or lock rings (e.g. DDC 53, 71, and 92 Series) retain cam bearing split shells.

> The smallest radius of a cam concentric with the camshaft centerline is known as the **base circle** or **inner base circle (IBC)**.

> The largest radial dimension from the camshaft centerline is known as **outer base circle (OBC)**.

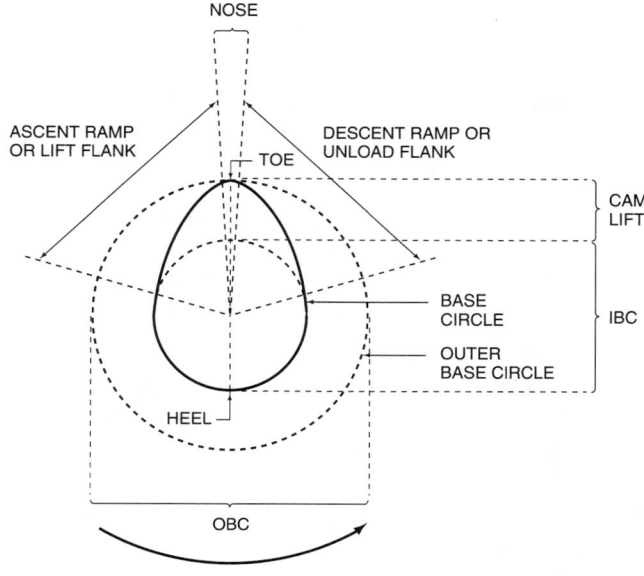

Figure 4-6 Camshaft geometry.

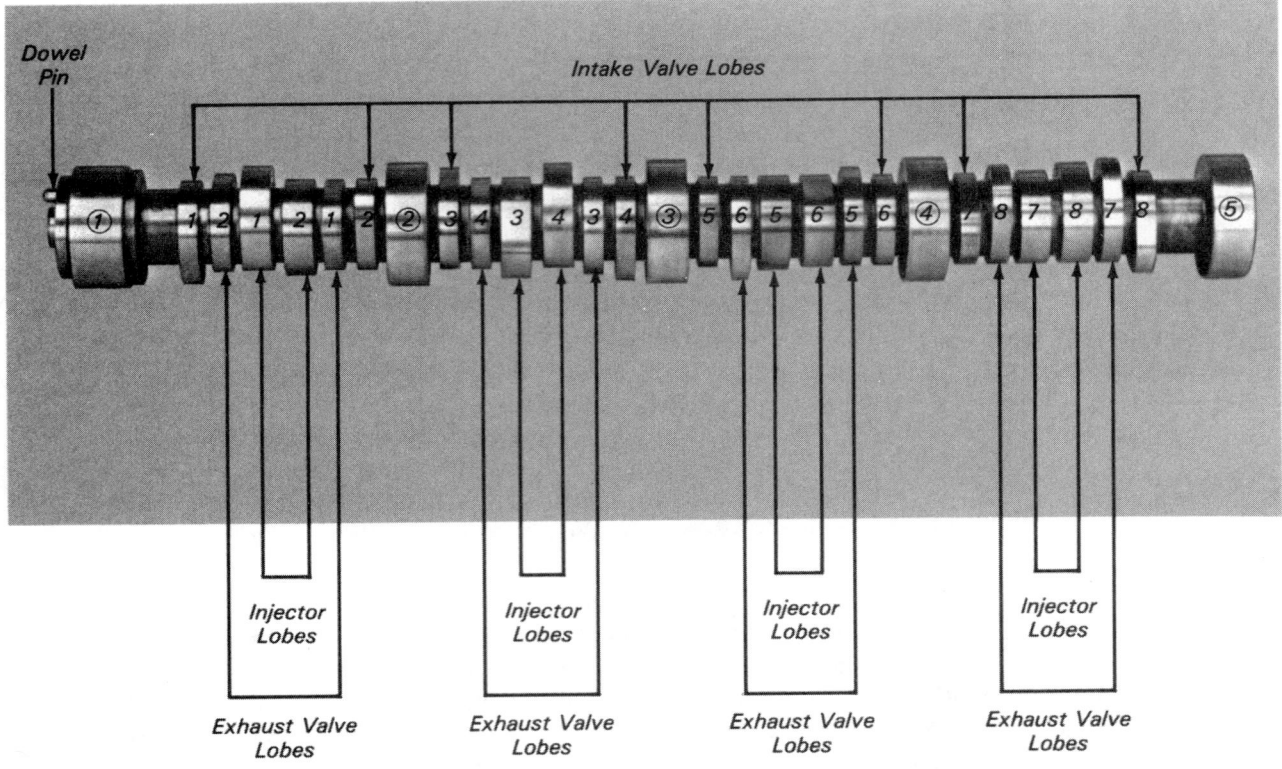

Dowel
Pin

Intake Valve Lobes

Injector Lobes

Injector Lobes

Injector Lobes

Injector Lobes

Exhaust Valve Lobes

Exhaust Valve Lobes

Exhaust Valve Lobes

Exhaust Valve Lobes

Figure 4-7 Camshaft configuration appearance when a center lobe is used to activate MUIs or EUIs. (Reprinted by permission of copyright owner Detroit Diesel Corporation, all rights reserved.)

Valve Train Components

Shop Manual Chapter 4, page 134.

Followers convert the rotary movement of the camshaft into linear motion.

Tappets, lifters, and cam followers usually are positioned to directly ride or at least be actuated by a cam profile.

The valve and injector trains (Figure 4-8) are responsible for transmitting the effects of cam geometry to action at the valve and injector assemblies. **Cam followers** convert the rotary movement of the camshaft into linear motion. This linear motion is converted to rotary movement once again at the rocker, which in its rocking action provides the force that opens valves and provides the pumping force for MUIs and EUIs. Rockers may be used in both cylinder-block-mounted camshaft engines and those with overhead camshafts.

Tappets, Lifters, and Cam Followers

Tappets, lifters, and cam followers (Figure 4-9) describe components that usually are positioned to directly ride or at least be actuated by a cam profile. The term **tappet** has a broader definition and is sometimes used to describe a *rocker lever.* Because North American engine and fuel system OEMs uses all three terms, the technician should be familiar with them. In this text, the term *tappet* generally will not be used to describe a rocker; the term *cam follower* will be used to describe a component that rides the cam profile, except when describing a specific OEM system. The function of cam followers is to reduce friction and evenly distribute the force imparted from the cam's profile to the train it is responsible for actuating. Diesel engines using cylinder-block-mounted camshafts use two categories of follower; those using overhead camshafts use either direct actuated rockers or roller-type cam followers.

Solid Lifters Solid lifters are manufactured from cast iron and middle alloy steels and are usually located in guide bores in the cylinder block that allow them to ride the cam profile over which they are positioned. Push rods are fitted to sockets in the lifter. The critical surface of a solid lifter is the face that directly contacts the cam profile. This face must be durable and may

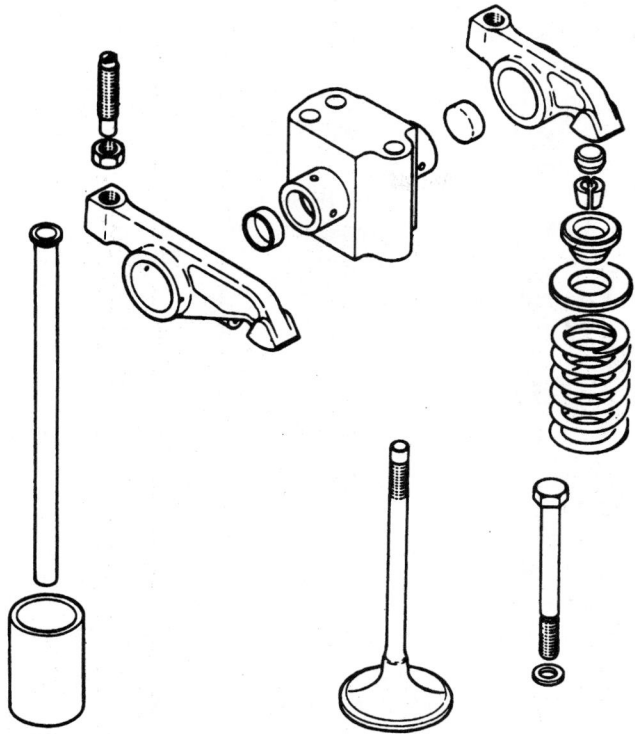

Figure 4-8 Typical valve train assembly. (Courtesy of Mack Trucks, Inc.)

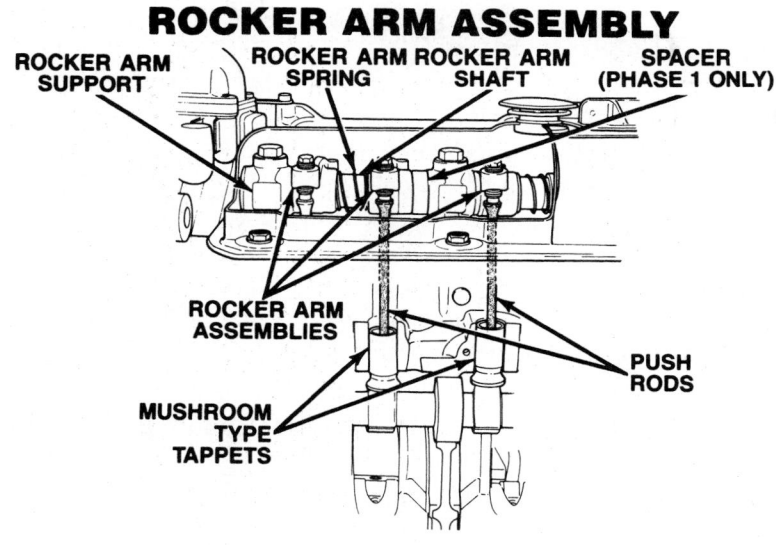

ROCKER ARM ASSEMBLY

ROCKER ARM
SUPPORT

ROCKER ARM
SPRING

ROCKER ARM
SHAFT

SPACER
(PHASE 1 ONLY)

ROCKER ARM
ASSEMBLIES

PUSH
RODS

MUSHROOM
TYPE
TAPPETS

Figure 4-9 Valve train with tappet, push rod, and rocker arm. (Courtesy of Freightliner Corporation)

either be chemically hardened with a toughened alloy or have a disc of special alloy steel molecularly bonded to the face. Solid lifters should be carefully examined at engine overhaul primarily for thrust face wear but also stem and socket wears. The guide bores in the cylinder block should also be measured using digital calipers or a telescoping gauge and micrometer. Sleeving lifter guide bores is a relatively simple procedure that involves boring to an interference fit on the new sleeve outside diameter. Always check that the lifters do not drag or cock in newly sleeved guide bores.

Roller-Type Cam Followers Roller-type cam follower assemblies generally consist of a roller supported by a pin mounted to a clevis. The clevis can either be cylindrical and mounted in a cylinder block guide bore or be a pivot arm fitted to either the cylinder block or cylinder head. In the case of some overhead camshaft designs, a roller-type cam follower assembly and the rocker arm are integral. An example is the overhead camshaft design used to actuate the DDC series 60, parallel port valve and EUI configuration. Roller-type cam followers distribute the loading they are subjected to better than solid types and, therefore, tend to outlast them. The rollers are usually chemically hard surfaced. Roller contact faces should be inspected for pitting, scoring, and indications of disintegration of the hard surfacing. The roller assembly should also be checked for axial and radial slop.

Push Rods and Tubes

Push tubes and push rods act as intermediaries in the train and are located between the cam followers and rocker assemblies.

Engines using in-block mounted camshafts require a means of transmitting the effects of cam action to the rocker assemblies in the cylinder head. **Push tubes** and **push rods** (Figure 4-9) act as intermediaries in the train and are located between the cam followers and rocker assemblies. Because they are subjected to shock loads that increase in proportion to engine rpm, they are manufactured from alloyed steels both to sustain these loads and minimize the mass of the train. Hollow push tubes are more commonly used than solid push rods, especially in cases where the camshaft is mounted low in the cylinder block and they are required to be fairly long. A typical push tube is a cylindrical, hollow shaft fitted with a solid ball or socket at either end. Balls and sockets form bearing surfaces at the follower and rocker. This is especially important at the rocker end as the rocker moves through an arc while the push tube moves linearly. Balls and sockets are usually chemically hard surfaced.

Hollow push tubes are used more often than push rods because the tubular shape provides nearly as high section modulus (shape characteristic related to rigidity) as a solid rod and much less weight. Push rods are cylindrical and solid. They are used mostly in applications that permit them to be short and relatively low in mass.

Rocker Arms

Rocker assemblies transmit camshaft motion to the valves. They are used in both in-block camshaft and overhead camshaft configurations.

Rocker assemblies (Figure 4-10) transmit camshaft motion to the valves. They are used in both in-block camshaft and overhead camshaft configurations. The rocker pivots on a rocker shaft. When a cam ramps off its base circle, it acts on the train that rides the cam profile and "rocks" the rocker arm, thereby actuating the components on the opposite side of the rocker arm (either valves or unit injectors). Lubricating oil is normally ducted through the rocker shaft to supply each rocker arm. Rocker arm ratio may be used to amplify the cam lift dimension in increased valve or unit injector plunger travel. This would require that the distance from the centerline of the rocker shaft pivot bore to the pushrod side be less than its distance to the valve or unit injector side. Rocker ratio expresses the mechanical advantage obtained over the actual cam lift dimension. A rocker ratio of 2:1 would convert a cam lift dimension of 1 in. into 2 in. of valve opening travel.

Rockers are normally supported on a shaft, which is supported by pedestals or brackets. The lubricating oil required to lubricate the rocker support bores and their bearings is provided through one or more of the support pedestals and transmitted through a bore machined into the rocker shaft. The rocker itself is usually supplied with oil, which is used to lubricate the actuating mechanism and the component(s) it actuates. The end of the rocker arm that actuates either valves, valve bridges, or injection pumps such as MUIs, EUIs, and PT injectors is known as the pallet end. The valve bridge or yoke is one means that can be used for a single rocker to actuate a pair of valves. The bridge is mounted on a guide pedestal shaft contacting the valve stem of the two valves it is to actuate. The pallet end of the rocker contacts the center point of the bridge directly above the guide pedestal.

Setting the clearance between the rocker pedestal and the bridge with thickness gauges normally sets valve lash in such an arrangement. Parallel port valve configurations usually require a more complex rocker mechanism, as do overhead cam actuated injection mechanisms. Often, the

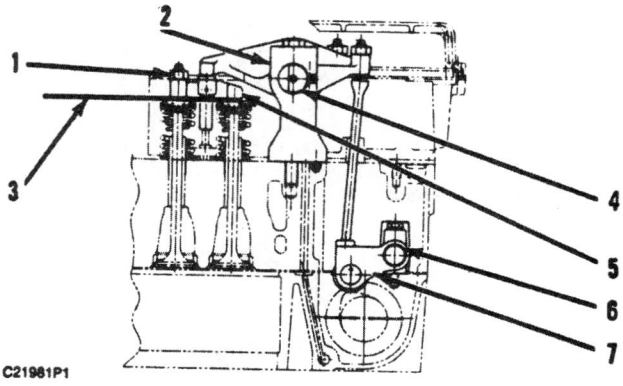

C21961P1

1. Adjusting screw and locknut
2. Rocker pedestal
3. Valve lash - valve stem to rocker pallet clearance
4. Rocker shaft
5. Valve bridge/yoke assembly
6. Cam follower pivot
7. Roller type cam follower/lifter

Figure 4-10 Valve train components. (Courtesy of Caterpillar Inc.)

exhaust valves, intake valves, and injection mechanism are each set on a different linear plane and are actuated from a single camshaft using rockers of different lengths.

Valves

Cylinder head valves (Figure 4-11) provide a means of timing and admitting air into the engine cylinders and timing and exhausting and gases. In all current medium and small bore North American engines, they are mechanically actuated. The timing and lift of engine cylinder valves is defined by the actuating cam geometry.

Shop Manual Chapter 4, page 135.

Valve Design and Materials

Cylinder head **valves** (Figure 4-12) are mushroom shaped, poppet type valves. They are fitted to guides in the cylinder head and loaded by a spring or springs to seal to a seat. A disc shaped, spring retainer locks the spring in position; split locks fitted to a peripheral groove at the top of the valve stem hold the spring retainer in position.

A BIT OF HISTORY

In July of 1892, Rudolf Diesel's first test engine contained only one poppet-style valve for both intake and exhaust. It had two conical seat heads, as found on some steam engines. Why he chose the dual-function valve and then made it even more complicated is not known. In operation, the intake air was drawn into the cylinder through ports in the exhaust pipe flange where the flange bolted to the cylinder head. Exhaust gases were expected to simply blast past these ports and out the exhaust pipe. This double-function valve design was changed after testing began.

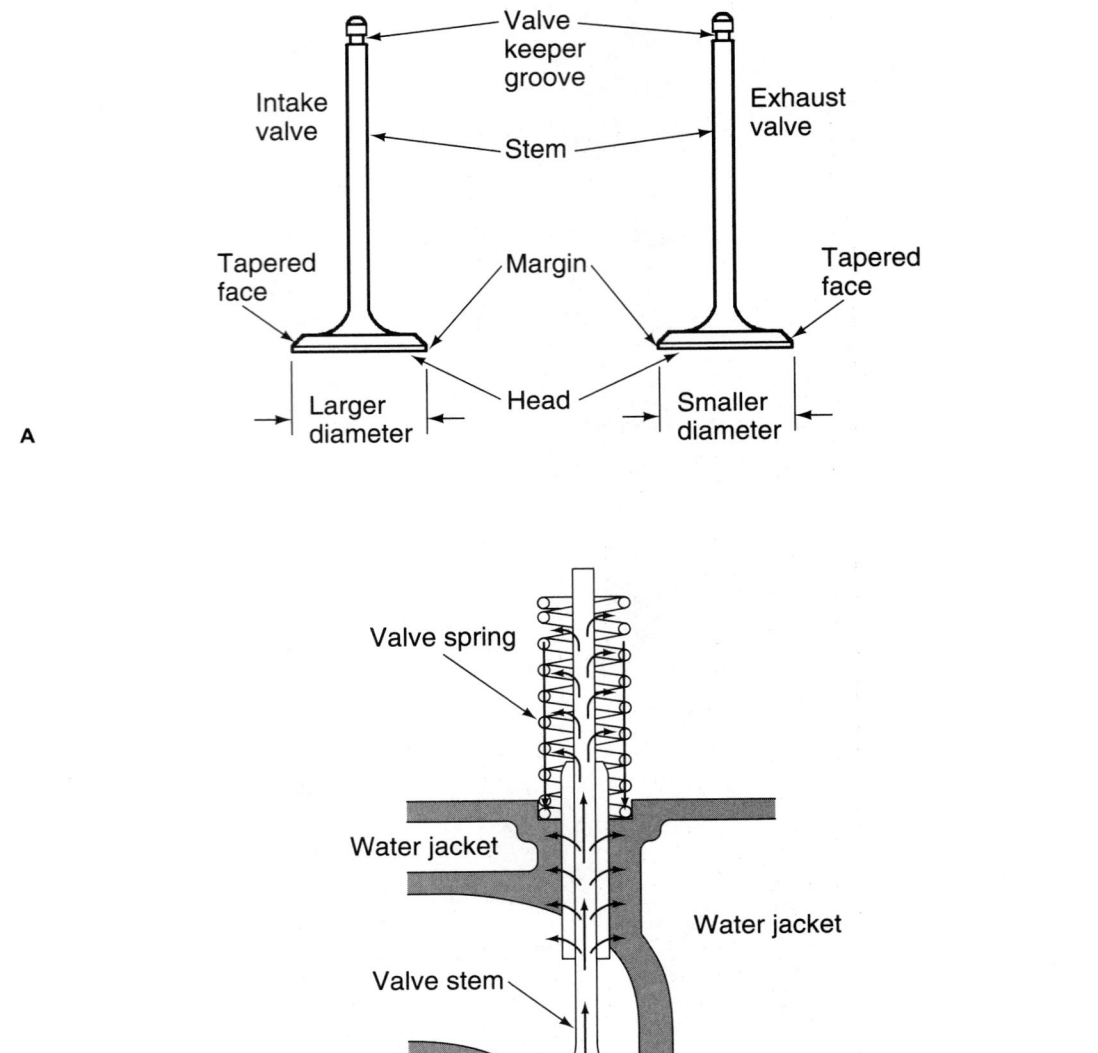

Figure 4-11 (A) Parts of a typical engine valve; (B) Valve assembly components.

Intake valves are manufactured from a variety of middle alloy steels. Valves are most vulnerable to high cylinder temperatures when they are in an open position (not seated). Therefore, intake valves do not have to sustain the high temperatures exhaust valves must. However, they are actuated at high speeds and must have some degree of flexibility. Running temperatures of exhaust valves are much higher than intake valves. When the exhaust valve is first opened during the latter portion of the power stroke, flame wrench has just taken place and cylinder temperatures are just past their peak. Only about 20% of the heat can be dissipated through the stem, so the exhaust valve must sustain high temperatures for the duration of the exhaust process until it seats.

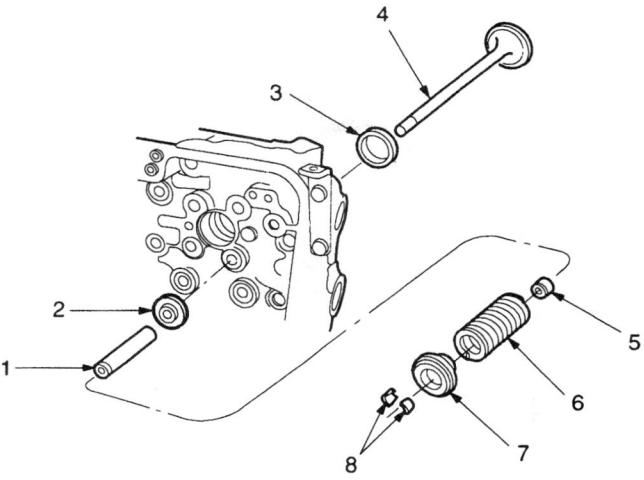

1.	Valve Guide	5.	Valve Stem Oil Seal
2.	Valve Spring Seat	6.	Valve Spring
3.	Valve Insert	7.	Valve Rotator
4.	Valve	8.	Valve Keepers

Figure 4-12 Valve assembly with spring and valve insert. (Reprinted by permission of copyright owner Detroit Diesel Corporation, all rights reserved.)

Exhaust valves are manufactured out of special ferrous alloys. The alloys may be clad at the head, and often contain chromium, nickel manganese, tungsten, cobalt, and molybdenum to improve flexibility, toughness, and heat resistance. Exhaust valves are cooled primarily by dissipating heat to the valve seats and, secondarily, by intake air at valve overlap and through the stem to the guide contact area.

Valve Operation

Current diesel engine valves rely on maximizing the seat contact area for cooling purposes; as a consequence, **interference angles** are seldom machined. An interference angle (Figure 4-13) requires that the valve be machined at an angle 0.5 to 1.0 degree more acute than the seat. This results in the valve seating with high unit pressures and inhibits the formation of carbon on the seat, while compromising the valve-to-seat contact area. A majority of current diesel engines use valve rotators to minimize carbon buildup on the seat and promote even wear. An interference angle is never machined to valves using valve rotators.

Valve rotators use a ratchet principal or a ball and coaxial spring to fractionally rotate the valve each time it is actuated. Valve rotation should be checked after assembly by marking an edge of the stem and then tapping the valve stem with a light nylon hammer a number of times. The valve should visibly rotate each time the stem is struck.

Valve Springs

A spring or pair of springs (Figure 4-14) seats cylinder head poppet valves. When a pair of springs is used, they are often oppositely wound to help cancel the coincidence of harmonies, which may contribute to valve flutter. Valve float can occur at higher engine rotational speeds when the valve spring force is insufficient; this may cause asynchronous (out of time) valve closures. Springs are critical components of the valve train assembly, and their importance is increased as engine speed increases and the real time periods within which they must seat the valves diminish.

A majority of current diesel engines use valve rotators to minimize carbon build up on the seat and promote even wear. An interference angle is never machined to valves using valve rotators

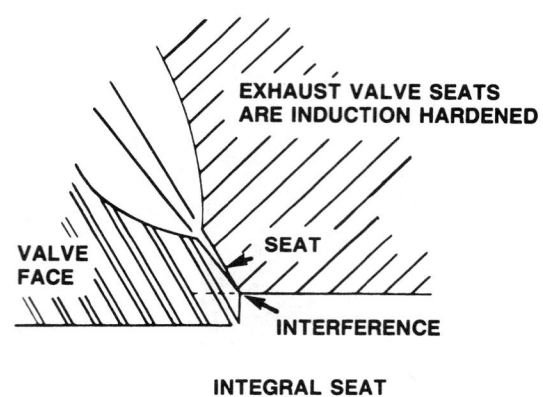

EXHAUST VALVE SEATS
ARE INDUCTION HARDENED

VALVE
FACE

SEAT

INTERFERENCE

INTEGRAL SEAT

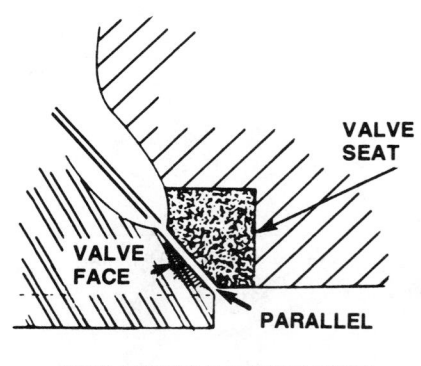

VALVE
SEAT

VALVE
FACE

PARALLEL

REPLACEABLE INSERT SEAT

Figure 4-13 Valve interference angle.

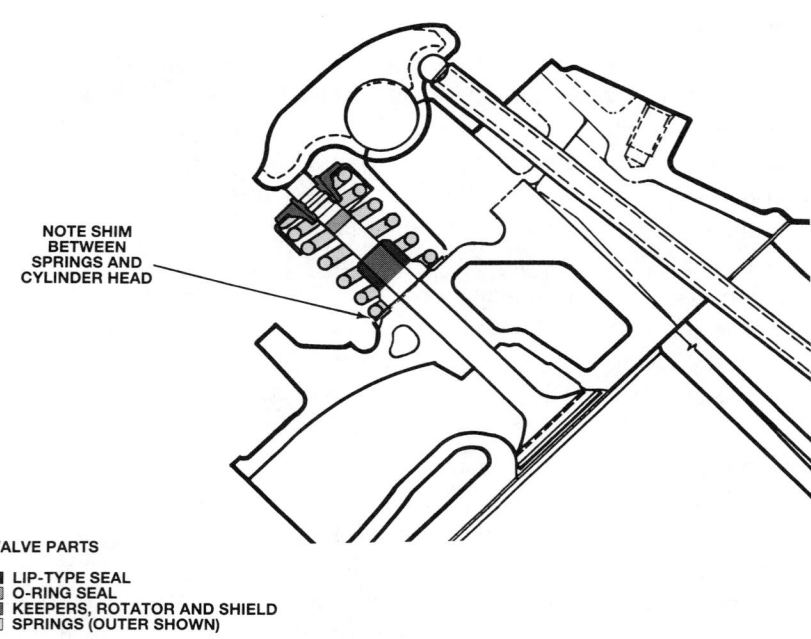

NOTE SHIM
BETWEEN
SPRINGS AND
CYLINDER HEAD

VALVE PARTS

■ LIP-TYPE SEAL
▨ O-RING SEAL
■ KEEPERS, ROTATOR AND SHIELD
☐ SPRINGS (OUTER SHOWN)

Figure 4-14 Valve spring. (Courtesy of General Motors Corporation, Service Operations)

Valve Bridges

Valve bridges or **yokes** (Figure 4-15) push down on two valves at a time. Most engines with four valves per cylinder use valve bridges. They do not usually need to be adjusted during a routine valve adjustment. The bridge must be properly supported to prevent the pedestal shaft from being damaged.

Shop Manual Chapter 4, page 140.

Valve Seat Inserts

Most current truck and bus diesel engines use **valve seat inserts** (Figure 4-16) rather than integral valve seats machined into the cylinder head. The primary advantages of valve seat inserts is that they can be manufactured from tough, temperature-resistant material and then easily replaced when the cylinder head is serviced. Some cast iron alloys toughened with nickel,

Shop Manual Chapter 4, page 127.

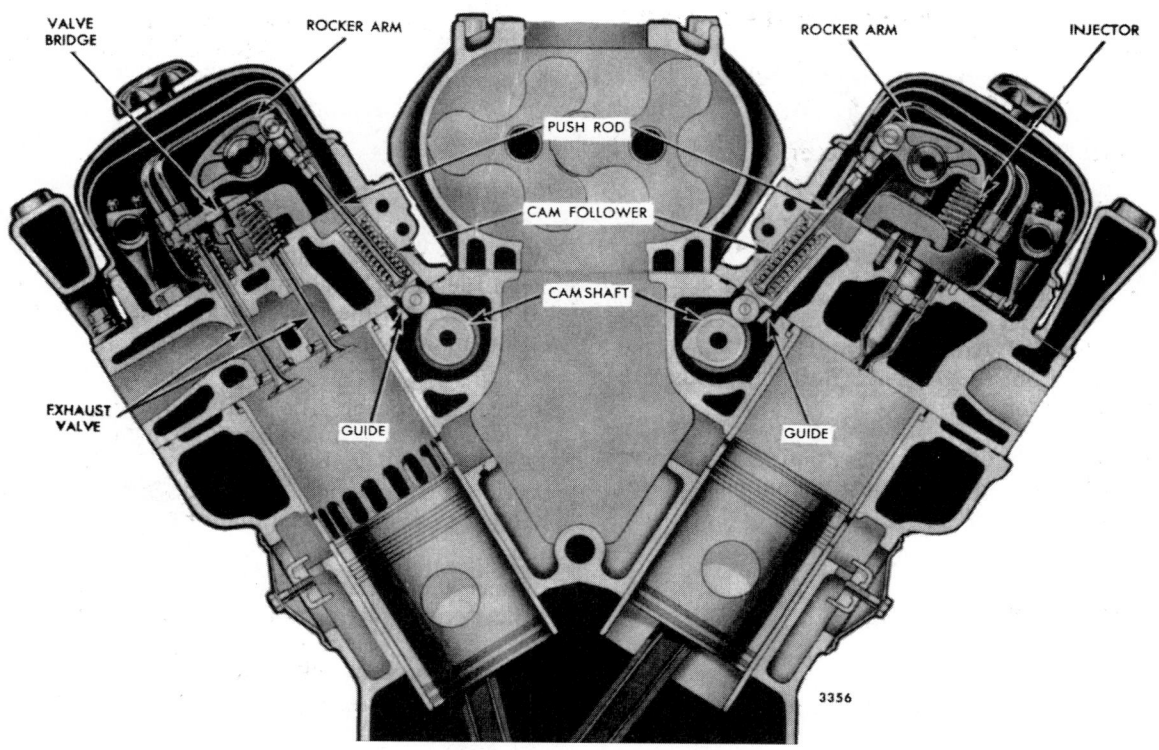

Figure 4-15 Valve bridge. (Courtesy of General Motors Corporation, Service Operations)

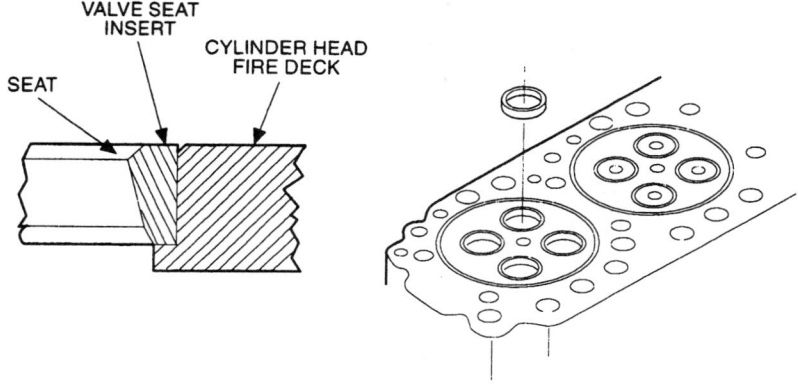

Figure 4-16 Valve seat insert. (Reprinted by permission of copyright owner Detroit Diesel Corporation, all rights reserved.)

chromium, and molybdenum are also used. However, the use of alloyed steels (some with Stellite cladding) is more common. Valve seat inserts are press-fit to a machined recess in the cylinder head. Sometimes they are staked to position after installation. Since most of the heat of the valve must be transferred from the valve to the seat, it is essential that the contact area of the seat and the cylinder head be maximized.

Timing Gears

Diesel engine timing gears (Figure 4-17) are normally located at the front of the longitudinally mounted engines that power most commercial trucks and the transverse mounted engines that power most buses. Timing gears are responsible for driving the camshaft and most of the engine accessories on a diesel-powered truck engine. The gear train is enclosed in a housing that permits engine oil lubrication of the rotating components. Most domestic medium-bore and large-bore, highway diesel engines do not use chains or belts as the drive medium for the camshaft and accessory drives within the timing gear housing. The timing gear train in some larger engines may be in the rear, or in both the front and the rear. The gear ratios will dictate the rotational speed. Timing gears are lubricated in two ways:

1. Splash: Gear teeth rotated through the lubricant in the oil sump will pick up oil and transfer it to other gears in the timing train before draining back to the sump.
2. Bearing spill: Oil used to lubricate the shaft support bearings spills to the timing gear housing where it is circulated before returning to the oil sump.

Shop
Manual
Chapter 4, page 144.

Timing gears are responsible for driving the camshaft and most of the engine accessories on a diesel powered truck engine

Timing Gear Construction

Diesel engine gears are cast or forged alloys that are heat tempered and then surface hardened by a flame, Nitriding, carbonizing, or induction hardening processes. The gear teeth are milled in manufacture to spur and helical designs. Combinations of both are used in the gear trains of some current engines. The noise produced by the gear train is a factor, and for this reason, helical cut gears tend to be more common. **Helical gears** have the advantage of increased tooth contact area, which lowers unit forces. The spur gear design offers the advantage of much lower thrust loads. The gears are commonly press fit to the shafts that they drive and are positioned on the shaft by means of keys and keyways.

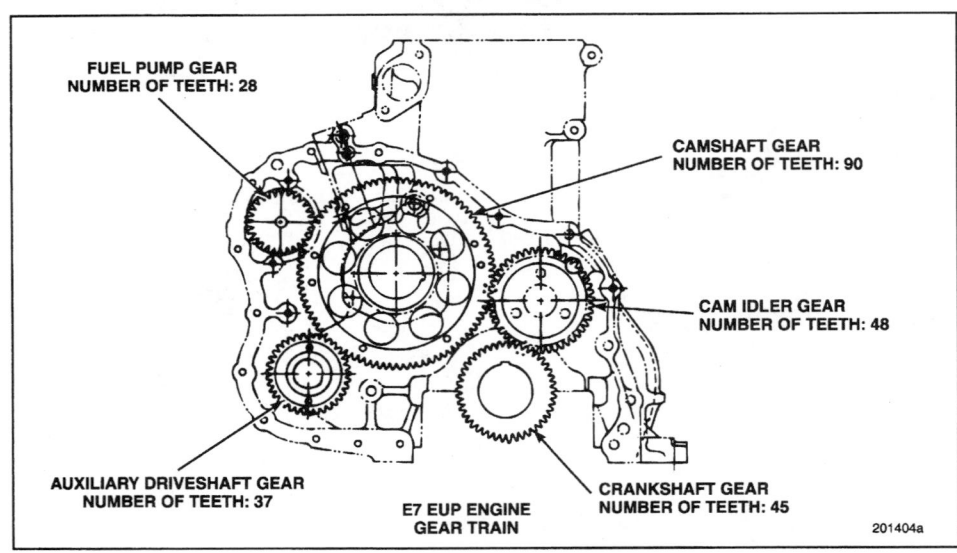

Figure 4-17 (A) Mack Trucks E-tech engine timing gear train (Courtesy of Mack Trucks, Inc.); (B) Caterpillar 3176 timing gear components (Courtesy of Caterpillar Inc.).

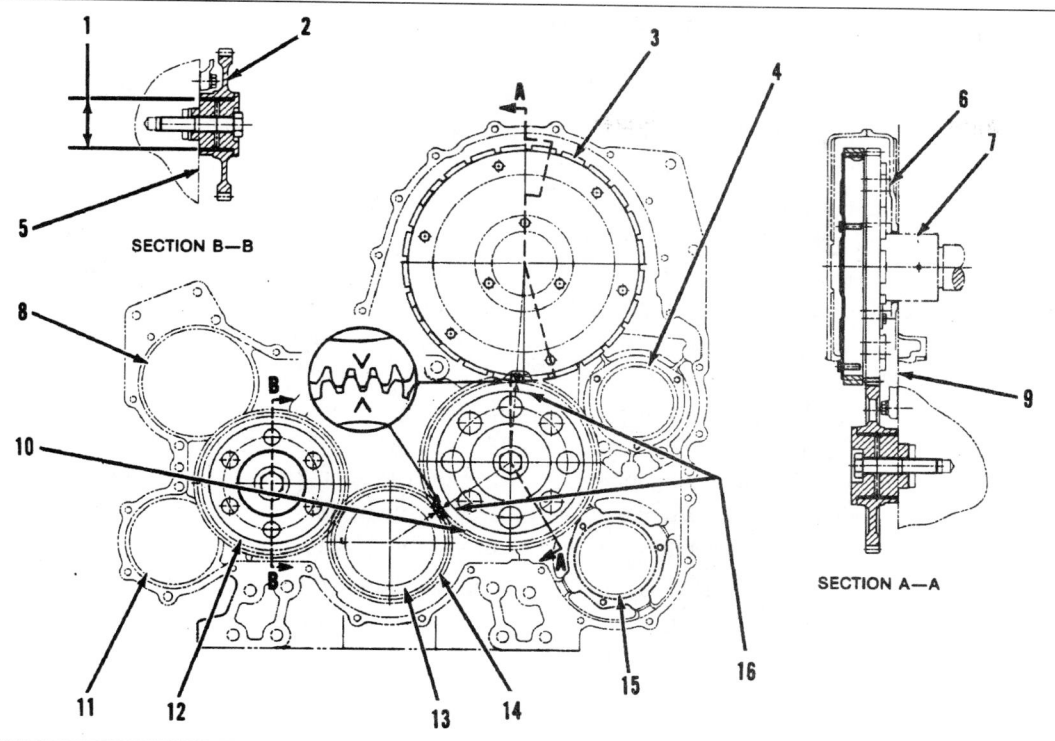

(1) Diameter of gear bore for bearing .... 60.163 ± 0.015 mm (2.3686 ± .0006 in)

(2) Gear assembly (Idler gear).

(3) Ring assembly. TDC occurs when the slot edge is in the center of the electronic control (speed sensor).

(4) Air compressor drive gear.

(5) Front face of the block.

(6) Front timing gear housing.

(7) Camshaft.

(8) Water pump drive gear

(9) Front face of block.

(10) Idler gear. Diameter of gear bore for bearing .... 74.452 ± 0.015 mm (2.9312 ± .0006 in) Diameter of shaft for idler gear .... 69.321 ± 0.020 mm (2.7292 ± .0008 in)

(11) Oil pump drive gear.

(12) Idler gear. Diameter of shaft for idler gear .... 55.047 ± 0.020 mm (2.1672 ± .0008 in)

(13) Crankshaft.

(14) Crankshaft gear. Heat gear to a maximum temperature of 316° C (601° F) for installation.

(15) Hydraulic pump drive gear.

(16) Align timing marks on idler gear with marks on crankshaft gear and camshaft gear as shown.

B

Figure 4-17 Continued

Summary

❏ The cylinder head houses the valve train assemblies, the injector mechanism, and the engine breathing passages. Torquing the cylinder in increments is designed to achieve the yield point of the cylinder head gasket components by evenly achieving the required clamping force.

❏ The rocker ratio can be used to provide the actuating cam with some mechanical advantage. Rocker assemblies are used with both block-mounted and overhead cam designs. Rocker assemblies provide a means of reversing the direction of linear movement of the push tube or follower, and in some cases providing a mechanical advantage.

❏ The cam base circle, or IBC, is that portion of the cam periphery with the smallest radial dimension. Cam OBC is that portion of the cam periphery with the largest radial dimension.

❏ Most diesel engine cam followers are of the solid or roller types (that is, they are not hydraulic).

Terms to Know

Base circle

Cam follower

Cam profile

Camshaft

Creep

Helical gear

Inner base circle (IBC)

Interference angle

Outer base circle

Push rod (Tube)

Rocker assemblies

❏ Most truck diesel engines with block-mounted camshafts use trains consisting of a follower assembly, push tubes, and a rocker; camshaft valve trains are adjusted with a lash factor to allow for expansion of the materials as the engine heats to operating temperature.

❏ Some injection pumping actuation trains are set at zero lash or even a slight load when the actuating cam profile is on its IBC.

❏ Cylinder head valves are used to aspirate (breathe) the engine cylinders. They are actuated by the cam geometry and time the air into and end gasses out of the engine cylinders. Exhaust valves are often manufactured out of more highly alloyed steels than intake valves as they must sustain much higher temperatures. Most diesel engines do not use an interference angle to seat valves because the seating contact surface area is compromised; valve rotators are widely used.

❏ Camshaft drive gears must be precisely timed with the crankshaft driven, engine gear train so that the events activated by the engine feedback assembly are synchronized with those in the engine. The camshaft drive gear is most often interference fit to the camshaft, positioned by a keyway. Camshaft gears are heat treated. When they are fitted to a camshaft, it is essential that they are heated evenly to a precise temperature, because overheating will cause premature failure.

❏ Camshafts may be rotated with or oppositely to the direction of engine rotation. Camshaft gears may use spur or helical cut gear teeth. Thrust loads are much higher when helical gear teeth are used.

Review Questions

Short Answer Essays

1. Describe the component parts of a cylinder head.

2. Outline the role of the cylinder head in the managing of engine functions.

3. Define component *creep* and *gasket yield*.

4. Describe the role of the rocker assembly.

5. Define *rocker ratio*.

6. Define the role of the camshaft in a typical diesel engine.

7. Describe the construction materials and methods of hardening diesel engine camshafts.

8. Identify the role the valve train components play in the running of an engine.

9. List the types of tappet/cam follower used in diesel engines.

10. Define the role of the cylinder head valves.

Fill-in-the-Blanks

1. The cylinder head houses the valve train assemblies, the _____ _____ and the engine breathing passages.

2. Torquing the cylinder in increments is designed to achieve the cylinder head gasket component _____ _____ point by evenly achieving the required clamping force.

3. _____ _____ can be used to provide the actuating cam with some mechanical advantage.

4. The cam _____ _____ circle, or IBC, is that portion of the cam periphery with the smallest radial dimension. Cam ____ is that portion of the cam periphery with the largest radial dimension.

5. Most truck diesel engines with block-mounted camshafts use trains consisting of a _____ _____, push tubes, and a _____.

6. Most valve trains are adjusted with a ____ _____ to allow for _____ of the materials as the engine heats to operating temperature.

7. Some injection pumping actuation trains are set at ____ _____ or even a slight load when the actuating cam profile is on its IBC.

8. Most diesel engines do not use an _____ _____ to seat valves because the seating contact surface area is compromised and the fact that valve rotators are widely used.

9. Camshaft drive gears must be _____ _____ with the crankshaft-driven, engine gear train so that the events activated by the engine feedback assembly are synchronized with those in the engine.

10. The camshaft drive gear is most often _____ ___ to the camshaft, positioned by a keyway.

ASE-Style Review Questions

1. *Technician A* says most 2-stroke cycle diesel engine cylinder heads use a reed valve design to maximize breathing efficiency. *Technician B* says 4-stroke cycle diesel engines use poppet valves in the cylinder heads. Who is right?
   A. A only
   B. B only
   C. Both A and B
   D. Neither A nor B

2. Which of these is the *most* likely reason cylinder head bolts are torqued progressively and in a pattern:
   A. Consistent clamping pressures
   B. Optimum heat transfer
   C. Even cylinder bolt loads
   D. To prevent leaks

3. Which of the following would be *least* likely considered part of the valve train?
   A. Push tubes
   B. Tappets
   C. Rockers
   D. Valve seats

4. Which of the following would *not* be considered to be part of the valve train?
   A. Push tubes
   B. Tappets
   C. Rockers
   D. Valve seats

5. Which direction must a camshaft be rotated on a diesel engine with a crankshaft that is rotated clockwise?
   A. Either clockwise or counterclockwise, depending on the engine
   B. Clockwise
   C. Counterclockwise
   D. Same direction

6. Which of the following is likely to cause a valve float condition?
   A. Engine lugdown
   B. Operating in the torque rise profile
   C. Engine overspeed
   D. Operating in the droop curve

7. The reason for using a pair of oppositely wound valve springs is to:
   A. Minimize valve dynamic flutter
   B. Double the closing force of two similarly wound valves
   C. Increase the valve closing velocity
   D. Diminish valve noise

8. *Technician A* says helical cut gears are often used in engine timing gear trains as they operate more quietly. *Technician B* says that shaft thrust loads are much higher when helical cut gears are used. Who is correct?
   A. A only
   B. B only
   C. Both A and B
   D. Neither A nor B

9. A camshaft on a 4-stroke cycle, truck diesel engine is driven at what speed in relation to the engine crankshaft speed?
   A. At 1/2 engine speed
   B. At engine speed
   C. At 2 times engine speed
   D. At 4 times engine speed

10. Most diesel engine cylinder heads are manufactured from:
    A. Cast aluminum alloy
    B. Cast iron alloys
    C. Composite fibers
    D. Magnesium alloy

Lubrication System

Upon completion and review of this chapter, you should be able to:

❏ Outline the function of the main components in a typical diesel engine lubrication circuit.

❏ List the required properties of heavy-duty engine oil.

❏ Define *hydrodynamic suspension* and describe how this principle is used in a typical diesel engine.

❏ Interpret API classifications and SAE viscosity grades.

❏ Identify the main components used in a truck or bus diesel engine lubricating system.

❏ Describe the two types of oil pump commonly used on diesel engines, and outline the operating principles of each.

❏ Describe the principle of operation of the pressure-regulating valve and its role in the lubrication circuit.

❏ Define *positive filtration*.

❏ Outline the differences between full-flow and bypass filters.

❏ Describe the operating principles of the various types of heavy duty oil filters and the relative operating efficiencies of each.

❏ Outline the objectives of the oil cooler in the lubrication circuit and its operating principles.

❏ Describe the two types of oil cooler used in current diesel engines.

Introduction

All medium- and large-bore diesel engines currently used on North American highways use a pressure lubrication system (Figure 5-1) to supply the bearings and all moving components with engine oil. Engine oil is the medium used in lubrication circuits, and it is specially formulated to fulfill the lubrication and service life requirements of diesel engines.

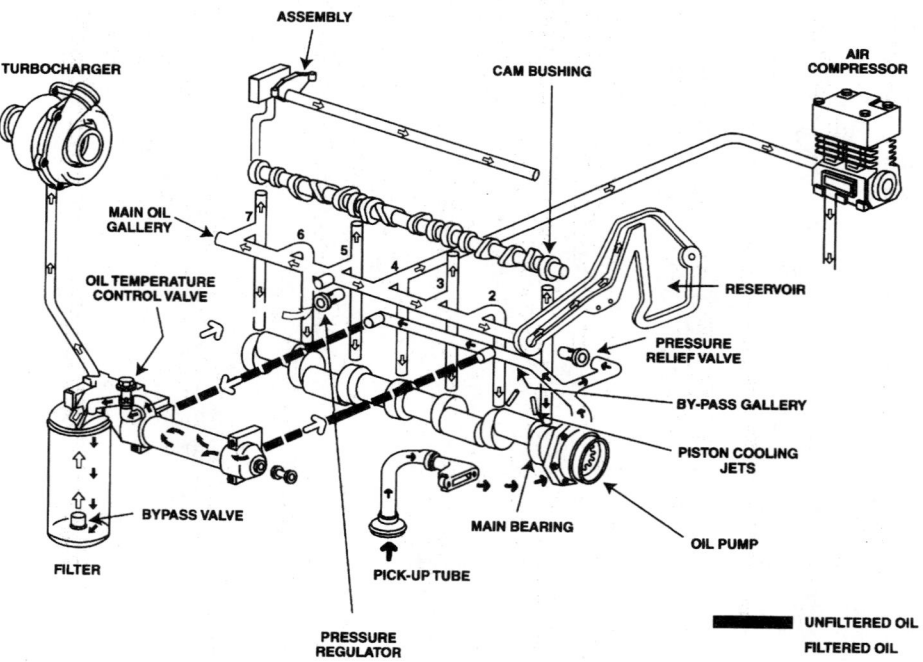

Figure 5-1 Navistar lubrication circuit. (Courtesy of Navistar International Engine Group)

WARNING: Engine lubricating oil temperatures run somewhat higher than engine coolant temperatures, and the actual oil temperature is probably a more reliable indicator of the true running temperature of the engine. If only the coolant temperature is monitored rather than the actual oil temperature, engine damage may result.

The basic components, as shown in Figure 5-2, of a pressure lubrication system are as follows:

Lubricant: A petroleum-based, liquid medium used to reduce friction, hydrodynamically support shafts, help seal pistons, and act as a cooling medium.

Sump: An oil storage space located at the bottom of the engine that encloses the crankcase; a remote located sump tank is called a **dry sump** system.

Pump: Responsible for moving the oil through the lubrication circuit to supply the system oil passages and bearings.

Filter: A system to remove particulates from the engine. Truck and bus diesel engine lubrication systems require multistage filtration systems.

Oil cooler: A heat exchanger that uses engine coolant as its medium. Heat picked up by the engine oil is transferred to the coolant from which it can be dissipated to atmosphere (Figures 5-1 and 5-2).

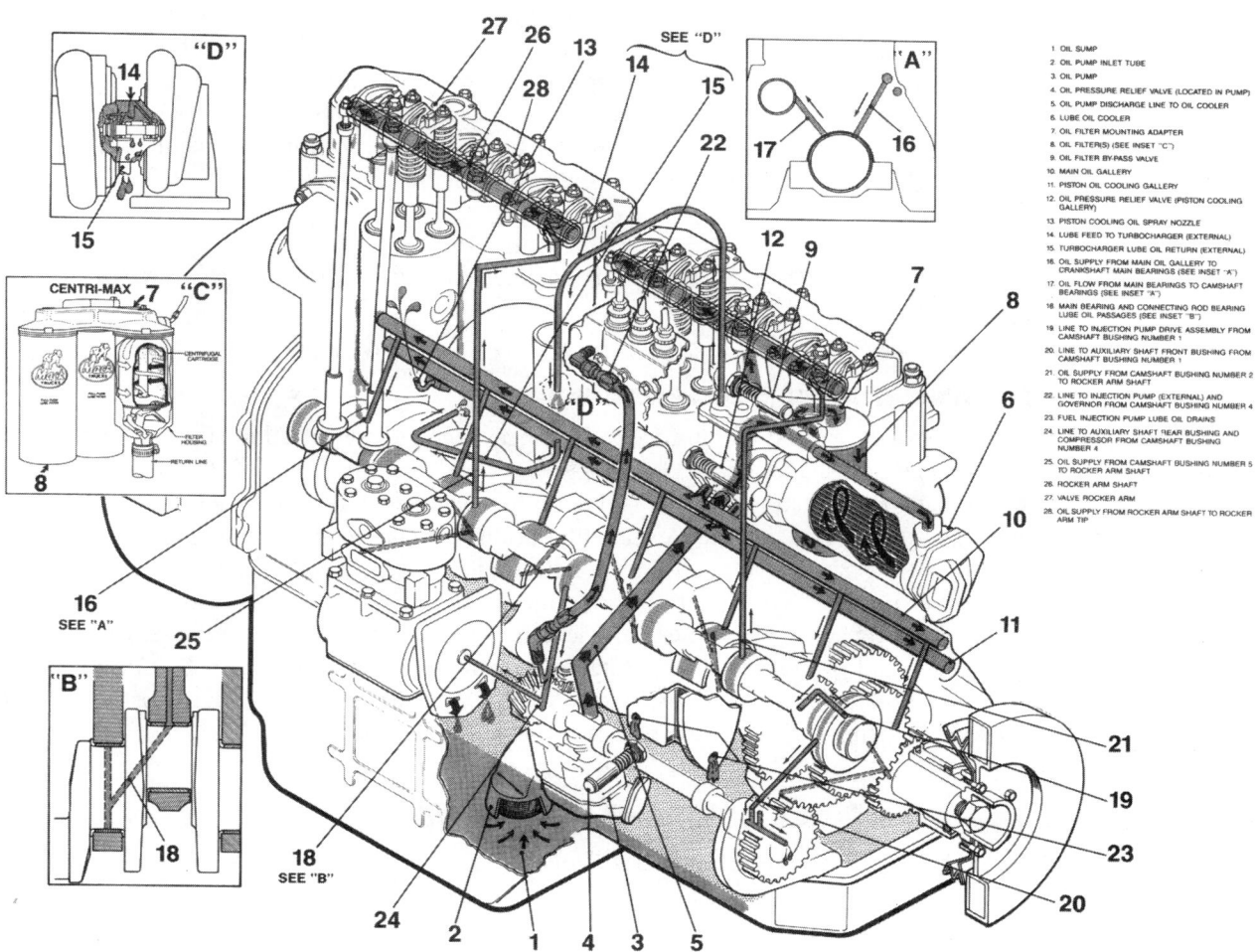

1. OIL SUMP
2. OIL PUMP INLET TUBE
3. OIL PUMP
4. OIL PRESSURE RELIEF VALVE (LOCATED IN PUMP)
5. OIL PUMP DISCHARGE LINE TO OIL COOLER
6. LUBE OIL COOLER
7. OIL FILTER MOUNTING ADAPTER
8. OIL FILTER(S) (SEE INSET "C")
9. OIL FILTER BY-PASS VALVE
10. MAIN OIL GALLERY
11. PISTON OIL COOLING GALLERY
12. OIL PRESSURE RELIEF VALVE (PISTON COOLING GALLERY)
13. PISTON COOLING OIL SPRAY NOZZLE
14. LUBE FEED TO TURBOCHARGER (EXTERNAL)
15. TURBOCHARGER LUBE OIL RETURN (EXTERNAL)
16. OIL SUPPLY FROM MAIN OIL GALLERY TO CRANKSHAFT MAIN BEARINGS (SEE INSET "A")
17. OIL FLOW FROM MAIN BEARINGS TO CAMSHAFT BEARINGS (SEE INSET "A")
18. MAIN BEARING AND CONNECTING ROD BEARING LUBE OIL PASSAGES (SEE INSET "B")
19. LINE TO INJECTION PUMP DRIVE ASSEMBLY FROM CAMSHAFT BUSHING NUMBER 1
20. LINE TO AUXILIARY SHAFT FRONT BUSHING FROM CAMSHAFT BUSHING NUMBER 1
21. OIL SUPPLY FROM CAMSHAFT BUSHING NUMBER 2 TO ROCKER ARM SHAFT
22. LINE TO INJECTION PUMP (EXTERNAL) AND GOVERNOR FROM CAMSHAFT BUSHING NUMBER 4
23. FUEL INJECTION PUMP LUBE OIL DRAINS
24. LINE TO AUXILIARY SHAFT REAR BUSHING AND COMPRESSOR FROM CAMSHAFT BUSHING NUMBER 4
25. OIL SUPPLY FROM CAMSHAFT BUSHING NUMBER 5 TO ROCKER ARM SHAFT
26. ROCKER ARM SHAFT
27. VALVE ROCKER ARM
28. OIL SUPPLY FROM ROCKER ARM SHAFT TO ROCKER ARM TIP

Figure 5-2 Mack Truck lubrication circuit. (Courtesy of Mack Trucks, Inc.)

Engine Lubricating Oil

The main functions of a diesel engine lubricating system are:

1. **Lubrication:** The primary task of engine oil is to minimize friction and act as a medium to support the **hydrodynamic suspension** of the crankshaft and camshaft.

2. **Sealant:** To enable the piston and ring assembly to seal compression and combustion gases from the crankcase.

3. **Coolant:** To dissipate the heat generated by combustion and friction to atmosphere via heat exchangers.

4. **Cleaning agent:** To eliminate the condensed by-products of combustion gases that end up in the engine crankcase and can combine to form harmful liquids (acids) and particulates (sludge).

5. **Oil film:** To maintain an oil film even when subjected to high thrust loads such as the compression loading of a piston/rod assembly.

Lubricating oils are **petroleum** products that are complex mixtures made up of many different fractions. Like gasoline and diesel fuel, engine lubricant is elementally about 85% carbon and 15% hydrogen. The fractional compounds are refined from petroleum and asphalt bases and subsequently mixed.

The theoretical action of engine oil is to form a film between moving surfaces so that any friction that results occurs in the oil itself: that is, it is **fluid friction**. Fluid friction generates considerably less heat than dry friction. The lubricating requirements of engine oil can be classified as *thick film lubrication* and *boundary lubrication* (thin film). **Thick film lubrication** occurs where mating tolerances between components are wide. Boundary lubrication is required where mating tolerances are narrow, such as when pressure is applied to one of the components. A breakdown of boundary lubrication will result in metal-to-metal contact. Quality engine oil should be capable of performing both thick film and boundary lubrication.

 CAUTION: Because engine lube oil is usually petroleum based it is flammable.

Hydrodynamic Suspension

Hydrodynamic suspension occurs when a shaft is rotated within friction bearings such as those used to support an engine crankshaft.

When a shaft is rotated within friction bearings, such as those used to support an engine crankshaft, a crescent-shaped gap (Figure 5-3) is formed on either side of the line of direct contact when that shaft is stationary. This is due to the clearance between the journal and the inside diameter of the friction bearing. A static film of oil prevents shaft-to-bearing contact when stationary. When the shaft is rotated and the bearing is charged with engine oil under pressure, a crescent-shaped wedge of lubricant is formed between the journal and its bearing. Oil is introduced to the bearing where the shaft clearance is greatest, usually at the top. This wedge of oil is driven ahead of the direction of rotation in a manner that permits the shaft to be floated on a bed of constantly changing, pressurized oil. This principle is used in most crankshaft and camshaft dynamics, and is known as hydrodynamic suspension.

Characteristics of Diesel Engine Oil

1. Good natural oxidation stability
2. High natural viscosity index
3. Careful product control to ensure high quality
4. Constant research for continuous product improvement
5. Extensive laboratory and field testing to ensure product performance

HYDRODYNAMIC LUBRICATION

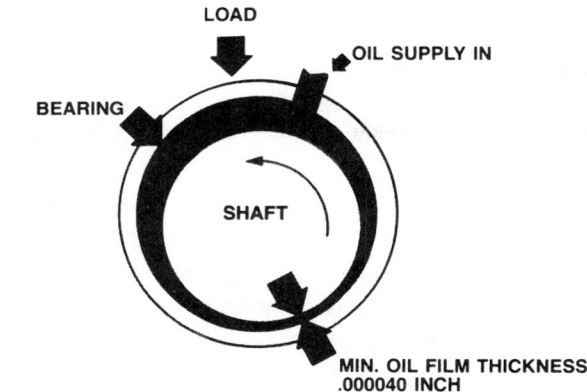

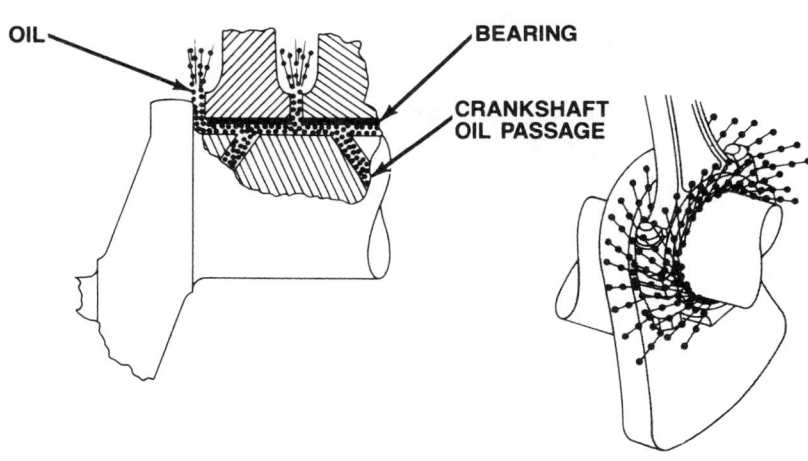

Figure 5-3: A: Hydrodynamic lubrication; B:Oil feed to crankshaft and connecting rods. (Courtesy of Freightliner Corporation)

Engine Oil Classification and Terminology

To fully understand the terms and codes used on the label of any engine oil one must learn about friction, wear, and lubrication. However, every technician should have a rudimentary understanding of the codes and terms used to describe engine oils. The following text introduces some of the basic language of lubricants.

Viscosity The **viscosity** rating of oil describes its resistance to flow. High viscosity oils have molecules with greater cohesion ability. However, properly defined, viscosity denotes resistance to sheer. When two moving components are separated by engine oil, the *lamina* (portion of the oil film closest to each metal surface) on each moving component should have the least fluid velocity, while the fluid in the center has the greatest fluid velocity. Sheer occurs when the lamina fluid velocity is such that it is no longer capable of adhering to the surface of the moving components.

The viscosity of oil is rated as an index.

Viscosity Index The **viscosity index** (VI) expresses the viscosity-to-temperature performance of lubricating oil. Temperature affects viscosity. Viscosity is considerably reduced when the temperature increases and, to a smaller extent, as fluid pressure increases. The greater the VI the less

The viscosity index determines the weight of the oil.

Shop Manual Chapter 5, page 163.

of an effect temperature will have on the actual viscosity. In other words, oils that show relatively small viscosity changes with changes in temperature have a high VI.

SAE Numbers and Viscosity The Society of Automotive Engineers (SAE) grades the viscosity of automotive engine oils. These gradings specify the temperature window within which the engine oil provides adequate sheer resistance under boundary lubrication conditions.

Multi-Viscosity Oils Multi-viscosity engine oils tend to be the lubricating oil of choice among the entire truck and bus engine OEMs. They have been around for many years, although initially some manufacturers took a cautious approach before recommending them. They have the advantage over straight grade oils of being able to provide proper lubrication to the engine over a wide temperature range. In other words, they have a relatively flat viscosity-to-temperature curve. These oils undergo special refining processes, and VI improvers are added. They possess good cold cranking characteristics and show comparatively small variations of viscosity over their nominal operating range.

The lubricity of oil comes from the nitrogen in the oil.

Lubricity Two oils with identical viscosity gradings can possess different **lubricity**. The lubricity of oil properly describes its flow characteristics. Lubricity is also affected by temperature. Hotter oils flow more readily, colder oils less readily. In comparing two engine oils, the one that has the lower frictional resistance to flow possesses the greater lubricity. In thick film fluid lubrication, flow friction is determined by the fluid's viscosity (that is, its resistance to shear). In thin film or boundary lubrication, flow friction is determined by the lubricity of the fluid. Sulfur compounds add to the lubricity of diesel engine oils.

Flash Point The flash point is the temperature at which a flammable liquid gives off enough vapor to ignite. The *fire point* of the same flammable liquid is usually about 10° C higher and is the temperature at which a flammable liquid gives off sufficient vapor for continuous combustion. The flash point value has some significance when assessing a diesel engine lubricating oil, because a large portion of the cylinder wall is swept by flame every other revolution in a four-stroke cycle diesel engine and every revolution in a two-stroke cycle diesel engine. However, actual cylinder wall temperatures are significantly lower than the temperatures of the combustion gases, and the oil is exposed to them for very short periods of time. Most diesel engine lubricating oils have flash points of 400° F (205° C) or higher.

Pour point is 5 degrees below the point at which the oil no longer flows.

Pour Point The temperature at which a lubricant begins to gel or simply ceases to flow is known as the *pour point*. Engine lubricating oils formulated for extreme cold weather operation have pour point additives or depressants, which act as "antifreeze." Pour point is an important engine oil specification, because of the tendency of many operators to use a multigrade engine oil with a viscosity nominally not suited for midwinter conditions in the northern part of the continent.

Inhibitors The term *inhibitors*, when applied to an engine lubricant, refers to the additives that protect the oil itself against corrosion, oxidation, and acidity. They make the oil less likely to partake in reactions with the contaminants in the crankcase, such as combustion by-products, moisture, and raw fuel.

Ash Ash in lubricant is mineral residue that results from oxide and sulfate incineration. High ash levels in oil are often the result of high temperature operation.

Film Strength Most mineral lubricants possess adequate film strength to prevent seizure and galling of contacting metal surfaces. However, high-speed, high output truck diesel engines usually contain additives to improve film strength and cohesiveness of the oil.

Detergents   Detergents are added to engine oils to prevent the formation of deposits on internal engine components. These function to keep soluble oxidation products from becoming insoluble. Polymeric detergents and amine compounds are used.

Dispersants   Dispersants are added to engine oils to help keep insoluble oxidation products in suspension and to prevent them from coagulating into sludge and deposits. When sludge and deposits do form in the crankcase, the capability of the dispersant has been exceeded. This is often an indication that the oil change interval should be reduced.

Synthetic Oils

Most diesel engine OEMs approve of the use of **synthetic oils** and often recommend them for severe duty applications, such as extreme cold weather operation, (providing the oil used meets their specifications). Synthetic lubricants are largely petrochemically derived. Synthesis is the process of creating compounds in a laboratory. Whereas conventional lubricants are obtained from crude oils by distillation or other refining methods, synthetics are manufactured, or synthesized, in chemical plants. Some examples are poly-alpha-olefins, diesters, polyesters, and silicone fluids. Only the very high price of synthetic lubricants limits their use in the trucking industry; most of the indicators suggest that they substantially outperform conventional lubricants. It should also be noted that, while most of the engine OEMs approve of the use of synthetic oils, they have not endorsed increased service intervals when they are used. Since synthetic oils often remain a bit of a mystery to the service technician, the following section will attempt to explain some of the terms used by the manufacturers of synthetic oils.

Poly-Alpha-Olefines (PAOs)

PAOs are the most widely used synthetics formulated for crankcase oil. These are all hydrocarbon structures that contain no sulfur, phosphorous, or metal residues. They are wax free and have low pour points, usually below −40° F (−40° C) and ranging to as low as −90° F. More important they have high thermal stability, meaning they can withstand high temperatures without decomposing. However, POAs are less resistant to oxidation than mineral oils. This requires that they contain antioxidants to withstand high temperature use and provide satisfactory service life without thickening and forming harmful acids. POAs have other disadvantages, such as a limited ability to dissolve necessary additives and, more serious in diesel engine applications, a tendency to shrink seals. Both problems can be overcome by the addition of a synthetic ester base fluid. Most PAO-based products contain from 5% to 20% diesters or polyesters. The applications are engine crankcase, gear lube, turbine lube, and compressor lube.

PAOs are used in fully synthetic and semisynthetic lubricating oil blends. The PAOs in these blends offer improved low temperature properties with lower volatility than in corresponding mineral oils. Poly-alpha-olefins cost between three and eight times what conventional mineral-based oils cost.

Diesters

Esters are synthesized by the reaction of an acid and an alkaline (the water formed as a by-product, must be removed). Diesters get their names from the diacids used to make them. Under some conditions, when water is present as a contaminant, the reverse reaction can occur and reformation of the parent alcohol and acids occurs. If strong acids are formed as a result of hydrolysis, corrosion can become a problem. Selecting the proper additive is crucial to avoid this problem.

Like PAOs, diesters do not contain sulfur, phosphorous, metals, or wax. They have super-low pour points, ranging from −60° to −80° F, and they exhibit good thermal stability. Also, like PAOs, they are susceptible to high temperature oxidation, and they can be made oxidatively stable with the right additives.

Diesters have excellent solvency, both for additives and potential deposit-forming substances. They are clean running lubricants and tend to dissolve varnish and sludge formed by mineral oils. Applications include crankcase lube additives, compressor lubes, misting oils, bearing lubes, and high-temperature hydraulic oils. Diesters are often used in small quantities with PAOs to provide neutral seal swell and adequate solvency for additives. Diesters cost five times the corresponding mineral oil grades.

Polyesters

Polyesters, like diesters, are formed by the reaction of an acid and an alcohol to form an ester and water. The term "poly" refers to a molecule that contains two or more alcohol fractions in its structure. Polyester characteristics are as follows:

1. Pour points from −20 to 95° F.
2. Viscosity indices from 120 to 160 (generally above 140).
3. Oxidatively more stable than diesters or PAOs.
4. Resist hydrolysis (chemical reaction with water, resulting in decomposition) better than diesters do.
5. Outstanding performance at both high and low temperatures.
6. Used in air and hot turbine systems.
7. Biodegrade much faster than mineral oils.
8. Cost ten to fourteen times what mineral oils in the same viscosity range cost.

Alkylbenzenes

Alkylbenzenes are used to improve the extreme low temperature properties of engine oils. Synthetic engine lubricants formulated for use in sub-Arctic and Arctic temperature conditions will often contain an alkylbenzenes component. As an ingredient in synthetic oil, its most important characteristic is its outstanding low temperature properties. They are used in the following applications: Arctic engine oils, compressors, transformers, and refrigeration systems.

Polyalkylene Glycols (PAGs)

Polyalkylene glycols that are synthesized from ethylene oxide are water-soluble and oil insoluble. When they are synthesized from propylene oxide, they are water insoluble and oil soluble. PAGs leave little residue after thermal decomposition, which makes them excellent carriers for graphite dispersions at very high temperatures. PAGs are most often used in hydraulic systems in which there is zero tolerance for leakage of petroleum-based fluids and fire resistance is crucial.

Phosphate Esters

Phosphate esters are used as synthetic base oils and also as anti-wear additives for both mineral and synthetic lubricants. They generally offer good lubricity.

Silicone Fluids

Silicone fluids have viscosity indices that are on the high end of the scale with values of 350 and above. They all contain synthetic and semisynthetic lubricants. They have other outstanding operating characteristics over a wide temperature range. These characteristics are as follows:

❑ Water repellant
❑ Resistant to chemical attack
❑ Compatible with a wide range of plastics and elastomers
❑ However, silicone fluids offer little wear protection for metal surfaces and this limits their use to hydrodynamic lubrication modes.

Lubrication System Components

Shop Manual Chapter 5, page 165.

Engine lubricant must be retained in a reservoir, pumped through the lubrication circuit, filtered, cooled, and have its pressure and temperature monitored. The group of components that perform these tasks are known as the lubricating circuit components. These components vary little from one engine to another, but OEM service literature should be consulted before servicing and reconditioning any components.

Sump Oil Pan

The sump (Figure 5-4) is a reservoir that is usually located at the base of the engine cylinder block enclosing the crankcase. Oil pans are manufactured from cast aluminum, stamped mild steel, and various plastics. The oil pan acts as the reservoir to collect the lubrication oil that drains to the crankcase. The oil pump pick-up can recycle it through the lubrication circuit. Because oil pans can act as a sort of boom box and amplify engine noise, they are usually designed to suppress noise. The oil pan also plays a role in dissipating lube oil heat into the atmosphere, but the effectiveness of this role obviously depends on the material it is manufactured from as aluminum will achieve this objective more effectively than a fiber reinforced plastic.

Dipsticks

The dipstick is a rigid band of hardened steel that is inserted into a round tube to extend into the oil sump. The oil level should be checked daily, so the dipstick is always accessible. In cab-over-engine (COE) chassis, the dipstick must be accessible without raising the cab, so it may be quite long.

 WARNING: To avoid over- or under-filling an engine with oil, it is crucial to use the correct dipstick for the engine.

To obtain an accurate oil level reading, be sure that the engine has been shut down for at least 5 minutes before reading the dipstick level.

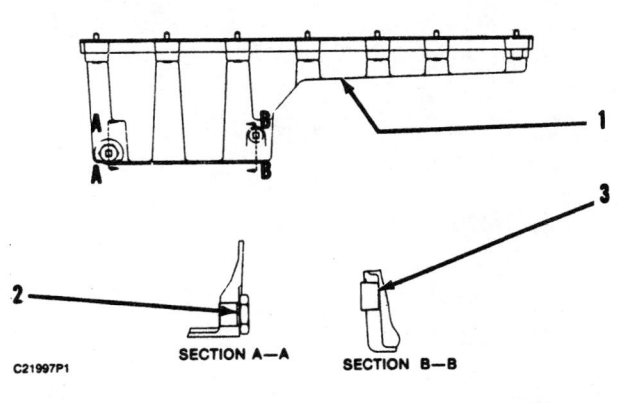

(1) Oil pan.
Oil pan is shown in rear sump mounting position.
An alternate front sump mounting position is available.

(2) Assemble gasket with seam against oil pan.

(3) Apply 9S3263 Thread Lock to plug threads, and tighten to a torque of .... 80 ± 11 N•m (60 ± 8 lb ft)

Figure 5-4 Oil pan. (Courtesy of Caterpillar Inc.)

Oil Pump

Engine oil pumps (Figure 5-5) are usually of the positive displacement type and have pumping capacities that greatly exceed the requirements of the engine. They are gear driven and usually located in the crankcase close to the oil they pump, although in some Cummins and Caterpillar applications they are external. Most oil pumps located in the crankcase are driven by a vertical shaft and pinion engaged with a drive gear on the camshaft. A pick-up is located close to, but not contacting, the base of the oil pan (see Figure 5-5). Two basic types of oil pump are used, external gear and Gerotor.

A BIT OF HISTORY

The original cylinder lubrication on Rudolf Diesel's engine occurred through grooved pieces as part of the piston skirt. These piston pieces dipped into the oil reservoir and splash lubricated the engine components including the bearings. Supercharging quickly proved this design unreliable because of the oil spray being transported with the air via the receiver into the combustion chamber. This design was replaced by a piston oil pump.

External Gear External gear pumps (Figure 5-6) consist of two meshed gears, one driving the other within a housing machined with an inlet (suction) port and outlet (charge) port. As the gears rotate, the teeth entrap inlet oil and force it outside between the gear teeth and the gear housing to the outlet port. Where the teeth mesh in the center, a seal is formed, which prevents any back flow of oil to the inlet. This is by far the most common design of oil pump on current engines. Gear-type engine oil pumps seldom malfunction; when they show evidence of wear, the underlying reason is usually contaminated engine lube.

Figure 5-5 Oil pump assembly.

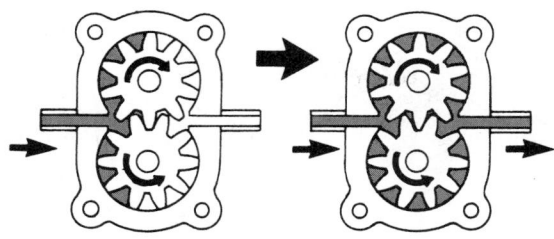

Figure 5-6 Gear-type oil pump. (Reprinted with permission by American Isuzu Motors, Inc.)

OIL PUMP ASSEMBLY

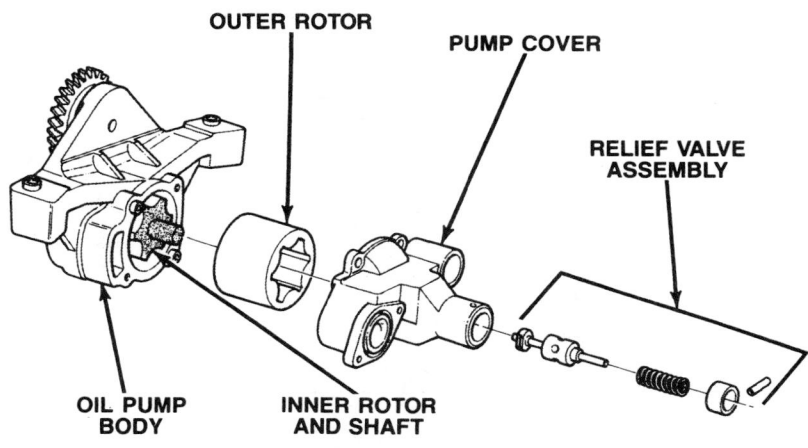

Figure 5-7 Gerotor oil pump. (Courtesy of Freightliner Corporation)

Gerotor **Gerotor** type oil pumps (Figure 5-7) use an internal crescent gear pumping principle. An internal impeller with external crescent vanes is rotated within an internal crescent gear, also known as a rotor ring. The inner rotor or impeller has one less lobe than the rotor ring. As the inner rotor is driven within the outer rotor, only one lobe is engaged at any given moment of operation. In this way, oil from the inlet port is picked up in the crescent formed between two lobes on the impeller. As the impeller rotates, oil is forced out through the outlet port as the lead lobe once again engages. The assembly is rotated within the Gerotor pump body.

Scavenge Pumps/Scavenge Pick-Ups Scavenge pumps are used in the crankcase of some off-highway trucks required to work on inclinations that could cause the oil pump to suck air. They are designed with a pick-up located at either end of the oil pan.

Pressure-Regulating Valves

Pressure-regulating valves (Figure 5-8) are responsible for defining the maximum system oil pressure. Most are adjustable. Typically, an oil pressure-regulating valve consists of a valve body with an inlet sealed with a spring-loaded, ball-check valve. Other types of poppet valve are also used, but the principle is the same. The regulating valve body is plumbed in parallel to the main oil pump discharge line. When oil pressure is sufficient to unseat the spring-loaded check ball, it

When testing the pressure regulating valve operation, always use a master gauge to accurately verify pressures.

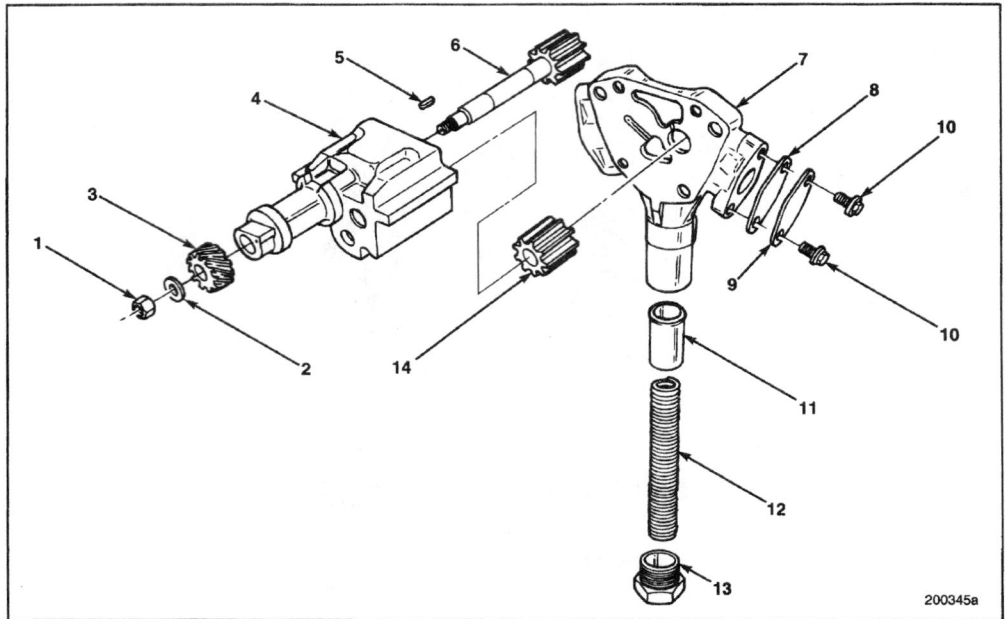

1. Self-locking nut	8. Gasket
2. Washer	9. Inlet flange plate
3. Drive gear	10. Capscrews
4. Housing	11. Plunger
5. Key	12. Relief valve spring
6. Pumping gear and shaft assembly	13. Relief valve cap
7. Oil pump housing cover	14. Idler gear

A

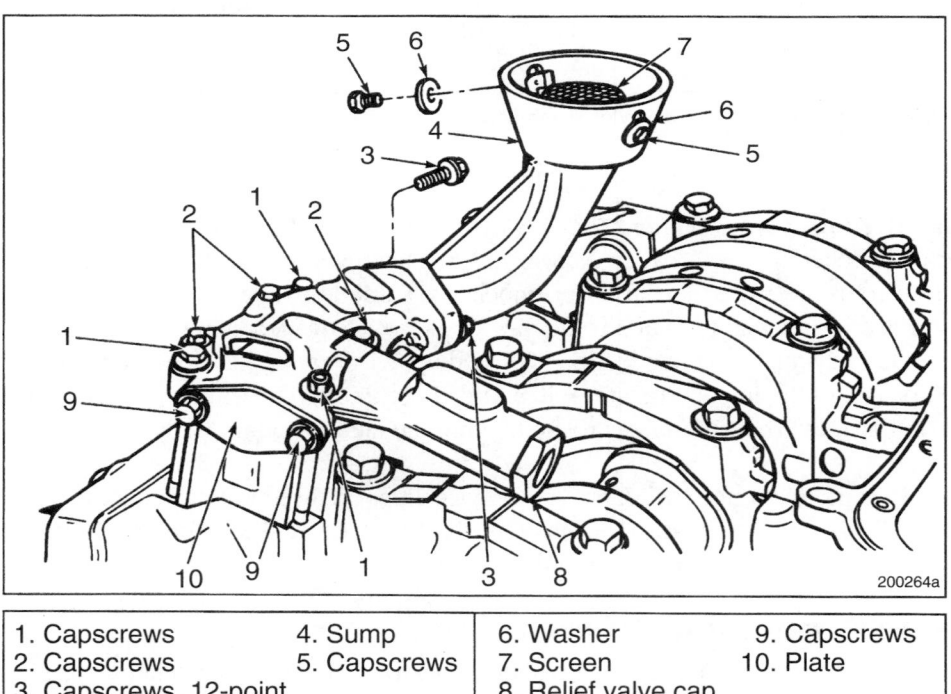

1. Capscrews	4. Sump	6. Washer	9. Capscrews
2. Capscrews	5. Capscrews	7. Screen	10. Plate
3. Capscrews, 12-point		8. Relief valve cap	

B

Figure 5-8 (A) Pressure regulator; (B) View of pressure regulator and oil pump pick-up. (Courtesy of Mack Trucks, Inc.)

unseals permitting oil to pass through the valve and spill to the oil sump. This action will cause the pressure in the oil pump discharge line to drop.

Filters

The role of oil filters (Figure 5-9) in the diesel engine lubrication system is to remove and hold contaminants while providing the least amount of flow restriction in the lubrication circuit. Filters use several different principles to accomplish this objective. The term **positive filtration** is used to describe a filter that operates by forcing all the fluid to be filtered through the filtering medium. Most engine oil filters use a positive filtration principle. Filters function at higher efficiencies when the engine oil is at operating temperature. Filters work to clean the engine oil by using the following methods:

Mechanical Straining   Mechanical straining is accomplished by forcing the lubricant through a filtering medium, which, if greatly enlarged, would have the appearance of a grid or matrix. The sizing of the grid apertures define the size of the particle that may be entrapped by the filtering medium. Most current diesel engine oil filters make use of mechanical straining in conjunction with other principles. Straining media include cotten fibers and rosin-impregnated paper that may be pleated to increase the effective area.

Absorbent Filtration   Absorbent filtration works by absorbing or sucking up engine contaminants as a sponge would. Effective absorbent filtering media include cotton pulp, mineral wool, wool yarn, and felt. These filters not only absorb coarse particles but may also remove insoluble oxidized particulate, moisture, and acids.

Adsorbent Filtration   Absorbent filters hold the molecules of dissolved substances or liquids to the surface of the filtering medium by adhesion. Adsorbent filtering media include charcoal, Fuller's earth, and chemically treated papers. Because adsorbent filters may act to remove oil additives, they are often used only where low-additive engine oils are specified.

The regulating pressure valve is adjusted by setting the tension of the spring (usually by shims or an adjusting screw). Some OEMs use color-coded springs to define the oil pressure values.

Figure 5-9 Oil filter configuration. (Courtesy of Mack Trucks, Inc.)

Filter Types and Efficiencies   Most current filters are spin-on disposable cartridges. Older engines may have permanent canisters enclosing a replaceable element. The canister is mounted to the filter pad with a long threaded shaft that extends through the length of the canister. They are seldom seen today, because most OEMs made conversion adapters so that disposable spin-on cartridges could replace them. Filters are categorized by the manner in which they are plumbed into the lubrication circuit.

Full-Flow Filters   The filter-mounting pad is usually plumbed into the lubrication circuit close to the oil pump outlet, and all of the oil exiting the pump is forced through the filter. The filtering medium is usually treated paper or cotton fiber. In most cases, these filters employ a mechanical straining principle, so the particles entrapped by the filtration media are those too large to pass through it. All **full-flow filter**s used on current truck and bus diesel engines use positive filtration and are rated to entrap particulates between 25 and 60 microns. Bypass valves located on full-flow filter mounting pad(s) protect the engine should a filter become plugged. In this event, the oil exiting the oil pump would be routed around the plugged filter to the lubrication circuit.

Bypass Filters   Bypass filters are used to complement the full-flow filters on current highway diesel engines. These are plumbed in parallel in the lubrication circuit, usually by porting them into the main engine oil gallery. They filter more slowly, but are rated to entrap particles down to 10 microns in size. Two types are used, Luberfiner and centrifugal.

Luberfiner filters are large canister-type filters designed to entrap much smaller particles than full-flow filters. They return the filtered lubricant directly to the oil sump. They may serve an additional role as an oil cooler, because they are often mounted in the airflow. Luberfiner filters are large volume filters that substantially increase the amount of engine oil required; this should be recalled when servicing the engine. A replaceable filter element is installed to the permanent canister. After a filter element is replaced, the canister should be purged of air after engine start-up.

Centrifugal filter filtration (or centrifuging) is used to entrap smaller particles than most full-flow filters, so they are usually of the bypass type. The filter consists of a canister within which a cylindrical rotor is supported on bearings. The filter is plumbed into the lubrication circuit, so that the rotor is charged with engine oil at the lube system pressure. It exits the rotor via two thrust jets angled to rotate the assembly at high velocity. The centrifuge forces the engine oil through a cylindrical filtering medium wrapped around the rotor. The filtering medium is usually a rosin-coated paper element. The filtered oil is returned to the oil sump.

 WARNING: Be sure to properly dispose of used oil filters in accordance with federal hazardous material disposal methods.

Filter Bypass Valves   Filter mounting pad **bypass valves** (Figure 5-10) operate in much the same way as the oil pressure regulating valves, except that their objective is to route the lubricant around a restricted full flow filter to prevent engine damage by oil starvation. When a filter by-pass valve is actuated and the check valve is unseated, it reroutes the oil directly to the lubrication circuit, instead of spilling the oil to the crankcase. This effectively shorts out the filter assembly.

Oil Coolers

The **oil coolers** operate in the same way as the coolant system radiator.

Oil coolers (Figure 5-11) are heat exchangers consisting of housing and a bundle (element/core) through which coolant is pumped and around which oil is circulated. Oil temperatures in diesel engines run higher than coolant temperatures, typically around 110° C (230° F). However, the engine coolant achieves its operating temperature more rapidly than the oil and plays a role in heat-

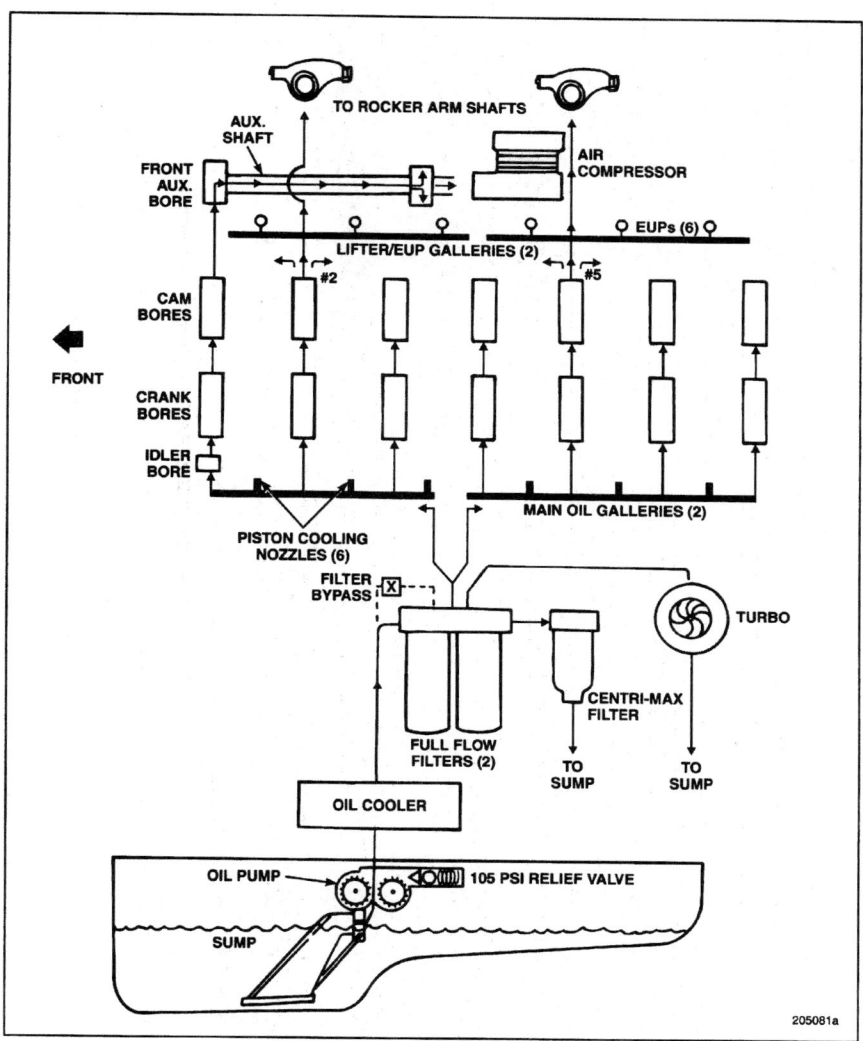

Figure 5-10 Oil flow schematic. (Courtesy of Mack Trucks, Inc.)

ing the oil to operating temperature in cold weather start-up/warm-up conditions. Two types of oil cooler are in current use, the bundle type and the plate type.

Bundle-Type Oil Cooler The bundle-type oil cooler (Figure 5-12) is the most common design. It usually consists of a cylindrical "bundle" of tubes, with headers at either end, enclosed in a housing. Engine coolant flows through the tubes, and the oil to be cooled is spiral circulated around the tubes by helical baffles. The assembly is designed so that the oil inlet is at the opposite end to the coolant inlet. This arrangement means that the engine oil at its hottest is first exposed to the coolant at its coolest, which slightly increases cooling efficiency.

Plate-Type Oil Cooler In the plate-type oil cooler (Figure 5-13), the oil circulates within a series of flat plates and the coolant flows around them within a housing assembly. They boast lower cooling efficiencies than bundle element coolers, but they are usually easier to clean and repair.

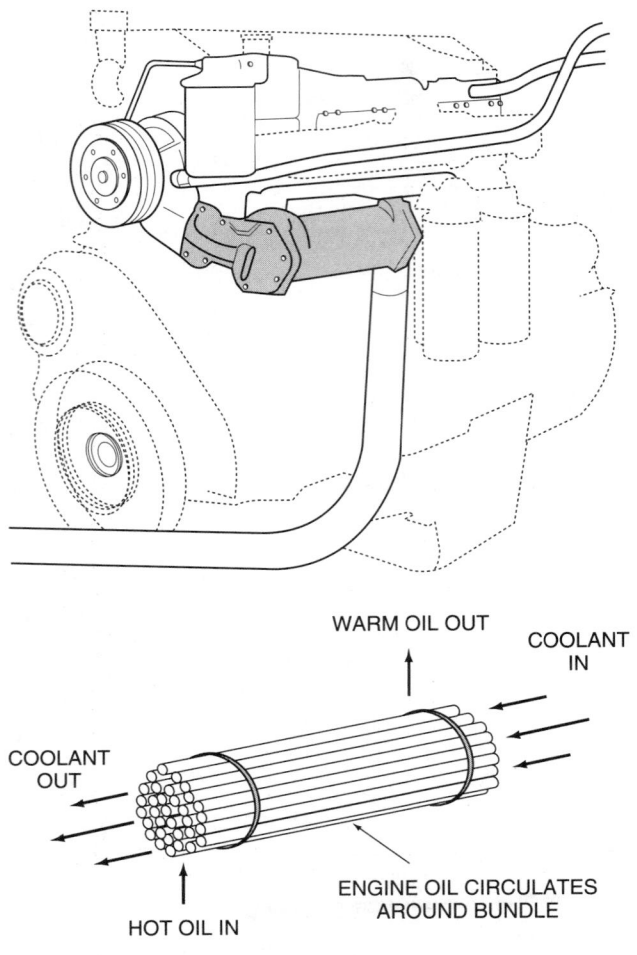

Figure 5-11 Typical oil cooler. (Courtesy of Mack Trucks, Inc.)

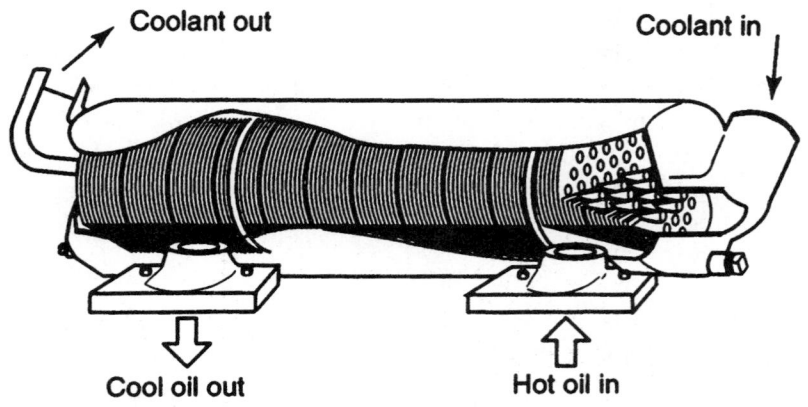

Figure 5-12 Bundle type oil cooler.

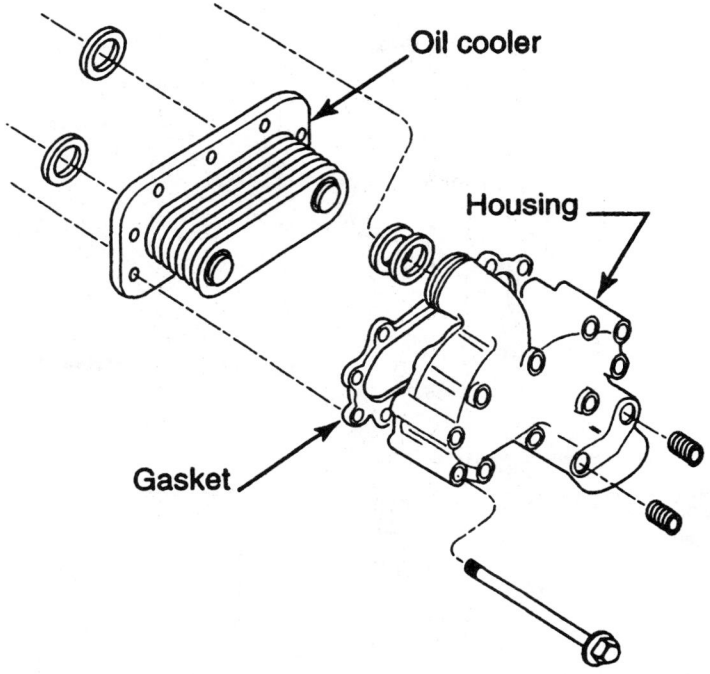

Figure 5-13 Plate-type oil cooler.

Oil Pressure Measurement

Shop Manual Chapter 5, page 174.

Of all the engine monitoring devices used on an engine, the *oil pressure* is one of the most critical.

 WARNING: In most cases, loss of engine oil pressure will cause immediate engine failure.

Back in the days when few of an engine's operating conditions were monitored and displayed to the operator, there was always a means of signaling a loss of oil pressure. Several types of sensor are used in today's engines.

Variable Capacitance (Pressure) Sensor

Most current truck and bus diesel engines use variable capacitance type sensors (Figure 5-14) that are managed electronically. The engine fuel management computer (ECM-electronic control modale) sends a reference voltage to the sensor. Oil pressure acts on a ceramic disc and moves it either closer or further away from a stationary steel disc varying the capacitance of the device and changing the reference voltage returned to the ECM. The ECM is responsible for outputting the signal that activates the dash display or gauge. It should be noted that engine oil pressure usually has to fall to dangerously low levels before programmed failure strategies are triggered.

Piezo Electric Sensor

Oil pressure can be measured using a piezo electric sensor. Certain crystals become electrically charged when exposed to pressure and produce a small voltage. The voltage increases as pres-

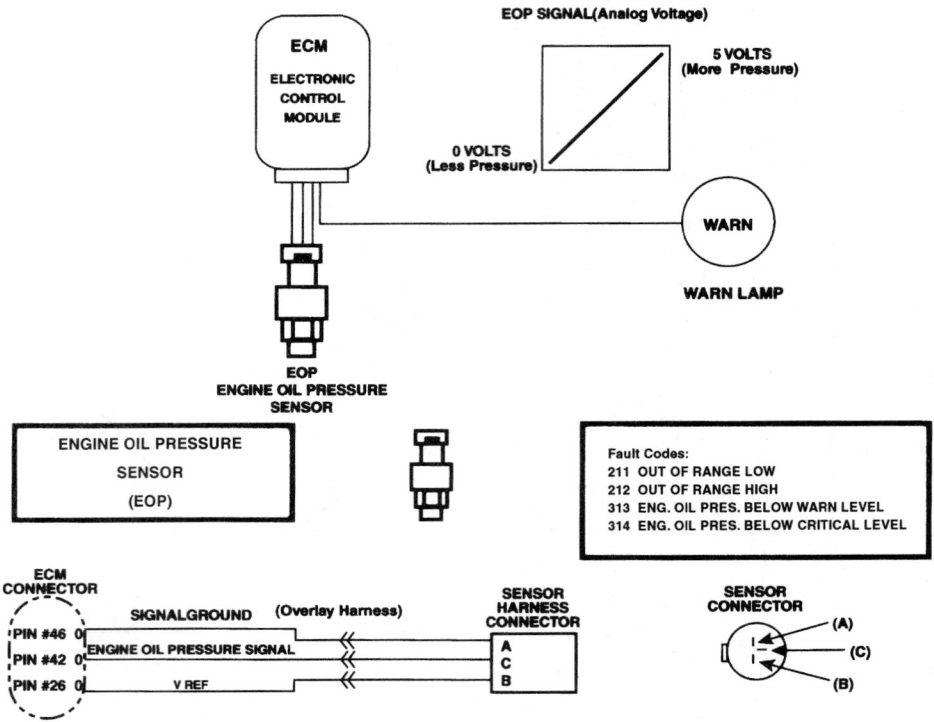

Figure 5-14 Navistar oil pressure sensing circuit. (Courtesy of Navistar International Engine Group)

sure increases. The signal is sent to the electronic control module (ECM) or a voltmeter-type display gauge.

Bourdon Gauge

A flexible, coiled, bourdon tube is filled with oil under pressure. The bourdon tube will attempt to uncoil and straighten incrementally as it is subjected to pressure. This action rotates a gear by means of a sector and pinion. A pointer is attached at the gear across a calibrated scale and provides a means of reading it. This is also known as a mechanical gauge.

Electrical Gauge

Engine oil pressure acts on a sending unit diaphragm, which in turn, moves a sliding wiper arm across a variable resistor, which incrementally grounds out a feed from the electrical gauge. The gauge is a simple armature-and-coil assembly that receives its feed from the vehicle ignition switch.

Shop Manual Chapter 5, page 174.

Oil Temperature Management

As EPA noxious emission standards become more stringent, there is an ever-greater requirement to manage the combustion temperatures, and engine oil temperatures are a factor of engine temperature. Whenever engine oil is used as hydraulic medium, its performance is dependent on its operating temperature to some extent. Figure 5-15 shows a Navistar temperature control circuit that enables the engine oil to bypass the oil cooler while the engine is warming up.

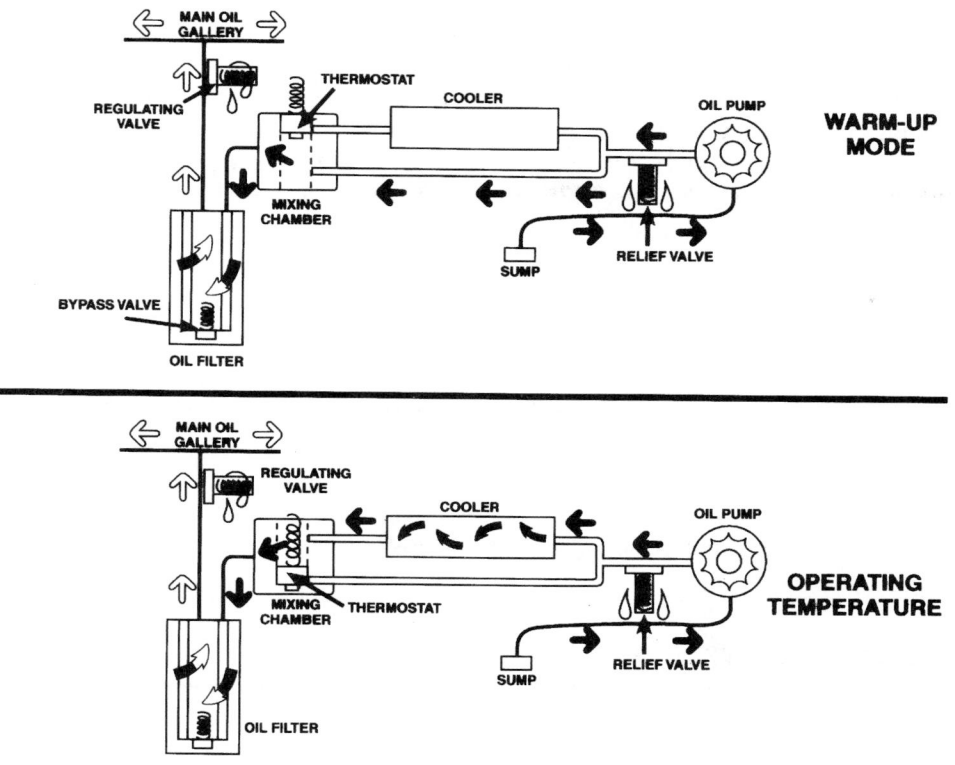

Figure 5-15 Navistar temperature control circuit. (Courtesy of Navistar International Engine Group)

Summary

❏ A diesel engine lubricant performs a number of roles, including minimizing friction, supporting hydrodynamic suspension, cooling, and cleaning. Good heavy duty diesel engine oil should have good oxidation stability, a high natural viscosity index, and proven field performance.

❏ Viscosity describes a fluid's resistance to sheer. Lubricity denotes the flow characteristics of a fluid. Lubricity in engine oils is affected by temperature, with hot oils tending to flow more readily than cold oils.

❏ Most diesel engine OEMs recommend the use of multigrade oils over straight grades and approve the use of synthetic engine oils, especially for operation in conditions of severe cold.

❏ The pour point of an engine oil is the temperature at which the oil starts to gel. Oils formulated for winter use have pour point depressant additives.

❏ Sludged oil is usually a result of oil degeneration caused by prolonged low-load, cold weather operation. Ash in used engine oils is mineral residue caused by oxide and sulfate incineration. High ash levels indicate high temperature operation. High oil sump levels can cause oil aeration.

❏ When interpreting API oil classifications, the S prefix denotes oil formulated for spark-ignited engines, and the C denotes oil formulated for compression-ignition engines. Most of the field research indicates that synthetic oils substantially outperform traditional oils. However, because they cost much more than traditional oils, they will probably not be extensively used by operators until engine OEMs endorse extended oil change intervals.

❏ It is important to maintain the correct engine oil level. The technician should be aware that the consequences of an excessively high oil level could be as severe as those for low sump levels.

❏ Positive displacement pumps of the external gear and Gerotor types are used in diesel engines. Most are of the external gear pump design. Oil pumps are designed to pump much

Terms to Know

API
Boundary lubrication
Bundle
Bypass filter
Bypass valve
Centrifugal filter
Dry sump
Fluid friction
Full flow filter
Gerotor
Hydrodynamic suspension
Lubrication
Lubricity
Oil cooler
Petroleum
Positive filtration
Synthetic oil
Thick film lubrication
Viscosity
Viscosity Index (VI)

greater volumes of oil than that required to lubricate the engine. An adjustable, pressure-regulating valve defines the peak system pressure.

❏ The filters used on a heavy duty lubrication system may be classified as full-flow or bypass, depending on how they are plumbed into the circuit. Full-flow filters are located in series between the oil pump and the lubrication circuit. **Bypass filters** are arranged in parallel and receive oil from an oil gallery and returning it directly to the oil sump. Filters are usually rated by their mechanical straining specification in microns, but they also filter by absorption and adsorption.

❏ Oil coolers are heat exchangers used on most heavy duty, highway diesel engines. The cooling medium used is engine coolant. The common type of oil cooler is the bundle, which consists of a housing within which coolant is pumped through a cylindrical "bundle" of copper tubes in one direction while engine oil is pumped around the tubes in the opposite direction.

❏ Engine oil pressure measurement in most modern engines is performed by a variable capacitance type sensor, which signals the management ECM, the dash gauge, or display. It is, therefore, an ECM output. Engine oil pressures usually have to fall to very low levels before programmed electronic failure strategies are triggered.

Review Questions

Short Answer Essays

1. Outline the function of the main components in a typical diesel engine lubrication circuit.

2. List the required properties of heavy duty engine oil.

3. Define *hydrodynamic suspension* and describe how this principle is used in a typical diesel engine.

4. List API classifications and SAE viscosity grades.

5. Describe the two types of oil pump commonly used on diesel engines, and outline the operating principles of each.

6. Describe the principle of operation of the pressure-regulating valve and its role in the lubrication circuit.

7. Define positive filtration.

8. Outline the differences between full-flow and bypass filters.

9. Describe the operating principles of the various types of heavy duty oil filters and the relative operating efficiencies of each.

10. Outline the objectives of the oil cooler in the lubrication circuit and its operating principles.

Fill-in-the-Blanks

1. Heavy duty diesel engine oil should have good _____ _____, a high natural viscosity index, and proven _____ _____.

2. Viscosity describes a fluid's resistance to _____.

3. Lubricity denotes the _____ characteristics of a fluid. Lubricity in engine oils is affected by _____.

4. The _____ _____ of engine oil is the temperature at which the oil starts to gel.

5. When interpreting API oil classifications, the __ prefix denotes oil formulated for spark-ignited engines, and the ___ denotes oil formulated for compression-ignition engines.

6. Oil pumps are designed to pump much _____ _____ of oil than that required to lubricate the engine. An adjustable, pressure-regulating valve defines the peak system pressure.

7. When an engine is overhauled, the __ _____ should be disassembled and checked for wear.

8. Full-flow filters are located in _____ between the oil pump and the lubrication circuit, while bypass filters are arranged in _____ and receive oil from an oil gallery and return it directly to the oil sump.

9. The common type of oil cooler is the bundle, which consists of a housing within which _____ is pumped through a cylindrical "bundle" of _____ _____ in one direction, while engine oil is pumped around the tubes in the opposite direction.

10. Engine oil pressure measurement in most modern engines is performed by a _____ _____ type sensor, which signals the management ____, the dash gauge, or display.

ASE-Style Review Questions

1. Which of the following SAE multigrade oils would likely be recommended by most highway diesel engine OEMs for North American summertime conditions?
   A. 5W-30 **C.** 20W-20
   B. 15W-40 **D.** 20W-50

2. Which of the following API classifications would indicate that the oil was formulated for a minimum emissions diesel engine?
   A. SF **C.** SG
   B. CC **D.** CG

3. Which of the following statements would usually correctly describe the relative temperatures of the coolant and lubricant when the engine is at operating temperature?
   A. Coolant temperatures are higher than engine oil temperatures.
   B. Engine oil temperatures are higher than coolant temperatures.
   C. Coolant and engine oil temperatures should be equal.
   D. Engine oil temperatures are lower than coolant temperatures.

4. Which type of oil pump is most commonly used in current highway diesel engines?
   A. External gear **C.** Centrifugal
   B. Plunger **D.** Vane

5. *Technician A* says the operating temperature of engine oil should be within 5° F of the engine coolant operating temperature. *Technician B* says engine oil temperatures are probably a more accurate means of assessing actual engine operating temperature than the coolant temperature reading. Who is right?
   A. A only **C.** Both A and B
   B. B only **D.** Neither A nor B

6. An auxiliary oil pump designed to feed oil to the lubrication circuit on a vehicle that operates at extreme working angles is known as a:
   A. Gerotor pump
   B. Scavenging pump
   C. Vane pump
   D. Emergency pump

7. Which of the following statements correctly describes *viscosity*?
   A. Resistance to flow
   B. Resistance to sheer
   C. Lubricity
   D. Breakdown resistance

8. *Technician A* says an oil cooler operates by having coolant pumped through the bundle tubes and oil circulated around the tubes. *Technician B* says a leaking oil cooler core must always be replaced. Who is right?
   A. A only **C.** Both A and B
   B. B only **D.** Neither A nor B

9. *Technician A* says 15W-40 is the most commonly used truck engine oil. *Technician B* says 15W-40 oil is not an appropriate all-season oil, because winter temperatures are often much lower than the SAE recommended ambient temperature window. Who is right?
   A. A only **C.** Both A and B
   B. B only **D.** Neither A nor B

10. *Technician A* says the device used to signal oil pressure values to the ECM on most electronically managed engines uses a variable capacitance principle. *Technician B* says the oil sump level can be signaled to the ECM using a thermistor. Who is right?
   A. A only **C.** Both A and B
   B. B only **D.** Neither A nor B

CHAPTER 6

Cooling System

Upon completion and review of this chapter, you should be able to:

❏ Define conduction, convection, and radiation.

❏ Describe the three types of coolant used in current highway diesel engines and the relative merits and disadvantages of each.

❏ Explain the properties of a heavy duty antifreeze and supplemental cooling additive (SCA) package.

❏ Explain the differences in the types of heavy duty radiator in use including downflow, crossflow, and counterflow.

❏ Outline the functions of a cooling system radiator cap.

❏ Explain the role of the water pump in the system and the principle of operation of a centrifugal pump.

❏ Define the role of the coolant filters and filter servicing requirements.

❏ List the types of temperature gauges used in current highway diesel engines and describe how their signals are output.

❏ Describe how a coolant level warning indicator operates.

❏ Explain the role and operation of the thermostat in the cooling system, and list the different types of thermostats in use.

❏ Define the roles played by the shutters and engine cooling fans in controlling airflow through the engine compartment.

❏ Explain the operating principles of a thermatic type, viscous fan hub.

Introduction

Cooling systems serve to dissipate a percentage of engine **rejected heat**. Rejected heat is that percentage of the potential heat energy of a fuel that the engine is unable to convert into useful kinetic energy (the energy of motion). Therefore, this heat must dissipate to the atmosphere either in the exhaust gas or indirectly through the engine cooling system. If an engine is operating at 40% thermal efficiency, 60% of the potential heat energy is rejected. Approximately half of the rejected heat is discharged in the exhaust gas, which leaves the engine cooling system responsible for dissipating the other half to atmosphere. This task is complicated by the extremes of the North American climate, because it is necessary to manage a consistent engine operating temperature at all engine speeds and loads to ensure optimum performance and minimum emissions. Because liquid cooling systems are universal in North American truck and bus applications, only they will be addressed in this section. The Deutz engine company of Germany manufactures air-cooled engines in the 200–500 BHP power range, but in North America their engines are generally found only in agricultural and off-highway applications.

 WARNING: Do not operate a diesel engine with straight antifreeze. The poor heat transfer qualities of antifreeze can cause the engine to overheat.

Cooling System Functions

The functions of a diesel engine liquid cooling system are to:

❏ Absorb heat from engine components
❏ Transfer the absorbed heat by circulating the coolant
❏ Dissipate heat to the atmosphere by means of heat exchangers
❏ Manage engine operating temperatures

Combustion heat is dissipated by the cooling system to the atmosphere in 3 ways:

1. **Conduction:** The transfer of heat through solid matter, such as through the cast iron material of a cylinder block. The transfer of heat by a medium (coolant) does not involve movement of the medium (coolant) itself.

2. **Convection:** Transfer of heat by currents of gas or liquids, such as in the movement of ambient air through the radiator or engine compartment. The transfer of heat by a medium (coolant) involves movement of the medium (coolant) itself.

3. **Radiation:** Transfer of heat by means of heat rays and not requiring matter such as a fluid or solid. The turbine housing of a turbocharger radiates a considerable amount of heat.

Cooling systems (Figure 6-1) are both sealed and maintained under pressure. By confining a liquid under pressure, its boil point is increased. Most cooling systems are designed to manage coolant temperatures at just below their boil points. The chemistry of the antifreeze and its concentration in the coolant will define the actual boil point of a coolant. Most antifreeze doubles as an antiboil agent. When an engine is about to overheat, the coolant will first boil at the location within the system where the pressure is lowest. In most cases, boiling will occur first at the inlet (suction side) of the system water pump.

Coolant Flow

Coolant flow, as shown in Figures 6-1 and 6-2, proceeds as follows: Coolant is drawn form the bottom tank of the radiator by the water pump, which passes the coolant through the front of the engine cylinder block into the liner water jackets. About midway through the block, coolant enters the oil cooler and then back to the water pump. The coolant flowing through the cylinder water jackets around the liners enters the cylinder head through the head gasket. A head gasket

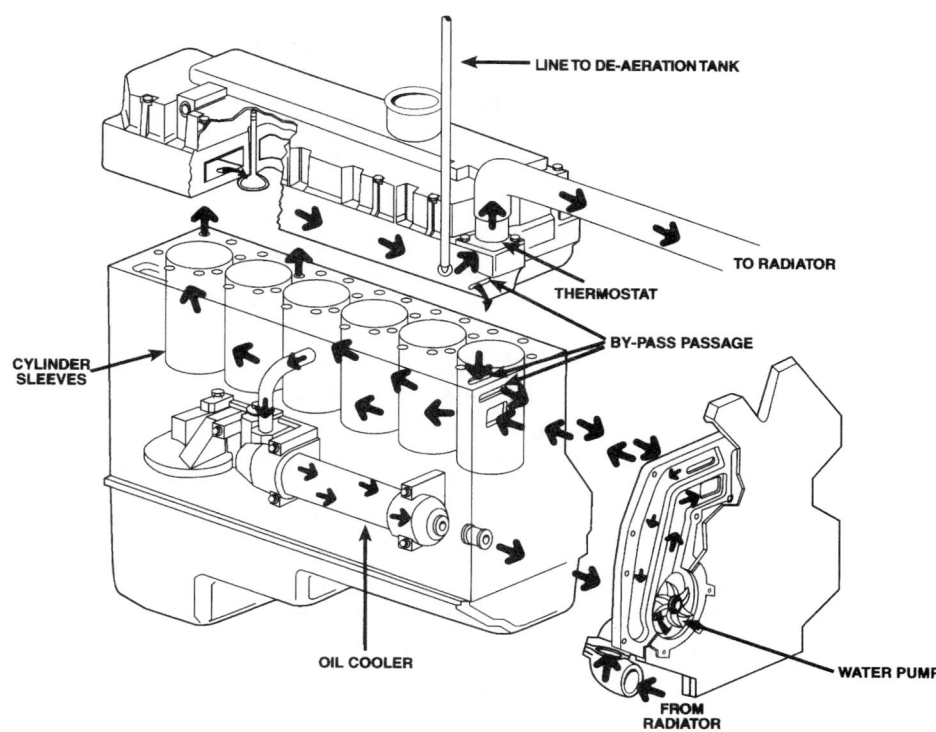

COOLANT FLOW

LINE TO DE-AERATION TANK

TO RADIATOR

THERMOSTAT

BY-PASS PASSAGE

CYLINDER SLEEVES

OIL COOLER

WATER PUMP

FROM RADIATOR

Figure 6-1 Diagram of a typical cooling system. (Courtesy of Navistar International Engine Group)

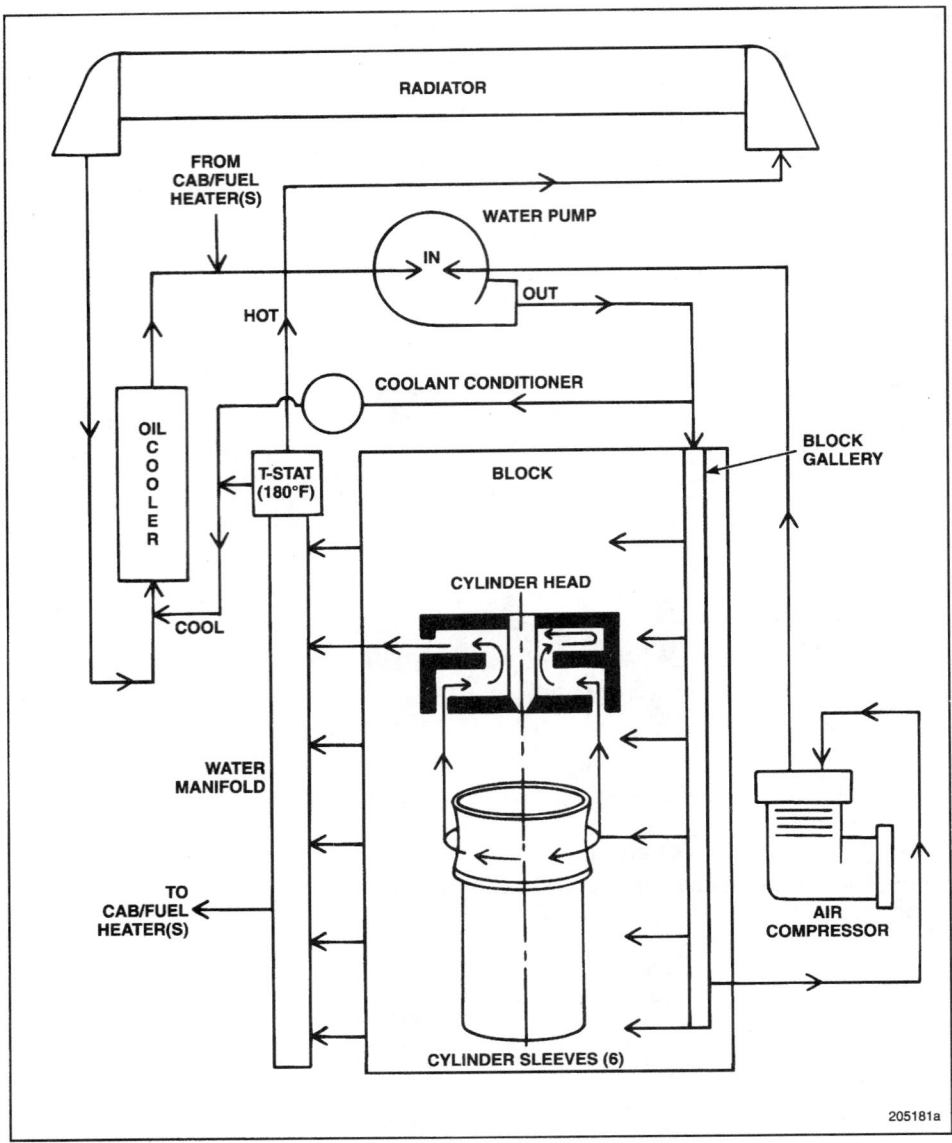

Figure 6-2 Coolant flow schematic. (Courtesy of Mack Trucks, Inc.)

prevents coolant from leaking into the cylinders and regulates coolant flow by the position and size of the holes within the gasket. Before reaching the thermostat, coolant flows through the cored passages in the cylinder head to cool the head. From the thermostat housing, the coolant either flows back to the water pump if the thermostat is in the closed position, or back to the radiator if the thermostat is open.

Engine Coolant

Water-based coolant is the medium used to absorb engine rejected heat, transfer that heat to a **heat exchanger,** and then dissipate it to the atmosphere. The coolant circulates through the engine water jacket to absorb the heat of combustion. Engine coolant is a mixture of water, antifreeze, and **supplemental cooling additives (SCA).** These additives are critical ingredients of the coolant mixture and consist of corrosion inhibitors, anti-foaming agents, and sodium silicates.

Shop Manual
Chapter 6, page 189.

Shop Manual
Chapter 6, page 189.

SCAs have been in use since the 1950s. The first types were chromate based, but these were phased out in the 1970s because of toxicity and, when mixed with ethylene glycol (EG), they formed chromium hydroxide, or green slime. The actual original equipment manufacturer's (OEM) recommended SCA package depends on whether wet or dry cylinder liners are used, the materials used in the cooling system components, and the fluid dynamics (high flow or low flow) within the cooling system.

If the only objective of diesel engine coolant were to act as a medium to transfer heat, pure water would accomplish this more efficiently than any currently used antifreeze mixture. However, water possesses inconvenient boil and freezing points, poor lubricating properties, and promotes oxidation and scaling activity. Almost all current truck and bus engines use a mixture of ethylene glycol, propylene glycol (PG), or carboxylate-type extended life antifreeze, plus water as coolant. Alcohol-based solutions are no longer used, because they evaporate at low temperatures. A properly formulated diesel engine coolant is always made up of the correct proportions of water, antifreeze, and supplemental coolant additives. When the antifreeze properties of the coolant are EG or PG based, the SCAs require monitoring and routine replenishing. **Extended life coolant (ELC)** is low maintenance in that the coolant life is six years with only one SCA charge required in that period.

A BIT OF HISTORY

In the early years of trucks, water was used as a coolant. The water worked well to remove the heat from the engine, but it would freeze, resulting in cracked blocks and liner failure. Several different substances were used to prevent freezing, including salt, calcium chloride, soda, sugar, honey, engine oil, and coal oil. The first successful antifreeze was made from wood and grain alcohol. However, these substances lowered the boiling point of the water and evaporated at higher temperatures. The first permanent antifreeze (meaning it would not evaporate) was glycerine based.

Water expands about 9% in volume as it freezes, and it can distort or fracture the containers it is housed in even when the container is a cast iron engine block. Water occupies the least volume when it is in the liquid state and close to its freezing point of 0° C or 32° F. As water is heated from a near freezing point to a near boiling point, it expands approximately 3%; a 50/50 mixture of water and ethylene glycol will expand even more (approximately 4%) through the same range. Consequently, cooling systems must be designed to accommodate the expansion and contraction of the cooling medium while it is in the liquid state. Most importantly, antifreeze is also antiboil.

The mixture of water, antifreeze, and SCAs that is referred to as engine coolant should perform the following:

1. Corrosion protection. Corrosion inhibitors in both the antifreeze and the SCA package protect the metals, plastics, and rubber compounds in the engine cooling system.
2. Freeze protection. The degree of freeze protection of the coolant is directly related to the proportion of antifreeze in the mixture.
3. Antiboil protection. The degree of antiboil protection of the coolant is directly related to the proportion of antifreeze in the mixture.
4. Antiscale protection. The antifreeze should contain antiscale additives that prevent hard water mineral deposits from adhering to the cooling system heat transfer surfaces.

Care should be taken when servicing coolant filters as some are charged with SCAs, and an excess of SCAs can cause coolant problems just as depleted SCAs.

pH is a chemical term for the amount of acidity or alkalinity in a substance. It is the concentration of hydrogen in the substance.

5. Acidity protection. A pH buffer is used to inhibit the formation of acids in the coolant, which would result in corrosion.

6. Antifoam protection. Antifoam agents prevent aeration of the coolant that could be caused by the action of pumping and flowing through the cooling circuit.

7. Antidispersant protection. This prevents insoluble matter from coagulating and plugging the cooling system passages.

Except for the antifreeze and antiboil characteristics of a glycol-based coolant mixture, the remaining protection additives deplete with engine operation and they must be evaluated and restored at appropriate maintenance intervals. Both propylene glycol and ethylene glycol are petrochemical products. Ethylene glycol was used as the standard antifreeze, but the Federal Clean Air Act and OSHA (Occupational Health and Safety Administration) have both come to regard EG as a toxic hazard. Propylene glycol in its virgin state is less toxic than EG, and for that reason it is gaining acceptance as a base antifreeze ingredient.

CAUTION: Leaks and spillage of both EG and PG should be regarded as dangerous to mammals (including humans) and plant life. The chemical characteristics of engine coolant change with use, and toxicity increases.

Ethylene glycol is derived from ethylene oxide, which is produced from ethylene, a basic petroleum fraction. **Propylene glycol** is derived from propylene oxide, which is produced from propylene, another basic petroleum fraction. When mixed in a solution with water, both PG and EG are described as aqueous. Extended life coolants claim to have lower toxicity, but OSHA has insufficient data to rule on PG and ELCs. Table 6-1 compares the freeze points of PG and EG.

A hydrometer is an instrument that measures the specific gravity of antifreeze.

Propylene glycol- and ethylene glycol-based coolants should never be mixed. The mixture itself will not cause any engine or cooling system problems, but it will be impossible to determine the antifreeze mixture strength with either a refractometer or a hydrometer. If a mixture of EG and PG is known to have taken place and the coolant cannot be immediately replaced, use a refractometer with an EG and a PG scale, and average the two readings. However, the cooling system should be drained and refilled with either aqueous PG or EG when practical to avoid problems later on. ELC is only sold premixed and is dyed a red color.

The correct instrument for reading the degree of antifreeze protection of a coolant is a refractometer.

WARNING: Only ELC premix should be added to the cooling system; in conditions of extreme cold, use ELC concentrate. ELC is incompatible with PG and EG. If you mix PG and EG with ELC, engine damage may result.

Table 6-1 EG and PG freeze points

Concentration of Antifreeze by % Volume	Freeze Point of Coolant			
	Ethylene Glycol		Propylene Glycol	
0 (Water only)	32° F	0° C	32° F	0° C
20%	16° F	−0° C	19° F	−7° C
30%	4° F	−16° C	10° F	−12° C
40%	−12° F	−24° C	−6° F	−21° C
50%	−34° F	−37° C	−27° F	−33° C
60%	−62° F	−52° C	−56° F	−49° C
80%	−57° F	−49° C	−71° F	−57°C
100%	−5° F	−22° C	−76° F	−60° C

High Silicate Antifreeze

High silicate concentrations are required to protect aluminum components exposed to the engine coolant. However, many OEMs require that low silicate coolant is used in their engines. High silicate and low silicate antifreeze should not be mixed. Generally, high silicate antifreeze should not be used except when required by the engine OEM. ELCs do not use silicates, nitrates, borates, phosphates, or amines to inhibit scaling. Instead they use a carboxylate base that, according to the manufacturer, significantly outperforms the complex chemical brew required of a EG or PG coolant.

Extended Life Coolants (ELC)

Extended life coolants promise a service life of 600,000 miles (960,000 km) or six years, with one additive recharge at 300,000 miles (480,000 km) or three years. This compares with a typical service life of two years during which up to 20 recharges of SCA would be required for conventional EG and PGs. ELCs are presently available only as a premixed solution to ensure that the water quality is at the required level. The pricing of ELCs by quantity is generally comparable with EG and PG, but because of reduced cooling system maintenance and extended service life, they will probably become the coolant of choice of the engine OEMs. No test kits are required to monitor the ELC SCA level. Notable among the advantages claimed for ELCs are:

1. Greatly extended service life (6 years, or 600,000 miles)
2. No inhibitor testing required
3. Improved water pump longevity due to much lower total dissolved solids (TDS) content (TDS are often abrasive)
4. Reduced hard water scaling
5. Improved cavitation protection
6. Improved corrosion protection
7. Improved heat transfer ability
8. No gelling problems (no silicates are used in ELC, which eliminates the problem of silicate drop-out, responsible for sludging EG, and PG)
9. Improved aluminum corrosion inhibitors
10. Better high temperature performance than EG or PG

ELCs may be used in engine cooling systems that have previously used either EG or PG. The EG and PG should be drained from the cooling system, which should then be flushed with clean water. The ELC can then be installed. If a coolant filter is used, replace the existing filter with an SCA-free filter. For midwinter Northern U.S. and Canadian operation, the ELC premix should be strengthened with concentrate; check with the ELC manufacturer. The Caterpillar Engine Company was the first to endorse the use of ELCs in their products, and many other engine OEMs will no doubt follow.

DexCool is the registered trademark of an extended life coolant.

Measuring the conductivity of the coolant solution in a test using a total dissolve solids (TDS) value, using the OEM service manual.

Cooling System Components

The components used to store, pump, condition, and manage engine coolant flow and temperature are known as the cooling system components. These components vary little from one diesel engine manufacturer to another.

Shop Manual Chapter 6, page 193.

 WARNING: When servicing and repairing cooling system components, always consult the appropriate service literature to avoid damage to expensive engine components.

Radiators

Radiators are heat exchangers. The engine power rating will usually dictate the frontal area of a radiator, typically 3–4 sq. in. per BHP unit. The cooling medium is ram air, that is, ambient air forced through the radiator core as the truck is driven downs the highway. Vehicle speed and ambient temperatures will obviously determine the radiator's efficiency as a heat exchanger. Fan shrouds improve airflow through the radiator core and the efficiency of the fan.

Radiator Materials and Construction　Most truck diesel engine radiators are currently fabricated from copper and brass components, but the use of aluminum and plastics is increasing. Radiators typically consist of bundled rows of round or elliptical tubes through which coolant flows. These tubes are connected to fins, which increase the sectional area of the tubes exposed to ram air. The tubes are usually are made of brass, and the fins are copper in truck applications, but OEMs are experimenting with aluminum, which is widely used in automobile radiator construction, because of its lower cost and lighter weight. Copper, brass (alloy of copper and zinc), and aluminum all contain high coefficients of heat transfer and are ideal as base material for radiators. Aluminum is more susceptible to corrosion than copper and brass, both from within (coolant breakdown, poor water quality) and outside (salt, both ambient and road salt). Plastics are widely used in the construction of radiator tanks, replacing bolted steel tanks. Plastic tanks are usually crimped onto the main radiator core, enclosing the headers.

All radiators are equipped with a drain valve (located at the lowest point in the assembly), inlet and outlet piping, and a filler opening sealed with a radiator cap. Most use a single pass, downflow principle, which requires that the cooling tubes run vertically from the top tank to the bottom tank. Radiators are classified by their flow characteristics. The following types are found in current truck applications:

Downflow Radiators (Figure 6-3)　Coolant enters the radiator through the top tank and flows to the bottom tank by means of vertical tubes that connect the upper and lower tanks. Downflow radiators are a single pass design, meaning that the coolant is routed from the top tank to the bottom and then exits.

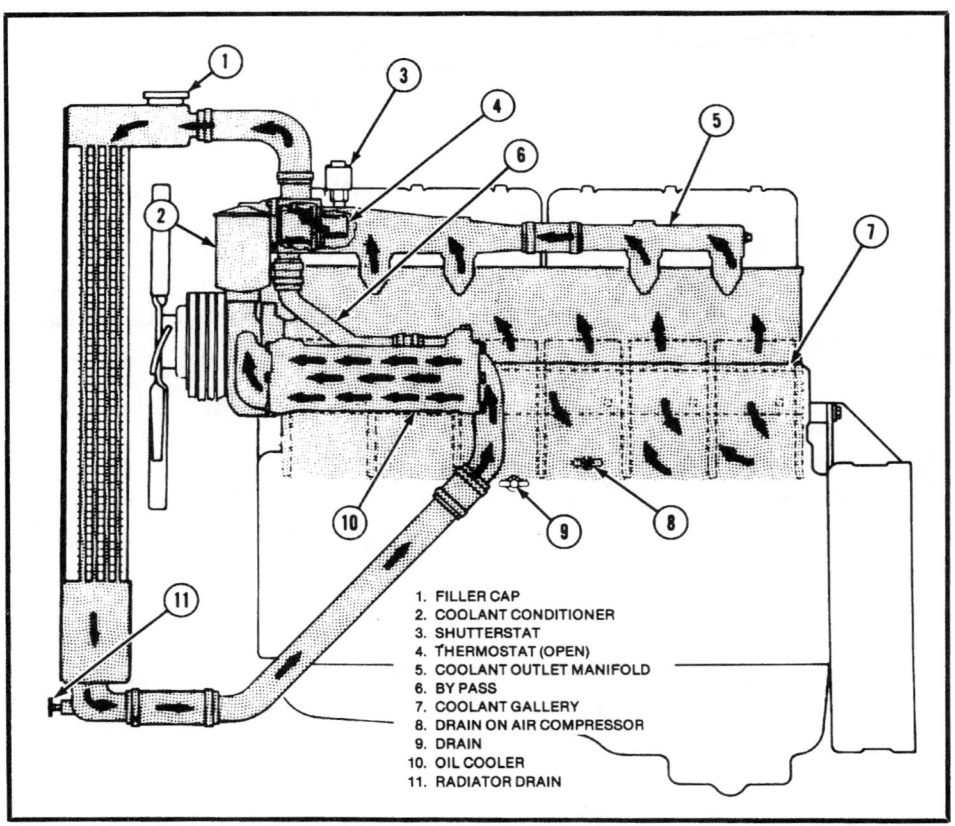

1. FILLER CAP
2. COOLANT CONDITIONER
3. SHUTTERSTAT
4. THERMOSTAT (OPEN)
5. COOLANT OUTLET MANIFOLD
6. BY PASS
7. COOLANT GALLERY
8. DRAIN ON AIR COMPRESSOR
9. DRAIN
10. OIL COOLER
11. RADIATOR DRAIN

Figure 6-3 Downflow radiator. (Courtesy of Mack Trucks, Inc.)

Crossflow Radiators (Figure 6-4)   The tanks are positioned on either side of the radiator core. The coolant enters through one of the side tanks and then flows through horizontal tubes to the tank at its opposite side. This design affords a lower profile than the downflow design, and it is used by chassis OEMs using a low-nose, aerodynamic truck design. Flow is single pass.

Counterflow Radiators   The coolant usually enters through a bottom tank that is divided into inlet and outlet sections. The coolant flows vertically upward from the inlet section of the bottom tank to the top tank, then downward to the outlet section of the bottom tank before being returned to the engine cooling circuit. Essentially, the cooling efficiencies of this design are improved mainly because the coolant is retained in the radiator for a longer period. The flow of coolant through the radiator is double pass. The design was popular when liquid-cooled, charge air heat exchangers were common.

Air in a cooling system can cause many problems. It severely compromises the coolant's ability to transfer heat, and it may promote corrosion. In extreme instances it can cause a cooling system to become air-bound. An air-bound system occurs when air is trapped in the inlet to the coolant pump and it effectively loses its prime. As a consequence, most cooling systems are designed to limit aeration problems. Some radiators have a divided top tank in which coolant enters the lower section from the cylinder head and the upper section from a standpipe. A baffle separates the two sections. This design tends to reduce cooling system aeration problems. Vent lines also help to deaerate the cooling system. The radiator top tank stores reserve coolant volume and accommodates the thermal expansion of coolant.

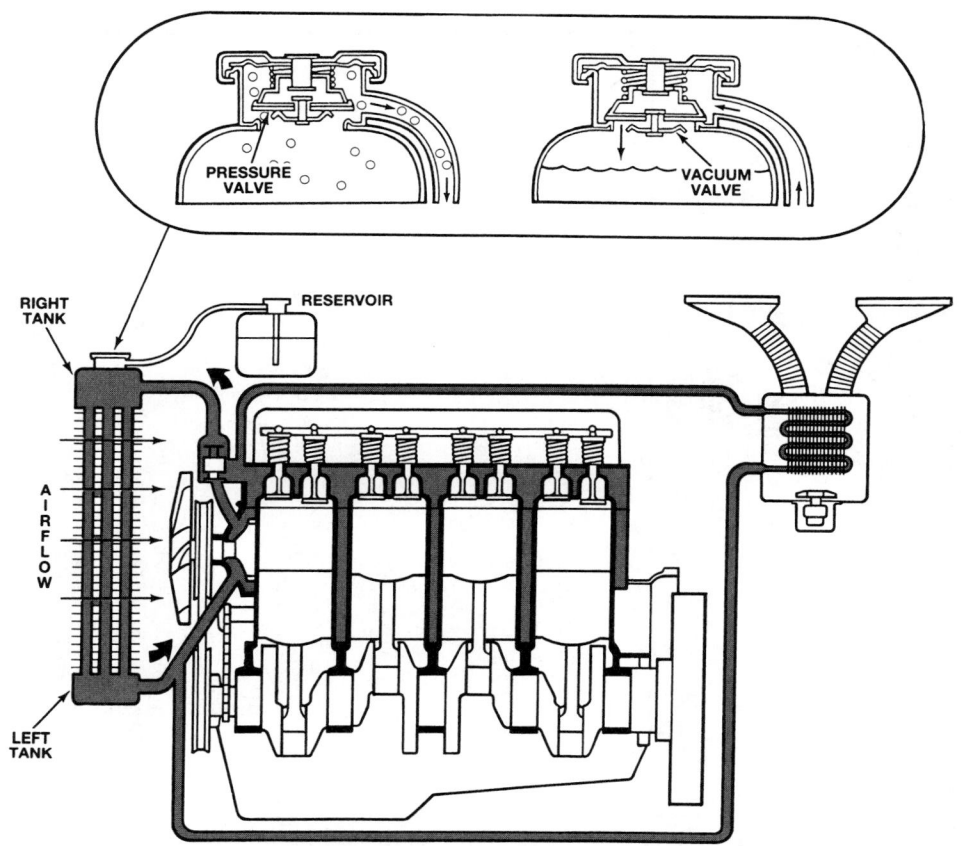

Figure 6-4 Crossflow radiator flow. (Courtesy of General Motors Corporation, Service Operations)

Radiator Cap

> **CAUTION:** Great care should be exercised when removing a radiator cap from the radiator. If the system is pressurized, hot coolant may escape from the filler neck with great force. Most filler necks are fitted with double cap lock stops to prevent the radiator cap from being removed in a single counterclockwise motion. If the radiator is still pressurized, the cap will jam on the intermediate stops. Never attempt to remove a radiator cap until the cooling system pressure is equalized.

Radiators are usually equipped with a pressure cap (Figure 6-4) designed to maintain a fixed operating pressure while the engine is running. This cap is also equipped with a vacuum valve to admit surge tank coolant (or air) into the cooling circuit (the upper radiator tank) when the engine is shut down to accommodate coolant thermal contraction. Radiator caps permit pressurization of a sealed cooling system. For each 1 psi (7 kPa) above atmospheric pressure, coolant boil point is raised by 3° F (1.67° C) at sea level. For every 1,000 feet of elevation, the boil point decreases by 1.25° F (0.5° C). System pressures will seldom be designed to exceed 25 psi (172 kPa); more typically they will range between 7 psi (50 kPa) and 15 psi (100 kPa).

Radiator caps are rated by the pressure required to overcome the cap spring pressures and unseat the seal. When this occurs, the coolant is routed to a surge tank. The surge tank coolant level will always be at its highest when the engine is running hottest. As the engine cools, the pressure within the cooling system drops. When it falls to a "vacuum" value of ±0.25 psi, the radiator cap vacuum valve is unseated, which allows coolant from the surge tank to be pulled back into the radiator.

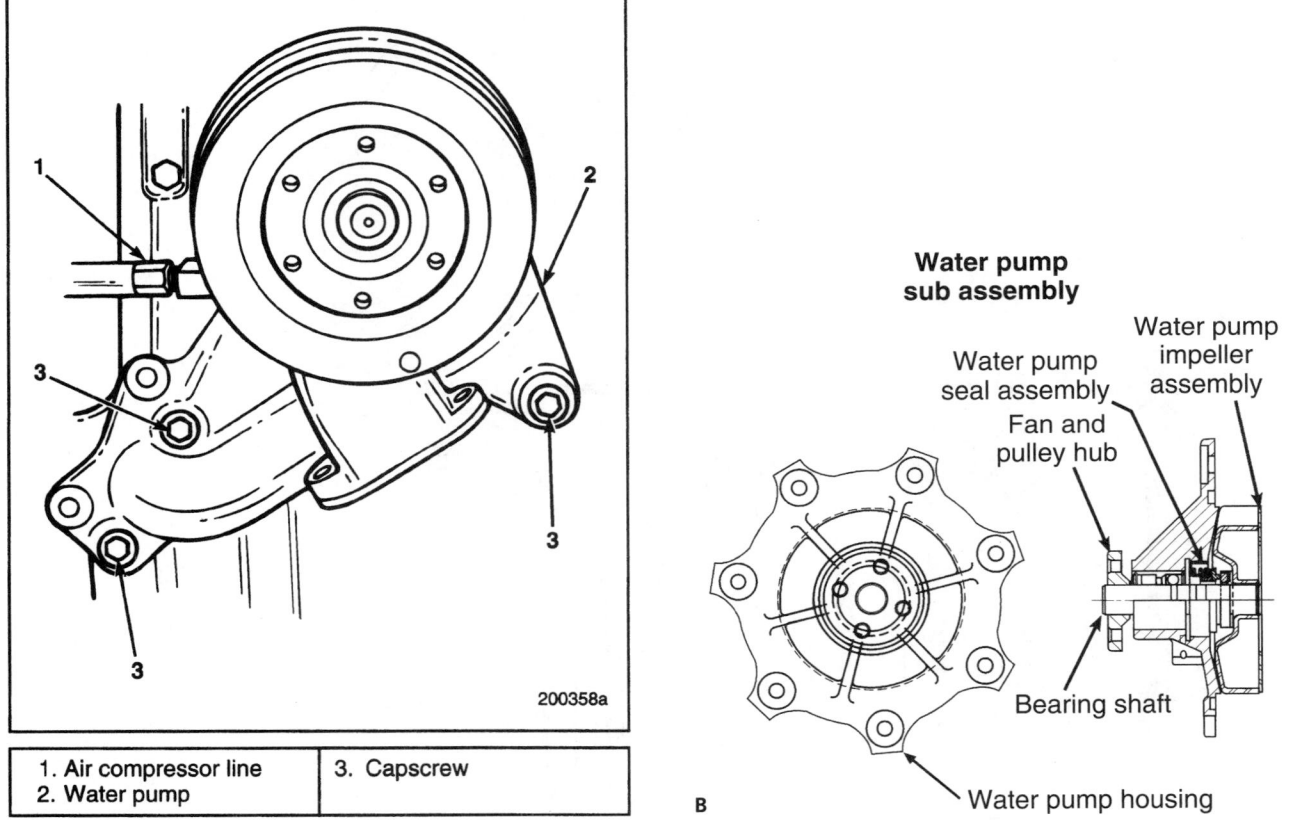

| 1. Air compressor line | 3. Capscrew |
| 2. Water pump | |

A

Water pump sub assembly

Water pump seal assembly

Fan and pulley hub

Water pump impeller assembly

Bearing shaft

Water pump housing

B

200358a

Figure 6-5 (A) Water pump, exterior view (Courtesy of Mack Trucks, Inc.); (B) Water pump sectional (Courtesy of Navistar International Engine Group).

Water Pumps

Water pumps (Figure 6-5) are usually nonpositive, centrifugal pumps driven directly by a gear or belts. When the engine rotates the coolant pump, an impeller is driven within the housing, creating low pressure at its inlet (usually located at or close to the center of the impeller). The impeller vanes throw the coolant outwards, and centrifugal force accelerates it into the spiraled pump housing and out toward the pump outlet. Because the cooling system pressure at the inlet to the coolant pump is at its lowest, boiling always occurs at this location first. This boiling will very rapidly accelerate the overheating condition, as the pump impeller will be acting on vapor. Water pumps are the main reason that engine coolants should have some lubricating properties, because they are vulnerable to abrasion damage when the coolant total dissolved solids levels are high.

Filters

Coolant filters (Figure 6-6) are usually of the spin-on cartridge type connected in parallel to coolant flow. Corrosion inhibitors may also be packaged within the OEM coolant filter, a good reason to avoid over-servicing. When coolant filters must be changed, check the type of shutoff mechanism used: some are automatic, others have manual shutoff valves. New filters do not require priming. Flow-through the filter is consistent with that of most other engine filters in that the coolant enters the canister through outer ports and exits through a central single port. Because some coolant filters are loaded with the SCA charge, be sure the correct filter is installed. Observe both the OEM recommendation and the type of coolant used in the cooling system. Some coolant filters are equipped with a zinc electrode to negate the electrolytic effect of the coolant, although these are more likely to be found in marine applications.

Coolant Temperature Sensors Indicating Circuit and Gauges

Shop Manual Chapter 6, page 202.

Coolant temperature is often a primary reference to the engine management software when factoring timing and air-fuel ratio parameters on electronically managed engine systems; thus the temperature display to the operator is secondary in importance. All current electronically managed engines can be programmed to default to failure strategies, which may include engine shutdown based on the coolant temperature readings. Thermistors (Figure 6-7) are most often used to sense the temperature of the

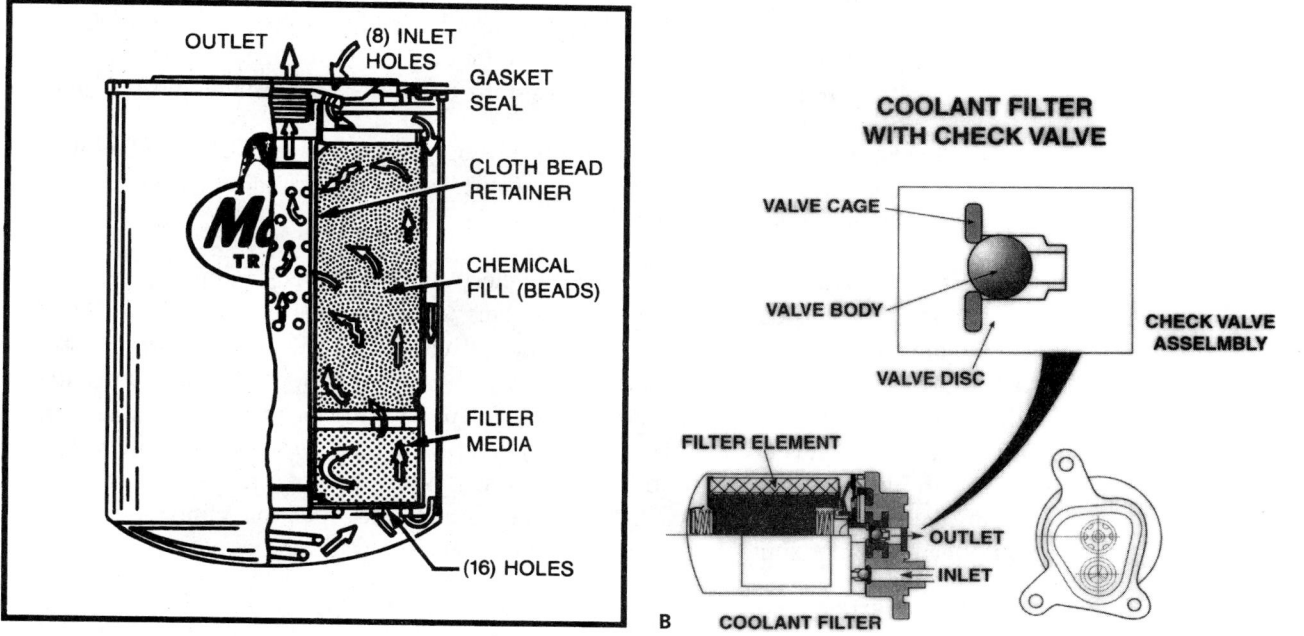

Figure 6-6 (A) Coolant filter (Courtesy of Mack Trucks, Inc.); (B) Coolant filter with mounting pad (Courtesy of Navistar International Engine Group).

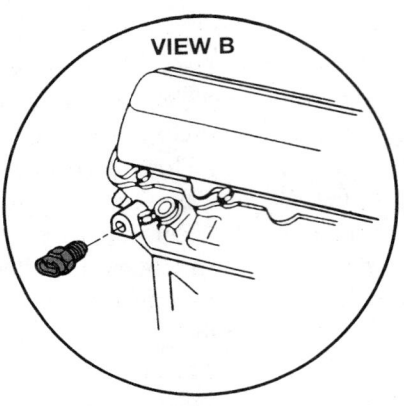

Figure 6-7 Coolant temperature sensor. (Courtesy of General Motors Corporation, Service Operations)

cooling system as well as other engine fluid temperatures including ambient air, boost air, and lubricating oil. The following methods are used to sense coolant temperature.

Thermistors Thermistors are solid state, semi-conducting devices whose internal resistance varies with temperature change. They are supplied with a specific reference voltage and output a signal based on temperature. Negative temperature coefficient (NTC) thermistors tend to be more commonly used. The internal resistance in a NTC thermistor decreases as temperature rises. Positive temperature coefficient (PTC) thermistors function in the opposite manner: the internal resistance increases as temperature rises. Thermistors are commonly used to read coolant temperature on electronically controlled engines. The reference voltage is an ECM output. The thermistor returns an electrical signal representative of the coolant temperature to the ECM. This signal is used in processing to help factor fueling logic and as an output for the dash digital display or gauge.

Electric Sensors Electric sensors use a bimetal arm in conjunction with a resistor fed with an modulated electrical signal from the temperature gauge. When the bimetal arm is heated, the greater linear expansion of one of the bimetal strips will cause it to bend one way; as it cools and contracts, it will bend in the opposite direction. Connected to the bimetal strip is a wiper, which short circuits the current flow through the resistor to ground, thereby altering the gauge value.

Expansion Sensing Gauge An expansion sensing gauge consists of a tube filled with a liquid that expands as it is heated. As it expands, it activates the gauge indicator needle. These gauges are not used much in today's engines.

Coolant Level Indicators Most current electronically managed engines have radiators equipped with low coolant level warning systems, as shown in Figure 6-8 for a GM 6.5L diesel engine. Most operate using the same principles. The ECM outputs a signal to a probe (or sensor), which grounds through the coolant. The probe is usually located in the top radiator tank. When the probe fails to ground through the coolant, a low coolant level warning is generated. The outcome depends on how the ECM has been programmed (this is a customer data program option). In most cases, a lag (from 5 to 12 seconds) is required before the ECM resorts to a programmed failure strategy, which this may be to alert the operator, ramp down to a default rpm/engine load, or shutdown the engine after a suitable warning period.

Thermostats

Thermostats (Figure 6-9) function as a type of automatic valve that will sense changes in engine temperature and regulate coolant flow to maintain an optimum engine operating temperature. To function effectively a thermostat must:

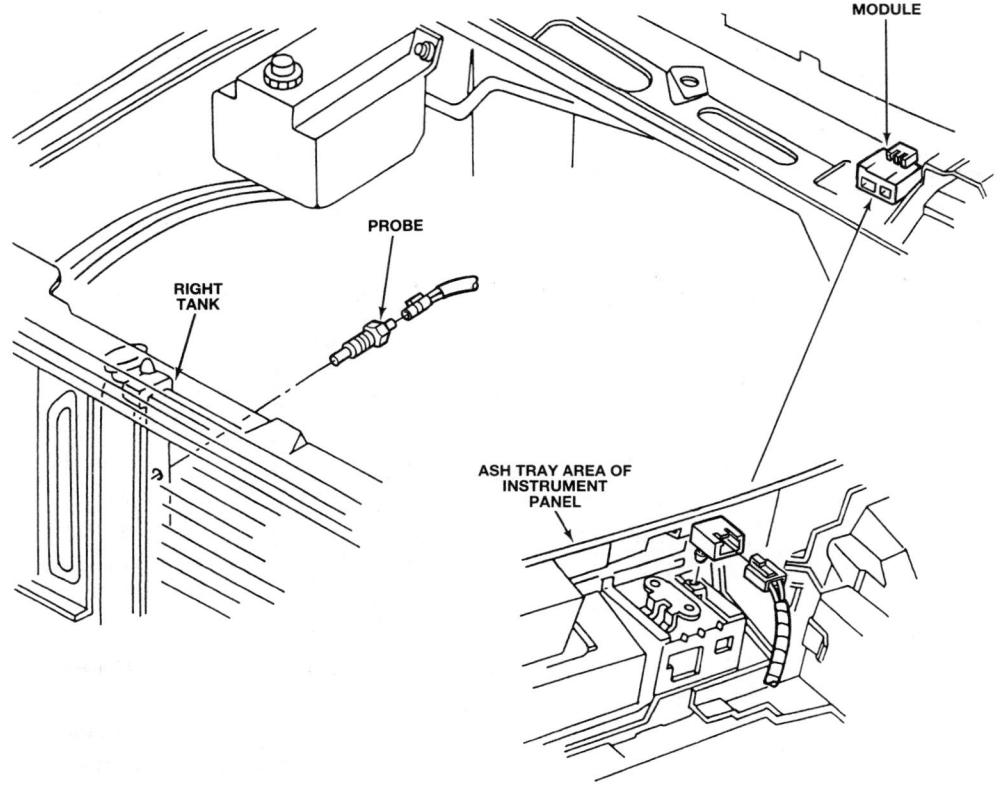

Figure 6-8 Low coolant warning system. (Courtesy of General Motors Corporation, Service Operations)

1. Start to open at a specified temperature
2. Be fully open at a set number of degrees above the specified temperature
3. Define a flow area through the thermostat in the fully open position
4. Permit zero coolant flow or a defined small quantity of flow when in the fully closed position

The cooling system thermostat is normally located either in the coolant manifold or in a housing attached to the manifold. Its primary function is to permit rapid warm-up of the engine. When the engine has attained its normal operating temperature, the thermostat will open and permit coolant circulation. As the thermostat defines the flow area for circulating the coolant, there may be more than one area. A heat-sensing element actuates a piston that is attached to the seal cylinder. Figure 6-10 shows that when the engine is cold, coolant is routed to the coolant pump to be recirculated through the engine. When the engine heats to operating temperature, the seal cylinder gates off the passage to the coolant pump and routes the coolant to the radiator. The heat-sensing element consists of a hydrocarbon or wax pellet into which the actuating shaft of the thermostat is immersed. As the hydrocarbon or wax medium expands, the actuating shaft is forced outwards in the pellet, opening the thermostat. Thermostats can be full blocking or partial blocking.

Top Bypass Thermostat   The top bypass thermostat simultaneously controls the flow of coolant to the radiator and the bypass circuit. During engine warm up, all of the engine coolant is directed to flow through the bypass circuit. As the temperature rises to operating temperature, the thermostat begins to open and coolant flow is routed to the radiator, increasing incrementally with the rise in temperature.

Running without a thermostat is illegal and will produce not only increased noxious emissions but also poor fuel economy and a host of other engine problems.

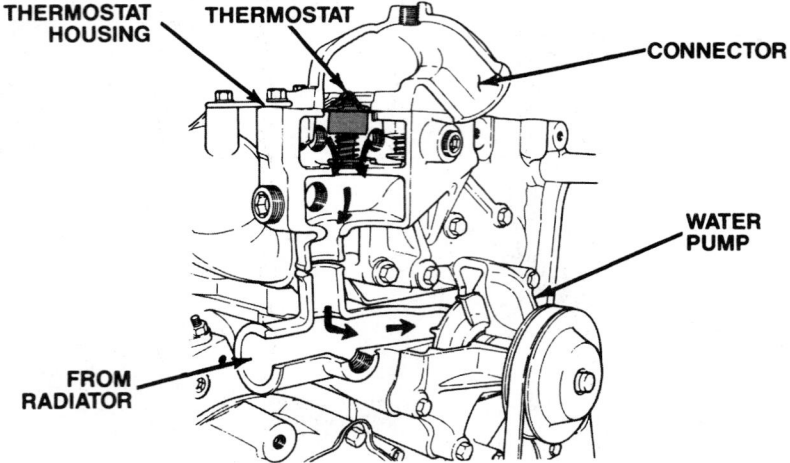

**COLD OPERATION
THERMOSTAT CLOSED**

THERMOSTAT
HOUSING

THERMOSTAT

CONNECTOR

WATER
PUMP

FROM
RADIATOR

A

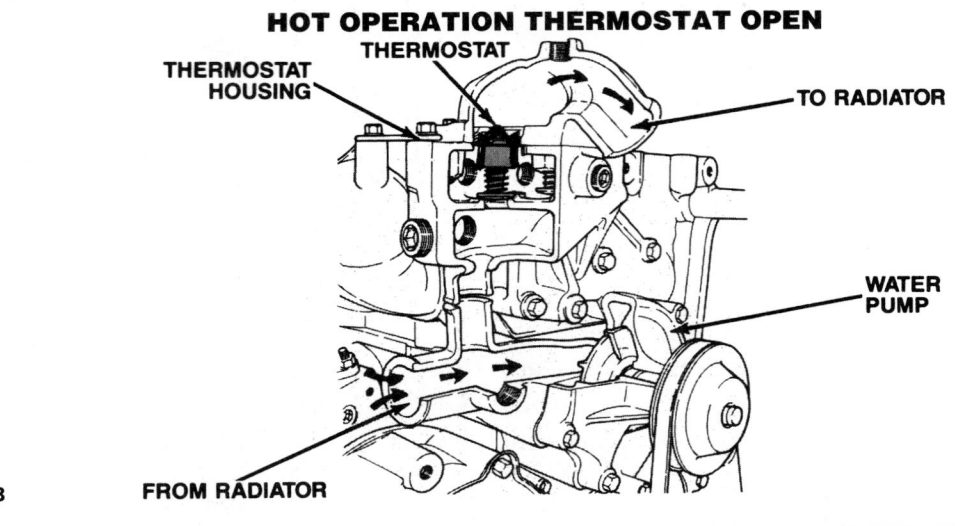

HOT OPERATION THERMOSTAT OPEN

THERMOSTAT

THERMOSTAT
HOUSING

TO RADIATOR

WATER
PUMP

B   FROM RADIATOR

Figure 6-9 (A) Cold-operating thermostat; (B) Hot-operating thermostat. (Courtesy of Freightliner Corporation)

Poppet or Choke Thermostats   Poppet type thermostats control the flow of coolant to the radiator only, and the bypass circuit is open continuously. Flow to the radiator is discharged through the top of the thermostat valve.

Side Bypass, or Partial Blocking, Thermostat   The side bypass thermostat functions similarity to the poppet type. It has a circular sleeve below the valve which moves with the valve, as it opens. This serves to partially block the bypass circuit and direct most of the flow to the radiator.

Vented and Unvented Thermostats   Vented thermostats have a small orifice in the valve itself or a notch in the seat that usually must be positioned in an upright position on installation. The function of the vent orifice is to help deareate the coolant by routing air bubbles out

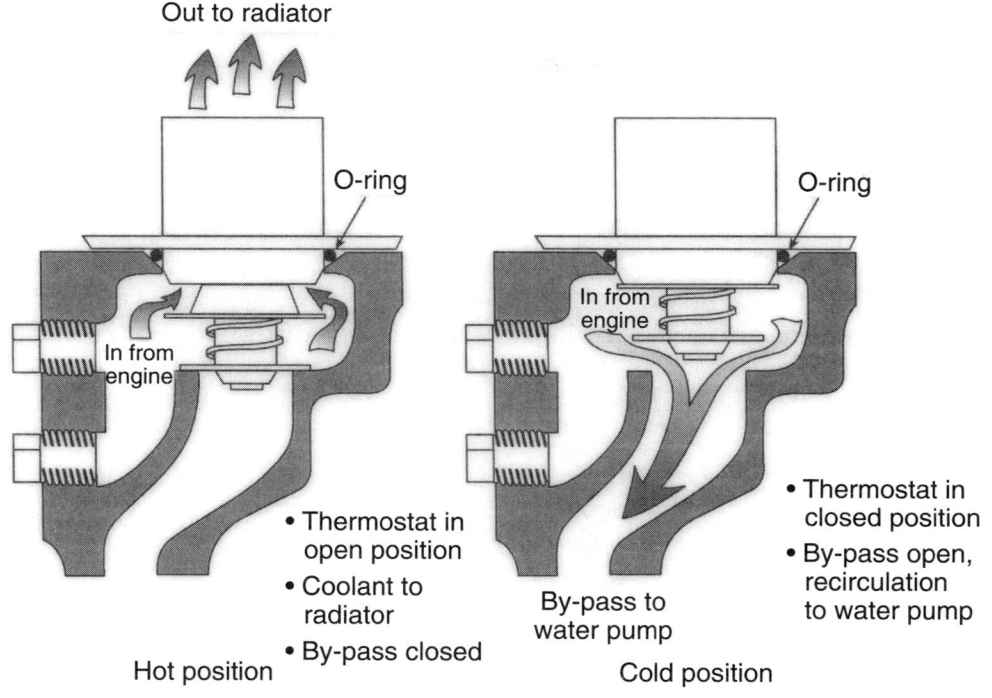

THERMOSTAT OPERATION

Out to radiator

O-ring

O-ring

In from engine

In from engine

- Thermostat in open position
- Coolant to radiator
- By-pass closed

By-pass to water pump

- Thermostat in closed position
- By-pass open, recirculation to water pump

Hot position

Cold position

Figure 6-10 Thermostat operation. (Courtesy of Navistar International Engine Group)

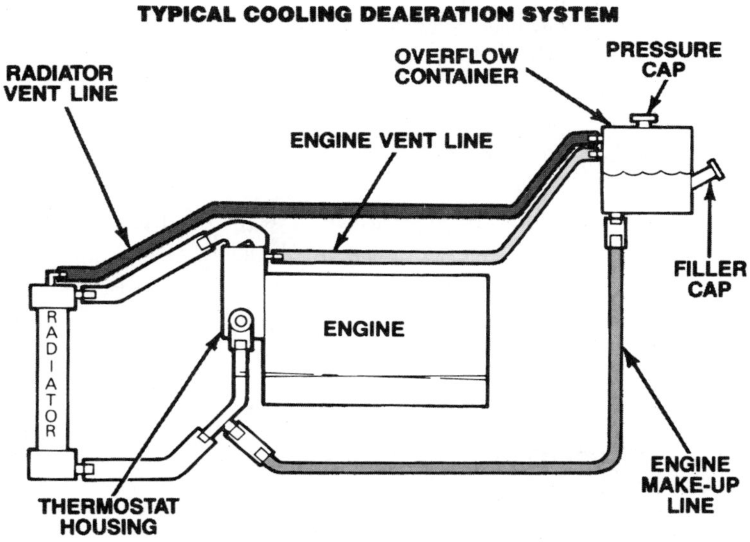

TYPICAL COOLING DEAERATION SYSTEM

RADIATOR VENT LINE

OVERFLOW CONTAINER

PRESSURE CAP

ENGINE VENT LINE

RADIATOR

ENGINE

FILLER CAP

THERMOSTAT HOUSING

ENGINE MAKE-UP LINE

Figure 6-11 Typical cooling dearation system. (Courtesy of Freightliner Corporation)

of the bypass circuit. Positive dearation type systems (Figure 6-11) usually require unvented thermostats.

Bypass Circuit The term bypass circuit describes the routing of the coolant before the thermostat opens, that is, through the engine cylinder block and head. The flow of bypass coolant permits rapid engine warm-up to the required operating temperature.

Shutters

Shutters control the airflow through the radiator and into the engine compartment. A shutterstat manages the system and is usually located in the coolant manifold. Shutters are sets of louver-like slats that pivot on shafts and are interconnected to rotate in unison from a fully open to a closed position in much the same manner as a set of venetian blinds. The shutterstat (Figure 6-12) is a temperature-actuated control mechanism. It receives a feed of system air pressure, which it will allow to pass through until the coolant temperature reaches a predetermined switching value denoted on the shutterstat. The shutter assembly is mounted on the radiator and is usually spring loaded to the opened position. When air from the shutterstat is delivered to the shutter cylinder, the plunger extends to actuate a lever and close the shutters. When the management system fails, no air is available at the shutter cylinder and the shutters will be held opened by spring force. During engine warm-up, the shutters are normally closed once there is sufficient air pressure in the system.

Cooling Fans

OEMs use two basic operating principles of engine compartment cooling requiring the use of either suction and pusher fans. Suction fans (Figure 6-13) pull outside air into the engine compartment while pusher fans do the opposite and push heated air out. Highway vehicles, which receive ram air assistance, most commonly use suction fans. In fact, depending on the variables of engine size, radiator design, and vehicle application, ram air is often sufficient to perform cooling 95% of the operating time. Fan design is important, and fans should draw the least possible amount of engine power; 6 BHP is typical of modern fiberglass fans, but twice that may be required to rotate a large steel fan. Consequently, many OEMs have adopted fiberglass fan blades over steel or aluminum. Many current designs use flexible pitch fan blades, which alter a fan's efficiency proportionally with its driven speed. This permits efficient fan operation at low engine speeds. Fan assemblies must be precisely balanced; an out-of-balance fan (a small fragment missing from one blade is sufficient) will unbalance the engine driving it. This can be severe enough to promote torsional stress failures of the crankshaft.

Because a fan assembly draws engine power, most current truck engines use lightweight temperature controlled fans. A **fanstat** activates most cooling fans. Pressure engaged clutch fans of the on-off type are used, as well as thermomodulated, viscous drive fans (Figure 6-14).

On-Off Fan Hubs On-off fan hubs use air, oil pressure, or electrically actuated clutches controlled by a fanstat or the engine management system. Vehicle air conditioning systems may also control fan hub cycles independently from the ECM using electric-over-pneumatic switching. A fanstat is usually located in the water manifold. It is a temperature triggered switch that can either

Winter fronts, which are installed on the hood grill to limit the amount of ram air driven through the engine compartment, can load a fan off its axis, creating an unbalancing effect whenever the fan is engaged. A winter front should not be necessary on most engines produced after 1990. When one is installed, be sure that the truck chassis OEM approves it. Never completely close a winter front.

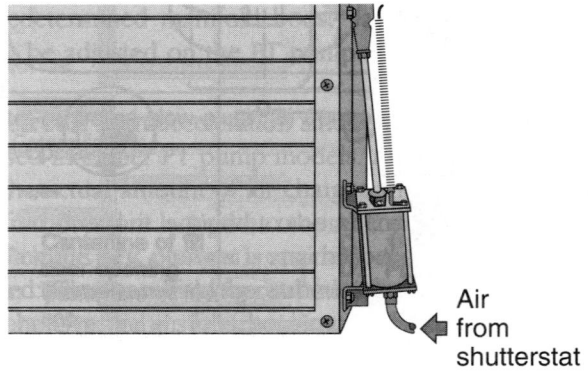

Air from shutterstat

Figure 6-12 Shutter with shutterstat. Air-operated shutters are closed by air and opened by springs. (Courtesy of General Motors Corporation, Service Operations)

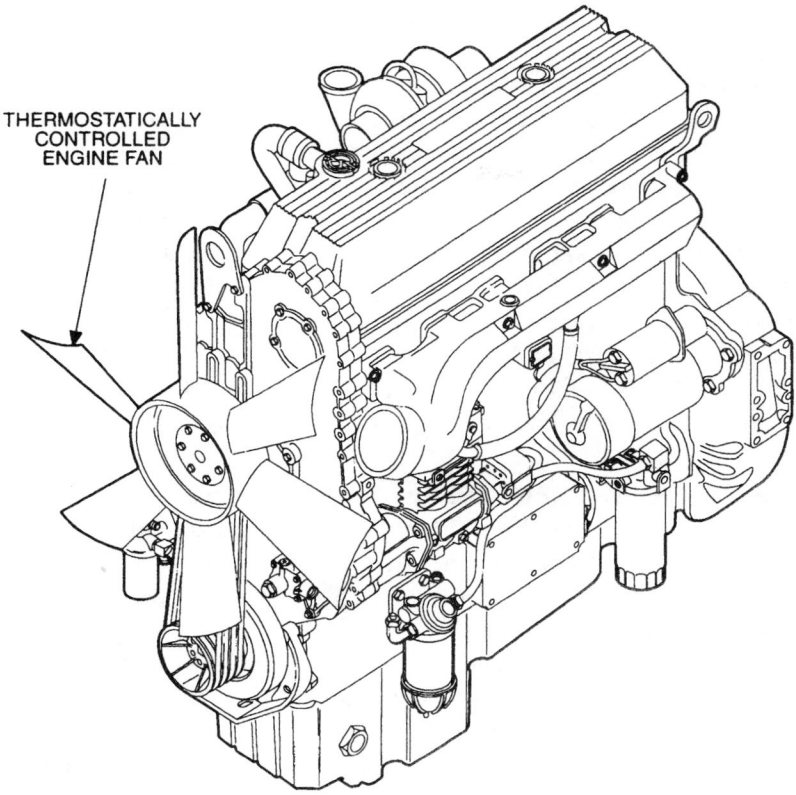

THERMOSTATICALLY
CONTROLLED
ENGINE FAN

Figure 6-13 Engine cooling fan. (Reprinted by permission of copyright owner Detroit Diesel Corporation, all rights reserved.)

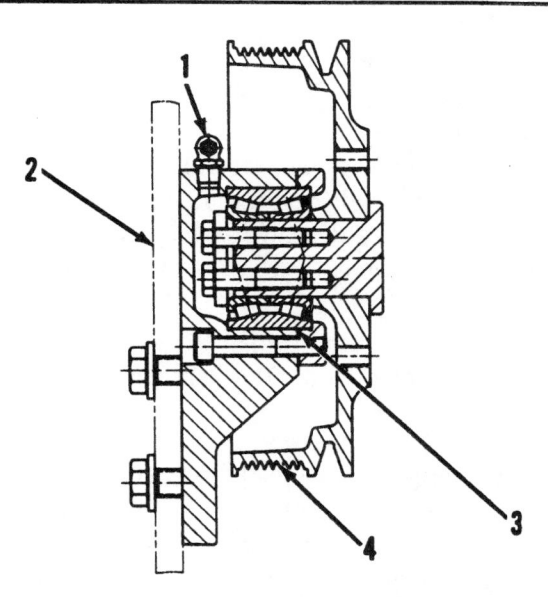

(1) Fitting. Fill seal cavity with 2S3230 Bearing Lubricant.

(2) Fan drive mounting plate.

(3) O-ring seal.

(4) Poly-vee belt pulley.

Figure 6-14 Poly V fan drive. (Courtesy of Caterpillar Inc.)

lock up or freewheel the fan hub using an electrical signal or chassis system air pressure as the medium to control the clutch. Most on-off fan hubs are spring loaded to the engage mode, and the pneumatic or electrical signal disengages the fan clutch to permit it to freewheel. This provides a locked fan hub in the event of control circuit malfunction. When an air pressure engaged, spring-released fan hub is used, the fan hub must be mechanically locked in the event of control circuit failure. On-off types of fan hubs use the least amount of engine power because the fan hub is off 90–95% of the time. Typically the coolant temperature is below the trigger value or turn-on point. However, the requirements of the vehicle air conditioning system may reduce the off time. Bendix, Rockford, Horton, Facet, Kysor/Cadillac, Eaton, Evans, and Schwitzer all manufacture air-operated fan clutches.

When engine oil is used as the hub drive coupling medium (as in some bus applications), the fanstat controls the flow of oil to the hub assembly. At the fanstat nominal value, oil pressure directed to the fan hub acts as a fluid coupling so there is always a percentage of slippage.

Thermatic Viscous Drive Fan Hubs/Thermo-Modulated Fans   **Thermatic viscous drive fan hubs** (Figure 6-15) are integral fan drive units with no external controls. They use silicone fluid as a drive medium between the drive hub and the fan drive plate. The hub assembly consists of three main subcomponents: a drive hub (input section), a driven fan drive plate (output section), and the control mechanism. There is no mechanical connection between the drive hub and the driven member. During minimum slip operation, torque is transmitted through the internal friction of the silicone drive fluid in the working chamber that couples the input and output sections of the drive. A wiper attached to the driven member continually wipes the fluid, and centrifugal force returns it to a supply chamber where an open valve cycles it back to the working chamber. As ambient temperature drops, a bimetal temperature sensing strip contracts to close the valve that supplies the silicone drive medium to the working chamber, and the fluid is trapped in the supply chamber. An advantage of the viscous drive fan hub is that it is capable of infinitely variable drive efficiencies, depending on the temperature of the bimetal strip.

Fan Shrouds   Fan shrouds are usually molded fiber devices that are bolted to the radiator assembly and may partially enclose the fan. They provide a small measure of safety if the fan is engaged when the engine is running. They play an important role in shaping the air flow through the engine compartment. A missing or damaged shroud can result in temperature management problems. In hot weather conditions, a defective or missing shroud can dramatically lower fan efficiencies, so they should be inspected at each preventative maintenance inspection.

The tension on V drive fan assemblies or poly V belts should be adjusted using a belt tension gauge to avoid bearing and slippage problems.

Fan Belts and Pulleys   Fan pulleys use external V or poly V grooves and internal bearings of the roller, taper roller, and bushing types. The belt tension should be adjusted using a belt tensioner.

▲ **WARNING:** A belt that is too tight will excessively load the bearings, which will shorten both bearing and belt life. If the belt is too loose, slippage will occur, which can destroy the belt even more rapidly than a belt that is too tight. Belts should be inspected periodically as part of a preventive maintenance routine. Replace any belts that are glazed, cracked, or nicked. The cost of replacing the belts will be much less than the cost of a breakdown should a belt fail in service.

Water Manifolds

The water manifold acts as a sort of main artery for the cooling system. It is usually a cast iron or aluminum assembly fitted to the cylinder heads and may house the thermostat(s). Figure 6-16 shows typical water manifold.

VIEW A

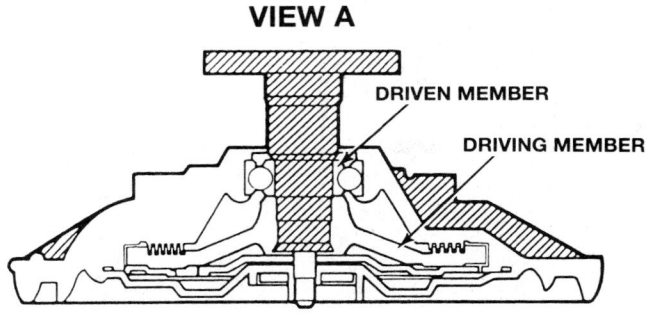

DRIVEN MEMBER

DRIVING MEMBER

VIEW B

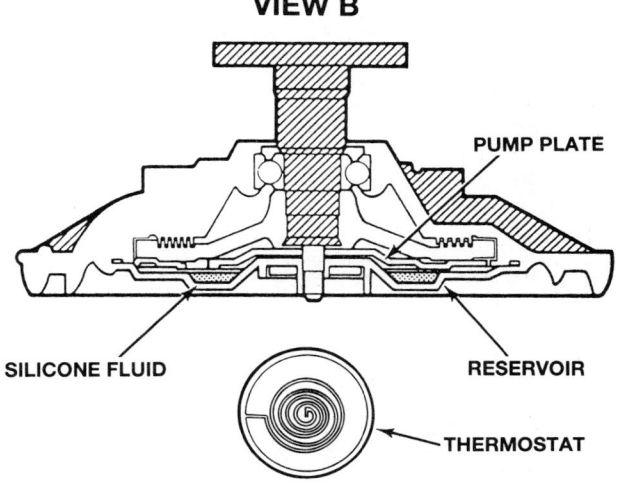

PUMP PLATE

SILICONE FLUID

RESERVOIR

THERMOSTAT

VIEW A

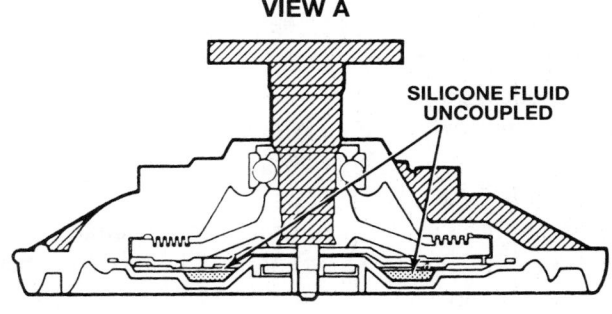

SILICONE FLUID UNCOUPLED

VIEW B

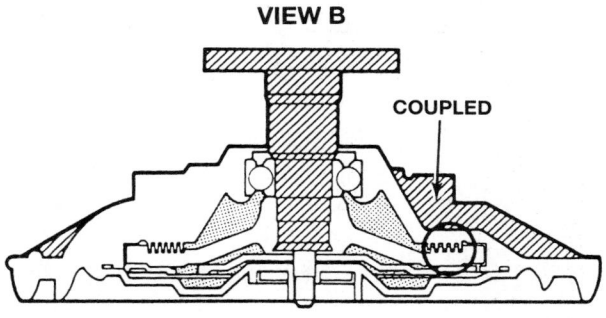

COUPLED

Figure 6-15 (A) Cooling fan viscous drive parts; (B) Cooling fan viscous drive operation. (Courtesy of General Motors Corporation, Service Operations)

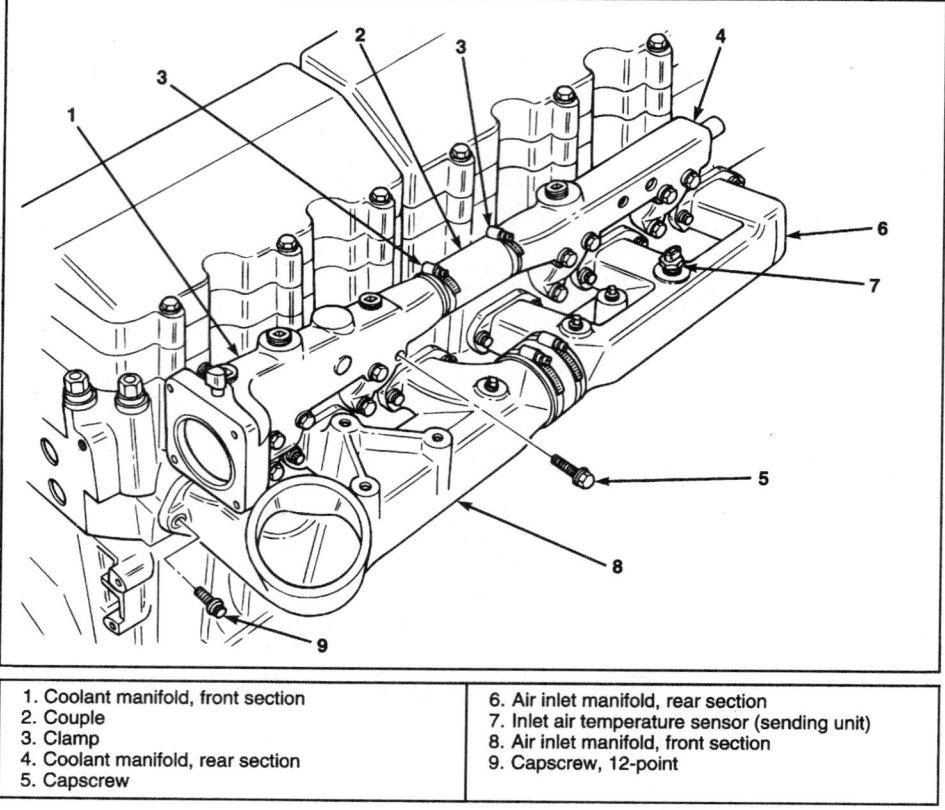

1. Coolant manifold, front section	6. Air inlet manifold, rear section
2. Couple	7. Inlet air temperature sensor (sending unit)
3. Clamp	8. Air inlet manifold, front section
4. Coolant manifold, rear section	9. Capscrew, 12-point
5. Capscrew	

Figure 6-16 Typical diesel coolant water manifold. (Courtesy of Mack Trucks, Inc.)

Cooling System Management

Testing a cooling system for external leaks is performed using a hand-actuated, pressure testing kit consisting of a pump, pressure gauge, and a variety of fill neck and radiator cap adapters.

Shop Manual
Chapter 6, page 203.

Most truck engine cooling systems are managed with the objective of opening the thermostat first, followed by the shutterstat, after which the fan will engage. Because the fan consumes power (around 6 hp at rated speed), most OEMs go with the premise that the fan should be driven as little as possible. However, some will elect to open shutters prior to the opening of the thermostat to enable the shutterstat full control of engine temperature and avoid the sudden changes of engine compartment temperature caused by cycling the shutters. Thermatic fans sense underhood or engine compartment temperature. An under-hood temperature of 155° F is generally equivalent to an engine coolant temperature of 190° F.

OEM Temperature Management Recommendations

Each engine OEM will manage its engines at different temperatures. Running an engine outside the recommended temperature window can result in greatly increased emissions, even when managed by an ECM. When engines are run cool, hydrocarbon emissions increase, fuel efficiency drops, and engine wear rates increase dramatically. When engines are run above their normal operating temperature window, NO_x (oxides of nitrogen) emissions increase and engine wear rates greatly increase.

Caterpillar Caterpillar engines use thermostats with start-to-open temperatures of 180° F and 190° F. Normal engine operating temperatures range from 160° F to 210° F. The fully open temperature of both thermostats is 15° F higher than the start-to-open value. Thermostatically engaged fans should be fully engaged at 205° F. All the temperature values are those of the engine coolant outlet. All engines equipped with air-to-air automatic coolers (ATAAC) have 190° F thermostats.

Cummins Cummins engines are designed to maintain the coolant at 170° F to 190° F, with outside limits of 158° F to 212° F. The company recommends the use of two types of shutter systems: an on-off type that opens at 185° F and closes at 178° F, and a modulating type that should be fully open at 185° F.

Two types of fan controls are used, categorized by whether they sense engine coolant temperature or engine compartment temperature. The coolant-temperature-sensing fan should be fully engaged at 200° F, while the air-temperature-sensing version should be at minimum slip at 195° F. These normally use thermatic modulating fan hub drives. Maximum coolant temperature of 212° F will trigger the overheat alert in most Cummins engines.

Detroit Diesel Corporation DDC engines are managed to operate at temperatures between 160° F and 210° F measured at the top radiator tank. DDC cooling systems use positive dearation and nonvented thermostats to manage the engine coolant temperature. When thermatic-type, viscous drive fans are used, these engage at an engine compartment temperature of 150° F. Either 180° F or 190° F thermostats are used, and both have fully open values 17° F higher than the rated values. The shutters on some DDC engines may be triggered to open before the engine thermostat opens. DDC fan hubs fully engage at 200° F with 180° F thermostats and at 205° F with 190° F thermostats.

Mack Trucks Mack Trucks recommends that coolant operating temperatures run between 170° F and 210° F, with an outside maximum of 225° F with a 180° F vented thermostat. The shutters are held open during normal engine operation, which permits the thermostat to control engine temperature. Mack prefers the use of a viscous fan clutch (thermatic fan) that is calibrated for constant airflow through the engine compartment. Mack fan hubs are designed to fully engage at 195° F to 200° F when they are of the on-off type. Current engines with CMCAC (chassis mount charge air coolers) use viscous drive, thermatic fans designed to operate at minimum slip at 175° F; non-CMCAC engines operate at minimum slip at 155° F. The objective is to permit the thermostat to fully control the engine temperature.

Navistar Navistar cooling systems use a rapid-warm, positive aeration system. Nonvented thermostats have start-to-open values of 180° F to 192° F with fully open values about 15° F higher. Shutters are set to open at 5° F above thermostat-opening temperature. Thermatic, viscous drive fan hubs are set for minimum slip 15° F higher than thermostat opening temperature.

Volvo Volvo heavy duty trucks use a fairly standard cooling system management sequence beginning with the thermostat opening at 190° F and on-off fan clutches engaging at 200° F. When shutters are used, they are set to open at the thermostat-opening temperature.

Summary

❏ Approximately 50% of the rejected heat of combustion is transferred to the engine cooling system, which is responsible for transferring it to atmosphere. The cooling system uses the principles of conduction, convection and radiation to transfer heat from the coolant to the atmosphere. A diesel engine cooling system has four main function: to absorb combustion heat, to transfer the heat using coolant-to-heat exchangers, to dissipate the heat from the heat exchangers to the atmosphere, and to manage engine operating temperatures.

❏ The main components of a diesel engine cooling system are the water jacket, coolant, a coolant pump, a radiator, thermostat(s), filter(s), shutters, and temperature-sensing circuit and fan assembly. Water expands in volume both as it freezes and as it approaches its boil point. The engine cooling system must accommodate this change in volume. Engine coolant is a mixture of water, antifreeze, and supplemental coolant additives.

Terms to Know

Antifreeze

Conduction

Convection

Coolant filters

Counterflow radiator

Crossflow radiator

Double pass radiator

Downflow radiator

Extended life coolant (ELC)

**Terms-To-Know,
Continued**

Ethylene glycol (EG)

Fanstat

Heat exchanger

Propylene glycol (PG)

Radiation

Radiator

Rejected heat

Supplemental cooling
system additives (SCA)

Shutterstat

Single pass radiator

Total dissolved solids
(TDS)

Thermatic viscous drive
fan or hub

Thermistor

Thermostat

Water pump

❏ Three types of diesel engine coolant are in current use: ethylene glycol (EG), propylene glycol (PG), and extended life coolant (ELC). A properly formulated diesel engine coolant should protect against freezing, boiling, corrosion, scaling, and foaming in addition to inhibiting acid build-up. The truck and bus industries are moving away from the use of EG based coolant, which has higher toxicity, to safer PG and ELCs.

❏ Supplemental coolant additives (SCAs) are vital components of engine coolant. The SCA levels must be routinely tested in EG and PG coolants, as they deplete in service. ELC is claimed to have a service life of 600,000 miles, or 6 years, with a single SCA recharge required.

❏ When possible, the radiator is located in the airflow at the front of the chassis to optimize ram air cooling effect. Most radiators are of the single-pass, downflow type.

❏ Radiator caps are equipped with a pressure valve to define the cooling system operating pressure and a vacuum valve to prevent hose collapse as the system cools and pressure drops. As altitude increases, the boil point of the coolant decreases.

❏ Coolant pumps are either belt or gear driven and are of the nonpositive, centrifugal type. They are lubricated by the coolant and tend to be vulnerable to high total dssolved solids (TDS) levels, which are abrasive.

❏ The commonly used coolant temperature sensor on today's electronically managed engines is the thermistor, which is a temperature sensitive, variable resistor. Coolant level sensors ground a reference signal into the coolant in the top radiator tank and trigger an alert when the ground circuit is broken for a programmed period.

❏ Thermostats are used to manage the engine temperature precisely to ensure optimum performance, fuel economy, and minimal noxious emissions. Thermostats route the coolant through the bypass circuit to permit rapid engine warm-up. The thermostat opens at the engine operating temperature and directs the coolant flow through the radiator.

❏ Shutters control the airflow through the engine compartment by means of louver-like slats that open and close like venetian blinds. Shutters are controlled by a shutterstat, which is located in the water manifold and uses chassis system air pressure to open and close them. They are usually designed to fail in the fully open position.

❏ Most engine compartment fans are temperature controlled, either directly on the basis of coolant temperature measured by a fanstat or indirectly based on engine compartment temperature. Flex blade, fiberglass fans are designed to alter their pitch based on their driven rotational speed, permitting higher fan efficiencies at low rpm. Viscous-type, thermatic fans sense under-hood temperatures and are driven by a fluid coupling designed to produce minimal slip at their nominal operating temperature.

Review Questions

Short Answer Essays

1. Define conduction, convection, and radiation.

2. Describe the three types of coolant used in current highway diesel engines and the relative merits and disadvantages of each.

3. Describe the differences in the types of heavy duty radiator in use including downflow, crossflow, and counterflow.

4. State the functions of a cooling system radiator cap.

5. Identify the different types of thermostats in use and describe their principles of operation.

6. Describe the role of the coolant pump in the system and the principle of operation of a centrifugal pump.

7. Define the role of the coolant filters and filter servicing requirements.

8. List the types of temperature gauge used in current highway diesel engines and describe how their signals are output.

9. Define the roles played by the shutters and engine fan in controlling airflow through the engine compartment.

10. Outline the operating principles of a thermatic-type, viscous fan hub.

Fill-In-the-Blanks

1. Mixing coolant solutions should be performed in a container _____ of the engine cooling system and then added.

2. Supplemental coolant additives are a vital component of engine coolant. The SCA levels must be _____ _____ in EG and PG coolants as they deplete in service.

3. _____ is claimed to have a service life of 600,000 miles, or 6 years, during which a single SCA recharge is required.

4. Where possible, the radiator is located in the airflow at the _____ of the chassis to optimize ram air cooling effect.

5. Thermostats route the coolant through the _____ _____ to permit rapid engine warm-up.

6. The thermostat opens at engine operating temperature and directs the coolant flow through the _____.

7. Shutters control the airflow through the _____ _____, by means of louver-like slats that open and close like venetian blinds. Shutters are controlled by a _____, which is located in the water manifold and uses chassis system air pressure to open and close them. They are usually designed to fail in the _____ _____ position.

8. Most engine compartment fans are _____ controlled, either directly on the basis of coolant temperature measured by a fanstat or indirectly based on engine compartment temperature.

9. Flex blade, fiberglass fans are designed to alter their _____ based on their driven rotational speed, permitting higher fan efficiencies at low rpm.

10. Viscous-type, thermatic fans sense _____ _____ temperatures and are driven by a fluid coupling designed to produce minimum slip at their nominal operating temperature.

ASE-Style Review Questions

1. Which of these diesel engine coolants is *most* likely considered toxic?
   A. EG
   B. PG
   C. Water
   D. ELC

2. Which of the following cooling mediums would transfer heat most efficiently?
   A. EG
   B. PG
   C. Pure water
   D. ELC

3. Which of these conditions will *most* likely cause wet liner cavitation?
   A. Aerated coolant
   B. Combustion gas leakage
   C. Air in the radiator
   D. Vapor bubble collapse

4. Which would be the warmest portion of a typical downflow type radiator when the engine is at operating temperature?
   A. The top tank
   B. The surge tank
   C. The bottom tank
   D. The center of the core

5. The temperature rating of a thermatic-type, viscous fan hub is a nominal 155° F. At approximately what equivalent coolant temperature will this produce minimum slip drive as temperatures rise?
   A. 155° F
   B. 165° F
   C. 190° F
   D. 225° F

6. In a typical diesel powered highway truck, which of the following statements is correct?
   A. Coolant temperatures run cooler than lube oil temperatures.
   B. Coolant temperatures run warmer than lube oil temperatures.
   C. Coolant temperatures should be equal to lube oil temperatures.
   D. Lube oil temperature is the same as lube oil pressure.

7. What is the pump operation type in a coolant water pump?
   A. Positive displacement
   B. Centrifugal
   C. Constant volume
   D. Gear type

8. Which of the following controls the drive efficiency of a thermatic-type, viscous drive fan hub?
   A. Bimetal strip
   B. Viscosity of the silicone drive medium
   C. Fanstat
   D. Solenoid

9. *Technician A* states that some ELC has an approved service life of up to 6 years. *Technician B* states that water should never be added to ELC. Who is correct?
   A. A only
   B. B only
   C. Both A and B
   D. Neither A nor B

10. *Technician A* says the ELC should always be mixed at more than 40% concentration with distilled water. *Technician B* says a mixture of 50/50 EG has better antifreeze protection than pure EG. Who is correct?
    A. A only
    B. B only
    C. Both A and B
    D. Neither A nor B

Air Induction and Exhaust System and Engine Brakes

Upon completion and review of this chapter, you should be able to:

❏ Identify the intake and exhaust system components that comprise the engine breathing circuit.

❏ Describe how intake air is routed to the engine cylinders and exhaust gases are routed to the tailpipe.

❏ Outline the principle of operation of a Roots blower and its function on a two-stoke cycle engine.

❏ Identify the main subcomponents on a truck diesel engine turbocharger and the role they perform; explain the operation of the turbocharger.

❏ Define the role of a charge air cooler and the relative efficiencies of each type.

❏ Relate the parallel port and crossflow valve configurations and valve seat angle to breathing efficiency and cylinder gas dynamics.

❏ Define the term *tuned* in relation to intake and exhaust manifold design.

❏ Define thermocouple pyrometer and outline its use on a diesel engine.

❏ Explain the role and operating principles of the exhaust muffler (silencer), and identify the two ways mufflers reduce combustion noise in an engine.

❏ Describe the operating principles of a truck catalytic converter.

❏ Identify the different types of cold engine starting aids used in a diesel engine.

❏ Identify and differentiate between the three types of engine brakes used on a diesel engine, and explain their operating principles and the relative advantages and disadvantages of each.

Introduction

Turbochargers are used on almost all current commercial vehicle diesel engines. Information that could be handled separately under the headings "intake systems" and "exhaust systems" will be dealt with in this section on engine breathing to facilitate the study of the component common to intake and exhaust systems. The turbocharger charges the intake manifold at pressures above atmospheric. Turbocharged engines are commonly described as having manifold boost. The manifold-boosted compression ignition (CI) engine cylinder is *charged* with air rather than induced. In a naturally aspirated engine, air or an air-fuel mixture is drawn into the engine cylinder by the atmospheric pressure created in the engine cylinder on the intake stroke of the piston. This air or air-fuel mixture is therefore induced into the cylinder. The correct term to describe the components responsible for delivering breathing air in naturally aspirated engines is *induction circuit*. The preferred term for describing the breathing air delivery components in an engine with manifold boost is *air intake circuit*.

The function of the air intake system (Figure 7-1) in a diesel engine is to supply a charge of clean, cool air to the engine cylinders for combustion, cooling, and scavenging. Diesel engines are designed for lean burn operation. Essentially, this means that the air charged to the engine cylinder always substantially exceeds that required to oxidize the fuel. Expectations of engine longevity have lengthened dramatically during the past decade, and this has altered the design of air intake system components.

The function of the exhaust system is to minimize both engine noise and noxious emissions while restricting the exhaust gas flow to a minimum degree. The turbocharger is a component common to both intake and exhaust systems. Turbochargers allow the exhaust system to recapture

In a **naturally aspirated** engine, air or an air-fuel mixture is drawn into the engine cylinder by the atmospheric pressure created in the engine cylinder on the intake stroke of the piston.

The function of the exhaust system is to minimize both engine noise and noxious emissions while restricting the exhaust gas flow to a minimum degree.

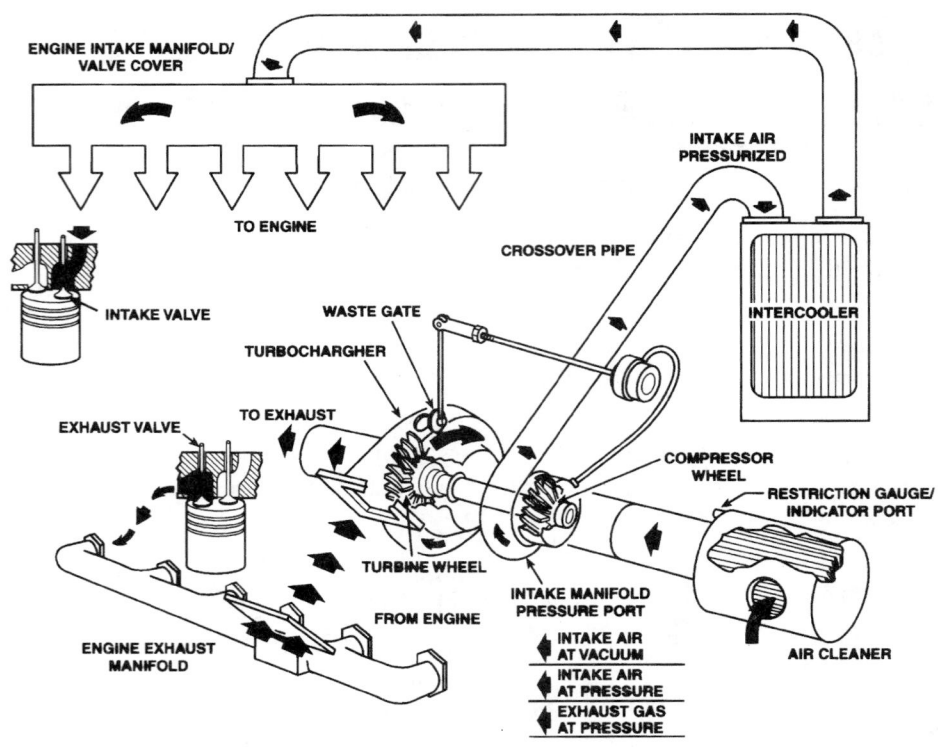

Figure 7-1 Engine breathing schematic. (Courtesy of Navistar International Engine Group)

Turbochargers allow
the exhaust system to
recapture some of the
engine-rejected heat to
pressurize the air
delivered to the engine
cylinders.

**Shop
Manual
Chapter 7, page 218.**

some of the engine **rejected heat** to pressurize the air delivered to the engine cylinders. Manifold-boosted diesel engine breathing requires that the gas dynamics of the intake and exhaust systems work to complement each other, especially during the very critical period of valve overlap. Diesel engine exhaust systems seldom have much of the post-combustion, emission control apparatus found on many spark ignition (SI) engines. Some manufacturers use oxidation catalysts, but noxious emission control is largely the responsibility of the engine management computer.

The function of engine braking systems is to supplement the truck's main braking system to safely stop the vehicle. Engine brakes are widely used on North American truck diesel engines. A large percentage of vehicle braking requires application pressures of 20 psi or less in a typical air brake system. In other words, less than 20% of brake system's potential is being used. The objective of engine compression brakes is to relieve the vehicle braking system of some of the light duty braking, especially in instances that require prolonged light duty braking, such as on an extensive downgrade. This chapter will contain an explanation of the operation of several different engine braking systems. Most of these braking systems are integral with the engine's exhaust system.

Air Intake System Components

The air intake system components are those responsible for delivering ambient air to the engine's cylinders. This air will always be filtered and in most truck diesel engines it will be pressurized well above atmospheric pressure values. Because the act of pressurizing the air charge increases its temperature, this reduces its density. To counter this, most turbo-boosted engines use some form of **heat exchanger** to cool the boost air before it is directed into the engine cylinders. Because the turbocharger (Figure 7-2) is driven by exhaust gas heat and serves to boost the intake charge, it is considered as both an intake and exhaust system component.

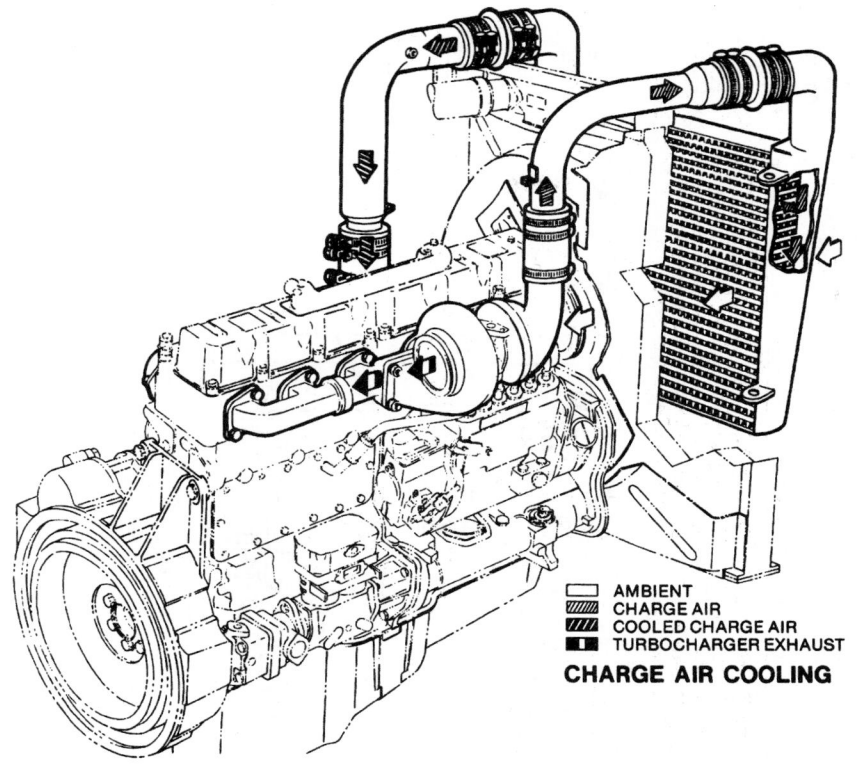

Figure 7-2 Airflow through a typical diesel air system. (Courtesy of Mack Trucks, Inc.)

4-Stroke Cycle Engine Breathing Components

Intake System Components (Figure 7-3)

1. Air cleaner system consisting of a main filter and a precleaner
2. Intake ducting/piping
3. Turbocharger and turbocharger controls
4. Intercooler, aftercooler, or charge air cooling circuit
5. Intake manifold
6. Valve porting and intake tract design

Exhaust System Components (Figure 7-4)

1. Valve configuration
2. Exhaust manifold
3. Turbocharger and turbocharger controls
4. Exhaust piping
5. Pyrometer
6. Engine silencer
7. Catalytic converter

Two-Stroke Cycle Engine Breathing Components

Intake System Components

1. Air cleaner system consisting of a main filter and a precleaner
2. Intake ducting/piping
3. Turbocharger and turbocharger controls (on some applications)

4. Roots blower
5 Charge air cooling circuit
6. Intake manifold
7. Air box
8. Liner intake ports

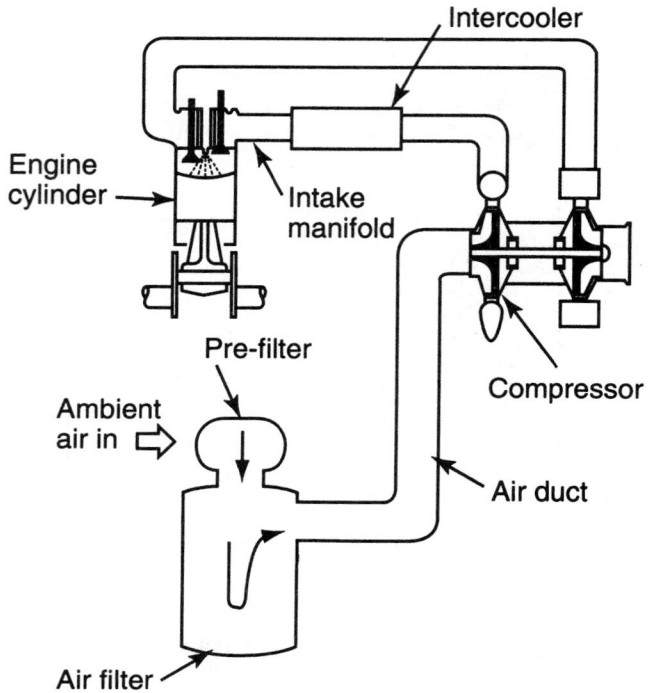

Figure 7-3 Basic air induction components. (Reprinted by permission of copyright owner Detroit Diesel Corporation, all rights reserved.)

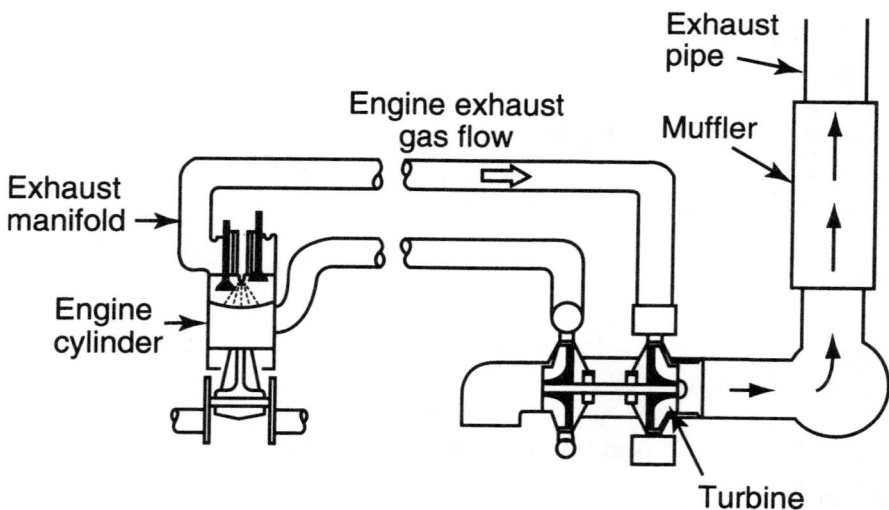

Figure 7-4 Basic exhaust components. (Reprinted by permission of copyright owner Detroit Diesel Corporation, all rights reserved.)

Exhaust System Components

1. Exhaust valves
2. Turbocharger
3. Engine silencer
4. Catalytic converter (some engines)
5. Exhaust piping

Air Cleaners

The function of the air cleaner system on a highway diesel engine is to filter airborne particulates from the air that will be delivered to the engine cylinders. Airborne dirt can be highly abrasive, and when it finds its way past the air cleaner system, it can destroy an engine in a very short period.

Precleaners   Precleaners (Figure 7-5) are required in trucks operating in dusty conditions or in North American winter conditions, especially where highways are salted. They may triple the service life of the main engine air cleaner. Located in front of the main air cleaner and sometimes designed to receive ram air, precleaners generate a cyclonic airflow with the objective of separating heavier particulates by centrifugal force from the air entering the main cleaner. Depending on the design of the air cleaner, the particulate removed from the air stream can either be discharged or collected in a dust bowl.

> ⚠️ **WARNING:** Trucks operating in conditions where they are subjected to high levels of larger airborne matter, such as ash and grain chaff, may use precleaner screens. However, these are vulnerable to plugging, so they should only be used in conjunction with inlet restriction gauges.

Dry Positive Filters   Dry **positive filtration** air cleaners are used in all contemporary North American commercial diesel engines. Because they use a positive filtration principle, all of the air entering the intake system must pass through the filtering media, which are usually resin-impregnated, pleated paper elements. A perforated steel mesh that provides the element with a limited amount of structural integrity surrounds the filtering medium. If a visible dent is evident in the surrounding mesh, the filter will probably not seal in the canister and should be replaced. Filtering efficiencies are high throughout the speed and load range of the engine, usually better

Precleaners are required in trucks operating in dusty conditions or in North American winter conditions.

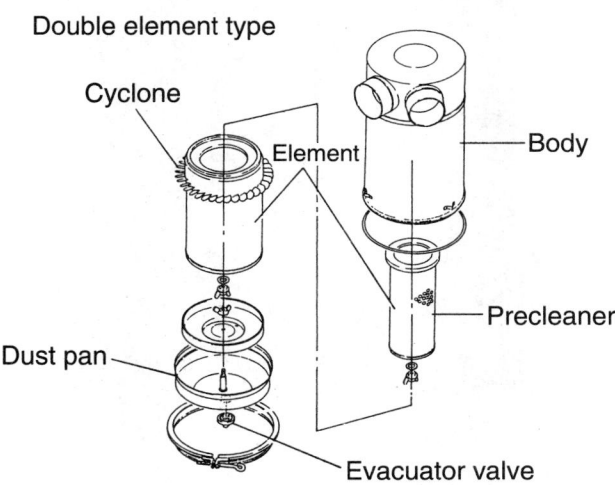

Figure 7-5 Precleaner. (Reprinted with permission by American Isuzu Motors, Inc.)

than 99.5%, and are highest just before replacement is required. Some dry element filters are two stage and may eliminate the need for a precleaner. These induce a vortex flow and use centrifugal force to separate heavier particulates, which are then discharged by an ejector valve. Dry paper element filters are designed to last for as long as 12 months in a line haul application, they should not be serviced unless the inlet restriction specification exceeds the OEM maximum specification or a used oil analysis report indicates that dirt may be bypassing the air cleaner. On-board inlet restriction gauges are not accurate instruments.

Inlet Restriction Gauges **Inlet restriction gauges** are resettable gauges mounted either on the filter canister filter canister or remotely in the vehicle dash. They provide readout in inches of water vacuum in the same way the manometer does. Although these instruments are not noted for their accuracy, they provide an indication of when filter service is required.

Oil Bath Filters/Viscous Impingement Oil Filters Low operating efficiencies have made these types of non-positive filter oil bath filters (Figure 7-6) of the past. They consist of a mesh filled canister with a sump filled with engine oil: air flow is cycloned through the canister and acts to wet down a cylindrical mesh to which dirt particles will attach themselves. This principle works effectively to remove larger particles from the air stream and has higher efficiencies when induced at the rated speed when airflow is highest. They possess low filtering efficiencies at low load, low speed operation. Replace the sump oil and wash down the mesh with solvent service oil bath filters.

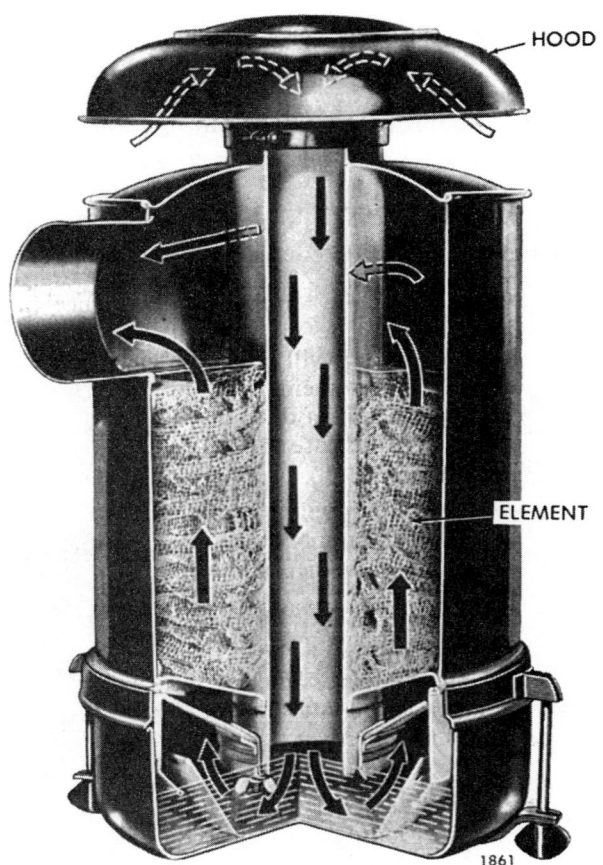

Figure 7-6 Oil bath air cleaner. (Reprinted by permission of copyright owner Detroit Diesel Corporation, all rights reserved.)

Roots Blowers

The **Roots blower** (Figure 7-7), also known as a supercharger, is a gear-driven, *positive displacement* air pump used to **scavenge** the engine cylinders in Detroit Diesel Corporation (DDC) two-stroke cycle engines. It produces relatively low peak pressures compared to those produced by turbochargers. As a scavenging component, the primary function of the Roots blower is to displace end gases from the engine cylinders. It may also produce a small amount of manifold boost. All DDC two-stroke cycle engines used in automotive applications (that is, highway trucks and buses) use Roots blowers for cylinder scavenging. These engines may also have a turbocharger to produce manifold boost. Under conditions of high engine loading, and therefore peak turbocharger efficiency, the turbocharger may be capable of performing cylinder scavenging. In some nonautomotive applications of DDC two-stroke cycle engines, the required scavenging may be performed exclusively by turbocharger(s) but only where engine loading is required to be consistently high, such as in a stationary generator, or genset.

Roots blowers are used as scavenging pumps because of their positive displacement pumping principle. They pump air to an air box, which in turn supplies air to the cylinders by means of intake ports in the cylinder liner. A positive displacement pump will displace a constant slug volume per cycle. This means that the Roots blower will unload a constant slug volume of air per rotation; the volume of air charged to the air box directly depends on the driven speed. Compared to a turbocharger, the Roots blower has somewhat higher efficiencies as a pump when engine rpm and loads are low. Turbochargers are driven by rejected heat energy and therefore tend to have low efficiency when engine loads are low. In addition, there is a lag between power demand and turbo response.

Roots Blower Construction

The DDC Roots blower housing is manufactured from aluminum alloy casting. Within this housing is a pair of intermeshing, spiral-fluted rotors supported by bearings located at either end of the assembly. One of the rotors receives gear-driven input, and it drives the other. Each rotor has three spiraled lobes timed to intermesh so one spiral rotor turns clockwise and the other turns counterclockwise. The rotor drive may be on either side, depending on the model. The rpm ratio of the blower to that of the engine also varies and will depend on factors such as whether a turbocharger and intercooler are used. The fuel pump and water pump are also be driven by the blower drive shaft (the water pump at the front and the fuel pump at the rear).

Lubrication   The DDC Roots blower is supported by pressure-lubricated roller bearings, which ensure that the rotors have close clearance in operation but no contact. As a consequence, the

Shop
Manual
Chapter 7, page 222.

The Roots blower is a gear driven, positive displacement air pump used to scavenge the engine cylinders in DDC two stroke cycle engines. It produces relatively low peak pressures compared to those produced by turbochargers.

Shop
Manual
Chapter 7, page 222.

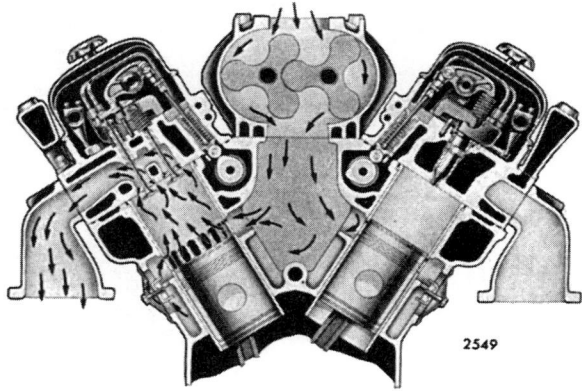

2549

Figure 7-7 Roots blower, cross-section. (Reprinted by permission of copyright owner Detroit Diesel Corporation, all rights reserved.)

rotors themselves require no lubrication, because in normal operation, they have no metal-to-metal contact. Lip-type seals prevent the engine oil from bypassing the rotor shaft and being pumped into the engine. Evidence of engine oil on the rotor lobes is an indication that a blower shaft seal is failing.

The oil to lubricate the blower gears, bearings, governor, and fuel pump drive is ported from the main oil gallery and will drain back to the crankcase via ducting. On in-line engines, oil drain-off from the rocker housing drains to the blower housing and plates and lubricates the moving components aided by a slinger.

Turbochargers

A turbocharger is an exhaust gas driven, centrifugal pump that "recycles" some of the rejected heat from the engines cylinders.

Turbochargers (Figure 7-8) are almost universally used on North American medium and large bore, highway Diesel engines. By definition, a **turbocharger** is an exhaust gas driven, centrifugal pump that "recycles" some of the rejected heat from the engines cylinders. Turbochargers may be driven to speeds exceeding 200,000 rpm in certain race car engine applications. In most highway applications of the technology, the turbocharger is used to deliver a pressurized charge of air to the engine's cylinders; in short, it increases the oxygen density in the air charge. In larger, off-highway diesel engines, the exhaust-gas–driven **turbine** may be used with a fluid coupling to drive reduction gearing connected to the engine crankshaft. In this instance, the turbocharger assists in driving the crankshaft.

Principles of Operation

A **turbocharger** (Figure 7-9) is an exhaust gas driven air pump consisting of a turbine wheel on the exhaust side and a compressor **impeller** (wheel) on the intake side. Both wheels are mounted on a common shaft. This shaft is floated (hydrodynamically suspended) on friction bearings supplied with pressurized lube oil. The turbine wheel is subject to engine exhaust gas energy (heat) and is driven within a turbine housing through which the exhaust is routed. The compressor impeller is enclosed in separate compressor housing and acts on intake system air, pumping it through the charge side of the intake system. The exhaust gas driving the turbine and the intake air compressed by the impeller do not come into contact. Basic operation begins with exhaust gases entering the turbine housing (Figure 7-10) causing the turbine wheel and shaft to rotate. As the exhaust gases spin the turbine wheel and shaft, the compressor impeller (wheel)

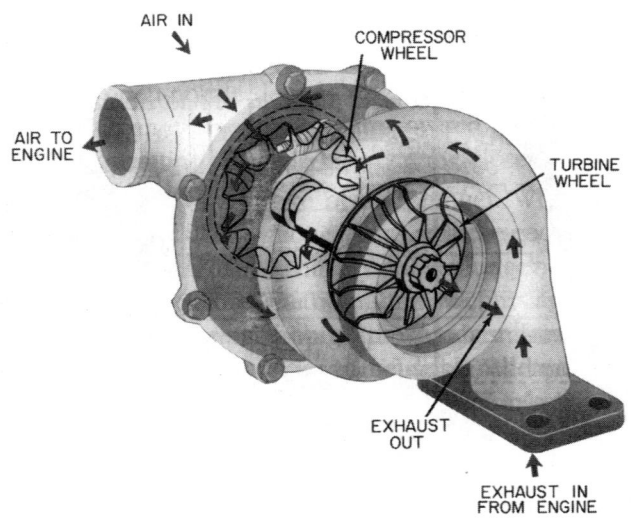

Figure 7-8 Turbocharger. (Courtesy of Schwitzer Corporation)

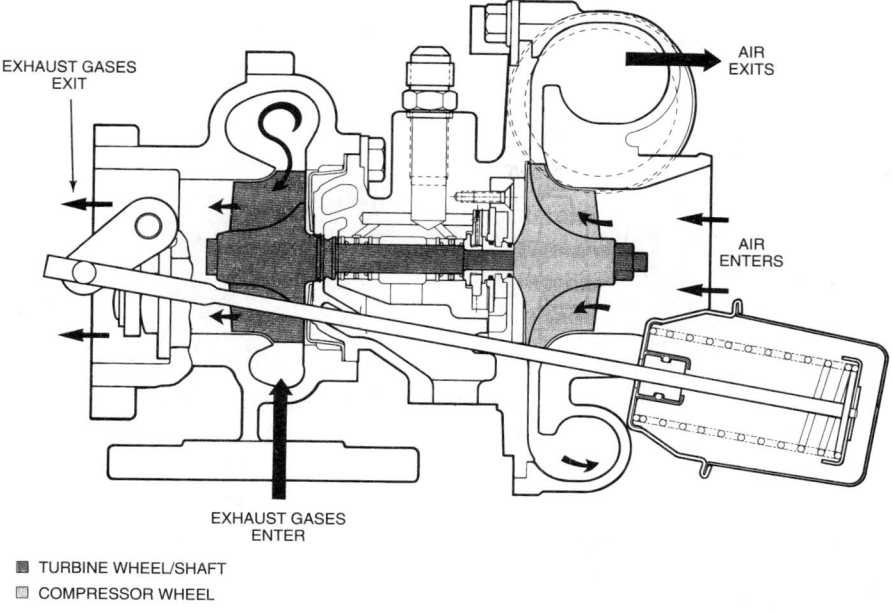

EXHAUST GASES
EXIT

AIR
EXITS

AIR
ENTERS

EXHAUST GASES
ENTER

■ TURBINE WHEEL/SHAFT
□ COMPRESSOR WHEEL

Figure 7-9 Turbocharger operation. (Courtesy of General Motors Corporation, Service Operations)

also spins. The action of the compressor impeller increases the pressure, using gas diffusion and the flow of the air intake charge, and is dependent on exhaust gas flow. A diffuser is shown in Figure 7-10.

The greater the engine rejected heat value (this increases somewhat proportionally with engine output), the more exhaust gas heat energy and the more the gases acting on the turbine

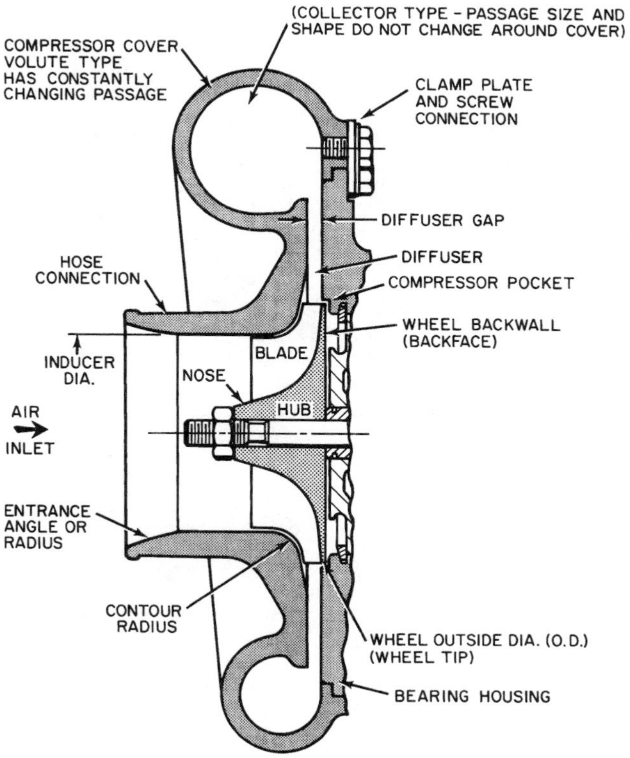

Diffusion is the process of slowing down the gas without turbulence, so the velocity energy is converted to pressure energy. The gas molecules are mixed together by having the molecules of one gas spread throughout the molecules of another gas.

COMPRESSOR COVER
VOLUTE TYPE
HAS CONSTANTLY
CHANGING PASSAGE

(COLLECTOR TYPE – PASSAGE SIZE AND
SHAPE DO NOT CHANGE AROUND COVER)

CLAMP PLATE
AND SCREW
CONNECTION

DIFFUSER GAP

DIFFUSER

COMPRESSOR POCKET

WHEEL BACKWALL
(BACKFACE)

HOSE
CONNECTION

INDUCER
DIA.

BLADE

NOSE

HUB

AIR
INLET

ENTRANCE
ANGLE OR
RADIUS

CONTOUR
RADIUS

WHEEL OUTSIDE DIA. (O.D.)
(WHEEL TIP)

BEARING HOUSING

Figure 7-10 Turbocharger diffuser. (Courtesy of Schwitzer Corporation)

wheel vanes will expand, so the greater the turbine rotational speed. The exhaust gas is routed to flow into a *volute* (a snail-shaped area), then exits and expands on the turbine vanes (Figure 7-11). A nozzle may be used instead of a volute, but the principle is the same. The volute flow area or the nozzle ring size helps to determine the maximum rotor speed. It is important to emphasize that turbocharger rotational speed is factored mainly by *exhaust gas heat* and not exhaust gas pressure.

Turbochargers are designed for optimum performance when the engine is fully loaded and operating at an rpm in the torque rise profile (in high torque rise engines, turbo optimum efficiency occurs close to peak torque or close to the lowest rpm in the torque curve). If the load demanded of an engine is reduced but the rpm maintained, the weight of the exhaust gas remains fairly constant (as does manifold pressure), but the reduced exhaust gas heat results in considerably reduced turbine speeds. The geometry of the turbocharger will always determine the specific engine rotational speed at which the turbocharger operates at peak efficiency. Filtered intake air is admitted to the compressor housing, pulled in by the impeller (compressor wheel), which is attached to the turbo shaft and driven by the turbine. This accelerates the air to high velocities. High-velocity air flows radially outward to a diffuser. The **diffuser** is designed to convert the **kinetic energy** (energy of motion) of the intake air into pressure. The diffuser may be either a volute type, with an increasing volume in the sectional area of the compressor housing, or a blade type. Blade type diffusers have higher efficiencies.

Shop Manual
Chapter 7, page 224.

Construction

Turbochargers (Figure 7-12) used in truck diesel engine applications wind out at peak rpm of up to 130,000 rpm, commonly perform at 70,000 rpm, and must sustain temperatures well over 1,200° F (650° C). For this reason, materials used in turbocharger construction must be able to

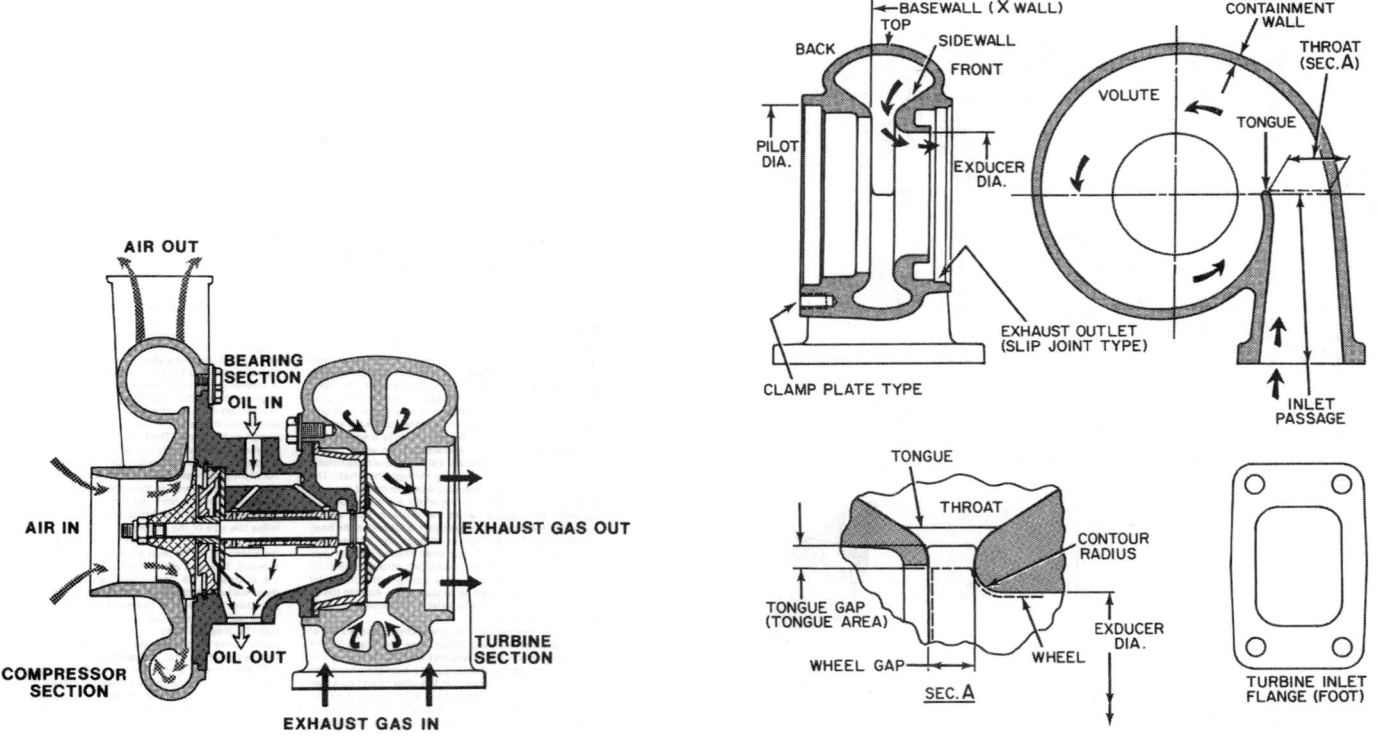

Figure 7-11 (A) Turbocharger, sectional view; (B) Turbocharger volute/nozzle. (Courtesy of Schwitzer Corporation)

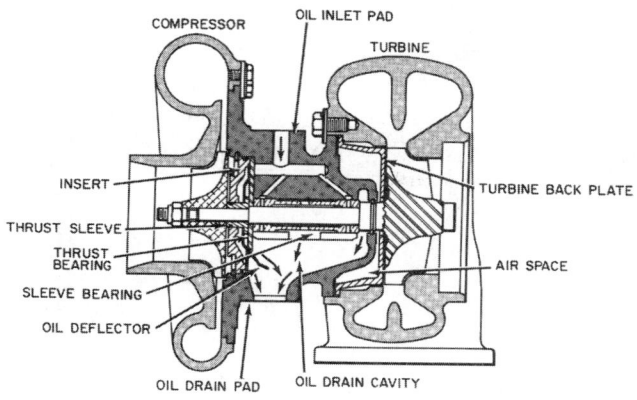

Figure 7-12 Turbocharger bearing and thrust sleeve. (Courtesy of Schwitzer Corporation)

withstand high temperatures and centrifugal forces. The compressor components are often manufactured from aluminum alloys. Compressor housings must be able to sustain rotor burst and are manufactured out of austenite (a particular crystalline structure of iron and carbon) or Ni-resist cast iron, which has high hot strength and resistance to oxidation. The turbines are manufactured from nickel or cobalt steels and ceramics. The turbo shaft is usually alloy steel to which the turbine is either welded or bolted and the impeller is bolted. Sometimes the shaft, turbine, impeller, and bearing assembly are manufactured in a cartridge for ease of replacement. When a turbocharger is rebuilt, the shaft assembly is replaced as a unit and the hub, turbine, and compressor housing are reused. All current turbochargers use floating friction bearings. Turbocharger bearings are designed to rotate in operation and do so at about one third shaft speed with a film of oil on either side of the bearing. A thrust bearing defines the axial play of the turbine shaft, and a pair of piston rings seals gas pressure on both the turbine and compressor housings.

■ **CAUTION:** When running turbocharger equipped diesel engine without the air intake, be sure to install an intake guard over the turbocharger inlet to avoid personal injury from the exposed rotating impeller.

Turbocharger Lubrication

The oil that pressure feeds the bearings (Figure 7-13) spills to an oil drain cavity and then drains to the crankcase by means of a return hose (Figure 7-14). The lubricating oil required for the bearings also plays a major role in cooling the turbocharger components.

Turbocharger Gas Flow

Turbochargers are often classified by the manner in which they are supplied with exhaust gas and whether turbocharger output can be managed. Gas flow into the turbine housing is radial and with an axial outflow. Airflow through the impeller housing is axial inflow and radial outflow. In the simplest turbocharger design, the turbine housing intake tract, or throat, is undivided, meaning that all the cylinders feed gas into a single passage. In a double-flow design (Figure 7-15), the turbine throat is divided and each half feeds one half of the turbine wheel circumference. Another variation is the twin flow, where two passages are fed from a common throat for the full rotation of the turbine wheel. V configured engines commonly use either the double- or twin-flow turbine housing design.

Common Manifold Turbocharging A single pipe manifold is flange mated to the exhaust tract of each engine cylinder, with the turbocharger turbine housing flange mounted to the center of the assembly so that the exhaust gas is routed through the turbine housing. This manifold

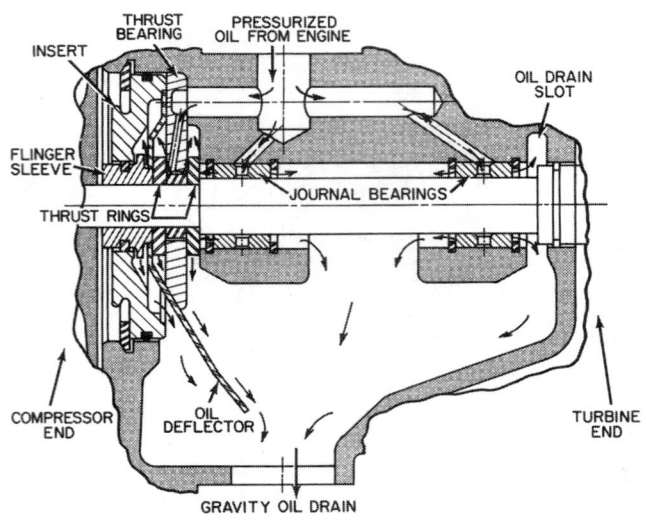

Figure 7-13 Oil flow through the turbocharger. (Courtesy of Schwitzer Corporation)

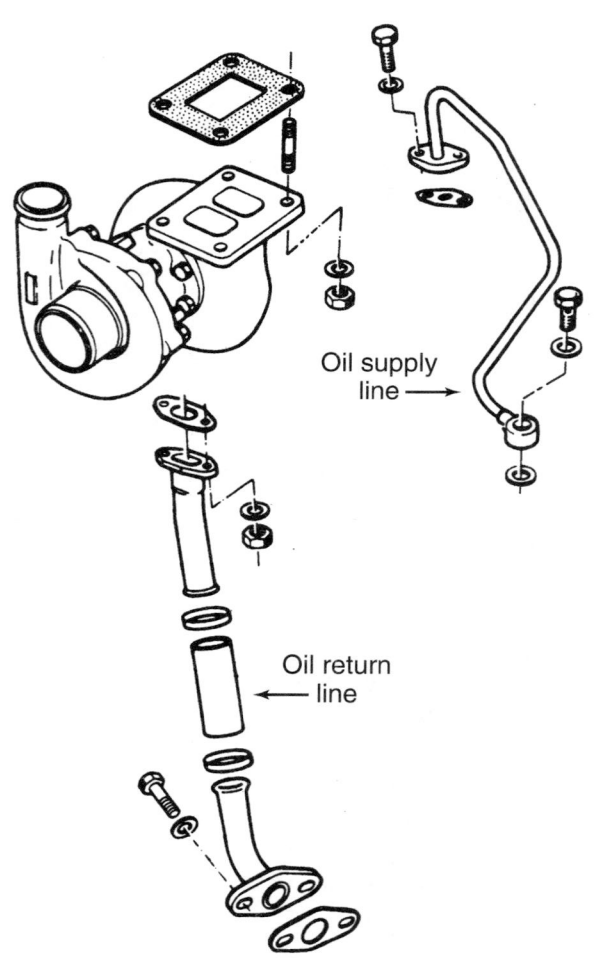

Figure 7-14 Turbocharger lubrication lines. (Courtesy of Mack Trucks, Inc.)

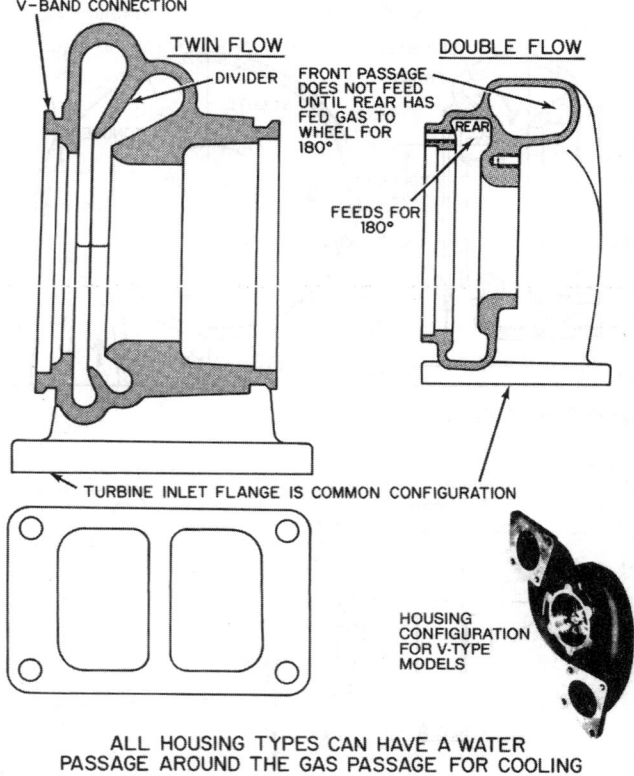

Figure 7-15 Double flow turbocharger design. (Courtesy of Schwitzer Corporation)

design creates certain problems with the gas dynamics and produces inconsistencies in the amount of effective heat per exhaust slug discharged into the manifold. Each exhaust slug volume discharge from cylinders #1 and #6 on an in-line 6 configured engine must travel considerably further before they enter the turbine housing tract than those discharged at cylinders #3 and #4 located at the center of the engine.

Pulse-Tuned Exhaust Manifold   Many current diesel engine OEMs use pulse-tuned exhaust manifolds that use geometrically tuned pipes to direct the exhaust gas from each cylinder almost directly into the inlet tract of the turbine housing. This minimizes gas flow interference and exhaust slug pressure variables that reduce the efficiency of common manifold systems. Pulse-tuned exhaust manifolds tend to diminish turbo lag (turbocharger response time) and increase performance at low engine load.

Turbocharger Geometry and Performance

Unlike most other engines, turbocharger efficiencies in highway truck diesel engines tend to be designed to be highest at peak torque, rather than at rated speed. A highway diesel engine described as high *torque rise* has an increase in torque as engine rpm decreases with a relatively constant power curve. Turbochargers used with high torque rise engines produce an approximately constant rate of flow per minute from the peak torque rpm through to the rated speed rpm. Therefore, the mass flow of air per cycle will be greatest at the peak torque speed. A turbocharger that is matched to the engine to produce optimum efficiency at peak torque will produce lower efficiencies at higher rpm. Also, as engine speed increases, the real time available for charging the cylinders with air and injecting fuel diminishes, providing the turbo with a self-regulating over-speed capability. The torque curve of a turbocharged engine quickly falls off in the lower speed range as the compressor efficiency plunges; fuel economy is also adversely affected.

Turbocharger performance depends on actual exhaust gas heat values and the gas flows dynamics, so to a certain extent they are self-regulating with respect to air requirements of the engine. Many truck diesel engines use constant geometry turbochargers that are precisely matched to the engine's requirements and do not require any type of external controls. Variable geometry turbochargers are managed externally by using controls, such as wastegating or nozzle ring control. Wastegating is a means of routing all or a percentage of the exhaust gas to bypass the turbine directly to the exhaust system. Figure 7-16 shows the wastegate in the closed position, which allows maximum boost. Figure 7-17 shows the wastegate in the open position, which allows exhaust gases to bypass the turbine wheel and directly enter the exhaust system. Some truck engine OEMs use wastegates to improve low load, low speed performance and to meet noxious emissions regulations. Managing manifold boost is obviously more critical in SI engines to avoid preignition.

Two-Stage or Series Turbocharging   Series turbocharging is not used in any current highway truck engines. Series turbocharging requires use of two turbochargers in series. The terms primary and secondary turbochargers (or low/high-pressure turbochargers) designate the roles played by the two units used. Cummins (in their NTC-475) last used this technology in early 1980s.

Paralleling   Paralleling describes the use of multiple turbochargers to charge each bank of a V configuration engine. Parallel turbocharger configurations are not used on any current truck diesel engines, but they are used extensively in larger, off-highway applications.

Compounding   The word "compounding" is often incorrectly used to describe two-stage turbocharging. The turbine is connected to a fluid coupling, which in turn is connected to the reduction gearing, and the output shaft is connected to the engine crankshaft. In other words, the turbocharger directly helps to drive the crankshaft. Compounding is not used in any current North American truck or bus engines.

A substrate is a catalytically inert material onto which active catalysts are coated.

Wastegating is a means of routing all or a percentage of the exhaust gas to bypass the turbine directly to the exhaust system.

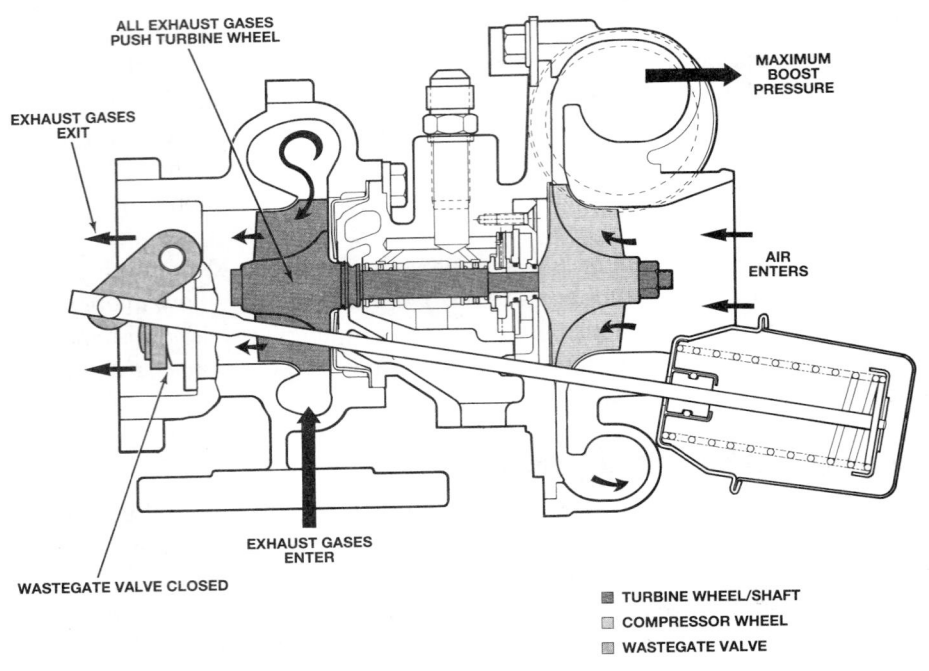

Figure 7-16 Wastegate closed. (Courtesy of General Motors Corporation, Service Operations)

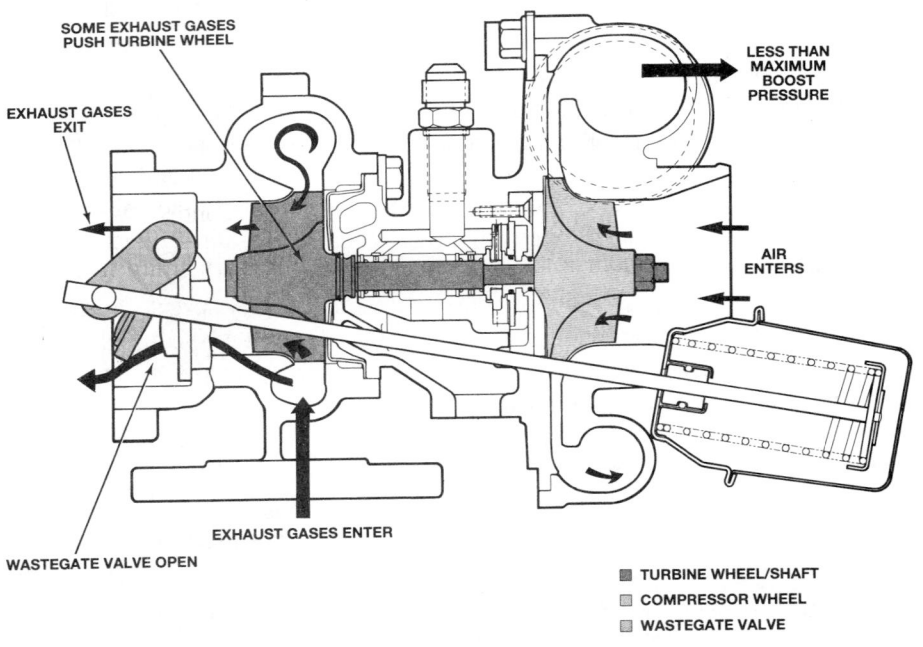

Figure 7-17 Wastegate open. (Courtesy of General Motors Corporation, Service Operations)

Charge Air Heat Exchangers

The act of compressing the intake air charge by the turbocharger typically produced air temperatures of 260° F (125° C) at a 68° F (20° C) ambient temperature and proportionally more at higher ambient temperatures. The objective of charge air coolers, whatever their cooling medium, is to cool the air pressurized by the turbocharger as much as possible while maintaining the pressure. As intake air temperature increases, air density diminishes and the oxygen charge in the cylinder is reduced. The result is lower power and higher cylinder temperatures as the proportion of rejected heat increases. To minimize this loss of power potential, the boosted air charge is cooled using one of several types of heat exchanger. As a rule, a 1° C increase in intake air temperature will produce a 2° C increase in exhaust gas temperature. It should also be noted that because the performance of charge air coolers greatly influences combustion temperatures and the density of the cylinder air charge, they are an integral component in the vehicle's noxious emission control system. Most North American boosted diesel engines use some type of heat exchanger to cool intake air.

Air-to-Air Cooling

Air-to-air coolers have the appearance of a coolant radiator and are often chassis mounted in front of the radiator (Figure 7-18). As the vehicle moves down the highway, ambient air is forced through the fins and element tubing. Ram air is therefore the cooling medium. Cooling efficiencies are highest when the vehicle is travelling at higher speeds. They are least efficient when the airflow through the cooler is minimal, and while the engine fan may assist, air to air-cooling may not be suitable for construction applications. Under optimal conditions, air-to-air heat exchangers have better efficiency ratios than any liquid-cooled heat exchangers. (Optimal conditions means maximal airflow through the element, as when the vehicle is travelling at full speed down the highway.) Thermal efficiency is increased by the air-to-air method of charge air cooling. If liquid-cooled heat exchangers are used, actual intake air temperatures are reduced to values only slightly higher than the coolant temperature values.

Air-to-air coolers are also referred to as charge-air coolers (CAC).

Aftercoolers and Intercoolers

Heat exchangers that use liquid engine coolant as the cooling medium are called either aftercoolers or intercoolers. The terms are synonymous; the one used merely reflects the OEM's preference. Boosted air is forced through an element containing tubing through which coolant from the engine cooling system is pumped. Cooling efficiencies tend to be lower than with air-to-air heat exchangers because of the heat of the cooling medium. They continue to be used in applications where low airflow through the engine housing precludes the use of an air-to-air exchanger. On Detroit Diesel two-stroke cycle engines, an intercooler is placed between the air discharge side of each turbocharger and the air inlet side of the engine blower. In this case, the intercooler reduces the temperature of the compressed air leaving the turbocharger before it

Aftercoolers and intercoolers are normally referred to as heat exchangers that use liquid engine coolant as the cooling medium. The two terms are synonymous and the one used merely indicates the OEM preference.

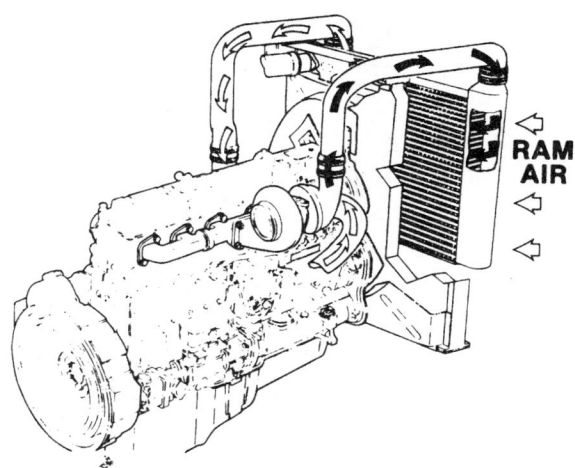

RAM AIR

Figure 7-18 Air-to-air charge cooler. (Courtesy of Mack Trucks, Inc.)

reaches the blower. Aftercoolers are located between the blower and the cylinders in the engine block. The aftercooler cools the air going into the engine after it passes through both the turbocharger and the blower.

Tip Turbine Heat Exchangers

Tip turbine heat exchangers (Figure 7-19) use both ambient air and engine coolant as cooling media. A main duct to a heat exchanger through which it is forced before entering the engine cylinders delivers boosted air. Engine coolant is circulated through tubing within the heat exchanger. However, a small amount of boosted air exiting the impeller housing of the turbocharger is ported off to drive a tip turbine. The tip turbine is an air-driven centrifugal pump that forces fresh, filtered ambient air through the heat exchanger element. This air does not come into direct contact with the charge air, but acts to cool both it and the engine coolant. Tip turbine, charge air cooling is popular in construction and stop/start applications where there is not sufficient airflow through the engine housing to use an air-to-air exchanger.

Jacket Water Aftercoolers

Jacket water aftercoolers (JWAC), as shown in Figure 7-20, use a split design intake manifold that contains the aftercooler core. The water pump pumps engine coolant through the aftercooler core. Coolant entering the aftercooler is approximately 190° F (88° C), and coolant exiting is at

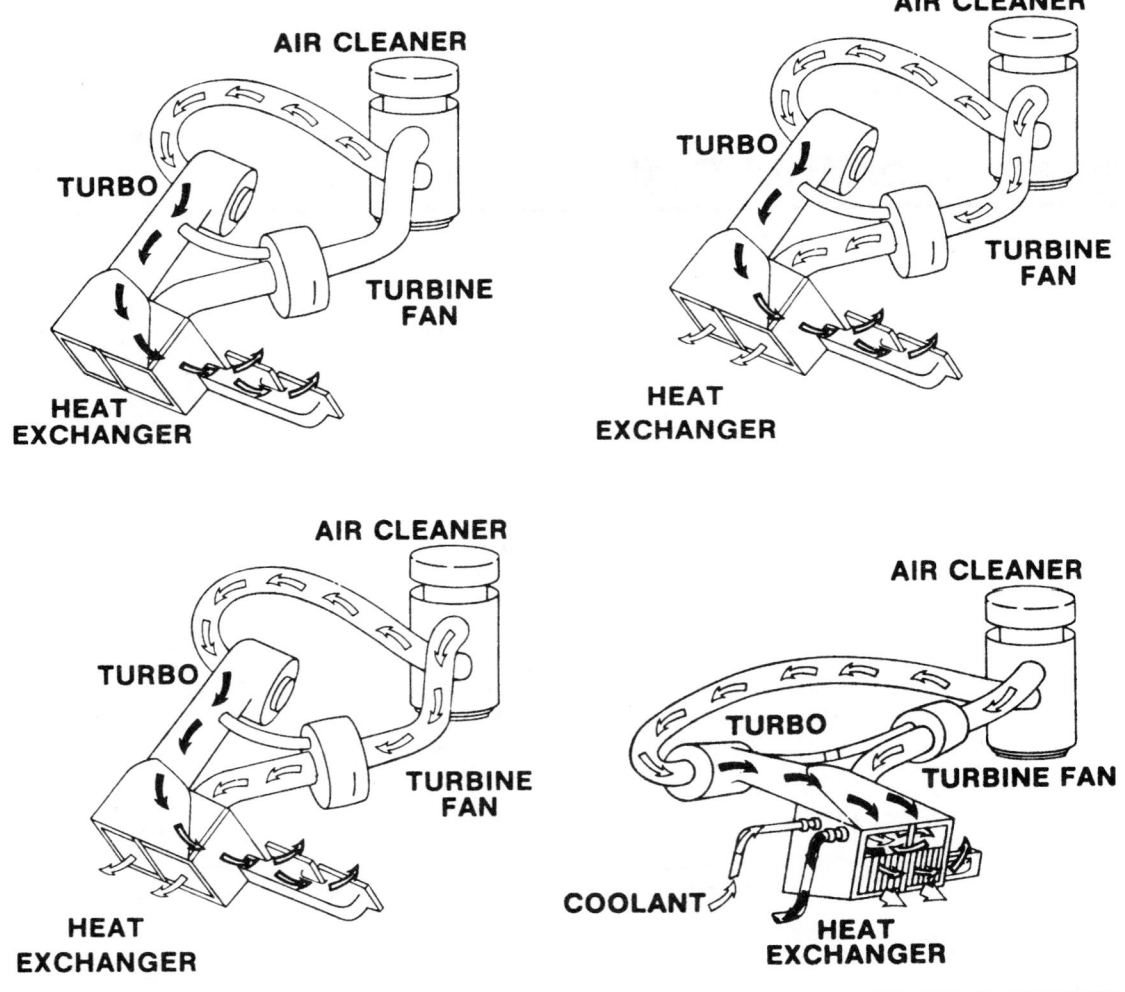

Figure 7-19 TIP turbine operation. (Courtesy of Mack Trucks, Inc.)

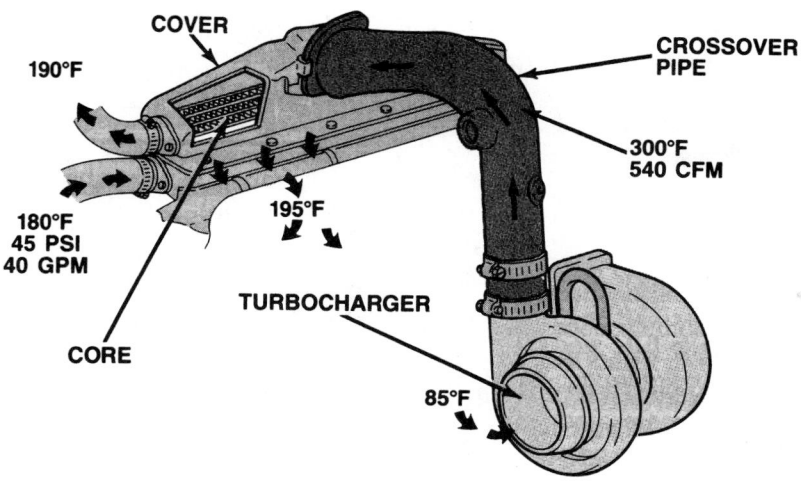

COVER

190°F

CROSSOVER
PIPE

300°F
540 CFM

180°F
45 PSI
40 GPM

195°F

TURBOCHARGER

CORE

85°F

Figure 7-20 Jacket water aftercooler (JWAC). (Courtesy of Freightliner Corporation)

approximately 180° F (82° C). Incoming air moving from the compressor side of the turbocharger through the intake manifold is at approximately 300° F (148° C) before going through the after-cooler core. This action reduces the air temperature to approximately 195° F (91° C), thereby increasing its density.

Intake Manifold Design

In a compression ignition (CI) engine, the intake manifold (Figure 7-21) is required to deliver air only to the engine cylinders. Because of the almost universal use of turbochargers on diesel engines, intake manifold design is generally less complex than those for naturally aspirated, gasoline-fueled engines. This usually means that the runners that extend from the plenum are of unequal lengths; this does not notably compromise engine breathing. A tuned intake manifold is one in which the shape and length of each runner are similar and designed to establish optimal gas dynamics for engine breathing. A single box manifold fed off the turbocharger compressor pipe can often meet a boosted engine's breathing requirements. Intake manifolds can either be wet (coolant ports) or dry; the latter is generally used in truck and bus applications. Materials used are aluminum alloy or cast iron, but some OEMs are experimenting with plastics and carbon-based fibers.

Where a tuned intake manifold is used in a boosted diesel engine, it is designed to work with a tuned exhaust manifold and establish ideal gas dynamics for effective scavenging during valve overlap. In full authority, electronically managed engines, a thermistor is located in the intake manifold; this thermistor is often a primary reference for determining air-to-fuel ratio. It is sometimes referred to as a manifold air temperature, or MAT, sensor.

Diesel engines cannot retain engine vacuum like a gasoline engine because they do not have a throttle to close off the intake manifold.

Valve Design, Configuration, and Breathing

Most current diesel engines use multivalve configurations consisting of two inlets and two exhausts valves. Parallel port configurations can enhance breathing efficiencies because each valve is responsible for an equal amount of gas flow without crossflow interference. Parallel port valve configuration enhances both cylinder charging and scavenging, but requires a more complex camshaft assembly (or dual camshafts). To some extent, it compromises gas velocity and swirl effect. In some instances this is desirable. In older, high-swirl engines, fuel sometimes was thrown

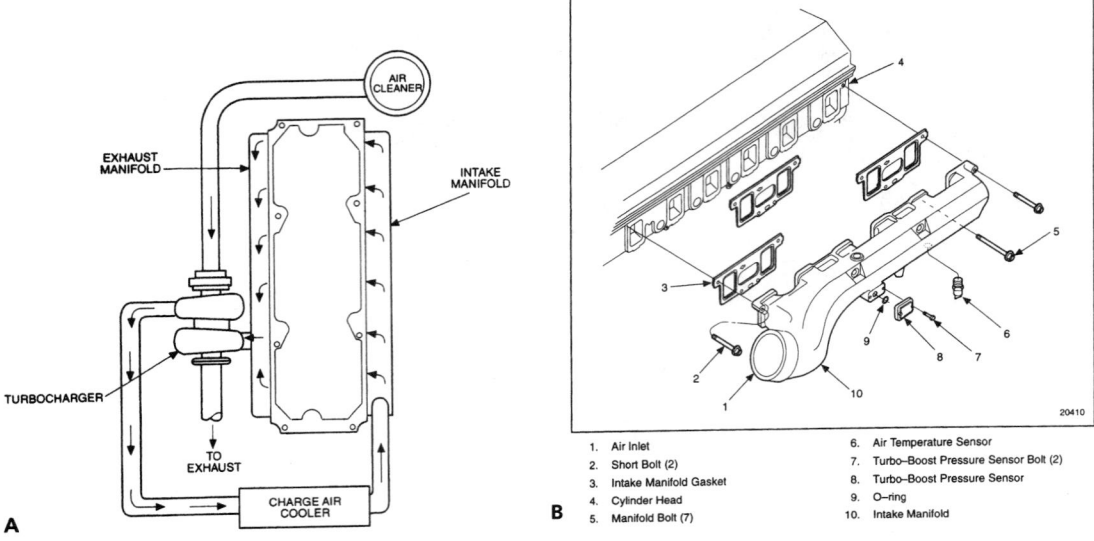

Figure 7-21 (A) Air intake system schematic; (B) Intake manifold and related parts. (Reprinted by permission of copyright owner Detroit Diesel Corporation, all rights reserved.)

into the cylinder wall causing it to condense and only partially combust. Many newer engines use quiescent cylinder gas dynamics (lower swirl) and much higher injection pressures to prevent fuel condensing in the cylinder. Valve seat angle will also affect cylinder breathing. Valve seats are usually cut and machined at either 30 degrees or 45 degrees.

Gas flow is generally about 20% greater using a 30-degree valve seat angle and the same lift as a 45-degree valve seat angle. However, a 45-degree seat is more often used because of the greater seating force and distortion resistance.

Exhaust System Components

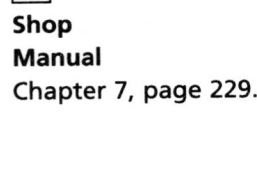

Shop Manual Chapter 7, page 229.

The exhaust system (Figure 7-22) is required to perform the following:

- ❏ Assist cylinder scavenging
- ❏ Minimize engine noise
- ❏ Minimize engine noxious gas emission
- ❏ Exhaust heat, noise and end gases *safely* to atmosphere.

Exhaust Manifold

The term "tuned" is used to denote any exhaust system designed with at least a little consideration for the exhaust gas flow.

The exhaust manifold collects cylinder end gases and delivers them to the turbocharger. Exhaust manifolds (Figure 7-22B) are usually manufactured in single or multiple sections of cast iron. Most diesel engine exhaust manifolds are "tuned" to deliver to the turbocharger exhaust gas with minimal and balanced flow resistance from each cylinder, such as in the pulse-tuned air manifold discussed earlier. The term "tuned" is used to denote any exhaust system designed with at least a little consideration for the exhaust gas flow. If the exhaust manifold and piping are properly designed, as each slug of cylinder exhaust is discharged it will not collide with that from another cylinder but instead be timed to unload into its tailstream. The pulsed manifolds discussed earlier in this section are an example of a tuned exhaust system. Exhaust back-pressure factors in a diesel engine are the turbocharger turbine assembly, the catalytic converter (if fitted), and the engine silencer. Exhaust manifold gaskets in truck and bus engine applications are usually of the embossed steel type, though occasionally fiber gaskets are used; they are usually installed dry.

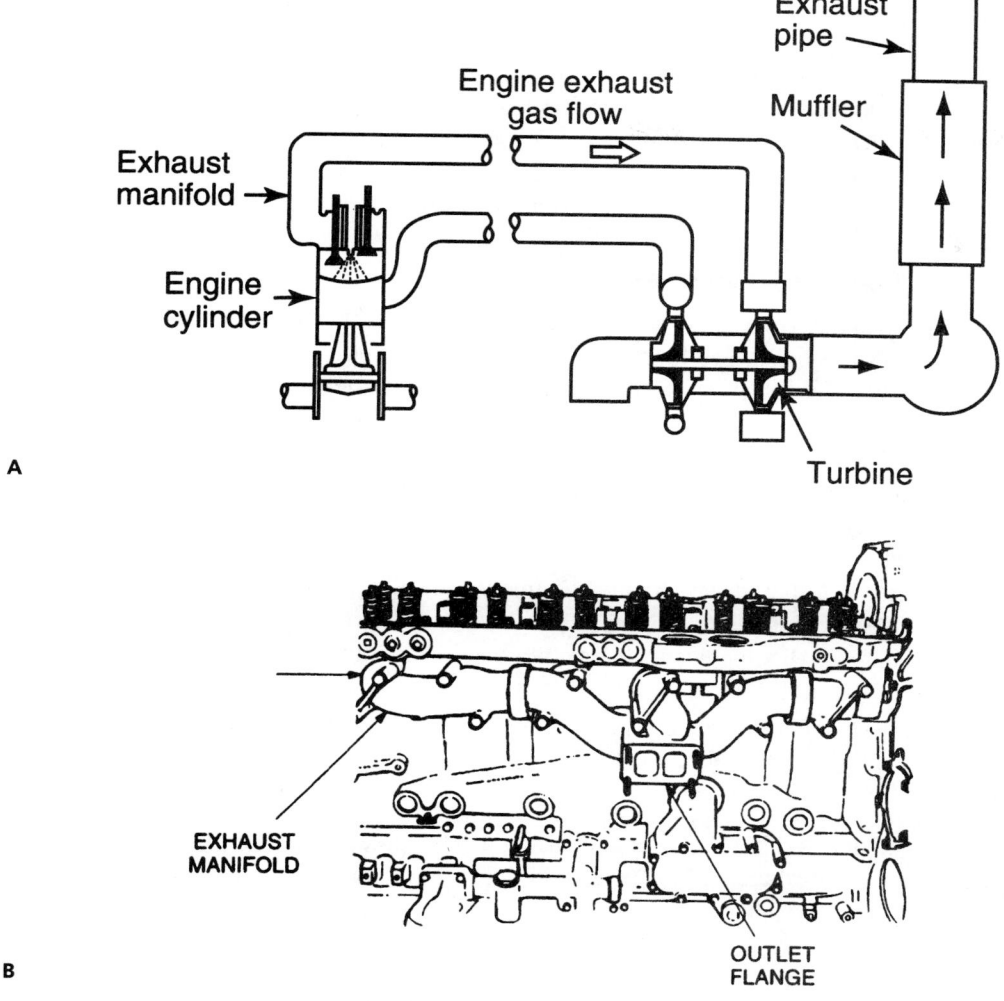

Figure 7-22 (A) Basic exhaust components; (B) Exhaust manifold. (Reprinted by permission of copyright owner Detroit Diesel Corporation, all rights reserved.)

Pyrometer

Normally found at a specified distance down line from the turbocharger, a pyrometer can provide the driver with engine loading information useful for shifting. Thermocouple pyrometers are used. This consists of two dissimilar insulated wires (often pure iron and constantin [55% copper and 45% nickel]) connected at each end to form a continuous circuit. The two junctions are known as the hot end, located where temperature reading is required, and the reference end, which is connected to a millivolt meter. Whenever the junctions are at different temperatures, there will be current flow; the voltage will increase with a greater difference in temperature. The practice of locating one pyrometer at the exhaust exit to each cylinder to indicate cylinder balance is not a current one in truck and bus engines.

Exhaust Piping

In order to protect from corrosion, galvanized steel has been used for mild steel piping, flex piping, and clamps. However, it is becoming more common to use stainless steels, which provide the exhaust system components with at least double the service life.

Shop Manual
Chapter 7, page 231.

Stainless steel band clamps have been used for many years and are a effective means of sealing flex pipe to straight pipe, because they yield to shape at installation; they are deformed at installation and should not be reused. The function of exhaust piping is to collect the engine exhaust gases unloaded by the turbine housing, deliver them first to the engine silencer/catalytic converter, and then to the atmosphere clear of the tractor cab and trailer structure. If rain caps are used on vertical exit pipes, these should be mounted transversely to open away from the chassis. They should never be mounted longitudinally, facing either forwards or rearwards.

Engine Mufflers (Silencers)

The firing pulses in engine cylinders are sonic nodes that produce sound and antinodes that respond to the sound and produce noise. The function of the silencer or muffler is to lower engine noise to legally acceptable levels. Essentially, this requires the scrambling of the frequency of the sonic nodes and antinodes produced by engine firing pulsations. Before rigorous sonic emission standards were developed in the 1990s, the turbocharger and piping could be designed to meet minimal standards. This is no longer the case. However, pressure waves caused by the unloading of each slug of exhaust gas into the piping can also have an amplifying effect on engine sound, and the turbocharger is capable of smoothing these pressure waves and enabling a simpler muffler design.

Muffler volume in today's truck engine systems is typically in the region of five times total exhaust slug volume. Exhaust slug volume is always somewhat greater than swept volume in manifold-boosted engines. Engine silencers use two basic principles to achieve their objective of dampening sound:

❑ Resonation: Resonation requires reflecting sound waves back towards the source, thereby multiplying the number of sound emission points. Separate chambers connected by offset pipes as well as baffles are used to achieve the objective of scrambling the frequency of the engine's firing pulses.

❑ Sound absorption: Exhaust gases pass through a perforated pipe enclosed in a canister filled with sound-absorbing material. Sound absorption involves converting sonic energy into heat by friction, and the efficiency of these devices will depend on the packing density of the sound-absorbing media and the pipe perforation geometry. The media must be resistant to pressure pulsations and mechanical vibration. Mineral and metal wool are used, most often basalt wool. Sound-absorption mufflers generally have improved flow resistance characteristics over comparable resonator-principle mufflers.

Truck and bus mufflers use one or a combination of the above principles to meet sonic emission requirements.

Catalytic Converters

For the most part, truck and bus engines meet noxious gas emission requirements by carefully managing the combustion process using comprehensive monitoring and an on-board computer. A few OEMs use **catalytic converter**s (Figure 7-23) on current engines, but this is necessarily a single-stage device incorporating just an oxidizing catalyst.

A catalyst is a substance that increases the rate of a chemical reaction without itself undergoing a chemical change.

A **catalyst** is a substance that increases the rate of a chemical reaction without itself undergoing a chemical change. Diesel vehicles also require control of particulate matter emissions. Control of carbon monoxide (CO) and hydrocarbons (HC) is straightforward, using the standard platinum group, metal-based catalyst; but control of oxidized nitrogen compounds (NO_x) is difficult because of the oxidizing nature of the exhaust. Catalytic particulate matter is equally difficult to control since it contains ash from fuel and oil combustion, carbon, and absorbed HC from the incomplete combustion of this fuel and oil. Extensive studies of exhaust filters (particulate traps) have shown that they oxidize and trap particulate matter with or without a catalyst. Yet, these filters suffer from clogging due to particulate matter buildup that requires complex filter regeneration procedures.

An automobile catalytic converter is normally described as a two-stage, three-way converter. The two stages refer to the objectives of the catalytic converter. When cylinder end gas

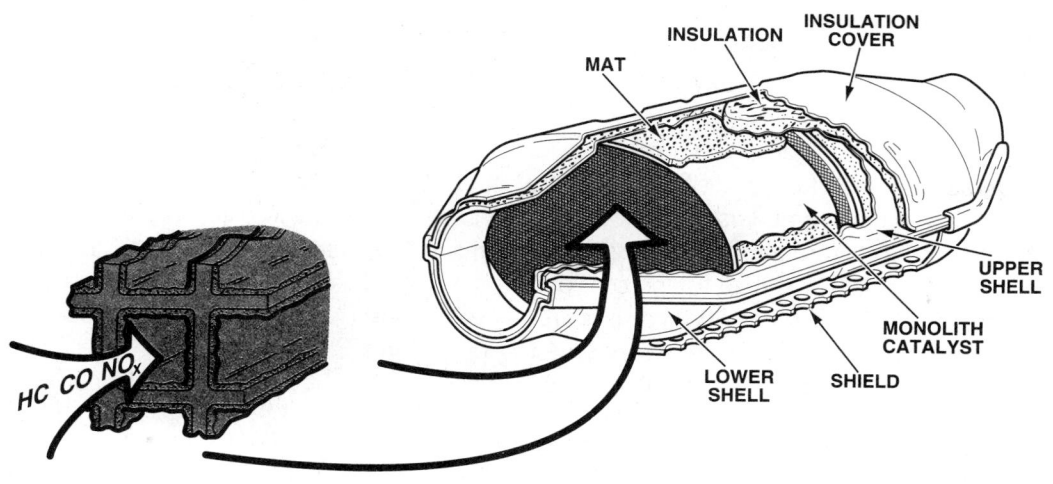

Figure 7-23 Catalytic converter. (Courtesy of General Motors Corporation, Service Operations)

contains HC and CO, the *oxidizing* stage of the catalytic converter attempts to oxidize these to H_2O and CO_2. Where oxidized nitrogen compounds (known collectively as NO_x) are present in the exhaust gas, a *reduction* stage attempts to reduce these to elemental nitrogen and oxygen. The term "three-way" denotes the catalysts.

Platinum and palladium are both oxidation catalysts; rhodium is a reduction catalyst. All three are classified as noble metals and are sensitive to lead contamination. Oxidizing catalysts enable catalytic afterburning, which is sometimes assisted by the introduction of fresh air into the reaction. Peak temperatures run in the range of 800° C–1,000° C. The commonly used reduction catalyst is rhodium. When oxidized nitrogen compounds have been formed, rhodium endeavors to reduce them to elemental oxygen and nitrogen. However, rhodium only functions as a reduction catalyst when the cylinder burn is managed close to the fuel-to-air ratio of 14.7 part air to 1 part fuel. This makes it of little use in lean burning diesel engines. Because of the lean burn characteristic of diesel engines, NO_x emission tends to be a problem, so the introduction of multistage catalytic converters may not be too far away.

The noble metal catalysts are thinly coated on aluminum oxide or granulate monolith substrate providing optimum utilization of catalyst surface area combined with minimal flow resistance. Catalytic converters operate efficiently only at operating temperatures and do little to limit HC emissions at start-up and warm-up. Computer-controlled variable timing and minimal fueling attempt to achieve this. At present, only CI truck and bus engines use oxidation catalysts.

Significant advancement in diesel engine technology (mostly electronic control) have reduced engine-out emissions; flow-through oxidation catalysts can now provide a level of particulate control that offers an alternative to filter (particulate trap) use. This matter is primarily the soot portion and the soluble organic fraction (SOF) or volatile organic fraction (VOF). Recent studies of flow-through oxidation catalysts for control of CO, HC, and SOF focused on the use of platinum-group catalysts, that is, platinum (Pt) and palladium (Pd). When hydrogen gas is sprayed on a platinum screen it combines with oxygen in the air to form water. This same principle works in an **oxidation catalyst**. Catalysts based on platinum and palladium are generally required due to the low exhaust temperatures in a diesel engine.

A substrate is a catalytically inert material onto which active catalysts are coated.

Other Means Used to Control Noxious Emissions

Particulate traps are used by some OEMs. They are essentially soot filters that use a ceramic filtration medium. Reduction catalysts used on automotive SI engines function as such only when air-fuel ratio AFR is stoichiometric or rich. As CI engines operate with excess air (to minimize particulate emission and maximize fuel economy), a reduction stage catalyst would not function. It should be noted that effective cooling of intake air lowers combustion temperatures, making it less likely that the nitrogen in the air is oxidized during the combustion process.

Opacity Meters   An opacity meter (opacimeter) is a light extinction, exhaust smoke testing tool that has become the aftermarket field-monitoring tool of choice for jurisdictions wishing to enforce smoke emission standards. The tool is fitted to the exhaust stack and consists of a light emitter that projects through the exhaust gas at a sensor. The percentage of emitted light that is picked up at the sensor is read as "percent opacity."

Smoke opacity readings of 5% or less are not easily observed. When using an opacity meter it is important to meticulously observe the test procedures if any credence is to be given to the results. The following lists some EPA opacity maximum readings on a Bosch opacimeter:

Idle speed	10%
High idle speed	15%
Maximum puff	50% (transient response)
Rated speed/load	20%
Engine lug	25%
Motoring	25%
Full acceleration	30%

Cold Starting Aids

Shop Manual Chapter 7, page 231.

Air Intake Preheater

Some medium duty diesel engines use an electric air heater grid located in the air inlet casting. This device preheats the intake air for improved starting during cold weather, and it also reduces white smoke after start-up. The system uses a control module, which is located on the left rear of the cylinder head on a Caterpillar 3116 engine, as shown in Figure 7-24. This module senses engine oil pressure, coolant temperature, and elapsed engine run time. The module activates a solenoid that turns the heater grid on and off. When the heater grid is on, an indicator light illuminates to alert the driver.

Glow Plug Systems

Some medium duty diesel engines use glow plugs (GP) (Figure 7-25) that are controlled by the electronic control module (ECM) or the powertrain control module (PCM), as shown in Figure 7-26. Figure 7-26A shows an ECM/PCM controlled glow plug system; Figure 7-26B shows a manual system controlled by an analog controller. In the ECM/PCM, a glow plug relay control is used to energize the glow plugs for assisting cold engine start-up. Engine oil temperature, battery positive voltage (B+), and barometric pressure (BARO) are used by the PCM to calculate glow plug on-time and the length of the duty cycle. On-time normally varies between 10 and 120 seconds. With colder oil temperatures and lower barometric pressures, the plugs are on longer. If battery voltage is abnormally high, the duty cycle is shortened to extend plug life. The glow plug relay will cycle on and off repeatedly only when there is a system high-voltage condition greater than 16 volts.

Ether Injection Systems

A fluid starting aid, such as the ether injection system shown in Figure 7-27, is designed to inject a highly volatile fluid into the air intake system to assist ignition of the fuel at low ambient temperatures. It consists of a pump and nozzle for injecting the fluid into the air intake and a suitable container for the fluid. The fluid is contained in suitable capsules to facilitate handling. The starting aid consists of a cylindrical capsule container fitted with a screw cap. Inside the container is a sliding plunger, like a piercing shaft. From the capsule container, a tube leads from the con-

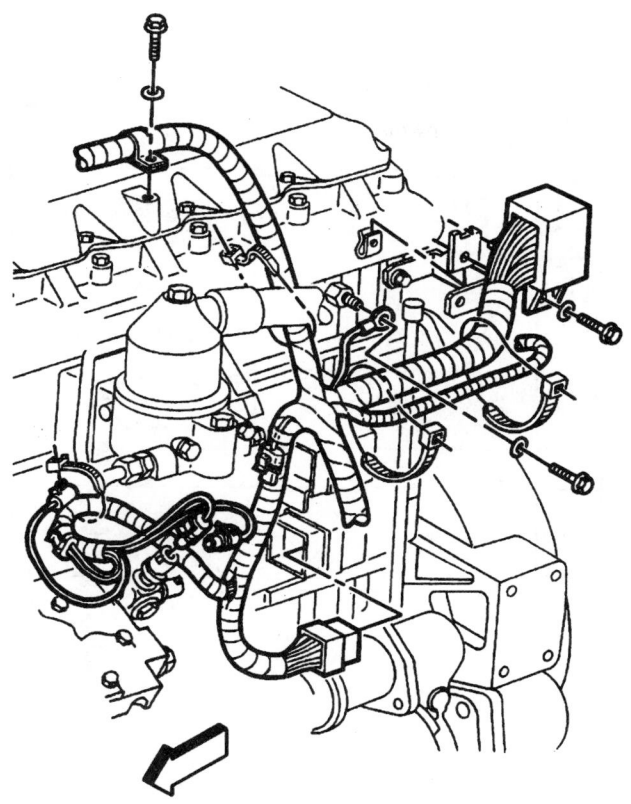

Figure 7-24 Air intake preheater module. (Courtesy of General Motors Corporation, Service Operations)

VIEW A

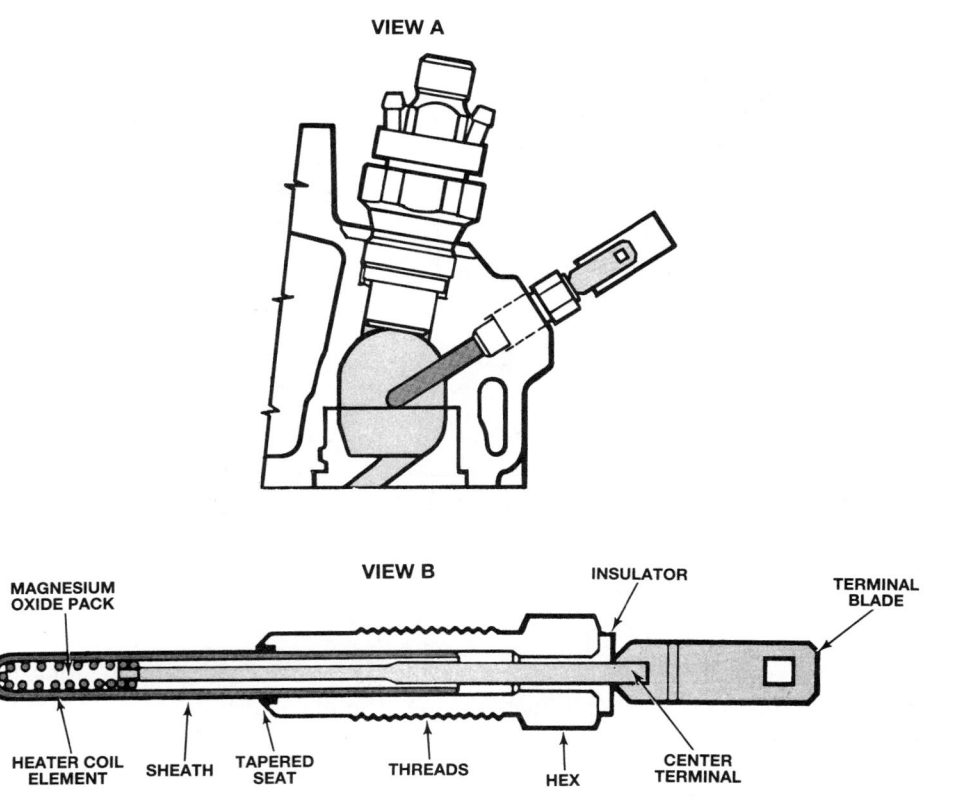

VIEW B

MAGNESIUM
OXIDE PACK

INSULATOR

TERMINAL
BLADE

HEATER COIL
ELEMENT

SHEATH

TAPERED
SEAT

THREADS

HEX

CENTER
TERMINAL

Figure 7-25 Glow plug. (Courtesy of General Motors Corporation, Service Operations)

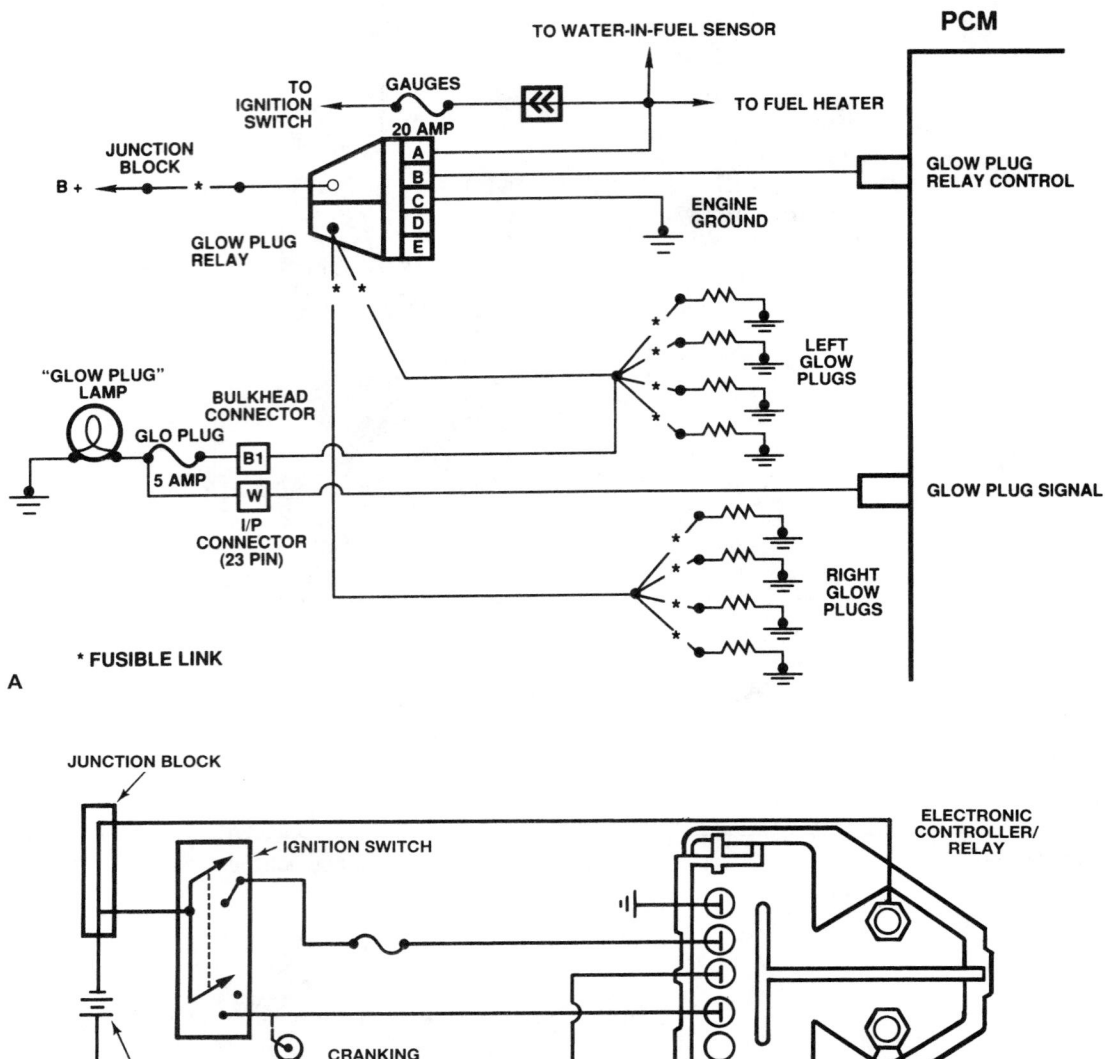

Figure 7-26 (A) Computer controlled glow plugs; (B) Controller controlled glow plugs. (Courtesy of General Motors Corporation, Service Operations)

tainer to a hand-operated pump, and another tube leads from the pump to an atomizing nozzle threaded into a tapped hole in the air inlet housing. The pump may be mounted on the instrument panel or in some other convenient location. The capsule container must be mounted in a vertical position away from such high heat areas as the exhaust manifold, muffler, etc., and should not be located under a hood or in a cab. The atomizing nozzle is screwed into a tapped hole in the air inlet housing.

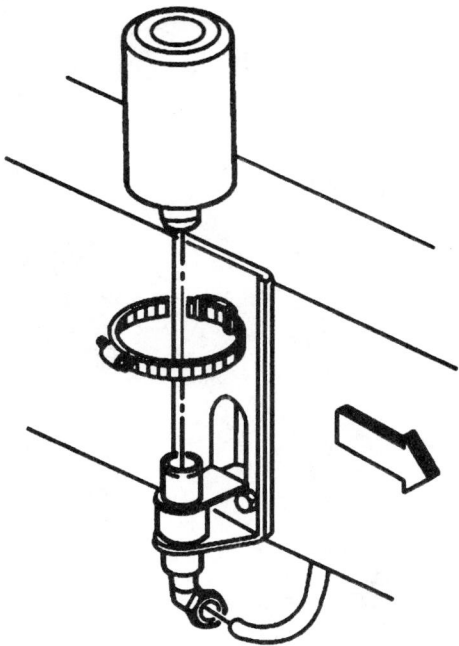

Figure 7-27 Ether injection system. (Courtesy of General Motors Corporation, Service Operations)

Crankcase Ventilation

**Shop
Manual**
Chapter 7, page 233.

A diesel engine has the maximal crankcase pressure potential at idle because at slow speeds there is more time for the air to leak past the rings. This time allows this blow-by gas pressure to gather in the crankcase. Many engines use a wire mesh breather in the oil fill cap or road draft tube that vents these blow-by gases. Some road draft breathers use a relief valve assembly.

Some medium duty diesel engines use crankcase ventilation systems (Figure 7-28) to ventilate the gases of combustion or blow-by. These systems use the open areas inside the engine cylinder crankcase and cylinder head (s) to transfer the gases coming from combustion blow-by into the intake manifold. At some location a type of crankcase ventilation valve is used to regulate the entry of these gases into the intake manifold to be re-burned in the engine. One OEM refers to this valve as a crankcase depression regulator (CDR), because it regulates the depressurization of the crankcase by engine intake suction.

The CDR valve (Figure 7-29) operates by engine vacuum or pressure differential. Inside the CDR valve a spring holds open a valve plate that connects the body to a flexible diaphragm. The valve plate can restrict the outlet passage to the air inlet duct when the pressure differential pulls it closed against the spring force. At idle, the CDR valve is fully open because diesel engines do not produce significant vacuum (pressure differential), and the airflow past the CDR outlet passage is not great enough to close the valve. The valve should be fully open (Figure 7-29A) because the crankcase pressure potential at idle is approximately 1 in. of water (at slow speeds there is more time for the air to leak past the rings). At higher engine speeds, the valve closes to provide more restriction (see Figure 7-29B). This action prevents the movement of oil vapors into the intake manifold by limiting the crankcase vacuum (measured at 2,000 RPM to be from 2 to 5 in. of water).

 WARNING: If the CDR valve were not restricted, it would attempt to depressurize the crankcase and suck all of the oil out of it.

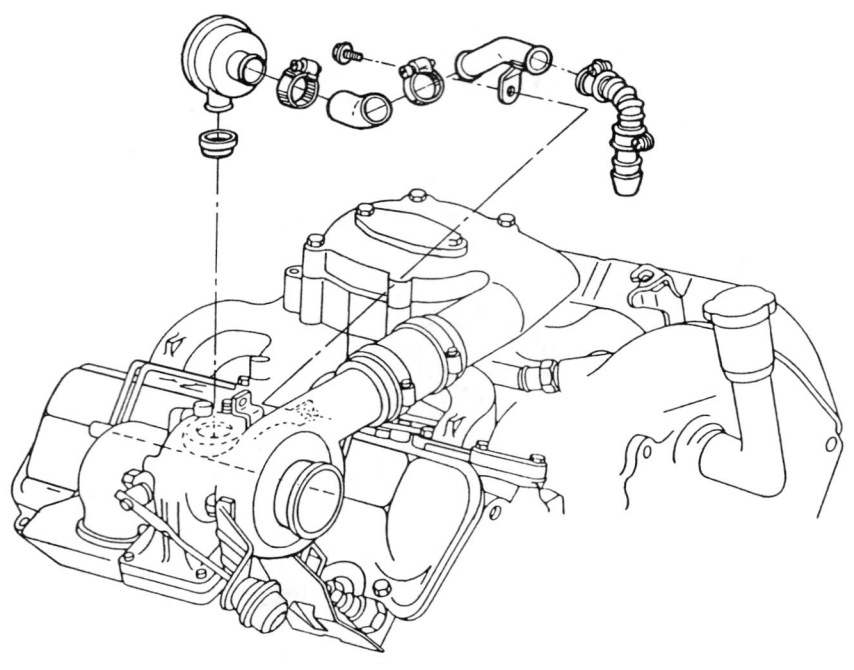

Figure 7-28 Crankcase ventilation system. (Courtesy of General Motors Corporation, Service Operations)

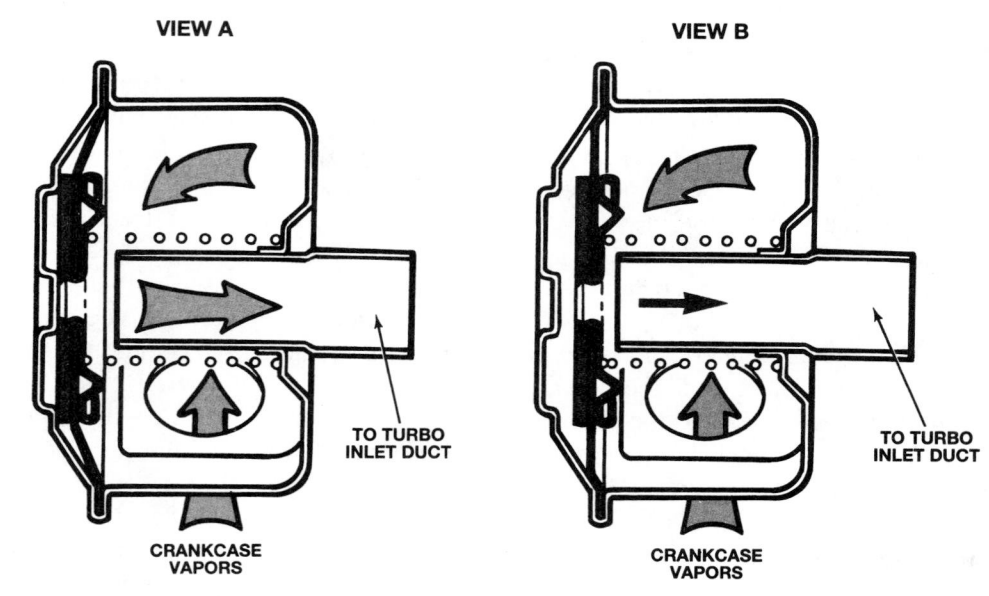

Figure 7-29 CDR valve. (Courtesy of General Motors Corporation, Service Operations)

Engine Brakes

**Shop
Manual**
Chapter 7, page 235.

Engine compression brakes are widely used on North American truck diesel engines. A large percentage of vehicle braking requires application pressures of 20 psi or less in a typical air brake system. In other words, less than one-fifth of the brake system's potential is used. The objective of engine compression brakes is to relieve the vehicle braking system of some of the light duty braking, especially in instances that require prolonged braking such as on an extensive downgrade. A truck air braking system converts the kinetic energy (energy of motion) of the vehicle to heat energy. This takes place at the brake foundation assembly, and its mechanisms are drums (or rotors), brake shoes (or calipers), and an actuating apparatus that converts the potential energy of compressed air into mechanical force.

When the brakes are applied, the friction facings on the brake shoes (or calipers) are loaded into a rotating drum or caliper and the resulting resistance, or friction, retards the movement. Because the friction produces heat, the brake system must be designed to sustain high temperatures and dissipate braking heat rapidly. The coefficient of friction, the ratio of the sliding force to the normal force, describes the aggressiveness of friction materials. It is desirable to maintain a consistent coefficient of friction in foundation brake components for obvious reasons. However, the coefficient of friction is altered by changes in temperature and the addition of a lubricant, such as water. When foundation brake systems are subjected to the high use, they dissipate a large amount of heat to the atmosphere. Failure to dissipate this heat can greatly increase foundation brake temperatures and will cause a reduction in the coefficient of friction of the critical friction surfaces at the shoe and drum, and expansion of the drum dimension.

> The coefficient of friction is the amount of force to move one body when it is in contact with another body. It is the ratio of sliding (surface) friction to the normal or total downward force.

⚠️ **WARNING:** When a tractor-trailer combination has to run downhill for prolonged periods, excessively high foundation brake temperatures result even though brake application pressures are relatively light. This causes a reduction in braking efficiency at a time when it is most needed, a condition known as brake fade.

A supplementary braking system such as an engine compression brake, or driveline **retarder,** can serve to support the main vehicle braking system, enhance peak braking performance and (in cases where engine braking is managed by the ECM), reduce some of the driver-actuated braking.

A BIT OF HISTORY

Clessie Cummins invented the internal engine compression brake after he had officially retired from his position as CEO of the company that bears his name. The Jacobs Manufacturing Company, perhaps better known for their drill chucks, has manufactured engine compression brakes since the first Cummins' engine brake was introduced in the late 1950s. They manufacture an engine brake for all of the major North American diesel engines and have established universal brand identification. Many truck drivers refer to the Jacobs Brakes (Figure 7-30) as a **"Jake brake,"** and this name typically refers to any engine compression brake.

> People in the trucking industry use the term "Jake brake" to refer to any internal engine compression brake and indeed, this manufacturer dominates the market place, designing their engine brakes for all the major OEMs. However, there are other manufacturers of similar devices and more evidence of this can be seen in imported truck engines.

Trucks may also use hydraulic retarders that may be located anywhere in the driveline, but they are usually an integral part of the engine or transmission. Hydraulic driveline retarders use a torque absorption principle that can be compared to that used in a torque converter. Electric driveline retarders (Figure 7-31) are also currently available and Jacobs markets one of them. The Caterpillar BrakeSaver®, which is coupled to the engine, will be covered in this chapter because it is an engine component rather than a transmission or driveline component.

The term "retarder" can be applied to any component or system that effects braking action, that is, retards or slows motion. The primary vehicle braking system is seldom referred to as a vehicle retarder system, although engine compression brakes and driveline brakes are commonly described as retarders. Compression and driveline brakes supplement the primary braking

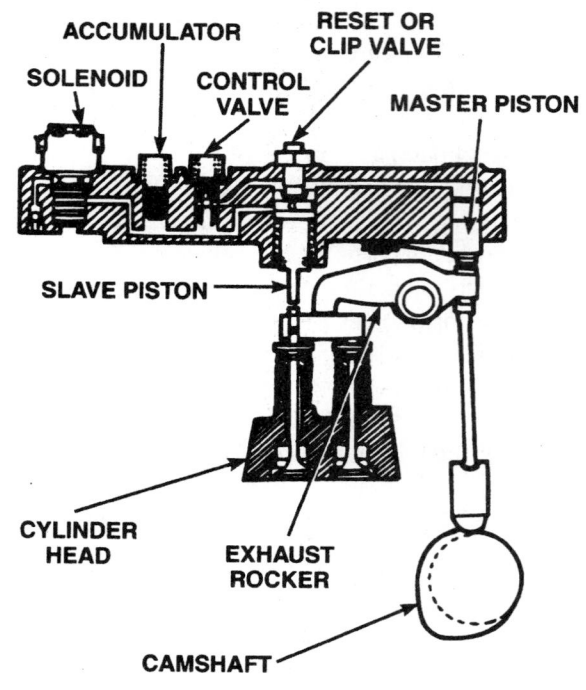

Figure 7-30 Exploded view of Jacobs engine brake. (Courtesy of Mack Trucks, Inc.)

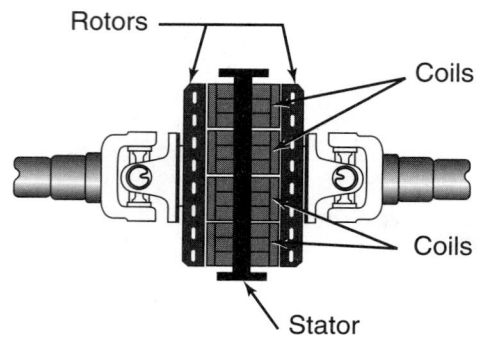

Figure 7-31 Driveline retarder components. (Courtesy of Jacobs Vehicle Systems™)

system. The retarding powerflow is confined to those axles directly linked to the drivetrain (that is, through the wheels on the drive axles). The objective of the engine brake system is identical to that of the main vehicle braking system, namely, to convert the energy of vehicle motion to heat energy. In the engine compression brake, kinetic energy (that of vehicle motion) is converted first to the potential energy of compressed air and then released into the exhaust system as heat.

The term engine brake can be used to describe retarder devices that use three distinct operating principles:

1. Internal engine compression brakes
2. External engine compression brakes
3. Hydraulic engine brakes or retarders

Internal Engine Compression Brakes

In the four-stroke cycle diesel engine, the piston is required both to perform work and receive work. On its upwards travel during the compression stroke it compresses the cylinder air charge (performs work), following which it is forced through its downstroke by expanding cylinder gases (receives work). The **internal compression brake** operates by making the piston perform the work of compressing the air charge on the compression stroke and then negating (canceling) the power stroke by releasing the compressed cylinder gases to the exhaust system (gas blowdown) somewhere around TDC. This has the effect of reversing the engine function, converting it from an energy-producing pump to an energy-absorbing compressor. All engine compression brakes use this principle of operation, although the mechanisms used to actuate, time, and control the brakes vary. The braking powerflow of the vehicle begins at the drive axle wheels, extends through the transmissions and driveshafts, and continues through the engine powertrain. The kinetic energy is converted to pressure; the potential energy of the compressed air is then dumped into the exhaust system. Braking efficiencies are highest when engine rpm is highest.

External Engine Compression Brake

The operating principle of the **external compression brake** is not very different from the internal compression brake. External engine compression brakes are also known as **exhaust brakes** (Figure 7-32). They consist of a housing located downstream from the turbine housing discharge in the exhaust system. Within the housing is a valve that, when actuated, chokes off the exhaust discharge. This restricts the exhaust gas flow and once again reverses the role of the engine, converting it into an energy-absorbing pump. However, in the internal engine compression brake, the effective pumping stroke is the compression stroke, while in the exhaust brake, the effective retarding stroke becomes the exhaust stroke. Because of this, as each succeeding exhaust slug is unloaded into the exhaust manifold, the pressure rises and braking efficiency increases. Retarding efficiency is to some extent diminished by pressure loss to the intake during valve overlap. Operation of the exhaust brake is managed so that the engine is not fueled during braking and, as with the internal engine compression brake, braking efficiencies are highest when the engine rpm is in its higher range.

Hydraulic Engine Brakes

Many types of electric and hydraulic driveline brakes exist, but these are more appropriately dealt with when studying transmission and driveline components. However, the Caterpillar BrakeSaver (Figure 7-33) is a hydraulic retarder that is coupled to the rear of the engine and uses engine lubrication oil as its medium. A rotor is coupled directly to the engine crankshaft and will rotate at any time the engine is running. The rotor is turned within the BrakeSaver housing, which is coupled to the flywheel housing and to which the transmission is mounted. The engine flywheel is bolted through to the crankshaft, but because of the BrakeSaver, the starter motor must crank the engine by means of a ring gear on a ring gear plate mounted to the crankshaft behind the brake saver rotor. The crankshaft therefore drives the rotor between the BrakeSaver housing and a stator. When the BrakeSaver is actuated, the housing is charged with pressurized engine oil. Vane pockets on the rotor means that when the BrakeSaver is charged with oil, the rotor encounters

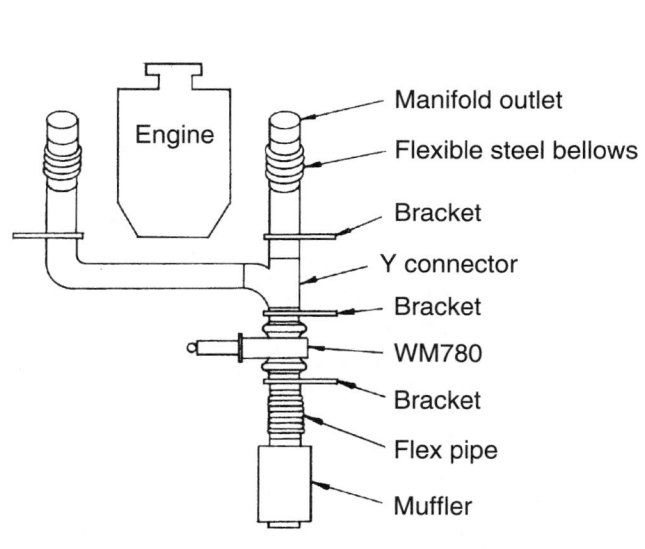

Figure 7-32 Layout of an exhaust brake on a V-engine. (Courtesy of Williams)

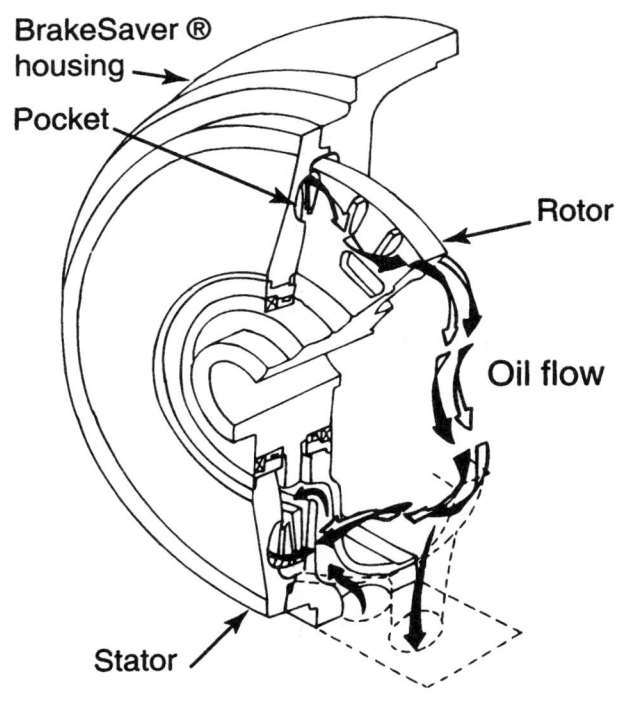

Figure 7-33 Caterpillar BrakeSaver. (Courtesy of Caterpillar Inc.)

fluid resistance defined by the stator geometry and the rotor rotational speed. Once again, retarding efficiencies are greatest when rotational speed is greatest. The oil that acts as the BrakeSaver hydraulic medium is engine oil, supplied from the engine sump by a section of the oil pump dedicated to changing the BrakeSaver.

**Shop
Manual
Chapter 7, page 235.**

Engine Brake Operation and Control Circuits

Engine brakes typically use electric control switches, which means that they may be actuated either directly by the driver or by the engine management ECM in "smart" cruise applications. When engine brakes are used in hydromechanically managed engines, the switching of the engine brake is usually electrical, but the retarding effect must be actuated either hydraulically or pneumatically. In such engines, a control circuit requires that a series of switches be closed before engine braking can be effected. The first of this series of switches is the driver control switch, which can be proportional depending on the system. The next is located at the clutch; this must be necessarily fully engaged to operate engine braking. The final switch in the series (others may be installed) is to ensure that the accelerator is not depressed. It is usually a fuel pump microswitch. This switch ensures that engine fueling is at zero (current engines) or at least at a minimum (in older engines) to limit the quantity of raw fuel dumped into the exhaust system. Current regulations require that no uncombusted fuel be discharged into the exhaust system. However, engine brakes used in many current engines are managed or monitored by the engine electronics. When an engine brake is electrically switched to the on position, the result depends on the type of engine brake used.

Internal/External Engine Braking Systems

**Shop
Manual
Chapter 7, page 239.**

Williams Exhaust Brakes

Williams exhaust brakes are classified as external engine compression brakes. Control of the brake is electric over pneumatic. The electrical circuit (Figure 7-34) required to actuate the engine brake consists of three switches, all of which must be closed: a control switch (dash mounted), a clutch switch (clutch must be fully engaged), and an accelerator switch (accelerator must be at zero travel). The sliding gate exhaust brake uses a pneumatically activated gate actuated by chassis system pressure. The air supply to close the gate is controlled by an electrically switched pilot valve. An aperture in the sliding gate permits minimal flow through the brake gate during engine braking. The butterfly valve version operates similarly.

**Shop
Manual
Chapter 7, page 235.**

Jacobs Compression Brakes

The Jacobs family of engine compression brakes (Figure 7-35) is classified as internal compression brakes. People in the trucking industry use the term "Jake brake" to refer to any internal engine compression brake and, indeed, this manufacturer dominates the market place by designing their engine brakes for all the major OEMs. However, there are other manufacturers of similar devices, seen especially in imported truck engines. Jacobs also manufacture driveline retarders.

**Shop
Manual
Chapter 7, page 237.**

As an internal engine compression brake, a Jacobs brake operates by having the piston perform the work of compressing the cylinder air charge on the compression stroke, and then negating the power stroke by opening the exhaust valves somewhere around TDC and releasing the cylinder compressed air to the exhaust. Jacobs engine brakes use electric switching of a hydraulic actuating circuit, so they adapt readily to electronic management, becoming an integral component in "smart" cruise control systems.

All Jacobs brakes hydraulically actuate the opening of the exhaust valve(s) in the engine cylinder performing the braking. The manner in which this hydraulic circuit is timed and actuated varies with the engine. However, the mechanical force required to actuate the hydraulic engine camshaft always provides circuit. This means that the timing of the exhaust valve opening

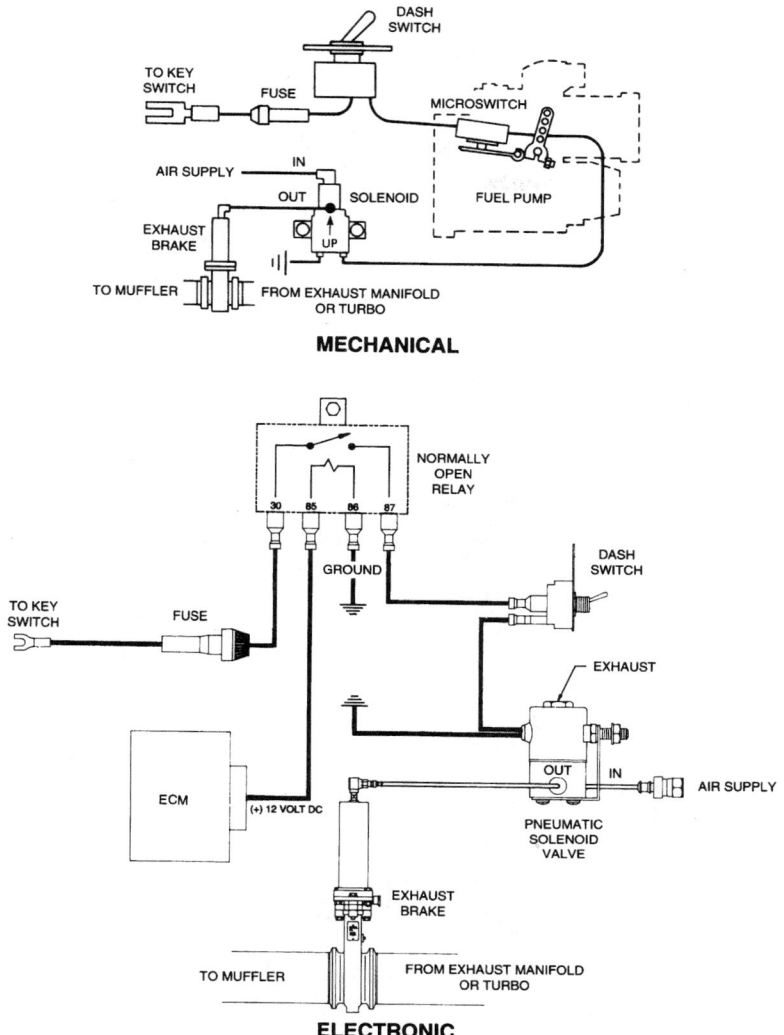

Figure 7-34 Control and actuation circuits on a Williams mechanical and ECM controlled exhaust brake. (Courtesy of Williams)

will also be governed by cam profile. The Jacobs brake was first designed for Cummins PT engines, so its operation on this engine will be described first.

The Jacobs brake (Figure 7-36) on a Cummins PT engine is managed electrically from a circuit that requires that switches at the clutch, throttle arm on the PT pump, and control switch on the dash all be closed. This energizes a three-way solenoid valve located in the rocker housing that charges the Jacobs hydraulic circuit with pressurized engine lubricating oil. The solenoid acts as an electrically controlled pilot valve that merely controls the flow of oil into the hydraulic circuit of the engine brake controls. This oil flows through the hold-down control piston and when the PT injector rocker arm loads the plunger into the cup, oil from the hold-down piston control valve is trapped between it and the hold-down piston preventing the PT injector plunger from lifting and therefore metering and pumping. This oil will remain entrapped in this circuit until the three-way solenoid valve is de-energized. When the three-way solenoid is energized, oil also flows to the slave piston control valve and feeds a circuit that has a master piston located directly above the push tube end of the injector rocker and a slave piston located at the exhaust valve bridge. The oil pressure first loads the master piston contacting the adjusting screw of the injector

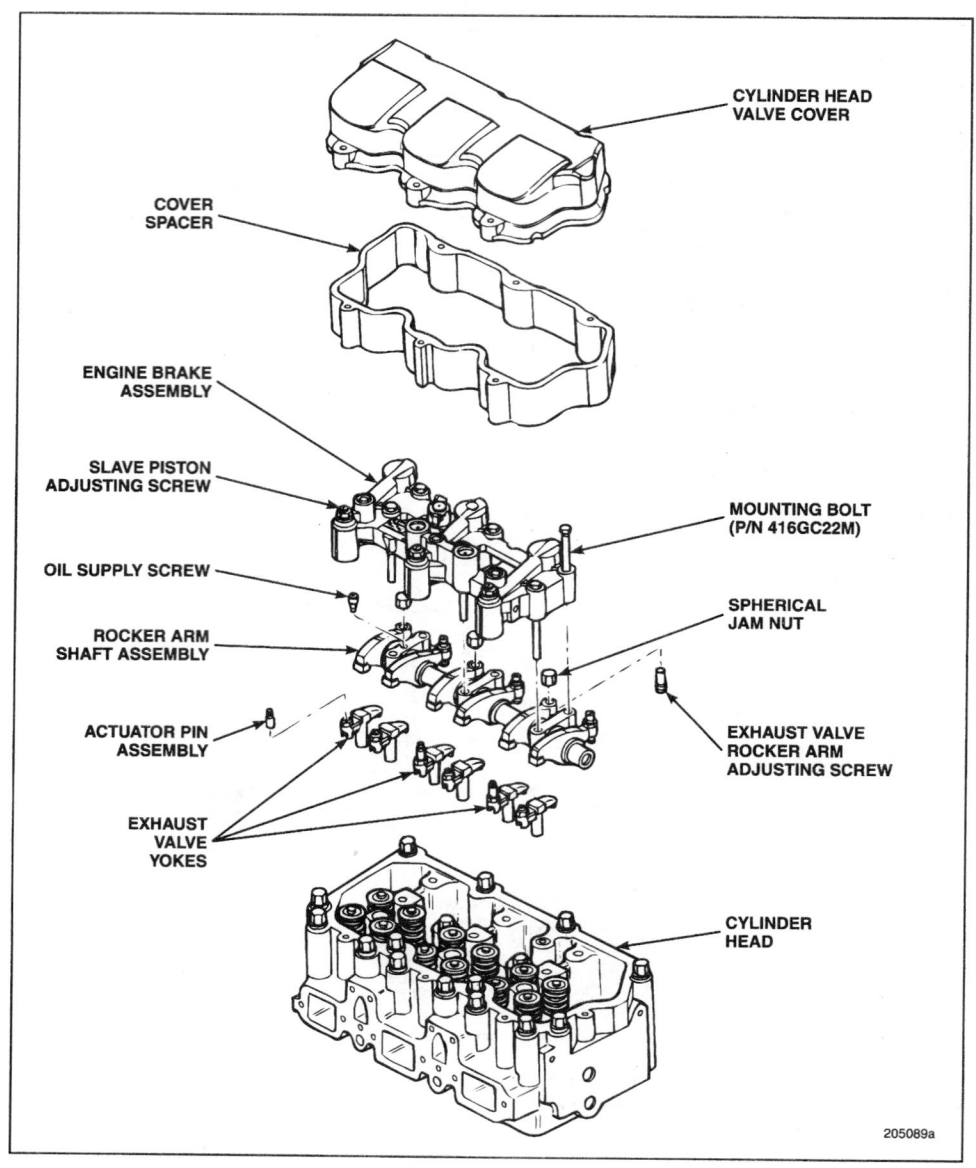

Figure 7-35 Jacobs brake on a Mack E-Tech engine. (Courtesy of Mack Trucks, Inc.)

rocker. When the injector rocker travels upward (actuated by cam profile), the oil is trapped in the circuit by the slave control piston and the oil pressure is driven upwards and acts on the slave piston located over the exhaust valves. The slave piston acts on the exhaust valve bridges opening them and causing gas blowdown to the exhaust manifold.

A version of the Jacobs brake is available for most North American built engines, and the principles used to actuate each are similar on each. For instance, on engines using mechanical unit injectors (MUIs) and electronic unit injectors (EUIs), both of which have cam-actuated pumping strokes, the Jacobs brake hydraulic timing and actuation is almost identical to that described for the Cummins PT engine. On Caterpillar and Mack engines using hydraulic injector nozzles, valve train movement must trigger the Jacobs brake hydraulic timing/actuation over another cylinder under the same cylinder head.

Though they may mistakenly be referred to as Jake brakes, there are a number of other internal engine compression brakes available on engines marketed in North America, most of them

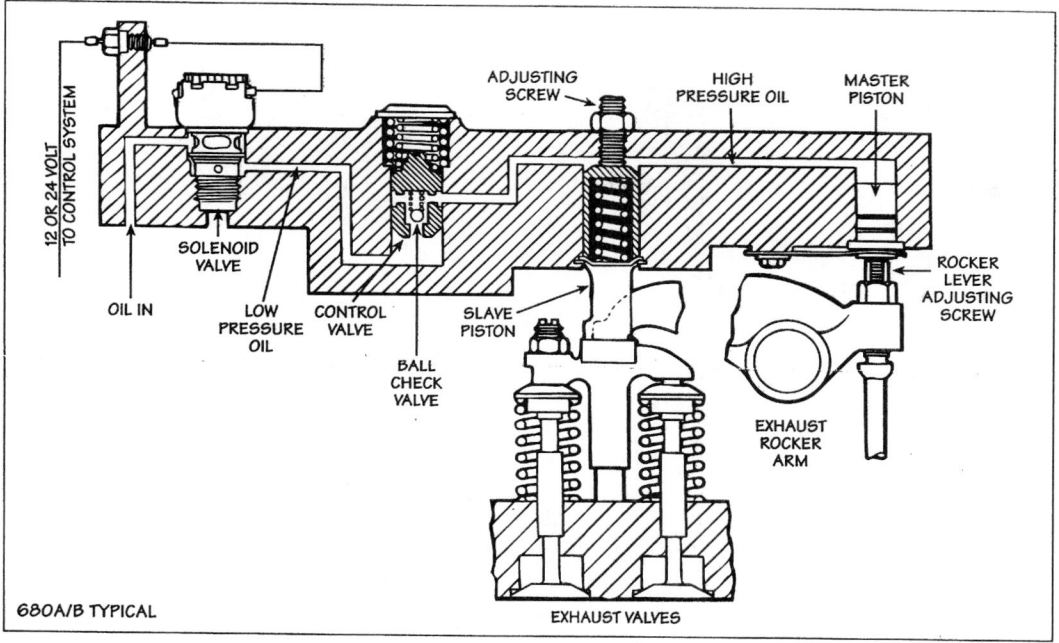

680A/B TYPICAL   EXHAUST VALVES

The blowdown of compressed air to atmospheric pressure prevents the return of energy to the engine piston on the expansion stroke, the effect being a net energy loss since the work done in compressing the cylinder charge is not returned during the expansion process.

Exhaust blowdown of the braking cylinder is accomplished by utilizing the pushrod motion of an exhaust valve of another cylinder during its normal exhaust cycle as follows:

1. Energizing the solenoid valve permits engine lube oil to flow under pressure through the control valve to both the master piston and the slave piston.

2. Oil pressure causes the master piston to move down, coming to rest on the corresponding exhaust rocker arm adjusting screw.

3. The exhaust rocker pushrod begins upward travel (as in normal exhaust cycle), forcing the master piston upward and directing high pressure oil to the slave piston of the braking cylinder. The ball check in the control valve traps high pressure oil in the master/slave piston circuit.

4. The slave piston (under the influence of the high pressure oil) moves down, momentarily opening the exhaust valves while the engine piston is near its top dead center position, releasing compressed cylinder air to the exhaust manifold.

5. Compressed air escapes to the atmosphere, completing a compression braking cycle.

The level of engine braking is controlled by using the solenoid to turn each housing ON or OFF. Figure 4 shows the relationships between master pistons, slave pistons and control valves within the housing.

Figure 7-36 Jacobs brake system schematic. (Courtesy of Jacobs Vehicle Systems™)

manufactured offshore. The Cummins C Brake was manufactured for a short time and was designed exclusively for Cummins engines. Cummins has since reverted to using Jacobs products once again, and the brakes known as C Brakes are currently manufactured by Jacobs. Volvo also manufactures their own compression brake for their VDE-12 engine. Pacific Diesel Brake (PACBRAKE) manufactures engine brakes for some contemporary engines, and Mercedes Benz also make a compression brake available on their medium-bore diesel engines.

Mack Dynatard

Mack Trucks used the Dynatard brake (Figure 7-37), until 1966, when the company decided to discontinue its manufacture and adopt engine brakes manufactured by Jacobs. The Dynatard is classified as an internal engine compression brake. It is important that the truck diesel technician understand something about the Dynatard brake, because all Mack engines have an engine brake camshaft whether they are equipped with an engine brake or not. This can lead to some confusion

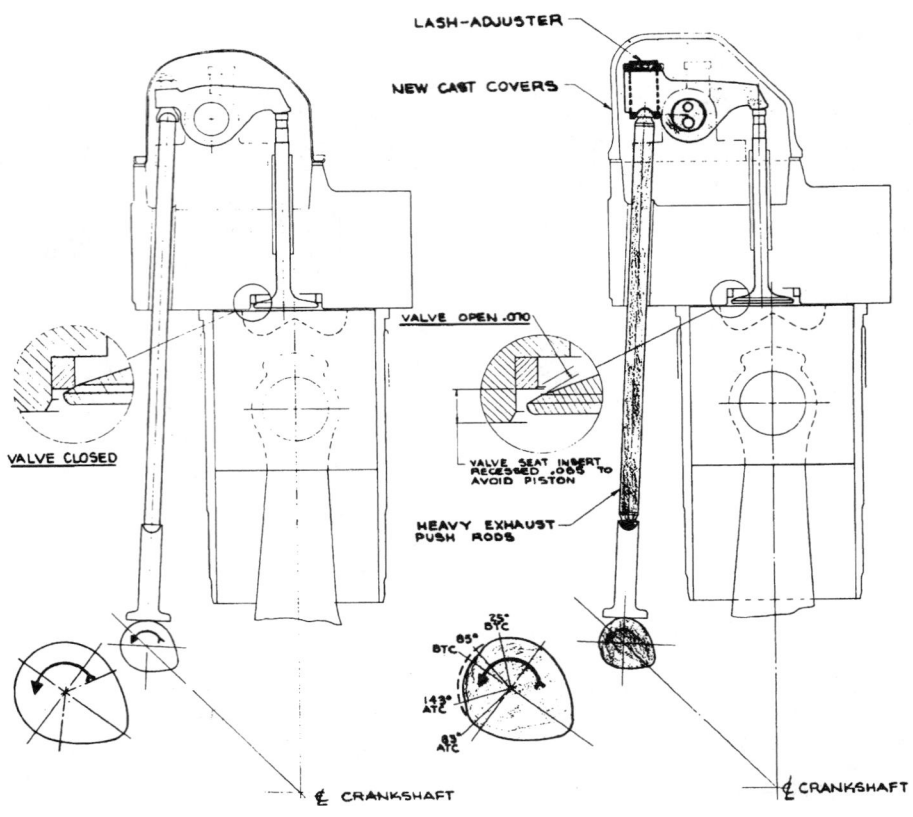

Figure 7-37 Section view of Mack Dynatard compression brake. (Courtesy of Mack Trucks, Inc.)

when adjusting the valves. Although at first glance the exhaust cam profile of the Mack engine camshaft appears to be of the inner base circle design, it is machined with a ramp between the base circle and the nose. This ramp is critical in the operation of the Dynatard brake.

The operating principle is that of all internal engine brakes: under braking, the piston compresses the air charge in the cylinder; just prior to the point at which the power stroke normally begins, the exhaust valves are opened for exhaust blowdown. The electrical control circuit of the Dynatard is almost identical to that of the Jacobs compression brake. It consists of a control switch, clutch switch, and a governor switch that is closed when the fuel pump rack is in its no fuel position. When the three electric control switches are closed, the Dynatard solenoid (Figure 7-38A) located on the rocker shaft, is energized. The solenoid armature is integral with a valve (Figure 7-38B) and, when energized, the engine oil that is normally circulated through the entire rocker assembly is rerouted to control hydraulic tappets (hydraulic lash adjusters) located at the push-tube end of each exhaust rocker.

The Dynatard tappets are continually charged with engine oil when the engine is running, and within the tappet, the oil pressure acts on both sides of the control piston. Under these conditions, the lower spring of the tappet has sufficient force to hold the control piston (top piston), thereby unseating the ball valve. This permits oil to flow in and out of the socket piston (lower piston) chamber and allows the socket piston to collapse when subjected to force by the exhaust push tube. When Dynatard is activated, the solenoid cuts the oil supply to the upper chamber of the Dynatard tappet and permits the oil to go below the control piston. To overcome the upper spring forcing the ball valve to seat: as the ball valve is seating, the lower spring forces the socket piston downward, filling the chamber behind it with oil, which is trapped the moment the ball seats. This forms a hydraulic lock and expands the tappet. The net result is that the exhaust valve train is effectively lengthened when the exhaust follower contacts the Dynatard ramp between

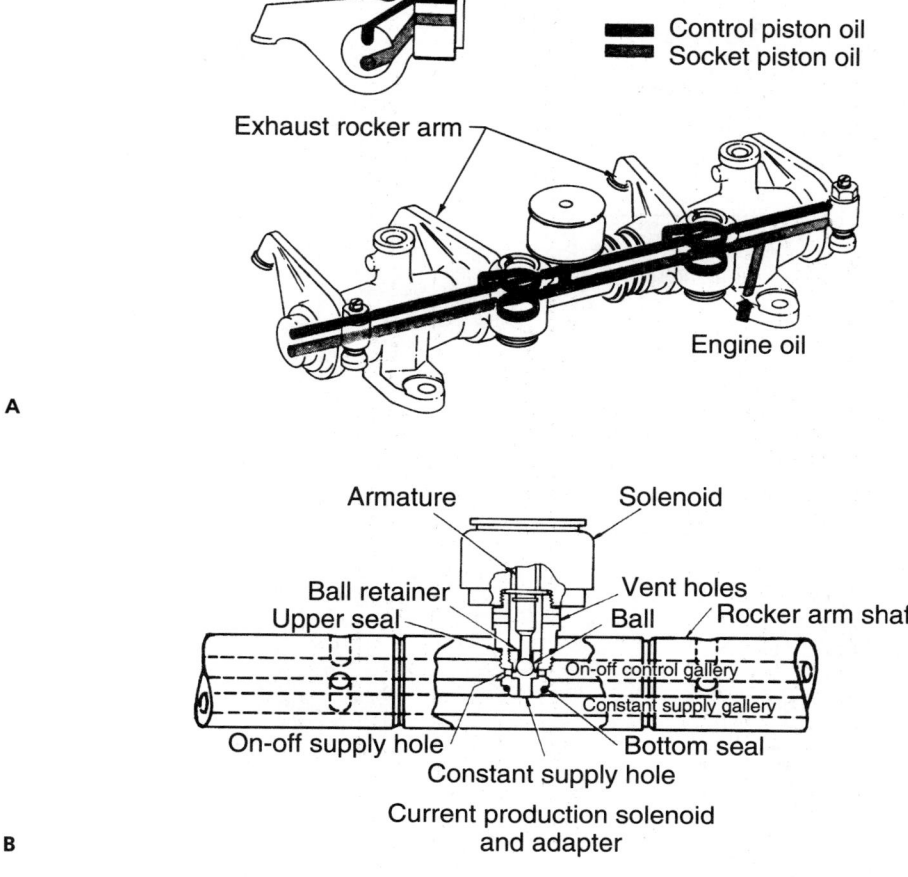

Control piston oil
Socket piston oil

Exhaust rocker arm

Engine oil

A

Armature   Solenoid

Ball retainer   Vent holes
Upper seal   Ball   Rocker arm shaft
On-off control gallery
Constant supply gallery
On-off supply hole
Constant supply hole   Bottom seal

Current production solenoid
and adapter

B

Figure 7-38 (A) Dynatard valve rocker assembly oil circuit; (B) Dynatard control valve and solenoid. (Courtesy of Mack Trucks, Inc.)

the base circle and the normal valve actuating ramp on the cam profile. The exhaust train is actuated just enough to open the exhaust valves at the moment where the exhaust gas blowdown is required.

In Dynatard-equipped engines, the exhaust valves are recessed in their seat in the cylinder head. This prevents the valves contacting the piston when Dynatard is actuated. The plumbing in the rocker shaft provides the constant oil supply to the socket pistons of the Dynatard tappets through the lower passage; the interrupt circuit supplying the control pistons is the upper passage.

Caterpillar BrakeSaver

The following description is of the Caterpillar BrakeSaver (Figure 7-39) on a hydromechanical 3406 engine. This engine-mounted hydraulic retarder system also lends itself to electronic controls with minor changes. The BrakeSaver rotor (Figure 7-40) is coupled directly to the rear of the crankshaft and driven within the BrakeSaver housing. When the BrakeSaver is not applied, the housing is not charged with pressurized engine oil, and the rotor can turn unimpeded in the housing. When the BrakeSaver is applied, the housing is charged with pressurized engine oil, and the rotor's ability to turn is moderated by fluid resistance and the pocket or vane geometry; this produces the retarding effect. The BrakeSaver offers modulated retarding. The control circuit is electrically or manually controlled and pneumatically actuated. Chassis system air pressure is

**Shop
Manual
Chapter 7, page 238.**

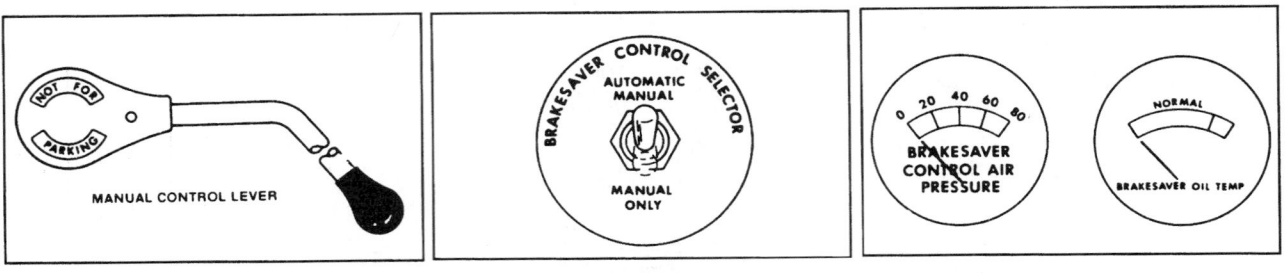

Figure 7-39 Caterpillar BrakeSaver controls. (Courtesy of Caterpillar Inc.)

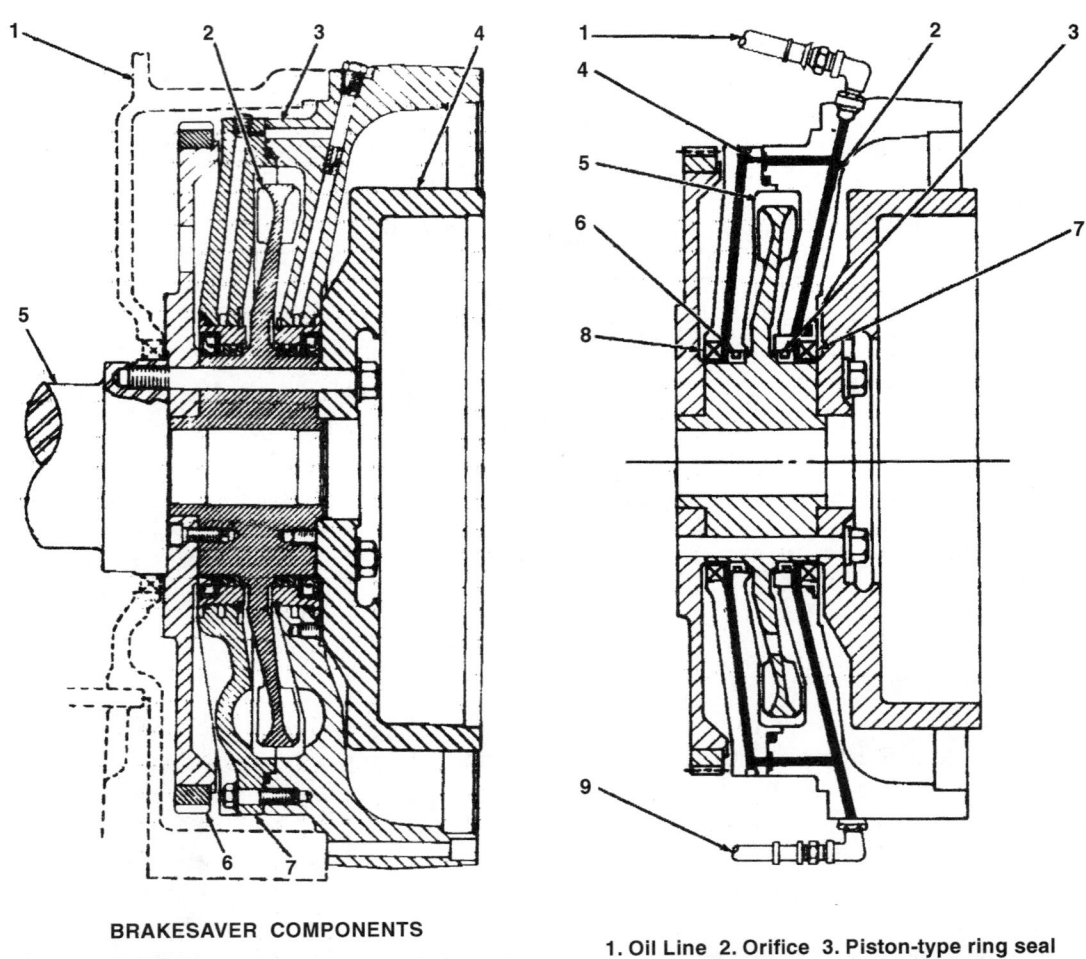

BRAKESAVER COMPONENTS

1. Flywheel housing 2. Rotor 3. BrakeSaver housing
4. Flywheel 5. Crankshaft flange 6. Ring gear plate
7. Stator

1. Oil Line 2. Orifice 3. Piston-type ring seal
4. Orifice 5. Chamber 6. Piston ring seal
7. Lip-type seal 8. Lip-type seal 9. Oil line

Figure 7-40 BrakeSaver, sectional view. (Courtesy of Caterpillar Inc.)

reduced by a pressure-reducing valve, which drops the pressure to a maximum value of 50 psi (345 kPa). This feeds the manual control valve and the solenoid control valve. Both the automatic solenoid valve and the manual control valve use air to control the oil control valve mounted on the flywheel. The function of the oil control valve is to meter the flow of oil to the BrakeSaver housing.

Caterpillar Control Valves The automatic control circuit requires an electrical circuit consisting of the ignition switch, mode selection switch, clutch switch, and accelerator switch. When the automatic option is toggled on the mode selection switch (manual/automatic), the BrakeSaver is actuated whenever the accelerator is released and the clutch and accelerator switches are closed. The automatic control circuit is on/off; that is, modulating the retarding effort of the BrakeSaver is not possible. The manual control valve, which is usually mounted on the steering column, meters the air delivered to the oil control valve. This controls the flow of engine oil to the BrakeSaver housing (and thus the braking effort). A pneumatic, two-way check valve prioritized delivery of the actuation signal at the highest pressure to the oil control valve.

Oil Flow Engine oil is the hydraulic medium used by the BrakeSaver. It is supplied to the oil control valve by the engine oil pump, which is designed with a front and a rear section. The front section functions to feed the engine lubrication circuit, while the rear section exclusively feeds the BrakeSaver oil control valve. BrakeSaver equipped engines require large capacity oil sumps. When pneumatically actuated by either the automatic or manual control valve, the oil control valve routes oil to the BrakeSaver housing at a high rate of flow and at a pressure value of ±70 psi (480 kPa). Because of the pumping action within the BrakeSaver, pressures at the outlet increase 50% over the charge pressure. When the BrakeSaver is not actuated, a small amount of oil is pumped through the housing to lubricate the seals. Oil exiting the housing is routed directly to the oil cooler.

Summary

❏ Most current trucks use a dry, positive filter system. Trucks operated in environments with airborne particulates such as grain chaff dust or construction and road dust should use some kind of precleaner to extend the service life of the air cleaner element. Dry, positive-type filters function at optimal efficiency just before their service life is completed. It is important not to overservice air filters, because every time the canister is opened, some dust will find its way downstream from the filter assembly. Many filters will last for up to a year in a line-haul application.

❏ Charge air/heat exchangers cool the turboboosted air charge and therefore increase its density. The result is more oxygen molecules in the intake charge delivered to the cylinder. Air-to-air charges air coolers boast higher cooling efficiencies than the liquid medium coolers. Yet they must have adequate airflow, and this makes them ideal for use in highway applications but not in high load, low road speed applications.

❏ Engine valve configuration affects both the cylinder breathing efficiency and the cylinder gas dynamics. Assuming identical lift, valve seats that are cut at a 45 degree angle produce higher flow restriction and higher seating force than valves cut at a 30 degree angle.

❏ Turbochargers represent an exhaust backpressure factor, but their objective is to recapture some of the engine-rejected heat by using this heat to pressurize the intake charge to the cylinders. The heat in the exhaust gas drives turbochargers; so the more heat, the faster the turbine speeds. Turbochargers in truck diesel engine applications may wind out at up to 130,000 rpm. Constant geometry turbochargers are more commonly used in truck diesel engine applications but some OEMs are using variable geometry turbochargers to increase the efficient operating range and reduce the duration of turbo lag.

❏ Engine silencers use resonation and sound absorption principles to alter the frequency of the sound emitted from the engine; most use a combination of both principles. The exhaust silencer volume on truck diesel engines is typically five times the total cylinder exhaust slug volume through a complete cycle.

Terms to Know

BrakeSaver

Catalyst

Catalytic converter

Charge-air cooling (CAC)

Diffuser

Exhaust brake

External compression brake

Heat exchanger

Impeller

Inlet restriction gauge

Internal compression brake

Jacket water aftercoolers (JWAC)

Jake brake

Kinetic energy

Oxidation catalyst

Positive filtration

Rejected heat

Retarder

Roots blower

Scavenge

Turbine

Turbocharger

❏ A catalyst is a substance that enables a chemical reaction without itself undergoing any change. A catalytic converter found on a truck diesel engine is of the single stage, oxidizing type. Catalytic converters used on diesel engines are vulnerable to plugging up when an engine is operated for prolonged periods under light or idle loads. An opacity meter (or opacimeter) is a light extinction tester that operates by aiming a light beam at a sensor that records the amount of light sensed. Jurisdictions that enforce smoking limits use the exhaust smoke opacity limits defined by the EPA.

❏ Some medium duty diesel engines use an electric air heater grid, glow plugs, or an ether starting system to aid in starting a cold engine.

❏ Engine brakes are designed to complement the main vehicle brake system, not replace it. An engine compression brake's ability to absorb power is low compared to the brake capacity of the vehicle. However, a large percentage of vehicle braking uses less than one-fifth of the total vehicle capacity, so engine brakes can greatly extend the life of the vehicle foundation brakes.

❏ An electrical circuit that requires a series of closed switches controls most engine brakes. At minimum, the control switch, a clutch switch, and an accelerator or governor switch must be closed. Internal compression brakes use the electrical control circuit to manage the brake's hydraulic circuit.

❏ The operating principle of an internal engine compression brake is to make the piston perform the work of compressing the cylinder air charge and then negate or cancel the power stroke by releasing the cylinder charge by opening the exhaust valves at TDC on the compression stroke. This changes the engine's role from that of a power-producing pump to that of a power-absorbing pump. The mechanical force used to time and actuate the hydraulic circuit of a Jacobs type internal engine compression brake is rocker movement. This may be the movement of the PT, MUI or EUI rocker, or in cases where hydraulic injectors are used, movement of an exhaust valves rocker over a different cylinder.

❏ The engine exhaust brake operates by choking down the exhaust flow. The retarding stroke of the piston is therefore the exhaust stroke. Exhaust brakes are managed electrically and actuated pneumatically.

❏ The BrakeSaver uses the principle of a torque converter in reverse to absorb energy and retard vehicle speed. It uses engine oil as its retarding medium. The engine oil is provided to the hydraulic brake from a larger capacity oil sump by a dedicated section of the oil pump. All engine retarders operate at optimal efficiency when engine rotational speeds are highest.

Review Questions

Short Answer Essays

1. Identify the intake and exhaust system components that comprise the engine breathing circuit.

2. Describe how intake air is routed to the engine cylinders and exhaust gases are routed to the tailpipe.

3. Define positive filtration.

4. Outline the principle of operation of a Roots blower and its function on a two-stoke cycle engine.

5. Identify the main subcomponents on a truck diesel engine turbocharger and the role they perform.

6. Outline the operating principles of an exhaust-gas–driven centrifugal turbocharger.

7. Define the role of a charge air cooler and the relative efficiencies of each type.

8. Describe the operating principles of a truck catalytic converter.

9. Identify the three types of engine brakes used in trucks.

10. Describe the operating principles of each type of engine brake and the relative advantages and disadvantages of each.

Fill-In-the-Blanks

1. Trucks operated in environments with airborne particulates such as grain chaff dust, or construction and road dust should use some kind of _____ to extend the air cleaner element service life.

2. Charge air heat exchangers cool the _____ _____ air charge and therefore increase its _____. The result is more oxygen molecules in the intake charge delivered to the cylinder.

3. Air-to-air charge air coolers boast higher cooling efficiencies than the _____ _____ _____. Yet they must have adequate airflow. This makes them ideal for use in highway applications but not in high load, low speed road speed applications.

4. Engine valve configuration affects both the cylinder _____ _____ and the cylinder gas dynamics. Parallel port valve configurations generally produce better and more balanced cylinder breathing efficiency but also produce lower swirl cylinder dynamics.

5. _____ represent an exhaust backpressure factor, but their objective is to recapture some of the engine-rejected heat by using and pressurizing the intake charge to the cylinders.

6. Sound is transmitted in sonic waves producing _____ and _____. An exhaust silencer operates by _____ (altering the frequency) these nodes and antinodes and thereby altering its nature.

7. Engine silencers use _____ and _____ _____ principles to alter the frequency of the sound emitted from the engine. Most use a combination of both principles. The exhaust silencer volume on truck diesel engines is typically _____ times the total cylinder exhaust slug volume through a complete cycle.

8. Some medium duty diesel engines use an electrical air heater grid, glow plugs, or an ether starting system to aid in _____ a cold engine.

9. The mechanical force used to time and actuate the hydraulic circuit of a Jacobs type _____ _____ _____ brake is rocker movement.

10. The engine exhaust brake operates by _____ down the exhaust flow. The retarding stroke of the piston is therefore the exhaust stroke.

ASE-Style Review Questions

1. Which of the following types of filter have the highest filtering efficiencies?
   A. Centrifugal precleaners
   B. Oil bath
   C. Dry, positive
   D. 5-micron filters

2. Which of the following best describes the catalytic converter fitted to a diesel engine?
   A. Two-stage, three-way
   B. Single-stage, oxidizing
   C. Single-stage, reduction
   D. Palladium oxidation catalyst

3. Which of the following components plays the largest role in minimizing NO_x emission in a highway diesel engine?
   A. Catalytic converter
   B. Air-to-air charge air cooler
   C. Particulate trap
   D. Turbocharger

4. In current electronically managed highway diesel engines, which of the following is a primary reference to the ECM in determining the air-fuel ratio?
   A. Pyrometer
   B. Intake manifold thermistor
   C. Inlet restriction sensor
   D. Engine speed sensor

5. Which of the following correctly describes the opacimeter used to measure exhaust smoke emission?
   A. Light extinction tester
   B. Filtration tester
   C. Constant volume sampling tester
   D. Flame ionization tester

6. Removing an engine coolant thermostat would likely produce an increase in which of the following noxious emission levels?
   A. NO_x
   B. HC
   C. Sulfur dioxide
   D. Carbon dioxide

7. The constant geometry turbocharger used with a high torque rise highway diesel engine is usually designed to produce optimal efficiency at which rpm?
   A. Rated
   B. Top engine limit (TEL)
   C. Peak torque
   D. High idle

8. Muffler volume in a truck diesel engine would typically be equivalent to:
   A. Engine displacement volume
   B. Five times the total exhaust slug volume
   C. Air cleaner volume
   D. Ten times the total exhaust slug volume

9. Which type of engine brake uses the compression stroke of the piston as its retarding stroke?
   A. Internal compression brake
   B. External compression brake
   C. Hydraulic retarder
   D. Electrical retarder

10. Which type of engine brake uses the exhaust stroke of the piston as its retarding stroke?
    A. Internal compression brake
    B. External compression brake
    C. Hydraulic retarder
    D. Electrical retarder

Diesel Fuel, Fuel Subsystems, and Injector Nozzles

Upon completion and review of this chapter, you should be able to:

❏ Explain the following fuel terms: specific gravity, viscosity, and cetane rating; determine how they affect diesel engine driveability diagnosis.

❏ Define cloud point and pour point, and state their importance in cold weather diesel engine operation.

❏ Define the role of primary and secondary fuel filters, and explain how a water separator works.

❏ Define the principles of operation of a transfer or charge pump.

❏ Differentiate between the hydromechanical injection operating principles of timing, pressurization metering, atomization, and distribution.

❏ Explain atomization and the droplet sizing factors required for a direct-injected diesel engine.

❏ Identify three types of hydraulic injector nozzles and the primary subcomponents of a nozzle assembly.

❏ Describe the hydraulic principles of operation of poppet, pintle, and multi-orifice nozzles.

Introduction

It is important, if not essential, that the diesel engine technician have a fundamental understanding of fuel chemistry and combustion. This knowledge can prove to be a useful tool when troubleshooting engine malfunctions. This chapter contains information concerning diesel fuel and fuel subsystems, including fuel tanks, lift pumps, filters, and lines. The fuel subsystem is best defined as the group of components responsible for fuel storage and its transfer to the injection pumping apparatus. While injection-pumping mechanisms differ greatly from manufacturer to manufacturer, the fuel subsystems that supply them tend to have much in common. Fuel tanks, filters, water separators, transfer pumps, fuel heaters, and their interconnecting plumbing will be examined. The chapter will also cover in detail the operation of mechanical fuel injection systems including pump/line/nozzle (PLN) systems and unit injector/governor systems.

A BIT OF HISTORY

The relatively low cost of diesel fuel compared to gasoline spurred the adoption of the diesel engine in the automotive industry. Because of their superior efficiency and improved operating costs, diesel engines far outnumbered gasoline engines by 1939 in heavy commercial road vehicles. This progress continued after World War II into the lighter truck applications (7 tons and below) and to the expanding tractor and light generator set markets.

Petroleum Formation

The elemental components of petroleum, carbon and hydrogen, originated in the organic materials that made up the primordial Earth. Single-celled aquatic plants and algae were abundant 570 million years ago. Rapid burial of these organisms preserved them in sedimentary rock.

Through a series of biological, physical, and chemical changes, they eventually became what we know as petroleum. The formation of petroleum requires heat, pressure, and the passage of time. Petroleum is a fossil fuel, an unrenewable energy source with limited reserves. It has never been replicated in a commercial laboratory.

The hydrocarbons in crude petroleum are classified into four groups: paraffins, olefins, naphthalenes, and aromatics. Crude oils are loosely classified by their content, which will predicate exactly what can be extracted from them in refining processes. There are three types: asphalt, paraffin, and mixed.

The area of the Earth's crust in which a crude oil is formed plays a role in defining its properties. A minimum temperature of 120° F (50° C) is required to allow the natural formation of crude petroleum. Oils formed at lower temperatures tend to be heavier and located in the upper strata of the oil window. The oil window is the area of the Earth's crust in which petroleum oils can be formed. It extends from a depth of 1,500 m (5,000 ft.) to 6,500 m (20,000 ft.). Oil formed in the deeper portion of the oil window at temperatures up to 35° F (175° C) tends to be lighter and commercially more valuable. Petroleum located at depths below the oil window is usually in the form of natural gas.

Oil may also be obtained from oil shales and oil sands. In these, organic matter has been reduced to kerogen, an intermediate stage between the organic matter and oil. Oil shale and oil sands originate from the fine-grained sand deposits below ancient lakes and oceans. The kerogens in oil shales and sands can be processed to produce around 100 L (27 gal.) per ton when heated to boiling point temperatures. Both Canada and the United States have large reserves of oil shale and sands. However, because of the energy required to process them to petroleum oils, at this moment in time it has proven more economical to import petroleum shortfalls.

Refining Petroleum

As shown in Figure 8-1, the refining of crude petroleum begins with two processes; vaporization and condensation. A crude oil is made up of many different hydrocarbon compounds, each with different characteristics. Each will have distinct boil and condensation temperatures permitting relatively easy separation. These are known as *fractions*. Boiled off vapors are passed through a distilling column.

The distilling column is a vertical cylinder containing a series of stacked vertical trays. Rising vapors bubble upward through holes in the trays and through any liquid that has already condensed. As these vapors rise they cool, condensing on the trays. Condensed liquid is permitted to spill from the trays to external containers. Each batch of liquid drawn off from the distillation column is called a cut or fraction.

The fractions that are drawn off at high temperatures are called *heavy fractions*, and those at low temperatures, *light fractions*. Straight-run distillation will therefore grade the fractions extracted from crude petroleum by cut points or boil temperature. The gas taken off the top of the distilling column (or tower) is natural gas, and in descending order of boil temperature, gasoline, naphtha, kerosene, light gas oil, and heavy gas oil. The liquid remaining at the bottom is known as residuum. After removal from the column, the cuts are purified.

Thermal and Catalytic Cracking

Most highway fuels are a brew of fractions carefully blended by the refiners. Racing fuels (gasoline type) are highly complex petroleum brews made from formulas that are sometimes kept secret from competitors. The gasoline and diesel fuel purchased at the pumps is a blend of fractions with a range of volatility designed to produce balanced combustion. The higher, or lighter, fractions predominate in gasoline; although a diesel fuel possesses some lighter fractions it will also contain heavier fractions with higher lubricity and calorific value than those found in a gasoline will. The lighter fractions are more highly valued than the heavier ones.

A problem inherent in the distilling of crude petroleum is that it produces too few of the lighter, more desirable fractions. Cracking is a process by which heavier fuel oils are chemically modified by dividing its heavy molecules into smaller light molecules. Two types of cracking are used. *Thermal cracking* involves subjecting crude petroleum to high temperatures and pressures in a cylindrical tower. *Catalytic cracking* is a more efficient method of accomplishing the same

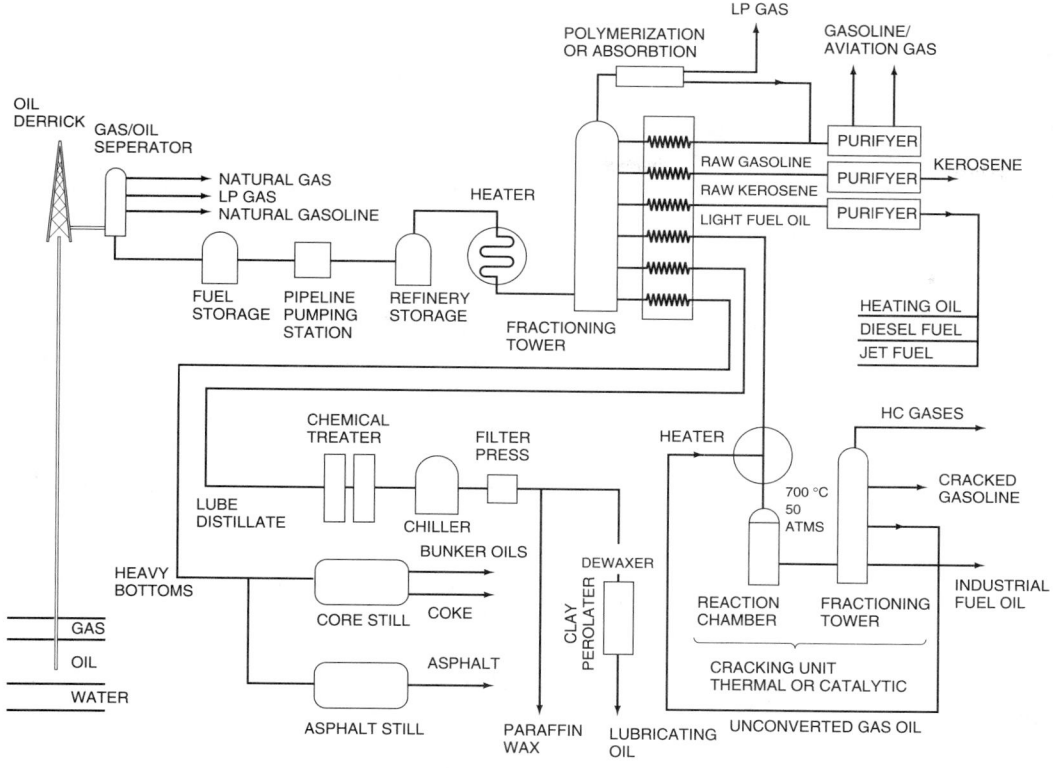

Figure 8-1 The petroleum refining process.

objective. A cracking tower is a reaction chamber, which is pumped full of cracking stock (low-grade crude) and catalysts. The contents of the cracking tower are then subjected to pressure and heat. The cracking products are drawn off from the reaction chamber and separated in a fractioner by density. The fractioner residuum (what is left in the catalytic cracker after the cracking process) is called cycle oil. This is sometimes run through a catalytic cracker again.

Hydrocracking (Figure 8-2) is a catalytic cracking process undertaken in the presence of hydrogen. It is used to produce the higher volatility, lighter fractions required in gasoline and light

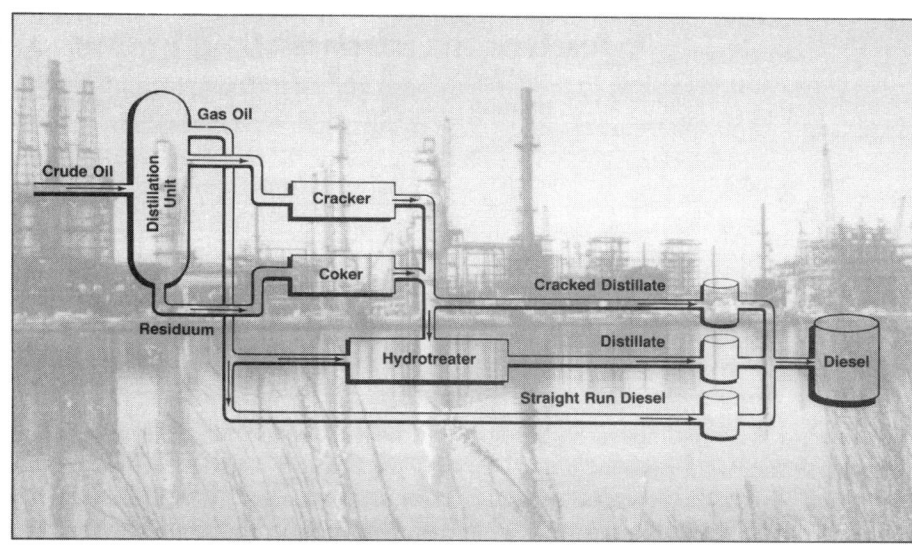

Figure 8-2 Diesel fuel blending. (Reprinted with permission from Chevron Products Company)

distillates such as diesel fuel. Before fuels are blended, their composite fractions must be purified. This generally refers to the removal of salt, sulfur, and water impurities.

Fuel Terminology

AFR   The air-to-fuel ratio is the actual ratio of air to fuel in a combustion reaction. AFR is distinct from stoichiometric, which describes the proportions of reactants required to complete the combustion reaction. It is normally expressed by mass (weight).

Ash   Diesel fuels normally contain a certain quantity of suspended solids or soluble metallic compounds, such as sodium and vanadium. The fuel's ash content can affect injector, fuel pump, and any engine components subjected to high temperatures, such as piston rings, exhaust valves, and turbochargers.

ASTM   The American Society for Testing and Materials is the organization that classifies diesel fuel (and other fuels) to a standard.

Boil Point   The temperature at which a liquid vaporizes is the boil point. When applied to liquid hydrocarbon fuels, it becomes a measure of volatility.

Calorific value   The potential heat energy of a fuel is measured in BTUs (British thermal units), joules or calories.

Catalyst   A catalyst is a substance that enables a chemical reaction without itself undergoing any change.

Cetane Number (CN)   The **cetane number** is a measure of the ignition quality of a diesel fuel (this is defined in some detail later in this section).

Cloud Point   The **cloud point** is the temperature at which the normal paraffin in a fuel becomes less soluble and begins to precipitate as wax crystals and make the fuel appear cloudy. Cloud point exceeds the pour point by 5° F (3° C) to 25° F (15° C).

CNG   Compressed natural gas (*see* "Natural gas").

Crude Oil   This is raw petroleum and consists of a mixture of many kinds of hydrocarbon compounds of differing molecular weights and small quantities of organic compounds such as sulfur. Crude oil is distilled and cracked in the refining process to produce residual oils, distillates, and fractions that are blended to manufacture fuel, oil, and tar.

Diesel Fuel   The text will use this term to describe distillate petroleum compounds and fractions formulated for use in highway compression-ignition (CI) engines. They are generally composed of fractions from the paraffin (the most volatile range used in a diesel fuel), naphthalene, and aromatic (the least volatile range found in a diesel fuel) series of crude oil fractions and are graded by the ASTM. Highway vehicles use diesel fuel that has a distillation range between 300° F (150° C) and 550° F (290° C) and specific gravity values ranging from 0.78 to 0.86.

Distillate   This term is sometimes used to describe diesel fuel formulated for on-highway use. The term diesel fuel is preferred.

Fire Point   The fire point is temperature at which a liquid hydrocarbon fuel evaporates sufficient flammable vapor to burn continuously in air. Fire point generally exceeds flashpoint by about 10° C in hydrocarbon fuels.

Flame Front   The flame front is the forward boundary of the reacting zone in the cylinder combustion.

Flame Propagation   Flame propagation is the way in which a fuel combusts inside the engine cylinder as determined by the manner the flame front spreads. It is dependant on cylinder gas dynamics, the actual AFR, temperature, and fuel chemistry.

Flashpoint   The flashpoint is the temperature at which a liquid hydrocarbon fuel evaporates sufficient flammable vapor to *momentarily* ignite in the presence of a flame. Flashpoint does not affect engine performance. It is specified because it relates to the hazards in handling and storing fuels. Number 2D diesel fuel has a flashpoint of 125° F; gasoline's flashpoint is −40° F.

Fractions   A fraction is a portion of a mixture separated by distillation or a cracking procedure, such as hydrocracking or catalytic cracking. Most fuels are carefully balanced brews of combustible petroleum fractions with a range of volatility. Each fraction possesses distinct characteristics.

Gasoline   Gasolines are the group of liquid petroleum fuels blended for use in spark-ignition (SI) engines. The actual composition of gasoline varies according the crude oil source, refining processes, and blend requirements. Typically, gasoline has a volatility range that extends from 90° F (35° C) to 400° F (210° C) and a specific gravity that ranges from 0.70 to 0.78.

Heat Value   The heating of combustion of a fuel expressed in British thermal units (Btu) per pound or per gallon is a measure of the amount of energy available to produce work. One Btu is the amount of heat necessary to raise the temperature of 1 lb. of water 1° F. In general, a diesel fuel having a higher volumetric heating value (Btu gal.) will produce more power or provide better fuel economy than a fuel of lower heating value. An important property of a diesel fuel is the heat energy it releases during combustion. This value is used to reckon the thermal efficiency of an engine's ability to produce power. The gross heat of combustion at a constant volume is determined by a *bomb calorimeter test*, which measures the amount of heat released by burning a known quantity of fuel as outlined by ASTM standard D240. Heating values may be expressed as:

> Joules per kilogram (J/kg)
> Joules per liter (J/L)
> Calories per gram (cal/g)
> Calories per liter (cal/L)
> BTUs per pound (Btu/lb.)
> BTUs per gallon (Btu/gal.)

Ignition Point (Auto Ignition)   This term applies to the event in which a combustible mixture under certain conditions of density, pressure, and temperature ignites without a flame or spark. For diesel fuel, the ignition point occurs at approximately 558° F (292° C).

Kerosene   This petroleum product is made up of heavier fractions than gasoline, and it is widely used in heating oil and jet fuel. Kerosene typically has a distillation range between 300° F (150° C) and 510° F (270° C) and specific gravity values ranging from 0.78 to 0.85.

Lubricity   This is literally the oiliness of a substance. This term denotes a fuel's ability to lubricate the moving parts of the mechanical fuel injection pump, nozzles, and unit injectors.

Microorganism Growth   Airborne bacteria and fungi commonly enter vehicle and storage tanks through the venting systems. When water is present in the bottom of the tank, bacteria may reside in it and feed off the fuel hydrocarbons. The metabolic waste from such microorganisms is acidic and may corrode fuel injection components. Fungal growth can plug fuel filters when pumped through the fuel subsystem. It is good practice to fill truck on-board tanks before parking the vehicle to minimize water condensation problems; draining tanks daily also helps. However, when the problem is at the storage tank, chemical treatment of the fuel is required.

Moisture   Most diesel fuels contain some moisture, either dissolved in the fuel as a liquid or suspended (dispersed) in the fuel as a liquid or as solid ice crystals. Such small amounts of moisture pass harmlessly through the fuel injection system. The amount is about 1 gal. in 10,000 miles. Yet, large amount of water cannot be suspended in the fuel and will settle out in a second layer. Free and dispersed moisture does not have to be solid. Liquid water can be freely dispersed in the diesel fuel, especially if the fuel contains a dispersent additive or emulsifier.

Natural Gas   Natural gas is the gaseous product of petroleum that is either suspended above liquid crude oil or dissolved in it, in which case it becomes the first product to be separated in the distillation process. Natural gas is composed primarily of methane (CH_4) with a lesser amount of ethane (C_2H_6), propane (C_3H_8), and butane (C_4H_{10}). Propane and butane are extracted from natural gas and stored as liquids under pressure. Both are used as an automotive fuel and are usually known as liquefied petroleum gas, or LPG.

Octane Rating   Octane rating is a measure of the antiknock quality of a fuel, usually a gasoline. *Knock* in an engine depends on complex combustion reaction phenomena and on the engine design. However, as a method of classifying gasoline, the American Society of Testing and Materials has standardized two methods, the motor method and the research method. Both relate the antiknock performance of a test gasoline to an actual fuel. Two primary reference fuels are used: isooctane, with ideal anti-knock characteristics and assigned an octane number of 100 and heptane, with poor antiknock characteristics and assigned an octane number of 0. A mixture of these two primary reference, pure hydrocarbon (HC) fuels is then used as the means for grading the actual performance of a gasoline. A mixture of 90% isooctane and 10% heptane would theoretically have identical antiknock characteristics as a gasoline sold at the pumps as 90 octane.

The actual test conditions used in the motor and research methods differ. The motor method testing operates at higher speed and inlet mixture temperatures than the research method. The research method is generally the better indicator of fuel antiknock quality for engines operating at full throttle and low speeds; the motor method is the better indicator at full throttle and high speeds. Federal regulations in the U.S. and Canada require the posting of the average of the research octane number (RON) and the motor octane number (MON) at dispensing pumps. This is expressed (R + M)/2.

A gasoline's tendency to knock can be decreased by the addition of antiknock additives, which are metalloorganic compounds that precipitate in the burn and slow the combustion rate. Tetraethyl lead (TEL) and tetramethyl lead (TML) were used until legislated out of use. Currently, potassium and oxygenated compounds are used as octane boosters. Manganese compounds are still used in some Canadian gasoline to increase octane number, but these are not permitted in the United States. Finally, when the octane rating of a gasoline is greater than 100, it is based on the milliliters of tetraethyl lead required to be added to the reference fuel isooctane to produce the improved rating. In North America, where leaded fuel is still available, it has a pump octane rating of 88. Unleaded fuels are available at the pumps with (R+M)/2 of generally between 87 and 91. Higher ratings are used for performance and specialty applications.

Oxidation Sability   The products of oxidizing stored diesel fuel can result in deposits, filter plugging, and lacquering of fuel injection equipment. Antioxidants in the fuel inhibit the condition.

Photochemical Smog   Photochemical smog results from the photochemical reaction of hydrocarbons and oxides of nitrogen (NO_x) with sunlight in the lower atmosphere. It manifests itself as a brownish haze and may cause reduced visibility, plant damage, eye irritation, and respiratory distress. A photochemical reaction occurs when a physical substance absorbs visible, infrared, or ultraviolet radiation.

Photochemical smog is primarily a problem in urban areas with ample sunshine and low air movement. Of these, the metropolis of greater Los Angeles, California, is most notable having an urban population of 11.5 million inhabitants (Compton's Encyclopaedia) effectively walled-in on its east side by the San Gabriel mountains, limiting air movement that is generally inclined to move from west to east.

Pour Point   As the temperature is lowered below the cloud point, more and more of the paraffinic compounds of the fuel crystallize. As they do, they begin to combine to form a solid gel-like structure, which cannot be poured or pumped. This crystalline structure holds the fuel in suspension, much the same way a sponge holds water. The **pour point** of a fuel is generally considered to be 5° above the temperature at which the fuel will no longer flow. Pour point is generally 5° F to 25° F (3° C to 15° C) below cloud point (#2D pour point is −45° F to +5° F). Pour point depressant additives are required for extreme cold weather operation in #1D and #2D diesel fuel. Pour point depressants have no effect on cloud point.

Specific Gravity   The specific gravity of a liquid is the weight of a volume of the liquid compared to the weight of the same volume of water at 60° F. Diesel fuel, for example, is heavier than gasoline, so it will have a higher specific gravity. The specific gravity of a petroleum-base fuel is a direct measure of its heating value (calorific value). At 60° F (15° C), a diesel fuel measured to a specific gravity of 0.85 would release 139.50 Btu/gal. when combusted. The importance of specific gravity to diesel fuel is that it must be heavy enough to achieve adequate fuel spray penetration into the combustion chamber. If the specific gravity is too high, all the force of combustion would be put on one small area of the piston rather than having an equal force across the dome.

Stoichiometric Ratio   The stoichiometric ratio is an expression of the exact ratio of the reactants required for a chemical reaction to take place. In the internal combustion engine, the specific stoichiometric ratio of a fuel at sea level depends on the chemistry of the fuel and not on the conditions of combustion. It is a ratio of masses.

Sulfur Content   On recommendation of the American Petroleum Institute (API), the EPA effected a maximum sulfur content of 0.05% by weight effective in October 1993 in the U.S. and October 1994 in Canada. An effect of the 0.05% fuel required for on-highway use in North America has resulted in fuels with inadequate lubricity even though they achieve the required viscosity. Low sulfur fuels have compromised the service life of both hydromechanical and electronic diesel fuel pumping apparatus. As yet, the fuel parameters that relate to this low lubricity fuel have not been identified, but some individual fuel injection pump manufacturers are recommending the use of lubricity enhancing additives with some of their systems. Fuels that have a high sulfur content will also be high in various nitrogen compounds and the elements of good lubricity come from the nitrogen.

Viscosity   **Viscosity** is a measure of a liquid's resistance to sheer, a value that generally decreases as temperature increases. It also influences the resistance to flow and therefore its fluidity. The fluidity of a liquid is graded by seconds Saybolt universal (SSU), which is a measure of its ability to flow through a defined flow area against time at 100° F (39° C). The SSU rating is used for comparing the viscosity of diesel fuels. For example, #1D is typically 34.4 SSU, #2D is typically 40 SSU, and #4 is 125 SSU (preheating is required). Viscosity and volatility in diesel fuels are closely associated. Fuel viscosity directly affects the longevity of the fuel injection pump.

Temperature affects fuel viscosity. The warmer the fuel, the less resistance there is to flow. Since viscosity influences the size of the fuel droplets, it governs the degree of **atomization** and fuel spray penetration. These are major factors in obtaining the mixing of the fuel and air that is essential to complete combustion. The viscosity of diesel fuel directly affects the spray pattern of the fuel into the combustion chamber. Fuel with *high viscosity* results in large droplets that are hard to burn, cause high injection pressures, and increase the ignition delay period due to a longer mixing period. Also, the large droplets may strike the relatively cold cylinder wall and fail to burn. Fuel with a *low viscosity* will spray in a fine, easily burned mist that can retard injection timing, also resulting in ignition delay. The fuel spray will not travel across the combustion chamber, resulting in poor mixing, incomplete combustion, power loss, and smoke. Viscosity affects injection pump internal leakage and lubrication along with injector lubrication and atomization. Lower viscosity fuels are thinner and therefore will cause a higher rate of internal pump and nozzle leakage. This action causes injector dribble that can also lead to power loss and black smoke.

Volatility   **Volatility** is the tendency of a liquid to vaporize. The volatility rating of a fuel is more important in spark-ignition (SI) engines because it will determine the vapor-to-air ratio at the time of ignition. To illustrate this fact, if both diesel fuel and gasoline are dripped onto a heated steel plate, the gasoline will evaporate without igniting, whereas the diesel fuel will not evaporate immediately but will burn with a flame when it reaches a certain temperature.

Diesel Fuel Characteristics

Because we have become accustomed to readily available, high quality, uncontaminated fuels in North America, we are often slow to attribute a fuel system or engine problem to the fuel being used. Filling a vehicle tank with contaminated or poor quality (low-grade) fuel can create performance conditions that last in the vehicle fuel system well beyond the fuel that caused them. When this happens, the cause is usually fuel that was retained for prolonged periods in a storage tank (most fuel degrades in contact with air). As North American trucking activity extends into Central and South America and fuel is purchased in those countries, the chances of fuel contamination are going to increase.

Cetane Number (CN) (CCI = calculated cetane index)

Cetane is a measure of the ignition quality of a Diesel fuel. A high **cetane number (CN)** indicates good ignition quality (short delay period) and a low cetane number indicates poor ignition quality (long delay period). Cetane is a colorless liquid hydrocarbon that has excellent ignition qualities. Cetane is rated at 100. Cetane number by American Society of Testing and Materials (ASTM) definition is the percentage by volume of a test fuel consisting of *cetane* (ideal ignition quality, CN 100) mixed with *heptamethylnonane* (poor ignition quality, CN 0) required matching the fuel to be classified. The percentage of cetane mixed with heptamethylnonane (that does not ignite) to produce a similar ignition quality of the fuel being tested is the cetane number rating. So a mixture of 45% cetane with 55% heptamethylnonane would have a CN of 45. The CN directly determines the ignition delay phase of the diesel combustion cycle, and it is not uncommon for diesel fuel refiners to adjust CN seasonally. However increasing CN generally reduces fuel density and therefore, fuel mileage. As the CN of a diesel fuel increases, its ignition temperature decreases.

The number scale of a cetane number represents blends of two pure hydrocarbon reference fuels. To get a cetane number rating, a fuel is compared with cetane, a colorless, liquid paraffin hydrocarbon that has excellent ignition qualities. The higher the cetane number of a diesel fuel, the shorter the time lag from when the fuel first enters the combustion chamber until it ignites. The exact rating is determined by mixing the cetane with a chemical called *alphamethylnapthalene,* which does not ignite and is rated at zero. The percentage of cetane mixed with alphamethylnapthalene to produce a similar ignition quality of the fuel bearing tested is the cetane number rating. The cetane number equals the percentage n-cetane. Sometimes the hydrocarbon *heptamethylnonane* (HMN) is used as the non-ignition base. It has very low ignition quality and represents a cetane number of 15. The formula is:

$$\text{cetane number} = \%\ n\text{-cetane} + .15\%\ HMN.$$

For example, 30% cetane and 70% cetane yields a cetane number of 40.5.

The engine used in cetane number determinations is a standardized ASTM single-cylinder, variable–compression-ratio engine with special loading and accessory equipment and instrumentation. The engine, the operating conditions, and the test procedure are standardized under ASTM Method D613.

The specified operating conditions include the following requirements: engine speed at 600 rpm, a coolant temperature 212° F, an intake air temperature of 150° F, an injection timing of 13 degrees BTDC and an injection pressure of 1500 psi. With the engine operating under the above conditions on the fuel whose cetane number is to be determined, the compression ratio is varied until combustion starts at top dead center. With the start of fuel injection timed at 13 degrees BTDC, and with the combustion timed to start at TDC, an ignition delay period of 13 degrees (2.4 milliseconds at 900 rpm) is produced. The above procedure is then repeated using reference fuel blends until the unknown fuel is bracketed between two reference blends differing by not more than 5 cetane numbers. Each time a reference fuel blend is tried, the compression ratio is adjusted to give the same 13-degree delay period and the compression ratio is noted. However, today's cetane numbers are indexed using a calculated cetane index (CCI) computer averaging system based on the chemical properties of the fuel, which is a much less expensive and less accurate process. Once the technician knows the cetane numbers of the bracketing blends and the compression ratios required by the fixed delay period for both the reference blends and the sample fuel, the cetane number of the sample can be calculated.

When a test engine is not available for determining cetane numbers, the calculated cetane index (astm d976) is a useful tool for estimating the cetane number of a diesel fuel. The formula, which is based on api gravity and the mid-boiling point (temperature for 50% evaporation), is particularly applicable to straight run fuels, catalytically cracked stocks, and blends of the two, and is suitable thus for most diesel fuels. The calculated cetane index is not applicable to fuels containing additives for raising the cetane number.

The cetane number of a diesel fuel depends primarily on its hydrocarbon composition. The cetane requirement depends on design and operating conditions. Cetane requirement increases as the compression temperature is reduced by such variables as low ambient temperature, low water jacket temperature, low compression pressure, or light load operation. High cetane numbers (Figure 8-3) are desirable for low temperature starting and to prevent roughness and knock. The higher the cetane number of a diesel fuel, the shorter the time lag before the fuel in the combustion chamber ignites. A fuel with a high cetane number may cause BLACK SMOKE under high-load operation. The short delay causes raw fuel to be sprayed into an established flame, which produces soot and black smoke. Engines equipped with glow plugs are less sensitive to cetane ratings.

> The higher the cetane number of a diesel fuel, the shorter the time lag from when the fuel first enters the combustion chamber until it ignites.

If the cetane number is too low, a long ignition delay may result and the engine may start hard (with white smoke and misfire). When the accumulated fuel does ignite, the rate of energy release is so great that it causes roughness and diesel knock. Since most of the fuel is injected during the long ignition delay period, a large amount of fuel can cause a fast pressure rise resulting in diesel knock. Long delay periods can have a positive side because more time is available for mixing.

Ignition Accelerators

Ignition accelerators (cetane improvers) are fuel additives that increase the CN value of a fuel. They are used as a scientifically preferable option to adding an alcohol-based fuel conditioner to tanks. An ignition accelerator should be added to fuel only after testing and on the recommendation of the fuel supplier. Cyclohexanol nitrate, when added to fuel at a 0.2% concentration will raise the CN by 7 points. It should always be added to a known quantity of fuel, which usually means adding it to a full tank, either onboard or ground storage. Other alkyl nitrates are also used as ignition accelerators.

CETANE NUMBER VS. IGNITION DELAY

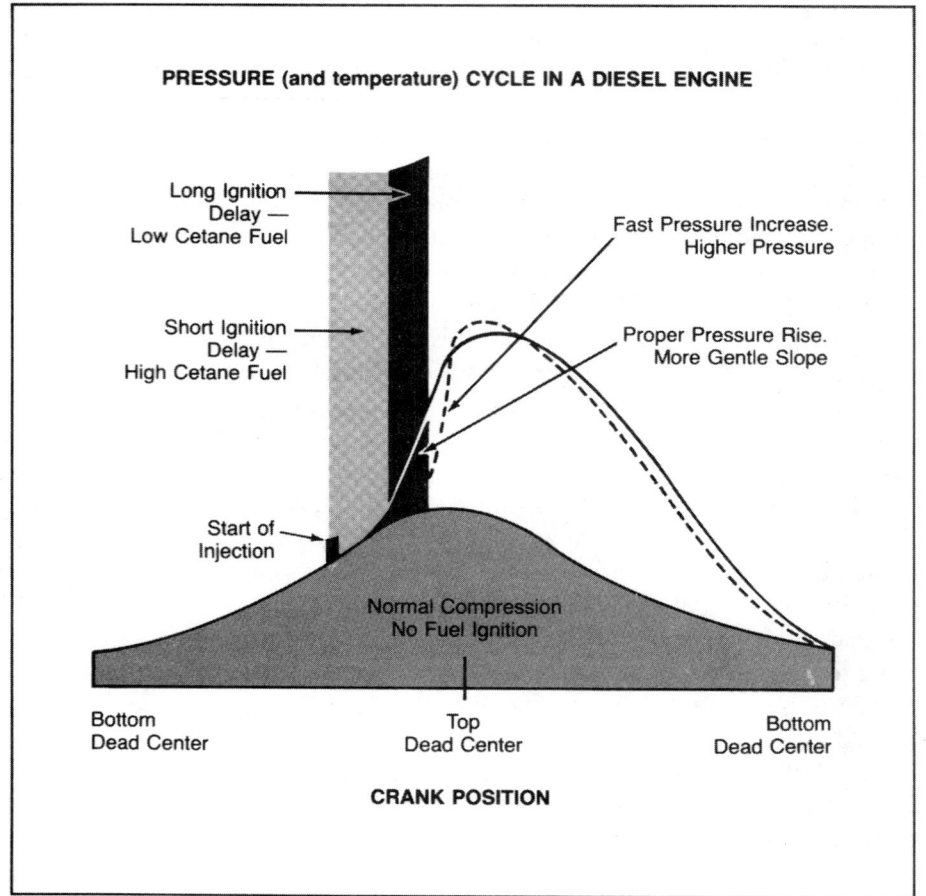

PRESSURE (and temperature) CYCLE IN A DIESEL ENGINE

Long Ignition Delay — Low Cetane Fuel

Fast Pressure Increase. Higher Pressure

Short Ignition Delay — High Cetane Fuel

Proper Pressure Rise. More Gentle Slope

Start of Injection

Normal Compression No Fuel Ignition

Bottom Dead Center

Top Dead Center

Bottom Dead Center

CRANK POSITION

Figure 8-3 Cetane number versus ignition delay. (Reprinted with permission from Chevron Products Company)

Engine performance factors that are influenced by the ignition quality of the fuel are cold starting, engine warm-up, combustion roughness, and acceleration and exhaust smoke density. The ignition quality requirement of an engine is governed largely by engine operating speeds, atmospheric temperature, and altitude. Increasing the ignition quality over the required level will have the effect of advancing the ignition timing, but this will not generally improve engine performance and can compromise engine life.

Both volume and weight bases have significance to the engine designer and equipment user, as diesel fuel is normally purchased by volume while fuel consumption is expressed by weight as kg/kW or lb./bhp. The power and economy of diesel engines cannot be directly correlated to the volatility of the fuel used, although in general, less volatile fuels have higher heating values, and more volatile fuels have better start-up/warm-up performance.

Heating Value

The heating value of a fuel (Figure 8-4) directly relates to its density though it is not the only determining factor of its heating value. The specific gravity of highway grade 1D and 2D fuels at 15° C with a CN value of 40 or better varies from 0.870 to 0.780 with high CN rating corresponding with lower specific gravity. Perhaps more important, especially to fleet operators, is the fact that #1D fuel generally possesses a lower calorific value than #2D fuel. In practice, this results in better fuel to mileage figures for #2D fuels over #1D, and it is this fact that has made it the fuel of choice among truck fleet operations.

HEATING VALUE RELATIONSHIPS

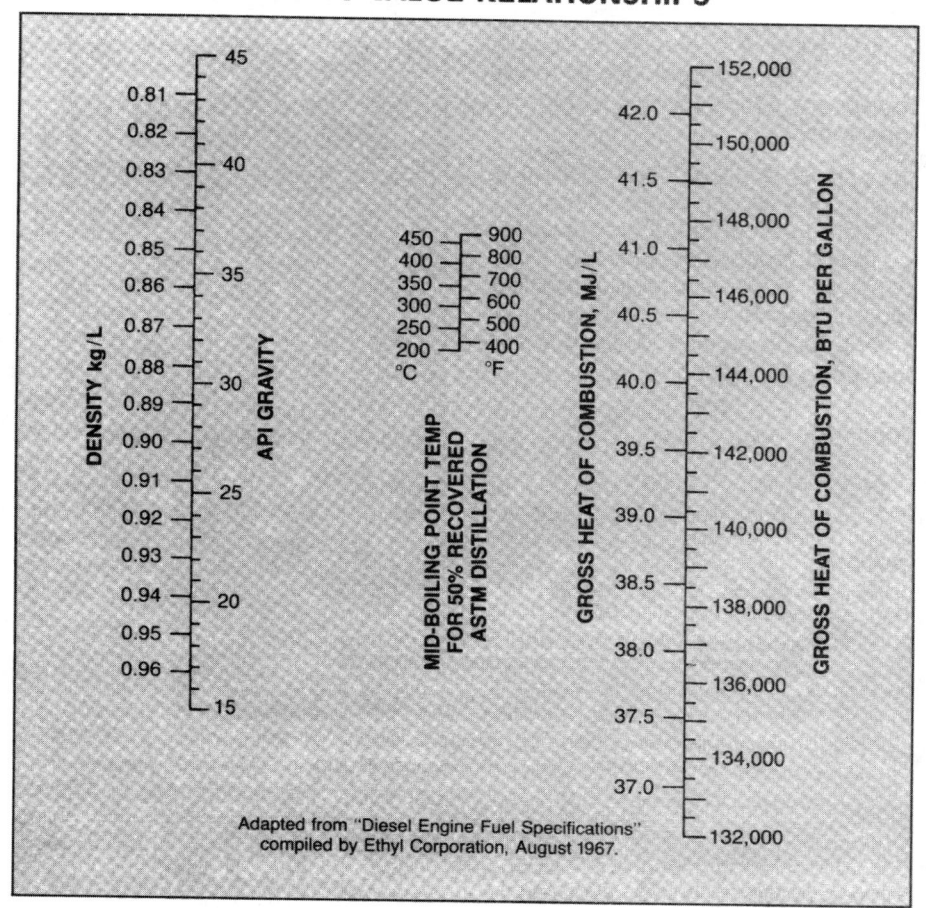

Figure 8-4 Heating value relationships. (Reprinted with permission from Chevron Products Company)

Low Sulfur and Ultra Low Sulfur Fuels (ULS)

Sulfur is present in most crude petroleum and is more prominent in the heavier residual fractions from the refining process. When combusted, sulfur in diesel fuel is oxidized to form sulfur dioxide (SO_2) which combines with hydrogen to form sulfuric acid (H_2SO_4).

WARNING: Sulfur dioxide produced from the combustion of diesel fuel is recognized as environmentally hazardous, and the sulfur content of diesel fuels has progressively been legislated to lower levels in recent years.

Today, low sulfur fuel is classified as diesel fuel containing 0.05% sulfur or less. All #1D and #2D diesel fuel currently sold in North America at the pumps as *clear* fuel is low sulfur. Low sulfur fuel has been in use since 1993, and because of a documented problem with its lubricating properties (or lubricity), it has caused failures in certain hydromechanical fuel injection pumps and even some reported premature failures in newer electronically managed fuel systems. In some cases, engine and fuel system manufacturers have recommended the use of additives that specifically enhance the lubricity of the fuel. When using a lubricity-enhancing additive, be sure that only the OEM recommended product is added to the tank in the correct proportions. This procedure should be undertaken only when the vehicle tanks are full, that is, when the quantity of fuel to be treated is known exactly.

Ultra low sulfur (ULS) fuels are currently being tested for possible use in North America. They are classified as fuels that contain 0.005% sulfur or less. ULS fuel is available in some Western European countries where it is used voluntarily by operators.

Fuel Conditioners

There are no SAE or ASTM standards that pertain to diesel fuel conditioners. Although most engine OEMs disapprove of their use, they are commonly sold in their dealerships (a measure of the profitability of the products rather than their worth). Diesel fuel conditioners are a vague mixture of cetane improvers, cleansing additives, and pour point depressants in an alcohol base. They are usually priced well beyond the cost of their ingredients. In a freeze-up, the addition of methyl hydrate is the fastest way out of the problem, and in those circumstances the negatives of putting water in solution with the alcohol are understandably overlooked. The addition of methyl hydrate to the tank under these circumstances is arguably preferable to using a fuel conditioner because the methyl hydrate is known to specifically address the problem. Similarly, when the objective is to improve the CN value of the fuel, adding a cetane improver in a measured quantity is a more scientific and cheaper strategy than adding a fuel conditioner. Most diesel fuels are carefully balanced by the refiner and the use of additives should be avoided whenever possible. Under circumstances such as the effect of reduced lubricity of 0.05 low sulfur fuel on some fuel systems, the OEM will recommend an additive to specifically manage the lubricity problem. Aftermarket fuel conditioners generally create more problems than they solve, but there are occasions when they have to be used. The technician should recognize the problems that can be caused by fuel conditioners and use them only when there is no other option. Diesel fuel conditioners should always be used according to the manufacturer's recommendations. Figure 8-5 shows the diesel additive classification on the left and a typical fuel specification for #2D on the right.

Cerium Dioxide (CeO$_2$)

Cerium dioxide is an in-combustion catalyst, which tends to increase the reactiveness of an HC fuel in the process of being oxidized, enabling more complete combustion to occur at lower temperatures. This can help to reduce both HC and NO$_x$, plus increase fuel mileage. The technology has been around for over 40 years, and research suggests that it works at least to some extent. The typical set-up involves passing intake air (or a portion of it) through a canister containing cerium dioxide. Minute particles of cerium dioxide are thereby metered to the engine cylinders through the intake system. In combustion, the cerium dioxide gives up one of its oxygen molecules, which participates in the oxidation reaction while also acting as an combustion catalyst. Cerium dioxide induction claims the following:

- ❏ Increased fuel to mileage economy
- ❏ Reduced HC emission
- ❏ Greatly reduced NO$_x$ emission
- ❏ Prolonged engine lube life
- ❏ Increased power output

Fuel SubSystems

In the fuel system shown in Figure 8-6, there is a clear divide between the suction side and the charge side, represented by the fuel transfer pump. Most fuel subsystems are of this type, and the terms "suction circuit" and "charge circuit" are used to describe each. A **primary filter** is most often located on the suction side of the transfer pump, while the **secondary filter** is located on its charge side. However, there are some fuel systems, notably the Cummins PT, where all movement of fuel through the fuel subsystem is under suction. When such a fuel system uses multiple filters, the terms primary and secondary tend not to be used. This section will describe fuel subsystems and troubleshooting methodology. The characteristics of particular fuel subsystems will be described in sections dealing with proprietary systems.

DIESEL ADDITIVE CLASSIFICATION

CLASSIFICATION	TYPE	FUNCTION
Contaminant Control		
Biocides	Boron compounds, ethers of ethylene glycol, quaternary amine compounds	Inhibit growth of bacteria and fungi — prevent filter clogging
Demulsifiers and Dehazers	Surface-active materials which increase water/oil separation	Improve separation of water and prevent haze
Rust and Corrosion Inhibitors	Organic acids, amines and amine phosphates	Prevent rust and corrosion in fuel systems, pipelines, and storage facilities
Fuel Stability		
Metal Detectors	Chelating agents	Inhibit gum formation
Oxidation Inhibitors	Alkyl amines	Minimize oxidation, gum, and precipitate formation
Dispersants	Polymeric amine surfactants	Prevents agglomeration and disperses residue
Engine Performance		
Detergents	Polyglycols, polyether amines	Prevent injector deposits Increase injector life
Dispersants	Basic nitrogen polymeric amine surfactants	Peptize injector deposits Increase filter life
Cetane Improvers	Alkyl nitrates	Increase Cetane Number
Smoke Suppressants	Overbased barium compounds	Minimize exhaust smoke
Fuel Handling		
Pour Point Depressants	Polymeric compounds	Reduce pour point and improve low-temperature fluidity properties
Cloud Point Depressants	Polymeric compounds	Reduce cloud point and improve low temperature filterability
De-icers	Low molecular weight alcohols	Reduce freezing point of small amounts of water to prevent fuel line plugging
Anti-Foam	Silicone and non-silicone surfactants	Minimize the formation of fuel foam

A

PROPERTY	SPECIFICATIONS	ASTM METHOD
Aromatics, % volume	35% Maximum	D 1319
Distillation, 90% Point, °C (°F)	378°C (640°F) Maximum	D 86
Flashpoint, °C (°F)	52°C (125°F) Minimum or Legal	D 93
Cetane Number	40 Minimum	D 613
Viscosity, cSt* @ 40°C (104°F)	1.9 – 4.1	D 445
Sediment and Water, Volume %	0.05 Maximum	D 1796
Ash, Weight %	0.01 Maximum	D 482
Total Sulfur, Weight %	0.05 Maximum	D 2622
Copper Strip Corrosion @ 100°C (212°F)	No. 3 Maximum	D 130
Ramsbottom Carbon Residue on 10% Residium, %	0.35 Maximum	D 524
Cloud Point, °F	See Note 1	D 2500

*Kinematic viscosity, centistokes.

Note 1: Cloud point no higher than –12°C (10°F) above the tenth percentile minimum ambient temperature for the area use. Tenth percentile minimum ambient temperature for the U.S. are shown in appendices of ASTM Standard D975-81, Diesel Fuel Oils. They are also included in this Recommended Practice here by permission of the American Society for Testing and Materials.

B

Figure 8-5 (A) Diesel additives; (B) Fuel specification. (Reprinted with permission from Chevron Products Company)

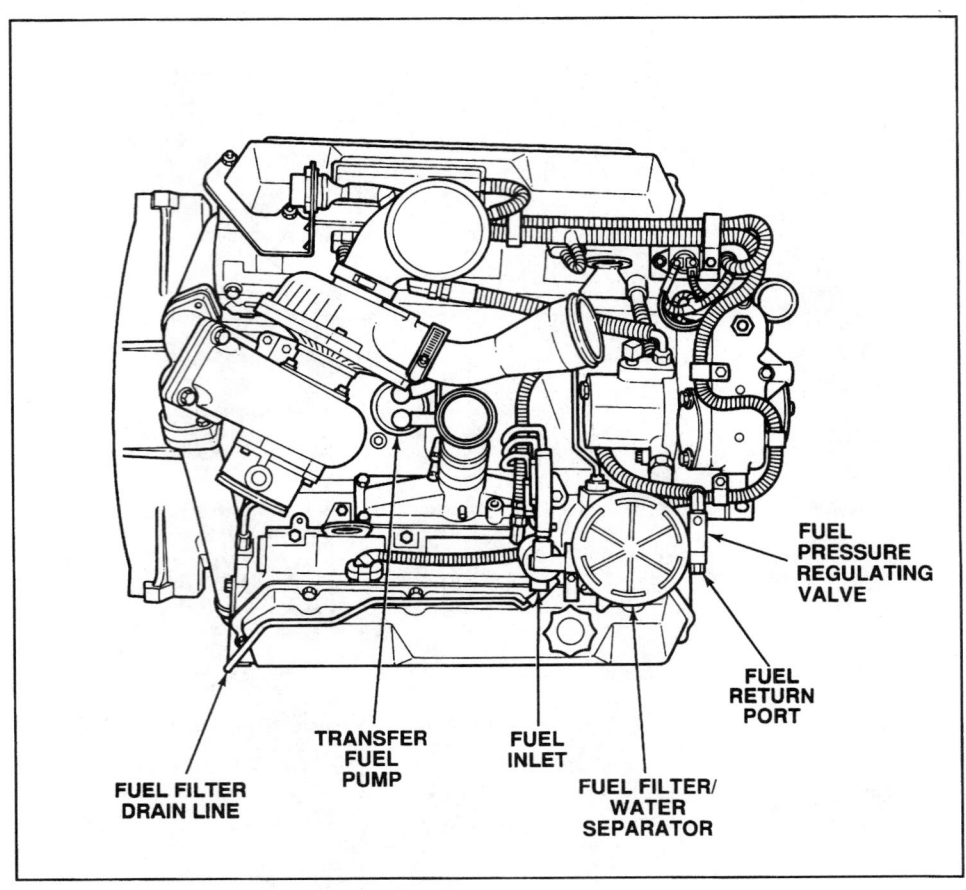

A

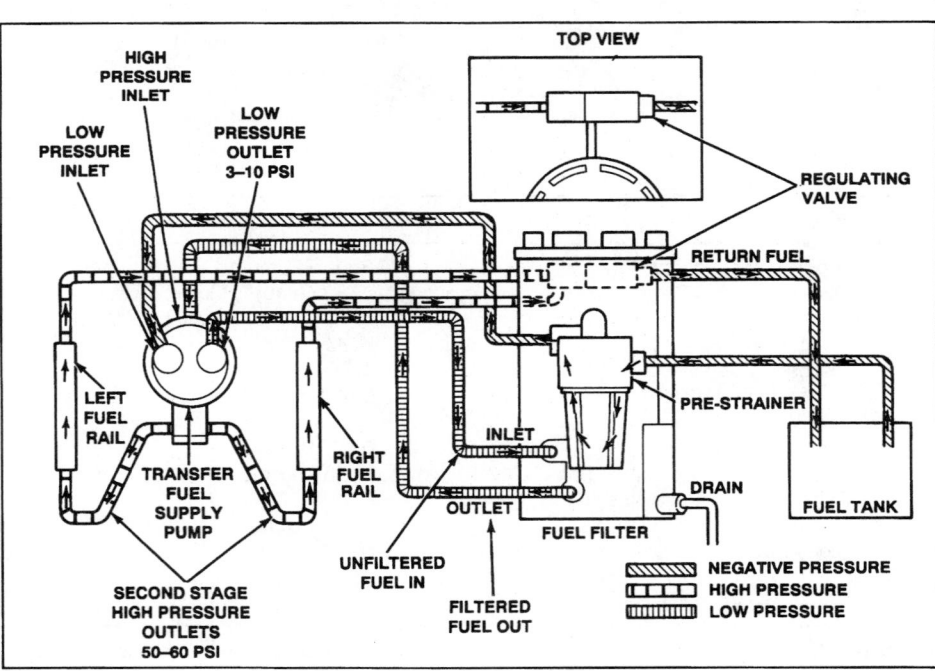

B

Figure 8-6 (A) Location of the fuel subsystems (Courtesy of TMC RP304B); (B) Subsystem schematic (Courtesy of Navistar International Engine Group).

Fuel Tanks

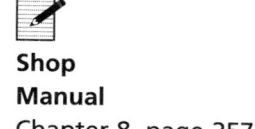

**Shop
Manual**
Chapter 8, page 257.

Fuel is stored on commercial vehicles in fuel tanks or reservoirs. Many diesel fuel management systems are designed to pump much greater quantities of fuel through the system than that required for actually fueling the engine. For example, at full throttle, 50% of the fuel pumped is returned to the tank.

> ■ **CAUTION:** When refueling a tank, never mix diesel fuel and gasoline in the same fuel tank. The combination of their different vapor pressures creates a potential volatile solution that could cause an explosion.

The amount of excess fuel varies from a minimal amount to values exceeding 60% of pumped fuel. The fuel is used to lubricate and cool high-pressure injection components, especially those exposed directly to the extreme temperatures of engine cylinders. As a cooling medium, the fuel transfers heat from the injection devices to the fuel tank, which means the fuel tank functions as a heat exchanger. A vehicle fuel tank will function most effectively as a heat exchanger if it possesses the following characteristics:

1. *Located in the airflow:* Truck fuel tanks tend to be mounted in cradle brackets (Figure 8-7) bolted to the outboard side of laded frame rails. This ensures fairly good airflow around the tank into which heat removed from the cylinder head by the fuel can be dissipated.

2. *Cylindrically shaped:* A cylindrically shaped vessel helps maximize the surface area of the tank exposed to the airflow. This shape also has a higher section modulus (relates the shape of a vessel or beam to rigidity) than a rectangular shaped vessel, permitting thinner wall thickness.

3. *Aluminum construction:* The coefficient of heat transfer of aluminum is high, which enables heat to be transferred efficiently from the fuel to atmosphere. Aluminum is less susceptible than steel to water corrosion at the base of the tank and is much lighter.

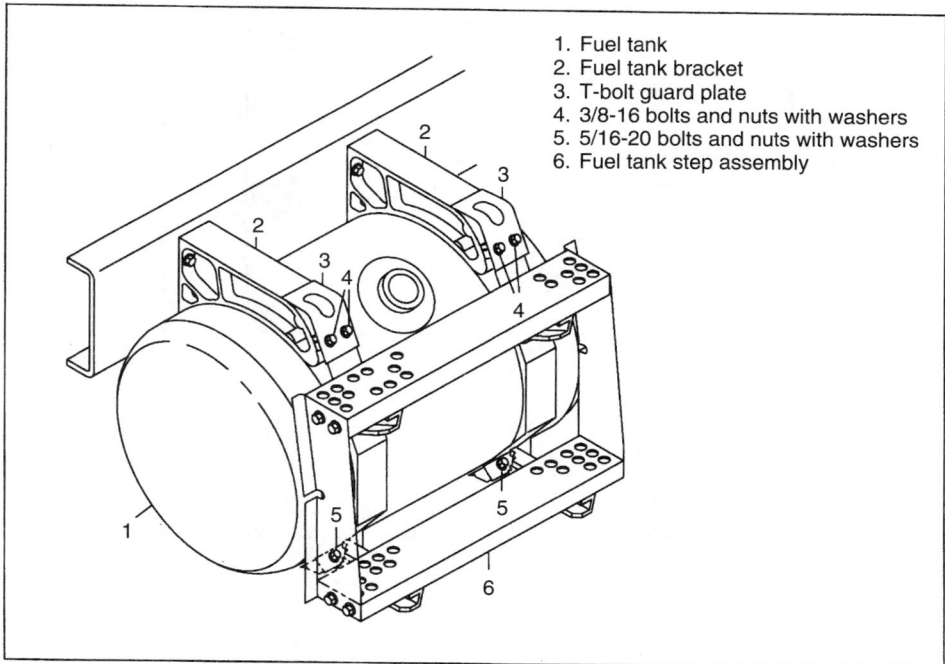

1. Fuel tank
2. Fuel tank bracket
3. T-bolt guard plate
4. 3/8-16 bolts and nuts with washers
5. 5/16-20 bolts and nuts with washers
6. Fuel tank step assembly

Figure 8-7 Typical fuel tank. (Courtesy of Freightliner Corporation)

4. *Maintained 25% full or better:* In a fuel system that circulates fuel through the system at a high rate, such as the Cummins PT system, the fuel can heat up to temperatures where its lubricity is compromised when the tanks are near empty. It is good practice in such systems to maintain the tank level at better than 25% full.

Fuel Tank Cap

Figure 8-8 shows a typical fuel tank cap. Light duty diesel applications use a fuel cap with a two-way check valve (vacuum and pressure). The pressure portion allows air to escape during the day when the tank heats up. In the event of a rollover, this valve prevents spillage. Under pressure, no more than 2 psi will exist in the fuel tank. The vacuum portion of the valve must allow air to enter the tank to push the fuel to the lift pump and replace the fuel used by the engine. A vacuum or pressure differential of no more than 1 in. Hg opens the vacuum valve. A slight hissing sound will occur when a fuel cap of this type is removed. Diesel fuel caps are unique to diesel fuel systems, so never use a fuel cap calibrated for gasoline, because the pressures are higher and will cause diesel driveability problems.

> Diesel fuel caps are calibrated with a pressure valve opening of 2 psi and a gasoline fuel cap has a pressure valve that opens at 4 psi.

Fuel Tank Sending Units

Most commercial truck fuel subsystems use remote (from the tank) fuel transfer pumps and not assemblies that incorporate the sending unit and a transfer pump. Figure 8-9A shows a typical sending unit. The fuel-sending unit is an integral assembly flange fitted to the tank. It consists of a float and arm connected to a variable resistor control current flow to a cab gauge proportionally with the fuel tank level.

> A crossover pipe connects a pair of cylindrical fuel tanks, saddle mounted on the outside of opposing frame rails.

Some dual fuel tank trucks, where the tanks are connected underneath by a crossover pipe may have a single sending unit located in one of the tanks. It is generally preferable to locate a sending unit in each tank and provide a dash gauge for each, providing the operator with some advance warning of a crossover pipe restriction. Fuel sending unit problems can be diagnosed with a digital multimeter (DMM) in resistance mode, by moving the float arm through its arc and observing readings.

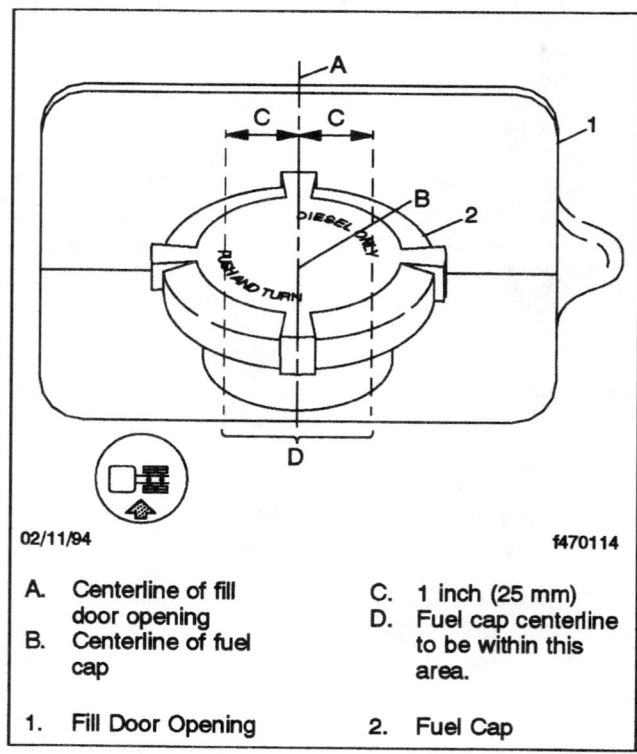

02/11/94   f470114

A. Centerline of fill   C. 1 inch (25 mm)
   door opening   D. Fuel cap centerline
B. Centerline of fuel   to be within this
   cap   area.

1. Fill Door Opening   2. Fuel Cap

Figure 8-8 Fuel cap. (Courtesy of Freightliner Corporation)

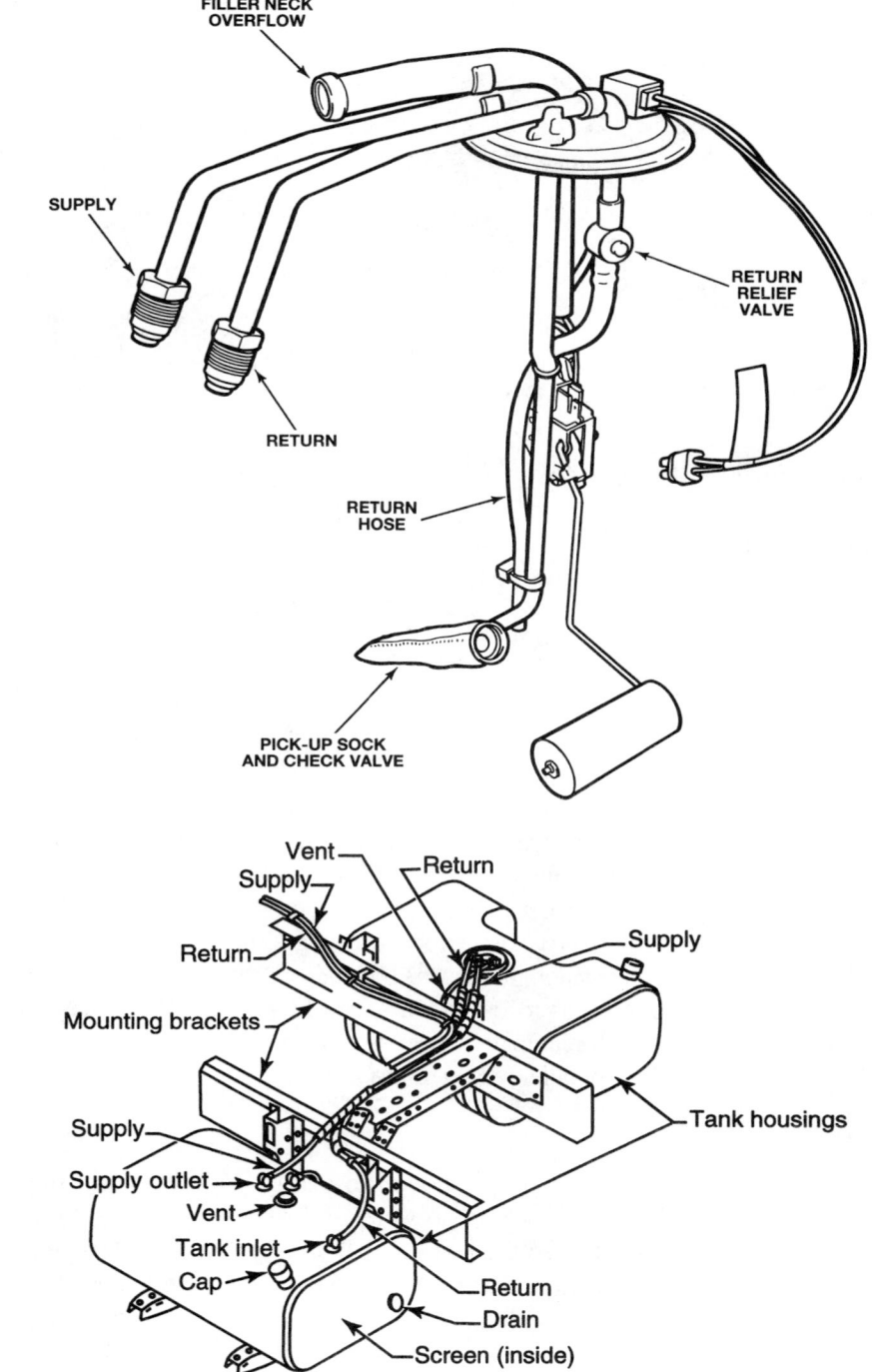

Figure 8-9 (A) Sending unit (Courtesy of General Motors Corporation, Service Operations); (B) Fuel tank parts (Courtesy of Navistar International Engine Group); (C) Duel tanks with no crossover pipe (Courtesy of Freightliner Corporation).

Dual Tanks

Most heavy duty trucks use multiple fuel tanks (Figure 8-9B) (usually two) to increase on-board fuel quantity and therefore range and also to evenly distribute fuel weight. It should be noted that 100 gal. of a typical diesel fuel weighs around 850 lb. To maintain even weight distribution as fuel is consumed, a single pick-up tube (Figure 8-9B) is located in one of the tanks, which are

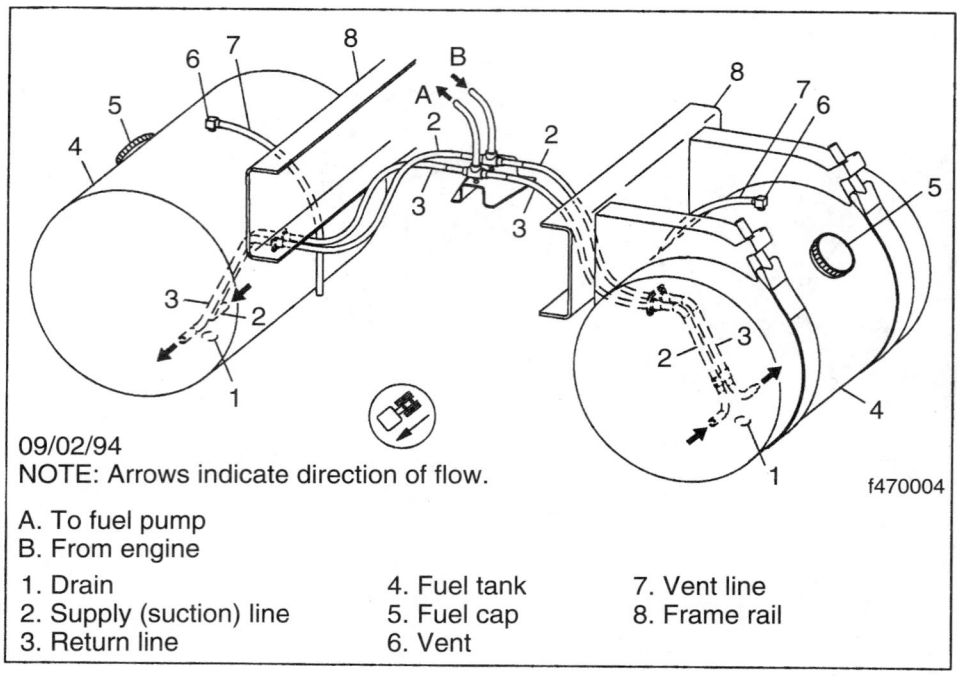

09/02/94
NOTE: Arrows indicate direction of flow.

A. To fuel pump
B. From engine

1. Drain	4. Fuel tank	7. Vent line
2. Supply (suction) line	5. Fuel cap	8. Frame rail
3. Return line	6. Vent	

C

f470004

Figure 8-9 Continued

connected by a crossover pipe. Assuming the tanks are of equal volume, fuel level is automatically equalized. Crossover pipes themselves can be the source of problems. The location of the pipe slung at the lowest point between the two fuel tanks makes them vulnerable to damage from road debris. The angle iron bracket that supports the crossover line provides little more than minimal protection. The crossover is also exposed to the airflow under the truck. Because of windchill factors in winter, any water present in the line will freeze. When crossover lines freeze up, alcohol (methyl hydrate) has to be added to the fuel tanks. Some OEMs have introduced dual fuel tank designs, which eliminate the crossover pipe (Figure 8-9C). In this configuration, the pick-up lines draw in a parallel arrangement from each tank.

Pick-Up Tubes

Fuel pick-up tubes (Figure 8-9, view C) are positioned so that they draw on fuel slightly above the base of the tank and thereby avoid picking up water and sediment. Pick-up tubes are quite often welded into the tank, in which case the tank must be replaced if the tubes fail. Fuel pick-up tubes seldom fail, but when they do it is usually caused by a metal fatigue crack at the neck. This results in no fuel being drawn out of the tank by the transfer pump whenever the fuel level is below the crack.

Venting

Currently, most jurisdictions in North America permit venting of diesel fuel tanks to the atmosphere. Therefore, as fuel is pumped out of on-board tanks it is replaced by ambient air drawn in through vent valves. Similarly, on refueling, fuel tank vapors are expelled to atmosphere. In hot weather conditions, some of the lighter fuel fractions may be boiled off, perhaps causing a slight reduction in the CN value. However, this boil-off seldom occurs at a rate sufficient to significantly compromise the fuel except in instances where a tank of fuel is retained for a prolonged period in extreme heat. Boiled-off fuel fractions from diesel fuel thus far have not been considered a noxious HC emission problem of significance by current emission legislation.

Fuel tank vents should be routinely inspected for restrictions and should be protected from ice build-up. A plugged fuel tank vent will rapidly shut down an engine, creating a suction side inlet restriction value the transfer pump will not be able to overcome.

Fuel Filters

Shop Manual Chapter 8, page 259.

Diesel fuel injection equipment is manufactured with minute clearances, and impurities in fuel, if not removed by the fuel subsystem, can cause premature failures. Most dirt found in fuel is a result of conditions in stationary fuel storage tanks, refueling practices, or improper fuel filter priming techniques by service technicians. The function of a fuel filter is to entrap particulate (fine sediment) in the diesel fuel. Although some current secondary filters will filter to the extent that water in its free state will not pass through the filtering media, a water separator is often used to remove water.

A typical fuel subsystem with a suction circuit and a charge circuit will employ a two-filter arrangement (Figure 8-10), one in each of the suction and charge circuits. Two basic types of fil-

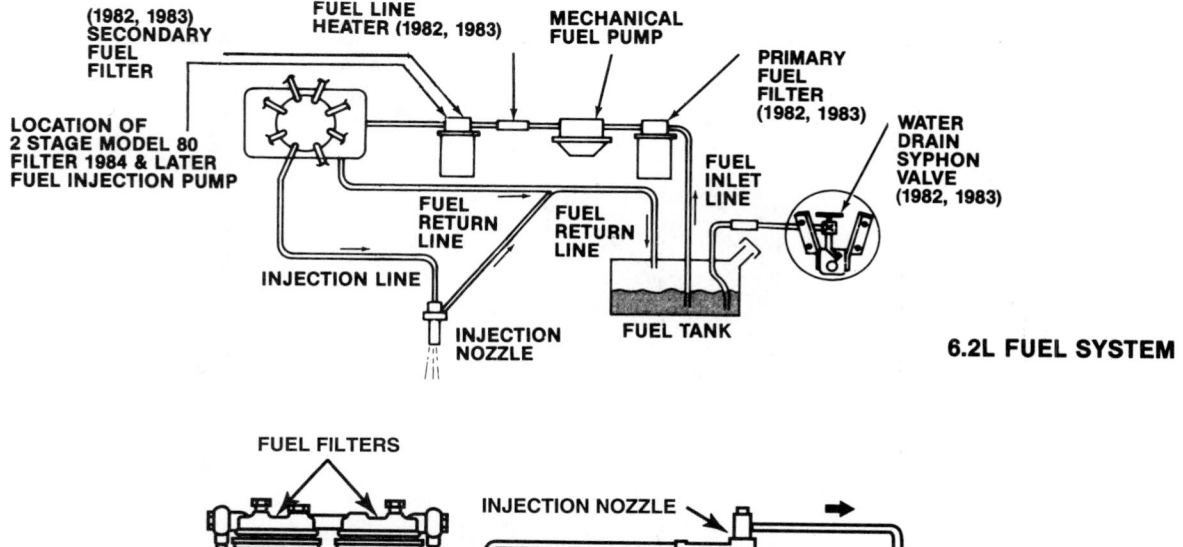

A

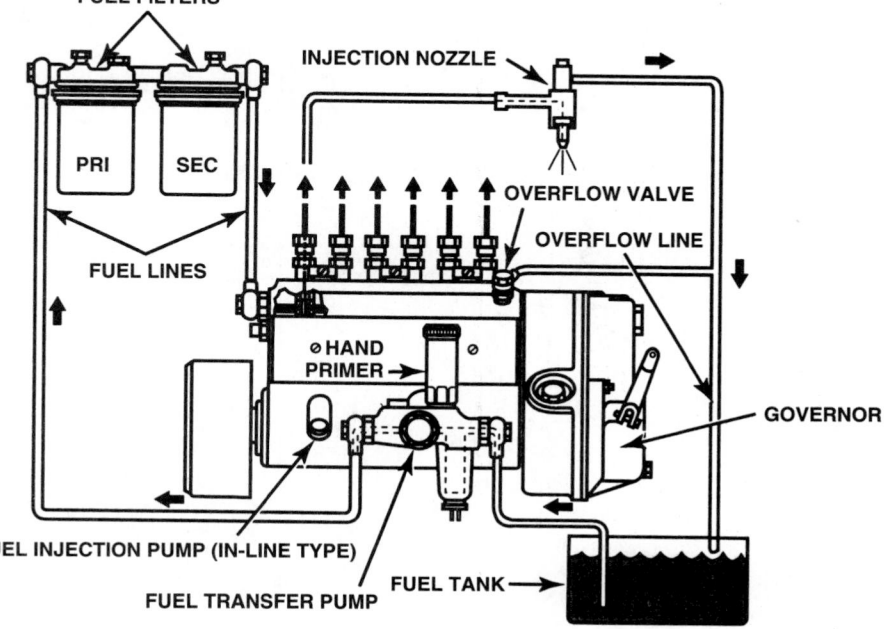

B

Figure 8-10 (A) Distributor-type diesel fuel system (Courtesy of General Motors Corporation, Service Operations); (B) Major parts of an in-line–type fuel system (Reprinted with permission from Robert Bosch Corporation).

ters are used; spin-on, disposable cartridge type and the canister and disposable element type. Spin-on filters are obviously easier to service and are the filter design of choice of most manufacturers.

Primary Filters

1μ (micron) is one millionth of a meter or 36 millionths (.000036) of an inch.

Shop Manual Chapter 8, page 260.

Primary filters (Figure 8-11) represent the first filtration stage in a typical two-stage filtering fuel subsystem. Primary filters are usually under suction, plumbed in series between the fuel tank and the fuel transfer pump. They are designed to entrap particles larger than 20–30μ (1μ = 1 millionth of a meter), depending on the fuel system, and achieve this using media ranging from cotton threaded fibers, synthetic fiber threads, and resin-impregnated paper.

Secondary Filters

Secondary filters (Figure 8-11) represent the second filtration stage in two-stage filtering. In a typical fuel subsystem, the transfer pump charges the secondary filter, and this enables use of more restrictive filtering media. The secondary filter therefore is normally located in series

Fuel filter
Two-stage box-type filter.

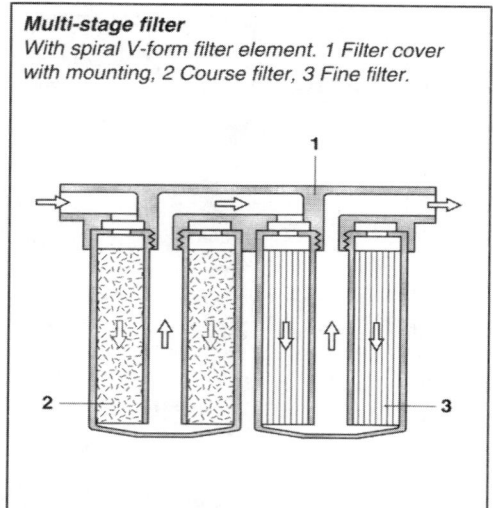

Multi-stage filter
With spiral V-form filter element. 1 Filter cover with mounting, 2 Course filter, 3 Fine filter.

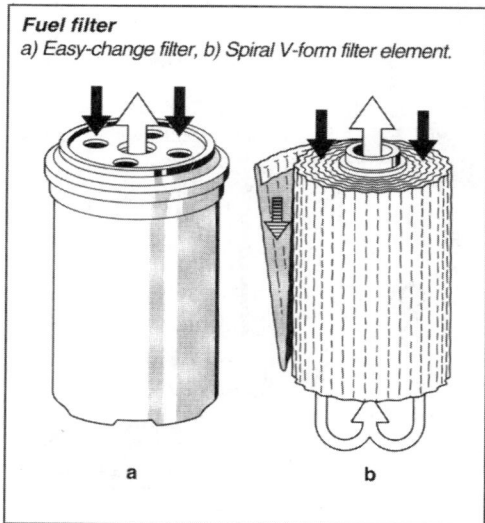

Fuel filter
a) Easy-change filter, b) Spiral V-form filter element.

Figure 8-11 Fuel filter types. (Reprinted with permission from Robert Bosch Corporation)

between the **transfer** (or charging) **pump** and the fuel injection apparatus. The transfer pump is responsible for pulling fuel from the fuel tank and charging the fuel injection components. In some diesel fuel subsystems using two-stage filtering, both a primary and secondary filter may be located on the same circuit, usually the charge circuit. In such cases, both filters are mounted on the same base pad with the primary filter feeding the secondary filter. Such an arrangement is more likely to be found in off-highway applications of diesel engines. Current secondary filters may entrap particles as small as 1µ but filtering efficiencies of 2–4µ are more common. Figure 8-12 shows a representation of a micron in comparison to other dimensions.

Water in its free or emulsified state will not be pumped through many of the current generation of fuel filters. This results in the filter plugging on water and shutting down the engine by starving it for fuel. Secondary filters use a variety of media, including chemically treated pleated papers and cotton filters.

In a fuel subsystem that is entirely under suction, such as the Cummins PT, the terms "primary" and "secondary" are not used to describe multiple filters when fitted to the circuit. As every filtering device used in the fuel subsystem is under suction, the inlet restriction specification is critical. If it exceeded the maximum, it would result in a loss of power caused by fuel starvation.

Dual-Element Filters

Figure 8-13 shows a dual-element filter that contains both a primary and secondary filter within one cartridge. Fuel coming from the lift pump flows into the bottom of the filter and then flows upward to the inside of the element and through the primary filter, which will only trap large particles. The fuel then flows out of the primary filter and down the outside of the element and through the secondary filter area where the filter element will trap particles larger than 10 µ. The fuel then goes to the injection pump. This filter is typically located on the pressure side of the lift pump.

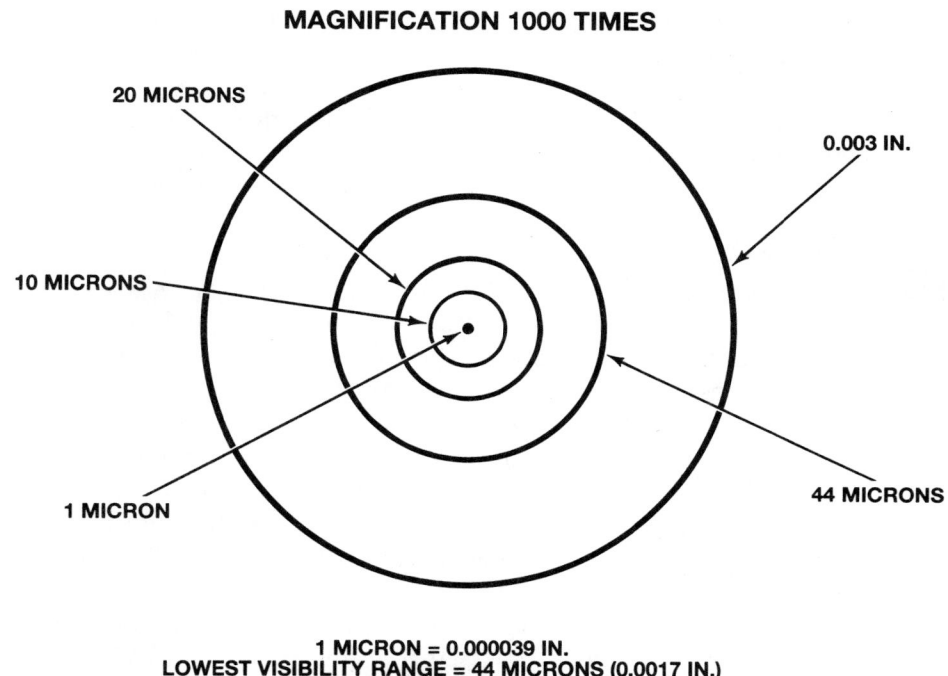

Figure 8-12 Micron particle sizes. (Courtesy of General Motors Corporation, Service Operations)

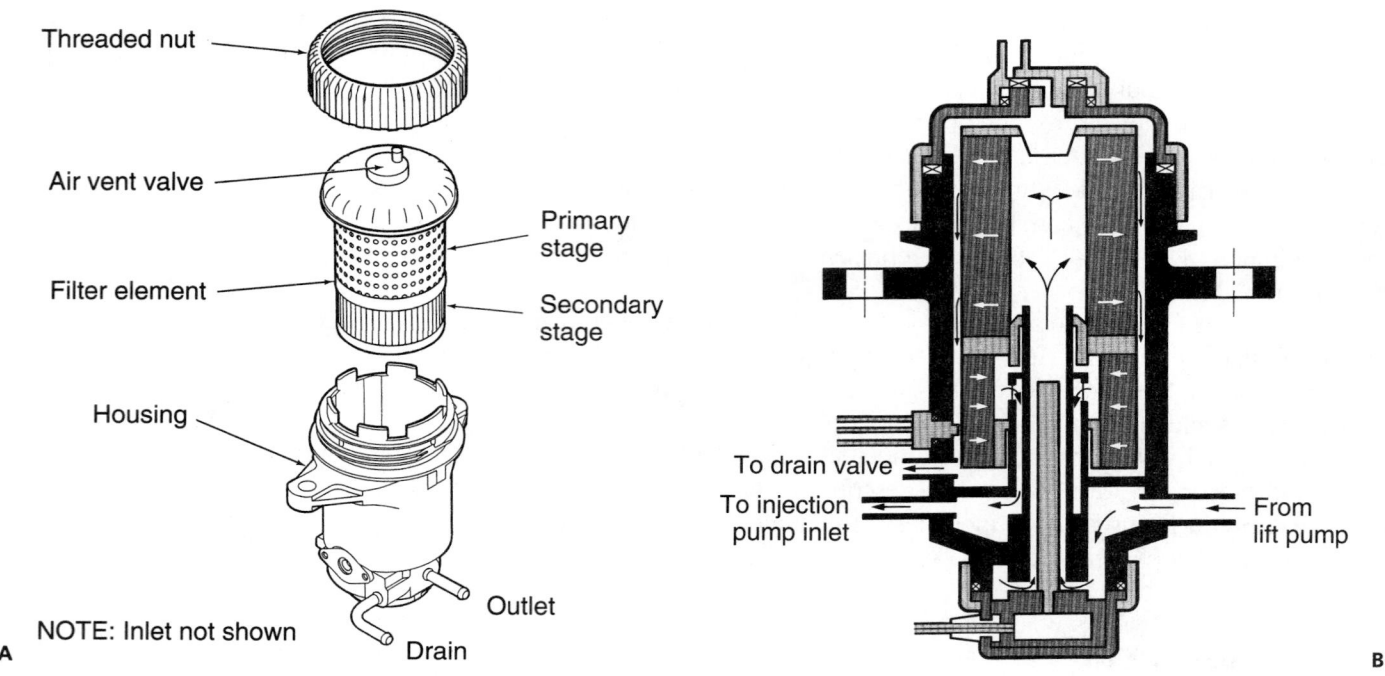

Figure 8-13 (A) Dual filter parts; (B) Fuel flow through a dual filter. (Courtesy of General Motors Corporation, Service Operations)

Water Separators

Most current diesel engine highway vehicles have a fuel subsystem with fairly sophisticated water removal devices. Water appears in diesel fuel in three forms, free state, emulsified, and semiabsorbed. Water in its free state will appear in large globules, and because it weighs more than diesel fuel, it will readily collect in puddles at the bottom of fuel tanks or storage containers.

Water emulsified in fuel appears in small droplets. Because these droplets are minute, they may be suspended for some time in the fuel before gravity takes them to the bottom of the fuel tank. Semiabsorbed water is usually water in solution with alcohol, a direct result of adding the de-icer/fuel conditioner methyl hydrate to fuel tanks to prevent winter freeze-up.

▲ **WARNING:** Water that is semiabsorbed in diesel fuel is in its most dangerous form because it may emulsify (the fine dispersion of one liquid into another) in the fuel injection system where it can seriously damage components. Water damages fuel systems for three reasons. It has lower lubricity than diesel fuel, it promotes corrosion, and its different physical properties affect pumping dynamics.

Diesel fuels are compressible at approximately 0.5% per 1000 psi.; water is less compressible at approximately 0.35% per 1,000 psi. A fuel injection pumping apparatus is engineered to pump diesel fuel, and if water (with its lower lubricity and compressibility) is pumped through the system, the resultant pressure rise can cause structural failures, especially at the sac/nozzle area of the fuel injectors.

Water separators have been used in diesel fuel systems for many years. These were mostly fairly crude devices that used gravity to separate the heavier water from the fuel. However, over the past two decades as both injection pumping pressures have steadily risen and the consumer's expectation of engine longevity has greatly increased, water separators have developed accordingly.

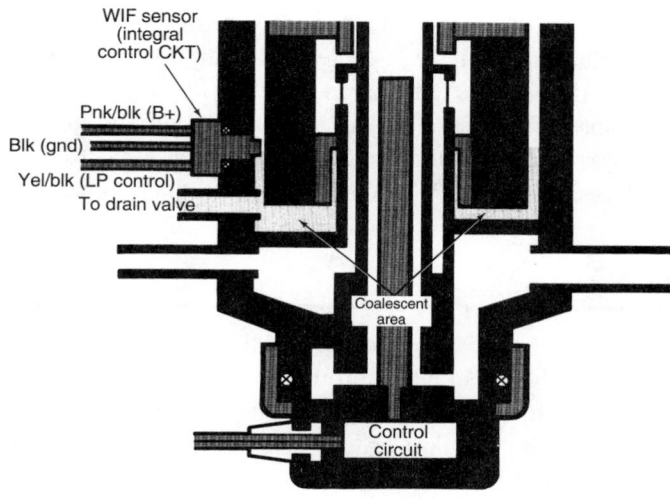

Figure 8-14 Water separator. (Courtesy of General Motors Corporation, Service Operations)

Often a water separator (Figure 8-14) will combine a primary filter and water separating mechanism into a single canister. Aftermarket suppliers, such as Racor, CR, Davco, and Dahl, manufacture many of these combination primary filter/water separators. These use a variety of means to separate and remove water in free and emulsified states; they will not remove water from fuel in its semiabsorbed state.

Water separators use combinations of several principles to separate and remove water from fuel. The first is gravity. Water (Figure 8-15) in its free state or emulsified water that has been coalesced (where small droplets come together to combine into larger droplets) will be pulled by gravity to the bottom of a reservoir or sump. Some water separators use a centrifuge to help separate both larger globules of water and emulsified water from fuel. The centrifuge subjects fuel passing through it to centrifugal force, throwing the heavier water to the sump walls where gravity can pull

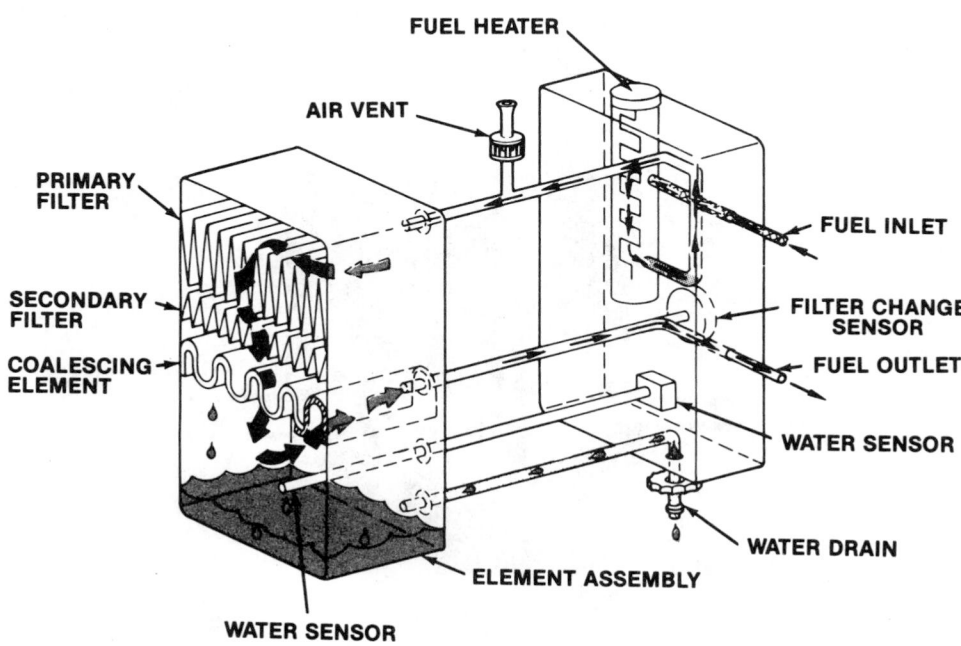

Figure 8-15 Filter flow schematic. (Courtesy of General Motors Corporation, Service Operations)

it into the sump drain. A centrifuge will act to separate particulate from the fuel in the same manner. Fuel that is directed through a fine resin-coated, pleated-paper medium will pass through the medium with greater ease than water. Water trapped by the filtering medium can collect and coalesce in large enough droplets to permit gravity to pull it down into the sump drain. In many cases, aftermarket water separators and fuel filters are designed to replace the fuel system OEM's primary filter; in others, the water separator works in conjunction with the primary filter.

Fuel Heaters

In recent years, more trucks have been equipped with **fuel heaters**, like the one shown in Figure 8-16. In fuel systems where fuel flows through the injection system circuitry at a rate much higher than that required for fueling the engine, constant filtering of fuel removes some of the wax (and therefore some of its lubricity) even when the appropriate seasonal pour-point depressants are present.

Pour-point depressants tend not to have too much effect on the cloud point of a fuel, which is the first stage of waxing. This is a condition to which #2D fuels are more prone than #1D fuels. However, there is some debate about the use of fuel heaters and the fuel system or engine manufacturer should always be consulted when fitting such a device. One engine manufacturer warns that its warranty is voided if electrical fuel heaters are used in its system. Fuel heaters allow the use of fuels at temperatures substantially below their fuel cloud point. There are two types of fuel preheaters in current use, which are described below.

Electrical Element Type Preheater

An electrical heating element (Figure 8-16) uses battery current to heat fuel in the subsystem. This type of fuel preheater offers a number of advantages, most notably that the heater can be energized before start-up so that cranking fuel is warmed up. Electrical element fuel heaters may be thermostatically managed so that fuel is heated only as much as required and not to a point that compromises some of its lubricating properties.

Engine Coolant Heat Exchanger Type Preheater

This type of fuel heater consists of a housing within which coolant is circulated in a bundle (heat exchanger core) and over which the fuel is passed. A disadvantage of this type is that the engine cooling system must be at operating temperature before the fuel can be heated.

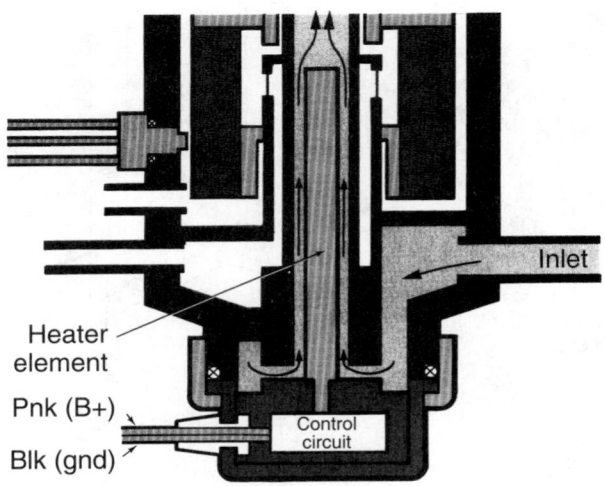

Figure 8-16 Fuel heater. (Courtesy of General Motors Corporation, Service Operations)

Some fuel heaters use both electrical heating elements and coolant medium heat exchangers to manage fuel temperature. Fuel temperature should not exceed 90° F (32° C). If fuel exceeds this temperature, its lubricating properties start to diminish reducing the service life of fuel injection components.

Fuel Charging (Transfer) Pumps

Shop Manual Chapter 8, page 265.

Fuel charging, or transfer, pumps (Figure 8-17) are positive displacement pumps driven directly or indirectly by the engine. A positive displacement pump displaces the same volume of fluid per cycle; therefore, the quantity of fuel pumped increases proportionately with rotational speed. Similarly, if a positive displacement pump unloads to a defined flow area, pressure rise is proportional with an increase in rpm. On most truck and bus fuel systems, charging/transfer pumps are of the plunger or gear types. These pumps are responsible for all movement of fuel through the fuel subsystem.

Fuel charging or transfer pumps are also called left or supply pumps.

Plunger-Type Pumps

Plunger-type pumps are often used with port-helix metering injection pumps. They are usually flange mounted to the injection pump cambox and driven by a dedicated cam on the pump camshaft. Single-acting and double-acting plungers may be used, with the latter type specified in higher output engines that require more fuel. A single-acting plunger pump (Figure 8-17) has a single reciprocating element with a single pump chamber and an inlet and outlet valve. Fuel is drawn into the pump chamber on the inboard stroke and pressurized on the outboard, or cam, stroke.

A double-acting pump (Figure 8-18) has twin chambers, each equipped with its own inlet and outlet valve. On the cam stroke, a two-way plunger charges the pump chamber while admitting fuel to the other. On the return stroke, the pump retraction spring reverses the process.

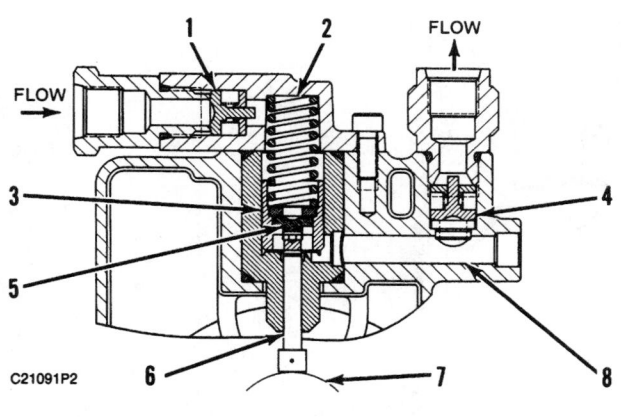

Fuel Transfer Pump
(1) Inlet check valve. (2) Spring. (3) Piston assembly. (4) Outlet check valve. (5) Piston check valve. (6) Tappet assembly. (7) Cam.
A (8) Passage.

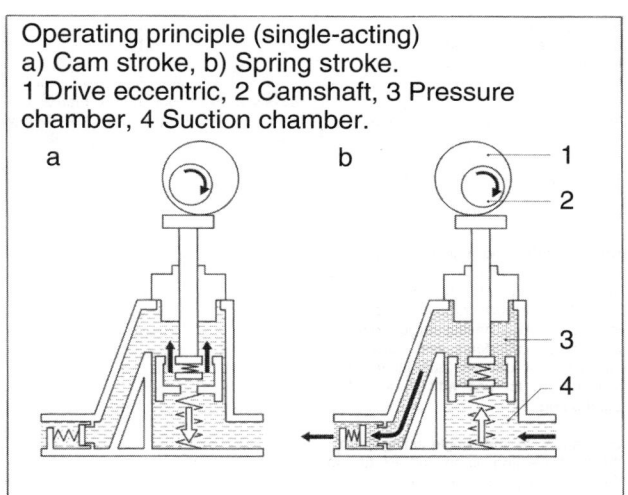

Operating principle (single-acting)
a) Cam stroke, b) Spring stroke.
1 Drive eccentric, 2 Camshaft, 3 Pressure chamber, 4 Suction chamber.

Figure 8-17 (A) Caterpillar single-acting transfer pump (Courtesy of Caterpillar Inc.); (B) Internal view of single-acting transfer pump (Reprinted with permission from Robert Bosch Corporation).

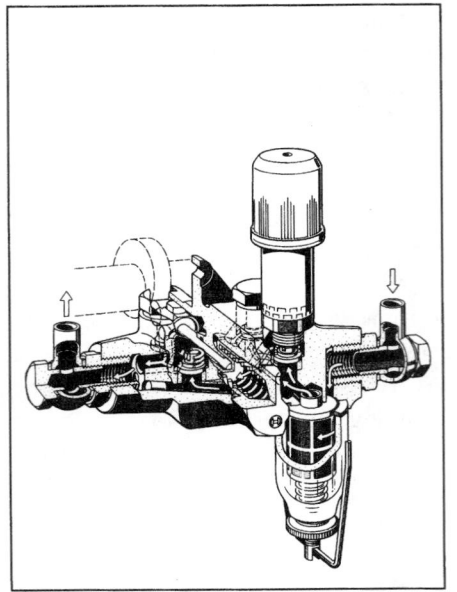

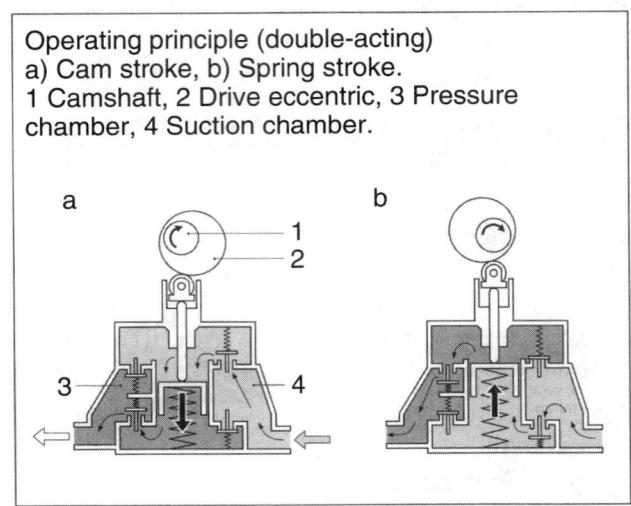

Operating principle (double-acting)
a) Cam stroke, b) Spring stroke.
1 Camshaft, 2 Drive eccentric, 3 Pressure chamber, 4 Suction chamber.

A

B

Figure 8-18 (A) Bosch double-acting transfer pump; (B) Internal view of double-acting transfer pump. (Reprinted with permission from Robert Bosch Corporation)

Hand Primer Pumps

A hand primer pump (Figure 8-19) can be a useful addition to the technician's tool kit, and it can be fitted to a fuel subsystem whenever prime is lost. It can be located on the fuel transfer pump body or on a filter-mounting pad. The function of a hand primer pump is to prime the fuel system whenever prime is lost. Typically it consists of a hand-actuated plunger and uses a single-

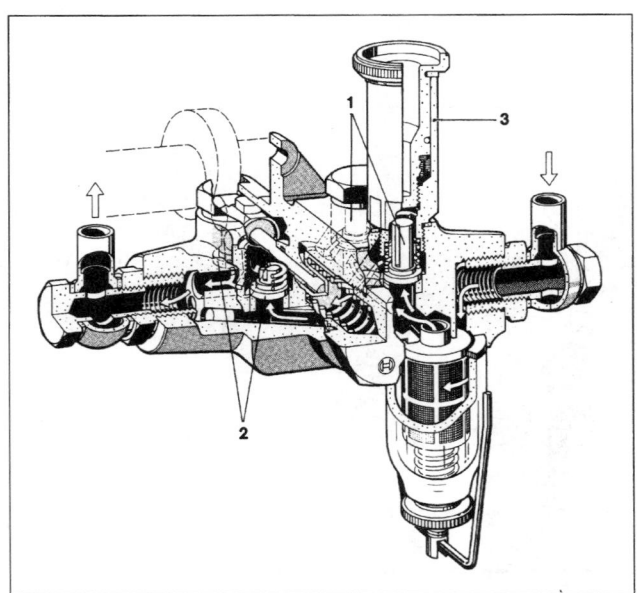

1. SUCTION SIDE CHECK VALVE
2. CHARGE SIDE CHECK VALVE
3. HAND PRIMER

Figure 8-19 Bosch charging pump with integral hand primer and primary filter. (Reprinted with permission from Robert Bosch Corporation)

acting pumping principle. On the outward stroke, the plunger exerts suction on the inlet side, drawing in a charge of fuel to the pump chamber. On the downward stroke, the inlet valve closes and fuel is discharged to the outlet. When using a hand primer pump, it is important to purge air downstream from the pump on its charge side. Some fuel subsystems mount a hand primer to the transfer pump housing. Some newer fuel systems have self-contained, electrical priming pumps that prime the system after servicing.

Gear Fuel Pumps

Gear pumps, as shown in Figure 8-20, are commonly used as transfer pumps. These are normally driven from an engine drive and are located wherever convenient. Gear pumps will usually have an integral relief valve, which will define the peak system charging pressure. Fuel injection systems designed to be charged at pressure values higher than typical tend to use gear type transfer pumps over cam-actuated plunger pumps. In instances where a gear pump feeds an injection system with no main filter in series, a filter mesh is sometimes incorporated to protect the injection pumping apparatus. When gear pumps are used, there is a small chance that gear teeth cuttings can be discharged into the system. Most full-authority, electronic management fuel systems use gear-type pumps.

Electrical Solenoid Fuel Pumps

Some light and medium duty diesel applications use an electrical solenoid fuel pump, as shown Figure 8-21A. It is typically located on the truck frame rail (Figure 8-21B).

Figure 8-21A, shows the operation of this fuel pump as follows: When the pump is at rest, a spring pushes the hollow plunger in the direction of the outlet (see Figure 8-21B). Both the inlet and outlet valves are closed, keeping fuel in the supply line from draining back to the fuel tank. When the pump has electrical power, a solenoid turns ON and pulls the hollow plunger

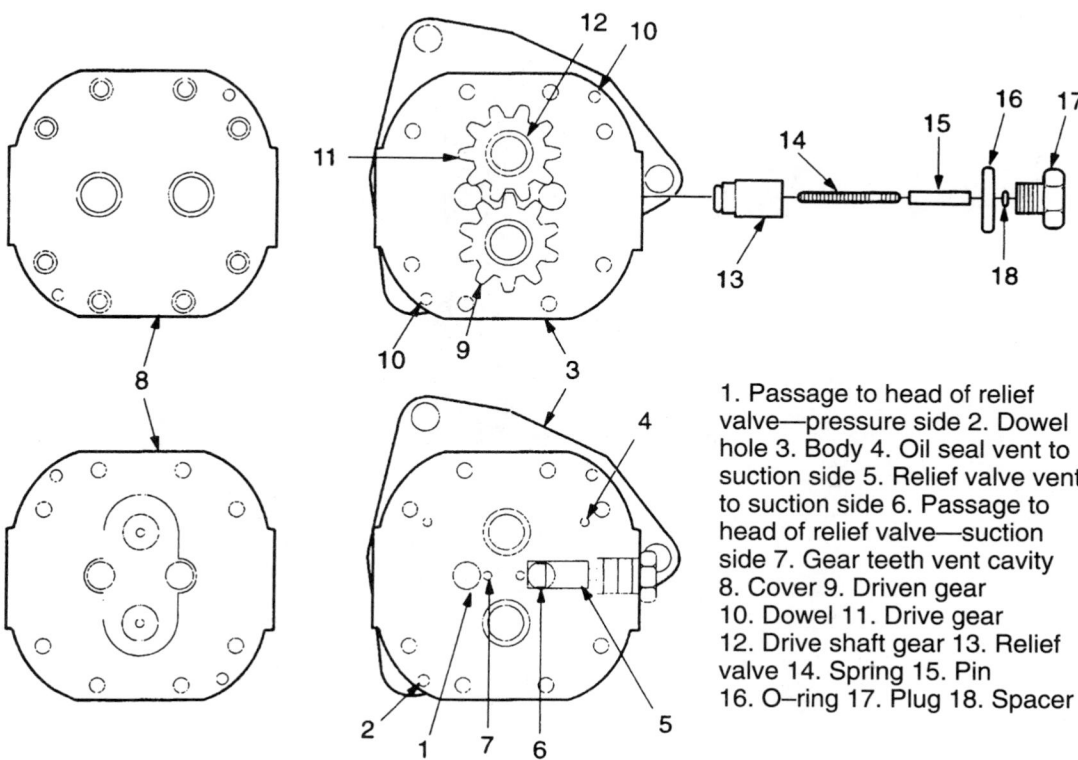

1. Passage to head of relief valve—pressure side 2. Dowel hole 3. Body 4. Oil seal vent to suction side 5. Relief valve vent to suction side 6. Passage to head of relief valve—suction side 7. Gear teeth vent cavity 8. Cover 9. Driven gear 10. Dowel 11. Drive gear 12. Drive shaft gear 13. Relief valve 14. Spring 15. Pin 16. O–ring 17. Plug 18. Spacer

Figure 8-20 Gear-type fuel pump. (Reprinted by permission of copyright owner Detroit Diesel Corporation, all rights reserved.)

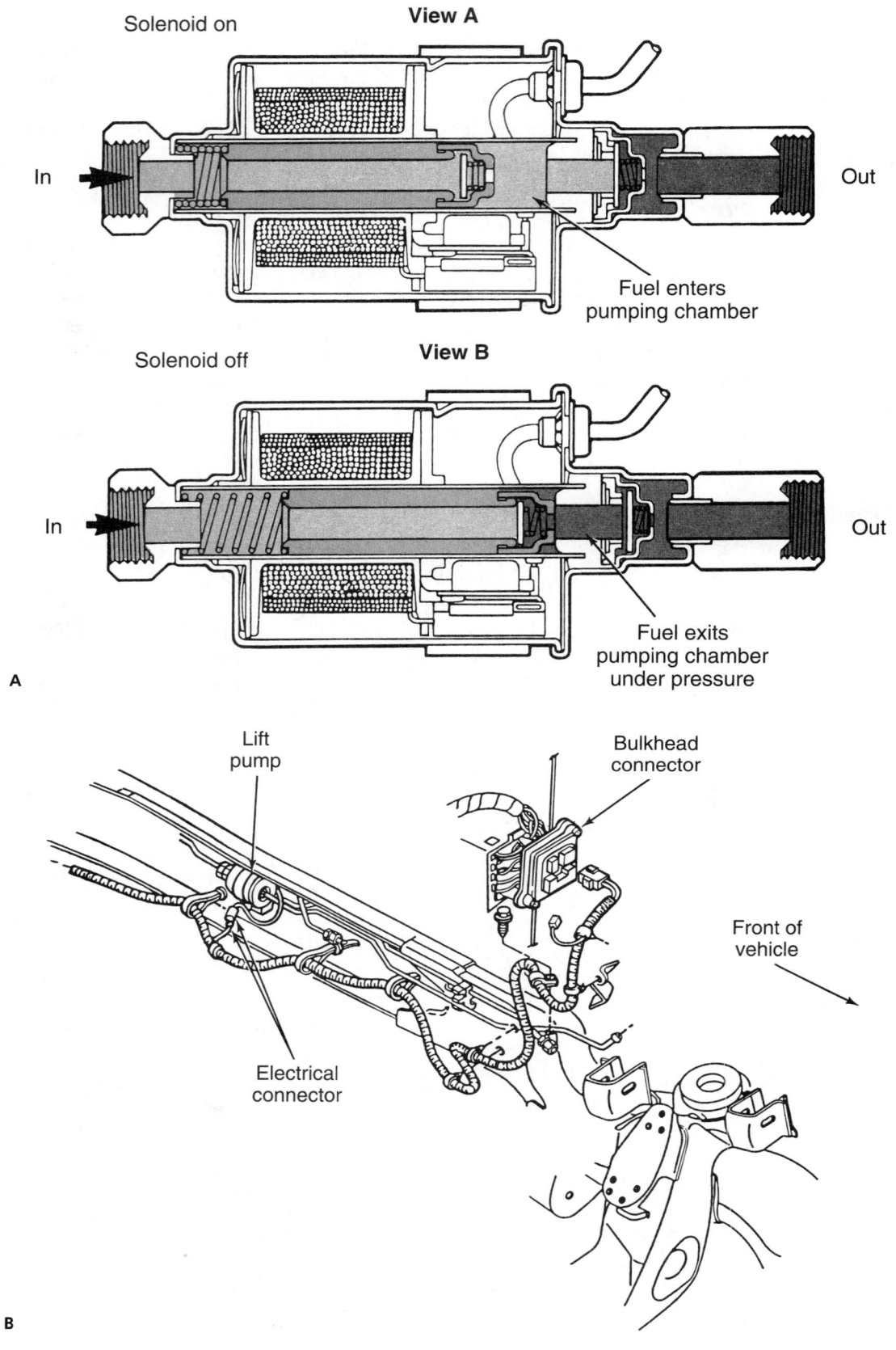

Solenoid on

View A

In

Out

Fuel enters
pumping chamber

Solenoid off

View B

In

Out

Fuel exits
pumping chamber
under pressure

A

Lift
pump

Bulkhead
connector

Front of
vehicle

Electrical
connector

B

Figure 8-21 (A) Electric transfer (lift) pump; (B) Lift pump location. (Courtesy of General Motors Corporation, Service Operations)

toward the inlet port against spring force (see Figure 8-21). This action causes the inlet valve to open and allow fuel to enter the pumping chamber. As the hollow plunger reaches full travel in the direction of the inlet port, the solenoid is turned off, and the spring force pushes the plunger toward the outlet port (Figure 8-21B). This action closes the inlet valve (causing fuel to be drawn from the fuel tank into the pump) and pressurizes the fuel in the pumping chamber (opening the outlet valve to allow fuel to travel to the fuel filter). As long as the fuel pump has electrical power, the solenoid on/off cycle causes the movement of the hollow plunger and valves necessary to deliver a fuel supply to the filter and injection pump.

Hydromechanical Injection Operating Principles

Before beginning the study of **hydromechanical** and electronic management of the diesel engine, a thorough understanding of engine components and principles of operation is required. Figure 8-22 shows a typical hydromechanical fuel system. In any *compression-ignition* (CI) engine, a power stroke will take place only if the fuel system is accurately phased to the engine and performs the following:

Timing

Fuel delivery **timing** is critical during all engine-operating phases. Typically, fuel will be injected into the engine cylinder slightly before the piston completes its compression stroke. Whereas older engines tended to have hard (nonvariable) timing values, newer engines have variable timing features that can be managed mechanically, hydraulically, or electronically. Variable timing is required in today's engines to produce optimal performance and minimal noxious emissions. Indirect injection (IDI) engines used in passenger cars and light trucks require timing delivery much advanced of that in direct injection (DI) engines because of the much larger droplet sizes emitted. The larger droplets emitted from IDI engines rely on turbulence in the prechamber to help reduce them to a smaller size suitable for combustion in the short amount of time available. DI engines with high nozzle opening pressure (NOP) values (smaller emitted droplet sizes) have injection timing typically around 20 degrees before top dead center (BTDC).

Pressurizing

The fuel system must be capable of pressurizing the fuel sufficiently to open the injector nozzles and deliver fuel to the engine cylinders correctly prepared for combustion. The means used to pressurize the fuel to the required injection pressures varies with the type of fuel system. In most cases, the fuel injection pumping apparatus is actuated mechanically, typically by a cam profile located either on the main engine camshaft or on a dedicated pump camshaft. The aggressiveness of the cam flank in conjunction with engine rpm will play a big role in defining peak system pressure values. In recent engines, pumping to injection pressures is by precisely controlled hydraulic pressure. Injector nozzle opening pressures in current engines range from 1,900 to 5,500 psi. Peak pressures in current CI engines can reach 30,000 psi.

Metering

Metering is the precise control of fuel quantity. The only factors that control the output of a diesel engine are the amount of fuel and when it is injected. Gasoline-fueled, spark-ignited engines define peak output by the amount of air that can be induced into the engine cylinders. The size of the throttle bore is the defining factor. Diesel engines operate with excess air; under any load or operating condition, there is always more air present than the minimum required to completely combust the fuel. Because of the excess air, most truck diesel engines given unlimited fuel will accelerate at a rate of up to 1,000 rpm per second. Diesel engines must have a means

Hydromechanical is a term used to describe a diesel engine that is managed without a computer.

Electronic management refers to any system that is managed by computer.

Metering is the precise control of fuel quantity.

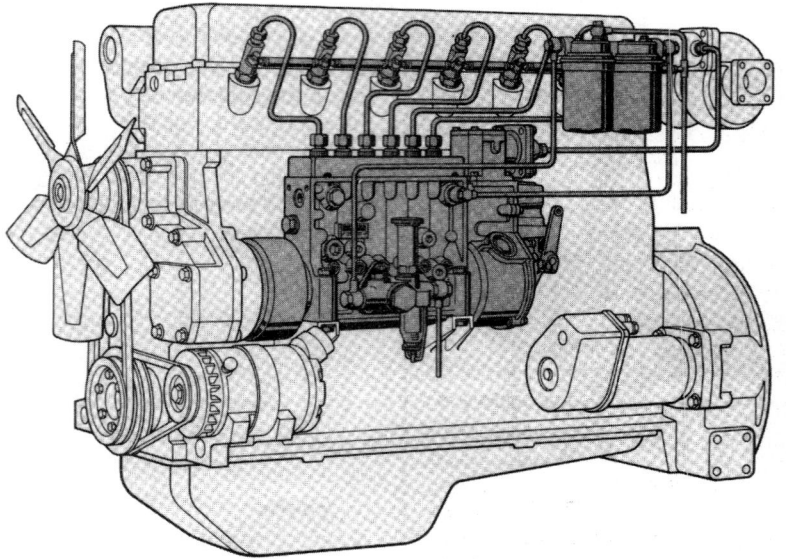

A

B

Figure 8-22 (A) Typical hydromechanical fuel system layout (Reprinted with permission from Robert Bosch Corporation); (B) Mack truck fuel schematic (Courtesy of Mack Trucks, Inc.).

Injection rate is the fuel quantity injected per crank angle degree.

of precisely metering fuel into the engine cylinders. Metering in today's engines is achieved by several distinct means. While diesel engine manufacturer's engines have many components and operating principles that are common, each fuel system that manages these engines tends to be distinct. Most current engines have fuel systems that are managed electronically (by a computer).

Atomization

In today's DI diesel engines, fuel must be atomized within a droplet size range between 10 and 100 μ. The reasons for the precise sizing parameters are dealt with in the section on combustion. The degree to which the fuel exiting the nozzle **orifice** is atomized depends on:

❑ The pressure of the fuel supplied to the nozzle orifice.
❑ The flow area, or sizing, of the nozzle orifice.

Atomized fuel is in the liquid state and, therefore, it must be vaporized and ignited before it can be combusted (oxidized) in the engine cylinder. The smaller the atomized droplets, the more rapidly they will respond to the heat within the cylinder to undergo the changes required.

Indirect injected engines use injector nozzles that provide a lesser degree of atomization (the emitted droplets are larger) and depend on prechamber turbulence to help reduce the emitted droplet sizing; that is, turbulence in the prechamber helps turn the fuel droplets emitted from the injector into smaller droplets. Because this takes time, fuel injection timing in IDI engines is significantly advanced compared with that in DI engines.

Distribution

In multicylinder engines, the fuel system must be phased (sequenced) to deliver the fuel to each engine cylinder at the correct time and in the correct firing order. Correct phasing is required to balance the engine output. If, in a given engine position, ignition is set to occur at 3 degrees BTDC, it must occur in that position in each cylinder in the engine.

In multicylinder engines, the fuel system must also dispense the fuel to the correct area of the combustion chamber so that the fuel droplets combust at the appropriate time (spray dispersion). The position of the fuel injector is critical in ensuring that the injected fuel droplets are directed to the correct area of the combustion chamber in DI engines. For instance, if an injector were seated on double-sealing washers (a not uncommon occurrence) the nozzle would be too high in the cylinder. The result would be to inject fuel above the combustion bowl on a Mexican hat piston crown, which could cause fuel to condense on the piston. Fuel exiting the nozzle orifice should be vaporized before it comes into contact with the cylinder hardware, so it does not condense. Fuel that condenses on the piston can cause erosion (pitting). Figure 8-23 shows the relationship between the cylinder and crank pressure.

Very different fuel delivery mechanisms have been developed since the late 1920s, and they have been fine-tuned to deliver the high thermal efficiencies and low noxious emissions required in diesel engines today.

Units of Pressure and Linear Measurement

While the tendency in North America is to use English unit values of measurement, most truck and bus engines and fuel systems engineered during the past two decades have used the S.I.

Atomization of the fuel delivered to the cylinder means preparing the fuel charge for combustion given the conditions within the cylinder and the time dimension it is to take place in.

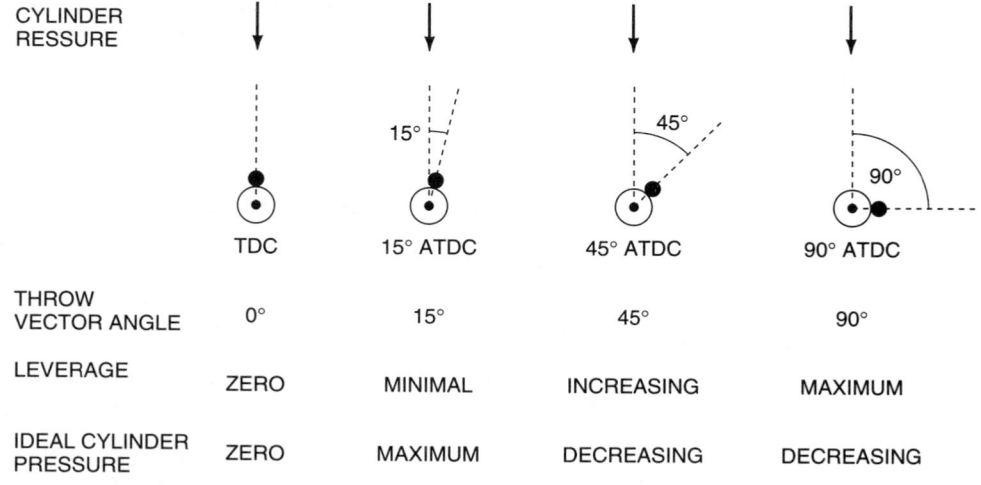

Figure 8-23 Relationship between cylinder pressure and crank pressure.

(Systeme internacionale d'unite) or metric system. Most North American–built diesel fuel injection equipment has always used the metric system. The contemporary diesel technician should be prepared to readily convert values from the English to the metric system and vice versa. It is ironic that the manufacturers who engineer machinery using the metric system then convert the specifications to English values for the technical service manual specifications. The trend today is to present specifications using both metric and English values. Manufacturers of fuel injection apparatus often express pressure values in units of atmospheric pressure known as atms. This has the advantage of being easily converted either to the metric or imperial systems. In Europe, units of *bar* (10^5 Newtons per square meter) are used to express units of atmosphere: 1 unit of bar is not precisely equivalent to 1 unit of atm. Here are some of units of measure that will be used in this book:

- ❏ 1 atm = 14.7 psi = 101.3 kPa (kilopascals)
- ❏ 200 atm = 2940 psi = 20.26 MPa (megapascals)
- ❏ 1 micron = 1 millionth of a meter
- ❏ 1 millimeter = 0.0394 inches
- ❏ 1 centimeter = 0.394 inches
- ❏ 1 meter = 39.4 inches

Shop Manual
Chapter 8, page 269.

Hydraulic Injector Nozzles

When the truck technician refers to an injector, the nozzle holder assembly is generally being described. The term also describes (somewhat incorrectly) components, such as mechanical unit injectors (MUI), electronic unit injectors (EUI), hydraulically actuated electronic unit injectors (HEUI), and other components that may incorporate a hydraulic nozzle but also perform the metering and timing functions. This text will refer to integral injector assemblies that define nozzle-opening pressure (NOP) and atomize fuel as *hydraulic injectors*. It should be noted that Cummins refers to these devices as mechanical injectors. Hydraulic injectors are normally classified by nozzle design. They are simple hydraulic switch mechanisms that atomize and inject fuel into the engine cylinders. There are three basic types of injector nozzles, two of which are effectively obsolete in medium and heavy duty truck and bus applications. However, both will be briefly described in this section. Poppet and pintle nozzles are best suited to indirect injected engine applications and therefore are not used in any current medium and heavy duty engines. **Multi-orifice nozzles** are used in most DI diesel engines using in-line, port helix metering, injection pumps. They are also used as an integral subcomponent of mechanically and electronically controlled unit injection systems.

All current, high-speed diesel engines found in highway applications use hydraulic nozzles with the exception with the open-nozzle designs used in Cummins hydromechanical and electronic common rail systems. Open-nozzle injectors are covered in later sections of this text. There are almost no similarities in operating principle between the closed hydraulic nozzles addressed in this chapter and open-nozzle injectors.

Nozzle Opening Pressure (NOP)

The term NOP can also mean nominal opening pressure (NOP).

An injector nozzle is a hydraulic switch. One of its primary functions is to define the pressure required to trigger its opening. The term "popping pressure" is also used to describe the opening pressure of a nozzle. The actual NOP value is defined by the mechanical spring tension of the injector spring. This spring tension loads the nozzle valve onto its seat and therefore determines the hydraulic pressure required to unseat the valve. Most injectors incorporate a means of adjusting the injector spring tension so that the NOP value can be set to specification. The spring tension adjustment mechanism is either shims or an adjusting screw and locknut. The NOP value is always one of the first performance specifications to be evaluated when testing injector nozzles on a bench test fixture (pop tester).

Poppet Nozzles

Caterpillar and General Motors used poppet nozzles (Figure 8-24) in the past on IDI engines with a precombustion chamber. They are perhaps the most simple of the hydraulic injector nozzles and as such were the cheapest to manufacture. Poppet nozzles are not easily reconditioned. The technician is normally required to bench test the nozzle for the correct NOP value and observation of spray pattern and either reject or accept it for continued service.

Poppet nozzles use an outward- or forward-opening valve principle. This means that all the fuel pumped to the injector ultimately ends up in the engine cylinder eliminating the necessity for the leak-off lines (return circuit) used with most other nozzle types. Nozzle opening pressure parameters generally range between 500 and 1,800 psi (35 and 125 atms). For instance, the Caterpillar poppet nozzles has a NOP of around 800 psi (55 atms).

Poppet nozzles operate as hydraulic switches to define the NOP and atomize the fuel. The nozzle spring loads the poppet valve onto its seat. Hydraulic line pressure (from the injection pump) acts on the sectional area of the seated upper portion of the poppet valve. When the injection pump delivers pressure sufficient to unseat the poppet valve, fuel passes around the poppet to exit from the nozzle's single orifice.

When a precombustion chamber design is used, the engine cylinder clearance volume is generally too large to generate compression temperatures in cold weather sufficiently high to ignite the fuel charge. Glow plugs were often designed into IDI engine systems as a cold weather starting aid.

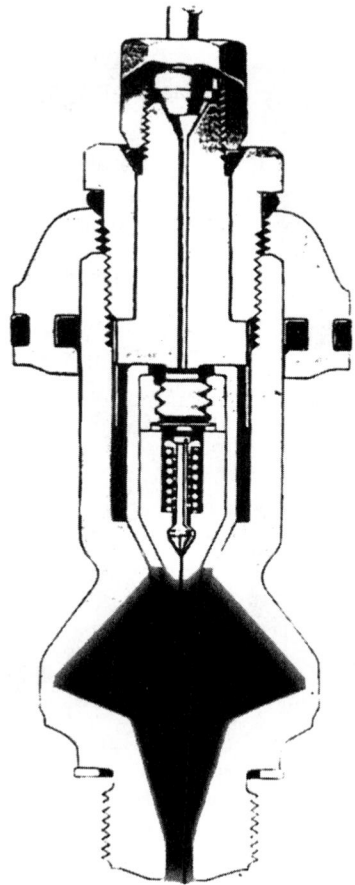

Figure 8-24 Poppet injector. (Courtesy of Caterpillar Inc.)

Pintle Nozzles

Like poppet nozzles, **pintle nozzles** (Figure 8-25) are used mainly in IDI engine applications and have seldom been used in truck and bus CI engines. The pintle nozzle body is designed with a single exit orifice. The valve pintle protrudes through this single orifice even when in the fully open position. This means that the fuel exiting the nozzle is forced around the pintle, producing a somewhat conical (cone-shaped) spray pattern. The valve body and pintle valve are lapped together in manufacture to a tolerance of around 2 μ which does not permit the components to be interchanged.

Pintle nozzles generally achieve better **atomization** than poppet nozzles; they have a much greater service life because they are easily reconditioned. Nozzle opening pressure parameter ranges between 1,470 and 2,000 psi (100 and 150 atm) with peak system pressures seldom exceeding 7,350 psi (500 atm).

Pintle Nozzle Operation   Fuel is delivered to the pintle injector (Figure 8-25) from the fuel injection pump by means of a high-pressure pipe. The nozzle valve is held in the closed position mechanically by spring pressure either acting directly on the pintle valve or transmitted by means of a spindle (a shaft that transmits spring force to the nozzle valve). Fuel is ducted through the injector assembly to the pressure chamber within the nozzle. This permits the line pressure to act on the sectional area of the pintle valve exposed to the pressure chamber. Whenever this pressure is sufficient to overcome the spring pressure that loads the valve on its seat, the pintle valve retracts. This action permits fuel to flow past the seat and exit through the injector orifice. Fuel

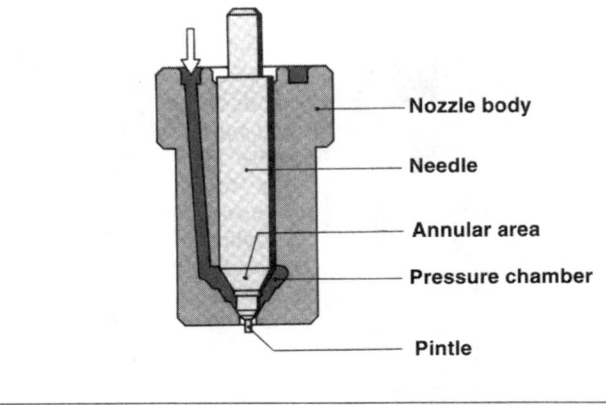

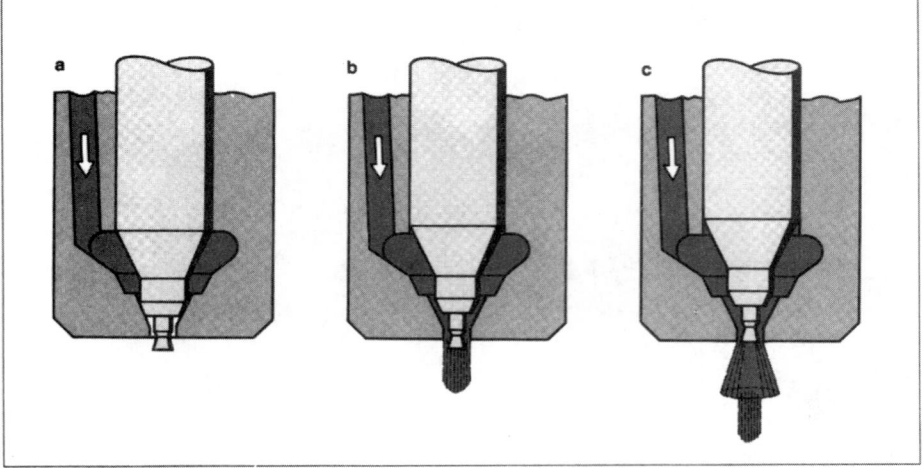

Figure 8-25 Pintle nozzle operation. (Reprinted with permission from Robert Bosch Corporation)

will continue to flow through the nozzle orifice as long as hydraulic pressure exceeds the mechanical force represented by the injector spring. The fuel injection pump determines the length of a fueling pulse, and when line pressure collapses, the spring reseats the nozzle valve, ending injection.

The pintle nozzle uses an inward opening valve principle requiring the use of leak-off lines or pipes to return fuel to the tank. The leakage takes place at the pintle valve-to-nozzle-body clearance and tends to increase as these matched components age.

The spray pattern emitted from a pintle nozzle depends on the pintle design. Throttling nozzles are designed to emit less fuel at the beginning of the injection pulse and may be identified by their conical-shaped pintle valves.

Orifice Nozzles

As stated in the introduction to this chapter, most current truck and bus diesel engines use closed hydraulic injector nozzles. These nozzles in almost all cases are of the multi-orifice type. Multi-orifice nozzles (Figure 8-26) appear in truck engines in traditional, pump-line-nozzle configurations, and they are an integral subcomponent in EUIs and HEUIs. They are also used with electronic unit pump (EUP) fueled engines. **Orifice** nozzles may have one or more orifices, almost

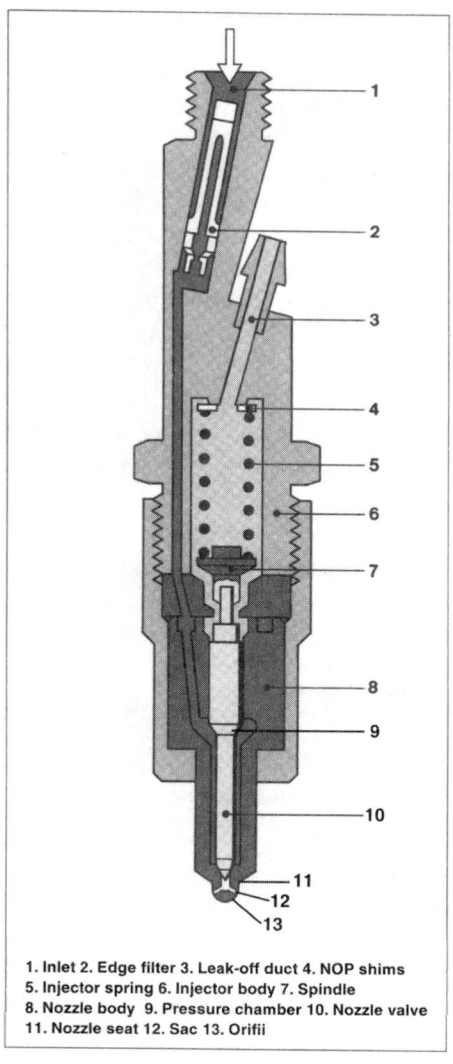

1. Inlet 2. Edge filter 3. Leak-off duct 4. NOP shims
5. Injector spring 6. Injector body 7. Spindle
8. Nozzle body 9. Pressure chamber 10. Nozzle valve
11. Nozzle seat 12. Sac 13. Orifii

Figure 8-26 Sectional view of a multi-orifice nozzle. (Reprinted with permission from Robert Bosch Corporation)

always the latter. Multi-orifice nozzles are required in most high-speed DI diesel engines because only these produce the necessary degree of atomization (droplets between 10 and 100 μ). Atomized droplets of 10 μ or less can result in erratic ignition or no ignition at all. Atomized droplet sizing 100 microns or more are too large to be completely combusted (oxidized) in the time dimension window of high-speed CI engines with rated speeds of 1,800–2,600 rpm. The time dimension within which combustion of the fuel must take place is defined by engine rpm. If an engine is run at 2,000 rpm, it takes exactly 0.030 seconds (or 30 msec) for a single revolution of the engine. This means that a piston travels from TDC to 90 ATDC (more or less the period available for combustion) in 7.5 msec. Top end engine speeds dictate the combustion or burn duration window and therefore the maximum atomized droplet sizing.

Orifice nozzles use an inward opening valve principle and that requires leak-off lines. The leakage occurs at the nozzle valve-to-nozzle-body clearance and is measured on the injector test bench as *back leakage*. NOP parameters in current engines range from 1,800–5,880 psi (120–400 atm) with peak system injection pressures ranging from two to ten times the NOP value. Multi-orifice nozzles used in hydromechanical engines typically have NOP in the region of 3000-psi (200 atm). The multi-orifice nozzles used in EUIs and HEUIs generally have NOP values closer to the higher range of the typical parameters, around 5,000 psi (350 atm).

Operation   The multi-orifice nozzle (Figure 8-26) is usually locked by a dowel or pin in the injector assembly to ensure that the spray pattern is directed to a specific location of the combustion chamber or, in the case of direct injection, to a specific location of the cylinder. Fuel from the injection pump element is delivered to the nozzle holder and then ducted to an annular recess in the upper nozzle valve body. Either single or multiple fuel ducts extend from the annular recess to the pressure chamber, meaning that the pressure here will be the same as that in the high-pressure pipe. The nozzle valve is loaded to a closed position on its seat, either directly by the injector spring or indirectly by spring pressure relayed to the nozzle valve by means of a spindle. This spring tension is adjustable by either a screw or by shims, and it will set the NOP value. When line pressure is driven upward by the injection pump, pressure will increase in the nozzle pressure chamber, and this pressure is sufficient to overcome the mechanical force of the injector spring. The nozzle valve retracts permitting the fuel to flow past the seat into the sac and exit the nozzle orifice. At center of the nozzle valve seat, a single duct connects to the nozzle sac. The sac is a spherical chamber into which the nozzle orifices are drilled. The sac hydraulically balances the fuel exiting the orifices, so that droplets start to exit from each orifice at approximately the same moment at the beginning of the fuel pulse. The injection pulse will continue for as long as the nozzle valve remains open. Depending on the specific means used to generate injection pressures, *maximum* fueling pulses can extend from 25 to 50 crank angle degrees. The fueling pulse ends when the metering/pumping element and the pressure in the nozzle assembly drops collapse the pressure to a value below NOP.

The emitted droplet sizing from the nozzle orifices is factored by pressure and flow area. The flow area is represented by the sizing of the nozzle orifices and therefore remains constant. However, the pressure values vary considerably, extending from a value lower than NOP up to the peak pressure value (the highest pressure attainable in a fuel injection system). As pressure increases, the droplet size will decrease. The more prolonged the injection pulse (that is, the longer the duration of the effective pump stroke), the higher the circuit pressure at injection pump port opening (end of delivery) and the smaller the atomized droplet sizing. The reduction in droplet sizing that occurs as the fueling pulse is extended is generally favorable for the complete combustion of the fuel, except for the short period of fueling that takes place *after* pump **effective stroke** completion and *before* nozzle valve closure. As the pressure collapses in the pump element and therefore the injection circuit, the emitted droplet sizing will increase until the moment the nozzle valve actually closes, so it is generally desirable for this pressure collapse to occur rapidly.

Nozzle Differential Ratio

The nozzle differential ratio describes the geometric relationship between the sectional area of the nozzle seat and that of the pressure chamber or valve shank. Nozzle valves are opened by hydraulic pressure acting on the sectional area of the valve subject to the pressure chamber. When this overcomes the spring pressure that loads them on their seat, they retract. However, the instant the nozzle valve unseats, hydraulic pressure is permitted to act over the whole sectional area of the nozzle valve. This area is the sectional area of both the seat and the pressure chamber.

When pressure rise begins in the injection line circuit prior to the opening of the nozzle valve, the hydraulic circuit that is being acted upon is closed (it is sealed at the nozzle seat). The instant the NOP value is achieved, the nozzle valve unseats and opens the circuit. This must result in a drop in line pressure. However, because at the moment the nozzle valve opens, that of the seat sectional area increases the sectional area over which the hydraulic pressure acts, less pressure is now required to hold the nozzle open. The nozzle differential ratio must be sufficient to prevent nozzle closure when this pressure drop occurs. After NOP, fuel passing around the seat fills the sac and pressure rise resumes because of the restriction represented by the minute sizing of the nozzle orifices. The nozzle differential ratio means that nozzle closure at the end of injection will occur at a value somewhat below the NOP value. It will also help define the specific residual line pressure value (pressure at which dead volume fuel is retained in the high-pressure pipe). More pressure is required to unseat a nozzle valve than the pressure required holding it off its seat. This is due to the nozzle differential ratio.

The term differential means a difference in sectional areas.

Nozzle Holders or Injectors

Injectors are simply mounting devices for hydraulic nozzles. They come in many different shapes and sizes for a variety of reasons; for instance, long-stem nozzles are easier to cool. Older hydraulic injectors were usually of the high-spring design. In high-spring injectors the spring was in the upper portion of the holder and relayed the spring tension to the nozzle value by means of a spindle. Spring tension was altered directly by an adjusting screw; setting the NOP value on the pop tester was a quick and easy procedure. A disadvantage of the high-spring design was at NOP. The nozzle valve slammed open until it was mechanically prevented from further inboard travel and because of its high opening velocity. The spindle was driven into the spring, which rebounded sufficiently to hammer the nozzle valve from its open position back into the seat, interrupting the pulse and affecting emitted droplet sizing. More recent hydraulic injector nozzles use a low-spring design that eliminates the spindle, thus reducing the mass of moving parts. This has minimized rebound interference of the injection pulse. However, low-spring injectors generally use shims acting on the spring to define the NOP value, which extends the time required to set the NOP value on the test bench.

Pencil-Type Injector Nozzles

Pencil-type injector nozzles are multi-orifice nozzles and share common operating principles with other multi-orifice nozzles with some small exceptions. They are seldom found in any current truck engine applications. Caterpillar has used them in some of their hydromechanical engines. The pencil nozzle assembly is cylindrical. Within the nozzle body, the nozzle valve extends through nearly the full height of the injector assembly, and the injector spring acts directly on top of the nozzle valve shaft. Adjustment is by means of an adjusting screw and locknut located at the top of the assembly. Some pencil nozzles have no leak-off lines. Fuel that bleeds by the nozzle valve during injection is accumulated in a chamber above the valve, and pressure equalization with line pressure occurs after nozzle closure. Pencil nozzles are most often damaged in the process of removal and service. When making any adjustments, use the recommended mounting fixture and observe the specified torque procedure.

VCO Nozzles

VCO (valve closes orifice) nozzles have eliminated the sac. The function of the sac in the nozzle is to provide balanced fuel dispersal, which is especially important in keeping the ignition lag time dimension consistent. However, at the completion of the injection pulse, the volume of fuel in the sac was essentially wasted fuel that added to hydrocarbon emission. At the instant of nozzle closure, the sac and the nozzle orifices would contain fuel that would be vaporized due to the heat of combustion but, at best, only partially combusted. Current injector nozzle designs have either substantially reduced the sac volume or eliminated it entirely. A true VCO nozzle has orifices that extend directly from the seat. VCO nozzles are most often used on electronically managed injection systems that use high NOP values, which can compensate somewhat for the compromising of the balanced fuel dispersal offered by nozzle sacs.

Terms to Know

Atomization

Cetane number

Cloud point

Delivery valve

Effective stroke

Fuel heater

Hydromechanical

Injection rate

Metering

Multi-orifice nozzle

Orifice

Pintle nozzle

Port closure

Port opening

Positive displacement

Pour point

Primary filter

Secondary filter

Timing

Transfer pump

Viscosity

Volatility

Water separator

Summary

❏ The ignition quality of a diesel fuel is rated by its cetane number (CN). Fuel deteriorates chemically and can be subject to microorganic contamination when stored for prolonged periods. Low sulfur fuels generally have lower lubricity and have produced failures with some of the older hydromechanical fuel systems and even some newer EUI systems.

❏ The typical fuel transfer system can be divided into a suction circuit and a charge circuit, separated by a transfer or charge pump. Some fuel systems may locate a transfer pump in the tank itself, while others, such as Cummins PT, may retain the entire fuel subsystem under suction.

❏ In the typical system, the primary filter is normally under suction. The secondary filter entraps smaller sized particulate than a primary filter and is subject to system charging pressure. Many current secondary filters will plug on water and shutdown the engine.

❏ Water may be found in fuel in three forms: free state, emulsified, and absorbed. Many fuel subsystems are equipped with a water separator designed to remove free state and emulsified water from fuel.

❏ Two types of fuel heaters are in current use: the electric element and coolant medium, heat exchanger types.

❏ Diesel fuel systems commonly use one of two different fuel transfer or charge pumps: reciprocating plunger pumps and gear pumps. They are both of the positive displacement type. Some fuel subsystems are equipped with a hand primer pump. The function of a hand primer pump is to purge air from the fuel subsystem.

❏ The fuel system must be timed to the engine and be capable of pressurizing the fuel to values up to 2000 atms. *Injection rate* refers to fuel injected per crank angle degree and is usually governed by the pump actuating mechanism geometry, usually a cam. The fuel system must atomize fuel to the precise amounts and deliver the fuel to the correct cylinder at the correct time to ensure balanced power output. Injector nozzle orifice sizing and the subjected pressure determine atomized droplet sizing.

❏ Injector nozzles are simple hydraulic switching mechanisms. Direct injected engines usually require the use of multi-orifice nozzles. Direct injected engines require atomized droplets within the 10–100 μ range.

Review Questions

Short Answer Essays

1. Explain the following fuel terms: specific gravity, viscosity, and cetane rating.

2. Explain how the cetane rating can affect diesel engine driveability.

3. Define the terms cloud point and pour point and their importance in cold weather Diesel engine operation.

4. Define the role of primary and secondary fuel filters.

5. Explain how a water separator works.

6. Define the principles of operation of a transfer or charge pump.

7. Differentiate between the hydromechanical injection operating principles of timing, pressurization metering, atomization, and distribution.

8. Explain atomization and the droplet sizing factors required for a direct injected diesel engine.

9. Identify three types of hydraulic injector nozzle and the primary sub-components of a nozzle assembly.

10. Describe the hydraulic principles of operation of poppet, pintle and multi-orifice nozzles.

Fill-In-the-Blanks

1. The ignition quality of a diesel fuel is rated by its _____ _____.

2. There is a correlation between ____ _____ and its heating value.

3. Fuel deteriorates chemically and can be subject to _____ contamination when stored for prolonged periods.

4. Low sulfur fuels generally have _____ _____ and have produced failures with some of the older hydromechanical fuel systems and even some newer EUI systems.

5. The fuel _____ is defined as the group of components responsible for fuel storage and its transfer to the injection pumping apparatus.

6. The typical fuel transfer system can be divided into a _____ circuit and a _____ circuit, separated by a transfer or charge pump.

7. _____ _____ refers to fuel injected per crank angle degree and is usually governed by the pump actuating mechanism geometry, usually a cam.

8. The fuel system must be capable of _____ fuel to the precise dimensions required of the specific fuel system and injector nozzle orifice sizing and the subjected pressure determines atomized droplet sizing.

9. The fuel system must deliver the fuel to the correct cylinder at the correct time to ensure _____ _____ _____.

10. The fuel system is responsible for managing engine _____ _____.

ASE-Style Review Questions

1. The temperature at which a diesel fuel begins to form paraffin wax crystals is known as:
   - A. Cloud point
   - B. Pour point
   - C. Flash point
   - D. Freeze point

2. What type of pumping principle does the typical hand primer pump use?
   - A. Single acting plunger
   - B. Double acting plunger
   - C. Rotary gear
   - D. Cam actuated, diaphragm

3. The precise control of injected fuel quantity is known as:
   - A. Timing
   - B. Metering
   - C. Atomization
   - D. Spray dispersion

4. Which of the following diesel fuel injection terms is used to describe the amount of fuel injected per crank angle degree?
   - A. Metering
   - B. Injection rate
   - C. Timing
   - D. Phasing

5. Which type of injection nozzle would be found on most direct injected truck diesel engines?
   - A. Poppet
   - B. Pintle
   - C. Throttling
   - D. Multi-orifice

6. At refueling on the completion of a trip fuel is steaming in the tanks. *Technician A* says this condition is a result of normal operation in some fuel systems. *Technician B* says such a condition can be a result of running the tanks low on fuel in a high flow fuel system. Who is correct?
   - A. A only
   - B. B only
   - C. Both A and B
   - D. Neither A nor B

7. *Technician A* says that diesel fuel can combine with air to form potentially explosive mixtures. *Technician B* says diesel fuel is likely to degrade more quickly in the heat of summer than in winter. Who is correct?
   - A. A only
   - B. B only
   - C. Both A and B
   - D. Neither A nor B

8. Which of the following factors will *most* likely to prevent secondary injections?
   - A. Pressure wave reflection
   - B. Nozzle differential ratio
   - C. The nozzle sac
   - D. The pressure chamber sectional area

9. Nozzle differential ratio is a ratio of:
   - A. Nozzle seat sectional area and total nozzle sectional area
   - B. The high pressure pipe sectional area and the orifice flow area
   - C. The nozzle pressure chamber sectional area and the high pressure pipe sectional area
   - D. The ratio of mechanical spring force and hydraulic pressure

10. Which type of injection nozzle would be found on most direct injected truck diesel engines?
    - A. Poppet
    - B. Pintle
    - C. Throttling
    - D. Multi-orifice

Mechanically Managed Diesel Fuel Systems

Upon completion and review of this chapter, you should be able to:

❏ Identify the major components of a typical in-line–type port-helix metering injection pump and explain its principles of operation.

❏ Identify the major components of a distributor-type injection pump and explain how the pump's principles of operation apply to these components.

❏ Define metering and the factors that control it.

❏ Explain the operation of injection pump peripherals, including aneroid devices, altitude compensators, and variable timing/timing advance mechanisms.

❏ Outline the principles of operation of a Detroit Diesel Corp. (DDC) mechanical unit injector (MUI).

❏ Describe how MUI effective stroke is varied to control injected fuel quantity.

❏ Describe the components that link the MUIs with the governor assembly.

❏ Outline the principles of operation of a Caterpillar mechanical unit injector system.

❏ Identify the major components in a Cummins PT-type metering injection pump and explain their operation.

❏ Explain the function and operation of a hydromechanical governor on a diesel engine.

Introduction

This chapter will cover in detail the operation of the various mechanical fuel injection pumps. These injection pumps include pump-line-nozzle (PLN) systems that use in-line type, distributor-type pumps, and the Cummins PT-type pump. One section will cover the mechanical governors used with in-line–type pumps. This chapter also covers the two **mechanical unit injector** systems.

A BIT OF HISTORY

The credit for the first commercially acceptable fuel injection pump goes to Robert Bosch who developed his prototype units in 1927 before World War II. From this invention and the experience gained from diesel engine submarines during World War I, the automotive, or high-speed, diesel engine evolved. In Germany, MAN and Daimler Benz produced the early models based on the marine engines as an initial start. Rapid improvements in design and adaptation by 1924 were followed by new interest in Switzerland, the United States, and England. By the late 1920s several engine manufacturers had produced successful prototype engines.

In-Line–Type Fuel Injection (Port-Helix) Systems

Robert Bosch's first high-pressure liquid fuel injection pump evolved into the fueling apparatus used by Caterpillar, Mack Trucks, Navistar, and other truck power plant manufacturers into the

Shop Manual
Chapter 9, page 287.

Many technicians refer to in-line port-helix pumps as line pumps.

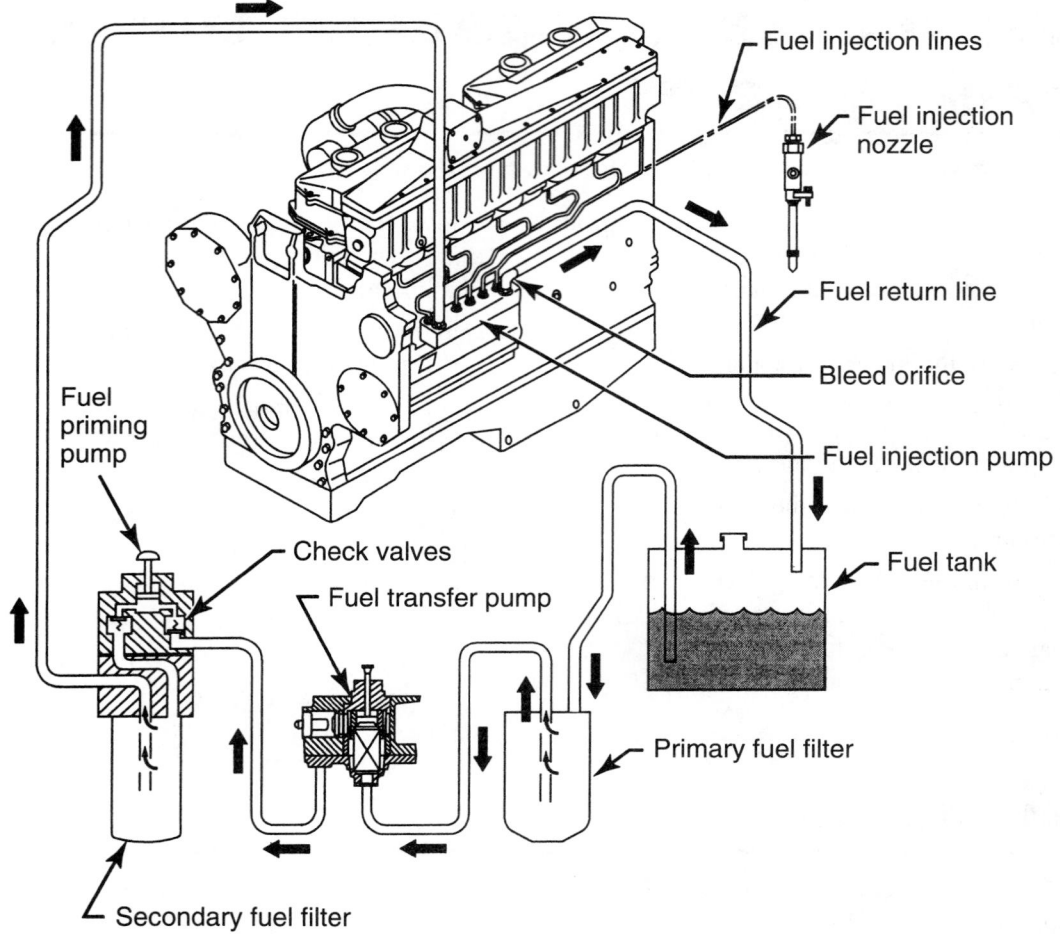

Figure 9-1 In-line–type pump fuel system. (Courtesy of Caterpillar Inc.)

1990s. The principles of pumping and metering a fuel charge have changed little since Bosch's first 1927 high-pressure injection pump. Figure 9-1 shows a typical in-line–type injection pump system. However, the port-helix (helix means scroll shaped) metering injection pump, emissions certified for the 1990's, will most likely be managed electronically rather than hydromechanically. The hard (not changeable) delivery-timing window, defined solely by the geometry of actuating cam profile and helix geometry, has given way to variable timing devices. This has permitted the technology of the in-line, port-helix metering pump to live on into the electronic age. In most cases, when the truck technician uses the term *pump-line-nozzle* (*PLN*), an in-line–type port-helix metering injection pump is being referred to. Despite the fact that these injection pumps are still in use at the present time, the days of using this technology in an environment of strict emission controls and simple, economical, electronically managed fuel systems are numbered. This section will describe the in-line–type port-helix metering pumps (Figure 9-2) engineered and manufactured by Bosch, Lucas Denso, Zexel, and Caterpillar and used by Caterpillar, Mack Trucks, Cummins, Navistar, Fuso, Hino, and others.

Technical Description

The typical port-helix metering pump (Figure 9-3) is used to fuel a truck or bus compression-ignition engine. It has an in-line configuration and is flange mounted to an engine accessory drive. It is driven through one complete rotation (360 degrees) per complete engine cycle (720 degrees). The internal pump components are housed in a frame constructed of cast aluminum,

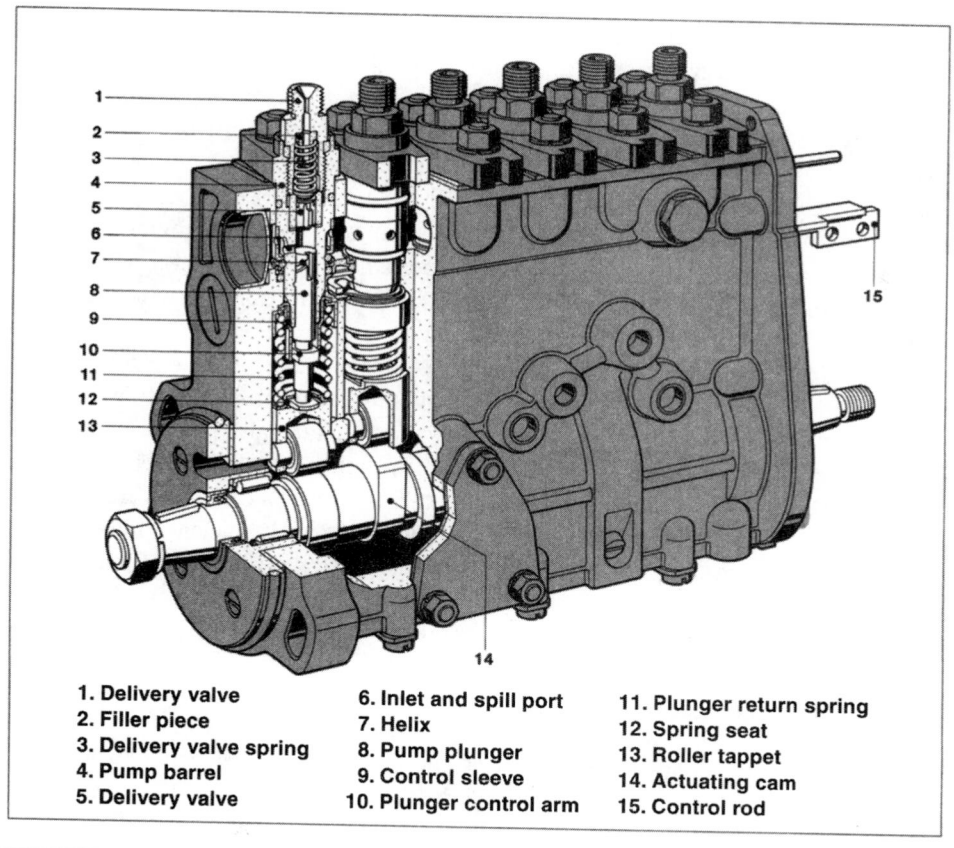

1. Delivery valve
2. Filler piece
3. Delivery valve spring
4. Pump barrel
5. Delivery valve
6. Inlet and spill port
7. Helix
8. Pump plunger
9. Control sleeve
10. Plunger control arm
11. Plunger return spring
12. Spring seat
13. Roller tappet
14. Actuating cam
15. Control rod

Figure 9-2 In-line–type port-helix pump cross-section. (Reprinted with permission from Robert Bosch Corporation)

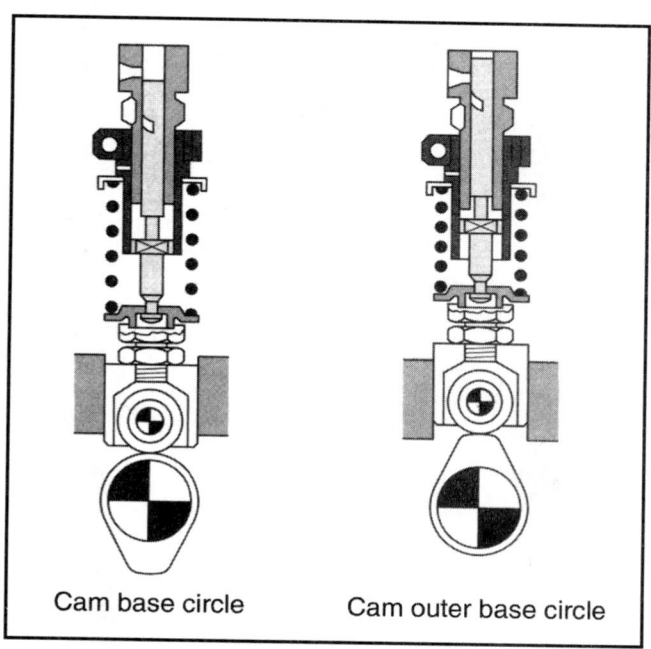

Cam base circle   Cam outer base circle

Figure 9-3 Actuating a port-helix–type pump element. (Reprinted with permission from Robert Bosch Corporation)

cast iron, or forged steel. The engine camshaft drives the injection pump by means of timed reduction gearing. The gear-driven pump drive plate is connected to the injection pump camshaft (the shaft fitted with eccentrics designed to actuate the pump elements), so rotating the pump drive plate rotates the pump camshaft. The camshaft is supported by main bearings and rotates within the injection pump **cambox**. The cambox is the lower portion of the injection pump, which houses the camshaft, tappets, and the integral oil sump.

In-line pump camshaft geometry provides one cam lobe for each engine cylinder. Riding each cam profile is a tappet assembly driving a pump element consisting of a plunger and a barrel. The barrel is stationary and drilled with two ports in its upper portion, which are exposed to the fuel-charging gallery.

Injection rate is defined as fuel quantity injected per crank angle degree; it is therefore determined by the geometry of the pump-element-actuating cam profile, as shown in Figure 9-4. Most port-helix metering injection pumps use a symmetrical cam profile whose periphery (outer surface) is mostly inner base circle. The plunger is held at the bottom of its travel for most of the cycle. Occasionally, asymmetrical cams and cams whose periphery is mostly outer base circle are used. The shaping of the cam contour will affect not only injection rate but also pump chamber breathing and cooling. A cam profile that is mostly outer base circle is sometimes known as a back-kick cam, which will preclude the engine from being started in reverse rota-

> The fuel quantity delivered to the combustion chamber is determined by the cam profile.

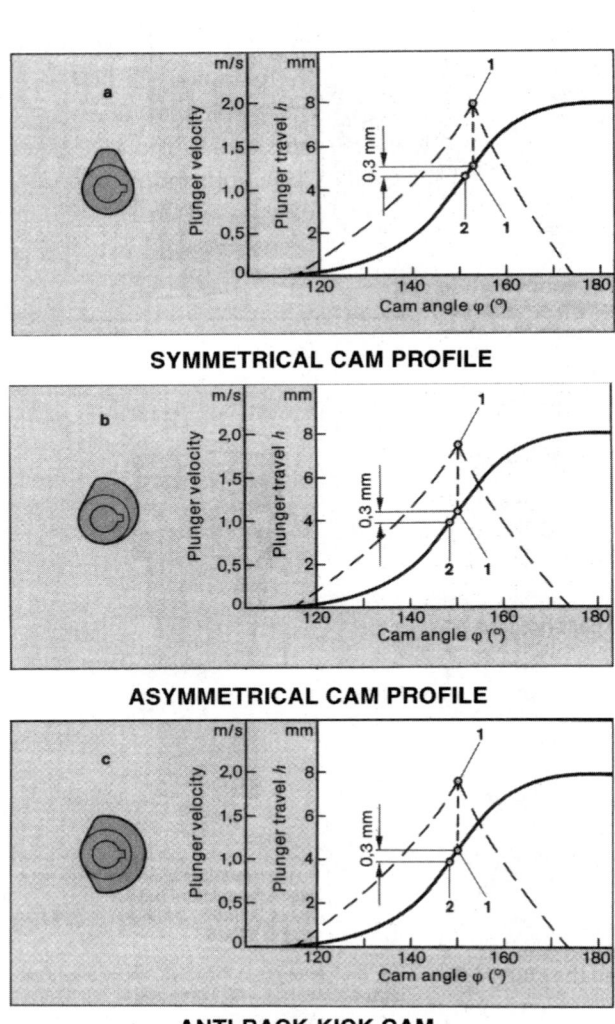

Figure 9-4 Actuating cam geometry. (Reprinted with permission from Robert Bosch Corporation)

tion. **Fuel rate** is a term usually used to define the actual rate of delivering fuel to an engine cylinder, so it is controlled by both cam profile and engine rpm.

The fuel gallery is charged with low-pressure fuel, typically between 1–5 atm (15–75 psi). This permits fuel to flow into and through the barrel ports when the plunger does not obstruct them. The plunger reciprocates within the barrel; it is loaded by spring pressure to ride its actuating cam profile. Therefore, actual plunger stroke is constant. Plunger-to-barrel tolerance is close; the components are lapped in manufacture to a tolerance of 2 to 4μ.

Fuel quantity to be delivered in each stroke is controlled by managing plunger **effective stroke** (Figure 9-5). The plunger is milled with a vertical slot or cross and center drillings and helical recesses. The function of the vertical slot or cross and center drillings is to maintain a constant conduit between the pumping chamber above the plunger and the helical recesses. In other words, whatever pressures exist in the pumping chamber must also exist in the helical recess. In this section, only plunger designs with a lower helix or helices will be discussed. Effective stroke describes the delivery stroke. The delivery stroke begins when the plunger is forced upwards by cam profile and the plunger leading edge (uppermost part of the plunger) traps off the spill port(s).

As the plunger rises through its stroke in the barrel, a rapid pressure rise occurs, ultimately creating the required injection pressures. The precise moment that effective stroke begins is known as **port closure** (Figure 9-6), when the inlet and spill ports close. Effective stroke is critically important to the diesel technician because its precise setting is used to control timing. It is of critical importance to the diesel technician because its precise setting is used to control timing. As pressure rises in the pump chamber, it acts first on a delivery valve, then on the fuel confined in the high-pressure pipe, transmitting fuel to the injector nozzle and finally delivering a fuel pulse to the engine cylinder. Effective stroke ends at **port opening** (Figure 9-7). This is the precise moment that the upward travel of the plunger exposes the helical recesses to the spill port. High-pressure fuel is spilled back to the charging gallery, causing a rapid collapse of pressure in the pump chamber, line, and nozzle. The injection pulse ceases when there is no longer sufficient pressure to hold the delivery and nozzle valves open. Port opening always occurs while the plunger is moving in an upward direction, that is, not at plunger top dead center (TDC) or beyond. This is required because the pressure in a port-helix pump element is designed to rise through the delivery stroke. This action produces smaller atomized droplets from the injector

The effective stroke begins at port closure and ends at port opening.

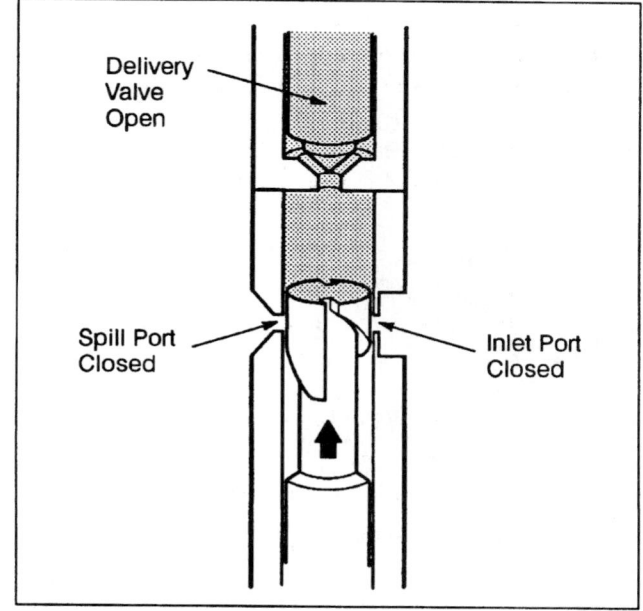

Figure 9-5 Effective stroke. (Courtesy of Mack Trucks, Inc.)

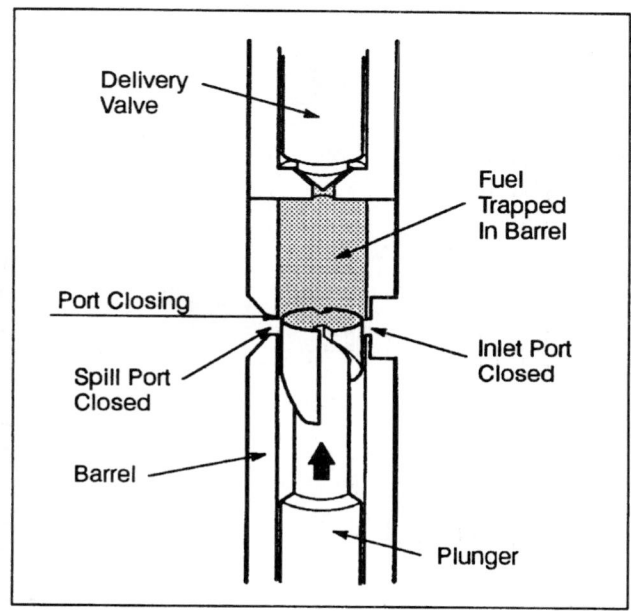

Figure 9-6 Port closure. (Courtesy of Mack Trucks, Inc.)

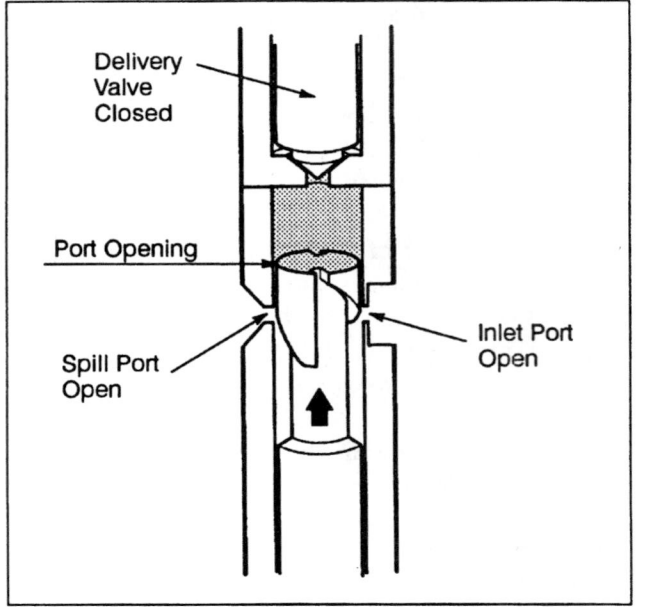

Figure 9-7 Port opening. (Courtesy of Mack Trucks, Inc.)

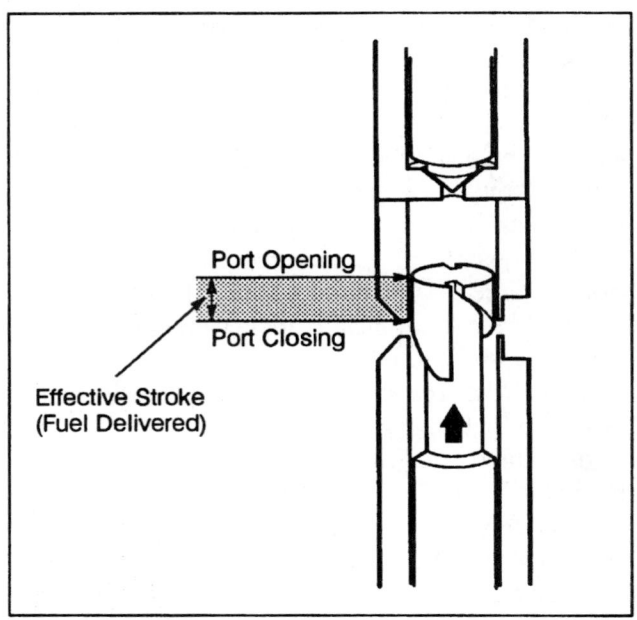

Figure 9-8 Plunger travel through the effective stroke. (Courtesy of Mack Trucks, Inc.)

toward the end of effective stroke. However, at the point of port opening, regardless of the length of the effective stroke, pump pressure should collapse as rapidly as possible and minimize the larger droplets emitted from the injector as pump pressure falls to a value below nozzle opening pressure (NOP).

The length of plunger effective stroke or the amount of fuel delivered (Figure 9-8) depends on the point at which the plunger helix vertically aligns with the spill port. This stroke lasts from the inlet port closure to spill port opening. A **control sleeve** (Figure 9-9) contains an integral

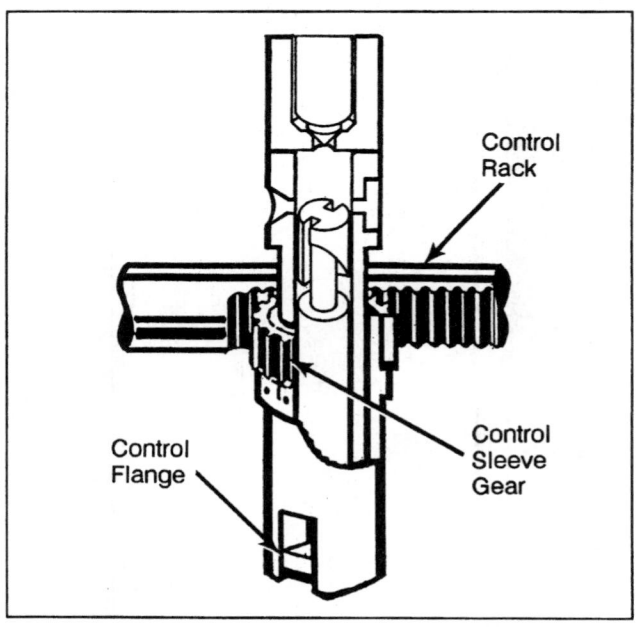

Figure 9-9 Control rack and sleeve gear. (Courtesy of Mack Trucks, Inc.)

gear lugged to the plunger that permits the plunger to be rotated while reciprocating. Rotation of the plunger in the bore of the barrel will change the registry of the plunger helix to the spill port. Therefore, plunger effective stroke depends entirely on the rotational position of the plunger. In multiple cylinder engines, the plungers move in unison to ensure balanced fueling at any given engine load.

The teeth of the control sleeves are meshed to a governor **control rack**, which, when moved linearly, rotates the plungers in unison. This means that in any linear position of the rack, all of the plungers will have identical points of register with their spill points, resulting in identical pump effective strokes. If this did not occur, the result would be an imbalance with retarded fueling, which would deliver different quantities of fuel to each cylinder.

Figure 9-10 shows the different control rack positions in relation to fuel delivery quantities. Engine shutdown is achieved by moving the control rack to the no-fuel position. The rotational position of the plungers is now such that the vertical slot will be in register with the spill port for the entirety of plunger travel. The plunger will merely displace fuel as it travels upward, with no pumping action possible. In other words, as the plunger is driven into the pump chamber, the fuel in the chamber will be squeezed back down the vertical slot to exit through the spill port and return to the charging gallery. Some plungers contain a starting notch that provides retarded timing for easier engine starting. When the rack is in the start position (that is, extended beyond the maximum operating position), the starting notch aligns with the fill port for filling. When the plunger is lifted at this position, maximum fuel will be delivered. Yet, it will occur later or retarded until the edge of the starting notch closes the fill port. Retarded timing aids in starting because the minimal advance will provide the best torque. Figure 9-11 shows the starting notch on the left and the cold start retarded timing effect on the right.

Most port-helix metering injection pumps use delivery valves (Figure 9-12) to isolate the high-pressure circuit that extends from the injection pump chamber to the seat of the nozzle valve. Fuel retained in the high-pressure pipes (pipes that connect the injection pump elements with hydraulic injectors) between pumping pulses is known as dead volume fuel. It is retained at a pressure somewhat below the nozzle opening pressure value. The high-speed hydraulic switching that occurs in the high-pressure circuit creates pressure wave reflections, which may have the effect of spiking (causing surges) line pressures. To ensure that these pressure spikes do not exceed the NOP value and cause secondary injections, the dead volume fuel is retained at about two-thirds of the NOP value; this is known as residual line pressure.

> Dead volume fuel is the fuel retained in the high-pressure pipes (pipes that connect the injection pump elements with hydraulic injectors) between pumping pulses.

> ■ **CAUTION:** Eye protection should be worn when working with or near the high-pressure fluids generated by in-line hydromechanical injector pumps. High-pressure atomized fuel is extremely dangerous, and no part of the body should contact it. Because diesel fuel oil is a known carcinogen (a cancer-causing agent), hands should always be washed after contact with it.

The delivery valve is held in its closed position on its seat by a spring and by the residual line pressure. This mechanical force is compounded when the residual line pressure acts on the sectional area represented by the delivery valve and establishes the pressure value that must be developed in the pump chamber before it is unseated. The delivery valve flutes provide a means of guiding the valve in its bore while permitting hydraulic access between the retraction collar (or retraction piston both terms are used) and the pump chamber (plunger and barrel assembly). The retraction collar seals the pump chamber from the dead volume fuel, which is retained at a higher pressure value. Consequently, when the delivery valve is first unseated, it is driven upward in its bore by rising pressure in the pump chamber. It acts as a plunger driving inward on the fuel retained in the high-pressure pipe (Figure 9-13).

When the retraction collar clears the delivery valve seat, fuel in the injection pump chamber and the high-pressure pipe unite. Rising pressure subsequently unseats the injector

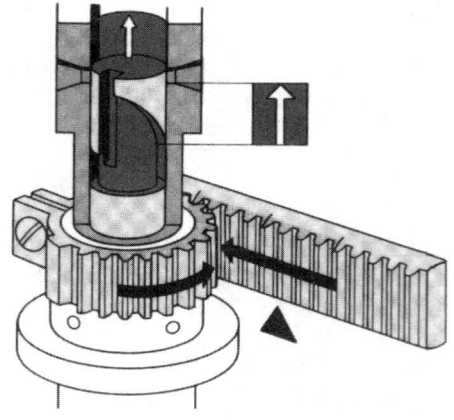

**FULL FUEL
MAXIMUM EFFECTIVE STROKE**

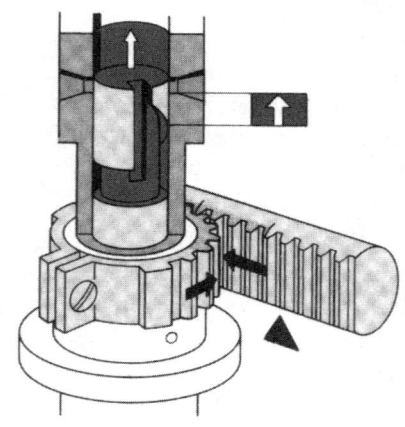

**MEDIUM FUEL
PARTIAL EFFECTIVE STROKE**

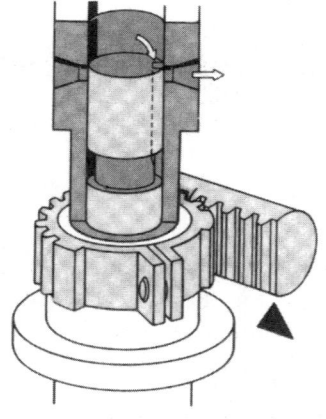

**NO FUEL ROTATIONAL POSITION
NO EFFECTIVE STROKE**

Figure 9-10 Rack position and relation to fuel delivery quantity. (Reprinted with permission from Robert Bosch Corporation)

Starting groove

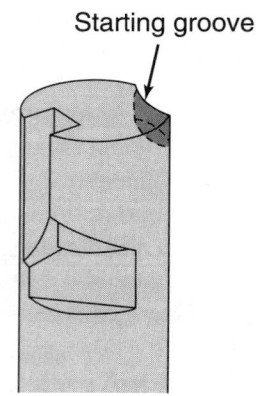

A

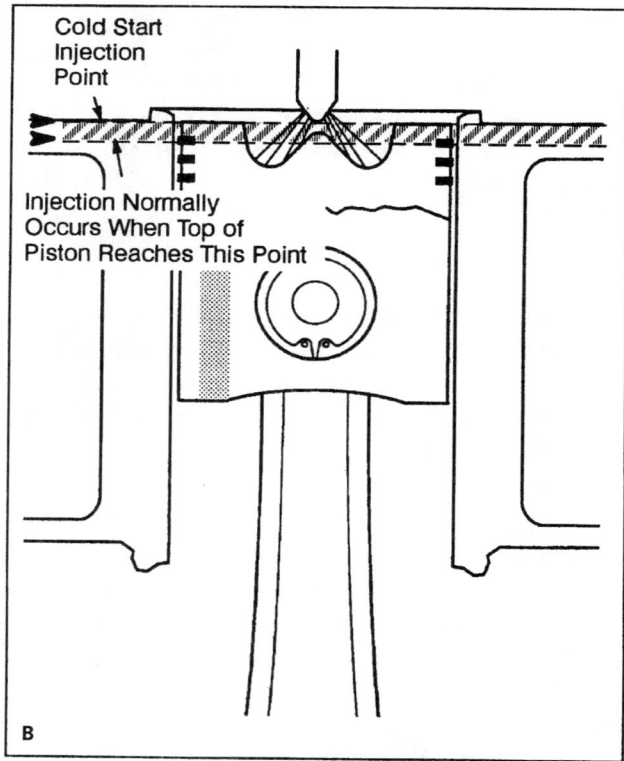

Cold Start Injection Point

Injection Normally Occurs When Top of Piston Reaches This Point

B

Figure 9-11 (A) Cold starting notch (Courtesy of Freightliner Corporation); (B) Cold start retarded timing injection (Courtesy of Mack Trucks, Inc.).

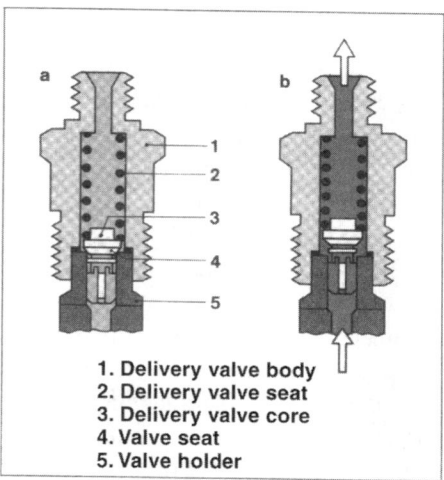

Figure 9-12 Typical delivery valve. (Reprinted with permission from Robert Bosch Corporation)

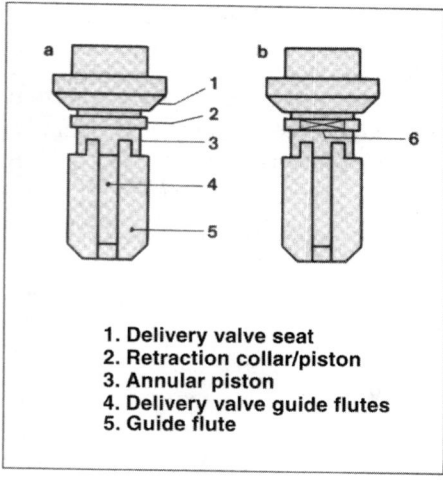

Figure 9-13 Delivery valve core terminology. (Reprinted with permission from Robert Bosch Corporation)

nozzle valve (NOP) and forces atomized fuel into the engine cylinder. At port opening, the effective pump stroke ends, beginning a rapid pressure collapse as fuel spills from the barrel spill port.

When there is insufficient pressure in the pump chamber to hold the nozzle valve in its open position, spring pressure overcomes hydraulic pressure and the valve seats, sealing the nozzle end of the high-pressure pipe. Almost simultaneously, the delivery valve begins to retract. The instant the retraction collar passes the delivery valve seat, it hydraulically seals the pump end of the high-pressure circuit. However, after sealing the pump end of the high-pressure circuit, the delivery valve must travel further before it seats, increasing the volume available for dead volume fuel storage. This causes a drop in line pressure and defines the residual line pressure value. The swept volume of the retraction collar increases the volume available for fuel storage in the high-pressure pipe. This volume of fuel is known as dead volume fuel and is retained at residual line pressure.

> ⚠ **WARNING:** The retraction collar swept volume is matched to the length of the high-pressure pipe to set achieve a precise residual line pressure value, so that if the pipe length is shortened, the pressure may increase enough to burst the pipe.

Most delivery valves are of the constant volume type, as described here. This rather more complex valve design helps minimize the effects of pressure wave reflection, which causes wear and cavitation.

Governors

Either a governor or rack actuator acts as the control mechanism for managing fueling. The only factor that determines the output of a diesel engine is the amount of fuel metered into its cylinders. Given unlimited fuel, a diesel engine is capable of accelerating at a rate of 1,000 rpm per second until self-destruction occurs. To prevent such an "engine overspeed" and to provide the engine with a measure of protection from abuse, a governor must precisely manage the fuel quantity delivered to the engine cylinders under all operating conditions. The fuel control mechanism on a port-helix metering injection pump is the rack, so the governor or rack actuator controls engine fueling by precise positioning of the rack. Governors are covered in detail at the end of this chapter.

Lubrication

The lower portion of the port-helix metering injection pump is lubricated by engine oil. In older versions, the injection pump lubricating circuit was often isolated from that of the engine, and the lube level would be checked and replenished through a dipstick located in the cambox. Contemporary injection pumps tend to be plumbed into the main engine lubricating circuit. The camshaft main bearings are often pressure lubed directly from an engine oil gallery; the remainder of the cam box components are splash lubricated from the oil held in the sump. The diesel fuel being pumped through the charging gallery lubricates the upper portion of the injection pump, so the lubricity of the diesel fuel is critical.

It is crucial for the operation of both the fuel injection pump and the engine it manages that the fuel used to lubricate the upper portion of the pump never comes into contact with the engine oil in the cambox. Plunger-to-barrel clearance is lapped in manufacture to a minute tolerance, and fuel that leaks by the recessed metering areas bleeds to an annular belt in the barrel. A duct connects the annular belt in the barrel to the charging gallery permitting bleed-by fuel to be routed there. This is known as a viscous seal. Trace leakage of a viscous seal can rapidly cause engine oil contamination and lead to lubricant breakdown. Viscous seals fail due to plunger side loading (single-helix design), prolonged usage wear, and fuel contaminants, especially water. Fuel that bleeds by the metering recess serves to guide the plunger true in the barrel bore, minimizing metal-to-metal contact. It should be remembered when studying the results of engine oil analysis that failure of the injection pump's viscous sealing ability is probably the least likely cause of fuel in the engine oil.

Timing Advance and Variable Timing Mechanisms

Older port-helix metering injection pumps were usually directly driven by reduction gearing from the engine camshaft gear. Such a system would dictate that the static timing value (port closure) occur at the same number of crank angle degrees before TDC regardless of engine load or speed when lower helix geometry is used. Fuel economy and emission control led to the development of mechanical advance mechanisms and, more recently, electronically managed, variable timing.

Mechanical timing advance mechanisms, as shown in Figure 9-14, are actuated by a set of flyweights and are eccentric CAM to advance the drive angle of the pump camshaft in relation to the pump drive gear, using a spiral gear on a shaft. The position of the fuel injection pump relative to that of the engine is advanced in direct proportion to the centrifugal force generated by the weight carrier. Port closure is therefore advanced as engine rpm increases. The extent of the advance offered by mechanical advance mechanisms can be as little as a 3 degree crank angle and seldom more than a 10 degree crank angle.

Electronically managed variable timing devices, such as those used by Mack Trucks and Caterpillar, employ a variable timing coupling mechanism between the pump drive gear and the pump camshaft. This intermediary was designed to establish a variable timing window of up to a 20 degree crank angle, managed by the engine management electronics. The static timing value (that is, port closure) specified is usually the most retarded parameter in the variable timing window. For example, an injection pump timed at 7 degrees BTDC with a 20 degree variable timing mechanism would permit the ECM to select any port closure value between 7 degrees BTDC and 27 degrees BTDC. Injection pump static timing (port closure) specifications of as little as 4 degrees BTDC are used, the reasoning being to limit combustion heat and therefore NO_x emission. Substantially retarded injection timing does not generally enhance either performance or fuel economy. In fact, the trade-off of retarded injection timing is an increase of hydrocarbon (HC) emission, but all emissions legislation is about keeping the noxious exhaust emissions within an acceptable window.

Aneroid Device

The aneroid device allows fueling changes at different altitudes.

By definition, an aneroid (Figure 9-15) is a low-pressure sensing device. In application, it is used on a turbocharged diesel engine to measure manifold boost and limit fueling until the boost pres-

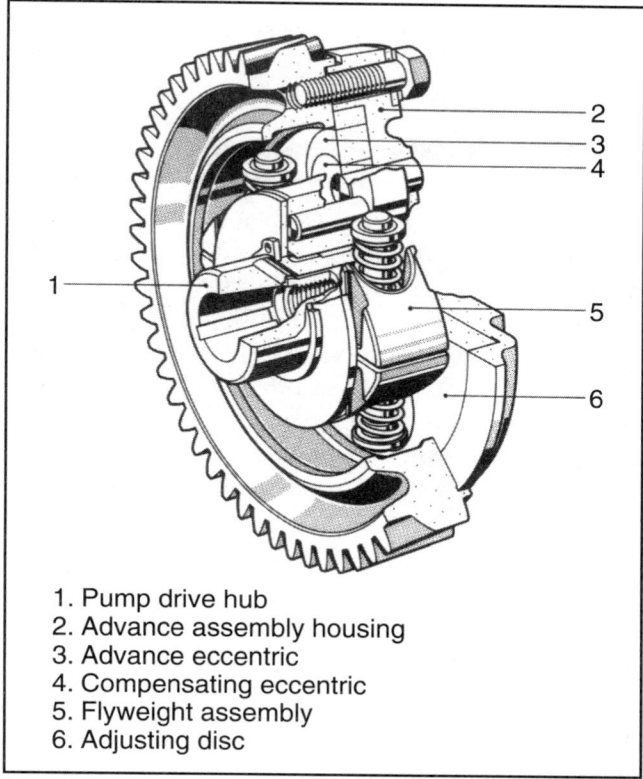

1. Pump drive hub
2. Advance assembly housing
3. Advance eccentric
4. Compensating eccentric
5. Flyweight assembly
6. Adjusting disc

Figure 9-14 Mechanical timing advance mechanism. (Reprinted with permission from Robert Bosch Corporation)

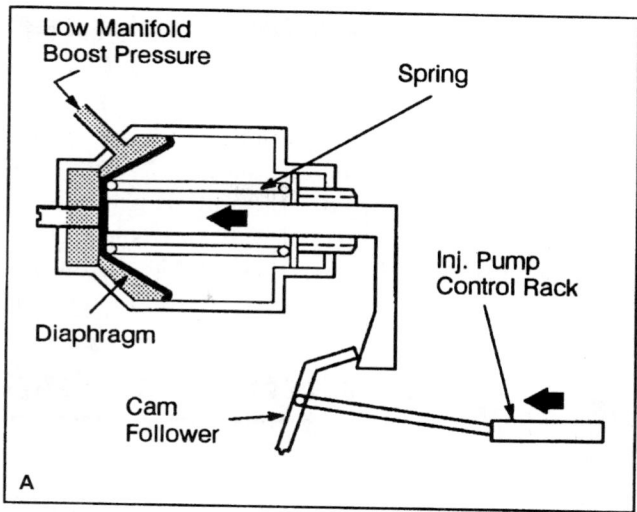

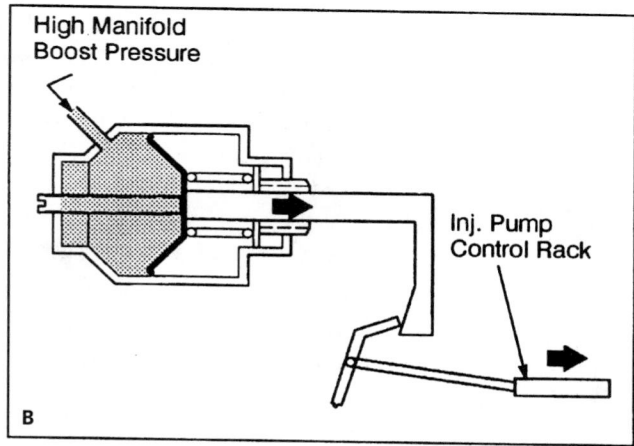

Figure 9-15 (A) Aneroid operation at low boost; (B) Aneroid operation at high boost. (Courtesy of Mack Trucks, Inc.)

sure achieves a predetermined value. Such devices are known variously as puff limiters, turbo boost sensors, AFC valves, and smoke limiters, and all seek to accomplish the same objective. They typically consist of a manifold within which is a diaphragm. Boost air is piped from the intake manifold to act on the diaphragm. Such devices are used on most current boosted (turbocharged) engines.

When an aneroid is used on an in-line pump, it is usually consists of a manifold, spring, and control rod. The manifold is fitted with a port, and a steel line connects it directly to the engine intake manifold. In this way, boost pressure is delivered to the aneroid manifold where it will act directly on the enclosed diaphragm. Attached to the diaphragm is a linkage connected either directly or indirectly to the fuel control mechanism (rack). A spring loads the diaphragm to a closed position, which will limit fueling by preventing the rack from moving into the full fuel position. When manifold boost acting on the diaphragm is sufficient to overcome the spring pressure, it acts on the linkage, permitting the rack full travel and thus maximum fueling.

Altitude Compensator

An altitude compensator device contains a barometric capsule that measures barometric pressure and down-rates engine power at higher altitudes to prevent overfueling. The devices are required when running at higher altitudes because the oxygen density in the air charge decreases with an increase in altitude, and unless the fuel system is aware of this, it effectively overfuels the engine. The critical altitude at which some measure of injected fuel quantity deration is required to accord with today's low emission engines is 1,000 ft. (303 meters) above sea level in altitudes. Older engines with altitude compensators usually did not derate the fuel system until an altitude of at least 2,000 ft. (606 meters) above sea level.

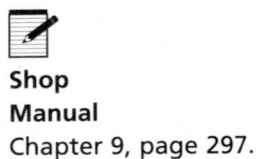

Shop Manual
Chapter 9, page 297.

Distributor-Type Fuel Injection Pumps

The distributor-type fuel injection pump pressurizes and distributes a metered amount of fuel to each cylinder nozzle at the proper time based on the calibrated needs of the engine. Figure 9-16 shows a distributor-type system. The use of a separate pumping element for each cylinder has been the general practice of the in-line pump. The distributor-type uses one pump barrel and a set of plungers to supply all cylinders in rotation. The pumping element operates more often (according to the number of cylinders), and is provided with a distributor or means of connecting the pump delivery to each of the injectors in turn. Some parts have surfaces with machining tolerances measured in microns and require handling by skilled technicians operating in a "clean room" with a ventilation system using temperature, humidity, and dust control.

Technical Description

Injection pump operation includes the functions of metering, pressurization/distribution, lubrication, and timing. The injection pump accomplishes its functions using fuel as a hydraulic fluid (Figures 9-17). Several different pressures, as shown in Figure 9-17, exist in the injection pump as it operates. During service, test gauges may be installed to diagnose the fuel system.

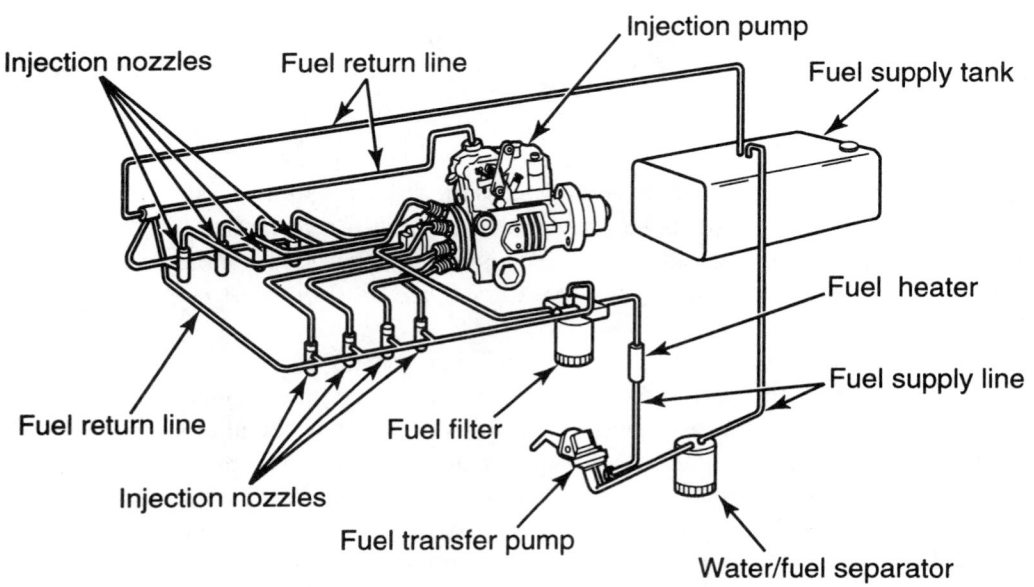

Figure 9-16 Distributor-type pump fuel system. (Courtesy of Freightliner Corporation)

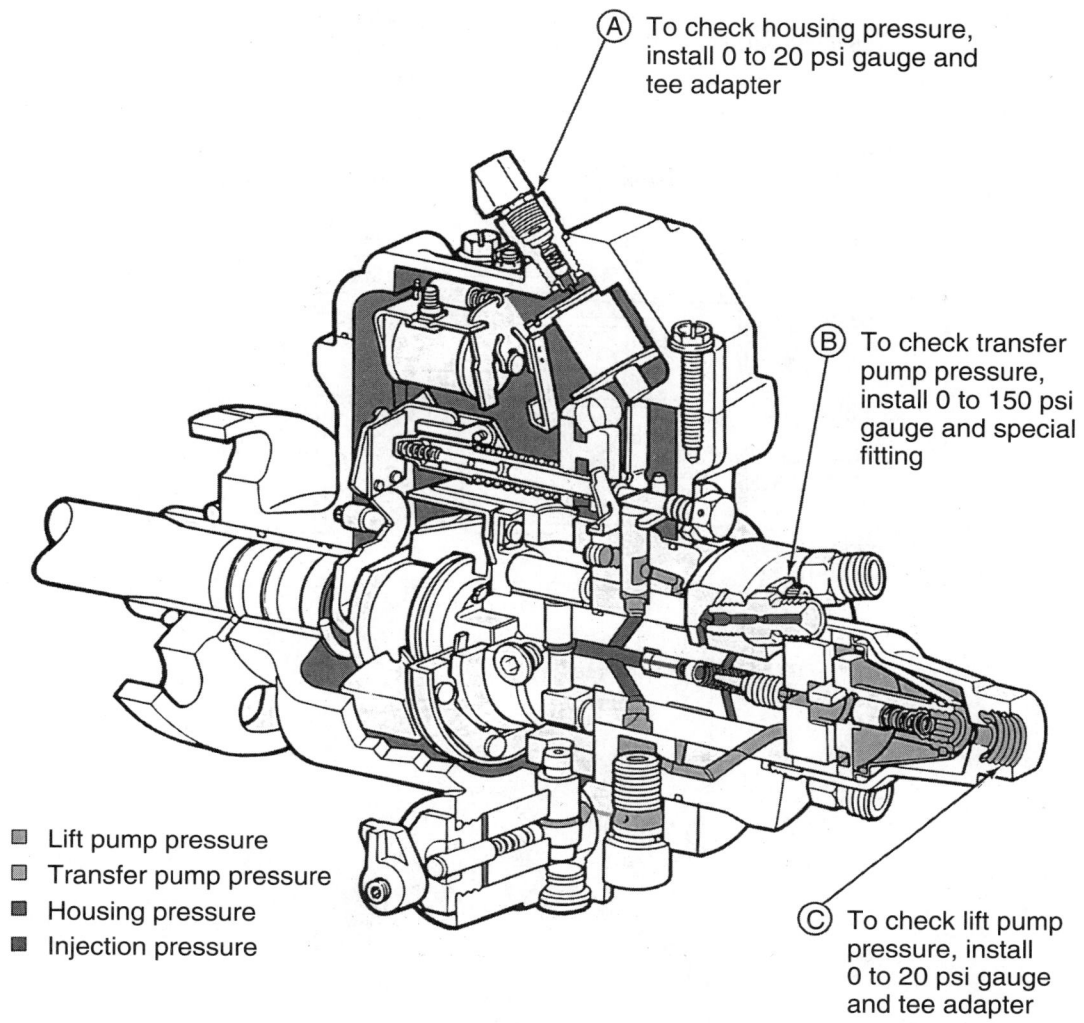

(A) To check housing pressure, install 0 to 20 psi gauge and tee adapter

(B) To check transfer pump pressure, install 0 to 150 psi gauge and special fitting

(C) To check lift pump pressure, install 0 to 20 psi gauge and tee adapter

- ▣ Lift pump pressure
- ▣ Transfer pump pressure
- ▣ Housing pressure
- ▣ Injection pressure

Note: no service check of injection pressure is allowed (pump damage may result)

Figure 9-17 Distributor-type pump cutaway with hydraulic pressures. (Stanadyne DB2) (Courtesy of General Motors Corporation, Service Operations)

Metering Fuel under lift pump pressure enters the inlet of the transfer pump (Figures 9-18A). The drive shaft rotates the rotor, which has slots in its end to operate the blades of the transfer pump inside a stationary cam ring. The transfer pump varies the pressure of the fuel, depending on the speed of the engine. At idle, the transfer pump outlet pressure is approximately 20–30 psi. At full engine speed, the transfer pump outlet pressure may be over 100 psi. A regulator valve controls the transfer pump outlet pressure and has an adjustment made during injection pump calibration. Fuel under transfer pump pressure travels through passages to several components, including a metering valve (Figures 9-18B). The metering valve directly affects the speed and power of the engine. The metering valve controls how much fuel under transfer pump pressure enters a circular charging passage in the high-pressure part of the injection pump. When the valve rotates in a clockwise direction, less fuel enters the charging passage. As the valve rotates in a counterclockwise direction, more fuel enters the charging passage.

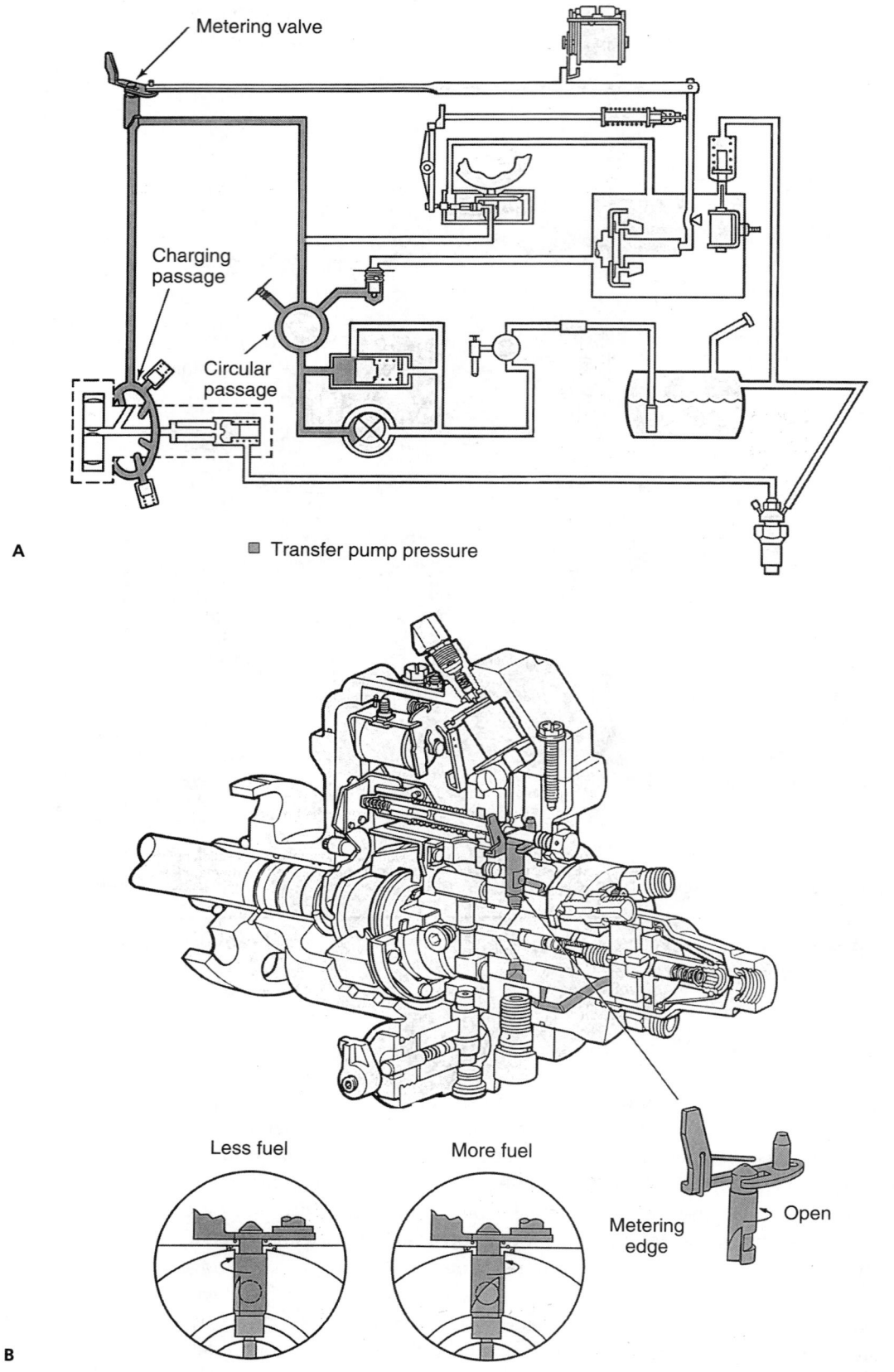

A

■ Transfer pump pressure

Less fuel

More fuel

Metering
edge

Open

B

Figure 9-18 (A) Metering valve flow schematic; (B) Metering valve position in pump cutaway. (Courtesy of General Motors Corporation, Service Operations)

A governor mechanism positions the metering valve by balancing the opposing forces of the throttle shaft position and the speed of the injection pump. The governor mechanism includes the parts shown in Figure 9-19. They are as follows: a metering valve, linkage connected to the metering valve, a governor arm connected to the linkage and pivoting on a pin in the injection pump housing, and two parts that contact the governor arm, the governor weight assembly and a min-max governor assembly. The governor weight assembly has a weight retainer that is mounted on the rotor and rotated by the drive shaft, and six weights that pivot further outward as injection pump speed increases (Figure 9-19A). A governor sleeve that is moved by the action of the weights in turn moves the bottom end of the governor arm. The min-max governor assembly has the parts shown in Figure 9-19B. It mounts in the injection pump housing by sliding on a guide stud and connects the throttle shaft to the upper end of the governor arm.

The min-max governor mechanism operates as follows: A fuel shut-off solenoid contacts the metering valve linkage when it is off, blocking fuel from entering the charging passage and

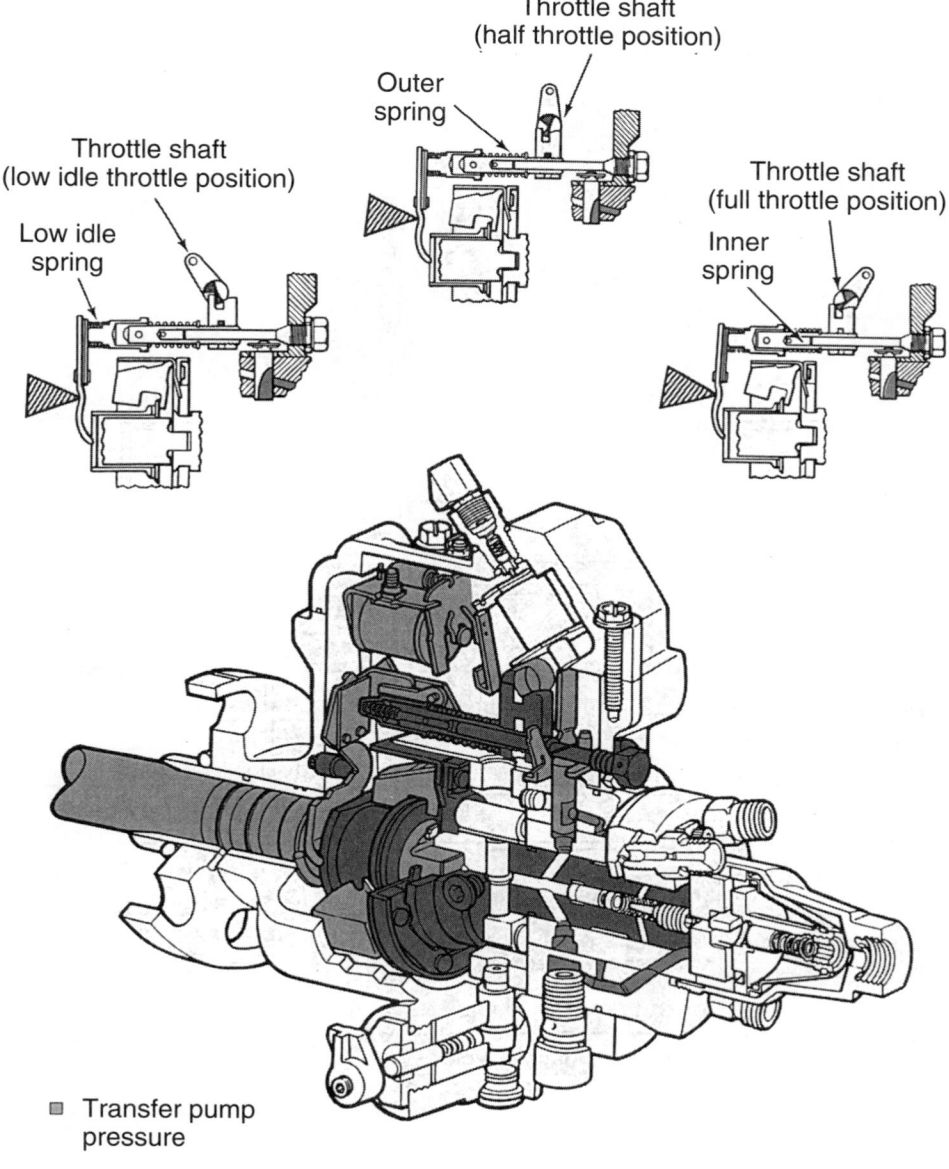

A   ▫ Transfer pump
        pressure

Figure 9-19 (A) Distributor-type pump governor operation; (B) Min-max limiting speed governor assembly. (Courtesy of General Motors Corporation, Service Operations)

253

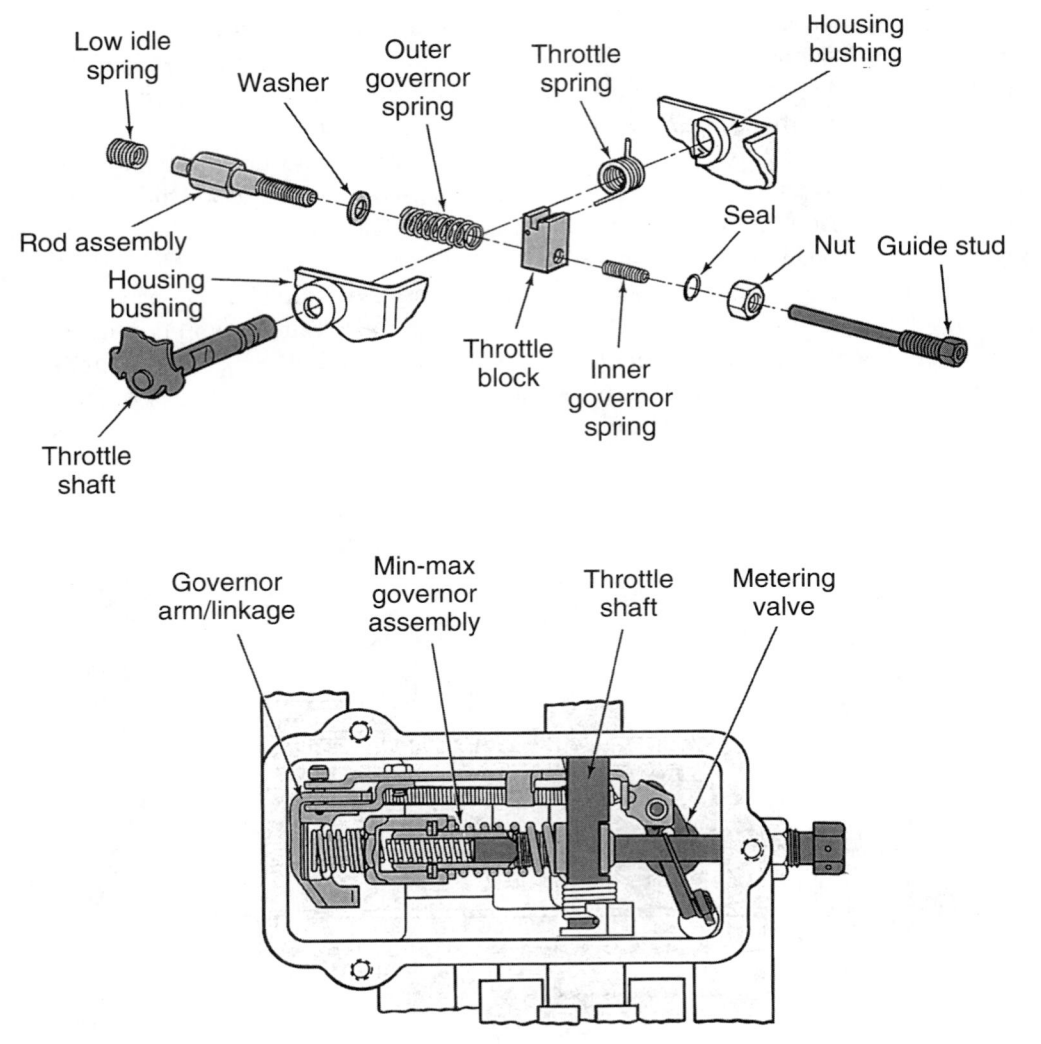

Low idle
spring

Washer

Outer
governor
spring

Throttle
spring

Housing
bushing

Rod assembly

Housing
bushing

Seal

Nut   Guide stud

Throttle
shaft

Throttle
block

Inner
governor
spring

Governor
arm/linkage

Min-max
governor
assembly

Throttle
shaft

Metering
valve

B

Figure 9-19 continued

stopping engine operation. At any engine speed, the metering valve is positioned by opposing forces. The low-idle throttle shaft position (min-speed) is maintained by the tension of the idle spring, which is part of the min-max governor. The max-speed is controlled by the force of the high-speed spring balanced against the force of the rotating weights.

Pressurizing and Distributing   The circular charging passage in the head of the injection pump has four, six, or eight ports that align in pairs with the two ports of the rotor pumping chamber. Metered fuel under transfer pump pressure travels through the charging passage and enters the rotor, pushing two pumping plungers outward as it fills the chamber (refer to Figure 9-20A). Two accumulators located in the transfer pressure annulus provide reserve pressure to help maintain a consistent pressure in the charging passage. Each pumping plunger contacts a shoe/roller assembly. The two shoe/roller assemblies contact the inner surface of a cam ring, which has eight lobes and valleys. During the charging of the pumping chamber, the valleys of the cam ring allow the pumping plungers and shoe/roller assemblies to move outward at a distance controlled by how much fuel fills the pumping chamber.

   As the injection pump rotor continues its rotation, the two ports of the pumping chamber are blocked from the charging passage (Figures 9-20B). At the same time, two of the cam ring lobes push the shoe/roller assemblies and pumping plungers inward, increasing fuel pressure in

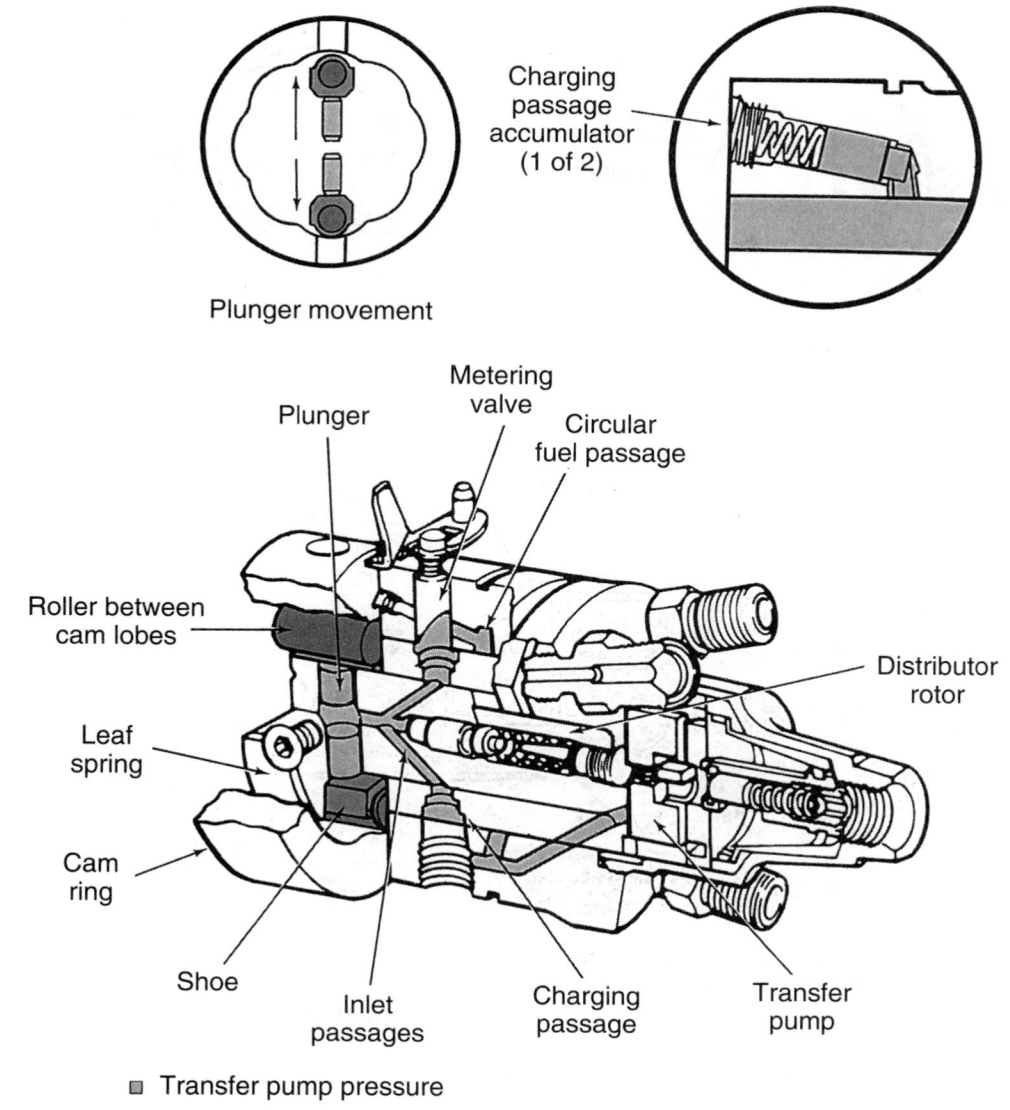

Plunger movement

Charging
passage
accumulator
(1 of 2)

Metering
valve

Plunger

Circular
fuel passage

Roller between
cam lobes

Distributor
rotor

Leaf
spring

Cam
ring

Shoe

Inlet
passages

Charging
passage

Transfer
pump

A   ▫ Transfer pump pressure

Figure 9-20 (A) Charge cycle; (B) Discharge cycle. (Courtesy of General Motors Corporation, Service Operations)

the pumping chamber to an amount approximately 100 times greater than the transfer pump pressure. When the fuel pressure in the pumping chamber rises, a delivery valve is pushed against spring force. The fuel then moves past the delivery valve to a discharge port of the rotor. When a port in the head connecting to the injection line and nozzle for a particular cylinder aligns with the rotor discharge port, fuel under a pressure wave exits the injection pump. The process of pressurizing and distributing fuel occurs four, six, or eight times in one revolution of the injection pump drive shaft and rotor.

Lubrication and Return   The outlet of the transfer pump connects to a threaded restrictor known as the vent wire assembly. This component causes fuel under transfer pump pressure to undergo a pressure decrease. It also vents the pump of air and uses a wire with hooked ends to assist in this task. The fuel passing through the vent wire assembly flows inside the pump housing to cool and lubricate most of the injection pump internal parts (Figure 9-21). An outlet port in the governor cover allows fuel to enter the return system and travel back to the fuel tank. The outlet port has a spring-loaded valve that regulates the flow of fuel from the injection

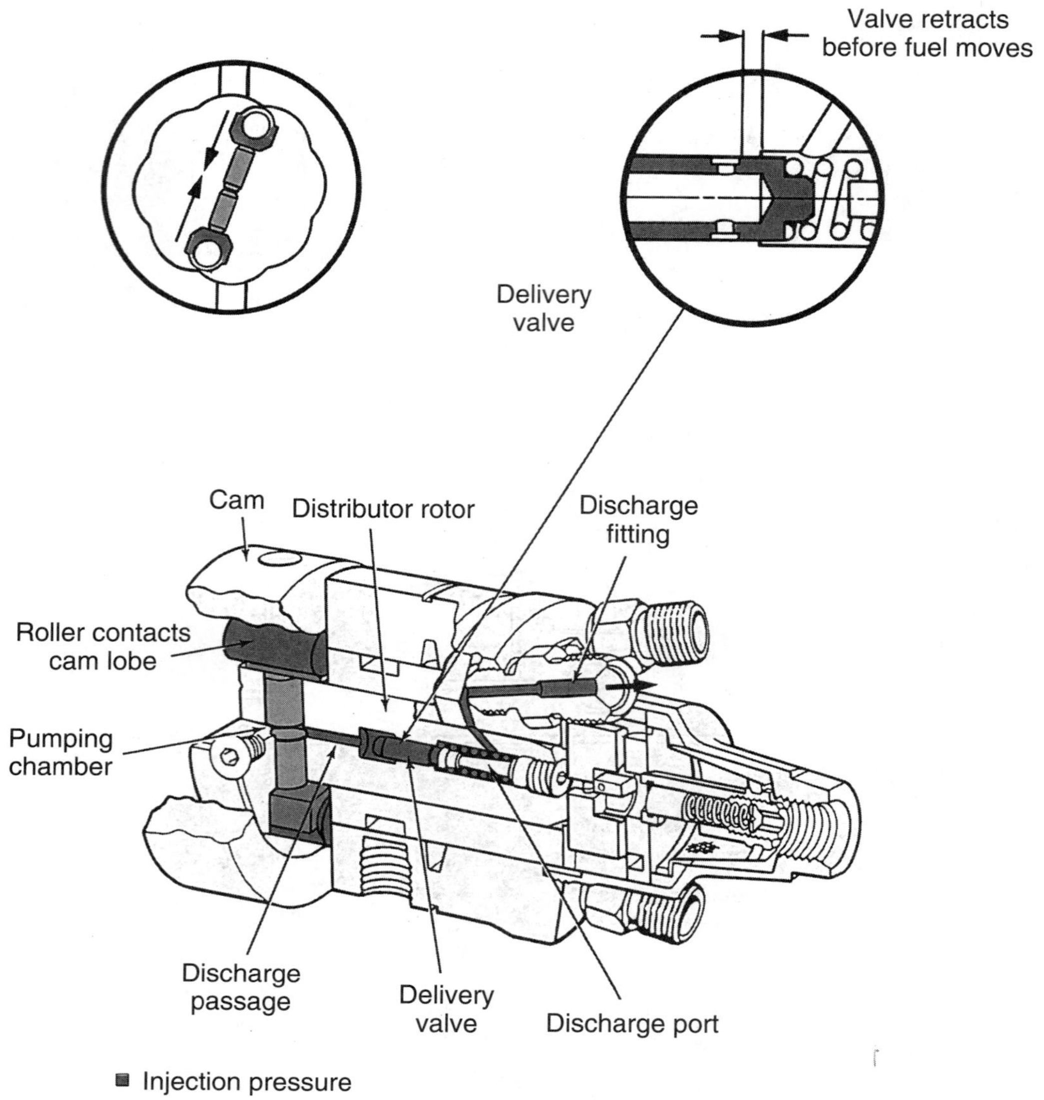

Valve retracts
before fuel moves

Delivery
valve

Cam   Distributor rotor   Discharge
                              fitting

Roller contacts
cam lobe

Pumping
chamber

Discharge
passage

Delivery
valve   Discharge port

B   ▪ Injection pressure

Figure 9-20 continued

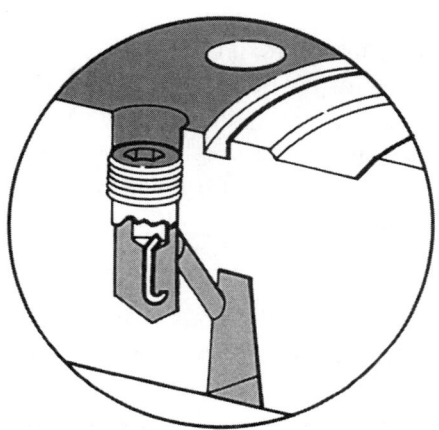

Figure 9-21 Distributor-type pump lubrication and vent-wire orifice. (Courtesy of General Motors Corporation, Service Operations)

pump housing into the return system. This valve, known as the housing pressure regulator, works with the vent wire assembly to provide a housing pressure of approximately 10 psi. The vent wire assembly has several sizes, based on the amount of return flow required. The selection of this component is a part of the injection pump calibration procedure. A solenoid in the governor cover uses a plunger to unseat the valve in the housing pressure regulator, causing housing pressure to drop to 0 psi. This component, known as the housing pressure cold advance (HPCA) solenoid, is on during cold engine operation to cause a change in fuel injection timing.

Timing   The cam ring that controls the pressurizing of fuel in the pumping chamber of the rotor can be rotated in the pump housing to change fuel injection timing. The automatic advance mechanism for the injection pump uses a sliding piston connected to the cam ring with a pin. As the piston slides in its housing bore, the cam ring rotates to change injection timing (Figure 9-22). One of the outlet passages of the transfer pump connects to the piston of the automatic advance mechanism, following a passage in the head of the injection pump, through a hollow head locating screw and into a housing passage. Fuel under transfer pump pressure pushes the piston, causing the injection timing to advance automatically in relation to engine speed.

Most distributor-type injection pumps have a pumping period with a variable beginning and a constant ending. That is, at minimal throttle positions the metering valve is open only a small amount, so the plungers move only a small amount. The rollers have to ride a great distance up the cam ramp before they can cause the plungers to pressurize the fuel. This causes retarded injection timing as compared to wide open throttle (WOT) position (maximum metering valve opening). At WOT, plunger pressurization begins as the rollers start up the cam ramp and peaks at the lobe. Therefore, timing must be advanced at light loads (part throttle) to compensate for this injection-timing lag. The throttle shaft has a face cam mounted on the outside of the injection pump that operates a servo valve inside the barrel of the advance piston (Figure 9-23). The face cam contacts a lever arm, which changes the spring force acting on the valve by pushing a plunger in contact with the spring. In this way, the timing of fuel delivery can be varied to correspond to engine requirements. The mechanical action of the throttle shaft, face cam, and rocker lever allows a more rapid advance of injection timing under light loads by permitting an unrestricted flow of fuel under transfer pump pressure to the advance end of the piston. Under heavier loads, the throttle shaft, face cam, and rocker lever increase the servo valve spring force, resulting in a restriction of fuel flow to the advance piston.

During warm engine operation, fuel under housing pressure pushes on the retard end of the advance mechanism piston, providing lubrication and a fluid cushion. When the HPCA

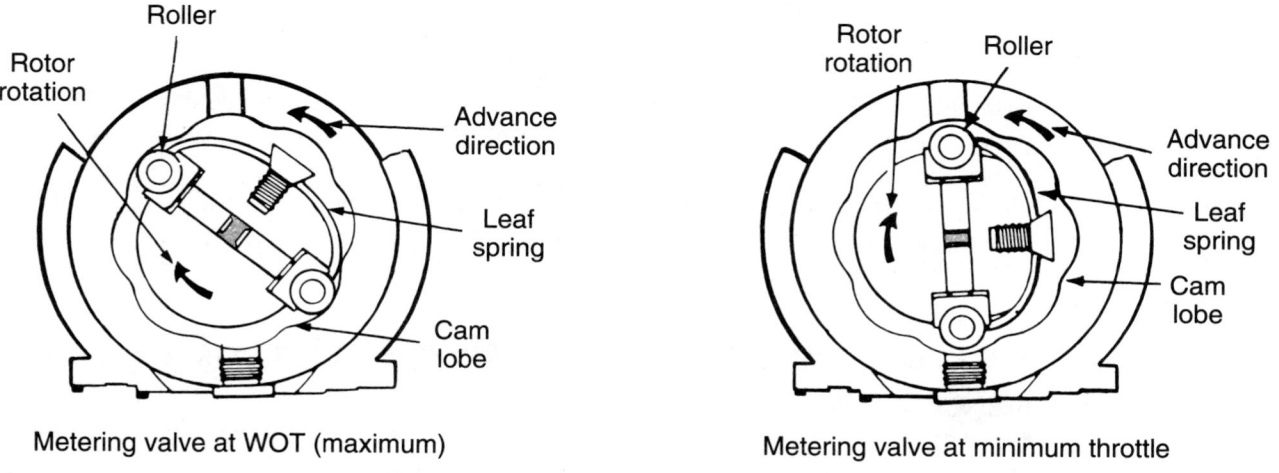

Metering valve at WOT (maximum)   Metering valve at minimum throttle

Figure 9-22 Cam positioning for timing. (Courtesy of General Motors Corporation, Service Operations)

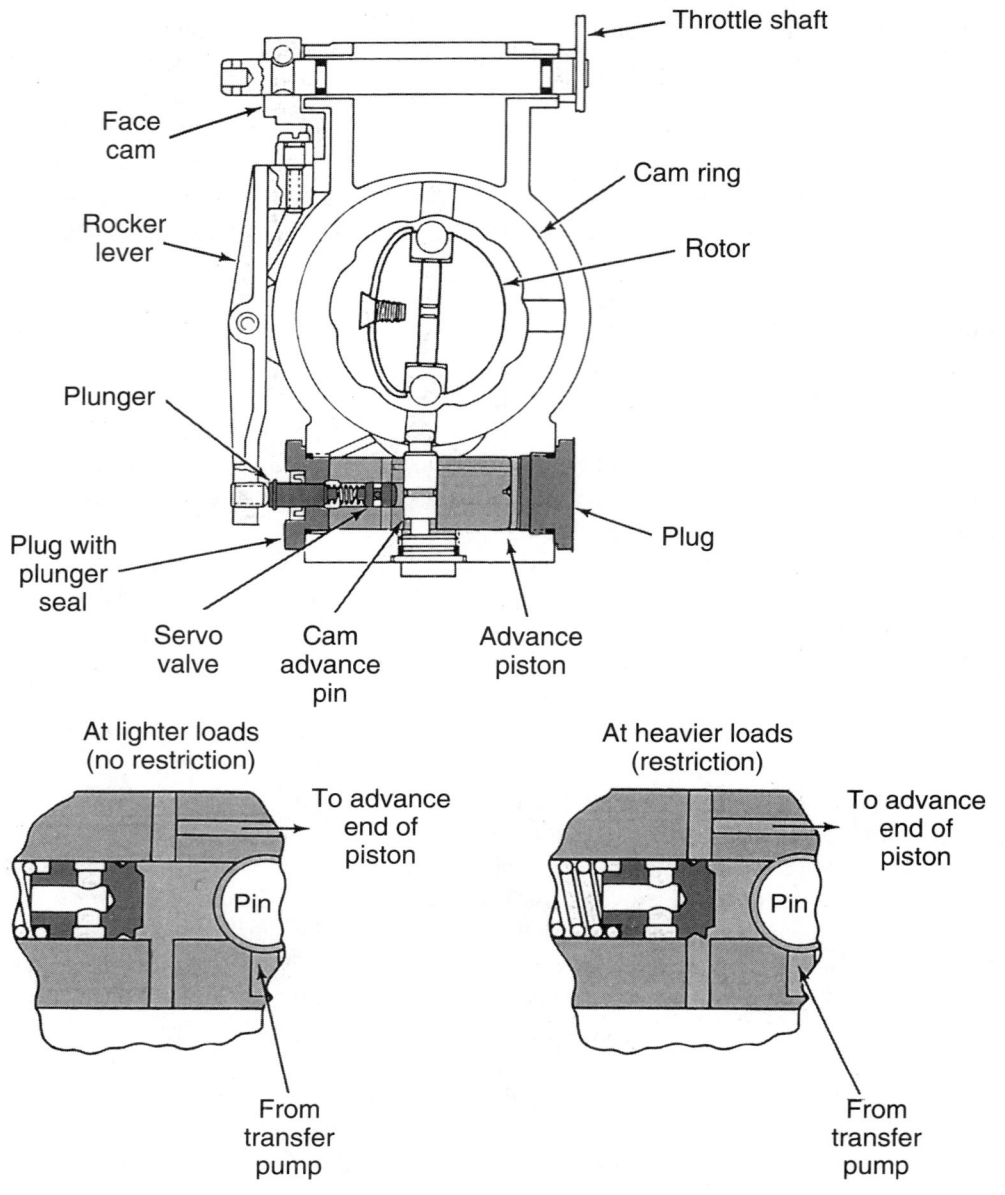

Figure 9-23 Distributor-type injection pump light load advance. (Courtesy of General Motors Corporation, Service Operations)

solenoid is on (during cold engine operation), a drop in housing pressure causes transfer pump pressure to push the advance piston farther. This results in smoother engine operation during warm-up and combines with the action of the fast idle solenoid to temporarily increase engine idle speed. If the HPCA solenoid is off during starting in cold temperatures, white exhaust smoke may result. If the HPCA solenoid is on during warm or hot engine operation, black exhaust during acceleration may result.

Detroit Diesel Corporation Mechanical Unit Injection System

The Detroit Diesel Corporation mechanical unit injection system has not been EPA certifiable for use on North American highways since 1991. However, it has been the bus and coach power plant of choice for nearly 50 years. Because of the tendency of transit corporations to recondition components many times over, these engines will survive for many more years. Detroit Diesel 2-stroke cycle engines enjoyed some popularity as a truck engine. This popularity peaked in the 1970s (I6-71, 8V-71 and 12V-71) and declined through the 1980s (8V-92). General Motors owned Detroit Diesel until 1987 when Penske Corporation purchased the division.

In an engine fueled by a mechanical unit injector fueled engine (Figure 9-24), each engine cylinder has its own unit injector. The injector is essentially a rocker-actuated, pumping and metering element combined with an hydraulic injector nozzle in a single unit. Each mechanical unit injector has fuel delivered to it at charging pressure, between 30 and 70 psi. (200 and 470 kPa) depending on engine speed.

Charging pressure is generated by a positive-displacement gear pump that is driven by the Roots blower drive shaft, and responsible for all movement of fuel through the fuel subsystem. A mechanical governor regulates fuel quantity and controls engine output by controlling the unit injector fuel racks.

Engine Identification

Figure 9-25 shows DDC engine identification. The following governor acronyms are used in truck and bus applications:

LS	Limiting speed
VS	Variable speed
DW	Double weight
SW	Single weight
TT	Torque tailored

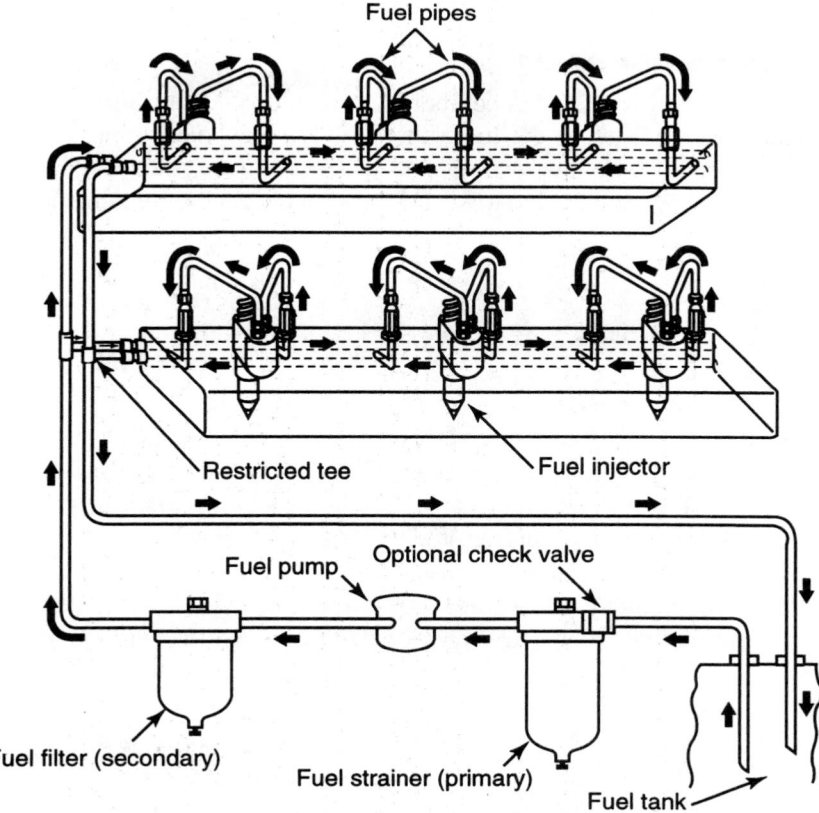

Figure 9-24 DDC unit injector fuel system. (Reprinted by permission of copyright owner Detroit Diesel Corporation, all rights reserved.)

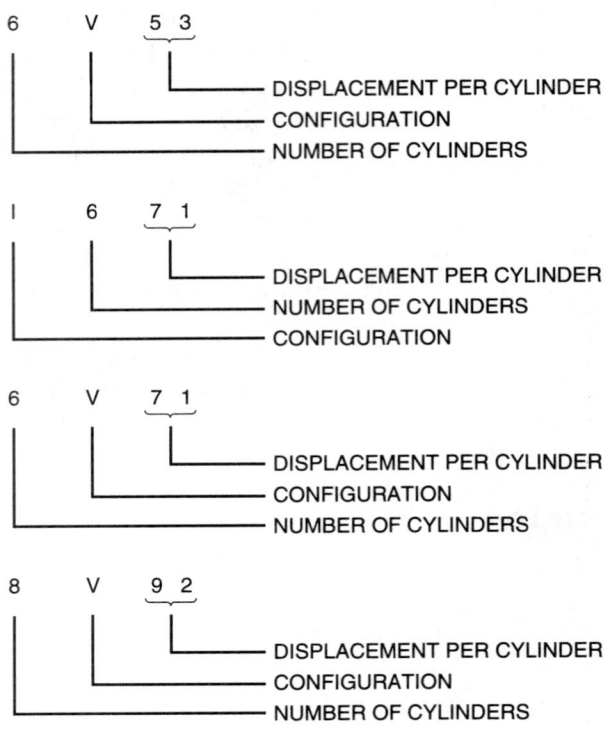

Figure 9-25 DDC engine identification. (Reprinted by permission of copyright owner Detroit Diesel Corporation, all rights reserved.)

Fuel System Components

The layout and most of the fuel system components are almost identical for 53, 71 and 92 series engines. The components listed in the fuel subsystem are for typical applications. Some differences may occur in specific chassis requirements of the fuel subsystem, such as in bus, coach, vocational, and truck applications, due to such factors as distance of the engine from fuel tanks and controls.

Fuel Filters   Most DDC two-stroke cycle engines with MUIs use a typical diesel fuel filter configuration consisting of two filters. One is under suction and the other is under charging pressure produced by the gear pump.

Primary Filters   Primary filters are located in series between the fuel tank and the engine-mounted gear pump. This filter is under suction. Older applications used non-disposable canisters within which was a disposable element. This element was surrounded by a cloth enclosure, and the flow of fuel was from the outside of element to the inside, with filtered fuel exiting the center of the assembly. Most newer engines use disposable spin-on cartridges. Older engines may be retrofit with filter mounting pads that use the spin-on cartridges for ease of servicing. The filter mounting pads are fitted with an inlet check valve to prevent fuel siphoning (drain back) to the tank.

Secondary Filters   Secondary filters are located in series between the gear pump and the inlet port in the cylinder head fuel manifold. Secondary filters on most current engines are disposable, spin-on filter cartridges, although the older canister and element filters are still in use. Current spin-on filter cartridges entrap particulate as small as 1µ and will inhibit water from passing through the filtering medium. The filters will become plugged and starve the engine for fuel.

260

Gear Pump   The gear pump is a traditionally configured pump. It is an engine-driven, positive-displacement pump responsible for charging the unit injectors with pressures between 30 and 70 psi (200 and 470 kPa). The terms transfer pump and charging pump are both used to describe the gear pump. It is driven by the Roots blower drive shaft.

Fuel Lines and Manifold   Hydraulic hose is used to connect the fuel tank with the filters and pump, and to deliver fuel to the fuel manifolds. In 53, 71, and 92 series engines the fuel manifolds are drillings within the cylinder head(s) that deliver fuel to the unit injectors at charging pressure and return fuel back to the tank(s). A restriction fitting is located at the exit port of each return fuel manifold. This defines the flow area that establishes the charging pressure window. A relief valve in the transfer pump prevents the charging pressure from exceeding 70 psi.

Fuel Pipes   Fuel (jumper) pipes connect the fuel manifolds in the head with the unit injectors; one charges the unit injector, the other returns fuel to the return manifold. Fuel is cycled through the system whenever the engine is running: in other words, all of the fuel delivered to the mechanical unit injector shown in Figure 9-26 is not used for fueling the engine. Fuel cycled through the MUI circuitry is used for cooling and lubricating the internal components of the assembly.

DDC Mechanical Unit Injectors (MUI)

A mechanical unit injector (Figure 9-27) combines a pumping element, a metering element, and a hydraulic injector nozzle in a single, cam-actuated unit. There are two types differentiated by

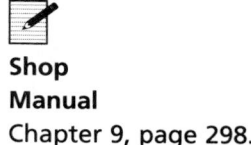

Shop
Manual
Chapter 9, page 298.

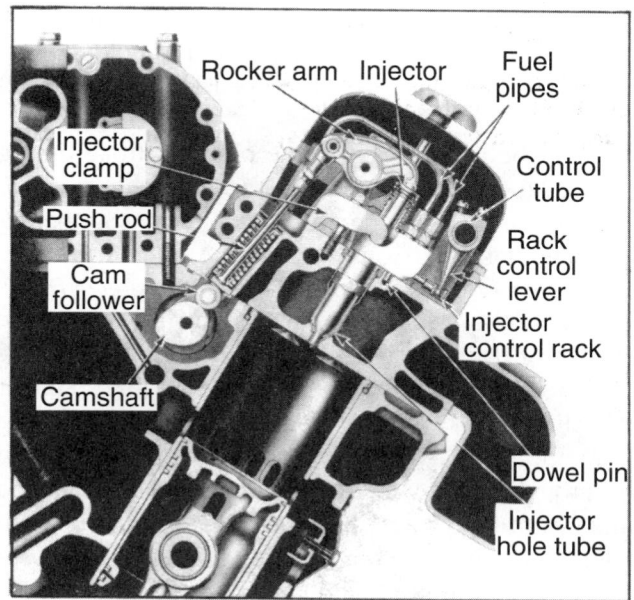

Figure 9-26 DDC engine cutaway. (Reprinted by permission of copyright owner Detroit Diesel Corporation, all rights reserved.)

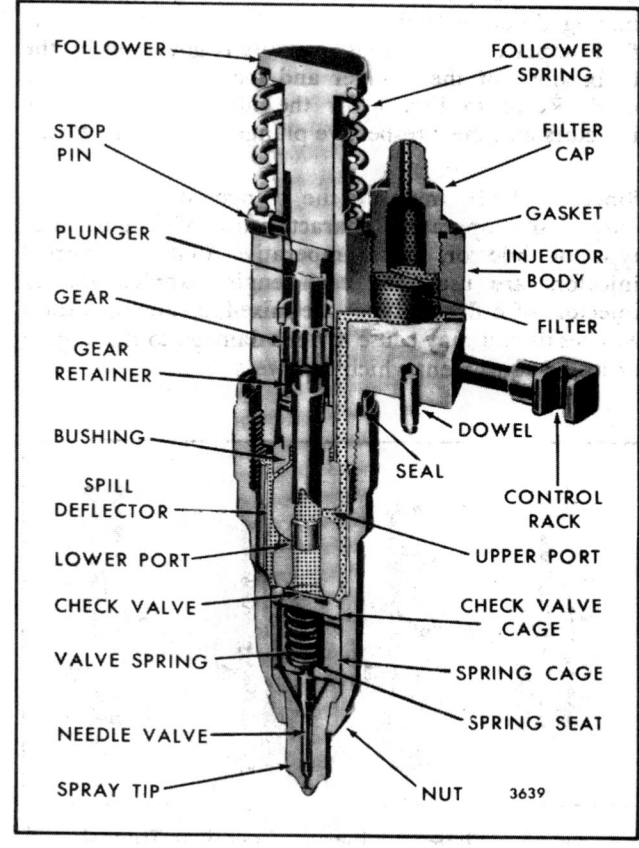

Figure 9-27 Sectional view of a DDC mechanical unit injector. (Reprinted by permission of copyright owner Detroit Diesel Corporation, all rights reserved.)

nozzle valve design, crown valve and needle valve. However, the crown valve type with NOP values of 800 psi can be considered obsolete and will not be discussed here. The needle valve type was introduced in the early 1970s and is almost universal in DDC mechanical engines in current use.

Design The plunger and bushing within the unit injector can be likened to the pumping element in an inline, port-helix metering pump. The bushing is stationary and machined with upper and lower ports, 180 degrees offset. The plunger reciprocates within the bushing. The engine camshaft actuates it by a cam profile that is mostly inner base circle and an injector train consisting of a follower, rocker, and pushrod. The plunger is milled (a machining process) with helical metering recesses and is center and cross drilled.

The unit injector follower is lug connected to the plunger, and the injector follower spring serves to load the plunger to ride its actuating cam profile. The teeth of a gear positioned over the upper portion of the plunger are meshed with the control rack, permitting the plunger to be rotated within the bushing when the rack is moved linearly. Surrounding the bushing is a spill deflector. This stellite alloy sleeve prevents high-velocity spilled fuel from eroding the injector body. Below the pumping element formed by the plunger and bushing is a needle valve assembly; ducting connects the pump chamber with the nozzle assembly. The needle valve is a simple multi-orifice, hydraulic injector nozzle. Spring pressure will determine NOP values, which will be within the range 2,200–3,400 psi.

MUI Operation For most of the cycle when the injector train is riding on the inner base circle of its actuating cam, the MUI plunger will be retracted by the injector spring, which also serves to load the injector actuating train. Fuel will flow into the MUI (Figure 9-28) from the supply jumper pipe, pass through the lower bushing port (charging the pump chamber), pass through the plunger center and cross drillings (charging the recessed metering helices), and exit at the upper bushing port. From the upper bushing port, the fuel flows through ducts to exit the MUI by means of return jumper pipes.

Actual plunger stroke is defined by cam profile geometry and therefore will not vary. At each rotation of the camshaft, each MUI plunger is driven through a single stroke that begins when the cam ramps off its base circle towards peak lift. Cam peak lift represents the maximum point of downward travel of the plunger into the bushing pump chamber. After peak lift, the cam ramps back toward cam base circle, and the injector spring lifts the plunger back into its retracted position.

The term effective stroke is used to describe that portion of the plunger stroke during which fuel is actually being pumped. The actual fuel quantity injected to the engine's cylinders is therefore the result of the MUI effective stroke. The length of the effective stroke depends on the rotational position of the plunger, which is controlled by the MUI rack. It depends on where the recessed metering helices register with the upper and lower ports machined into the bushing. For an effective stroke to occur, both bushing ports must be closed for some part of the downward stroke of the plunger. When the injector cam rotates off base circle, the cam ramp loads the injector train, and the rocker arm drives the plunger through its downward stroke within the bushing. As the plunger begins to move, its leading edge first closes the lower bushing port. As the plunger descends through its stroke, fuel in the pump chamber is displaced, passing through the plunger center and cross drillings and exiting the upper bushing port. Fuel continues to be displaced by the plunger until the helical edge of the plunger metering recess closes off the upper bushing port. As the lower bushing port has already been closed, the fuel is trapped in the pump chamber, and the effective stroke begins. Effective stroke, or pumping stroke, will continue until the lower edge of the metering recess is exposed to the lower bushing port. At this point, the high pressure fuel in the pump chamber spills, exiting the pump chamber through the now exposed lower bushing port.

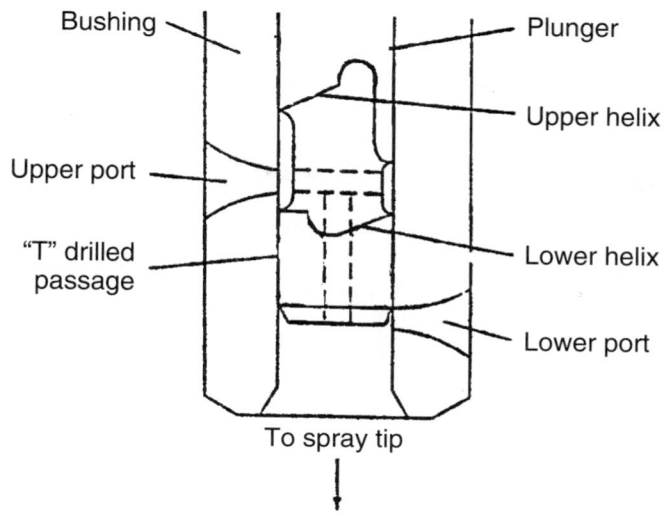

Bushing — | — Plunger

Upper port — | — Upper helix

"T" drilled passage — | — Lower helix

— Lower port

To spray tip

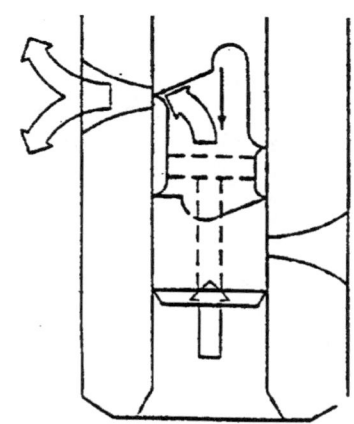

Injector Operation: The plunger descends, first closing off the lower port and then the upper. Before the upper port is shut off, fuel being displaced by the descending plunger may flow up through the "T" drilled hole in the plunger and escape through the upper port.

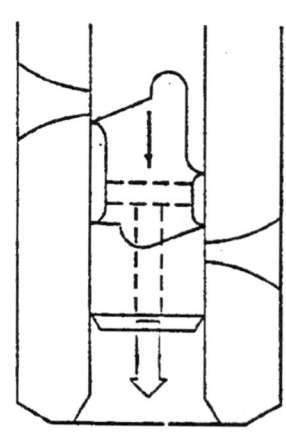

After the upper port has been shut off, fuel can no longer escape and is forced down by the plunger and sprays out the tip.

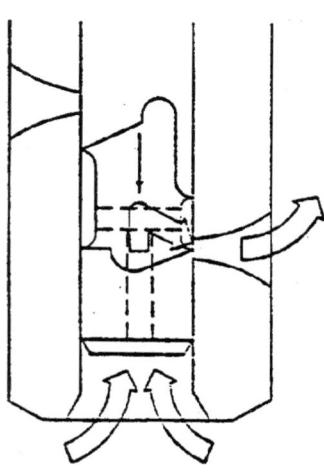

As the plunger continues to descend, it uncovers the lower port, so that fuel escapes and injection stops. Then the plunger returns to its original position and awaits the next injection cycle.

Figure 9-28 Effective stroke in a MUI with plunger variable beginning, variable ending geometry. (Reprinted by permission of copyright owner Detroit Diesel Corporation, all rights reserved.)

The length of the effective stroke depends on the rotational position of the plunger and the points at which the plunger metering recesses register with the upper and lower bushing ports. Effective strokes only take place when both bushing ports are closed off. To shut down the engine, the plunger rotational position must be such that one of the bushing ports is always exposed. The no-fuel position requires that the plunger be rotated to a position where the lower bushing port is exposed before the upper bushing is closed off. At no-fuel, the fuel in the MUI pump chamber will be displaced through the entire plunger stroke. As the plunger descends into the pump chamber, fuel will be forced up through the plunger center and cross drillings to exit at the bushing ports.

The geometry of the metering recesses/helices will determine the injection timing characteristics in the unit injector. The term "injection timing" is used to denote the points at which the

Injection timing determines when the fuel injection will begin and end.

injection pulse begins and ends. These points are usually specified in engine crank angle degrees. Depending on the application, these may be:

Variable beginning, variable ending of injection pulse

Constant beginning, variable ending of injection pulse

Variable beginning, constant ending of injection pulse

MUI Identification   A circular identification tag pressed into the unit injector body identifies the class number. Unit injectors with no line under the manufacturer name are those with the obsolete crown valve nozzles. A needle valve nozzle would have this line under the manufacturer's name (Table 9-1):

Table 9-1 MUI Identification Chart

OEM ID:	GM	Reliabilt	DDC	DDC
CODE:	N60	B60	N70	C70
NOZZLE TYPE:	Needle	Crown	Needle	Needle

In the first example above, the line under the manufacturer's name identifies the injector as having a needle valve nozzle. The letter codes indicates the plunger specific plunger geometry (for instance, "C" identifies the plunger as having variable beginning, variable ending timing characteristics. The number following the letter designation is the comparator bench specification for fuel output per 1,000 strokes, measured in cubic millimeters. The digits engraved on the nozzle valve identify the number of orifices, orifice sizing, and spray angle.

MUI Subcomponents   The technician should be familiar with the components that collectively make up the MUI assembly. In the following section the components are briefly described.

Follower Assembly   This assembly is the actuating mechanism for the MUI plunger. The follower is directly linked to the plunger by means of a slotted lug. A stop pin limits the upward travel of the plunger. The MUI spring surrounds the follower. Its function is to hold the follower (and therefore the plunger) in the raised position and to load the injector train.

Plunger   The plunger is the moving component of the pumping element. It reciprocates within the bushing, which is held stationary in the MUI body. The plunger is milled with helical-shaped metering recesses (for purposes of varying effective stroke) and center and cross drilled to provide a hydraulic connection between the pumping chamber below the plunger and the metering recesses. Plungers are lapped to bushings in manufacture. Lapped components are not interchangeable.

Bushing   The MUI bushing is a cylindrical housing within which the plunger reciprocates and with which it forms the pump element. Drilled with upper and lower ports, the bushings are positioned in the MUI by dowels.

Spill Deflector   The spill deflector is a cylindrical, stellite alloy sleeve that surrounds the bushing and prevents high-velocity fuel spilled from the MUI pump chamber at the completion of effective stroke from eroding the unit injector body.

Gear   The MUI gear is connected to the plunger by means of a flat machined into its inside bore and meshed to the control rack. The gear therefore rotates whenever the MUI rack is moved linearly. In this way, the gear provides a means of rotating the plunger (required to alter effective stroke) while it reciprocates.

Control Rack The teeth of each MUI control rack is meshed with the plunger gear so when it is moved linearly, the plunger is rotated. The control rack is manufactured with a clevis into which the foot of the control lever is placed on the control tube. The control levers that extend from the control tube determine the linear position of the control rack.

Check Valve In the event of the needle valve being unable to seat due to carboning or any other reason, the check valve functions to prevent cylinder gases from entering the MUI circuitry beyond the nozzle assembly.

Needle Valve Nozzle Assembly The DDC needle valve assembly uses all the same operating principles as any other multi-orifice nozzle assembly, as discussed earlier. As with any hydraulic injector nozzle, a MUI needle valve nozzle is held closed and seated by the nozzle valve spring. Fuel pressures developed in the MUI pump chamber below the plunger are ducted to act on the sectional area of the nozzle valve of the pressure chamber. When that pressure is sufficient to overcome the nozzle valve spring tension, the needle valve retracts, allowing high pressure fuel from the MUI pump chamber to pass around its seat to a sac and exit through the nozzle orifices, thus beginning the injection pulse. When the pump element effective stroke ends, fuel starts to spill from the pump chamber, exiting the lower bushing port. The moment there is insufficient fuel pressure acting on the sectional area of the needle valve to retract the valve spring, the spring closes, ending the injection pulse.

Injector to Governor Linkages The output of the mechanical unit injectors is governor controlled by means of a set of linkages that mechanically connect the governor with the MUI. Governor control rods extend from the governor differential lever (a twin-arm lever that pivots on a fulcrum) in the governor housing and connect to a control tube lever by means of a clevis and pin. Linear movement of the governor control rods is converted into rotary movement of the control tube, which runs lengthwise through the upper cylinder head. Extending from the control tube are the rack levers. When the control rods rotate the control tube, the rack levers rotate with it. The feet of the rack levers connect to the MUI fuel racks by means of a clevis, so rotary movement is converted to linear movement in order to position the racks. An adjusting screw (or pair of screws in older engines) adjusts the mechanical relationship between the MUI rack and the rack lever on the rack lever to control tube clamp.

The rack levers link the individual MUIs with the control tube. When the governor control rods rotate the control tube, the rack levers extending from the control tube will linearly move the unit injector racks.

⚠️ **WARNING:** A critical procedure in DDC engine tuneup is ensuring that the unit injectors are balanced, that is, that the point of register of the helices with the bushing ports in each unit injector in the engine is identical.

Correctly setting the control tube adjusting screw(s) achieves this. The adjusting screw(s) locate the radial position of the rack lever in relation to the MUI fuel rack.

Governor

A simple mechanical governor is used in most truck and bus applications. Governor components consist of a set of engine-driven flyweights, a thrust collar, bell crank, spring set, and differential lever. As in most mechanical governors, centrifugal force exacted by the flyweights attempts to diminish engine fueling, while spring force moderated by speed control lever position attempts to increase engine fueling. The thrust collar acts as an intermediary between the spring force defined by precise adjustment and moderated by position of the speed-control lever position, and the centrifugal force that directly correlates to engine speed. Extending from the governor housing are the control rods that link the MUI fuel control linkages with the governor assembly. Both limiting-speed and variable-speed governors are used.

Caterpillar Mechanical Unit Injector (MUI)

Technicians familiar with DDC MUI systems will require little introduction to the Caterpillar MUI system. However, there are some critical differences, and Caterpillar technical literature must be referenced when performing their version of the tune-up sequence. Most technicians will see the 3126 version of this engine with Caterpillar's HEUI fuel system, a computer-controlled, hydraulically actuated, unit injector system; this is covered in Chapter 11. The 3116 engines are available exclusively to General Motors Corporation at this time.

Fuel Subsystem Components

Fuel is pulled from the fuel tank (Figure 9-29) by a plunger-type transfer pump. A primary fuel filter may be fitted between the fuel tank and the transfer pump. The transfer pump is integral with the governor housing and charges a secondary or main fuel filter and the unit injector assemblies. The charge side of the transfer pump is pressure protected by an outlet check valve and a pressure relief valve. When present, the hand primer pump is located on the secondary filter-mounting pad. Its function is to purge air from the fuel subsystem responsible for charging the unit injectors. Fuel is delivered from the fuel subsystem to drilled passages in the cylinder heads, which intersect annular galleries around the cylindrical bore of the unit injectors.

The fuel transfer pump (Figure 9-30) is a cam-actuated plunger pump located in the front housing of the governor. Inlet and outlet check valves aspirate the pump chamber; these close when the engine is shutdown. Outlet pressure should test at a minimum of 200 kPa (30 psi) at high idle.

Caterpillar MUI Operation

The mechanical unit injectors (Figure 9-31A) are responsible for metering, pumping to injection pressure values, and atomizing fuel directly to the engine cylinders. In the MUI, the pump is actuated by a dedicated cam profile on the engine camshaft. The high-pressure pipe is eliminated, and the hydraulic injector nozzle is integral with the assembly. The mechanical unit injector is installed within a sleeve bored to the cylinder head. This is directly exposed to engine coolant passages.

The injector cam profile actuates a train consisting of a roller-type lifter, push rod, and rocker arm. The rocker arm acts on the tappet assembly of the mechanical unit injector using

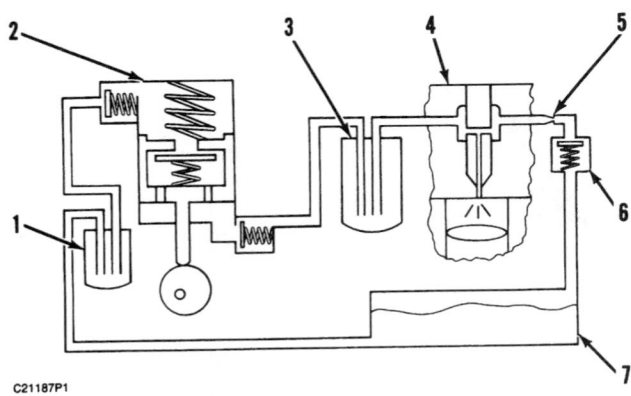

Fuel System Schematic
(1) Screen. (2) Fuel transfer pump (integral with governor).
(3) Main filter. (4) Cylinder head. (5) Pressure regulating orifice.
(6) Check Valve. (7) Fuel tank.

Figure 9-29 Caterpillar MUI fuel system schematic. (Courtesy of Caterpillar Inc.)

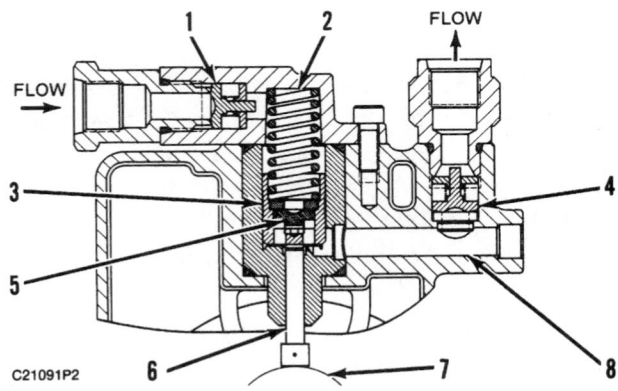

Fuel Transfer Pump
(1) Inlet check valve. (2) Spring. (3) Piston assembly. (4) Outlet check valve. (5) Piston check valve. (6) Tappet assembly. (7) Cam. (8) Passage.

Figure 9-30 Caterpillar plunger type fuel transfer pump. (Courtesy of Caterpillar Inc.)

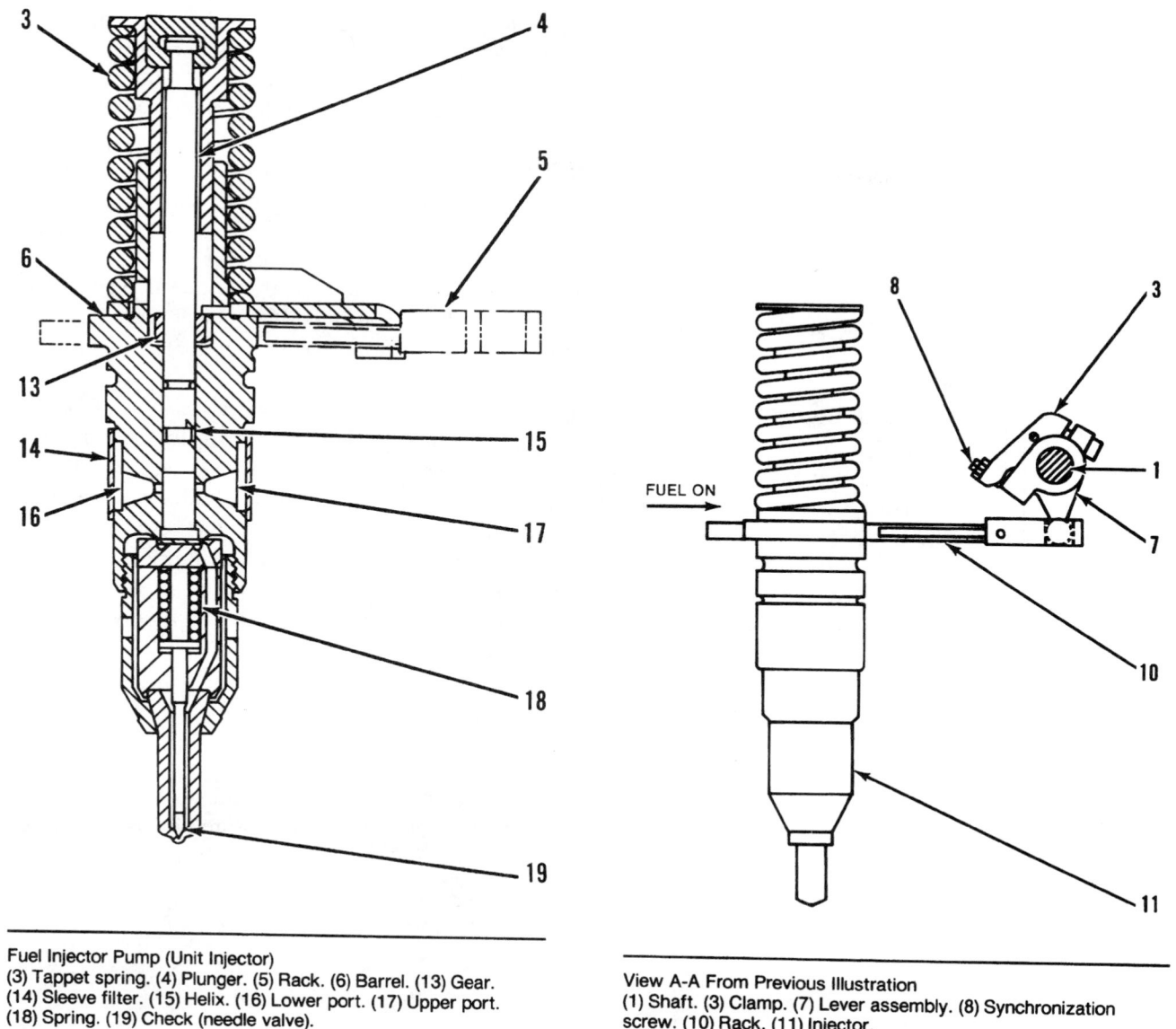

Fuel Injector Pump (Unit Injector)
(3) Tappet spring. (4) Plunger. (5) Rack. (6) Barrel. (13) Gear.
(14) Sleeve filter. (15) Helix. (16) Lower port. (17) Upper port.
A (18) Spring. (19) Check (needle valve).

View A-A From Previous Illustration
(1) Shaft. (3) Clamp. (7) Lever assembly. (8) Synchronization
screw. (10) Rack. (11) Injector.   B

Figure 9-31 (A) Caterpillar MUI; (B) MUI on-position. (Courtesy of Caterpillar Inc.)

a floating button as an intermediary. This prevents side loading of the tappet by the rocker arm. The plunger (Figure 9-31B) is linked to the tappet and reciprocates within a stationary barrel. It is milled with a helix that can be classified as a lower helix design, resulting in a constant beginning, variable ending of fuel delivery timing. With this design of helix, the beginning of the effective pumping stroke is determined by the angular location of the cam. The barrel is machined with an upper port and a lower port diametrically offset. Metering of fuel is accomplished by rotating the plunger in the barrel bore and altering the point of register of the spill ports with the helix. The plunger is rotated by means of a gear, which is tooth-meshed to the rack. The gear collars the plunger in a manner that permits plunger rotation while it reciprocates.

Effective Stroke   The plunger actual stroke is determined entirely by cam profile geometry and therefore remains constant. Control of fuel quantity depends on the rotational position of plunger, specifically the point of register of the helix with the upper barrel port. When the actuating cam ramps off its base circle, the plunger descends within the barrel and its leading edge closes off the upper barrel port. As the barrel continues to descend, it displaces fuel below it in the pump chamber, forcing it out of the lower barrel port. Effective stroke begins when the lower bushing port is closed by the plunger leading edge. Pressure rise occurs in the pump chamber below the plunger.

At nozzle opening pressure sufficient pressure rise has occurred to overcome the spring pressure that loads the needle valve on its seat causing it to retract. Fuel passes around the needle valve seat and is forced out through the nozzle orifices. Effective pumping stroke will continue as long as the plunger closes off both the upper and lower barrel ports. It ends at the moment the upper barrel port is exposed by the recessed helix milled into the plunger, permitting fuel to be spilled to the return gallery. The resultant pressure collapse will enable the injector nozzle spring to close and seat the needle valve, ending injection.

Fuel Rack Control   The governor is mechanically connected to the rack shaft assembly (Figure 9-32) by means of a link and lever assembly. When the governor demands fuel, the injector rack must move away from the injector body rotating the plunger counterclockwise (CCW). The governor output shaft therefore moves to rotate the shaft, causing the rack levers to move the injector racks simultaneously outboard. To reach the no-fuel position, the injector racks must be forced inboard, causing the plungers to move clockwise (CW). The fueling positions of the injector rack on Caterpillar mechanical unit injectors are opposite those in DDC engines. A torsion spring at each rack lever assembly allows the rack control linkage to return to a no-fuel position should one injector seize in a fuel-on position.

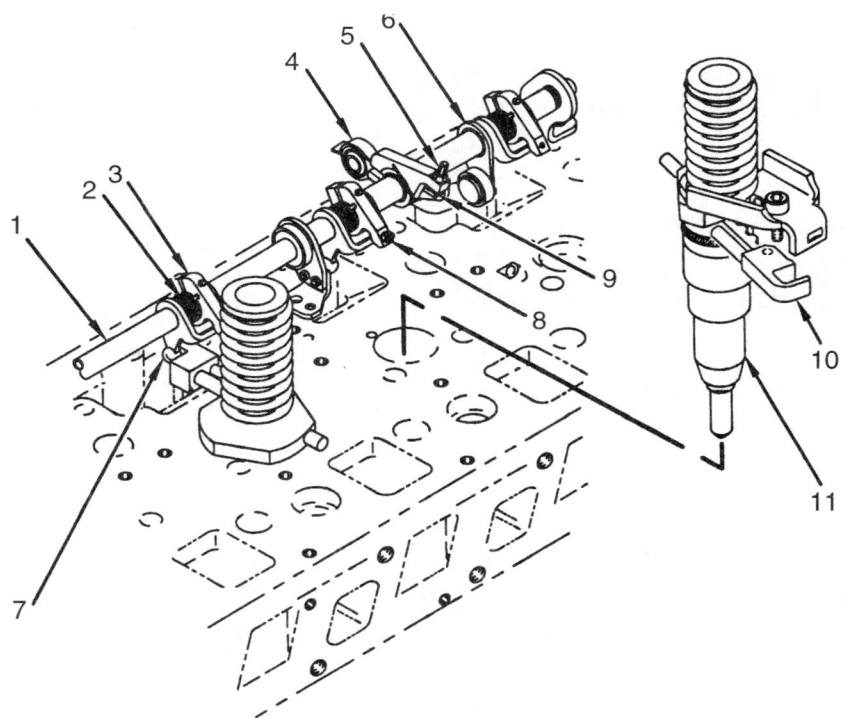

Injectors And Rack Control Linkage (Type I)
(1) Shaft. (2) Spring. (3) Clamp. (4) Link. (5) Fuel setting screw. (6) Lever assembly.
(7) Lever assembly. (8) Synchronization screw. (9) Clamp assembly. (10) Rack. (11) Injector.

Figure 9-32 MUI location and rack assembly. (Courtesy of Caterpillar Inc.)

Fuel Ratio Control Turbocharged engines use a fuel ratio control (FRC) to limit engine smoke during conditions of low boost. Essentially, the FRC limits fueling to the cylinders until there is sufficient air in them to combust it. Boost air is ported from the engine intake manifold to the FRC inlet port. At low boost, when the governor output shaft moves to the fuel-on position attempting to draw the unit injector racks outboard, the limit lever will contact the set screw on the FRC lever. As manifold boost increases, it acts against the FRC diaphragm and overcomes the FRC spring pressure. This permits the retainer shaft to move outwards, enabling the FRC lever and the limit lever to rotate CW and the rack shaft to rotate towards full fuel.

Fuel Shut-Off Solenoid The fuel shut-off solenoid is an energized-to-run (ETR) latching solenoid (a pilot valve that stays in the energized position even when not electrically energized and remains so until the circuit it controls is shut down). A spring-loaded plunger inside the solenoid acts on a lever assembly located in the front housing of the governor. This lever assembly forces the governor output shaft to the fuel-off position when the solenoid armature is released either electrically or manually. At start-up, the solenoid armature is energized to latch in the run position, permitting the governor output shaft to move to the fuel-on position. Engines with latching solenoids can be manually shut off by depressing the override button, but they cannot be manually latched in the run position for starting. Some engines have an ETR solenoid with no latching capability.

Governors

Six different governor types (Figure 9-33) are used on 3116, 3114, and 3126 engines. Each is identified sequentially by Roman numerals. Governors for Caterpillar mechanical unit injector fuelled engines will be covered only briefly in this section. Proprietary literature (SEN6454) and test equipment (IU7326 Governor Test Stand) are required to service and adjust these governors.

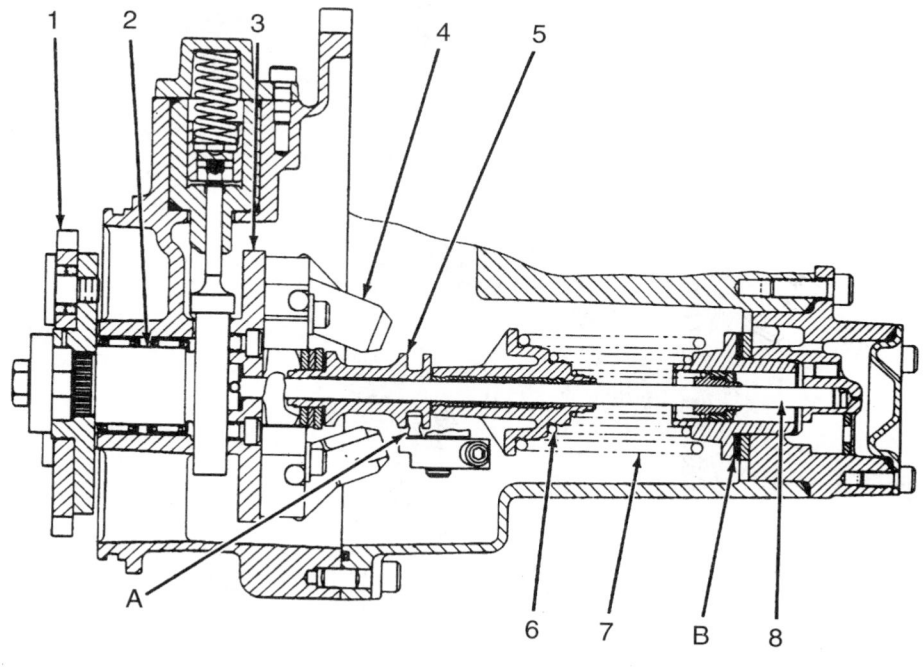

Governor
(1) Governor drive gear. (2) Shaft. (3) Flyweight carrier. (4) Flyweights. (5) Riser. (6) Low idle spring.
(7) High idle spring. (8) Shaft. (A) Pin. (B) Shims.

Figure 9-33 Sectional view of one type of Caterpillar governor. (Courtesy of Caterpillar Inc.)

Cummins PT Pump Fuel System

A BIT OF HISTORY

Until World War II, diesel engine manufacturers generally had been content to use port-helix metering fuel injection systems; there were a few exceptions, but most did not survive. However, with port-helix metering injection, a disproportionate percentage of the total engine cost must be allocated to the fuel management apparatus, and engine designers began to explore alternative systems. The Cummins Engine Company introduced their PT (pressure-time) fuel system in the early 1950s, and by 1970, PT-managed engines were the North American market sales leaders in the 200–500 bhp range. PT system production continued until the 1994 EPA emission legislation rendered the PT-managed engine ineligible for on-highway certification.

Common rail fuel systems are those in which a fuel pump supplies injectors from a common pipe or accumulator.

Unlike other hydromechanical fuel injection systems, the PT system (Figure 9-34) did not readily lend itself to electronic management. So Cummins developed their own full-authority electronic management system (CELECT) to replace the PT **common rail system**. It is important to stress that there are a number of similarities in terms of basic operating principles between the PT system and the several different electronic common-rail, open (valveless) nozzle systems that are currently being introduced to the market place.

PT System Theory

The power output of any diesel engine is essentially determined by the amount of fuel metered into its cylinders. The Cummins PT system manages fueling using a set of principles quite dis-

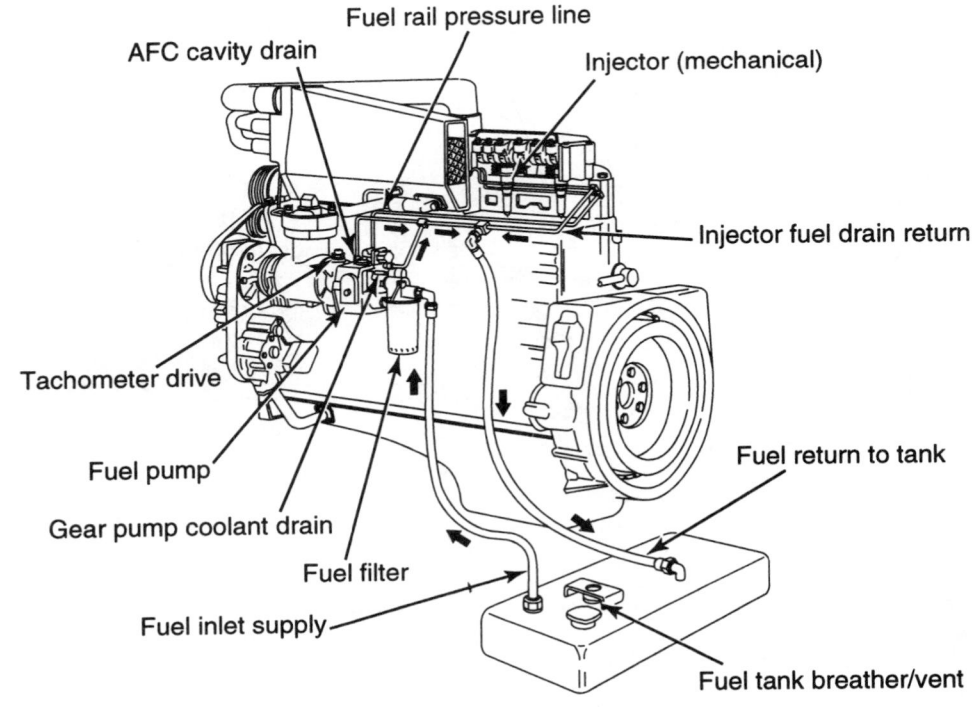

Figure 9-34 PT-type pump fuel system. (Courtesy of Cummins Engine Company)

tinct from those used in port-helix metering systems. According to Cummins, the metering of fuel in the PT system is based on the principle that a volume of liquid passing through a fixed orifice is directly proportional to the square root of the *pressure* and to the *time* of the flow, thus the term PT, pressure-time. The sequence of the two letters in the acronym PT is important, as Cummins also uses the TP, representing time-pressure. In each case, the first letter of the acronym denotes the control variable. In the case of the PT system, the control variable is pressure.

The time is regulated by the speed of the engine, since the injector plunger is cam driven. By varying the two elements of fuel pressure and time, the engine speed and horsepower are controlled. If the pressure is increased and time (rpm) is held constant, more fuel is injected into the cylinders. Likewise, lugging down the engine increases time and the pressure is held constant, so more fuel is delivered; thus, the system has a built-in torque rise. A simple hydraulic equation constructed to determine a volume of flow requires the following data: fluid pressure, flow time, and **flow area**.

Changing any of the above values in this hydraulic equation will change the outcome, which in this case determines a specific volume of flow. In the Cummins PT system, the critical flow area (the most restricted sectional area of a hydraulic circuit under flow) is designed to remain constant, so the variables in the hydraulic equations are:

Flow time: Variable, but dictated by engine speed, which has little relationship with fuel quantity required. The time period will diminish inversely with rpm increase.

Fluid pressure: Variable and precisely managed by the system to control engine output.

The critical component in terms of managing the fueling is the PT pump. It is best regarded as a flow-control device, not an injection pump. The PT pump feeds all the injectors by means of a common rail (that is, a single outlet pipe from the PT pump that supplies a rail gallery in the cylinder heads). The injectors are cylindrical and receive and return fuel through exterior annuli separated by O-rings. The PT injectors are mounted in parallel and the only way in which the rail fuel can pass into the cylinder head return gallery is by passing through the PT injectors. Therefore, their inlet orifices will collectively identify the flow area into which the PT pump unloads. This flow area remains constant (PT injectors are accurately sized on the test bench), so it can be said that the PT pump directly manages the rail pressure. Rail pressure values are used to manage engine output through the entire rpm and load ranges of the engine. Typically, the rail pressure ranges from 8–210 psi. Actual quantity of fuel to be metered per cycle will depend on the following five factors:

The PT pump fuel system is a common-rail system that maintains variable rail pressures to control fueling.

1. **Rail pressure:** The PT pump is a flow-control device responsible for managing rail pressure. Actual rail pressure will be factored by the quantity of fuel unloaded into the single rail pipe that supplies all of the PT injectors and the flow area as defined collectively by the PT injectors.

2. **Balance orifice sizing:** All of the fuel unloaded by the PT pump to the rail is routed through the PT injectors mounted in parallel. The **balance orifice** is the inlet orifice of each PT injector. Each balance orifice is accurately sized because collectively they will define the flow area for factoring the actual rail pressure value parameters. Although the term "balance orifice" will be used in this text, Cummins also use the term "calibrating orifice" to describe this precisely sized inlet to the PT injector.

3. **Engine speed:** Engine speed will define the real time dimension available for metering and pumping in the PT injector. It will also determine the PT gear pump (a subcomponent of the PT pump) pressure. This will increase proportionally with a rise in rpm, given a constant flow area, and accordingly will influence rail pressure.

4. **Metering orifice sizing:** The PT injector is essentially a pumping device responsible for converting rail pressure values to injection pressure values. The larger percentage of fuel entering the PT injector circulates (for purposes of lubing and cooling) and exits to the return gallery in the cylinder head, from which it is routed back to the fuel tank. The percentage of fuel to be used for fueling the engine is that forced through the metering orifice within the PT injector in the time period factored by engine rpm and the cam geometry (the PT injector pumping stroke is actuated by cam profile).

The PT injector-metering orifice defines the flow area for determining the actual volume of fuel metered to the engine cylinder.

5. **Injector timing:** The pumping action of the PT injector is actuated by a dedicated cam profile (located between the valve cams in the cylinder) and injector train, which consists of a cam follower, push tube, and rocker assembly. Timing adjustments are critical, and marginal inaccuracies can cause both fueling problems and physical damage to the injector and injector train.

The PT fuel system uses a common rail fueling principle. The pump is a flow-control device that manages rail pressure. It feeds PT injectors positioned in parallel and extending from the common rail. Fuel exiting the rail must pass through the PT injectors where a portion is metered for injection and the remainder routed to the return manifold. Fuel flows through a circuit in the PT injectors at any time the engine is running. The fuel used for metering is diverted from this circuit within the PT injector.

PT Injector

The collective restriction of the balance orifice in the rail (six in a six-cylinder engine) defines the flow area required to factor rail pressure.

The PT injector (Figure 9-35) is a rocker-actuated, pumping element responsible for converting rail pressures to injection pressures. It also defines a couple of critical flow areas essential for metering. It is cylindrically shaped and has two exterior annuli separated by O rings. When installed in the injector bore in the cylinder head, the lower annular area, in which the balance orifice is located, is subjected to rail pressure. In this way fuel enters the PT injector, circulates within it, and exits through an orifice in the upper exterior annulus, to be returned to the fuel tank.

The PT injector plunger (Figure 9-36) extends through the length of the body and is seated in the cup at the base of the assembly. The cup orifices are sealed only when the plunger is seated.

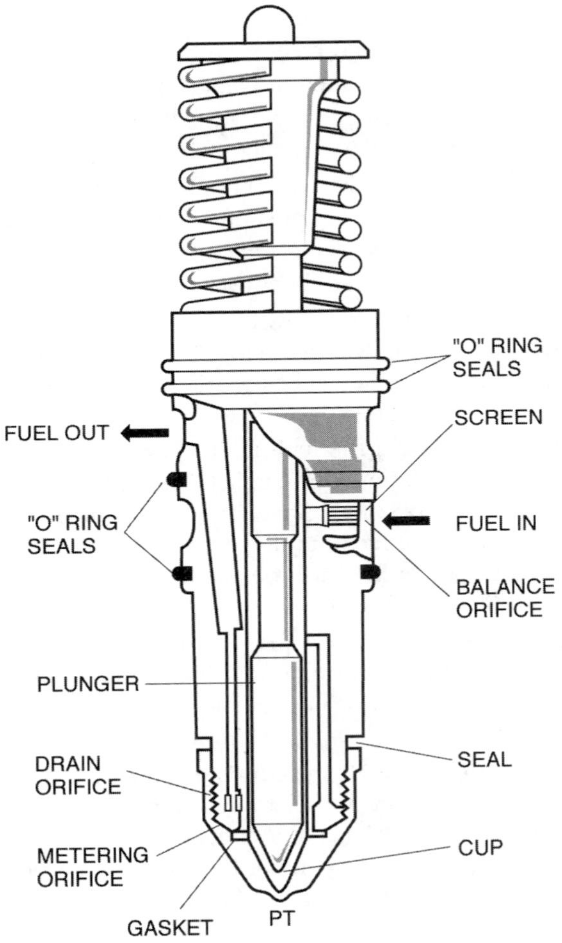

Figure 9-35 Older type Cummins PTC type injector. (Courtesy of Cummins Engine Company)

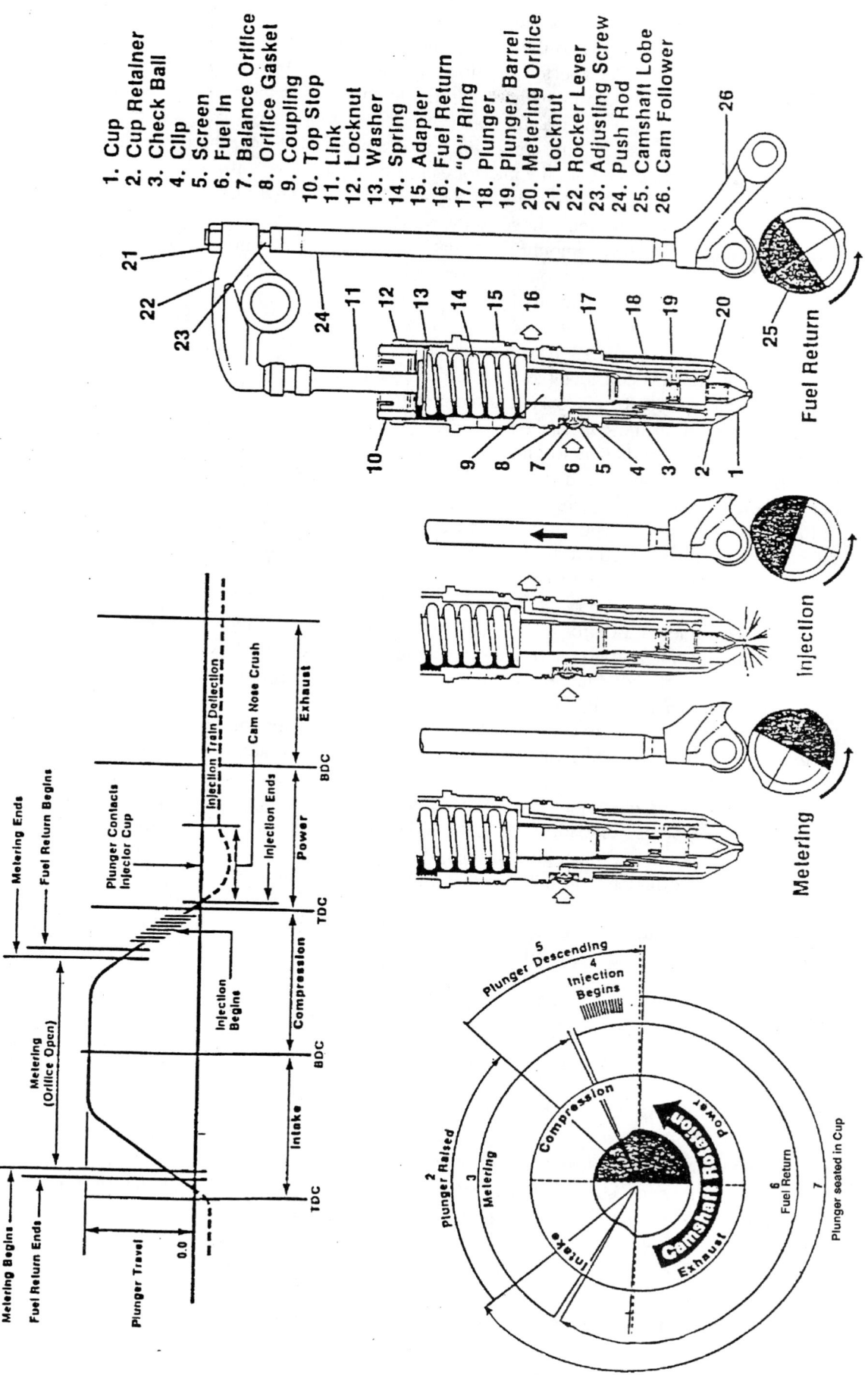

1. Cup
2. Cup Retainer
3. Check Ball
4. Clip
5. Screen
6. Fuel In
7. Balance Orifice
8. Orifice Gasket
9. Coupling
10. Top Stop
11. Link
12. Locknut
13. Washer
14. Spring
15. Adapter
16. Fuel Return
17. "O" Ring
18. Plunger
19. Plunger Barrel
20. Metering Orifice
21. Locknut
22. Rocker Lever
23. Adjusting Screw
24. Push Rod
25. Camshaft Lobe
26. Cam Follower

Fuel Return

Injection

Metering

Figure 9-36 The metering and injection cycle of a PT injector. (Courtesy of Cummins Engine Company)

Due to the geometric design of the injector cam lobe (more than half its periphery is outer base circle), the plunger is loaded into the cup for the larger portion of the cycle. When seated, fuel will circulate through the PT injector, the flow volume dictated by whatever value the rail pressure happens to be at a given moment. When the injector cam is moved to a rotational position to unload the injector train, the injector spring will lift the plunger off its seat in the cup. As the plunger is lifted, it exposes the metering orifice, permitting fuel to flow through it into the cup.

Metering will continue until the injector cam profile passes over the inner base circle of the cam profile onto the ramp toward outer base circle. This ramp drives the injector train (follower, push tube, and rocker) and the plunger downward into the cup, first purging air from the cup and then, acting on whatever amount of fuel is metered to pump it through the cup orifices. This action creates the required injection pressures as the fuel metered into the cup must be forced out through the cup orifices, directly into the engine cylinder.

PT injectors use an open nozzle design. This means that during metering, air from the engine cylinder (under compression) can seep into the cup. As engine compression pressures exceed rail pressure by a considerable margin, this could cause fuel flow reversal. To prevent this from occurring, a check ball is located at the metering orifice. This check ball is designed to seat when the pressure in the cup exceeds the metering pressure. Although this interferes with the metering process, it tends only to present a problem at low engine speeds. The minute sizing of the cup orifices means that at higher speeds, there is insufficient time during metering for cylinder pressures to seep to the injector cup. The check ball will also engage whenever a full charge of fuel is metered into the cup. In this instance, the plunger acts directly on the fuel metered into the cup before it has closed off the metering orifice. Once again, the check ball (or lock-off ball) will engage whenever cup pressure exceeds metering pressure, regardless of the reason.

Typically, only about 35% of the fuel exiting the PT pump rail pipe actually fuels the engine. The remainder is used for lubrication and, very importantly, taking heat away from the PT injectors. The specific percentage of the rail volume used to fuel the engine is known as the fuel rate. Cam profile will determine discharge rate. Cam profile, engine speed, and the sizing of the cup orifices will determine the actual droplet sizing.

It would be incorrect to describe PT injectors as unit injectors because they do not manage metering. Their function is to create the injection pressure and to atomize the fuel. The two flow areas they define are critical for determining engine fueling. For instance, if a set of low flow balance orifices were installed in a set of injectors, rail pressure parameters would go up; that is, the flow area the PT pumps unload to would be reduced, so the pressure would increase. Remember that all of the rail fuel must pass through the PT injector because there is no other connection between the cylinder head rail manifold and the return manifold.

PT Injector Components   The main components of PT injectors are identified below. Step timing control (STC) injectors have some additional components, which are described later in this chapter.

Body
- ❏ Provides mounting for other components
- ❏ Provides fuel routing within the body

Plunger and Cup
- ❏ Form the high-pressure pump element
- ❏ The cup provides space for fuel storage during metering
- ❏ The cup orifice sizing determines droplet sizing and spray dispersal characteristics
- ❏ The plunger and cup are lapped in manufacture, and therefore are not interchangeable

Plunger Spring
- ❏ Lifts plunger for metering
- ❏ Loads the injector train

Check Ball

❑ Prevents fuel flow reversal that could be caused by preinjection pressures or seepage of compression pressures to cup (The check ball, also known as a lock-off ball, is designed to isolate the cup pump chamber from the remainder of the PT injector circuitry at any time cup pressure exceeds metering pressure.)

Balance (Calibrating) Orifice Plug

❑ Provides a means of customizing fuel flow through the injector, which permits standardization of the injector body to a number of engines

❑ Balances injector sets to an engine by defining the flow area the rail unloads to (The balance orifice is sized by burnishing, a procedure that must be performed using Cummins service literature and a PT injector comparator.)

Metering Orifice

❑ Passage within the PT injector through which fuel must pass to enter the cup (It is exposed to the cup whenever the plunger is lifted, so when the plunger is lifted, some of the fuel that is flowing through the PT injector is diverted through the now-exposed metering orifice to flow into the cup. Defines the critical flow area that determines the actual quality of fuel to be injected.)

Plunger Link

❑ A short rod inserted into an aperture on the PT injector follower to help prevent side loading of the plunger (which moves linearly) by the actuating rocker arm (which moves radially). It is a surface-hardened device that should be replaced when evidence of surface-hardening failure is observed.

PTG-AFC Pump

Only the PTG-AFC (pressure-time and governor air-fuel control) pump will be discussed in this text. PTR pumps are not used in highway applications of PT technology, and the PTG-AFC pumps replaced PTG pumps many years ago. In a PTR pump, the "R" stands for manifold fuel pressure regulator, which regulates the maximum fuel pressure. In a PTG pump the "G" stands for governor, where the governor controls the maximal fuel pressure.

The PTG-AFC pump (Figure 9-37) has the following functions:

❑ Responsible for all movement of fuel in the system
❑ Manages rail pressure
❑ Provides limiting speed (LS) governing
❑ Limits fueling when manifold boost is low (internal aneroid)
❑ Provides for ignition key engine shutdown

The PT pump (Figure 9-37) cannot be described as an injection pump. It is a precise flow control device that manages rail pressures, the control factor in the Cummins PT fueling equation. The best way to study this pump is to analyze performance schematics and identify the subcomponents.

PT Pump Components In order to effectively correct any malfunctions in the PT system, the technician must understand the subcomponents of the PT pump and how they interact with each other. The technician should understand the role of each subcomponent in the circuit before proceeding to the next set of schematics.

Gear Pump The gear pump is located at the rear of the PT pump and is encased in a cast iron housing. It is responsible for all movement of fuel in the system. Therefore, the entire fuel subsystem is under suction. In early versions of the PT pump, the fuel filter mounting pad was integral with the pump. In contemporary versions, the filter pad is always remotely mounted. Remote positioning of the filter mounting pad facilitates troubleshooting the admission of air into the fuel

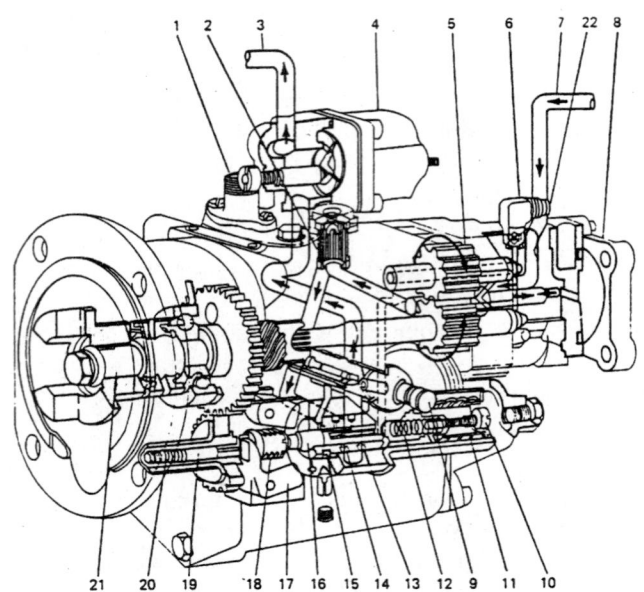

1. Tachometer shaft	12. Idle spring
2. Filter screen	13. Gear-pump pressure
3. Fuel to injectors	14. Fuel-manifold pressure
4. Shutdown valve	15. Idle pressure
5. Gear pump	16. Govenor plunger
6. Check-valve elbow	17. Govenor weights
7. Fuel from tank	18. Torque spring
8. Pulsation damper	19. Weight-assist plunger
9. Throttle shaft	20. Weight-assist spring
10. Idle adjusting screw	21. Main shaft
11. High-speed spring	22. Bleed line

Figure 9-37 PTG pump components. (Courtesy of Cummins Engine Company): 1, Tachometer drive shaft; 2, filter screen; 3, supply to injector rail; 4, shutdown solenoid; 5, gear pump; 6, check valve elbow; 7, fuel from tank; 8, pulsation damper; 9, throttle shaft; 10, idle-adjust screw; 11, high-speed spring; 12, idle spring; 13, gear pump pressure; 14, fuel manifold pressure; 15, idle pressure; 16, governor plunger; 17, governor flyweights; 18, torque spring; 19, weight assist plunger; 20, weight assist spring; 21, main shaft; 22, bleed line.

subsystem (a not uncommon problem) by permitting the insertion of a diagnostic sight glass in series between the gear pump and the filter pad. The diagnostic sight glass consists of clear plastic tubing in a section of hydraulic hose, with coupling fittings at either end, enabling it to be inserted into the fuel circuit. The gear pump is positive displacement; that is, it unloads a constant slug volume into the outlet through each cycle. So if it unloads to a defined flow area, the pressure rise will be proportional to its rotational speed. It is driven at pump-driven speed (in other words, at engine rpm). The pressure values produced by the pump always exceed rail pressure.

Pulsation Damper The gear pump produces pressure in pulses, that is, there is a surge each time a slug of fuel is unloaded by the gear teeth into the ducting that supplies the governor assembly. To smooth these pressure pulses, a pulsation damper is used. It consists of a steel disc supported on either side by a pair of O-rings within an aluminum housing that is mounted directly onto the gear pump housing. A fractured pulsation damper disc would produce fluctuating rail pressures, and cause surging of the engine rpm.

Internal Filter The internal filter is a cylindrical core consisting of a mesh and a magnet. It is located downstream from the gear pump, and its purpose is to minimize the possibility of a gear tooth cutting passing into the rest of the PT pump circuitry. The internal pump filter is in no way designed to substitute for the filters provided in the fuel sub-system. It can be replaced without disassembling the pump, but it is not considered part of routine service maintenance and should be left untouched.

Governor Assembly Figure 9-38   The governor assembly is driven at approximately two times the PT pump driven speed. This allows for smaller weights to be used. The governor employs all the principles of a typical mechanical governor. In any mechanical governor, centrifugal force generated by rotating flyweights attempts to diminish engine fueling at speeds beyond governor break (rated speed), while spring force attempts to increase engine fueling. At engine high-idle speed, the centrifugal force generated by the flyweights must be capable of overcoming any of the spring forces that are applied against it, to prevent engine overspeed. At any engine speed, the position of the governor plunger within the governor barrel will be determined by the balance between flyweight force acting on one end of the plunger and spring force acting on its opposite end. The governor plunger is machined with a recessed annulus, which is cross and center drilled. It is driven by a radial lug that rests on the flyweight feet and, therefore, rotates within the stationary governor barrel. As PT driven speed increases, the plunger is forced down the governor barrel. Fuel from the gear pump enters the governor barrel through the supply passage at any time the engine is running.

The supply passage is always exposed to the recessed annular area of the governor-plunger. The cross and center drillings hydraulically link the plunger annulus with the governor **button (idle spring plunger),** which is spring loaded to contact the inboard end of the plunger. Sizing of the governor button recess determines how much fuel pressure is required to overcome the spring pressure applied to the opposite side of the governor button. When the button separates from the plunger, the fuel will spill back into the bypass passage and recirculate through the pump. The smaller the governor button recess surface area, the more pressure is required in the governor barrel before button separation and spill occur. A governor button with a small recess has the effect of increasing supply pressure beyond the governor assembly.

Fuel is spilled to the bypass passage continually; the amount spilled will determine the actual pressure in the plunger annular recess. Fuel may also exit the plunger recess through the

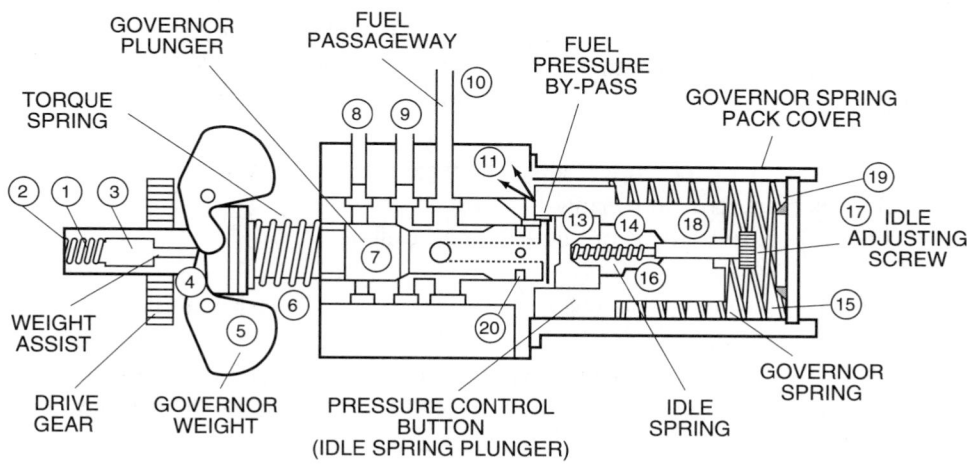

Figure 9-38 PTG governor. (Courtesy of Cummins Engine Company): 1, Weight assist spring; 2, weight assist spring shims; 3, weight assist plunger; 4, governor weight carrier; 5, governor flyweights; 6, torque control spring; 7, governor plunger; 8, idle passage; 9, main passage; 10, supply passage; 11, bypass passage; 12, idle plunger guide; 13, governor button (idle spring plunger); 14, idle spring; 15, governor spring; 16, idle spring seat washer; 17, idle spring adjusting screw; 18, idle screw retention spring; 19, governor shims.

idle passage and the main passage. Fuel exiting these passages will ultimately determine rail pressure. At idle speeds, both the idle and main passages are exposed to the plunger annulus. As the speed increases and centrifugal force drives the governor plunger inboard, the idle passage ceases to register with the plunger annulus. Through the torque rise profile, the torque control spring becomes a factor: this spring helps define torque rise duration. Beyond rated speed, the governor plunger is driven inboard sufficiently so that the flow to the main passage begins to diminish, reducing the flow potential to the rail.

As engine speed progresses through the droop curve (The droop curve is a graduated fuel deration that occurs as the engine rpm increases from the rated to the high idle speeds.) the flow area to the main passage continues to diminish. This means that most of the supply fuel passes through the plunger cross and center drillings, acts on the governor button, and spills to the bypass passage. At the end of the droop curve or high idle, transverse drillings in the governor plunger, called overspeed dump holes, register with the bypass passage, and most of the fuel entering the governor barrel is recirculated to the gear pump intake. The tension of the governor spring assemblies will determine how much centrifugal force (this will correlate to an rpm value) is required to drive the plunger to a position where the main passage is taken out of register with the plunger annular recess. The governor spring tension is set with shims and will define the high idle speed. The sizing of the governor button recess will moderate the actual supply pressure value and will affect engine fueling throughout the speed range of the engine. The governor button is loaded against the inboard end of the plunger by the idle spring. Final adjusting of this spring tension is performed dynamically and sets the engine idle speed.

Throttle Shaft Assembly   The throttle shaft assembly receives fuel from two passages in the governor assembly and supplies fuel to the AFC circuit and the no-air set screw. A throttle lever (accelerator lever) with a spring actuates it break-over. The throttle lever is clamped to the throttle shaft, which is set to move through an arc of 27 degrees ± 1 degree. Moving the throttle shaft will set a flow area from the main passage to the AFC circuit by setting the extent of register of the throttle shaft fuel orifice with the main passage. When the throttle lever travel is measured with a protractor and exceeds the required arc of 27 degrees ± 1 degree, the complaints usually involve low power, and it is an indication that the throttle arm stops have been tampered with.

When the throttle shaft is in the idle position, a small quantity of fuel flows through the throttle shaft fuel orifice; this fuel is defined as throttle leakage. The idle passage feeds an eccentric in the throttle assembly and bypasses the throttle fuel orifice. As the throttle shaft is rotated, its fuel passage is brought into full register with the main passage. A fuel adjusting screw within the throttle shaft bore sets the maximal flow area through the throttle fuel orifice. It is used to adjust rail pressure on the pump calibration stand.

No-Air Set Screw   The no-air set screw defines the maximal flow area from the throttle assembly to the rail before a predetermined manifold boost value (usually 15 psi [105 kPa]) has been achieved. It should only be adjusted on the PT pump calibration stand.

AFC Circuit   The **AFC circuit** is an acceleration smoke control mechanism that replaced the external aneroid device used in earlier PT pump models. All turbo-boosted engines must have fuel delivery modulated to the actual amount of air charged into the engine cylinders. In a Cummins PT engine, intake manifold pressure is piped to the AFC manifold on the PT pump, where it acts on the diaphragm to which the AFC plunger is attached. A spring acting on the AFC plunger loads it into its closed position. When manifold boost reaches a predetermined value (usually 15 psi), it acts on the diaphragm to overcome spring tension to drive the AFC plunger inboard, exposing the AFC ducting and increasing the fuel flow potential to the rail. Unlike most aneroid devices, which are on-off, the PTG-AFC gradually increases flow to the rail as manifold boost increases. The AFC circuit is usually fully open by the time 25 psi of manifold boost is attained, regardless of peak boost on the engine (which is usually around 35 psi).

Shutdown Solenoid The shutdown solenoid is an energized-to-run device that permits the engine to be shut down using the ignition key: it acts by gating the rail. When the solenoid is not energized, a spring loads a disc into a closed position that gates the supply from the AFC ducting to the rail pipe. When the solenoid is energized, the disc is pulled into the solenoid permitting flow. A manual override is incorporated in the device in the event of a chassis electrical failure. The shutdown solenoid override when turned CW displaces the shutdown disc by it forcing back against the spring, thus opening the rail circuit.

PT Pump Circuit Schematics

Probably the best way to understand the operation of a PT pump is to study the following sequence of schematics that show how every internal pump component responds to any given set of running conditions.

Shop Manual
Chapter 9, page 311.

Idle Speed (Figure 9-39)

1. Fuel is pulled through the fuel subsystem by the gear pump.
2. The pulsation damper smoothes pressure waves caused by the gear pump.
3. Fuel is forced through an internal filter consisting of a mesh and a magnet.

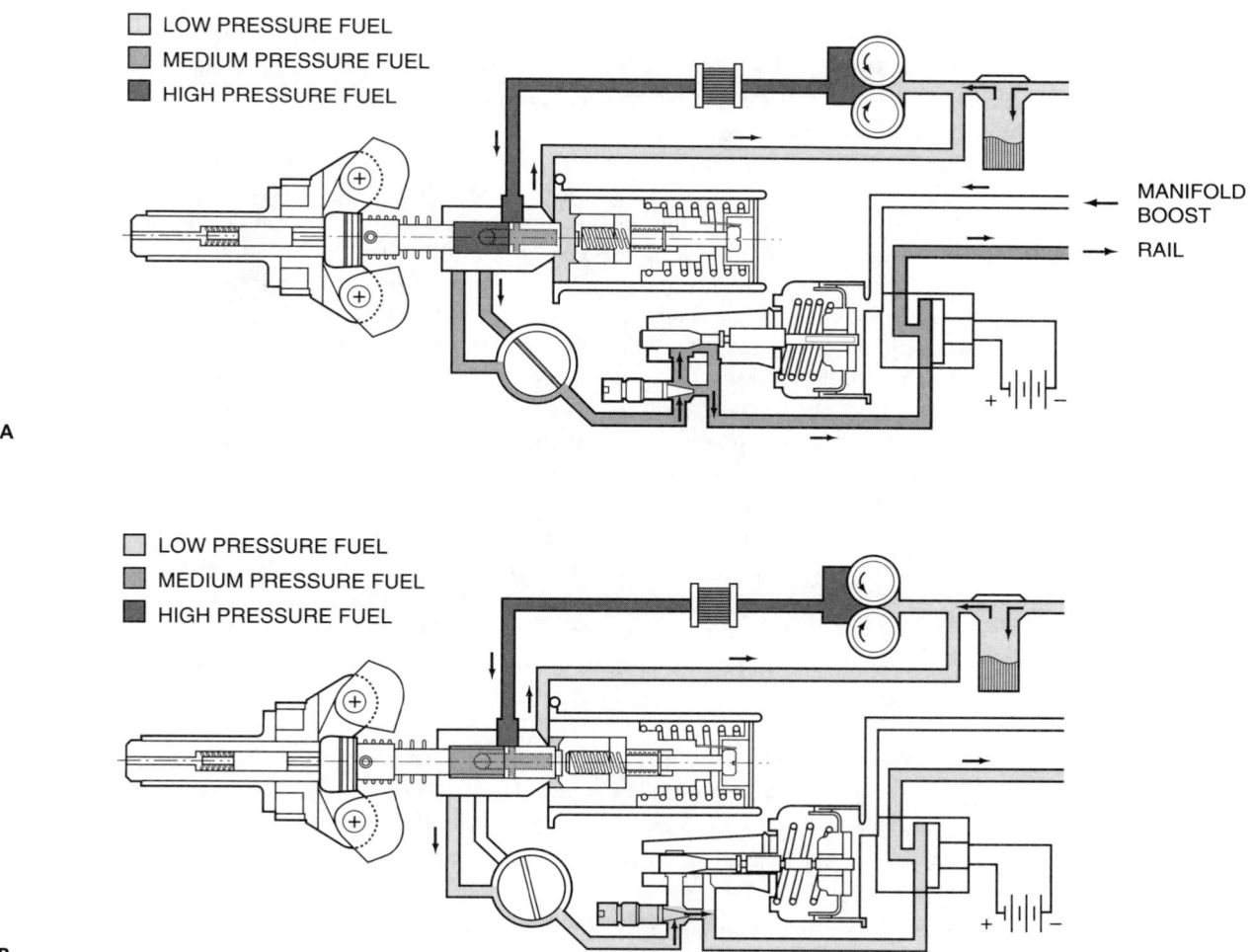

Figure 9-39 (A) PTG-AFC circuit at normal operation; (B) PTG-AFC circuit at idle speed. (Courtesy of Cummins Engine Company)

4. Fuel enters the governor assembly through the supply port in the governor barrel. The governor weights generate little centrifuge at idle speed and therefore the governor plunger (which rotates within the governor barrel) is in a position where both the idle and main ports are in register with the recessed annular area of the plunger. Fuel at supply pressure will exit through the idle and main ports. Because the governor plunger is center and cross-drilled, supply pressure will also act on the governor button, separating it, and recirculate fuel to the bypass circuit.

5. Fuel from the governor assembly arrives at the throttle assembly from the idle and main passages. However, the throttle is in the idle fuel position, which locates the throttle fuel orifice out of register with the main port. A small quantity of fuel is permitted to pass through the throttle fuel orifice and this is known as throttle leakage. Fuel from the idle passage is permitted to bypass the throttle shaft. Most of the idle fuel quantity bypasses the throttle shaft in this manner.

6. Fuel from the throttle assembly is ducted to the AFC circuitry. However, at idle speed there is little or no manifold boost to act on the AFC diaphragm; accordingly, the AFC plunger is in the closed position. The flow area defined by the no-air set screw setting is the total flow area feeding the rail pipe.

7. Fuel is ducted from the AFC circuitry and exits the PT pump through the solenoid to which the rail pipe is connected.

Normal Engine Operation with the Engine Anywhere within Torque Rise Profile (Figure 9-39)

Pressure within the defined flow area represented by the duct feeding the governor barrel will be higher than at idle because the positive-displacement gear pump is being rotated at a higher speed.

1-3. The steps are the same as in the preceding schematic.

4. Fuel enters the governor assembly through the governor barrel supply port and circulates in the annular recess of the governor plunger. However, because of the higher rpm and greater centrifugal force generated by the flyweights, the governor plunger has been driven inboard into the barrel sufficiently to take the idle passage out of register with the plunger-recessed annulus. Supply fuel exits the governor barrel through the main passage and also acts on the governor button, spilling fuel to the bypass for recirculation. This action moderates supply pressure. The torque spring is now a factor; this helps resist further inboard plunger travel toward a diminished fuel position.

5. Fuel from the governor is ducted to the throttle assembly by the main passage. As the engine is running somewhere within the torque rise profile, the throttle shaft is positioned so there is some degree of register of the fuel orifice with the main passage, defining a flow area.

6. Fuel flows from the throttle assembly to the AFC circuitry. Fuel flows up to and around the no-air set screw, but now there is sufficient manifold boost acting on the AFC diaphragm to have overcome the AFC spring and driven the AFC plunger inboard to permit flow through the AFC ducting. This increases the flow area to the rail.

7. Fuel from the AFC circuitry is ducted through the shutdown solenoid and out to the rail pipe.

High-Speed Governing (Figure 9-40)   High-speed governing occurs at governor break, that is, when engine operation is in the droop curve. This is a common running condition for Cummins PT engines in which the accelerator is held in the full-fuel position by the operator. Supply pressure is higher than in previous schematics because of the higher rpm, but otherwise the operating conditions are the same.

1-3. The steps are the same as in the preceding schematic.

4. Fuel enters the governor assembly through the governor barrel supply port, but now centrifugal force exacted by the flyweight has driven the governor plunger inboard to the extent that the main passage flow area is reduced. The fuel passes through

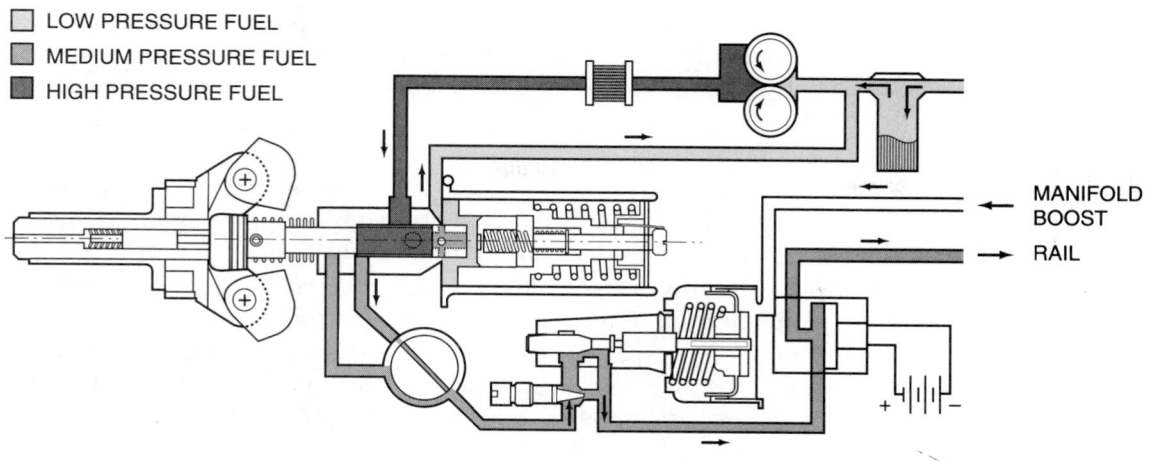

MANIFOLD
BOOST

RAIL

Figure 9-40 PTG-AFC circuit schematic at high-speed governing. (Courtesy of Cummins Engine Company)

the center and cross drillings and pushes on the governor button to enter the by-pass circuit, spilling more fuel.

5. For the engine to be run in this condition, the accelerator would necessarily be fully depressed, allowing the throttle shaft fuel orifice full register with the main passage.

6. Fuel from the throttle assembly flows to the AFC circuitry. However, until the rpm penetrates well into the droop curve, the engine will be sufficiently fueled to generate enough rejected heat to ensure that the turbocharger can maintain manifold boost in excess of 15 psi (105 kPa). This permits at least some flow through the AFC ducting.

Complete High-Speed Governing (Figure 9-41) In complete high-speed governing engine rpm has risen above high idle; that is, the engine is being run at an rpm beyond the droop curve. In most cases, this means that the engine is being driven by the drivetrain of the vehicle, a condition that would occur when running on a prolonged downhill. The schematic shows what is occurring in the PT pump circuitry when the engine speed exceeds the high idle speed. Supply pressure, which depends on rpm, is at its highest.

1-3. The steps are the same as in the preceding schematic.

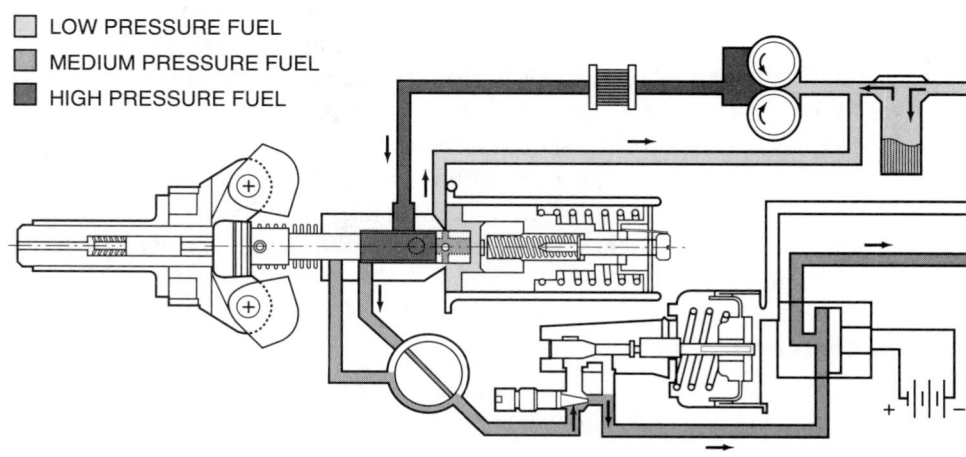

Figure 9-41 PTG-AFC circuit schematic at complete high-speed governing. (Courtesy of Cummins Engine Company)

4. Fuel enters the governor barrel through the supply port. In this running condition, the governor flyweights are maximally extended and have driven the governor plunger inboard against the governor spring (high-speed spring) to the extent that:

   A. The main passage has been entirely or almost entirely taken out of register with the recessed plunger annulus.

   B. The plunger has been driven inboard to the extent that a transverse drilling through the plunger has extended beyond the governor barrel, allowing most of the supply fuel to be spilled directly to the bypass.

   C. The rpm at which complete high-speed governing occurs is determined by the spring tension of the governor (high idle) spring, which is set by shims.

5. At complete high-speed governing, so little fuel is being injected into the engine's cylinders, that manifold boost will drop below the 15 psi required to hold the AFC ducting open. So all of the fuel that exits the PT pump to the rail must do so by passing around the no-air set screw.

Step Timing Control (STC)

A characteristic of PT, common rail injection is to have a variable beginning and constant ending of the injector fueling pulse. This translates into the end of injection always occurring in exactly the same engine location regardless of engine load or speed. The Cummins PT fueling window was originally engineered for optimal performance around the rated speed and load-operating zone. This meant that when the engine was operated at low speeds and lighter loads, the beginning of the injection pulse was effectively retarded. In fact, as the beginning of injection depends directly how much fuel has been metered into the cup, the lighter the engine load, the more retarded the beginning of injection becomes. In an effort to rectify this problem, Cummins introduced the step timing control (STC) system in 1986.

STC is actuated and managed hydraulically. It attempts to "advance" the beginning of injection whenever engine loading is light. Although the term "advance" will be used in describing the STC system, perhaps the objective of STC would more properly be described as getting the injection pulse within a normal range.

Governors

James Watt invented the mechanical governor with the objective of regulating the speed of his steam engines. Speed sensing was accomplished by driving a set of flyweights in a carrier at a speed proportional to engine speed. The flyweights pivoted in the carrier and were loaded into their most retracted position by a spring. As the carrier was rotated, the centrifugal force exacted from the flyweights would act against the applied spring force. The spring tension could then be set so that at a specified maximal speed, centrifugal force would overcome the spring force to act directly on a fuel control mechanism to limit fueling. Most truck and bus governors classified as hydromechanical use variants of Watt's governor.

The diesel technician should have a familiarity with this terminology, as it is used in performance testing and assessment. At the end of the chapter, there is a glossary of governor terminology, which gives more detailed explanations of terms used within the text than those found in the main glossary. It is recommended that these terms be referenced when studying this chapter.

Governor Types

Governors are classified by type, or operating principles. The mechanical governor is probably the most common on hydromechanically managed (engines with fuel systems managed without the use of computers). Every governor managing a truck or bus engine has the benefit of a driver who will regulate engine output according to the requirements of the trip being undertaken. The accelerator, therefore, represents an essential input to the governor. The governor must also know exactly how fast the engine is rotating in any given mode of operation; so another essential input to the governor is a means of precisely determining engine rpm. In vehicle engine governors, these two inputs are essential. A governor has control over how the engine is fueled. It limits fueling at the highest intended engine rpm to prevent overspeed, but it is also required to define the lowest no-load rpm of the engine to enable it to idle without any input from the driver's accelerator pedal. The governor must also provide a means of no-fueling the engine to shut it down. Most modern governors also have many other features such as the ability to provide extra fuel over the normal engine operating rpm (torque rise: identified to the operator by the recommended shift rpm), graduated fuel deration at the top engine speeds, and excess fuel/timing adjustments for start-up. Each governor type listed in this section is simply a means of grouping governors into categories. Not every mechanical governor is the same; some are very basic while others have an extensive array of optional features.

Mechanical Governors Speed sensing is by means of rotating flyweights (Figure 9-42) driven at a speed proportional to engine rpm. In a mechanical governor, centrifugal force generated by the rotating flyweights acts directly on the fuel control mechanism. The governor spring tension is set to oppose the centrifugal force and will define the engine top limit or high idle speed (the highest speed at which an engine is designed to run). The thrust collar (thrust washer or thrust sleeve) acts as an intermediary between the spring force and centrifugal force and is usually connected to the fuel control mechanism. Governor spring tension is usually designed to be variable and increases with accelerator pedal travel.

❏ Centrifugal force (produced by rotating the flyweights) acts on the thrust collar to attempt to decrease fueling.
❏ Spring force (often variable and increases with accelerator pedal travel) acts on the thrust collar to attempt to increase fueling.

Hydraulic Servo-Type Governor In a hydraulic, servo-type governor speed sensing is accomplished by rotating flyweights driven at a speed proportional to engine speed; opposing this force is the governor spring. However, the intermediary is a control valve that feeds oil (usually pressurized oil from the engine lubricating circuit) to a servo, which moves the fuel control mechanism. Therefore, the actual force used to move the fuel control mechanism is hydraulic. This design permits the use of smaller flyweights and is more sensitive to speed and load fluctuations.

Hydraulic Non-Servo Governor In a hydraulic non-servo governor, speed sensing is accomplished by using a positive-displacement, engine-driven transfer pump, which unloads to a defined (and constant) flow area so fuel pressure rises proportionally with engine speed. This type of governor is used in inlet metering fuel systems. Gating or diverting fuel from the inlet metering apparatus at a specified fuel pressure regulates speed.

Pneumatic A pneumatic governor is used on naturally aspirated diesel engines and therefore is seldom seen in truck and bus applications. Fueling is factored by manifold vacuum, which is regulated by a throttle valve controlled by an operator. This is common in older marine and agricultural applications.

Electronic Governing Electronic governing is commonly used to govern the truck and bus diesel engines of today. Engine speed sensing is by means a reluctor-type, inductive sensor, which signals an electronic control module (ECM) that plots a fueling profile and then effects it by switching actuators. Today's trucks and buses are drive-by-wire, meaning there is no direct mechanical connection between the accelerator pedal and the fuel control mechanism. In other words, the accelerator position is one of a number of sensor inputs to the ECM. Electronic governing is used by all the diesel engine electronic management systems including Detroit Diesel DDEC, Cummins CELECT and IS (Interact System), Caterpillar (PEEC, 3176, 3406E, C10, C12,

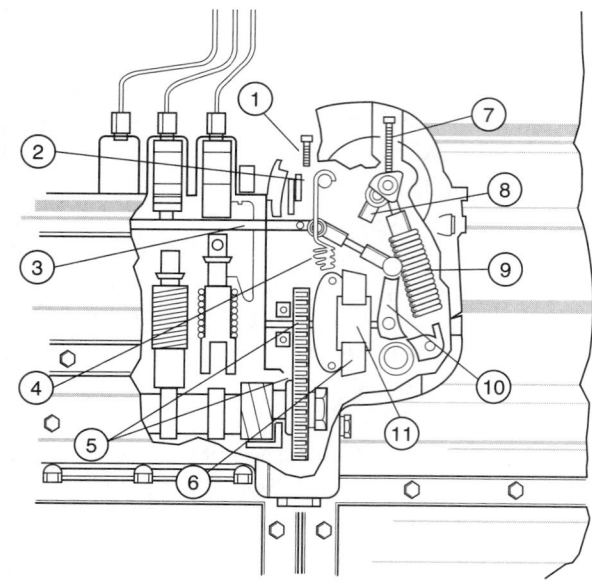

GOVERNOR CUTAWAY

1—Low idle adjusting screw. 2—Spring-loaded stop assembly.
3—Fuel rack. 4—Shut-off spring. 5—Gears.
6—Governor weight. 7—High idle adjusting screw.
8—Low idle stop lever. 9—Governor spring.
10—Terminal lever.11—Thrust bearing and sliding sleeve

A

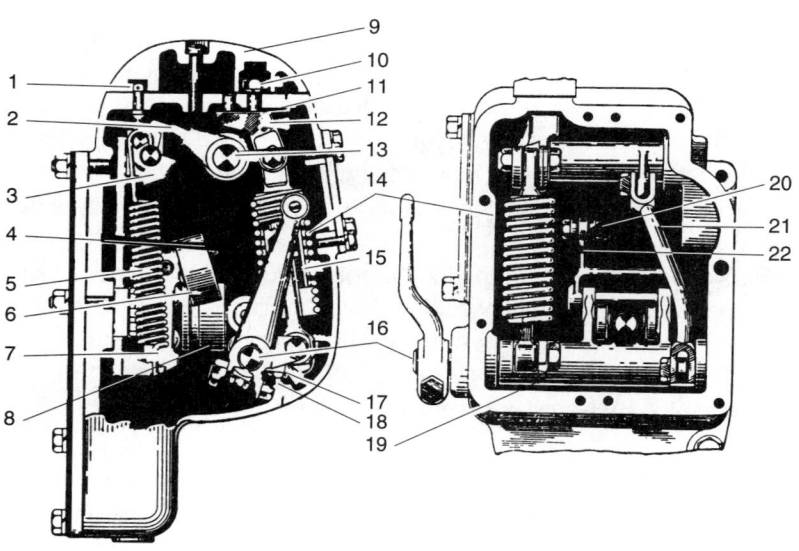

SECTIONAL VIEW OF GOVENOR
(SCHEMATIC)

1—Low idle adjusting screw. 2—Low idle stop lever. 3—Stop. 4—Small governor
weight. 5—Shut-off spring. 6— Large governor weight. 7—Spindle assembly.
8—Thrust bearing and sliding sleeve. 9—Top cover. 10—Capscrew. 11—Shims.
12—Anchor. 13—Stop lever shaft. 14—Governor spring. 15— Small spring.
16—Terminal lever shaft. 17—Throttle shaft control lever. 18—Setscrew
(earlier engines). 19—Terminal lever. 20—Link. 21—Rod. 22—Lever.

B

Figure 9-42 (A) Governor cutaway (Courtesy of Caterpillar Inc.); (B) DDC governor cutaway (Reprinted by permission of copyright owner Detroit Diesel Corporation, all rights reserved.).

HEUI), Mack Trucks V-MAC, and Navistar HEUI/etc. These systems will be discussed in some detail elsewhere in this text.

Governor Classification

The governor classifications found on contemporary truck and bus diesel engines fuel management systems are limiting speed, variable speed, and isochronous.

Limiting speed A **limiting speed (LS; min-max) governor** sets the engine idle speed, defines the high-idle speed, and permits fueling between those parameters to be controlled by an operator (driver). Limiting speed governors are the most common in commercial vehicle applications, and one of their advantages is to make the diesel engine respond to accelerator input in much the same manner as the spark-ignition (SI) engine responds to throttle control. A governor classified as limiting speed will, in most cases, provide excess start-up fuel, define a torque rise profile, define droop curve, and be capable of no-fueling the engine for shutdown. A mechanical limiting speed governor is sometimes known as an automotive governor.

Variable Speed Governor – VS A **variable speed (VS) governor** is known as an all-speed governor in the United Kingdom. It sets engine idle-speed and defines high idle and any speed in the intermediate range, depending on accelerator pedal position. A given amount of accelerator pedal travel will correspond to an engine rotational speed. As engine loading either increases or decreases, the governor will manage fueling to attempt to maintain that engine speed. Hydromechanical variable speed governors were common in many Mack Trucks and Caterpillar applications, among others, where power take-off (PTO) management was a consideration. PTO management is used when the engine drives auxiliary equipment. From the driver perspective, the VS governor takes a little getting used to. Most of today's electronic management systems can be toggled to either LS or VS mode. A governor classified as variable speed will usually provide excess start-up fuel, define a torque rise profile, define droop curve, and be capable of no-fueling the engine for shutdown.

Isochronous Governor Isochronous governing is only required when driving a generator in which application, the engine must respond instantly to load changes with zero **droop** (no rpm fluctuation when engine load changes) or the frequency will alter. However, the term is used to describe an option in diesel engine electronic management systems. In this instance, isochronous governing mode is used to manage PTO fueling while stationary, and one OEM uses the term to describe engine fueling at an electronically managed, default rpm (when critical input signals are lost).

Governor Terminology

All speed governor: A variable speed governor, but Mack Trucks and Bosch also use the term.

Deadband: The sensitivity of a governor. It is the speed window around set speed within which the governor makes no fueling correction.

Droop: A transient (of short duration) speed variation from set speed when engine load changes.

Droop curve: Expressed as a percentage of high idle speed. The droop curve is the descent in the fuel profile between rated load and speed (peak power) and the high idle speed. The droop curve is usually between 5% and 20% in hydromechanical highway diesel engines.

Governor cut-off: Speed at which the governor cuts off fueling.

Governor flight path: *See* **Droop curve**.

Governor spring: Force used to counter centrifugal force developed by the flyweights in a mechanical governor.

High Idle: (WOT. or top engine limit). Maximum no-load speed of an engine.

Hunting: Rhythmic change in engine speed often caused by unbalanced fuel delivery in multicylinder engines.

Hydromechanical governing: Refers to engines that are governed without the use of computers (ECM).

Idle: Any no-load running speed of an engine, but usually refer to low idle, the lowest speed the engine is designed to run at, usually with no input from the speed control mechanism.

Overrun: The inability of a governor to keep the engine speed below the high idle speed when it is rapidly accelerated.

Overspeed: Any speed above high idle.

Motoring: Running an engine at zero throttle, with chassis momentum driving engine.

Peak torque: The rpm at which the engine develops peak torque, often located at the base of the torque rise profile.

Rated speed: The rpm at which peak power is achieved from a diesel engine.

Road speed governing: Any governor system in which engine fueling is moderated by a predetermined road speed value.

Runout: *See* **Droop curve**.

Sensitivity: Ability to respond to maintain a set rpm without fluctuation as the load changes.

Speed drift: Engine speed rises above or below set speed, often in surges. Differentiated from **hunting** by the fact it is not rhythmic.

Stability: Ability to maintain a set rpm.

TEL: *See* **Top engine limit**.

Throttle: Airflow to the intake manifold control mechanism used in SI gasoline and diesel engines with pneumatic governors. The term is commonly used to describe the speed control/accelerator/fuel control mechanism in a diesel engine.

Thrust collar: In a mechanical governor, the intermediary between the centrifugal force exacted by the flyweights and the spring forces that oppose it. The thrust collar in a governor is usually connected to the fuel control mechanism.

Top engine limit (TEL): High idle, or the fastest no-load rpm that the engine is designed to run at.

Torque rise: The rpm range through which an engine can maintain near maximum torque. The engine's operating range.

Torque rise profile: A graphic representation of engine fueling on a fuel map or graph indicating the torque rise window. Often, but not necessarily, the profile begins at peak torque and ends at rated load speed.

Underrun: A governor's inability to maintain the engine low-idle speed when rpm is quickly dropped.

Work capacity: A measure of a mechanical governor's ability to produce the centrifugal force required to move the fuel control mechanism. The work capacity of a mechanical governor driven at engine rpm exceeds that of one driven at camshaft speed, assuming the same flyweight mass.

Wide-open throttle (WOT): Wide-open throttle. A term used mainly on SI automotive engines and less often on diesels meaning full fuel request.

Summary

Terms to Know

AFC (air-fuel control) circuit

Balance orifice

Button (idle spring plunger)

❑ The port-helix pump is driven through one full rotation (360 degrees per full effective cycle of the engine (720 degrees in a four-stroke cycle). Its camshaft is supported by main bearings and driven in the cambox, which also acts as a lubrication sump. Pump-element–actuating tappets are spring loaded to ride the cam profiles. Cam geometry dictates the pump element activity with a pump element dedicated to each engine cylinder, consisting of a stationary barrel and a reciprocating plunger. The plunger is milled in manufacture with a metering recess known as a helix or scroll. Plunger rotational position will determine the point of register of the barrel spill port and the helix. Plungers are rotated in unison by a toothed rack meshed to slotted control sleeves, which are themselves lugged to the plungers.

❏ Plunger effective stroke begins at port opening and ends at port closure. Delivery valves separate the pump elements from each high-pressure pipe and act to retain dead volume fuel at pressure values approximating two-thirds NOP. Delivery valves increase the volume available for dead volume fuel storage in the high-pressure pipe by the swept volume of the retraction collar.

❏ Injection rate is a term defined as fuel injected per crank angle degree. Most current port-helix metering injection pumps have a variable timing mechanism that acts as an intermediary between the pump drive gear (on the engine) and the pump camshaft coupling. In-line pumps often incorporate an aneroid device and a altitude compensator to prevent more fuel being injected to a engine cylinder than there is oxygen to burn it.

❏ The distributor-type pump uses one pump barrel and a set of plungers to supply all cylinders in rotation.

❏ Metering of fuel in a Detroit Diesel Corporation mechanical unit injector is accomplished by rotating the plunger within the bushing bore by means of a gear, which is toothmeshed to the MUI control rack. The DDC MUI plunger is milled with helices which, depending on helix geometry, may offer constant beginning, variable ending; variable beginning, constant ending; or variable beginning, variable ending, delivery timing characteristics.

❏ The Caterpillar MUI systems use a fuel ratio control device that limits fueling under engine operating conditions of low boost. The fuel shut-off device used by Caterpillar on their MUI system is an energized-to-run (ETR) latching solenoid. A hydraulic equation constructed to calculate a volume of flow is factored by the time of the flow, the flow area, and fluid pressure.

❏ The Cummins PT system uses the pressure variable in a simple hydraulic equation to manage fueling. It further uses a flow-control device and a common rail feed system to supply PT injectors mounted in parallel. The collective restriction of the balance orifices in the rail (six in a six-cylinder engine) defines the flow area required to factor rail pressure. The PT injector metering orifice defines the flow area for determining the actual volume of fuel metered to the engine cylinder. The PTG-AFC pump is a flow-control and governing device that manages rail pressure values to control engine fueling.

❏ Most hydrochemical governors sense engine speed change using rotating flyweights that rotate at a speed proportional to engine speed. On a variable speed governor, accelerator position commands a specific engine speed. As the engine load increases and decreases, the governor adjusts engine fueling to attempt to maintain that engine speed. In a basic mechanical governor system, centrifugal force exacted by flyweights will attempt to diminish engine fueling, while spring force (often moderated by operator demand) attempts to increase fueling. However, at a predetermined maximal engine speed value, the centrifugal force must always overcome the spring force.

Cambox
Common rail system
Control rack
Control sleeve
Droop
Effective stroke
Flow area
Flow control
Fuel rate
Governor spring
Injection rate
Isochronous
Limiting speed (LS) governor
Lower helix
Metering orifice
Mechanical unit injector (MUI)
Plungers
Port helix
Port closure
Port opening
Pressure-time (PT) fuel system
STC-step timing control
Upper helix
Variable speed (VS) governor

Review Questions

Short Answer Essays

1. Identify the major components of a typical port-helix metering injection pump.

2. Explain the principles of operation of an in-line, port-helix metering injection pump.

3. Identify the major components of a distributor-type injection pump, and explain how the pump's principles of operation apply to these components.

4. Explain how the pump element components interact to create injection pressures.

5. Explain the operation of injection pump peripherals, including aneroid devices, altitude compensators, and variable timing/timing advance mechanisms.

6. Outline the principles of operation of a DDC mechanical unit injector.

7. Describe how MUI effective stroke is varied to control injected fuel quantity.

8. Outline the principles of operation of a Caterpillar mechanical unit injector system.

9. Identify the major components in a Cummins PT-type metering injection pump and explain their operation.

10. Explain the function and operation of a hydromechanical governor on a diesel engine.

Fill-In-the-Blanks

1. A pump element consists of a stationary _____ and a reciprocating _____.

2. Plunger effective stroke begins at ____ _____ and ends at ____ _____.

3. _____ _____ is a term defined as fuel injected per crank angle degree.

4. Metering of fuel in a DDC MUI is accomplished by _____ the plunger within the bushing bore by means of a _____.

5. The Detroit Diesel MUI plunger is milled with _____ which, dependant on helix geometry, may offer constant beginning and a _____ ending.

6. The Caterpillar MUI systems uses a _____ _____ _____ device limits fueling under engine operating conditions of low boost.

7. The fuel shut off device used by Caterpillar on their MUI system is an _____ to run (ETR) latching solenoid.

8. The Cummins PT System uses a _____ _____ _____ and a common rail feed system to supply PT injectors mounted in parallel.

9. The Cummins PT System uses the _____ variable in a simple hydraulic equation to manage fueling.

10. In a basic _____ governor system, centrifugal force exacted by flyweights will attempt to diminish engine fueling while spring force often moderated by operator demand, attempts to increase fueling; however at a predetermined maximum engine speed value, the _____ force must always overcome the spring force.

ASE-Style Review Questions

1. What is the function of the flutes machined into a typical delivery valve core?
   A. Guide the valve in the body
   B. Permit the valve to seal before it seats
   C. Define the residual line pressure
   D. Define valve core travel in the valve body

2. The precise control of injected fuel quantity is known as:
   A. Timing
   B. Metering
   C. Atomization
   D. Spray dispersion

3. Which of the following diesel fuel injection terms is used to describe the amount of fuel injected per crank angle degree?
   A. Metering   C. Timing
   B. Injection rate   D. Phasing

4. Which of the following in-line fuel injection pump components would limit engine fueling at high altitudes?
   A. Barometric capsule
   B. Aneroid
   C. Governor
   D. Variable timing

5. *Technician A* says that when static timing a port-helix metering pump to an engine, the #1 element is must always timed to the #1 pump cylinder. *Technician B* says the OEM seldom recommends timing current port-helix metering pumps to engines with the spill timing method. Who is correct?

A. A only

B. B only

C. Both A and B

D. Neither A nor B

6. Effective stroke in a DDC mechanical unit injector begins when:

A. Both upper and lower bushing ports are exposed

B. Both upper and lower bushing ports are covered

C. The plunger is on its downstroke

D. The plunger is on its upstroke

7. When considering the actual quantity of fuel delivered to the cylinders of a PT-type pump fuelled engine, which of the following *most* likely represents the critical flow area?

A. Balance orifice

B. Metering orifice

C. No-air screw

D. AFC valve

8. Which type of hydromechanical governor would *most* likely be used to manage a highway truck or bus diesel engine?

A. Mechanical, limiting speed

B. Hydraulic, limiting speed

C. Mechanical, variable speed

D. Hydraulic, variable speed

9. What component is used on Caterpillar MUI systems to limit fueling under conditions of low manifold boost?

A. ETR solenoid

B. Aneroid

C. FRC device

D. AFC valve

10. *Technician A* says the phasing of an in-line, port-helix metering pump involves the correct spacing of port closure in each pump element through its effective cycle and is usually performed with a degree wheel (protractor) on the comparitor bench drive turret. *Technician B* says that the calibration of an in-line, port-helix metering injection pump, can be performed on the engine with the appropriate governor adjusting tools. Who is correct?

A. A only

B. B only

C. Both A and B

D. Neither A nor B

Electronically Managed Pump-Line-Nozzle (PLN) Systems

Upon completion and review of this chapter, you should be able to:

❏ Identify the major components and explain the operation of a distributor-type electronic injection pump system.

❏ Define the components and explain the operation of the Caterpillar PEEC fuel injection system on a 3406 engine and define why it is a partial authority system.

❏ Describe how Mack Trucks and Bosch adapted an in-line, port-helix metering injection pump for computerized management and control.

❏ Explain the operation of the MACK V-MAC II fuel injection systems and differentiate between V-MAC II and I.

❏ Define the terms PACE and PACER.

Introduction

This chapter will cover fundamentals of operation for distributor-type electronic pumps and limited authority in-line type pump systems with electronic governor control. This coverage will include component identification, pump metering, timing, electronic sensors, and microprocessor control. The section on in-line–type pumps will cover the electronic type governor and will not cover the mechanical operation of the in-line–type hydromechanical areas. See Chapter 9 for these specific areas. Specific coverage will extend to the following original equipment manufacturer (OEM) systems: Stanadyne DS, Caterpillar PEEC, and V-MAC I and II.

A BIT OF HISTORY

The Electronic engine control revolution for diesel engines began in 1985 when the Detroit Diesel Division (then part of General Motors) introduced the first DDEC (Detroit Diesel electronic control) system. Caterpillar followed in 1987 with its partial authority **PEEC** (programmable electronic engine control) system based on the Bosch in-line–type pump in the 3406 engine for California applications. In 1994, GM introduced the first distributor-type electronic pump system on the 6.5 L diesel engine.

Distributor-Type Electronic Injection Pump System

Stanadyne DS Pump

General Description (Figure 10-1)

1. Eight-cylinder, four-plunger rotary distributor pump.
2. Fuel delivery controlled by a single high-speed electronic spill control solenoid.
3. Constant beginning of injection/variable end of injection (spill) vs. mechanical pump's (DB2) variable beginning and constant ending.
4. Hydromechanical automatic timing advance electronically controlled by a stepper motor.

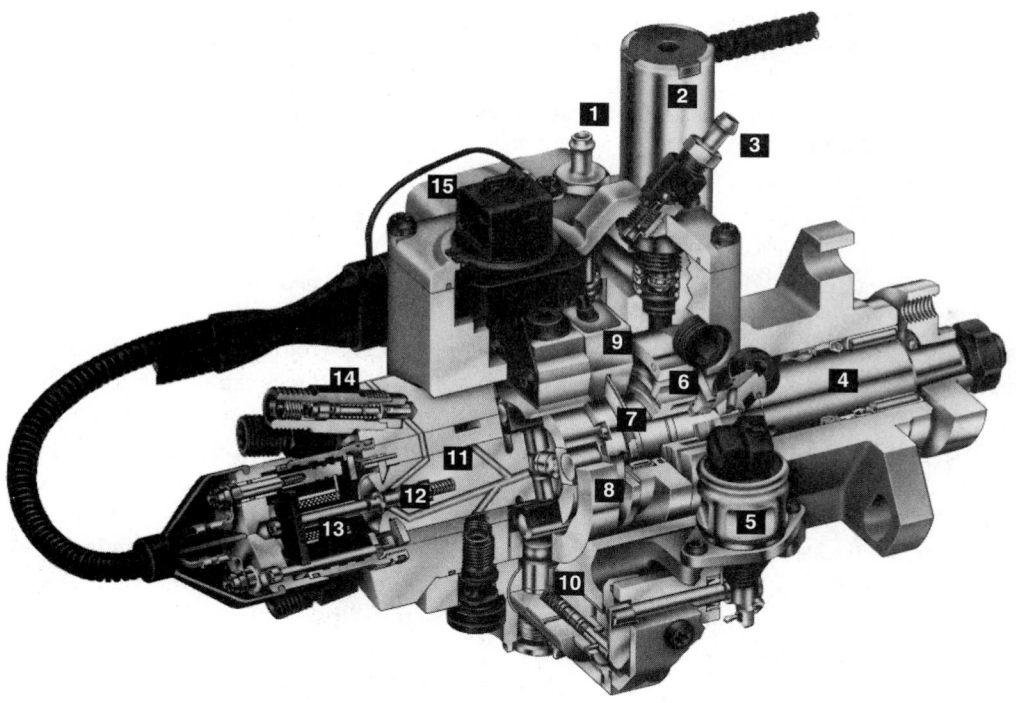

1. Fuel inlet fitting
2. Electric shutoff solenoid
3. Return line connector/housing pressure regulator
4. Heavy duty drive shaft
5. Advance stepper motor
6. Transfer pump
7. Datatrack disk
8. Cam ring
9. Encoder sensor
10. Servo advance mechanism
11. Distributor rotor
12. Poppet valve
13. Fuel control solenoid
14. Delivery valve and snubber components
15. Pump mounted driver (PMD)

Figure 10-1 Model DS fuel injection pump. (Courtesy of General Motors Corporation, Service Operations)

5. Integral vane-type transfer pump providing regulated pressure for charging and advances operation.

6. No mechanical governor (electronically controlled).

DS Pump Electronic Components

1. Electrical shut-off solenoid (ESO); a plunger-style electromagnetic solenoid that is energized to run (ETR).

2. Pump-mounted driver (PMD) or fuel solenoid driver:

A. Solid state high current switching device.

B. Receives inject commands from the powertrain control module (PCM) and provides a regulated 10 amp current to the fuel solenoid, which controls injection.

C. Sends poppet valve closure signal (injection pulse width) to the PCM.

3. Optical sensor: optical speed timing encoder (OSTE) (also known as the angular speed timing encoder, or ASTE):

A. Two-channel optical sensor mounted on the pump cam ring.

B. Reads a two-track **Datatrack disk** mounted concentric to the pump driveshaft.

C. High-resolution track (DTC 17 or P0370) has 512 notches that provide a 0.044 angular resolution. The PCM uses this track information to energize and de-energize the fuel control solenoid for accurate fuel delivery.

All of the fuel subsystems, such as transfer (lift) pump, lines, and filter, are the same as on the mechanical system covered in Chapter 8.

Shop Manual Chapter 10, page 365.

D. Low-resolution track or cam reference pulse (DTC 18 or P0251) has eight notches that provide an angular relationship between TDC and the cam ring for timing purposes.

   E. Contains a thermistor fuel temperature sensor for measuring fuel temperature (DTC 42 and 43).

4. Fuel control solenoid:

   A. High speed electromagnetic actuator, which, when pulsed, actuates a poppet valve (GM term: control valve) in the rotor to control fuel delivery.

   B. At the rated speed of 3,400 rpm solenoid pulses 227 times/second.

The housing pressure regulator is also known as the return line or back leak connector.

Additional Pump Features   The **housing pressure regulator** (return line connector), as shown in Figure 10-2, uses a two-stage positive seal regulator. This reduces drain back to the fuel tank. This regulator contains the same positive sealing poppet-valve–type regulator used on the DB2 pump (6.2 L/6.5 L mechanical system), in the first stage and a spring loaded glass ball check regulator with a notched seat in the second stage. They are both rated at 5 psi. On engine shut-down air could migrate through a porous return hose into the pump generating air-in-fuel symptoms through the notched seat. Residual line pressure is 600–800 psi. The first design single-stage, spring-loaded ball-check regulator with a notched seat was rated at 10 psi. The second design contains a positive sealing poppet-valve–type regulator rated at 2–4 psi and a spring-loaded, ball-check regulator with a notched seat rated at 6 psi.

Housing Pressure Values

Enging running   8–12 psi

Test bench   6–14 psi

Cranking   4 psi or lower

The energized-to-run (ETR) electric shut-off solenoid (ESO) (Figure 10-3) is located on the right side of the transfer pump and upstream of the transfer pump with the regulation downstream.

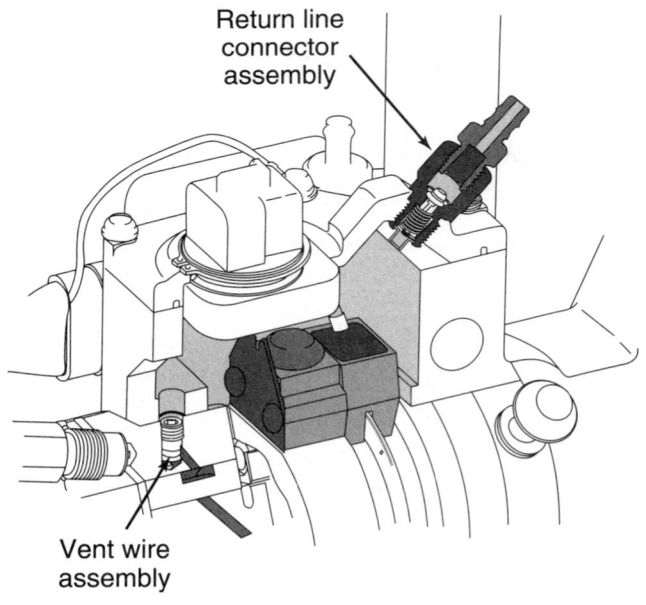

Figure 10-2 Return line connection and vent wire assembly. (Courtesy of General Motors Corporation, Service Operations)

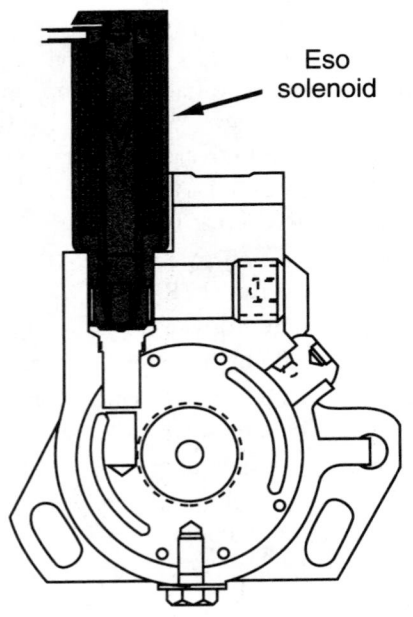

Figure 10-3 ESO electric shut-off solenoid. (Courtesy of General Motors Corporation, Service Operations)

DS Pump Component Inputs and Outputs to and from the PCM

Optical Sensor The angular/optical speed timing encoder (ASTE/OSTE) in Figure 10-4 is a two-channel **optical sensor** that is mounted on the pump cam ring. It reads a two-track Data-track disk mounted concentric to the pump driveshaft. This sensor provides rpm, cam and crank-shaft position, and injection duration to the powertrain control module. It sets either of two DTC 17 (P0370), high-resolution failure (512X signal) or DTC 18 (P0251) cam reference pulse error (8X signal). Figure 10-4B shows the optical sensor circuit.

Datatrack Disk The Datatrack disk (Figure 10-4) contains two separate tracks or channels. These tracks contain a series of notches or windows. The inner track or cam reference pulse (DTC 18 [P0251]) has eight notches (windows), which provide accurate individual cylinder references. This track is also referred to as the low-resolution track and can provide an angular relationship between the crankshaft position sensor (TDC) and pump cam ring for timing purposes. The optical sensor (ASTE) transmits these references to the PCM. A second (outer) track on the Datatrack disk

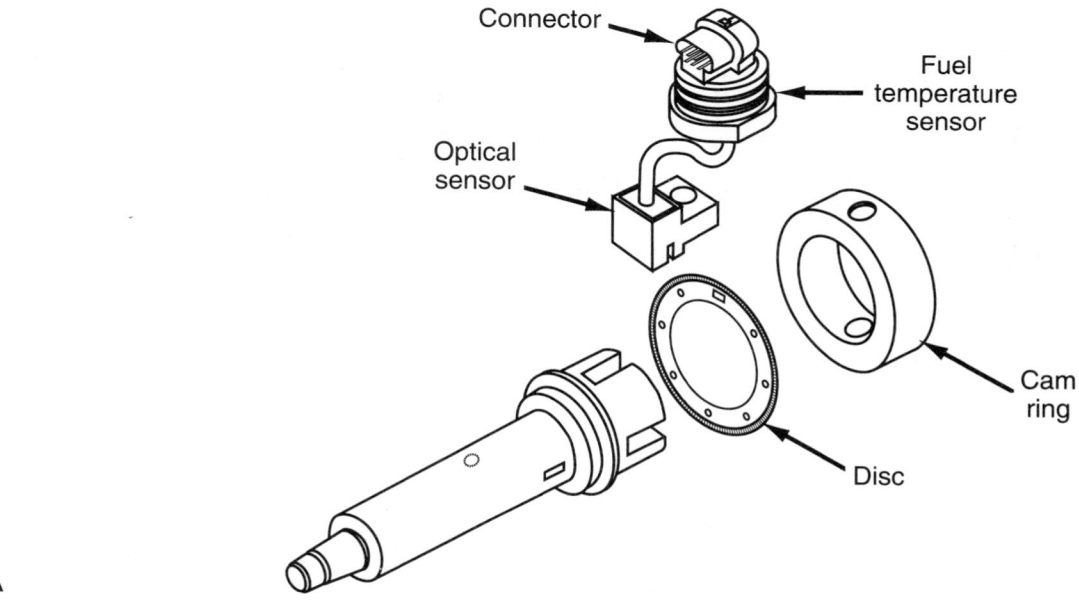

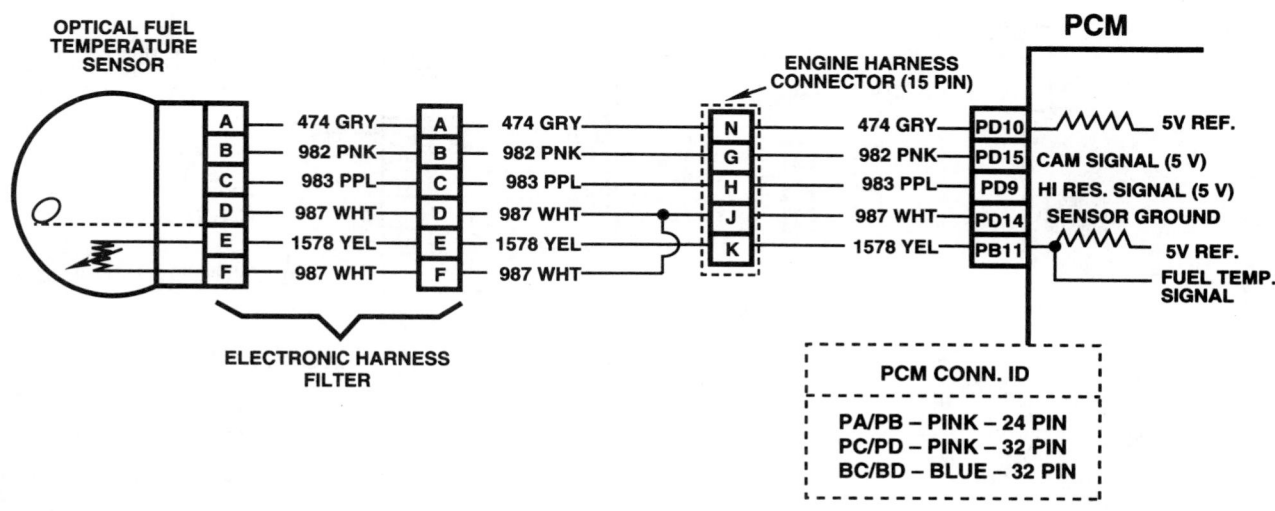

Figure 10-4 (A) Optical sensor and Datatrack disk; (B) optical sensor circuit. (Courtesy of General Motors Corporation, Service Operations)

contains 512 notches (windows) whose signals are sent to the PCM via the optical sensor (ASTE). The PCM multiplies these signals 16 times through an angular clock mechanism to provide 0.044 degree of angular resolution. This is called the high-resolution track (DTC 17 [P0370]). The information provided enables the PCM and the fuel solenoid driver (pump-mounted driver) to energize and de-energize the fuel control solenoid at precise intervals to achieve accurate control of fuel delivery at all throttle positions and vehicle speeds. The 0.044 figure is derived as follows: 512 notches times the 16× angular clock = 8,192; 360 degrees divided by 8,192 = 0.044 degrees.

Fuel Sensor　The fuel sensor (Figure 10-5) is a thermistor that senses fuel temperature. Resistance is inversely proportional to temperature. (As the temperature increases the resistance decreases). Low fuel temperature produces high resistance, and high fuel temperature produces low resistance. Failure of this sensor sets the following DTC 42 (P01820) fuel temperature low with high temperature indicated and DTC 43 (P0183) fuel temperature high with low temperature indicated.

Fuel Solenoid Driver or Pump-Mounted Driver (PMD)　The PMD (Figure 10-6), or fuel solenoid driver is a solid-state, high-current switching device that receives injection commands from

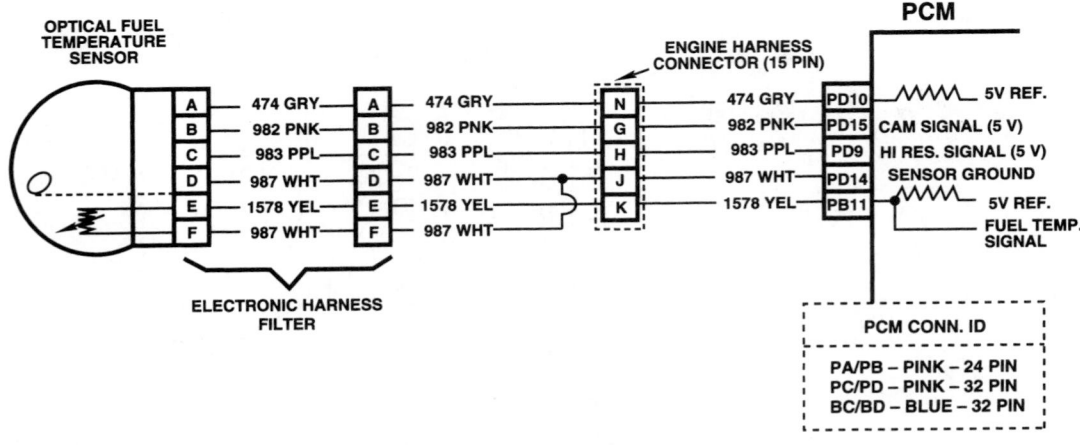

Figure 10-5 Fuel temperature sensor. (Courtesy of General Motors Corporation, Service Operations)

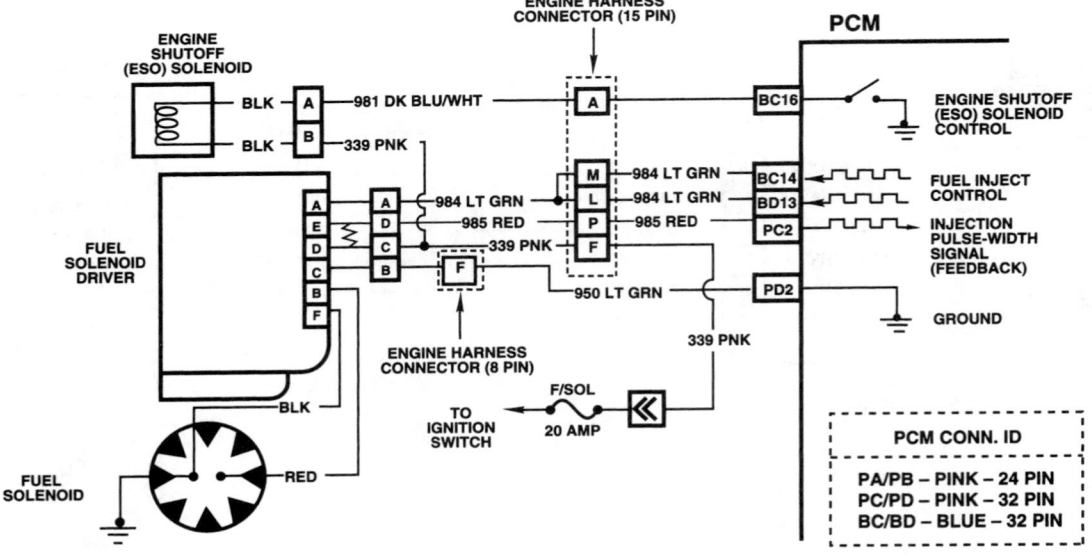

Figure 10-6 Fuel solenoid driver/PMD circuit. (Courtesy of General Motors Corporation, Service Operations)

the PCM and provides a regulated 10-amp current to the fuel solenoid, thereby controlling injection. It also sends a poppet valve closure signal (injection pulse width [DTC 35/P1216 and 36/P1217]) to the PCM. If the scan tool (DDR) shows the injection signal out of limits (default to 1.95 closure signal), check part of the engine harness.

DS Pump Operation

Fuel Flow Through the Pump Fuel enters (Figure 10-7) at the top of the pressure regulator and flows into the regulator bore above the regulator piston and out through the four ports in the regulator sleeve, and then through the inlet screen. With the ESO energized, fuel flows to the inlet side of the transfer pump. Fuel is pressurized as the pump rotates. The DS4 transfer pump pressure range is 14–170 psi. Transfer pump pressure at idle (approximately 600 rpm) is in the 45–70 psi range. Pressurized fuel flows through the transfer pump porting screw and to horizontal passage in the housing, to the head locking screw, and into the charge annulus in the hydraulic head.

Figure 10-7 also shows a spill accumulator that works like a shock absorber to absorb or snub possible high-pressure spikes. The purpose of the charge accumulator is to assist charging at high speeds to allow additional reserve fuel for high rate periods when time is a limiting factor.

Charging/Fill As the rotor turns, two of the eight rectangular charging ports in the head bore align with two rotor charging ports. Pressurized fuel then flows into the central (angled) bore in

The original intent of the charge accumulator was anticipating a direct injection application of the 6.5 L engine.

The eight rectangular charging ports in the head bore are located in the pump hydraulic head.

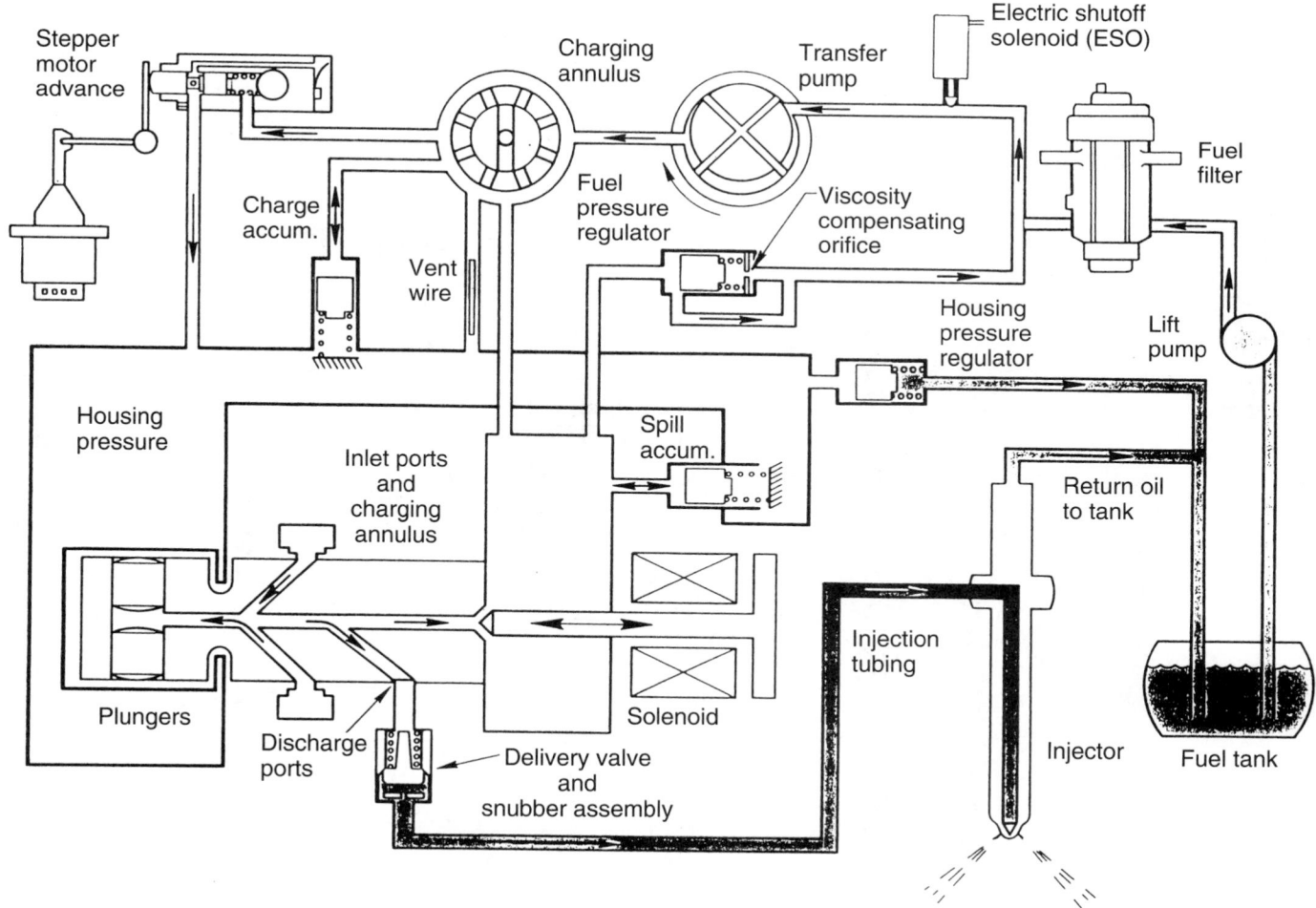

Figure 10-7 Fuel flow schematic. (Courtesy of General Motors Corporation, Service Operations)

the rotor and into the pumping chamber (Figure 10-8A). Two charging ports are used in the rotor to ensure the maximum charge. The poppet valve (what GM calls the control valve) is open (the fuel solenoid is de-energized) during the charging event, and pressurized fuel flows past the poppet valve to help assist in charging.

The four rollers ride down the cam lobe to the base circle of the cam, permitting the four plungers to move to the maximum extended position (base circle of the cam ring). The pumping chamber is completely filled during every charging event. The rotor discharge passages are not in registry during the charging event (Figure 10-8B).

Solenoid Energized/End of Fill The command to energize the fuel control solenoid is generated as follows (Figure 10-9): The optical sensor (ASTE) reads the high and low resolution tracks on the Datatrack disk and sends these signals to the PCM. The rising edge of the cam reference pulse (low-resolution track) sets several computer clocks in the PCM. The on-delay angle clock (ONDLYA) determines when the fuel control solenoid will be turned on. The ASTE also reads the high resolution track and when the predetermined number of counts for the on-delay angle is measured, a command to is sent to the fuel driver solenoid (PMD) to turn on the fuel control solenoid. The ONDLYA values are programmed to ensure that poppet valve closure occurs at the same point on the cam (approximately 8 degrees before the pumping ramp) throughout the speed range. As the speed increases, the ONDLYA value decreases (less time). This is done to always allow poppet valve closure to occur 8 degrees before the pumping ramp.

The term ONDLYA is computer machine language called a mneumonic.

The predetermined number of counts for the ONDLYA determines when fuel injection begins.

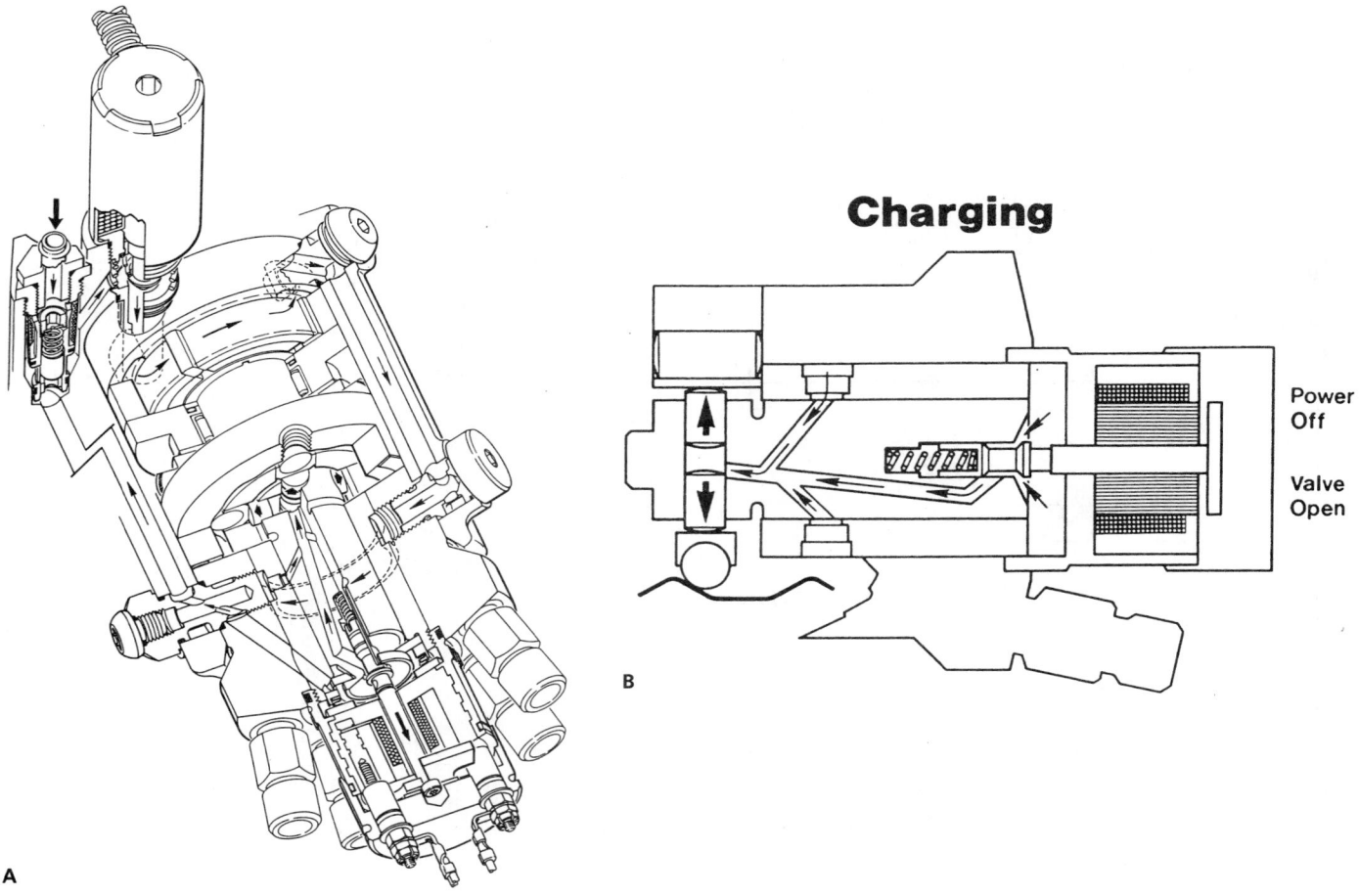

Figure 10-8 (A) Fuel charging/pump cutaway view; (B) charging cross section. (Courtesy of General Motors Corporation, Service Operations)

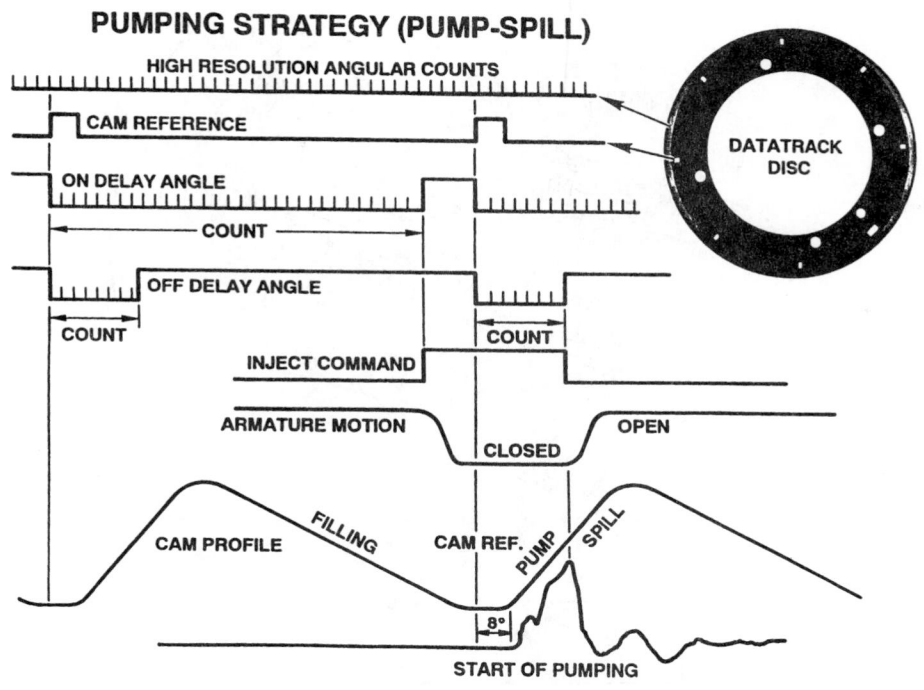

PUMPING STRATEGY (PUMP-SPILL)

Figure 10-9 Pumping strategy (Pump-Spill). (Courtesy of General Motors Corporation, Service Operations)

Start of Pumping The fuel solenoid is now "on" (Figure 10-10), the rotor continues rotation, and the charging (inlet) ports pass out of registry. The rotor discharge port aligns with one of the eight high-pressure discharge outlets prior to roller contact with the cam lobe. Rollers contact the cam lobes and climb the cam ramp, forcing the plungers inward. The fuel in the pumping chamber is pressurized and discharged through the central rotor passage and discharge port of the rotor, through drilling in the hydraulic head, and out to the discharge fitting (Figure 10-11). Peak injection pressure is 3,000–5,000-psi. Pressurized fuel lifts the delivery valve and snubber plate, located in the discharge fitting, and flows out through the high-pressure line to the injector. High injection pressure

The poppet valve always closes 8 degrees before the rollers reach the pumping ramp.

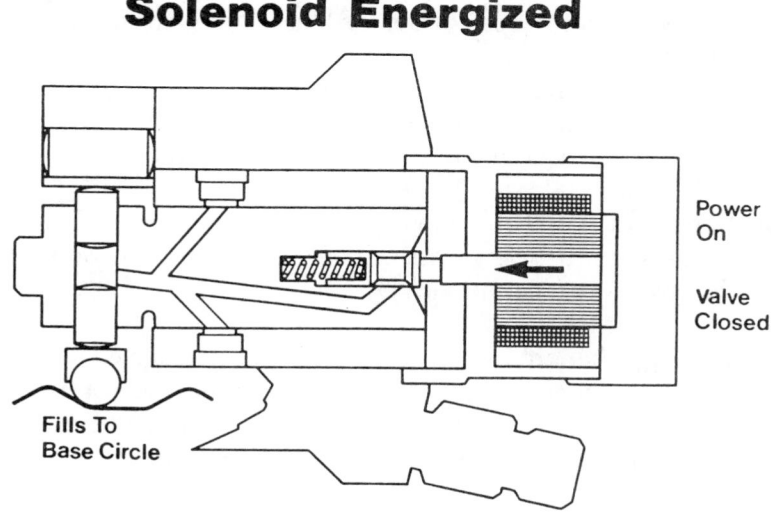

Solenoid Energized

Figure 10-10 Fuel solenoid energized. (Courtesy of General Motors Corporation, Service Operations)

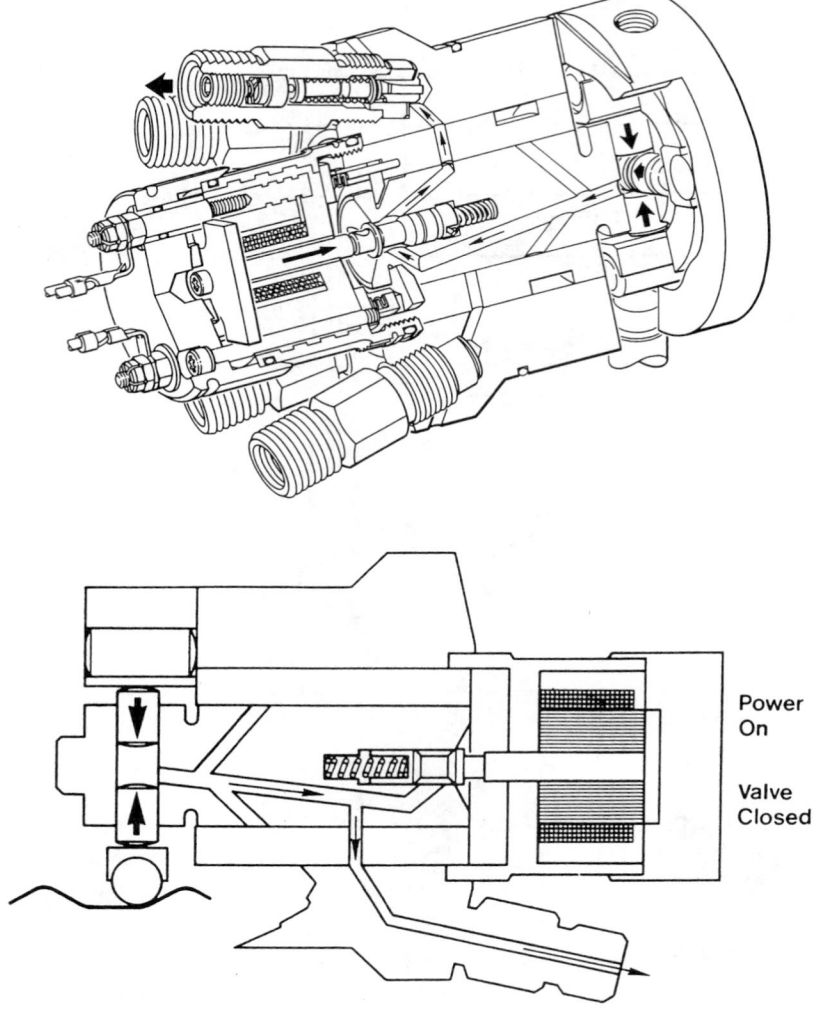

Figure 10-11 A: fuel start of pumping in the hydraulic head cutaway with fuel solenoid energized; B: start of pumping fuel solenoid energized. (Courtesy of General Motors Corporation, Service Operations)

lifts the needle valve in the injector and sprays atomized fuel into the engine's prechamber. During the pumping event, the delivery valve and snubber plate raise off their seats (Figure 10-12).

The retraction cuff of the delivery valve adds its volume to the spring chamber between the two valve seats. The triangular snubber plate (Figure 10-12) permits fuel flow around its edges; fuel, under injection pressure, is delivered to the high-pressure lines.

CAUTION: Although these pump are electronically controlled, they still generate very high pressures, so safety glasses should be worn when working with or near these injector pumps to avoid personal injury.

Spilling/End of Pumping, Solenoid De-Energized   The pump-mounted driver receives a signal from the PCM to terminate fuel delivery. PMD de-energizes the fuel solenoid (Figure 10-13) and the poppet valve opens. Cam rollers have not reached the end of the cam ramp. The discharge pressure spills into the cavity above the poppet valve seat. The pressure abruptly decays and pumping ends.

The delivery valve and snubber plate move rapidly back to their seats. Volume displaced by the delivery valve during the pumping event is now removed from the high-pressure line.

The pump mounted driver (PMD) is similar to a relay that carries high voltage. The PCM uses low voltage to activate the PMD similar to relay operation.

During retraction

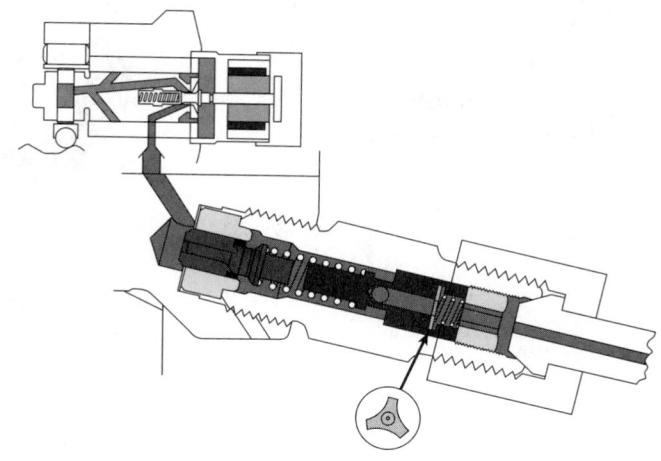

Figure 10-12 Injector nozzle opening as a result of fuel pressurization in the pump. (Courtesy of General Motors Corporation, Service Operations)

Spilling

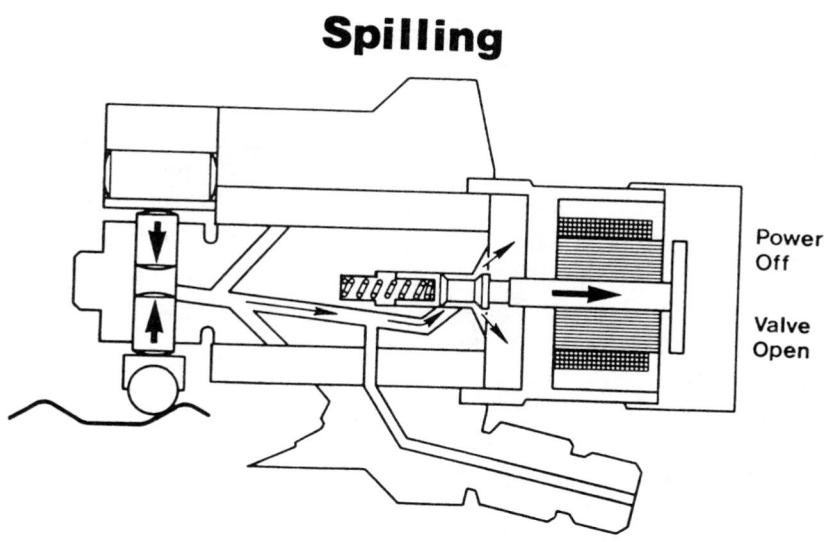

Power
Off

Valve
Open

Figure 10-13 Solenoid off and fuel spilling. (Courtesy of General Motors Corporation, Service Operations)

Pressure drops in the high-pressure line. The injector nozzle valve rapidly returns to its seat and fuel delivery ends. Reflected pressure waves generated in the high-pressure line are dampened (weakened) by the orifice in the snubber plate, as the pressure equalizes above and below the snubber valve. Residual line pressure in the line remains at a value lower than injection opening pressure. It often varies from one application to another, but is generally in the 600–800 psi range on both the DS and DB2 pumps.

The snubber plate orifice size line is 0.016 in. in the 5067 or 10225930 light duty, and 0.020 in. in the 5068 or 10225929 heavy duty; however, in the single pump (10225930 or 5067) now used, the size is 0.016 in.

Pumping Strategy   The optical sensor (ASTE/OSTE) reads the high and low-resolution tracks on the Datatrack disk and sends signals to the PCM (Figures 10-14). Rising edge of the cam

The inner track or cam reference pulse (DTC 18) has eight notches (windows), which provide accurate individual cylinder references. This track is also referred to as the low-resolution track and can provide an angular relationship between the crankshaft position sensor (TDC) and pump cam ring for timing purposes.

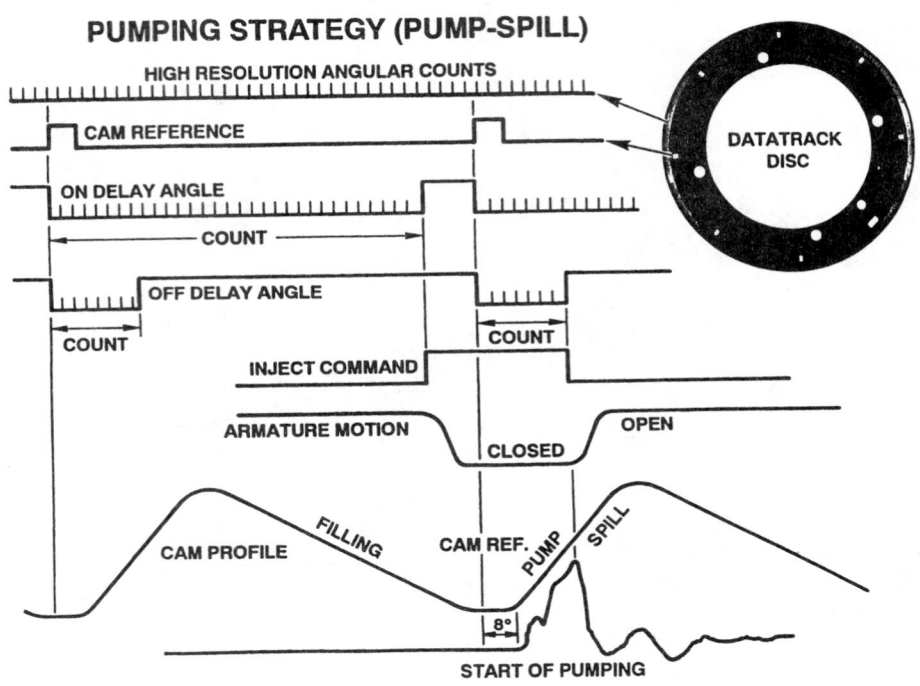

Figure 10-14 Delivery valve retraction after spilling with fuel solenoid off. (Courtesy of General Motors Corporation, Service Operations)

reference pulse (low-resolution track) sets several clocks in the PCM. The On Delay Angle (Mnemonic: ONDLYA) is determined when the injection command begins, and the off-delay angle (OFFDLYA) determines when the inject command ends (fuel solenoid off).

The optical sensor (ASTE or OSTE) counts the 512 notches on the high-resolution track. (Angular Resolution of 0.044 with 8,192 possible counts in a single revolution, e.g. 606 Counts = $0.044 \times 606 = 26.6$ degrees. This could represent an OFFDLYA angle.) When a predetermined number of counts for the ODDDLYA is measured, a command to turn off the fuel solenoid is sent to the PMD by the PCM.

OFFDLYA is calculated by the PCM based on sensor information (inputs), such as engine coolant temperature (ECT), accelerator pedal position (APP) module, fuel temperature, and optical sensor. The larger the OFFDLYA angle (number of counts), the greater the amount of fuel injected.

Injection Pump Timing

PCM Timing Control Inputs and Outputs The stepper-motor–actuated, dual-powered, automatic advance allows the start of pumping to happen earlier to offset intrinsic ignition delay (ID) relative to piston TDC (speed advance) and delays of fuel injection (line lag). The system advances or retards the start of pumping in response to engine load changes (that is, light loads require reduced fuel quantities and produce lower cylinder temperatures). Lower cylinder temperatures at light engine loads require additional advance, irrespective of speed, to provide stable ignition and combustion (light load advance).

Injection Timing Stepper Motor The stepper motor is located on the right side of the pump (Figure 10-15). The motor housing contains two coils that are controlled by voltage from the PCM through four circuits:

- ❏ CKT 564 for coil 1 low position
- ❏ CKT 565 for coil 1 high position
- ❏ CKT 566 for coil 2 low position
- ❏ CKT 567 for coil 2 high position

Shop Manual Chapter 10, page 368.

The best timing is where the minimum ID occurs (MBT = minimum advance best timing).

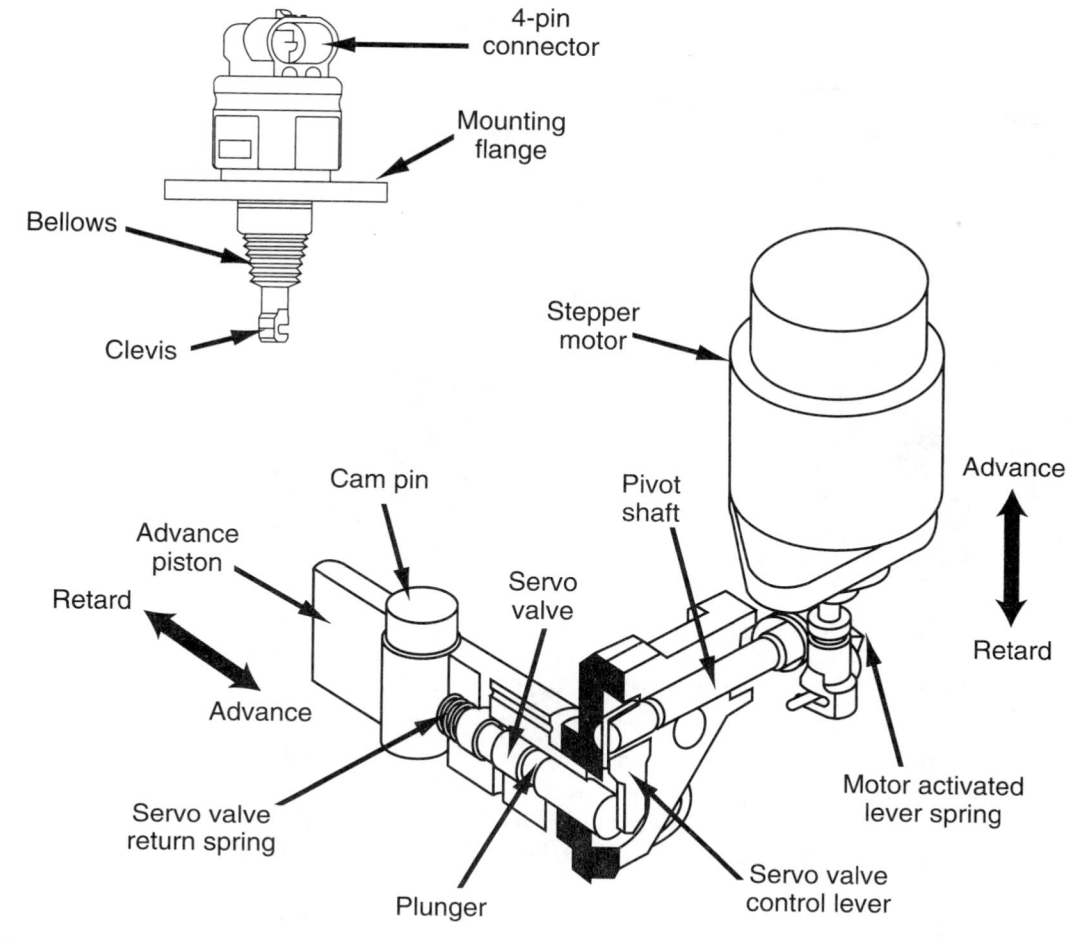

Figure 10-15 Injection timing stepper motor. (Courtesy of General Motors Corporation, Service Operations)

DTC 34 or P0216 sets when rpm is steady and the PCM detects injection timing that is a 5-degree difference between desired injection and measured injection timing. This electric stepper motor converts electrical impulses into discrete rotational movements. It is a threaded shaft in a worm gear that moves up and down at 12 steps per revolution.

❏ One step = 0.1 degree pump advance
❏ Stepper motor response time
❏ Cold: 50 steps/second (5 degree pump)
❏ Hot: 120 steps/second (11 degree pump)

During cold operation increased torque is required to move the advance components. As the step rate increases, the torque decreases.

Advance Piston Advance Mode The PCM signals the stepper motor to the advance position (Figure 10-16A) and retracts the arm in steps. The arm is linked to the advance plunger (a spring in the pack keeps it together), which allows the servo valve to shift to the right. The servo valve uncovers a port in the advance piston. Transfer pump pressure flows to the power side (left) of the advance piston. This pressure opens a reed valve permitting the fuel pressure to move the advance piston to the right and pivoting the cam ring on the cam pin rotating the cam ring in a direction opposite to pump rotation, thus advancing timing. The cam rollers contact the cam lobes sooner, and pumping starts earlier. Whenever pressure is fed to one side of the advance piston, the other side is being vented to housing pressure (advance piston movement is 0.022 in. per degree of movement).

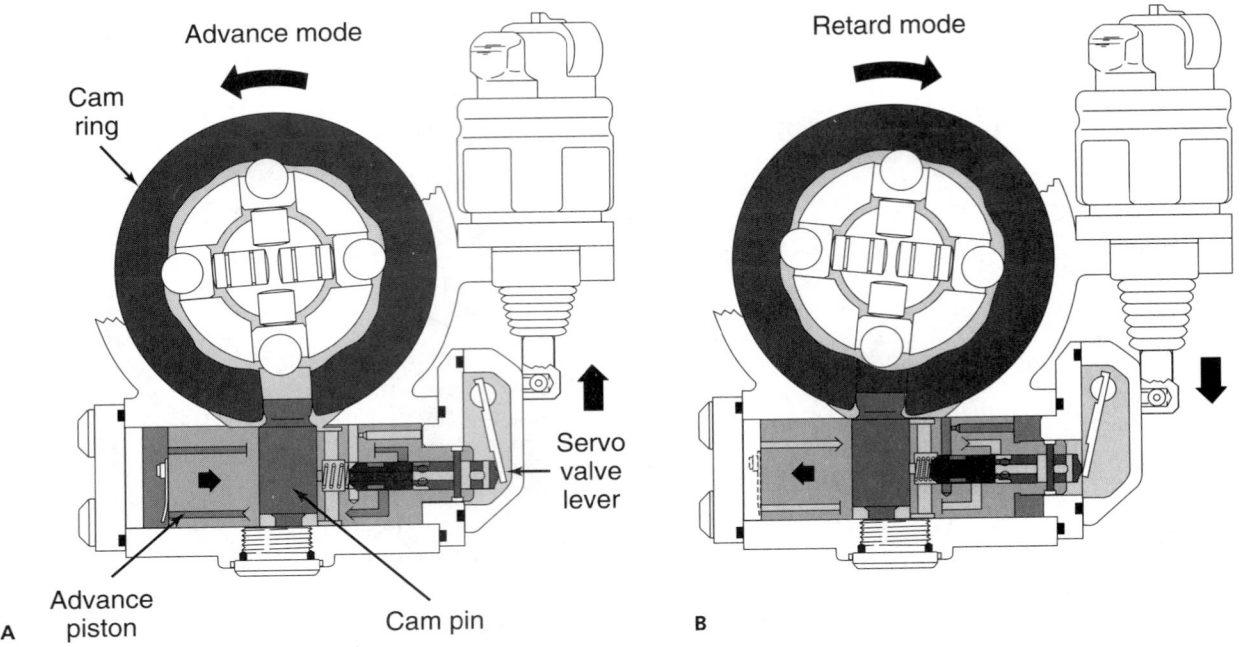

Advance mode

Cam
ring

Servo
valve
lever

Advance
piston

Cam pin

A

Retard mode

B

Figure 10-16 (A) Advance mode; (B) Retard mode. (Courtesy of General Motors Corporation, Service Operations)

Advance Piston Retard Mode   The PCM signals the stepper motor to the retard position (Figure 10-16B) and extends the arm downward in steps. This allows the servo valve to shift to the left. The servo valve covers the feed passage and opens the bleed passage. Fuel pressure flow stops to the power side (left) of the advance piston. The reed valve closes. Pressure behind the power side of the advance piston dissipates to the housing interior via the advance plunger and servo valve center bores. At the same time, the servo valve feeds transfer pump pressure to the stepper motor end of the advance piston. Fuel pressure moves the advance piston to the left, pivoting the cam ring on the cam pin and rotating the cam ring in the direction of pump rotation, retarding timing. The cam rollers contact the cam lobes later, and pumping starts later. During periods of constant speed and load a steady state advance operation exists. The servo valve remains fixed, the advance piston moves slightly in the retard direction, and pressure is simultaneously fed to both sides of the advance piston.

PCM Inputs and Outputs

**Shop
Manual
Chapter 10,
page 364.**

The powertrain control module contains the normal operating parameters for fuel delivery, timing, emission control, and transmission control.

The powertrain control module (PCM), or engine computer, provides the control and operation of this system. Internally the PCM is programmed with calibration information specific to the vehicle. This program tells the PCM what the normal operating parameters are for fuel delivery, timing, emission control, and transmission control, hence the name, powertrain control module. Externally, the PCM is hard wired to numerous sensors known as "inputs," as well as to solenoids, relays, and indicator lamps, known as "outputs." Figure 10-17A shows all of the PCM inputs and outputs. It further contains a fuel table for an engine over-temperature protection feature. The PCM constantly receives and interprets information from these inputs. It processes this information and compares it to the nominal or normal values with which it is programmed. The PCM then either sends or inhibits electrical responses to output devices in order to control fuel delivery, timing, and other emission control systems.

Engine Coolant Temperature (ECT) Sensor   The ECT sensor (Figure 10-17B) is a **thermistor** that changes its value based on temperature. Resistance is inversely proportional to temperature; as temperature increases, resistance decreases. Low coolant temperature produces high resistance and high coolant temperature produces low resistance.

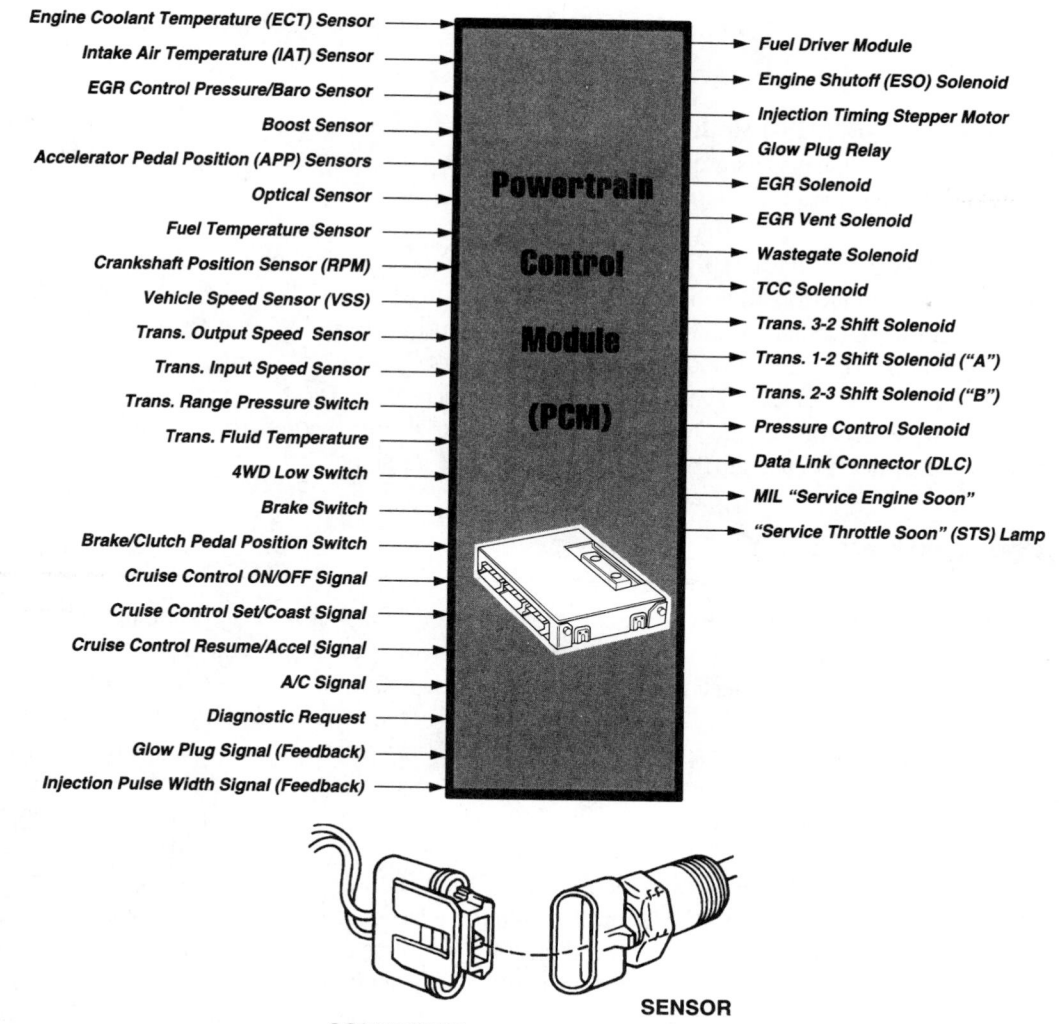

A

Engine Coolant Temperature (ECT) Sensor →
Intake Air Temperature (IAT) Sensor →
EGR Control Pressure/Baro Sensor →
Boost Sensor →
Accelerator Pedal Position (APP) Sensors →
Optical Sensor →
Fuel Temperature Sensor →
Crankshaft Position Sensor (RPM) →
Vehicle Speed Sensor (VSS) →
Trans. Output Speed Sensor →
Trans. Input Speed Sensor →
Trans. Range Pressure Switch →
Trans. Fluid Temperature →
4WD Low Switch →
Brake Switch →
Brake/Clutch Pedal Position Switch →
Cruise Control ON/OFF Signal →
Cruise Control Set/Coast Signal →
Cruise Control Resume/Accel Signal →
A/C Signal →
Diagnostic Request →
Glow Plug Signal (Feedback) →
Injection Pulse Width Signal (Feedback) →

Powertrain Control Module (PCM)

→ Fuel Driver Module
→ Engine Shutoff (ESO) Solenoid
→ Injection Timing Stepper Motor
→ Glow Plug Relay
→ EGR Solenoid
→ EGR Vent Solenoid
→ Wastegate Solenoid
→ TCC Solenoid
→ Trans. 3-2 Shift Solenoid
→ Trans. 1-2 Shift Solenoid ("A")
→ Trans. 2-3 Shift Solenoid ("B")
→ Pressure Control Solenoid
→ Data Link Connector (DLC)
→ MIL "Service Engine Soon"
→ "Service Throttle Soon" (STS) Lamp

CONNECTOR   SENSOR

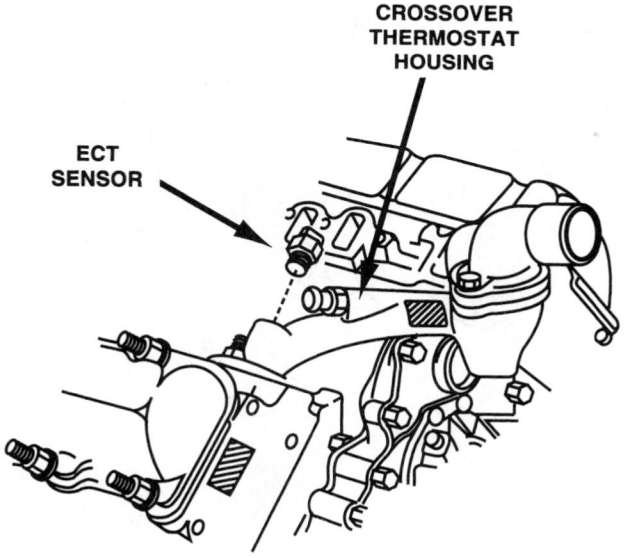

B

ECT SENSOR

CROSSOVER THERMOSTAT HOUSING

Figure 10-17 (A) Distributor-type electronic pump ECM inputs/outputs; (B) Engine coolant sensor (ECT). (Courtesy of General Motors Corporation, Service Operations)

Intake Air Temperature (IAT) Sensor This sensor is mounted on the intake manifold and works the same as the ECT.

Accelerator Position (APP) Module The drive-by-wire potentiometer variable resistor (Figure 10-18) controls fuel delivery, as requested by the driver through the accelerator pedal. Three sensors are used for redundancy.

Crankshaft Position Sensor (CKP) A Hall-effect style sensor (voltage is dependent on a magentic field) is mounted on the engine front cover (Figure 10-19). When the sprocket teeth do not align with the crankshaft position sensor, the north and south poles are close enough to generate a magnetic field, which produces 5 volts. When a tooth comes into alignment with the sensor, the magnetic field passes through the lower reluctance of the tooth instead of the Hall-effect device. This pulls sensor voltage to 0. This high-low variable provides crank positioning reference, with the optical sensor DATATRACK furnishing the #1 signal.

Shop Manual Chapter 10, page 369.

Caterpillar PEEC System

Caterpillar PEEC (programmable electronic engine control) on 3406 engines is appropriately classified as a partial authority, electronic management system (Figure 10-20). It is based around Caterpillar's new hydromechanical, scroll injection pump, so it is an adaptation of the pump-line-nozzle system used by Caterpillar for many years. The pumping apparatus of the new scroll pump remains the same as does the fuel control mechanism, the rack. The injection pump elements unload to hydraulic, multi-orifice injector nozzles directly to the engine's cylinders. However, the pump is driven through an electrically controlled timing advance unit, and the hydromechanical governor housing is replaced by rack actuator housing. Governing is fully electronic, an electronic control module (ECM) managed function. The system has many similarities to the Bosch PE7100/8500/RE30 injection management system used by Mack Trucks V-MAC. Because Caterpillar is in the process of phasing out the 3406 PEEC system, probably to focus on the 3406E engine with its full-authority electronics, the PEEC system will be covered in somewhat lesser detail than currently manufactured systems.

The Caterpillar PEEC system has monitoring and programming capabilities consistent with typical full-authority systems. Yet it is classified as a partial authority system, because an existing hydromechanical injection pump has been adapted for computerized control and, as such, control over the fueling pulse and timing is limited to a tighter hard window than in electronic unit injector (EUI) systems.

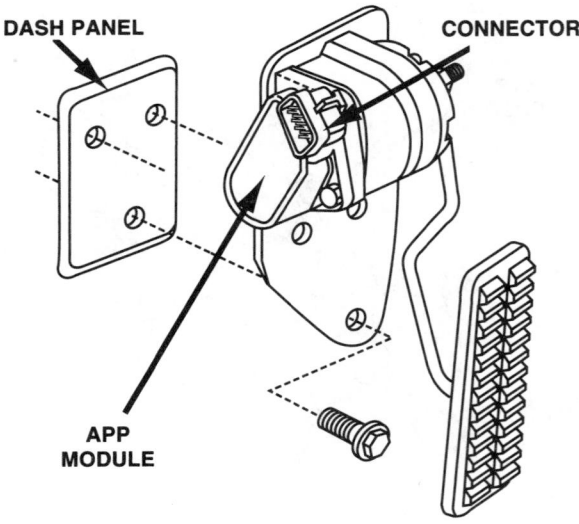

Figure 10-18 Accelerator pedal position module (APP). (Courtesy of General Motors Corporation, Service Operations)

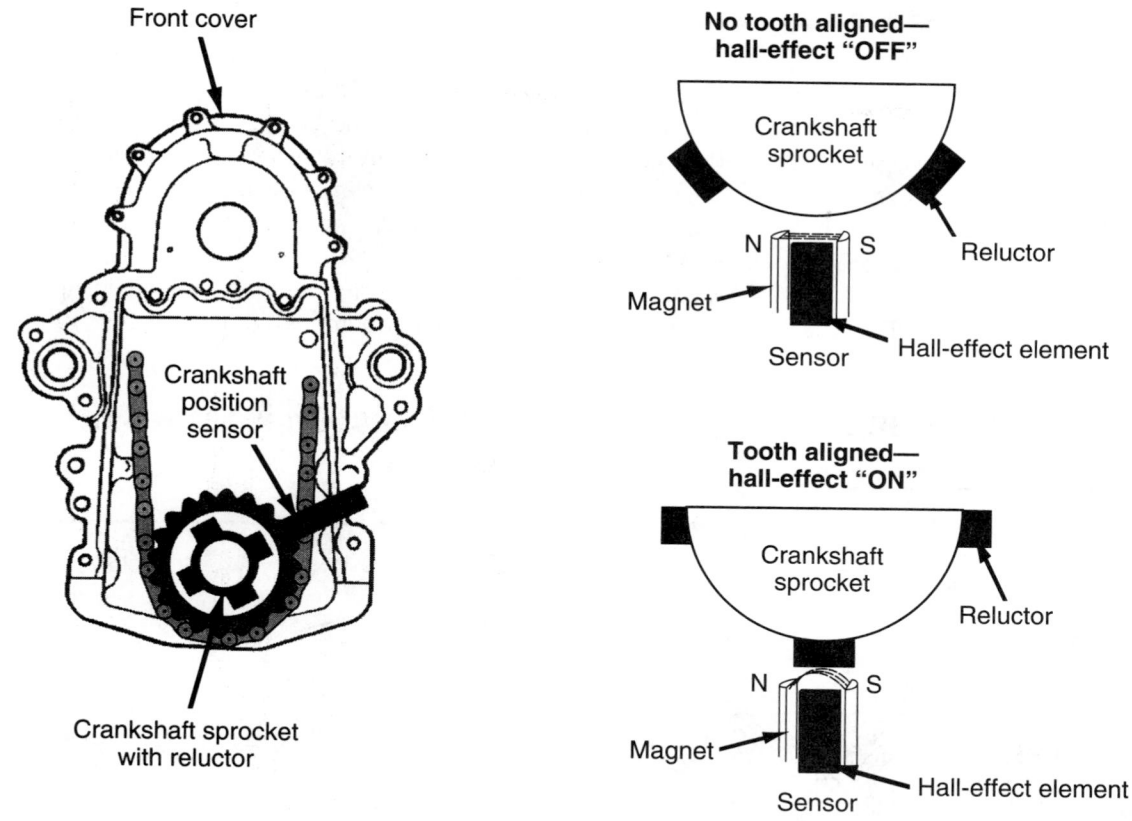

Figure 10-19 Crankshaft position sensor (CKP). (Courtesy of General Motors Corporation, Service Operations)

Fuel System Identification

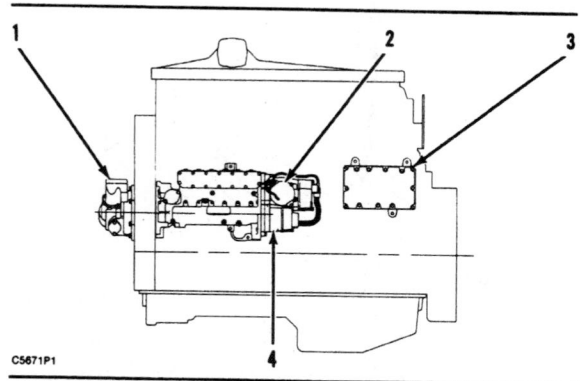

C5671P1

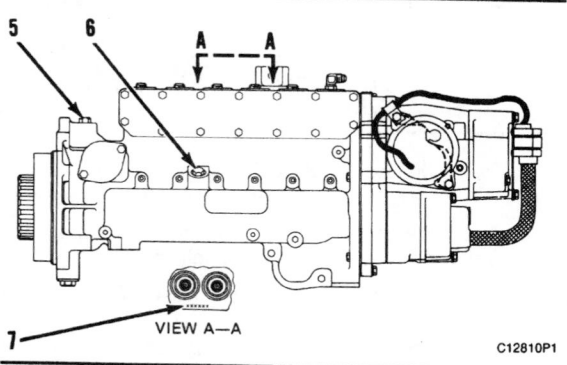

C12810P1

(1) Timing advance unit.

(2) Rack actuator.

(3) Control module.

(4) Transducer module.

(5) Location for rack centering pin.

(6) Location for timing pin.

(7) Location of stamped part number and serial number for FUEL INJECTION PUMP AND GOVERNOR GROUPS.

See FUEL SETTING AND RELATED INFORMATION FICHE for the correct fuel injection timing.

Injection sequence (firing order) 1,5,3,6,2,4

Rotation of Fuel Pump Camshaft (when seen from pump drive end) .... counterclockwise

Figure 10-20 PEEC component location. (Courtesy of Caterpillar Inc.)

System Layout

The 3406 PEEC engine uses a fuel subsystem that is almost exactly the same as in their hydromechanical version of the engine. A plunger-type transfer pump actuated by an eccentric on the injection pump camshaft pulls fuel from the fuel tank and charges it through a secondary filter to the charging gallery of the fuel injection pump. As an in-line port-helix metering injection pump fuels Caterpillar PEEC engines, knowledge of its operating principles is assumed. As a pump-line-nozzle system, the key functions of metering, pressurizing, and atomizing the fuel are unchanged in PEEC-managed engines from its hydromechanical relative. The management of the pump is essentially what has changed. Figure 10-21 shows a transfer pump and hydraulic injector used in the 3406 PEEC engine. A sectional view of the injection pump is shown in Figure 10-22.

Fuel Transfer Pump

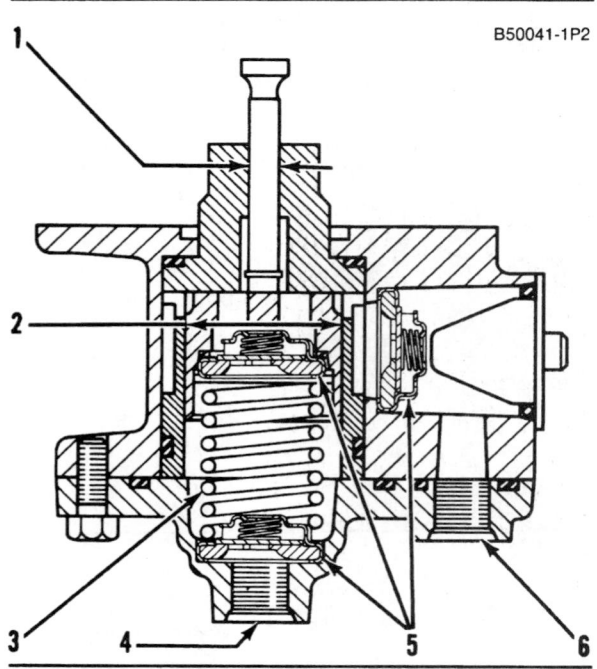

B50041-1P2

(1) Clearance between guide and
tappet .... 0.0025 to 0.0101 mm (.0001 to .0004 in.)

NOTE: The guide and tappet are a matched set. If one part is worn, a new guide and tappet assembly must be installed.

(2) Bore in sleeve for the piston .... 34.000 ± 0.015 mm
(1.3386 ± .0006 in.)

Diameter of piston .... 33.972 ± 0.008 mm
(1.3375 ± .0003 in.)

(3) 7W134 Spring:
Length under test force .... 33.67 mm (1.326 in.)
Test force .... 211 ± 11 N (49.2 ± 2.5 lb.)
Free length after test .... 56.1 mm (2.21 in.)
, Outside diameter .... 27.36 mm (1.08 in.)

(4) Fuel inlet port.

(5) Check valves.

(6) Fuel outlet port.

Fuel Injection Nozzle And Adapter

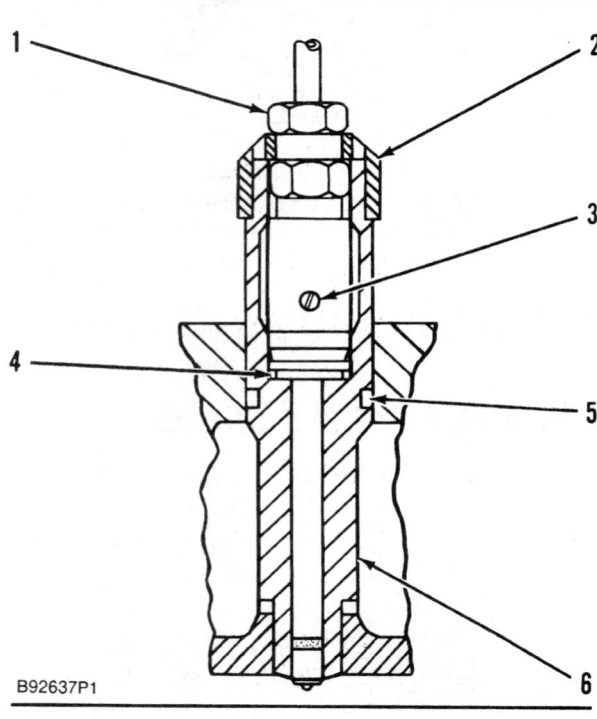

B92637P1

(1) Torque for fuel injection line nut .... 40 ± 7 N•m
(30 ± 5 lb.ft.)

(2) Torque for retainer .... 48 ± 7 N•m (35 ± 5 lb.ft.)

(3) Torque for bleed screw .... 2.2 ± 0.8 N•m (20 ± 7 lb.in.)

(4) Washer. Make sure the correct washer is used when the nozzle assembly is installed. Only copper washers are to be used with this adapter.

(5) Put liquid soap on rubber O-ring and bore in head before assembly.

(6) Put 5P3931 Anti-Seize Compound on the threads of the adapter before installation. Torque for the adapter .... 205 ± 14 N•m (150 ± 10 lb.ft.)

See Testing And Adjusting section, for fuel injection nozzle test procedure.

Figure 10-21 PEEC fuel transfer pump and injection nozzle adapter. (Courtesy of Caterpillar Inc.)

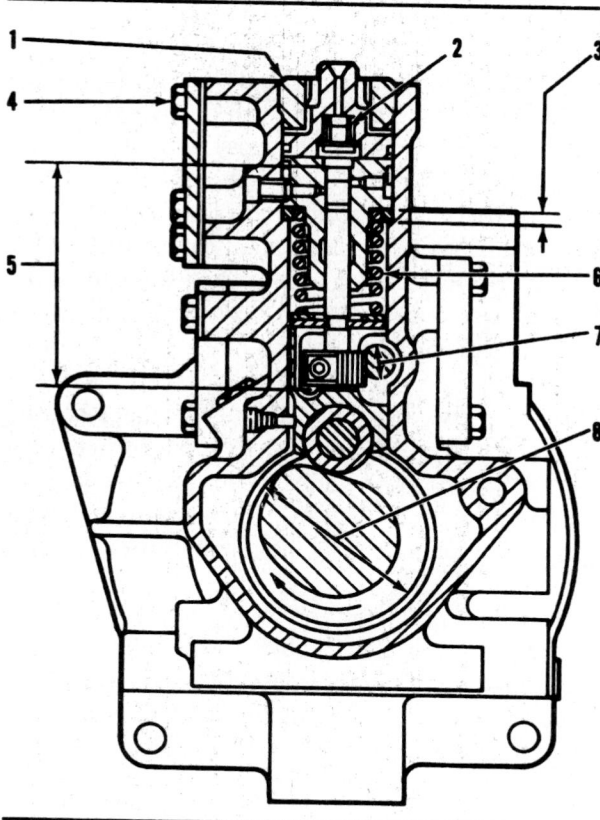

1) CHECK VALVE BUSHING
2) REVERSE FLOW CHECK VALVE
3) SPACER
4) COVER PLATE
5) PUMP PLUNGER
6) TAPPET SPRING
7) FUEL CONTROL RACK
8) ACTUATING CAMSHAFT

Figure 10-22 Injection pump cutaway. (Courtesy of Caterpillar Inc.)

In the PEEC system a wiring harness connects the following:

1. Rack actuator housing
2. Timing advance unit
3. Sensor/monitoring circuit
4. Command circuit
5. Electronic control module (ECM)
6. Transducer module

Rack Actuator Housing The new scroll injection pump hardware is essentially the same as that on nonelectronic 3406 engines. The pumping elements unload to high-pressure pipes connected to hydraulic injector nozzle assemblies in the cylinder heads. However, in PEEC, the **electronic control module** manages the governing function, and the scroll pump's mechanical governing unit is replaced by **rack actuator** housing. This unit uses a series of sensors to signal operational conditions to the ECM and actuators to effect ECM computations.

The rack actuator used by Caterpillar is electronically controlled but hydraulically actuated. A brushless torque motor (BTM) is a proportional rotary solenoid that is ECM controlled. The

A BTM (brushless torque motor) is a proportional rotary solenoid that is ECM controlled.

BTM controls a double-acting hydraulic servo that uses engine oil ported from the transducer module as its medium. The hydraulic servo effects the linear movement of the fuel-control rack, which protrudes into the rack actuator housing. The BTM is spring biased to the fuel shutoff position. When the BTM is energized by the ECM switch command, it moves the servo, which in turn acts to move the rack hydraulically. The ECM feeds the BTM with pulsed voltage values from ranging from 0–3.6 V.

Within the rack actuator housing are a number of sensors. A linear position sensor is responsible for continually signaling precise rack position data to the ECM; it is electromagnetically coupled to the rack. Moving the rack varies magnetic field strength, and thus, the percentage of a reference voltage value returned to the ECM (0.3–5.25 V). Actual rack position defines injected fuel quantity so this data is critical and must be accurate. This device is known as the **rack position sensor.**

An engine speed sensor is also located in the rack actuator housing. This is a simple AC pulse generator sensor is based on the reluctor wheel/chopper wheel principle. It is mounted on the rear of the injection pump camshaft. A stationary magnetic pick-up sensor signals rotational speed data back to the ECM in the form of an AC voltage pulse. Figure 10-23 shows a sectional view of the rack actuator housing that contains the timing sensor, solenoid, and rack actuator.

Timing Advance Unit   The timing advance unit is used in place of the mechanical (centrifugal) advance unit that was speed sensitive only. The electronic timing advance unit (Figure 10-24) is ECM controlled and is perhaps more accurately described as a *variable* timing advance unit. Effectively, this unit is capable of advancing static timing (port closure) through a much larger window (27 degree crank) than the former centrifugal advance units (9 degree crank). The static tim-

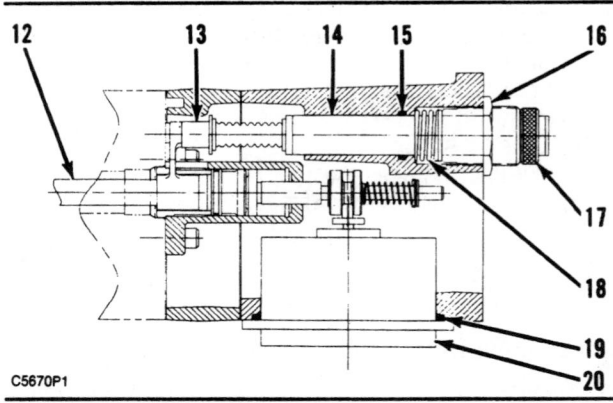

Top View Of Governor Housing

(12) Fuel rack.

(13) Magnet.

(14) Rack position sensor. See Testing and Adjusting section for correct installation.

(15) O-ring seal.

(16) Nut. Tighten to a torque .... 55 ± 7 N•m (41 ± 5 lb.ft.)

(17) Adjusting collar.

(18) Spring.

(19) O-ring seal.

(20) Rack solenoid (proportional rotary solenoid).

Figure 10-23 Timing sensor and solenoid and rack actuator. (Courtesy of Caterpillar Inc.)

Governor And Fuel Pump Drive Group (Timing Advance Unit)

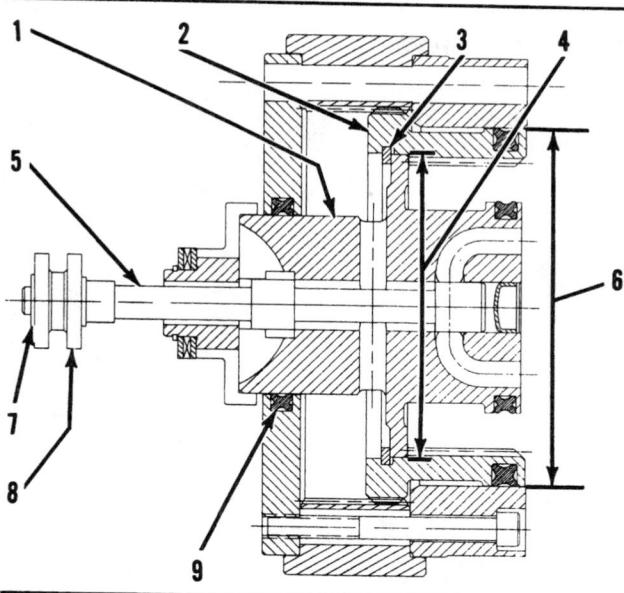

(1) Body assembly.

(2) Carrier.

(3) Internal retaining ring.

(4) Body assembly and carrier are pressed together.
Body outside diameter .... 86.252 ± 0.013 mm
(3.3957 ± .0005 in.)
Carrier inside diameter .... 86.220 ± 0.013 mm
(3.3945 ± .0005 in.)

(5) Spool valve.

(6) Guide ring diameter .... 102.03 ± 0.05 mm
(4.017 ± .002 in.)

(7) External retaining ring.

(8) Sleeve assembly.

(9) Typical quad seal.

Figure 10-24 Overhead view of the rack actuator housing. (Courtesy of Caterpillar Inc.)

ing value is therefore the most retarded port closure location the ECM may select when plotting injection timing.

To advance injection timing, Caterpillar uses the same BTM used in the rack actuator housing to control a double-acting solenoid that uses engine oil to effect movement of a helical splined pump drive sleeve within a cylindrical drive carrier. This advances the pump rotary position relative to engine position. The timing advance unit is, therefore, an intermediary between the pump drive gear on the engine and the injection pump drive plate. Figure 10-25 shows a sectional view of the timing sensor and BTM timing actuator.

Sensoring/Monitoring Circuit In terms of sensoring and monitoring engine and chassis functions, the Caterpillar PEEC system differs little from other electronically managed systems except

Fuel Rack Controls

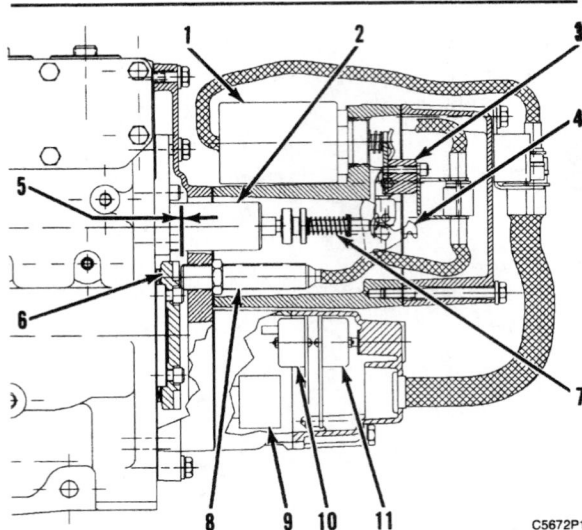

Side View Of Governor And Fuel Injection Pump Group

(1) Shutoff solenoid (energized to run).
Tighten to .... 70 ± 10 N•m. (50 ± 7 lb.ft.).

(2) Fuel rack servo valve.

(3) Shutoff lever group.

(4) Manual shutoff. Shut off override shaft and lever assembly.

(5) Clearance between cam retainer (6) and speed sensor (8) .... .76 ± .15 mm (.030 ± .006 in.)

NOTE: This distance is set by turning speed sensor into threads until magnet is against the gear tooth while the engine is stopped. Now, back speed sensor out ½ turn ± 30° and tighten the nut to a torque of .... 13 ± 2 N•m (10 ± 1 lb.ft.)

(6) Cam retainer.

(7) 1N3801 Spring.
Length under test force .... 26.33 mm (1.037 in.)
Test force .... 4.45 ± 0.22 N (1.000 ± .05 lb.)
Free length after test .... 49.29 mm (1.940 in.)
Outside diameter .... 11.32 mm (.446 in.)

(8) Speed sensor.

(9) Oil pressure sensor.

(10) Inlet air pressure sensor.

(11) Boost pressure sensor.

Timing Sensor And Solenoid

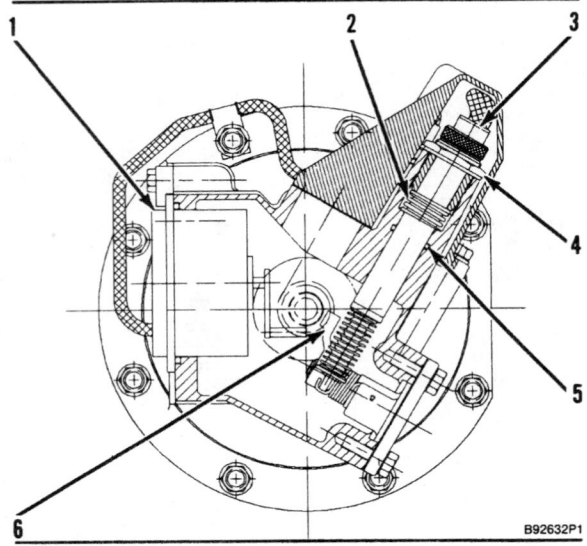

(1) Timing solenoid (proportional rotary solenoid).

(2) 4B4909 Spring.
Length under test force .... 11.906 mm (.4688 in.)
Test force .... 67 ± 4 N (15 ± 1 lb.)
Free length after test .... 19.050 mm (.7500 in.)
Outside diameter .... 20.638 mm (.8125 in.)

(3) Sensor. See Testing and Adjusting section for correct installation.

(4) Nut. Tighten the nut to a torque of .... 55 ± 7 N•m (41 ± 5 lb.ft.)

(5) O-ring seal.

(6) Bell crank.

Figure 10-25 PEEC timing advance unit. (Courtesy of Caterpillar Inc.)

that, as one of the first introduced, it lacks the comprehensiveness of current systems. Primary sensors are the rack input circuit (rack position), engine speed, boost pressure, timing advance, inlet air pressure, oil pressure, coolant temperature, and vehicle speed. Inlet air pressure, boost pressure, and oil pressure sensors are located in the transducer module. Reference voltage is 5.0 ± .25 V.

Command Circuit Caterpillar PEEC (Figure 10-26) incorporates a cruise control system that is dash switched and in which the upper and lower speed limits are ECM programmed. The upper limit cannot exceed the programmed road speed limit value. The Caterpillar throttle position sen-

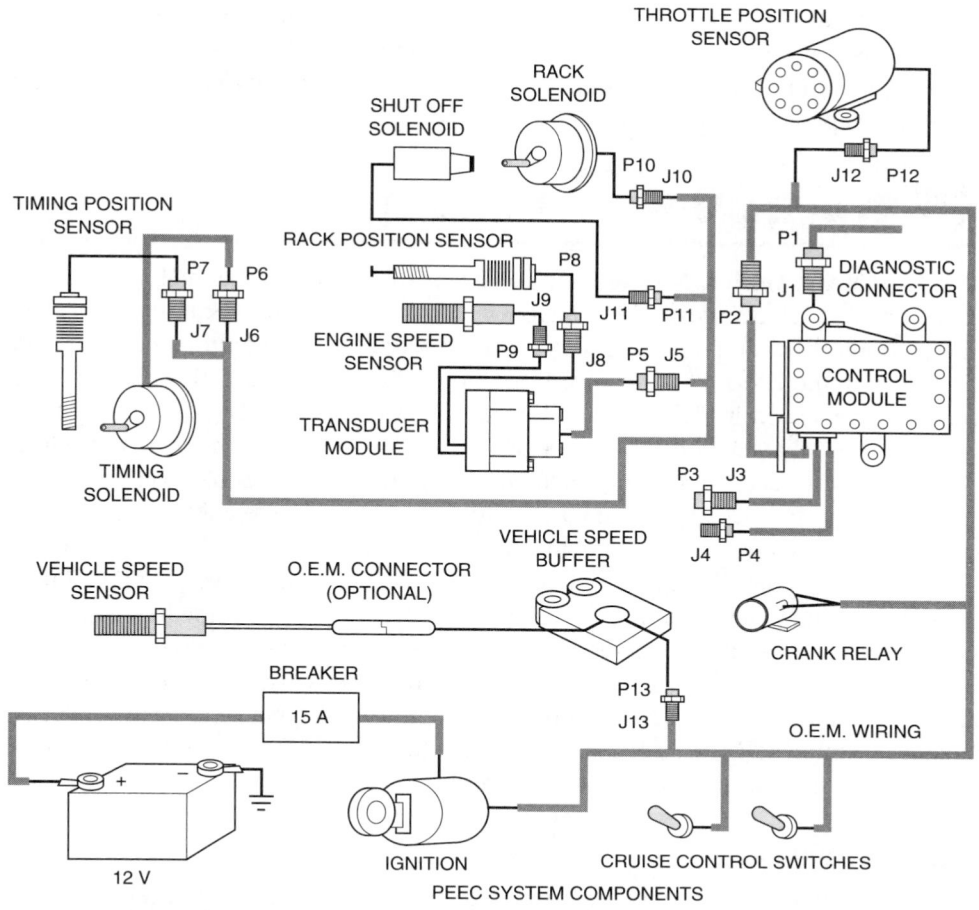

Figure 10-26 PEEC electrical components and rack control. (Courtesy of Caterpillar Inc.)

sor (TPS) is distinct from those found in other current truck drive-by-wire systems in that it uses a variable reluctance principle to generate a pulse width modulated (PWM) or digital signal to the ECM rather than a potentiometer-derived analog signal.

▲ **WARNING:** This TPS should be properly adjusted using Caterpillar PEEC SENR 3479-03, and it is especially important that the pedal linkage stop is set so that the TPS roll pin stop is not loaded with foot pressure.

ECM The ECM is a combined microprocessor and actuator assembly that incorporates switching for Caterpillar PEEC primary outputs—the rack actuator and timing advance BTM. Fuel is routed through the base of the ECM housing for purposes of cooling the electronic components. The ECM is plumbed downstream from the fuel transfer pump and ahead of the secondary filter-mounting pad. The ECM will draw a maximum of 7.5 amperes, and receives chassis voltage at 12.0 V. A minimum of 9.0 V is required when cranking and 11.0 V when running; the system is protected to accommodate transient surges to 28 V.

Check customer data programming with ECAP (Caterpillar's electronic control analyzer programmer), DDT (digital diagnostic tool) or other EST.

▲ **WARNING:** A removable (earlier models) or integral (current models) personality module (PROM/EEPROM) card accommodates data required to make the engine specific to chassis and customer requirements; the system is reprogrammable with both customer and Caterpillar data. Failure to recognize earlier and late model personality modules can result in engine damage.

Transducer Module   The transducer module is located in the lower portion of the rack actuator housing. It is responsible for converting pressure signals to electrical signals to be returned to the ECM. Manifold boost and inlet air and oil pressure valves are monitored in the transducer module. It also supplies oil to the rack actuator servo.

Mack Trucks V-MAC I and II

Shop Manual Chapter 10, page 370.

Mack Trucks **V-MAC (Vehicle Management and Control)** versions I and II are both classified as a partial authority, electronic engine (chassis) management systems, because the system is built around a hydromechanical, port-helix metering injection pump, adapted for computer control. It was introduced in the late 1980s and was used until 1998, when V-MAC III replaced it. The fuel management system used in V-MAC I and II is distinct from the V-MAC III, E-Tech system that is discussed in a later chapter in this textbook. With the two early versions of V-MAC electronics, Mack Trucks chose to use a Bosch port-helix metering, high-pressure injection pump and manage it electronically. The actual metering and pumping apparatus differs little from that found on any previous Mack hydromechanical engine. The essential differences of V-MAC II and I focus on the means used to control injected fuel quantity and beginning of delivery timing. V-MAC electronics also comprehensively manages vehicle and chassis systems, plots the fueling logic to meet a variety of chassis and running conditions, and additionally permits a wide range of customer and proprietary (Mack) programming options. Mack Trucks was also the first of the truck OEMs to make the portable PC (personal computer) their primary diagnostic, system reader, and reprogramming instrument. In this section, it will be assumed that the reader has a sound knowledge of in-line, port-helix metering injection pumps.

System Overview

The V-MAC I system layout (Figure 10-27) incorporates two modules bussed (electronically connected) to each other to manage engine fueling, diagnostic, and chassis functions. The fuel injection control (FIC) module was essentially the Bosch electronic diesel control (EDC) module used on other electronic applications of the PE 7000/8000 series of port-helix metering, injection pumps (used by John Deere, Volvo, and other OEMs). The FIC module sensed and switched most of those functions directly associated with engine fueling. The FIC module was bussed to the V-MAC module, which "managed" the FIC module from the broader perspective of the specific chassis application of the engine, performance, noxious emissions management, and command inputs (such as PTO and cruise controls). V-MAC II eliminated the separate FIC module, incorporating its functions in a single V-MAC II module, currently manufactured by Motorola. Both versions of V-MAC had the V-MAC module(s) located in the cab dash compartment on the passenger side of the vehicle. In the case of V-MAC I, the FIC module was piggybacked on top of the V-MAC module.

V-MAC modules can be accessed for fault code display, reading system parameters, and customer and proprietary data programming via the standard six-pin ATA/SAE serial communications connector (SAE J1708/J1939) located under the dash on the driver's side of the vehicle. Active system faults can also be read by blinking them through the dash "electronic malfunction" light.

While a ProLink 9000 reader programmer can be used to read system parameters and active and historical codes plus perform most customer data programming, Mack Trucks prefer technicians to use a PC. Mack Trucks has developed a software package that simplifies sequential troubleshooting plotting, making it user friendly for those with moderate computer skills.

A PC must be used to effect Mack data reprogramming. This means downloading a mainframe-based file to diskette, reprogramming of the on-board modules from that diskette, and subsequent uploading of a verification file to the mainframe. This procedure has become more user friendly over time. It is backed up by technical support literature and help lines and is unlikely to intimidate the even the first-time user.

Shop Manual Chapter 10, page 371.

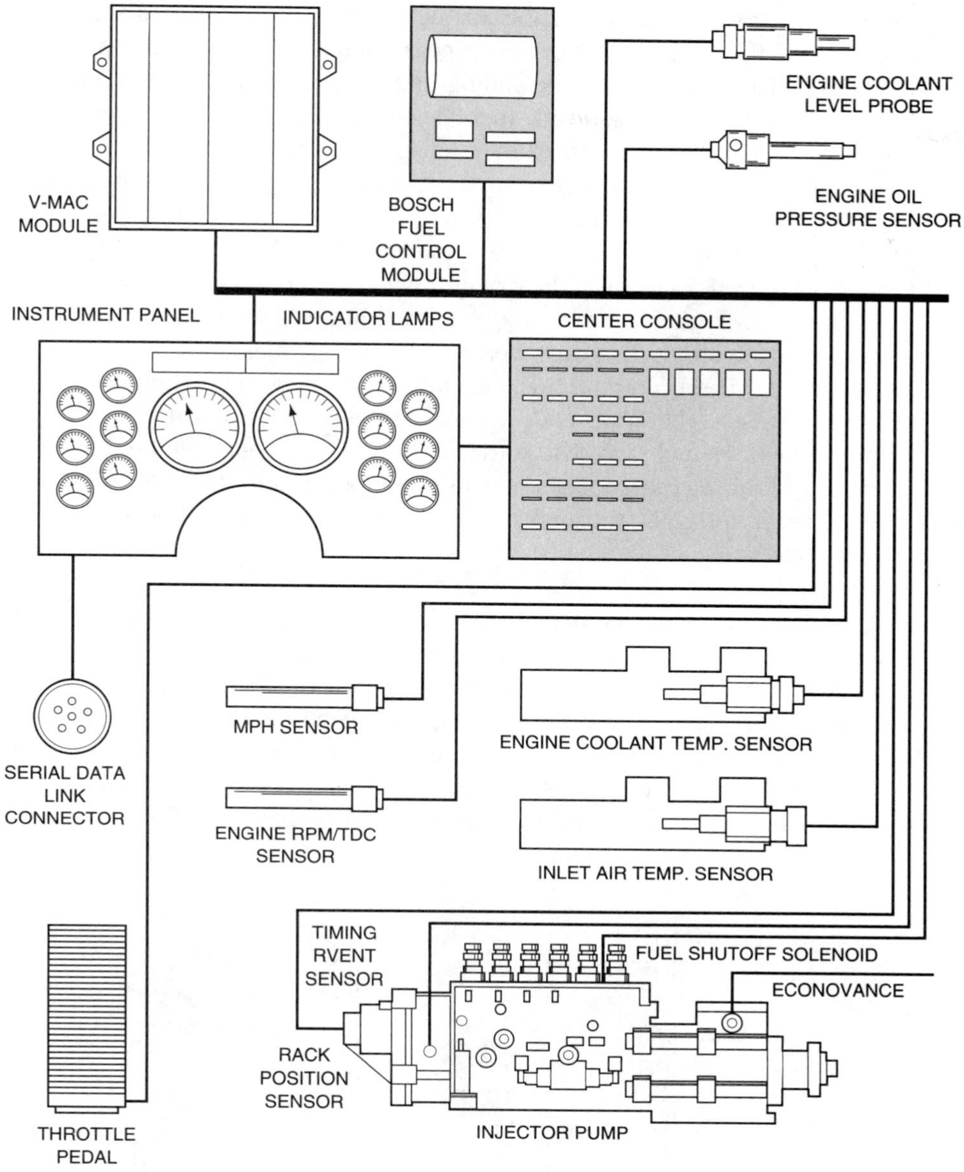

ENGINE COOLANT LEVEL PROBE

ENGINE OIL PRESSURE SENSOR

V-MAC MODULE

BOSCH FUEL CONTROL MODULE

INSTRUMENT PANEL　　INDICATOR LAMPS　　CENTER CONSOLE

SERIAL DATA LINK CONNECTOR

MPH SENSOR

ENGINE COOLANT TEMP. SENSOR

ENGINE RPM/TDC SENSOR

INLET AIR TEMP. SENSOR

TIMING RVENT SENSOR

FUEL SHUTOFF SOLENOID

ECONOVANCE

RACK POSITION SENSOR

INJECTOR PUMP

THROTTLE PEDAL

Figure 10-27 V-MAC I system overview. (Courtesy of Caterpillar Inc.)

WARNING: To effectively work with V-MAC systems (as with any other current diesel electronic management system), OEM training is essential.

Bosch PE7100/PE8500 Injection Pump

These pumps are essentially "beefed-up" versions of the Bosch PE3000 series injection pump. Both can be managed with hydromechanical governors. The PE7100 was used with the first V-MAC I systems. The injection pump itself was designed to produce peak injection pressures of 1050 atm (106.4 MPa or 15,500 psi) using 10 mm plungers with lower helix geometry and a cam lift of 12 mm. The pumping elements were fitted with funnel spill deflectors and a reinforced housing. The more recent version, PE8500, is closely related to the PE7100 in most of its critical specifications but is capable of peak injection pressures of 1150 atm (116.5 MPa or 16,900 psi). Both pumps are driven off the engine timing gear train at camshaft speed and use an electronically controlled, electric-over-hydraulic variable timing device, called **Econovance**. It is located between the pump and the engine accessory drive gear. An injection pump with plungers of the lower helix design will have a constant beginning and variable ending of the fuel pulse, the length of which would be measured in crank angle degrees. Using a variable timing device "softens" the hard timing parameters represented by plunger and helix geometry by giving the ECM a window within which it can select effective stroke. Therefore, the constant beginning, variable ending timing characteristic of the typical lower helix geometry is adapted into a variable beginning, variable ending of the delivery pulse, entirely managed by ECM processing logic. All V-MAC system injection pumps use Econovance; the extent to which the port closure value can be moderated depends on the application. This will be described in detail later in this section.

The function of the hydromechanical governor is to manage fueling on the basis of inputted fuel demand (accelerator/speed control lever), manifold boost (puff-limiter) and engine rotational speed (sensed centrifugally). The ECM sends a signal to the rack actuator that uses a rack actuator mechanism to move the fuel rack. The software in the ECM contains the governor parameters. The rack actuator housing incorporates sensors that input data to the ECM (V-MAC module) and a switched output from the ECM, capable of precisely controlling rack position and, therefore, injected fuel quantity.

RE24/30 Rack Actuator Housing (Figure 10-28)   It was stated previously that, although its location positions it at the rear of the fuel injection pump, the rack actuator housing should not be confused with a governor. The governor is a self-contained unit that after calibration by the technician sets hard parameters (rigid values) for fueling the engine through its performance range. The V-MAC I and II rack actuator housing is a unit that contains mechanisms both for reporting data to the system module(s) (that is, sensors) and for effecting the fueling commands that are the result of ECM processing logic. These fueling commands are the result of the fuzzy logic com-

> The V-MAC I and II rack actuator housing is a unit that contains mechanisms both for reporting data to the system module(s) (that is, sensors) and for effecting the fueling commands that are the result of ECM processing logic.

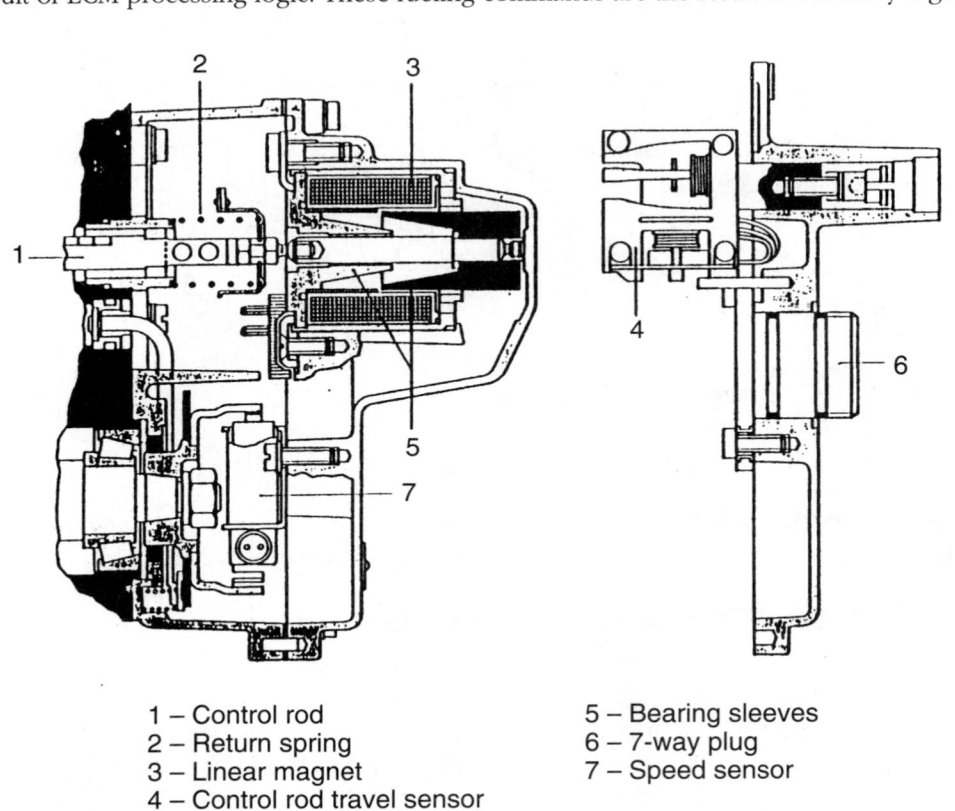

1 – Control rod
2 – Return spring
3 – Linear magnet
4 – Control rod travel sensor
5 – Bearing sleeves
6 – 7-way plug
7 – Speed sensor

Figure 10-28 Sectional view of Bosch RE30 rack housing actuator. (Reprinted with permission from Robert Bosch Corporation)

putations (that is, based on multiple inputs and variables); therefore, the fueling parameters can always be described as "soft" or non-rigid. Truly, the software programmed for the electronic control unit (ECU) is the "governor" in a V-MAC management system. (Mack Trucks uses the term electronic control unit rather than electronic control module.) The rack actuator housing is not as a rule, field-repaired but its operation must be understood by the technician seeking to effectively troubleshoot V-MAC systems. Subcomponent testing is outlined in Mack technical support literature and detailed in Robert Bosch diagnostic literature.

The rack actuator housing consists of a rack actuator mechanism controlled electrically by a V-MAC II module or by an FIC module in V-MAC I. Within the housing there are also several critical sensors that input engine data to the ECU(s).

Rack Actuator   The function of the rack in an in-line, port-helix metering injection pump is to rotate the plungers in unison within their barrels when it is moved linearly. When plungers with lower helices are used, port closure (static timing) will determine the beginning of the fueling pulse at a constant timing location of the engine, while port opening will set the end of the pulse, which will vary according to the amount of fuel to be pumped.

The fueling window will therefore be defined precisely by where the plunger helices register with the barrel spill ports. Plunger rotation is effected by linearly moving the rack, which is tooth meshed to plunger control sleeves. In Bosch rack actuator housings, the control rack extends into the rack actuator housing and is attached to the rack actuator mechanism. The rack actuator can be electrically defined as a linear magnet or proportional solenoid. The rack assembly is spring loaded to the rack no-fuel position. This no-fuel rack position rotationally positions all the pump plungers so their vertical slots register with their barrel spill ports throughout their cam-actuated stroke. This plunger position permits them to do no more than displace fuel as the plungers are actuated. When the rack actuator is energized by V-MAC module switching, the rack moves linearly in opposition to this spring pressure, thereby altering the rotational position of the plungers to increase plunger effective stroke. As the amount of current flowed through the rack actuator coil is increased, the further inboard the rack is forced. The further inboard the rack is forced against the spring pressure, the length of the plunger effective stroke increases, and so does the quantity of fuel pumped per cycle. The rack retraction spring would therefore be maximally compressed in the full-fuel position when the rack is forced to its fully inboard position by the rack actuator proportional solenoid. The rack actuator (Figure 10-29) is

The rack actuator is also known as the fuel control actuator.

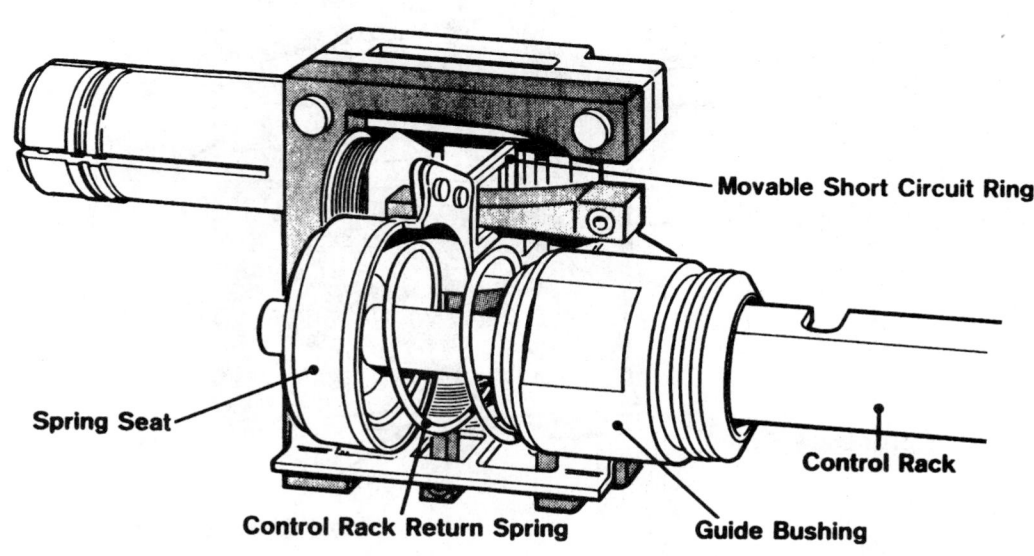

Figure 10-29 Rack actuator assembly components. (Reprinted with permission from Robert Bosch Corporation)

an ECU module switched output and, therefore, returns no data back to the module itself. It simply responds to command signals.

Rack Travel Sensor   The function of the **rack travel sensor** (Figure 10-30) is to report exact rack position to the electronic control module. It does so several times per second. The assembly is mounted in a laminated iron core and consists of a measuring coil, which is energized with reference voltage (5 V). Attached to the rack is a short circuit ring and protruding from the fixed measuring coil is an iron bar.

The short circuit ring attached to the rack is designed to slide over the iron bar without physically contacting it. As the fuel control rack is moved linearly by the rack actuator, the short circuit ring will either move closer to or further away from the measuring coil, thereby varying the electromagnetic field spread by the coil and the voltage signal returned to the V-MAC module. This component is necessarily sensitive and required to be precise. The ECM requires exact rack position data (Figure 10-31).

Reference Coil   All electromagnets are temperature sensitive, so in order to accurately interpret the voltage signal returned to the V-MAC module by the rack travel sensor, a precise "thermometer" is required to verify the signal. The **reference coil** is located close to the rack travel sensor coil (Figure 10-30). It is wound identically to the rack travel sensor coil. When it is energized, this coil provides a fixed magnetic field. The signal is sent to the electronic control module (Figure 10-32) by the rack travel sensor. The reference coil will be modulated by the actual temperature conditions in the rack actuator housing. When the rack is in a certain linear position, the signal outputted by the rack position sensor in a cold engine will be different from an engine that is at running temperatures. The reference coil is used to enable the ECM to accurately interpret the signal from the rack actuator. For instance, as temperature in the rack actuator housing rises, so does the temperature in both electromagnets used in the rack position sensor and the reference coil. The result of the temperature rise is that resistance and current draw increase

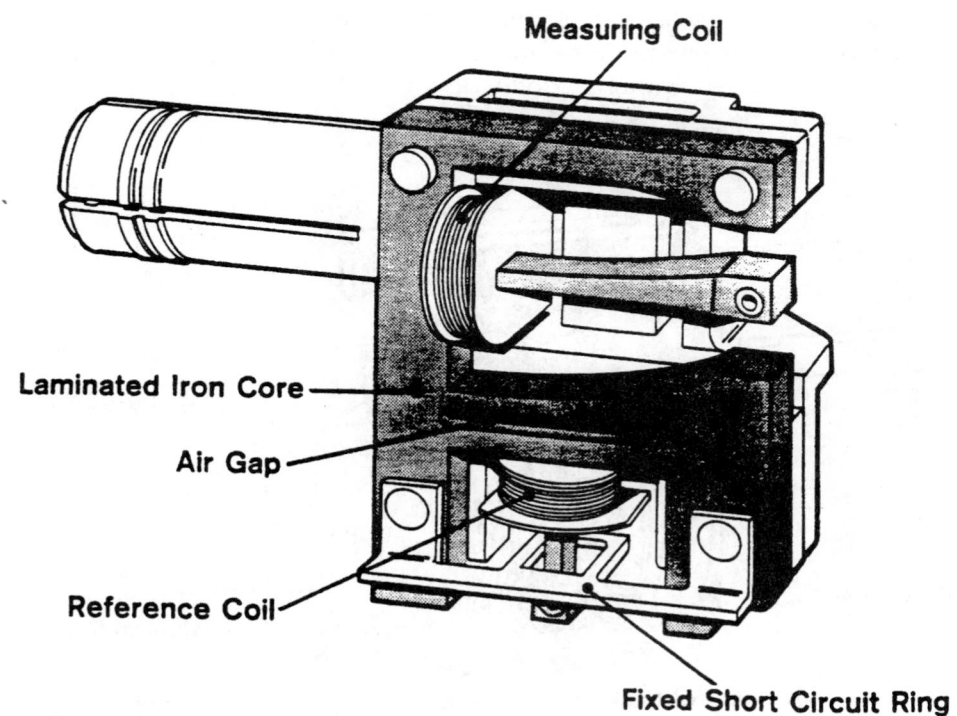

Figure 10-30 Rack travel sensor components. (Reprinted with permission from Robert Bosch Corporation)

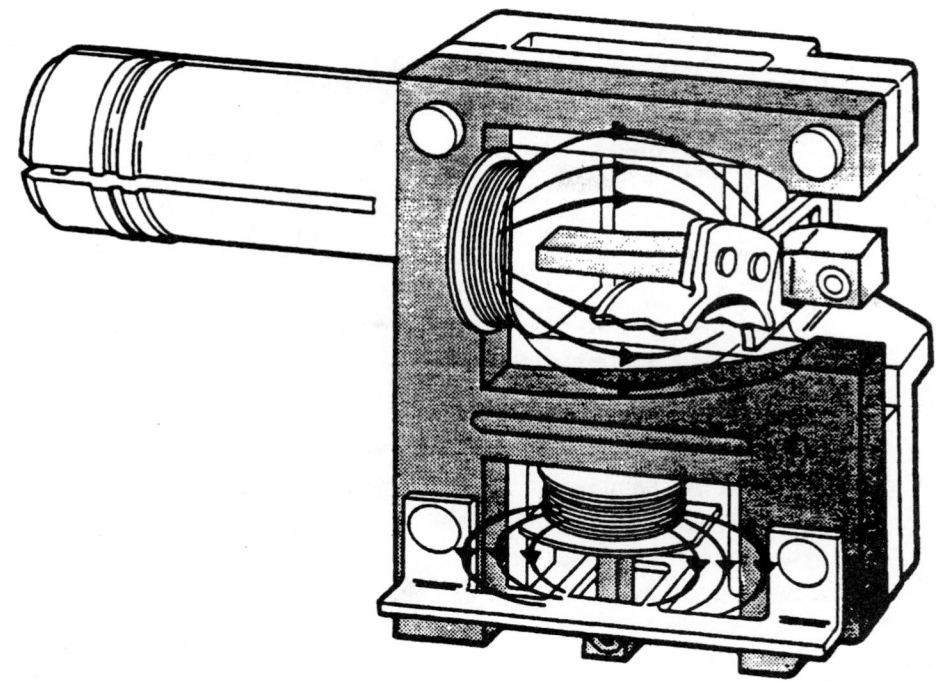

Figure 10-31 Rack travel sensor at max fuel. (Reprinted with permission from Robert Bosch Corporation)

in each coil winding, and the signal returned to the ECM varies. The opposite occurs when the temperature decreases. The signal returned by the reference coil is varied only by temperature change. By comparing the signals from both electromagnets, the ECM is able to accurately evaluate the voltage signal returned from the measuring coil on the rack position sensor and, thereby, determine exact rack position.

Pulse Wheel   The pulse wheel is the only rotating component in the rack actuator housing. The pulse wheel is a toothed impeller located at the rear of the injection pump camshaft and driven within the rack actuator housing. It is a sintered steel component whose total indicated runout (TIR) deviation must not exceed 0.03 mm. Straightening of impeller teeth should never be attempted and the impeller should be replaced if measured or observed to be defective. The impeller or pulse wheel (tone wheel) is responsible for two inputs to the V-MAC module. The first is engine speed (it is the primary reference for engine rpm) and the next is engine location reported by the timing event marker (TEM). There are 16 protruding teeth on the pulse wheel impeller that cut through the rpm sensor's magnetic field to produce a voltage signal to the ECM that increases proportionally with an increase in rotational speed. The TEM sensor consists of single notch that cuts the magnetic field of the TEM sensor, producing a signal that works on the same principle as the rpm sensor, but whose input is used by the ECM to locate precise engine position. The TEM notch is also used for the electronic static timing of the pump to the engine (this procedure is covered later in this chapter). Both these sensors can be electrically classified as induction pulse generators, which feed a small AC analog signal to the ECM.

Econovance

The PE7100 or PE8500 injection pump is driven off the engine crankshaft gearing at camshaft speed on Mack engines. The Econovance unit serves as an intermediary between the engine driven pump drive gear and the injection pump drive. Econovance provides an advance on the static timing, port closure location of an 8 degree, 10 degree or 20 degree crank angle, depending on the engine model. Current versions of V-MAC tend to use a 20 degree Econovance, which

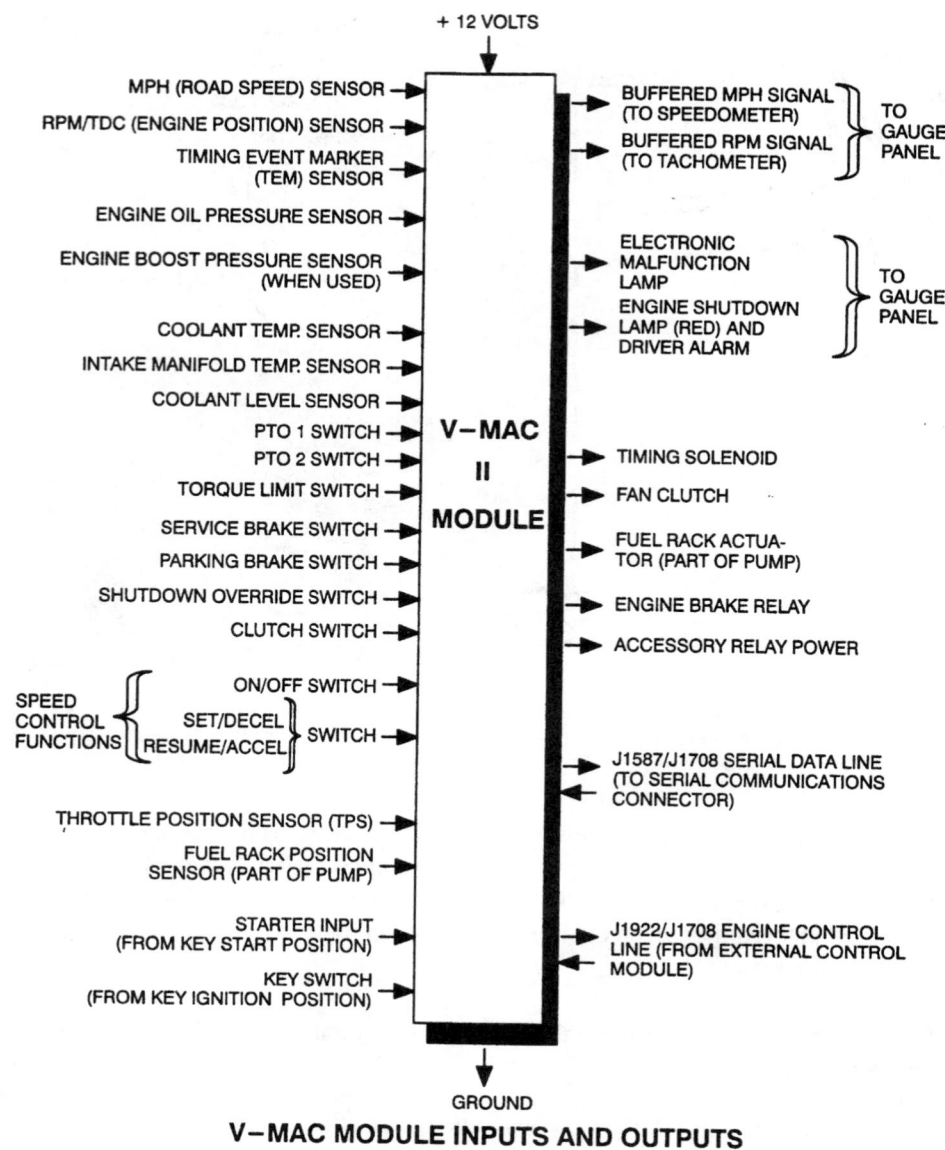

+ 12 VOLTS

MPH (ROAD SPEED) SENSOR →

RPM/TDC (ENGINE POSITION) SENSOR →

TIMING EVENT MARKER (TEM) SENSOR →

ENGINE OIL PRESSURE SENSOR →

ENGINE BOOST PRESSURE SENSOR (WHEN USED) →

COOLANT TEMP. SENSOR →

INTAKE MANIFOLD TEMP. SENSOR →

COOLANT LEVEL SENSOR →

PTO 1 SWITCH →

PTO 2 SWITCH →

TORQUE LIMIT SWITCH →

SERVICE BRAKE SWITCH →

PARKING BRAKE SWITCH →

SHUTDOWN OVERRIDE SWITCH →

CLUTCH SWITCH →

ON/OFF SWITCH →

SPEED CONTROL FUNCTIONS { SET/DECEL RESUME/ACCEL } SWITCH →

THROTTLE POSITION SENSOR (TPS) →

FUEL RACK POSITION SENSOR (PART OF PUMP) →

STARTER INPUT (FROM KEY START POSITION) →

KEY SWITCH (FROM KEY IGNITION POSITION) →

V–MAC II MODULE

→ BUFFERED MPH SIGNAL (TO SPEEDOMETER)
→ BUFFERED RPM SIGNAL (TO TACHOMETER) } TO GAUGE PANEL

→ ELECTRONIC MALFUNCTION LAMP
→ ENGINE SHUTDOWN LAMP (RED) AND DRIVER ALARM } TO GAUGE PANEL

→ TIMING SOLENOID

→ FAN CLUTCH

→ FUEL RACK ACTUATOR (PART OF PUMP)

→ ENGINE BRAKE RELAY

→ ACCESSORY RELAY POWER

⇄ J1587/J1708 SERIAL DATA LINE (TO SERIAL COMMUNICATIONS CONNECTOR)

⇄ J1922/J1708 ENGINE CONTROL LINE (FROM EXTERNAL CONTROL MODULE)

↓ **GROUND**

V–MAC MODULE INPUTS AND OUTPUTS

Figure 10-32 V-MAC ECU inputs and outputs. (Courtesy of Caterpillar Inc.)

is perhaps better described as a variable timing. Econovance establishes a port closure timing window within which the ECM can select port closure. For example, if static timing was set at 7 degrees before top dead center (BTDC) on a V-MAC engine with a 20 degree Econovance, V-MAC could select port closure occurrence at any value between 7 degrees BTDC and 27 degrees BTDC.

Econovance alters the position of the injection pump camshaft relative to that of the engine. The V-MAC module controls the device electronically. The engine's injection pump drive gear rotates a hub within which is a sliding sleeve machined with a helical spline. This is spring loaded to a nonadvanced position. To advance timing, engine oil used as hydraulic medium acts to move the sliding sleeve along the axis of rotation of the helical splines. The following components are critical in actuating Econovance.

(1) Proportional Solenoid The proportional solenoid is used as the timing actuator. It is a linear magnet, which can be smoothly moved to any position within its range of travel by varying current flow through the coil. The V-MAC module switches the proportional solenoid, and its function is to control the Econovance spool valve.

(2) Hydraulic Spool Valve This controls oil flow (engine lube) to the Econovance sliding sleeve. The hydraulic spool valve is directly controlled by the proportional solenoid, and it establishes the extent of advance by managing oil flow. Actual engine oil pressure values will not affect its operation (so long as the oil pressure is above the V-MAC failure value) because the operation of Econovance is flow dependent. The ECM monitors the advance location of the injection pump by reading the TEM signal.

Econovance Operation Static timing values as retarded as 4 degrees BTDC seem to make little sense mechanically, because the firing pulse and cylinder pressure rise would appear to be completely out of synchronization with the crank throw vector angles and thus inefficiently load the engine power train. However, it should be remembered that static timing the engine at 4 degrees BTDC with a 20 degree Econovance simply provides the ECU with the option of selecting a port closure value anywhere from 4–24 degrees BTDC and an actual 4 degree BTDC port closure setting would seldom be used. Emission control in truck diesel engines is often a case of staying within a narrow range of cylinder temperature values, in which the consequences of being slightly under optimum value would be excess hydrocarbon emissions and the consequences of being slightly over would be higher NO_x emissions. Diesel engine designers cannot use reduction catalysts such as rhodium in the exhaust piping because they will not function with lean burn diesel combustion, where excess air substantially exceeds the stoichiometric requirement. When the V-MAC programming opts for port closure to occur at 4 degrees BTDC, the reason is to reduce NO_x emission. When injection timing is this retarded, there is almost certainly an adverse effect on performance and fuel efficiency.

V-MAC Electronics

V-MAC incorporates comprehensive monitoring, engine management, customer and proprietary data programming, and fault and data retention consistent with a full option, partial authority management system. In terms of operating principles, V-MAC is closely related to Caterpillar PEEC system on 3406 engines. The V-MAC sensor system is as comprehensive as most of the full authority systems. The primary sensors on which the V-MAC software relies to plot fuel ratio and timing logic are the two temperature thermistors, engine coolant and ambient air. Figure 10-32 shows the V-MAC II ECU and inputs and outputs.

Pace and Pacer

PACE and PACER are Cummins PT, partial authority computer control systems. A partial authority engine management system is one in which a hydromechanical fuel system is adapted to computer controls. The section on partial authority engine management should be consulted to understand the operating principles of ECM managed engines. A brief account is provided here.

The PACE/PACER management systems adapt the PT system for some electronic controls but leaves most the key components of the system operating as if there were no computer controls. The primary feature of the PACE/PACER system is road speed governing. In fact, PACE and PACER are not true acronyms but instead refer to the road speed governing feature. A major disadvantage of the hydromechanical PT system in truck applications was the tendency of drivers to operate the engine in the governor droop curve rpm (that is, drive the vehicle with foot to the floor, resulting in an engine speed somewhere above the rated speed and below high idle). When an engine is run in the droop curve, it will usually run at lower than optimum efficiency and produce higher levels of noxious emissions. The PACE/PACER essentially governs the operator out of the droop curve.

The PT fuel system remains largely intact. PACE/PACER is mastered by a PT control module (PTCM) supplied with inputs from a transmission tailshaft-mounted vehicle speed sensor

(VSS), a brake switch sensor, clutch switch sensor, and engine position sensor. The PTCM processes input data and controls fueling by means of an electronic fuel control valve assembly mounted on top of the PT pump. This unit is equipped with a throttle bypass valve positioned in parallel with the PT throttle shaft, which is capable of routing fuel alternately (bypassing the rail) when energized. Its function is solely to limit fueling; the PTCM can thereby define a maximum road speed or a set power takeoff (PTO) mode speed.

Terms to Know

CEL – check engine light

Datatrack disk

Electronic control module (ECM)

Econovance

Housing pressure regulator

Multiplex

PEEC

Optical Sensor

Powertrain control module (PCM)

Rack actuator

Rack position sensor

Rack travel sensor

Reference coil

Thermistor

Transducer module

V-MAC – vehicle management and control

VSL – variable speed limit

VSG – variable speed governor

VSS – vehicle speed sensor

V-Ref – reference voltage

Summary

❏ ECM-controlled distributor-type injection pumps control fuel delivery by a single high-speed electronic spill control solenoid. They use a constant beginning of injection, variable end of injection (spill). The ECM or PCM controls the hydromechanical automatic timing advance electronically using a stepper motor. An integral vane-type transfer pump provides regulated pressure for charging and advances operation.

❏ The Caterpillar 3406B PEEC is a partial authority engine management system that has adapted a hydromechanical new scroll pump for electronic management by replacing the integral governor with a computer-controlled rack actuator housing.

❏ Caterpillar PEEC has an input circuit incorporating monitoring and command sensors. The primary ECM outputs are the rack actuator and the timing advance mechanism. The ECM controls rack position by switching a BTM, which controls engine oil flow to the rack servo. Variable timing is also controlled by the ECM by means of a BTM, which uses engine oil as a medium to alter the position of the fuel pump camshaft relative to that of the engine crankshaft.

❏ V-MAC I and II are classified as partial authority engine management systems and both use a Bosch hydromechanical, port-helix metering injection pump adapted for electronic control.

❏ The primary MACK V-Mac ECU outputs are the rack actuator and Econovance. The rack actuator is a linear magnet used by V-MAC to precisely control rack position.

❏ The PACE/PACER system is a very early example of diesel electronic management. The standard PT system is left largely intact. In short, PACE/PACER is a hydromechanically controlled PT fuel system with electronic governing of road speed, cruise control, and PTO operation. Should the PACE/PACER system fail, the PT system continues to operate but at slightly reduce power.

Review Questions

Short Answer Essays

1. Identify the major components and explain the operation of a distributor-type electronic injection pump system used on a GM 6.5 L diesel engine.

2. Define the components and explain the operation of the Caterpillar PEEC fuel injection system on a 3406 engine and explain why it is a partial authority system.

3. Describe how Mack Trucks and Bosch adapted an in-line, port-helix metering injection pump for computerized management and control.

4. Explain the operation of the MACK V-MAC II fuel injection systems and differentiate between V-MAC II and I.

5. Explain the operation of the optical sensor used on the DS4 electronic distributor-type pump.

6. Explain the two-stage housing pressure regulator used on the Stanadyne DS4 pump.

7. Define the Cummins terms PACE and PACER.

8. Explain the function of the rack travel sensor used on V-MAC systems.

9. Explain the operation of the reference coil used on the V-MAC II system.

10. Define the brushless torque motor used on the Caterpillar PEEC system.

Fill-In-the-Blanks

1. Electronic control module (ECM) controlled distributor-type injection pumps control fuel delivery by a single high-speed electronic _____ _____ _____.

2. Distributor-type electronic solenoid injection pumps use a constant beginning of injection and a _____ end of injection (spill).

3. The ECM or PCM controls the hydromechanical automatic timing advance electronically using a _____ _____.

4. An integral vane-type transfer pump provides _____ pressure for charging and advances operation.

5. The Caterpillar 3406B PEEC is a _____ _____ engine management system.

6. The PEEC system adapted _____ _____ _____ for electronic management by replacing the integral governor with a computer-controlled, rack actuator housing.

7. Caterpillar PEEC has an input circuit incorporating _____ and _____ sensors.

8. The primary PEEC ECM outputs are the _____ _____ and the timing advance mechanism.

9. V-MAC I and II are classified as _____ _____ engine management systems.

10. Both the V-MAC I and II systems use a Bosch hydromechanical, _____ _____ _____ injection pump adapted for electronic control.

ASE-Style Review Questions

1. What controls actual port closure timing in any given moment of operation in a Caterpillar PEEC managed engine?
   A. Centrifugal force
   B. The ECM
   C. Engine speed
   D. The oil pressure value

2. Which of the following sensors is responsible for signaling engine position data to the V-MAC module?
   A. rpm/TDC sensor
   B. TEM sensor
   C. CTS
   D. OPS

3. Which of the following components inputs data to the PEEC ECM that defines the metered fuel quantity in any given moment of operation?
   A. Rack actuator
   B. BTM
   C. Rack position sensor
   D. Throttle position sensor

4. What is the reason for routing subsystem fuel through the base of the ECM housing?
   A. Cool ECM components
   B. Lubricate the ECM components
   C. Filter the fuel
   D. Heat the fuel

5. Which of the following components inputs data to the V-MAC module that defines metered fuel quantity in any given moment of operation?
   A. Rack actuator
   B. Rack travel sensor
   C. Reference coil
   D. Timing event marker

6. Which is the moving component on the Bosch rack position sensor assembly?
   A. Reference coil
   B. Short circuit ring
   C. Proportional solenoid
   D. Pulse wheel

7. Which two of the following sensors is responsible for signaling engine position data to the V-MAC module?
   A. RPM/TDC sensor
   B. TEM sensor
   C. CTS
   D. OPS

8. Which dash diagnostic light is used to flash active DS distributor-type electronic system fault codes?
   A. Oil pressure warning
   B. Service engine soon (SEL)
   C. Electronic malfunction
   D. Check engine light (CEL)

9. What is the V-Ref value in the DS distributor-type electronic engine management system?
   A. 5 volts
   B. 12 volts
   C. 24 volts
   D. 36 volts

10. What force is used to move the rack to rack to a fuel on position on a 3406B PEEC injection pump?
    A. Mechanical
    B. Pneumatic
    C. Electrical
    D. Hydraulic

Electronic Unit Injector (EUI) and Common Rail Diesel Fuel Systems

Upon completion and review of this chapter, you should be able to:

❏ Identify the components and explain the operation and programming of the Detroit Diesel Corporation (DDC) Detroit Diesel electronic controls (DDEC) system.

❏ Explain the principles of operation of the DDC EUI and describe the importance of programming EUI calibration data to the ECM.

❏ Identify the components and explain the layout and operation of the Caterpillar full authority, advanced electronic engine management system (ADEMS).

❏ Identify the major components and explain the principles of operation of the Cummins CELECT system and the differences between CELECT and CELECT Plus.

❏ Identify the major components and system layout and explain the principles of operation of the electronic unit pump (EUP) on the DDC series 55 and Mack Trucks E7 engines.

❏ Identify the four primary subsystems that manage a hydraulic electronic unit injector (HEUI) engine's output.

❏ Identify the major components and the four primary subsystems of the HEUI common rail systems and explain their operation.

❏ Identify the components and explain the operation of the Cummins HPI-TP electronic common rail fuel system on a Signature 600 engine.

Introduction

This chapter will cover electronically managed diesel fuel systems that use electronic unit injectors systems and common rail pressure electronic metering systems. This coverage will include component identification, metering, electronic sensors, and microprocessor control used on electronic unit injectors and pumps (jerk pump) and electronic common rail systems. Specific coverage will extend to the following original equipment manufacturer (OEM) systems: Detroit Diesel DDEC, Cummins CELECT, Caterpillar ADEM, Navistar/Caterpillar hydraulic electronic unit injectors (HEUI), and the new Cummins HPI-TP used on the Signature 600 engine.

Detroit Diesel Electronic Controls (DDEC I-II-III-IV)

A BIT OF HISTORY

Detroit Diesel Corporation (DDC) was the first diesel engine manufacturer to offer an electronically managed diesel engine in the North American truck and bus markets. DDC electronics are described by the acronym **DDEC** that stands for **Detroit Diesel electronic controls**. DDEC I was introduced in 1985 and made generally available in 1987 on 2-stroke cycle, 92 series engines (displacement of 92 cubic inches per cylinder). DDEC I was a full authority management system with mechanically actuated (camshaft) **electronic unit injectors (EUI)** that were computer controlled by an electronic control module (ECM) located on the dash and an engine-mounted, injector driver unit called an electronic distributor unit (EDU). Injector drivers are ECM actuators responsible for effecting ECM plotted fueling commands by switching the EUI duty cycle.

In 1987, DDC launched their successful series 60 (in-line, six-cylinder engine available in 11.1 L. and 12.7 L. displacement) and series 50 (in-line four-cylinder version of the series 60 engine) range of 4-stroke cycle engines managed with DDEC II electronics. DDEC II combined the computer (ECM) and switching hardware (EDU) into a single, engine-mounted housing and enhanced the monitoring and output circuits. The current version of DDEC electronics is DDEC III, introduced at the beginning of 1995. It will be found on series 60, series 50, series 92, and a number of off-highway engines. DDEC III will be used as the basis of this description, but it should be noted that it does not differ significantly from either previous version. A feature of all electronics packages is that they are improved continually, and the earliest versions of each numeric distinction of DDEC are upgraded significantly almost as each month passes. For instance, DDEC III was introduced in 1993 and during 1995 was improved with features such as ECM-controlled ether cold start (with no driver action) and multiplex ability with transmission controllers. Figure 11-1 shows the system schematic.

System Overview

DDEC is a full authority, computerized engine management system that uses cam-actuated electronically controlled EUIs. DDEC III evolved from the earliest DDEC I version, which offered limited programmability and monitoring capability. DDC introduced DDEC IV in 1998. DDEC III and IV offer comprehensive engine monitoring, audit trails of self-diagnostics, other on-board electronic system interface capability, and customer and proprietary (DDC) data programming. The system can be divided as follows: fuel subsystem, ECM, input circuit

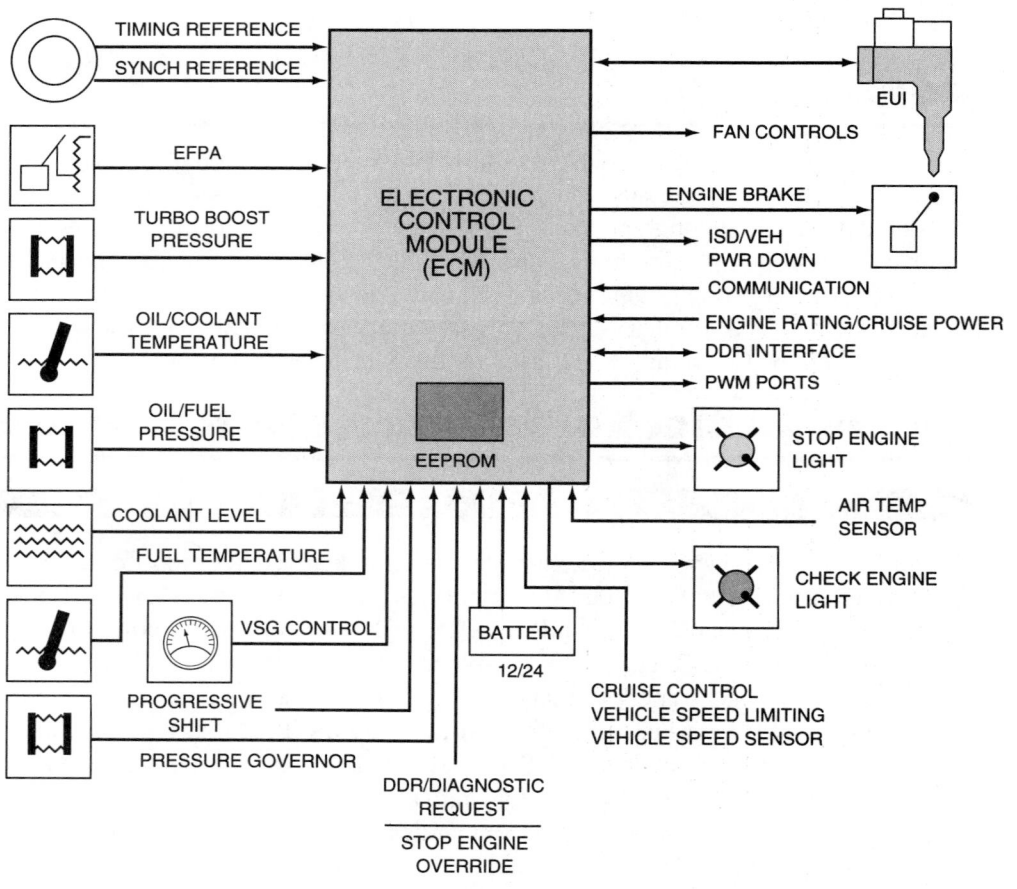

Figure 11-1 DDEC system layout. (Reprinted by permission of copyright owner Detroit Diesel Corporation, all rights reserved.)

(sensors), and output circuit (EUIs). DDEC III will be referenced in the context of a series 60 engine.

Fuel Subsystem

Movement of fuel from the fuel tank to the EUIs is accomplished in much the same manner as in a DDC hydromechanical engine. Fuel is drawn from the fuel tank by suction created by the gear pump; a primary filter is plumbed in series between the fuel tank and the transfer pump. Fuel discharged by the transfer pump is routed through a secondary filter, the ECM heat exchanger, and then to the fuel galleries located in the engine cylinder head. The Fuel Pro system, previously available as an option, is now standard on DDEC IV. This replaces the two-filter system with one extended service, filter separator system. The fuel galleries then charge the cylindrically shaped EUIs. Series 60 EUIs are manufactured with two external annular recesses in the cylindrical injector body, the lower of which is exposed to the charging gallery in the cylinder head. The upper annular recess in the EUI discharges fuel to the return gallery from which it is sent back to the fuel tank. In series 92 engines, fuel is transferred to and from the EUIs by means of jumper pipes in much the same manner as in the older hydromechanical unit injectors. Fuel is constantly circulated through the EUIs whenever the engine is running. Charging pressures are typically around 50 psi (3.3 atm, 334 kPa).

Electronic Control Module (ECM)

The DDEC I computer (Figure 11-2) was located within the vehicle cab, with the injector drivers (switching apparatus) or EDU (electronic distributor unit) mounted on the engine close to the EUIs. DDEC I rapidly evolved into DDEC II, so there are few examples of the former in service. DDEC II and III house the computer and injector drivers in a single unit, usually mounted on the engine. The physical size of the ECM housing has decreased in DDEC III systems despite having much greater computing capability and speed compared to DDEC II.

The DDEC II, III, and IV ECMs are microprocessor-driven management and switching modules. The single module therefore houses all of the microprocessing and output actuators. It is responsible for engine governing, fuel timing logic, self-diagnostics, comprehensive

Shop Manual Chapter 11, page 394.

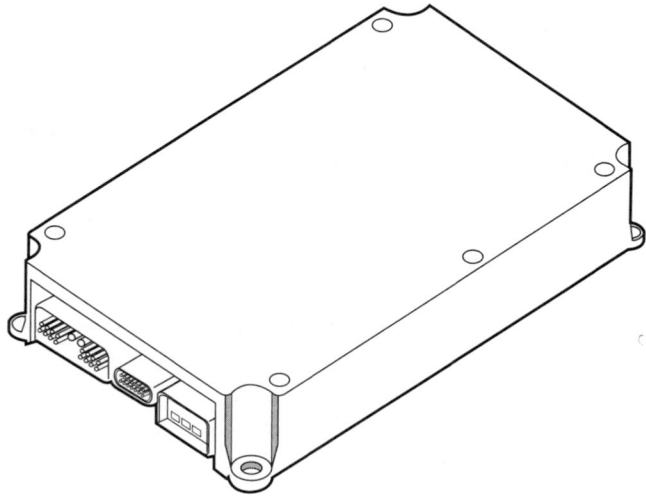

P/N 23513553 FOR 4 CYCLE (WITHOUT J1939)
P/N 23516923 FOR 2 CYCLE (WITH J1939)

Figure 11-2 DDEC III ECM. (Reprinted by permission of copyright owner Detroit Diesel Corporation, all rights reserved.)

system monitoring, audit trails (tattle-tale or covert writing of an electronically monitored event or condition to EEPROM), customer and DDC data programming. Each DDEC ECM can manage up to eight EUIs; in off-highway engines that have more than eight cylinders, multiple ECM units are required. In multi-ECM–managed engines (off-highway applications), one of the ECMs masters the engine. The master ECM receives all inputs and processes data, while the other(s) are referred to as receivers and are controlled by the master ECM. Receiver ECMs act as EUI actuators and basically effect EUI switching on command. However, in the event of master ECM communication failure, receiver ECMs can temporarily enable operation of each portion of the engine.

The DDEC ECM is programmed with engine management logic to fuel the engine through variable speeds and loads, while at the same time monitoring both engine and chassis operating conditions and continually performing self-diagnosis. DDEC III features EEPROM (electronically erasable programmable read only memory), which qualifies the master engine management program in ROM (read only memory) with application-specific detailing, such as engine governing, torque profile shaping, cold start logic, and engine protection strategy. The results of ECM arithmetic, logic and fetch and carry computations in the processing cycle are converted to actuator or control signals that switch the EUIs. Command pulses can be referred to as *duty cycle* or **pulse width (PW)** and are measured in milliseconds. Duty cycle or PW is the time period within the EUI cam actuated downstroke in which fuel is actually being pumped. **PW** is calculated and controlled by the ECM; it is effected by the injector driver unit of the ECM known as the EDU, which energizes the control solenoids of the EUIs.

DDEC ECMs are mounted on a heat sink–type heat exchanger through which diesel fuel in the fuel subsystem is routed and acts as a cooling medium. Fuel from the transfer pump is routed through this heat exchanger before charging the cylinder head fuel gallery. DDEC III ECMs can manage both 12- and 24-volt systems. They require either a direct or bus strip connection to the vehicle batteries. DDC states that DDEC III and IV ECMs have an operating voltage range of 11–32 V and will sustain a transient cranking low of 9 V. The DDEC III ECM is roughly half the size of the DDEC II ECM, processes data eight times as fast, and can retain seven times the memory. The DDEC IV ECM has about 50% more speed and memory capability than the DDEC III module.

Input Circuit—Sensors

As DDEC has evolved, the sensor circuit has become more comprehensive. The following is a description of DDEC III sensors.

Timing Reference Sensor   The **timing reference sensor (TRS)** (Figure 11-3) signals crank position to the ECM. It tells the ECM where each piston is in its cycle. An AC pulse generator is used. A 6-tooth pulse wheel (reluctor) is used on DDEC II and a 36-tooth pulse wheel is used on DDEC III and IV. In both DDEC II and III the pulse wheel is fitted to the front of the crankshaft.

Synchronous Reference Sensor   The **synchronous reference sensor (SRS)** (Figure 11-3) develops a signal to tell the ECM that the next TRS signal to be generated corresponds to cylinder #1 for that engine. It corroborates the TRS signal. An AC pulse generator is used. Both TRS and SRS must work for the engine to run; if either is nonfunctional in DDEC II, a code is logged after 10 seconds of cranking. In DDEC III, this code is produced immediately. The SRS consists of a single dowel that acts as a reluctor tooth located on the cam gear and a stationary magnetic pickup in series 60/50. On series 92 the SRS and TRS both pick up on the pulse wheel, located on the camgear.

Electronic Foot Pedal Assembly   In common with most truck/bus electronically managed engines, DDEC is drive by wire. The critical sensor in the **electronic foot pedal assembly (EFPA)**

The ECM controls EUI pulse width through the injector driver unit known as the EDU, which energizes the control solenoids of the EUIs.

The timing reference sensor (TRS) tells the ECM piston crank angle or where the piston is.

The synchronous reference sensor (SRS) identifies cylinder #1 and begins the sequence in firing order for the ECM to follow.

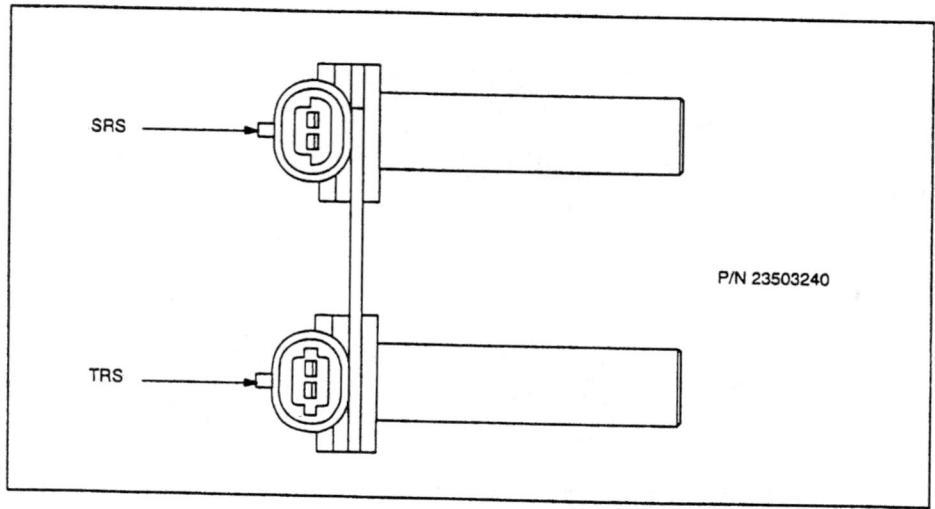

The SRS and TRS Sensors – 6V–92, 8V–92, and 12V–71 Engines

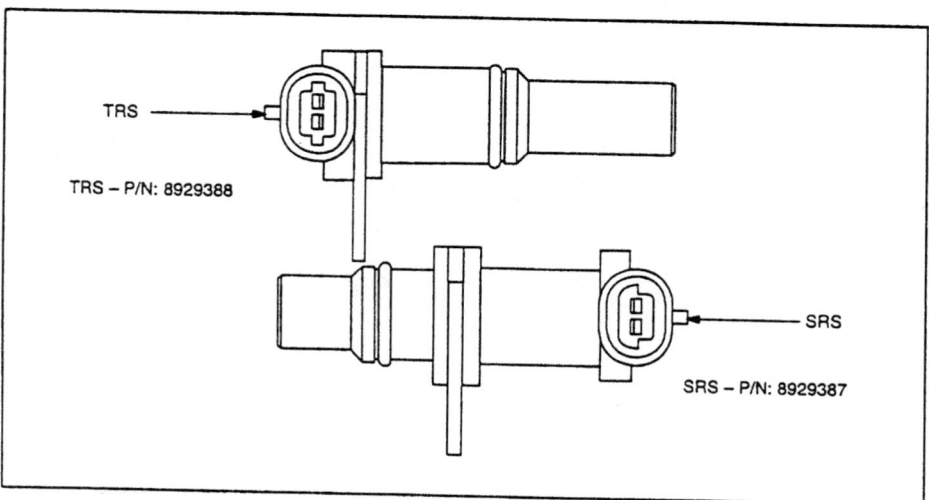

The SRS and TRS Sensors – Series 50 and Series 60 Engines

Figure 11-3 SRS and TRS sensors. (Reprinted by permission of copyright owner Detroit Diesel Corporation, all rights reserved.)

is the **throttle position sensor** (**TPS**) (Figure 11-4). The TPS receives reference voltage (V-Ref) of 5 V and returns a portion of it, proportional to pedal mechanical travel. The DDEC TPS is a three-terminal potentiometer or voltage divider; the resistance within the TPS that the V-Ref must overcome depends on wiper position on the TPS resistor. The actual voltage signal returned to the ECM is converted to counts, which can be read with an electronic service tool (EST). TPS counts should be read as follows:

Zero travel (idle)	100–130 TPS counts
Full travel	920–950 TPS counts
48 counts or less	Low voltage code generated
968 counts or higher	High voltage code generated

 WARNING: The chassis manufacturer that supplies the EFPA must use DDC specifications.

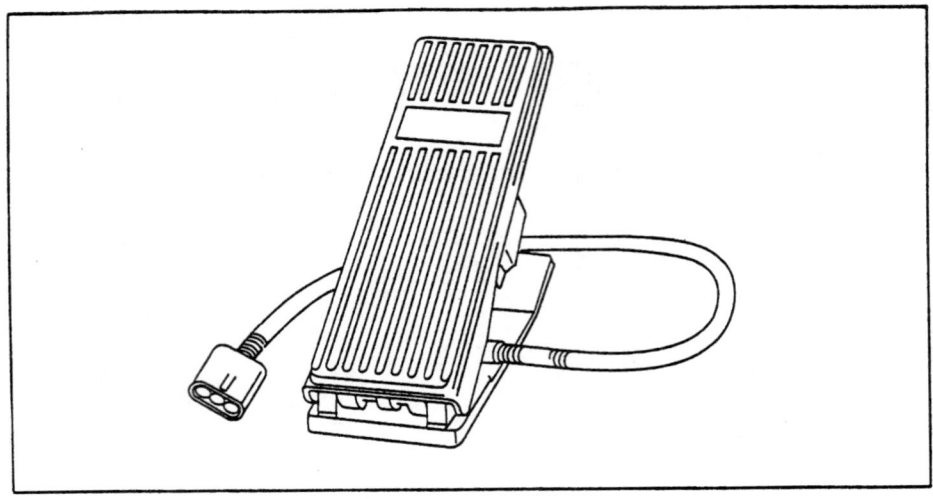

Electronic Foot Pedal Assembly (EFPA)

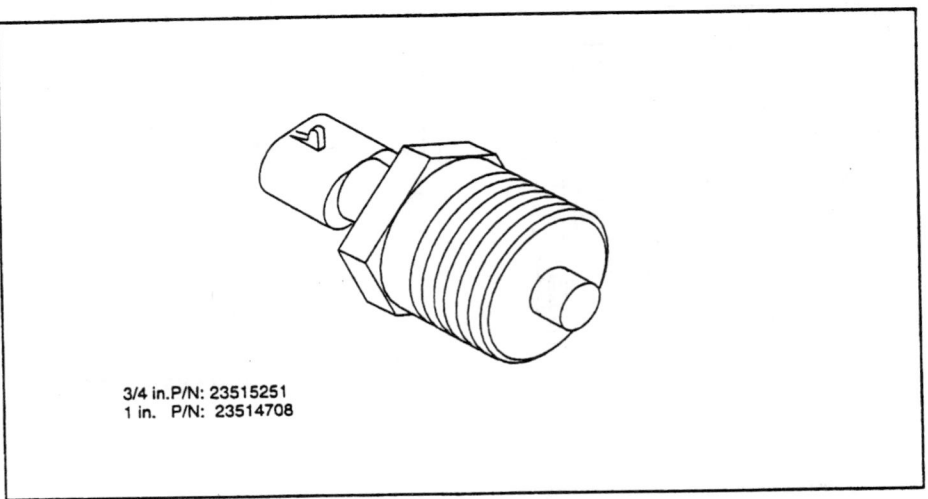

3/4 in. P/N: 23515251
1 in. P/N: 23514708

Coolant Temperature Sensor (Series 50 and Series 60)

Figure 11-4 EFPA and CTS sensors. (Reprinted by permission of copyright owner Detroit Diesel Corporation, all rights reserved.)

Turbo Boost Sensor   The DDEC turbo boost sensor (TBS) (Figure 11-5) is responsible for signaling the ECM with both manifold boost and barometric pressure data. The TBS uses a variable capacitance principle. Pressures from 0 to 45 psi (3 atm; 304 kPa) can be measured. The TBS is a DDC provided sensor.

TBS data is required by the ECM to plot air-fuel ratio (AFR). By monitoring atmospheric pressure, oxygen density is known, so ECM fuel quantity programming is automatically derated at altitude. Turbo boost pressure monitoring indicates actual engine load and prevents transient overfueling. Therefore, a sudden full-fuel demand at the accelerator pedal will not result in excess fuel being injected before there is sufficient air (supplied by the turbocharger) in the engine cylinders to completely burn it.

Fuel Pressure Sensor   The **fuel pressure sensor (FPS)** (Figure 11-5) is located on the charge side of the transfer pump. It is a variable capacitance–type sensor capable of signaling pressure values from 0 to 75 psi (5 atm; 505 kPa). The FPS is a DDC supplied sensor, which will warn the operator of a fuel subsystem problem such as a restricted filter.

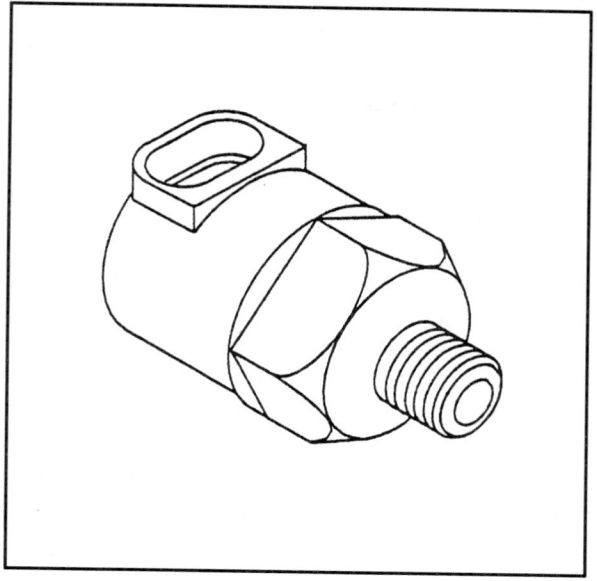

Oil, Fuel and Coolant Pressure Sensor

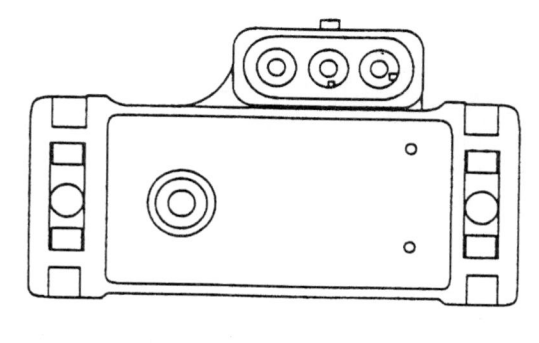

The Turbo Boost Sensor

Figure 11-5 Oil, fuel, and coolant pressure sensors and turbo boost sensor. (Reprinted by permission of copyright owner Detroit Diesel Corporation, all rights reserved.)

Oil Pressure Sensor (OPS) The OPS will trigger engine protection strategy if engine oil pressure drops below a specified value. It is a variable capacitance–type sensor capable of signaling values from 0 to 75 psi (5 atm; 505 kPa). The OPS is a DDC supplied sensor.

Coolant Pressure Sensor (CPS) The CPS is a variable capacitance type sensor used to warn the operator of low coolant pressure. It is a DDC supplied sensor.

Fuel Temperature Sensor (FTS) The FTS is a thermistor supplied with V-Ref of 5 V in which resistance decreases as the temperature increases. Therefore, it can be classified as a negative temperature co-efficient (NTC) type sensor. The voltage signal returned to the ECM increases with temperature rise. The FTS signal is required by the ECM to help factor the air-fuel ratio and log fuel consumption data. It is a DDC sensor.

Coolant Temperature Sensor (CTS) The DDEC CTS is also a thermistor (Figure 11-4) of the NTC type and will trigger engine protection strategy once a specific programmed limit is exceeded. It is a DDC sensor.

The coolant temperature sensor (CTS) is a thermistor that, when its cold, is a poor ground; when it gets hot, the resistance is reduced and the ground gets better.

Air Temperature Sensor (ATS) The DDEC ATS is another thermistor of the NTC type. Its signal input is critical to the ECM for calculating cold start fueling and timing, factoring AFR, and hot idle speed. It is a DDC sensor.

Oil Temperature Sensor (OTS) The DDC OTS is a thermistor of the NTC type and is used to corroborate data from CTS and ATS sensors and can also trigger engine protection strategy once the specified limit is exceeded. It is a DDC sensor.

Fire Truck Pump Pressure Sensor The fire truck pump pressure sensor (Figure 11-6) is used with DDEC III to master engine speed governing to produce a specific pump pressure value. It is of the variable capacitance type and is DDC supplied.

Vehicle Speed Sensor The **vehicle speed sensor** (**VSS**) is of the induction pulse generator type, usually a standard 36-tooth reluctor located on the transmission tailshaft. Although the VSS is optional, it is required to enable cruise control and vehicle speed limiting ECM functions, and it is incorporated in most DDEC applications. The chassis manufacturer (OEM) supplies the VSS.

Coolant Level Sensor (CLS) The CLS is a switch-type sensor consisting of an integrated resistor that grounds through the engine coolant and is located in the top radiator tank. If the ground is interrupted, engine protection strategy is initiated after a prescribed time period. The CLS is supplied by the OEM.

Multiplexing

Where DDEC III is required to interface with other chassis system ECMs, the J1708/J1587/1939 data communication links are used. This synergizes separate electronic system operation and avoids the need to duplicate vehicle sensors, such as the TPS and VSS.

Output Circuit

Output circuit devices are components that are controlled by the ECM. They are the components that affect the results of the programmed software and the processing computations. Any component that is switched or can be controlled by the ECM can be called an output.

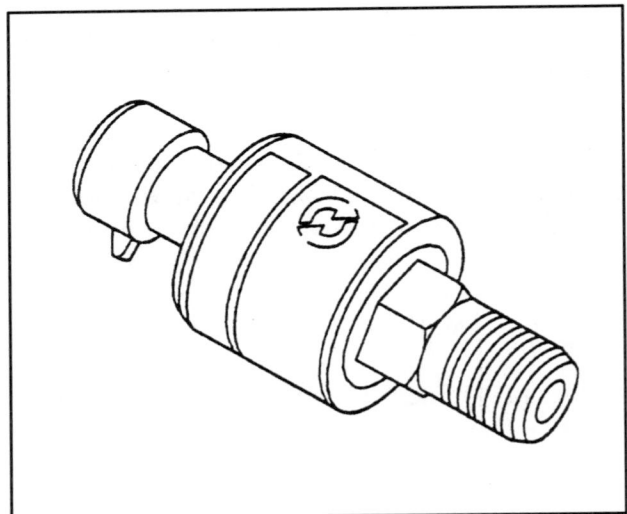

The Firetruck Pump Pressure Sensor

Figure 11-6 Fire truck pump pressure sensor. (Reprinted by permission of copyright owner Detroit Diesel Corporation, all rights reserved.)

Electronic Unit Injector (EUI)

DDEC EUIs (Figure 11-7) are integral pumping, metering, and atomizing devices. In common with the mechanical unit injectors (MUIs) used by DDC in the past, they are cam actuated but with that the similarity ceases. The EUI plunger has no cross and center drillings or helical recesses, but is a simple cylindrical plunger element that reciprocates within the pump stationary element or bushing. Actual plunger travel does not vary, as this is dictated by cam profile. The EUI tappet spring holds the plunger in a retracted position for most of the cycle until the actuating cam on the engine camshaft ramps toward the cam nose, and the EUI plunger is driven downward into the pump chamber.

Fuel is circulated through EUI (Figure 11-8) internal ducting whenever the engine is run at charging or supplies pressure (50 psi; 3.3 atm; 340 kPa). The EUI internal fuel circuit includes the pump chamber and circuitry that routes fuel through the control solenoid. Connected to the control solenoid armature is a poppet valve. Whenever the poppet valve is in the open position (that is, when the solenoid is not energized), fuel is permitted to pass through the solenoid ducting. The instant the ECM energizes the EUI solenoid, the poppet control valve closes, blocking the passage of fuel through the solenoid ducting. This action effectively traps the fuel in the pump chamber under the EUI plunger. A duct in the base of the pump chamber connects it with a hydraulic injector nozzle. When the required nozzle opening pressure (NOP) is achieved, the nozzle valve opens, fuel passes around the nozzle valve seat and exits the EUI through the nozzle orifii.

DDEC EUIs are designed for direct injection of a fuel charge to both two- and four-stroke cycle engines at NOPs typically around 5,000 psi (340 atm; 34.4 MPa) and possible peak pressures of over 25,000 psi (1700 atms; 172 MPa). The fueling window within which the EUI can switch an effective pumping stroke is dictated by cam profile, as the EUI plunger must be in the act of descending for this to occur. Plunger mechanical stroke is constant. Plunger effective stroke must occur within the mechanical stroke hard window, beginning when the EUI solenoid is energized and ending when it is switched open. DDEC electronic unit injectors are switched at 12 V.

DDEC III and IV EUIs are graded with a calibration code, which ranges from 00–99 which enables the ECM to be programmed with data concerning how a specific EUI flows fuel. The **diagnostic data reader (DDR)** is used to program this data to the ECM and enables highly accurate and balanced fueling of the engine.

To program the calibration code, the injector cylinder number would be selected first followed by entering the digital calibration code value. DDEC II calibration codes are alpha codes,

Shop Manual Chapter 11, page 398.

Plunger effective stroke on most EUIs is a hard mechanical window not controlled by the ECM.

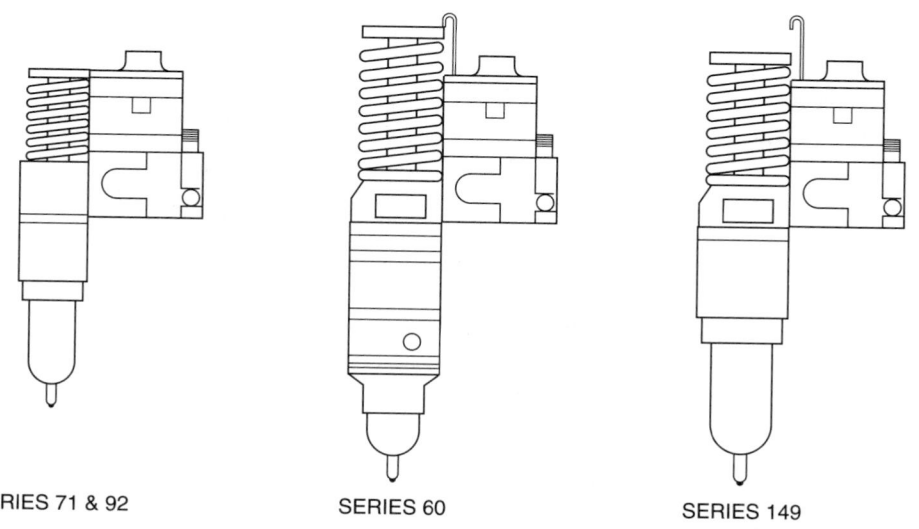

SERIES 71 & 92   SERIES 60   SERIES 149

Figure 11-7 DDEC EUI. (Reprinted by permission of copyright owner Detroit Diesel Corporation, all rights reserved.)

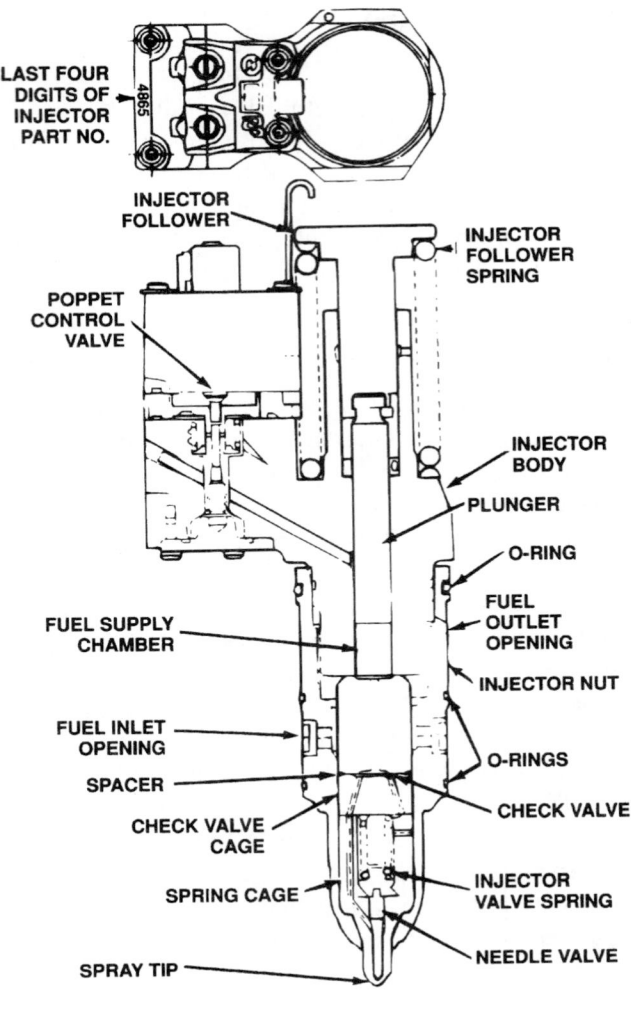

LAST FOUR DIGITS OF INJECTOR PART NO.

INJECTOR FOLLOWER

POPPET CONTROL VALVE

INJECTOR FOLLOWER SPRING

INJECTOR BODY

PLUNGER

O-RING

FUEL OUTLET OPENING

FUEL SUPPLY CHAMBER

INJECTOR NUT

FUEL INLET OPENING

O-RINGS

SPACER

CHECK VALVE

CHECK VALVE CAGE

SPRING CAGE

INJECTOR VALVE SPRING

NEEDLE VALVE

SPRAY TIP

Electronic Unit Injector Cross Section

Figure 11-8 Overhead and sectional view of DDEC EUI. (Reprinted by permission of copyright owner Detroit Diesel Corporation, all rights reserved.)

A, B, or C, with A denoting the least fuel flow and C the most fuel flow. This critical procedure helps the ECM balance cylinder fueling.

Injector Response Time DDEC EUIs are switched at 12 V, unlike other EUI systems that spike the application voltage to values around 100 V. The time lag between **beginning of energizing (BOE)**, or the instant that the EDU starts to flow current to the EUI, and **beginning of injection (BOI)** solenoid is variable. DDEC electronics are programmed to measure the response time between BOE and BOI by studying the actuation voltage wave of the previous two actuations of each EUI on a continual basis. Injector response time (IRT) (Figure 11-9) is essentially the time lag between the output of the actuating signal at the EDU and the moment the EUI poppet control valve actually closes. It is measured in milliseconds and read on the DDR display. DDEC is capable of adjusting to minor electrical circuit problems within a certain window to maintain balanced fueling and timing. It should be also noted that there is a fractional lag between EUI control valve closure (BOI) and the actual opening of the EUI hydraulic nozzle valve (NOP), which truly begins injection (Figure 11-10). Similarly, at the completion of the switched duty cycle or PW known as **ending of energizing (EOE)**, there is not an immediate ending of injected fuel.

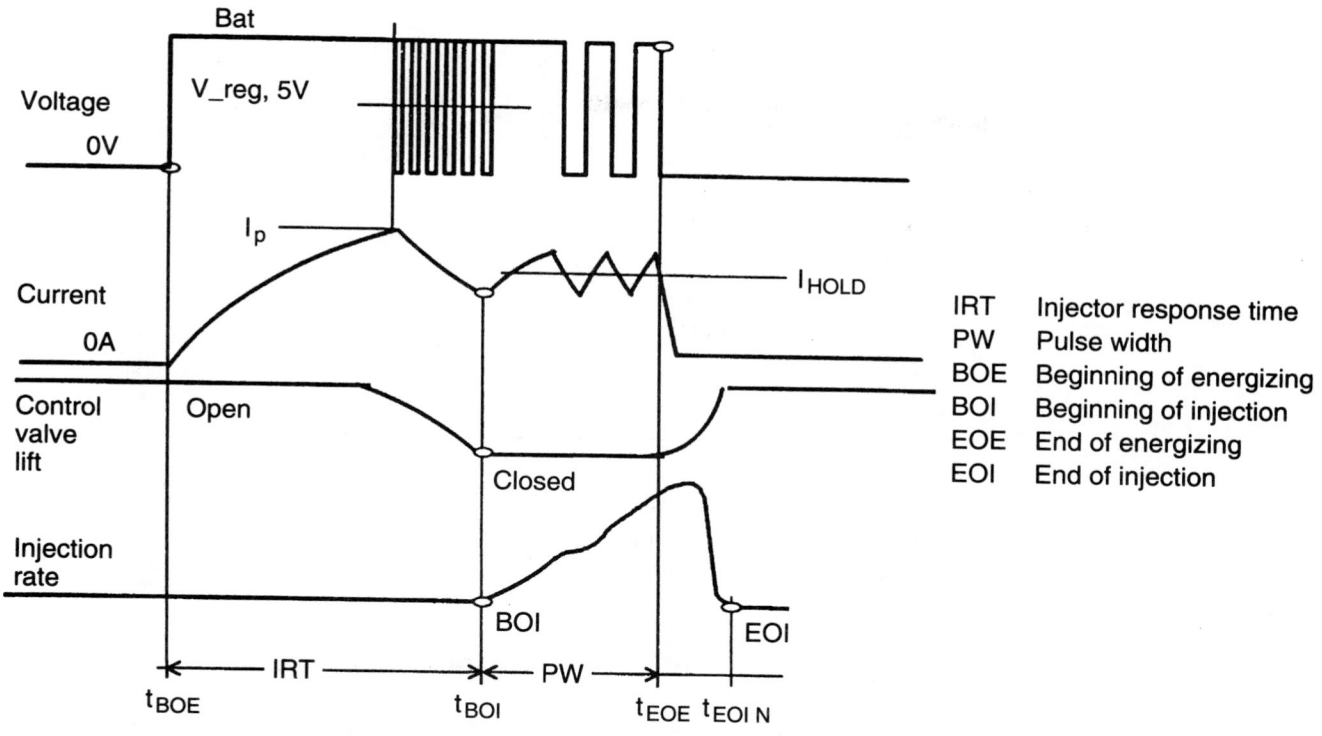

Figure 11-9 DDEC injection cycle graph. (Reprinted by permission of copyright owner Detroit Diesel Corporation, all rights reserved.)

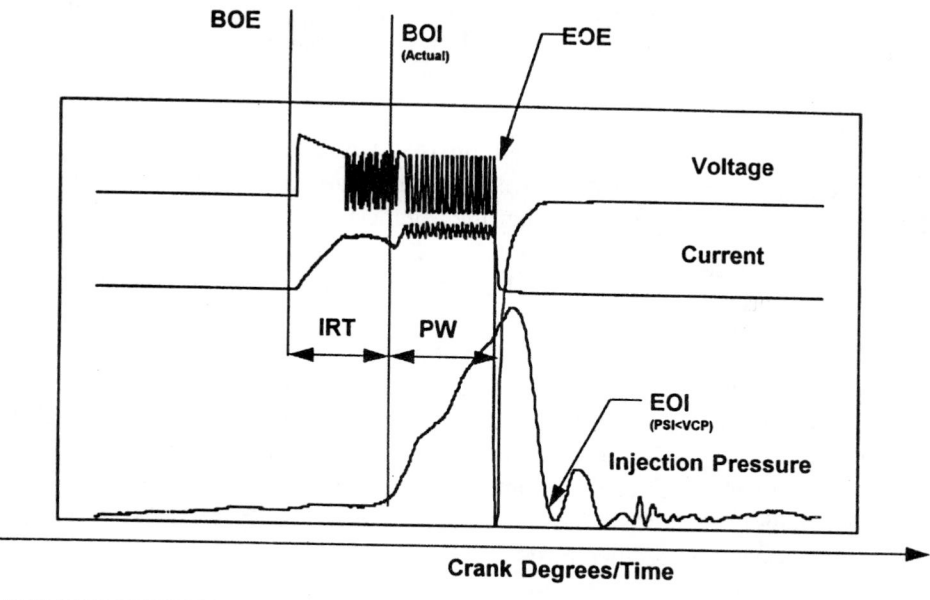

Figure 11-10 DDEC injection cycle electrical waveform to time graph. (Reprinted by permission of copyright owner Detroit Diesel Corporation, all rights reserved.)

Some fraction of time is required to collapse the pressure in the EUI pump chamber to the VCP (valve closing pressure) value, at which point the injection pulse truly ceases.

Figures 11-10 and 11-11 graphically represent what is occurring electrically and hydraulically through the EUI duty cycle. The primary DDEC output is EUI management, but DDEC will support an engine compression brake and bus into other chassis system ECMs.

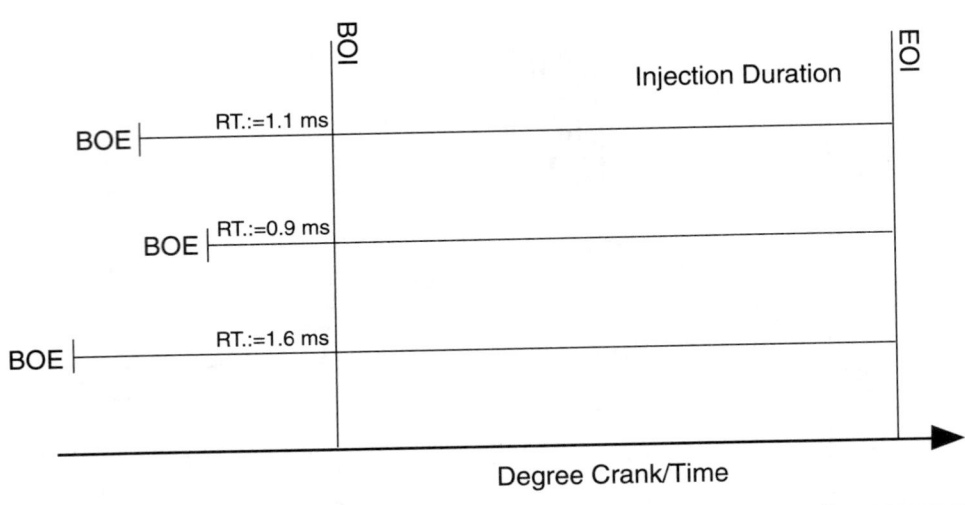

Figure 11-11 Response time effect on the injection cycle graph. (Reprinted by permission of copyright owner Detroit Diesel Corporation, all rights reserved.)

Pilot Injection DDEC III and IV electronics can manage **pilot injection**. This technology may be used as part of ECM cold start strategy where it is used to eliminate diesel knock and minimize start-up smoke emission. Pilot injection essentially breaks up the fueling pulse by switching the EUI at high speeds. Using this technology under cold start conditions would mean delivering a short pulse of fuel to the engine cylinder, calculating the moment of ignition, and at that point resuming the EUI fueling pulse. Diesel knock is a cold start detonation condition caused by delayed ignition and excess fuel in the cylinder at the point of ignition. Besides reducing engine wear, pilot injection also greatly lowers cold start emissions.

Governing

DDEC III offers both limiting speed governing (LSG) and variable speed governing (VSG). Though managed electronically, both LSG and VSG are designed to function similarly to mechanical LSG/VSG. The LSG option will fuel the engine based on percent throttle (pedal position) selected by the operator. The VSG option will hold engine speed at a specific requested value and attempt to maintain it as the load on the engine fluctuates. It is used for power take-off (PTO) operation and electronic cruise control.

The ECM is programmed with a detailed fuel map, which manages engine fuelling through both normal and transient abnormal conditions. Smoke control and cold start fuelling are managed by selecting optimum injection timing and injected fuel quantity based on such factors as ambient temperature, barometric pressure, and manifold boost and throttle demand. Cold start strategy on DDEC III may produce an automatic increase in engine idle speed and advanced injection timing to minimize cold operation white smoke emission. This would revert to normal as the engine reaches operating temperatures.

Cruise Control Cruise control can be operated in any gear when engine speed is above 1,100 rpm and providing the road speed is above 20 mph (32 kph). Cruise control can be programmed to master engine brake operation so that if the vehicle accelerates above set cruise speed on a downgrade, the engine brake will automatically actuate.

Progressive Shifting Progressive shifting is a feature designed to achieve better fuel economy by encouraging the driver to upshift before the engine reaches governed speed. When the speed of the engine is limited, the engine remains in the higher torque range (where the engine operates at better fuel efficiency) for a longer period of time.

Engine Protection DDEC III monitors a range of vehicle and chassis functions and is programmed with engine protection strategy that may derate or shut down the engine when a running condition could result in catastrophic failure. The operator is alerted to system problems by illuminating dash lights, the check engine light (CEL) and the stop engine light (SEL). The CEL is illuminated when a system fault is logged to alert the operator that DDEC III fault codes have been generated. The SEL is illuminated when the ECM detects a problem that could result in a more serious failure. DDEC III may also be programmed to shut down or ramp down to idle speed after such a problem has been logged. A shut-down override switch, known as a stop engine override (SEO) switch, will permit a temporary override of the shut-down command.

DDEC III and IV Comparison

Because vehicle management electronics are upgraded on a continual basis, it is difficult to pinpoint the significant changes that define a numeric graduation from one generation of electronics to the next. This section outlines some of the improvements that DDC highlights in DDEC IV when comparing it with DDEC III.

Fuel Economy Driver Incentive The vehicle owner can set a fuel economy goal and program a maximum vehicle speed. If the driver fails to meet the goal, nothing changes. However, for each 10-mpg increase over the fleet economy target, the maximum vehicle speed is increased by a speed value programmed by the owner. DDC claims that early testing of this feature can produce some dramatic results.

Low Gear Torque Limiting This permits the use of smaller transmission by limiting torque in the lower ratios and allowing higher engine torque output in the higher gears.

Jacobs Compression Brake DDEC IV permits three levels of braking, and DDC claims 29% more braking power with the DDEC IV Jake brake.

Fuel Pro Filter This filter has been available as an option but is now standard with DDEC IV. Fuel Pro eliminates the need for separate primary and secondary filters. It greatly extends filter service intervals and acts as a water separator. A Fuel Pro option is a thermostatically controlled fuel heater that prevents gelling during low temperature operation.

Optimized Idle System DDEC IV can be programmed to an optimized idle mode, which operates like a home heating thermostat when the vehicle is parked. The idea is to automatically stop and start the engine to:

❑ Maintain a comfortable cab/bunk temperature
❑ Keep the vehicle batteries charged
❑ Keep the engine warm

Ether Start System This feature was available in later versions of DDEC III. Ether start provides automatic ether injection during start-up. The ECM is programmed to use the ether start option only when the temperature conditions require its use. By eliminating driver-controlled ether injection, the chances of abusing an engine with ether are reduced. A low ether supply indicator is included with the package.

Data Analysis As the computing power of vehicle ECMs increases, more parameters are monitored, and the analyzed data can be used as an incentive (for example, for the driver to increase fuel economy). Data analysis can be enhanced by downloading the data to a personal computer (PC) with software designed to analyze every performance detail of the vehicle. Data analysis is useful to the service technician. First generation electronic systems identified what was malfunctioning in a system but offered few clues as to why the problem was occurring. Current

The check engine light (CEL) signals a fault or intermittent fault to the driver.

The service engine light (SEL) signals the driver that the engine needs immediate service.

troubleshooting packages can produce precise performance profiles that will tell the diagnostic technician exactly how the vehicle has been driven and handled, and the conditions at the exact moment of the logging of a trouble code. ProDriver (described earlier in this section) data can be downloaded by a fleet manager for analysis. With DDC software it can produce a profile of exactly how a driver operates a piece of equipment. ProDriver data may be accessed with ProDriver Reports software (requires MS Windows 95) or the RDI system. RDI is weatherproofed so it can be mounted in a convenient location such as a fuel island for hard wire downloads. ProDriver can now also be programmed for wireless extraction of data from the vehicle electronic system.

Diagnostic Link Software Diagnostic Link software is a MS Windows 95 troubleshooting package that includes a software-based service manual, and it can guide troubleshooting and view or modify vehicle speed settings, engine protection strategy and outcomes, cruise options, progressive shifting, fault codes, and just about any engine or trip data. Diagnostic Link software is designed with user-friendly architecture and is the preferred EST in late version DDEC systems.

Shop Manual Chapter 11, page 400.

Caterpillar ADEM

The Caterpillar "E" engines use electronically controlled, mechanically actuated unit injectors and as such can be defined as "full authority," electronically controlled engines. The first versions of these engines were based around the Lucas/CAV 100 EUI and were introduced to the truck market place as follows:

3176	1989	10.3 L. displacement
3406E	1994	14.6 L. displacement
3406E	1998	15.8 L. displacement (Cat EUI)
C10	1996	10 L. displacement
C12	1996	12 L. displacement

All engines in this series are metric engineered, six-cylinder engines designed for highway truck applications. The 3176 (Figure 11-12) was engineered for EUI fueling, whereas the 3406E is a re-engineered adaptation of the successful 3406PEEC and hydromechanical 3406 engines. The C10 and C12 are essentially an evolution of the 3176 engine platform and incorporate improvements learned from seven years of market experience. These engines use articulating pistons with a cast steel crown and forged aluminum skirt. Three rings are used: the top compression ring is of a barrel-faced, keystone design, the second is taper faced, and the oil control ring is double railed, profile ground with a coil spring expander. All three rings are cladded; the top with a plasma face coating, the other two are faced with chrome.

Engine breathing is turbo boosted with a Siamese plenum, canted configuration intake ports (two cylinders charged from one intake plenum) and a canted exhaust port configuration (one discharge port per cylinder), that locates both intake and exhaust manifolds on the right side of the engine. The engine cylinder liners use a mid-supported design that permits lower liner bore temperatures at the top ring upper stroke turnaround point. They are manufactured from molybdenum alloy cast iron with an induction-hardened interior wall. Charge air-cooling is air to air. This series of engines will support Jacobs's engine brakes and driveline retarders. The aluminum spacer block used with 3176 engine has been eliminated in C10 and C12 engines.

Caterpillar 15.8 L 3406E Engine Caterpillar introduced their 15.8 L 3406E engine in 1998 to claim a stake in the high brake power segment of the truck engine market. The engine is based on the 3406E platform. The bore is increased from 5.4–5.5 in. and the stroke from 6.5–6.75 in., but otherwise, the engine retains high level of common components with the smaller displacement model. It is currently available in brake power ratings of 575 BHP and 600 BHP.

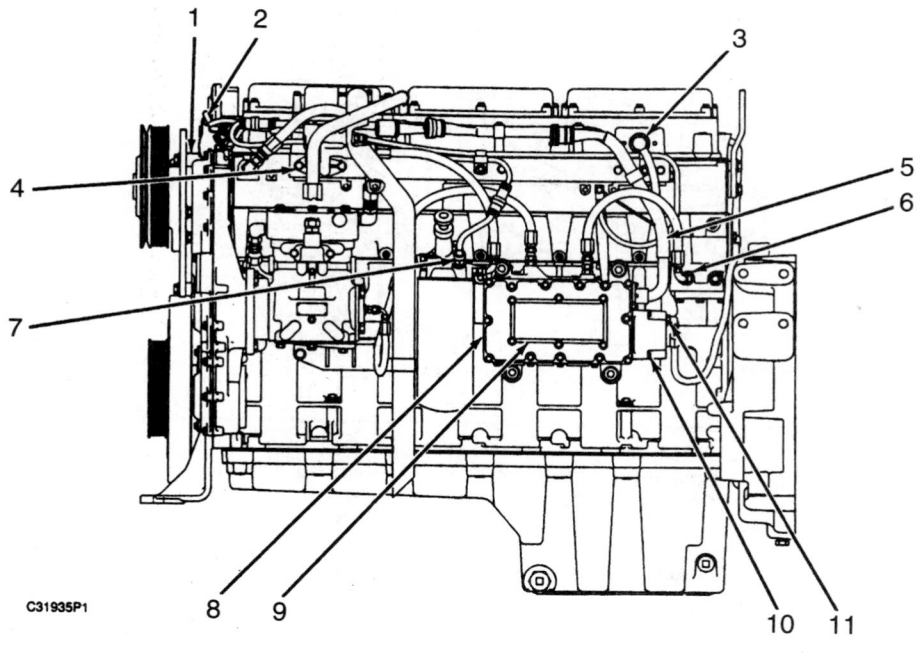

Electronic Control System Components
(1) Speed/timing sensor. (2) Coolant temperature sensor. (3) Injector enable circuit. (4) Fuel manifold. (5) Engine wiring harness. (6) Fuel transfer pump. (7) Fuel pressure sensor. (8) Electronic control module (ECM). (9) Personality module. (10) Transducer module. (11) Boost sensor.

Figure 11-12 Caterpillar ADEM electronic control system components. (Courtesy of Caterpillar Inc.)

ADEMS System Overview

Consistent with other full authority electronic engines, the mechanically actuated EUIs have an effective pumping stroke managed and switched by the ECM. However, because the actual EUI plunger stroke is cam actuated, the ECM is confined to the hard limit window represented by cam profile. The following represents an accounting of features and terminology that makes this series of Caterpillar engines distinct from other manufacturers' engines classified as full authority. Caterpillar engines with full authority management systems may be divided into the following subsystems for purposes of studying fueling control:

> The ECM control of fueling is confined to the hard limit window represented by cam profile, because the actual EUI plunger stroke is cam actuated.

1. Fuel subsystem
2. Electronic input circuit
3. ECM
4. Output circuit

Fuel Subsystem

The fuel subsystem (Figure 11-13A) incorporates those components that enable the transfer of fuel from the fuel tank to the EUIs. A fixed clearance gear pump is responsible for fuel movement through the supply circuit to the ECM. This gear pump is flange mounted to be driven by the camshaft through a pair of helical gears. It incorporates a check valve and pressure relief valve. A hand-actuated, plunger pump is located downstream from the gear pump and is used to prime the fuel subsystem; it is usually integral with the filter mount pad. Fuel from the fuel subsystem is delivered to the fuel supply and return manifolds. An adapter or siphon break prevents fuel drain off from this manifold when the engine is not running. While the engine is running, fuel is continually circulated through the system, passing through a heat exchanger to which the ECM is mounted and through the EUIs acting as coolant. System charging pressure is maintained at around 4 atm (60 psi; 400 kPa). Figure 11-13B shows the 3176 fuel schematic.

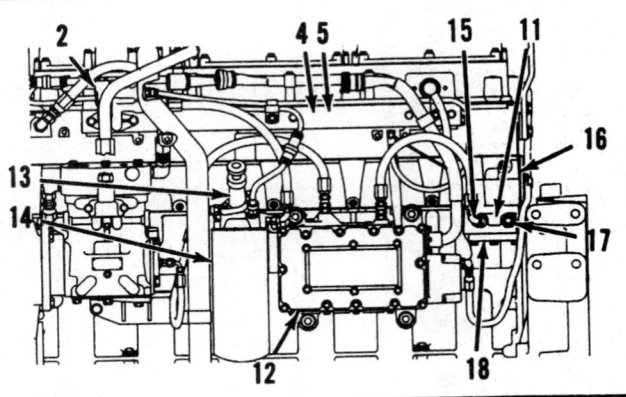

Fuel System Components
(2) Adapter (siphon break). (4) Fuel return manifold. (5) Fuel supply manifold. (11) Fuel transfer pump. (12) Electronic control module (ECM). (13) Fuel priming pump. (14) Fuel filter (remote mounted fuel filter as an attachment). (15) Fuel outlet (to ECM). (16) Spacer block. (17) Fuel inlet (from tank). (18) Cover assembly.

A

B

Fuel System Schematic
(1) Vent plug. (2) Adapter (siphon break). (3) Electronically controlled unit injectors. (4) Fuel manifold (return path). (5) Fuel manifold (supply path). (6) Drain plug. (7) Pressure regulating relief valve. (8) Fuel tank. (9) Check valve. (10) Pressure relief valve. (11) Fuel transfer pump. (12) Electronic control module (ECM). (13) Fuel priming pump. (14) Fuel filter (remote mounted fuel filter as an attachment).

Figure 11-13 (A) 3176 fuel system components; (B) 3176 fuel system schematic. (Courtesy of Caterpillar Inc.)

ADEMS

Caterpillar refers to the electronics package that manages their engines as ADEMS, or the advanced electronic engine management system. The first generation of ADEMS did not appear in any highway application Caterpillar engines. The second generation of ADEMS was the electronics package used to manage 3176 and 3406E engines beginning in 1993, and the third generation of ADEMS electronics is that used on their current engine lineup. The current ADEMS ECMs have 32-bit processors running at 24 MHz and dual 70-pin connectors that enable up to 140 inputs and outputs.

Input Circuit   Command and monitoring sensors and switches are consistent with other comparable full authority management systems with the exception of the throttle position sensor. The Caterpillar throttle position sensor, shown in Figure 11-14, is a variable capacitance device that outputs a constant frequency pulse width modulated (PWM) signal rather than the analog signal produced by the potentiometer based TPSs used by other manufacturers. Caterpillar states that this digital PWM signal overcomes errors that can be produced by analog signals due to pin-to-pin leakage or contamination in either the connectors or harnesses, an important factor in any drive-by-wire system. If the PWM signal is invalidated for whatever reason, the engine will only run at the idle speed value. The sensor is designed to have 30 degrees of active travel with an additional 5 degrees of undertravel and 10 degrees of overtravel for linkage tolerance. Two TPS types are used, remote mounted and pedal mounted. They both operate on a variable capacitance principle. The **personality module** must be programmed for the correct TPS. Current Caterpillar TPS units are pedal mounted and automatically recalibrate themselves. They are a Caterpillar (not OEM) supplied component. All the sensors are remote mounted from the ECM,

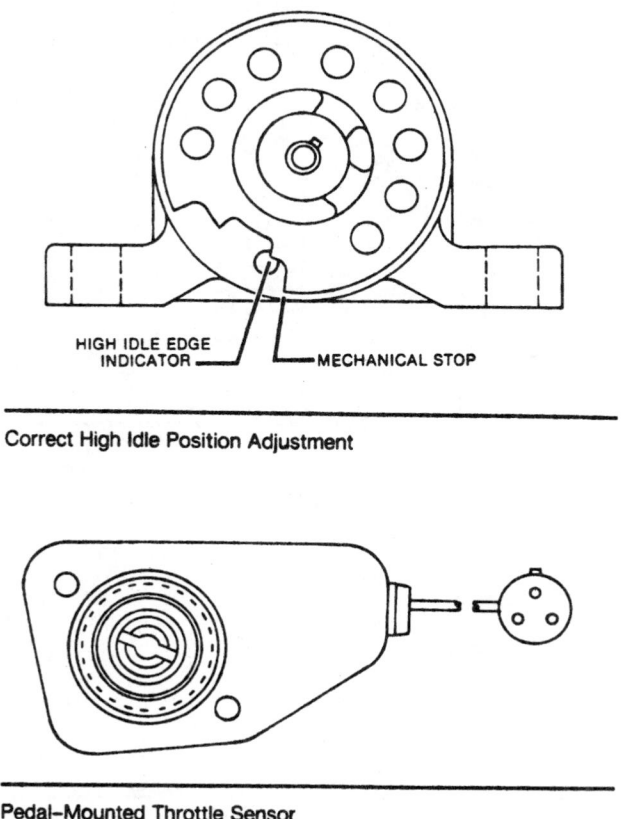

HIGH IDLE EDGE INDICATOR — — MECHANICAL STOP

Correct High Idle Position Adjustment

Pedal–Mounted Throttle Sensor

Figure 11-14 Caterpillar floor and pedal throttle sensor. (Courtesy of Caterpillar Inc.)

and the functions of the transducer module have been localized to eliminate the routing of hoses to the ECM.

The ECM manages the control of fuel delivery

ECM The ECM (Figure 11-15) is a typical microprocessor-driven management and switching system, which is mounted on the left side of the engine cylinder block. The ECM is responsible for engine governing, fuel timing logic, self-diagnostics, system monitoring, and tattletale logging; it is programmable with customer and proprietary (Caterpillar) data. Consistent with other systems of its kind, the Caterpillar ECM is resistant to radio frequency, electromagnetic interference and other low-level radiation. The ECM is responsible for supplying system reference voltages, powering up sensors, and integrally housing the injector drivers.

Bolted to older version ECMs was a personality module that was programmed with system and chassis operating parameters and contains chips that perform the programmable read-only memory (PROM) and electronically erasable programmable read-only memory (EEPROM) functions required by the system. Also attached to the older version ECMs is a "transducer module" responsible for transducing all system pressure values to electrical signals to signal to the ECM. Current version ECMs (Figure 11-16) have eliminated the attached personality and transducer modules and incorporated their functions in a single, more compact unit with greatly improved data processing power and speed.

Dual microprocessors are used in the current ECMs. The new units continue to be mounted on a fuel-cooled heat exchanger. Mounted internally is a flash memory chip containing the engines controls software. It is not removable but "flashing," or electronically reprogramming, the chip may make program changes. This is a version of what is generically described as EEPROM. Caterpillar ECMs have a nonvolatile RAM (NV-RAM) component for storing customer data; an internal battery backs this up, so battery disconnection or failure will not result in the loss of this data. Caterpillar electronic engine governing is designed to replicate the driver feel of a hydromechanical LS governor with mechanical accelerator linkage.

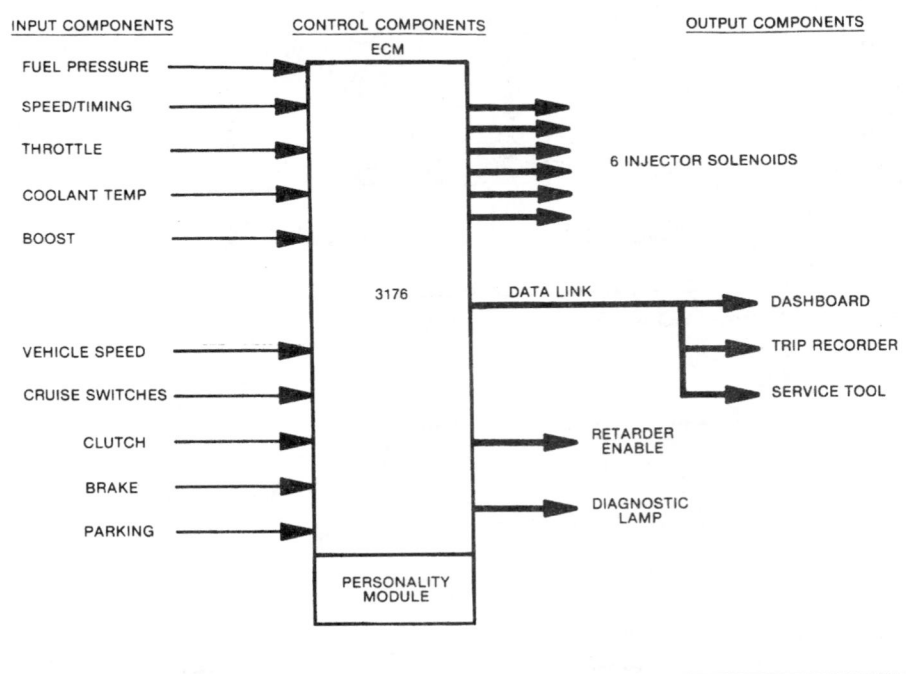

3176 Electronic Control System

Figure 11-15 3176 electronic control inputs and outputs. (Courtesy of Caterpillar Inc.)

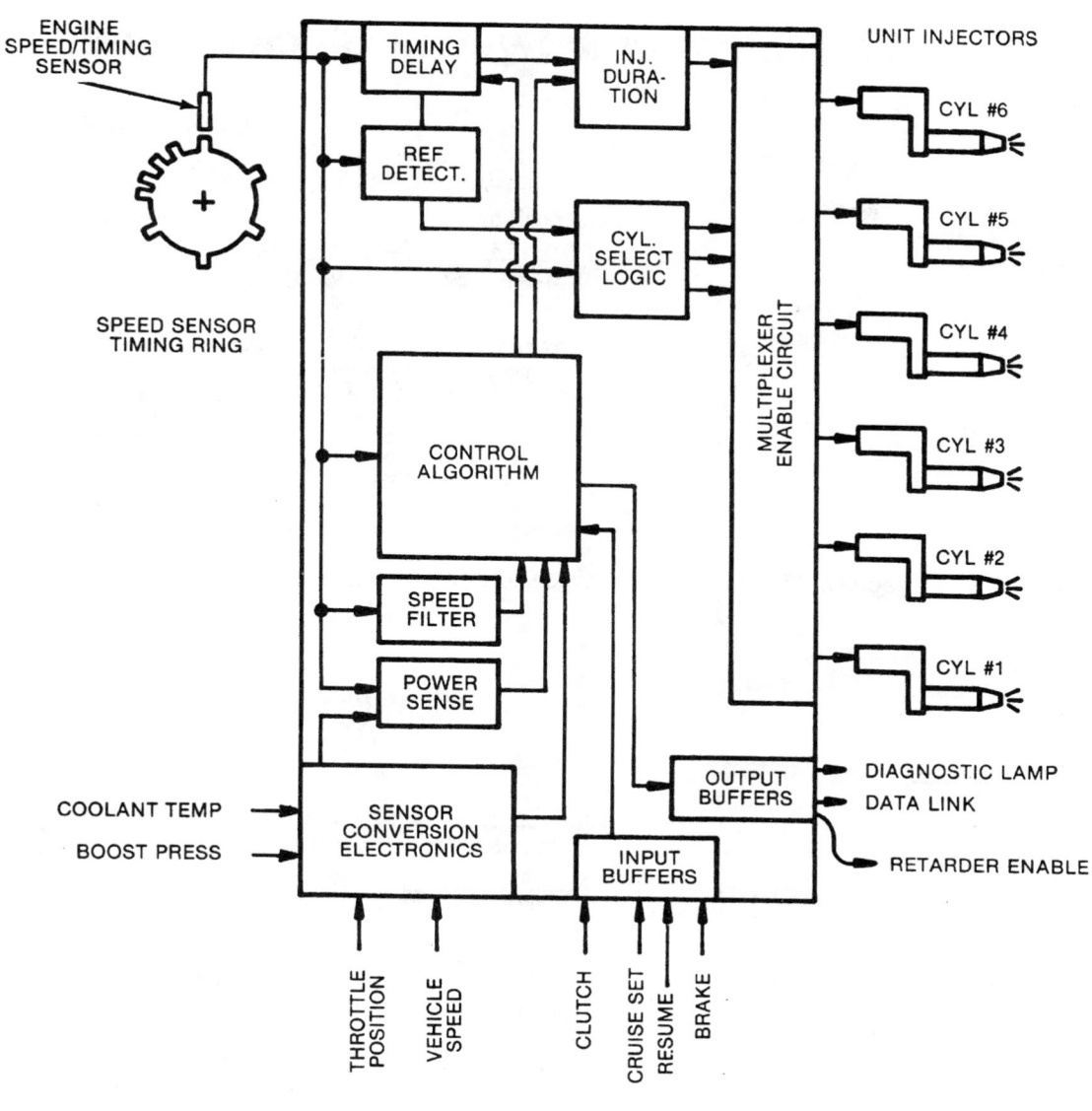

Fuel System Electronic Control Circuit

Figure 11-16 Caterpillar ECM processing mechanical drawing. (Courtesy of Caterpillar Inc.)

A comprehensive fuel map is written to the Caterpillar ECM. The ECM analyzes the data input from the command and monitoring sensors and, referencing the instructions programmed to its data retention media, plots a fueling profile. Start-up fueling strategy, failure modes, and all the system default values are programmed into the ECM software. The detailed fuel maps programmed into it govern the engine. Some Caterpillar engines have software that enables the cycling of cylinders during certain conditions, such as warm idling, with the objective of saving fuel and minimizing engine wear.

The ECM is programmed to run diagnostic tests on input and output circuits and identifies specific component and circuit faults using blink codes, reader programmers, or PCs. Most troubleshooting can be performed using the appropriate EST and a DMM. Figure 11-17 shows the circuit block diagram.

1994 3406E Electronic System Block Diagram

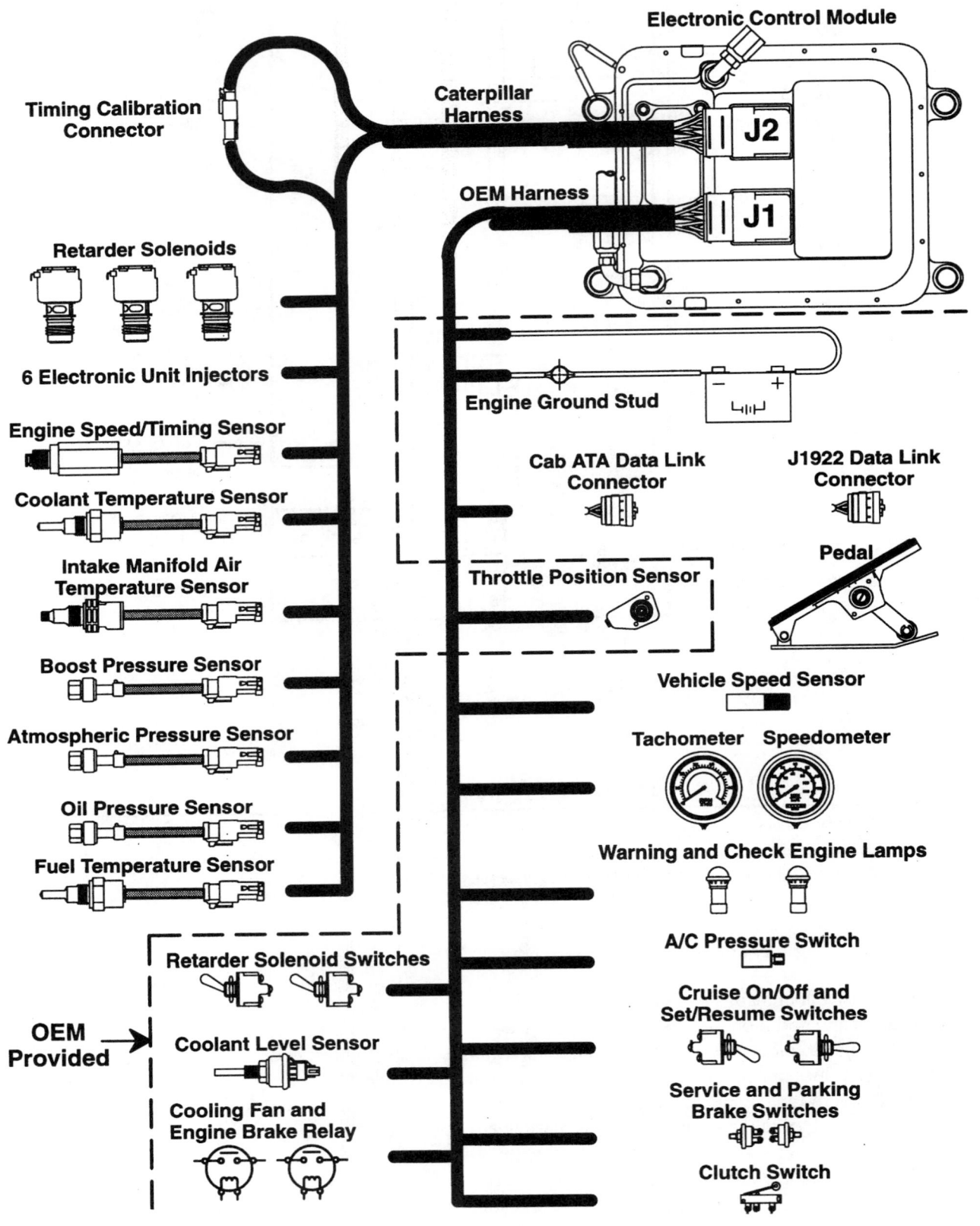

Figure 11-17 3406E electronic circuit block diagram. (Courtesy of Caterpillar Inc.)

342

Smart Options   Smart parameters or functions are those that will bend a hard programmed limit or value when it makes sense to do so. Current Caterpillar electronics feature some smart programming options. Some examples are given below.

Vehicle Speed Limit (VSL)   The hard programmed VSL can be optionally reprogrammed with soft limit VSL technology. Soft limit programming permits a cushion on either side of a hard limit value. Soft VSL modulates engine fueling when the truck is running over rolling terrain, permitting some latitude either side of the programmed vehicle speed to maximize fuel economy. Vehicle maximum speed can be programmed at a value under peak cruise speed. This feature encourages the driver to use the cruise control feature and eliminate some of the irrational variables encountered when a driver's foot is managing the accelerator pedal.

VSL can be programmed as a soft parameter. This means that instead of an abrupt fuel cutoff when the vehicle achieves the programmed VSL, the engine is fueled to keep the manifold boost up and gain some speed advantage on the upcoming hill. Soft VSL improves fuel economy and driver comfort.

Programmable Extended Droop   Governor droop can be programmed to extend up to 150 rpm above the top engine limit (TEL) or rated speed value. This reduces shifting frequency.

SoftCruise   SoftCruise is a smart cruise control option that "learns" running and terrain conditions, then modulates engine fueling to eliminate abrupt cutoffs in fueling and engine braking actuations that are characteristic of hard cruise programming. Soft cruise can manage the engine retarder if programmed to do so. Automatic brake activation reduces driver fatigue while also increasing the vehicle fuel economy.

Cooling Fan Control   The ECM constantly monitors the input from the temperature sensors monitoring the engine coolant, intake manifold, compression brake, and AC refrigerant system pressure to determine whether the cooling fan should be actuated. An ECM output switches and manages the engine compartment cooling fan, eliminating the requirement for such devices as fanstats. In addition, the ECM may engage the fan when the engine retarder high mode is selected to increase retarding effort.

PTO Ramp Rate   This can reduce the PTO work cycle time by enabling programming rotational speed from 50–1,000 rpm/sec. Engine speeds is therefore factored by PTO speed. The TPS is disabled when this option is in use to prevent PTO overspeed.

Gear-Down Protection   This programmable feature prevents engine overspeed conditions by limiting engine rpm in the higher gear ratios with the objective of ensuring that cruising speed can be attained only in the top gear.

Progressive Shifting   Progressive shifting can assist less experienced drivers of trucks by providing engine speed prompts during different vehicle speeds. Progressive shifting can make a significant difference to vehicle fuel economy.

Idle Shutdown Timer   In common with most other truck electronic management systems, Caterpillar offers an idle shutdown timer programmable from 1 minute to 24 hours. A driver-activated override feature can defeat the timer programming.

Trip Data Log   This standard feature permits monitoring and recording of the engine parameters by three methods:

  1. Lifetime totals
  2. Trip totals
  3. Instantaneous readings

An EST through the SAE/ATA (Society of Automotive Engineers/American Trucking Association) data link may access the data. The following parameters are monitored:

1. Engine running hours, ECM hours
2. Machine miles
3. Idle mode miles
4. PTO mode hours
5. Total fuel consumed
6. Idle fuel consumed
7. PTO fuel consumed
8. Average engine load factor

Output Circuit

Output circuit devices are those components used by a computer system to effect the results of processing into action. As the computer in question is an ECM managing the fueling of an engine, the primary output would concern the control of fuel delivery. Also categorized as outputs would be digital data displays in the truck dash and the downloading of any ECM retained data for analysis.

ECM Injector Drivers The injector drivers are integral with the ECM. They produce a spiked control signal to energize the EUIs to a voltage value of 100 V, obtained by using induction coils. EUI logic determines duty cycle, which is converted to effective EUI pumping stroke by the duration the solenoid is energized. This duration is measured in milliseconds.

Electronic Unit Injectors (EUI) The EUI (Figure 11-18A) used by Caterpillar on these engines was originally designed and built by Lucas/CAV and consequently could be found on other engines. However, the currently used EUIs (Figure 11-18B) on 3406E were designed specifically for the engine and are built in Caterpillar's Pontiac, Illinois, fuel systems plant where HEUI technology was engineered and is manufactured. An injector train consisting of rocker, push rod, and roller tappet and cam profile mechanically actuates Caterpillar EUIs. The ECM electronically controls the EUI; the injector driver mechanism (switching) is integral with the ECM.

EUI Operation The EUI is fitted to a cylindrical bore in the engine cylinder head. Figure 11-19 details EUI operation. Low-pressure fuel from the fuel supply manifold, via drilled passages in the cylinder head, enters the EUI through the fill port and charges the pumping chamber in the barrel of the pump element. The plunger and barrel form the EUI pump chamber. When the injector train actuating cam is ramped off inner base circle toward the cam nose, the EUI plunger is driven downward and acts on whatever amount of fuel has been metered into the pump chamber. As the plunger descends, its leading edge first closes off the fill passage and subsequently displaces fuel in the EUI pump chamber through the internal passages leading to the spill port. Effective stroke can occur only when the EUI solenoid is energized. This action moves the EUI control valve to close off the passage leading to the spill port, trapping fuel in the EUI pumping chamber and internal ducting. Located below the EUI pump chamber and connected by a passage is a hydraulic, multi-orifice injector nozzle. Pressure rise, caused by trapping fuel in the EUI pump chamber, acts on the nozzle valve pressure chamber; when the required NOP is attained, it unseats, beginning the injection pulse. After a momentary dip (at the instant the nozzle valve opens), the pressure will continue to rise after NOP, and peak injection pressures depend on the length of the duty cycle or pulse width. The effective stroke is ended when the ECM switches the EUI plunger open and pressure within the EUI is permitted to collapse.

EUI NOP values are typically around 5500 psi (37,000 kPa; 375 atm). By factoring the nozzle valve pressure differential ratio, closing pressure values are typically 3,700 psi (25,500 kPa; 252 atm). When the EUI's actuating cam has attained peak lift, the injector train is unloaded, and the

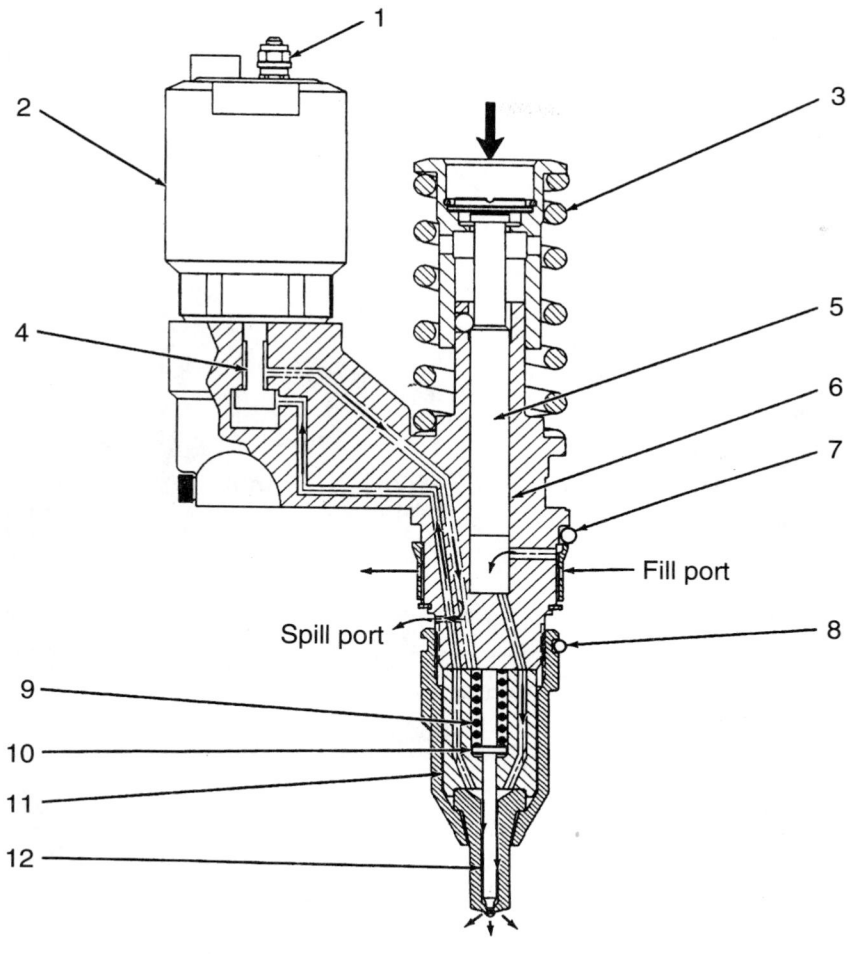

Electronically Controlled Unit Injector
(1) Solenoid connection (to the multiplex enable circuit). (2) Solenoid valve assembly.
(3) Spring. (4) Valve (shown in the closed position).
(5) Plunger. (6) Barrel. (7) Seal. (8) Seal. (9) Spring. (10) Spacer. (11) Body. (12) Check.

A

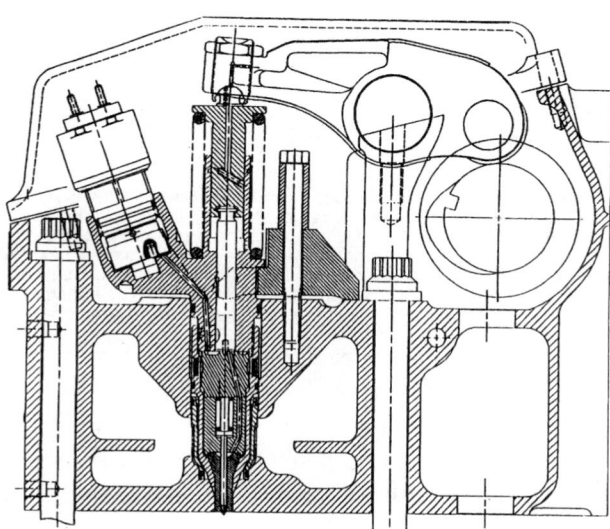

B

Figure 11-18 (A) sectional view of Lucas CAV EUI; (B) Cutaway showing 3406E EUI. (Courtesy of Caterpillar Inc.)

Operation of Electronic Unit Injector

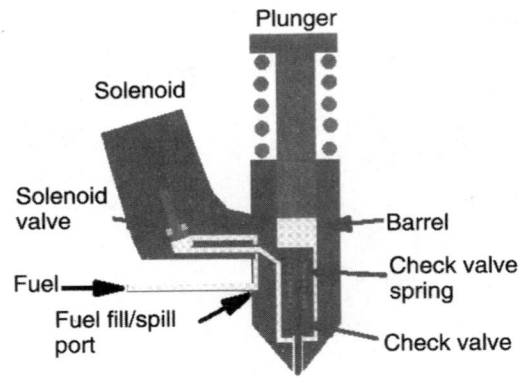

Without pressure applied to the plunger from the cam, a spring keeps the plunger retracted. Propelled by the new low pressure fuel transfer pump, fuel flows into the injector through the fill/spill port. From there it flows past the solenoid valve, down through the internal injector passages to the spring loaded check valve at the injector's tip and back up into the barrel. The pressure from the transfer pump is too low to unseat the spring loaded check valve at the injector's tip.

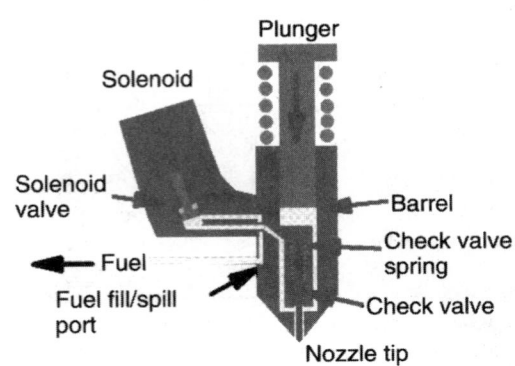

As the cam rotates, it starts to drive the plunger downward. Injection of the fuel may occur at any time after the plunger starts its downward travel. Until the ECM signals the start of injection, the displaced fuel is forced back out through the solenoid valve to the fill/spill port.

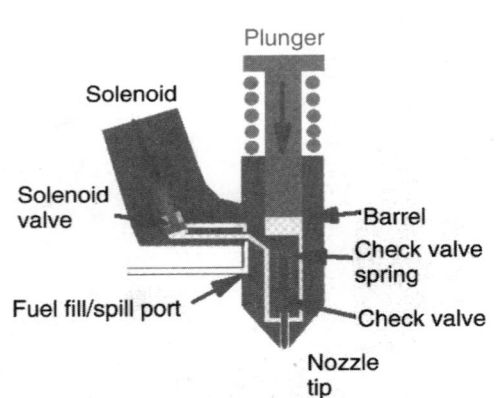

When the ECM signals the start of injection, the solenoid pulls the fuel valve closed, blocking the fuel's path to the fill/spill port. With this valve closed, pressure elevates at the injector tip to the 37 931 kPa (5500 psi) needed to unseat the spring loaded check valve. Once this valve is overcome, fuel is injected into the cylinder.

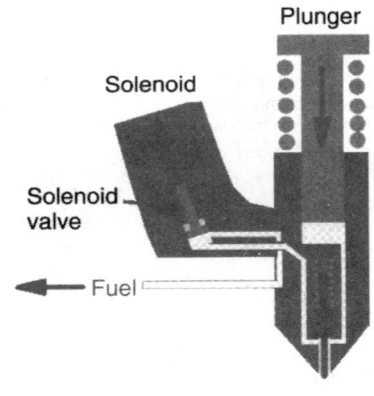

Fuel will continue to be injected until the ECM signals the solenoid to open the valve, allowing fuel to exit through the open valve and out the fill/spill port. The pressure at the injector tip immediately drops and the check valve snaps shut ending the injection cycle. The plunger will continue on its downward path however, displacing fuel through the open valve to the fuel manifold and back to the tank. This flow of fuel helps to cool the injector.

Figure 11-19 Operation of a Caterpillar ADEMS EUI sequence. (Courtesy of Caterpillar Inc.)

tappet ramps down the cam flank toward cam base circle. Simultaneously, the EUI spring lifts the plunger within the barrel, returning it to its retracted position and thereby exposing the fill port. Charging pressure fuel is then permitted to circulate throughout the EUI passages for purposes of cooling the assembly, exiting the spill port until the effective cycle is repeated. Caterpillar states that peak injection pressures in C-10 and C-12 engines reach values of 25,500 psi (1735 atm; 176 MPa).

CAUTION: Peak injection pressures may attain values of 28,000 psi (1905 atm; 193 MPa) in 3406E engines. Technicians need to be very cautious around pressures this high, or bodily injury may result.

Cummins Celect and Celect Plus Fuel System

Shop Manual Chapter 11, page 401.

Cummins introduced CELECT (Cummins electronic engine control), their full authority electronic management system, in 1991 on L10 and N14 engines, and later on their M11 engine when it was introduced in 1994 (Figure 11-20). CELECT Plus electronics were released into the marketplace in 1996 and are used to manage current M11 Plus and N14 Plus engines.

It should be noted that L10 (10 liter) and N14 (14 liter) engines could have PT, CELECT, or CELECT Plus fuel management, but each engines incorporates different hardware. CELECT Plus represents an enhancement of the CELECT engine management system by providing more comprehensive engine monitoring, faster processing speed, greater memory retention, and greater programmability. The CELECT Plus system will be used as the basis for this description of Cummins electronic management systems.

System Overview

CELECT Plus is a full authority, computerized engine management system that uses cam-actuated, electronically controlled CELECT injectors, which can be likened to the EUIs on other full authority systems. In fact, CELECT Plus has much in common with its competitor's full authority management systems. To properly understand this section, the reader should have a

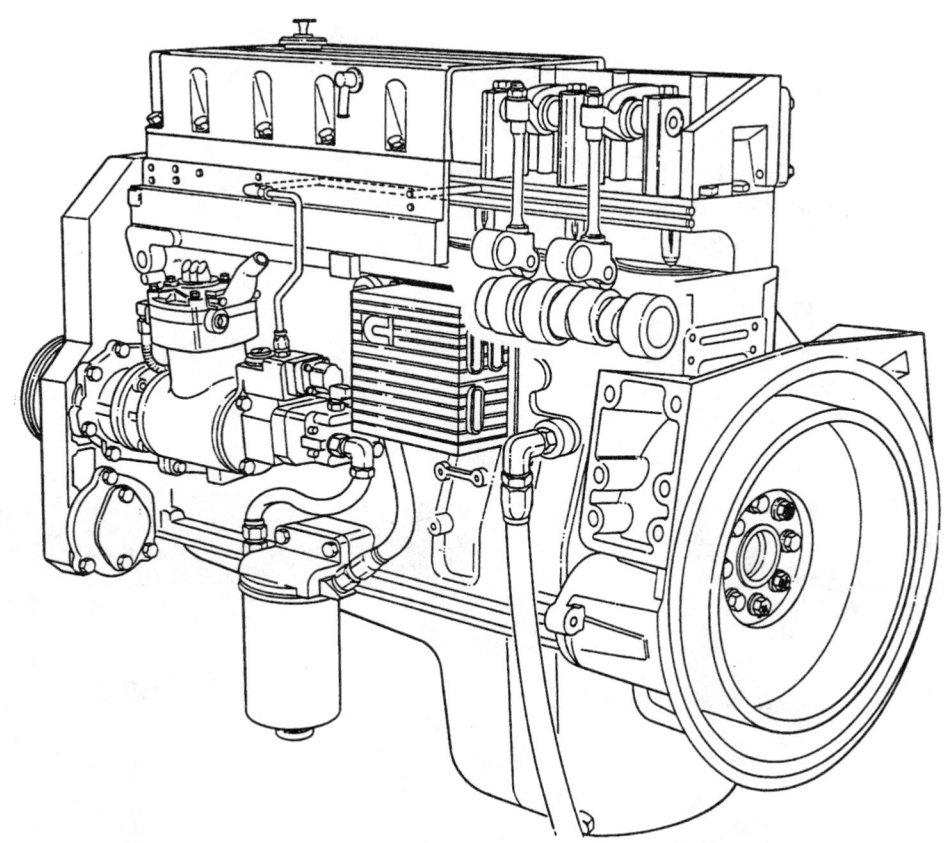

Figure 11-20 Location of key components on a Cummins CELECT fuel system. (Courtesy of Cummins Engine Company)

thorough understanding of hydromechanical fuel systems and basic vehicle computer operation covered elsewhere in this text.

The CELECT Plus system can be divided as follows:

1. Hydromechanical subsystem
2. Input circuit—sensors
3. ECM
4. Output Circuit—CELECT injectors, C-brake, etc.

Fuel Subsystem

A gear pump provides fuel movement in the fuel subsystem that is driven by an accessory drive that rotates both the air compressor and gear pump (Figure 11-21) at engine speed. The gear pump is flange mounted behind the compressor. It is fitted with a pressure regulator responsible for maintaining a 150 psi (10.2 atm; 1.03 MPa) charge pressure to the CELECT injectors. An electrical shutdown valve is positioned on top of the fuel pump at the fuel outlet. It is energized to run. The fuel pump is also fitted with a small filter consisting of a mesh and magnet to preclude a cutting from the gear pump entering the charge pipe.

Fuel is drawn from the fuel tank by the gear pump. It is pulled through the cooling plate (heat sink) on which the CELECT electronic control module is mounted. It circulates through this heat exchanger, helping to dissipate heat from the ECM. From the ECM cooling plate, fuel passes through the fuel filter, which has a high capacity 10 μ filtering medium. Therefore, both the ECM cooling plate and the filter are under suction. From the filter head, the fuel is piped to the gear pump, which pressurizes the fuel at a constant charge pressure of 150 psi (10.2 atm; 1.03 MPa). The fuel is then discharged to tubing that delivers it to the cylinder head charging galleries and makes it available to the CELECT injectors. After circulating through the CELECT injectors, fuel exits by means of a fuel drain and enters the cylinder head return gallery from which it is piped back to the fuel tank.

Input Circuit

Figure 11-22 details the CELECT Plus processing cycle. In the CELECT Plus electrical subsystem, components are connected to the ECM by a single wiring harness, integrating the separate sensor and actuator harnesses of the CELECT system. CELECT Plus sensors are defined by type; the electrical principles on which they are based are explained elsewhere in this text.

Engine Monitoring Sensors

1. Ambient pressure sensor (Piezo electric type) for altitude fuel deration
   A. M11, located in front of ECM
   B. N14, located behind ECM

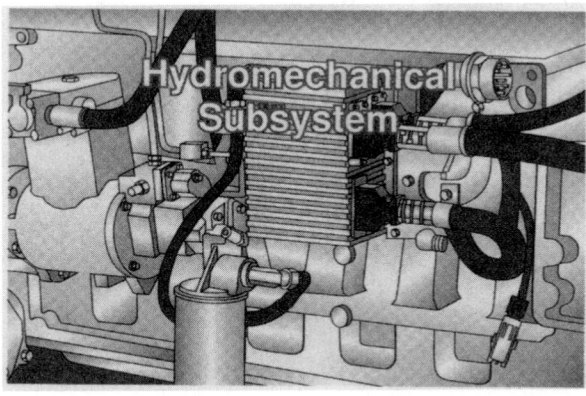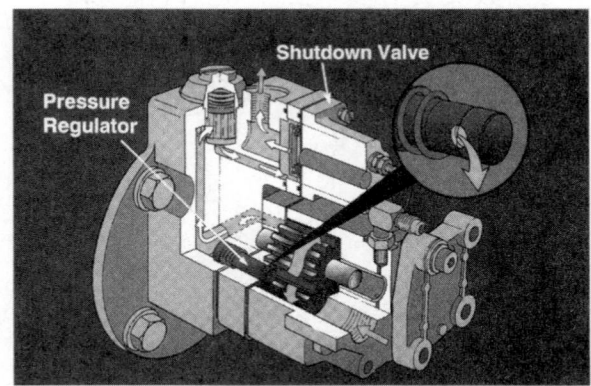

Figure 11-21 Hydromechanical fuel system of the CELECT system. (Courtesy of Cummins Engine Company)

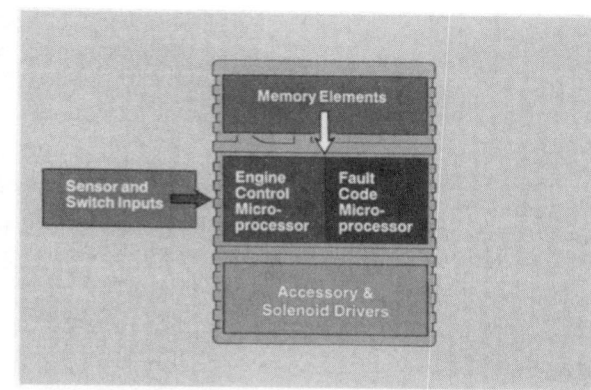
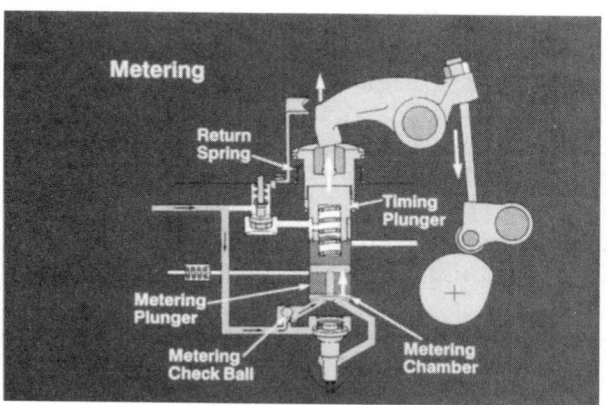
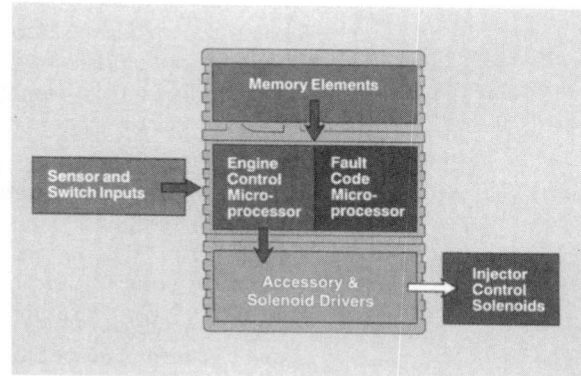
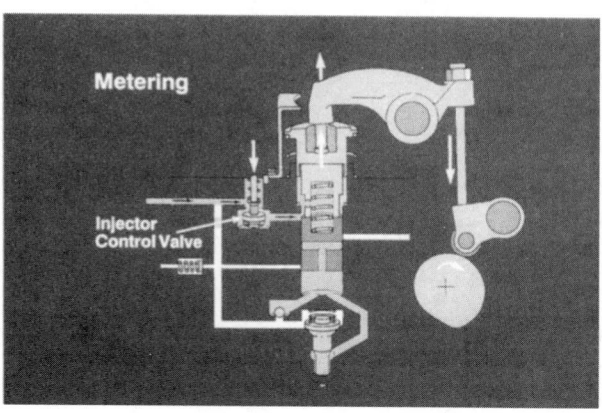
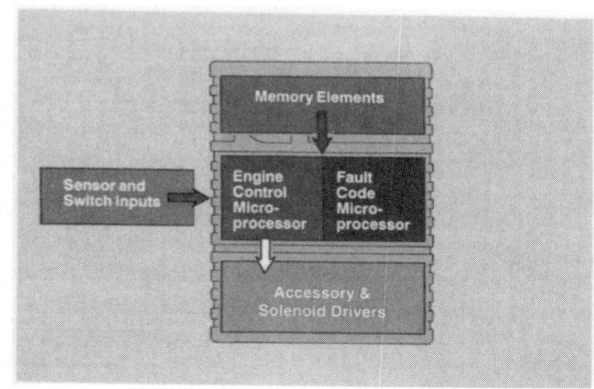
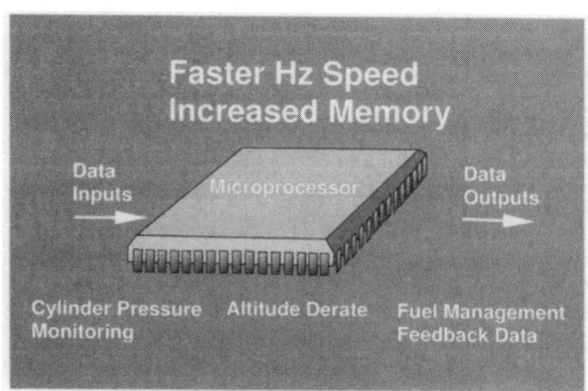
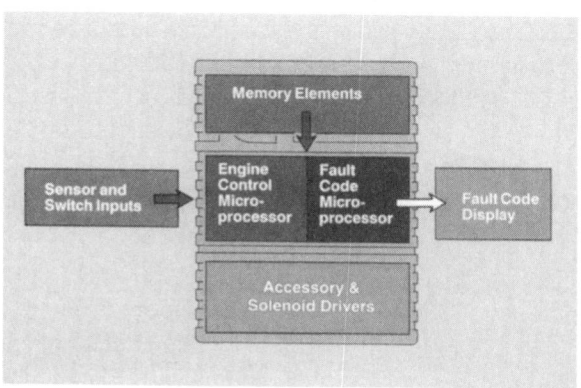

Figure 11-22 CELECT-PLUS processing cycle graphics. (Courtesy of Cummins Engine Company)

2. Oil pressure sensor (Piezo electric type), located in the main oil rifle in the block.

3. Engine position sensor (a magnetic pulse generator type sensor). The cam gear located pulse wheel has 24 evenly spaced recesses to feed back rotational speed data and a small dowel at TDC for engine position data.

4. Oil temperature sensor (thermistor), located in the main oil rifle in the block.

5. Intake manifold temperature sensor (thermistor), located centrally on the intake manifold.

6. Boost pressure sensor (Piezo electric type), located in the intake manifold.

7. Coolant temperature sensor (thermistor), located in the thermostat housing.

8. Coolant level sensor (a switch type sensor) that grounds through the engine coolant with dual probes to indicate high or low coolant levels OEM supplied.

Command Sensors and Switches

1. Throttle pedal assembly: OEM supplied accelerator pedal, which incorporates a 5 V throttle position sensor (potentiometer) and an idle validation switch (open-close circuit). In common with their competitor product, CELECT Plus is a drive-by-wire system with no mechanical backup. An accelerator interlock is optional.

2. Jacobs C-brake: A Cummins designed, Jacobs manufactured, engine compression brake managed by the CELECT ECM. The current version will permit engine braking down to 900 rpm (previously 1,500 rpm) and can be automatically actuated by the ECM in cruise control mode.

3. Cruise control: Managed by the CELECT Plus ECM with some advanced features:
   A. Governor tailoring (see section on governor tailoring)
   B. Cruise auto resume after shift
   C. OEM cruise switch configuring
   D. Automatic engine braking in cruise mode (The minimum cruise control speed limits the top vehicle speed while in cruise control mode and this cannot exceed the maximum vehicle speed setting.)

4. PTO control: Controls engine rpm at a constant speed selected by the operator (Either a cab or remote switch can be used, or PTO mode can be selected up to 6-mph [10-kph] vehicle speed.)

Electronic Control Module

The ECM receives and processes information from the various engine and chassis sensors and switches, plots fueling and timing commands, and outputs signals to the injector driver circuitry (switching apparatus). A more detailed accounting of ECM functions is covered elsewhere in this text. The CELECT Plus injector driver unit uses coils to operate the EUI solenoid voltage at 78 V and is the primary output of the ECM.

Consistent with other manufacturers, the CELECT Plus ECM performs comprehensive self-diagnostics, which can be read by *flash codes* or ESTs and can be programmed to adopt a variety of failure mode strategies. Cummins claim that CELECT Plus offers significant advantages over CELECT, including:

Step-up transformer type coils are used to increase the Cummins CELECT EUI solenoid voltage to 78 V.

1. Faster processing speed
2. Altitude fuel deration
3. Cylinder pressure monitoring
4. Faster, more accurate throttle response
5. Low idle clutch engagement
6. Multilevel programming security
7. Engine warm-up protection
8. Accelerator interlock (i.e., bus door, vehicle brake, or PTO switched)

Output Circuit

The primary ECM output function is control of the CELECT injectors and this operation will be described in detail. However, the Jacobs C-brake, PTO apparatus, and all ECM readouts can also be described as being part of the output circuit. The CELECT ECM drives output circuit functions through the actuator harness. A brief description of C-brake and PTO operation is given in the section on the input circuit.

Cummins states that the primary diagnostic and programming tool for use with CELECT Plus systems is the PC with MS Windows–driven, INSITE software and the appropriate serial links. Echeck and Compulink (proprietary ESTs) software packages will also be made available.

CELECT and CELECT Plus Injectors

The CELECT injector (Figure 11-23) is substantially different from the Cummins PT injector; probably the only real resemblance is that they both are cam actuated injection devices. CELECT and CELECT Plus injectors use similar operating principles, but the CELECT Plus injector will be referenced in the following description. CELECT injectors are cam-actuated, electronically controlled injector assemblies, which use multi-orifice, hydraulic injector nozzles and therefore have much in common with the EUIs used by their competitors. However CELECT injectors differ in metering principles.

CELECT injectors are fitted to cylindrical bores in the engine cylinder head. They circulate fuel at any time the engine is running and the gear supply pump is rotating. Fuel enters and

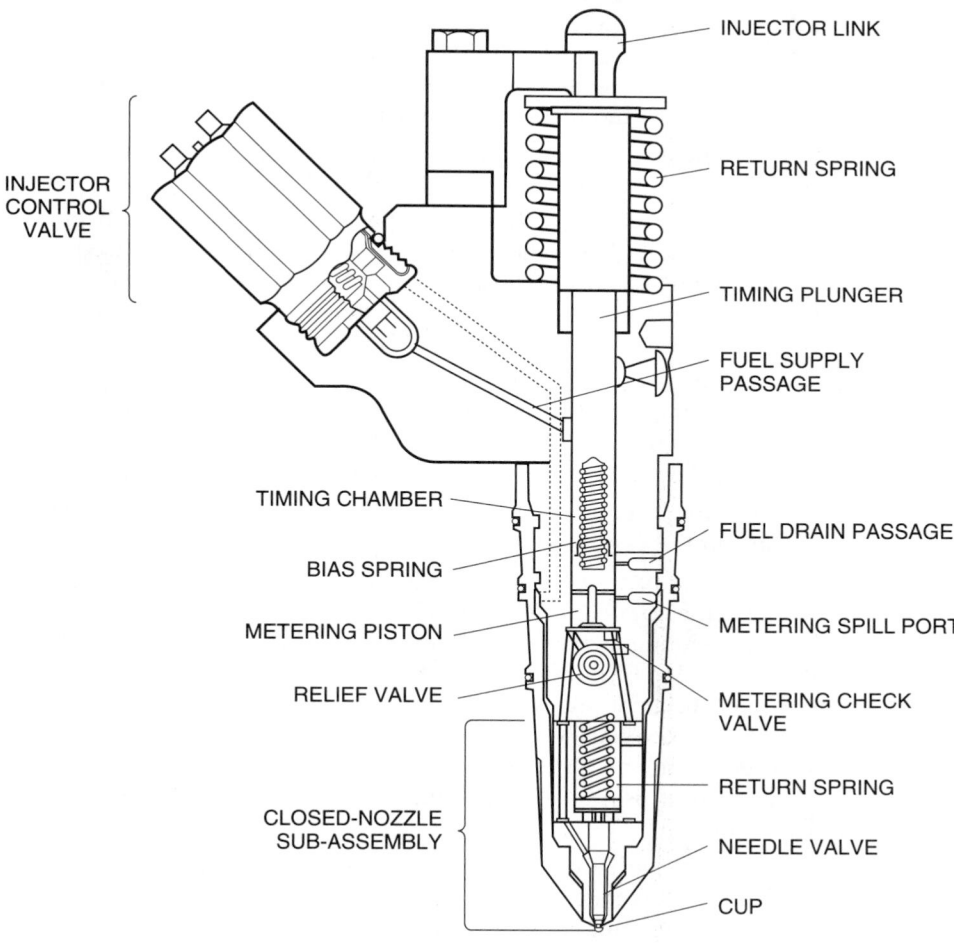

Figure 11-23 Sectional view of a CELECT injector. (Courtesy of Cummins Engine Company)

exits the injector by means of annular recesses sealed by rubber O-rings. The lower annular recess is exposed to the supply gallery in the cylinder head and receives fuel at charging pressure (150 psi; 10 atm; 1 MPa). Fuel passes from the lower annular recess to the internal injector circuitry, through the metering spill port, and is circulated through the injector and solenoid control valve ducting. It exits through a fuel drain in the upper external annulus. To understand the operation of this injector assembly it is necessary to follow the sequence in Figure 11-24, which matches the text below as numbered:

1. At the start of metering, both the metering plunger and the timing plunger are at their lower travel limit as their actuating cam is at peak lift. The injector control valve closes, actuated by a 78 V induction-coil–derived spike delivered from the CELECT injector driver unit.

2. As the cycle continues, the cam ramps off the nose toward base circle, unloading the injector train and permitting the timing plunger return spring to lift the timing plunger. This enables fuel to flow past the metering checkball into the metering chamber. This flow continues as long as the timing plunger is moving upward and the injector control valve is closed. Supply pressure acting on the bottom of the metering piston forces it to maintain contact with the timing plunger.

3. The ECM will determine the end of metering by switching the injector control valve to its open position. This action causes the metering checkball to seat, and permits fuel to pass around the injector control valve.

4. Fuel at supply pressure then flows into the timing chamber, stopping metering piston travel. The bias spring ensures that the metering plunger remains stationary, preventing it from drifting upwards as the timing plunger moves upward. This same force against the metering plunger results in sufficient fuel pressure below the metering piston to keep the metering checkball seated. The result is a precisely metered quantity of fuel in the metering chamber.

5. As the cycle continues, the injector train continues to ride toward cam inner base circle, permitting the timing chamber to fill with fuel.

6. Next, the injector cam passes over the inner base circle location and begins ramping toward outer base circle. This action loads the injector train and, consequently, the timing plunger begins its downstroke. Initially, the injector control valve remains open, allowing fuel to spill from the timing chamber reverse flowing it through the fuel supply passage.

7. The delivery sequence begins when the ECM switches the injector control valve to its closed position, trapping fuel in the timing chamber. This trapped fuel acts as a hydraulic link between the timing plunger and the metering plunger; this forces the metering plunger downward with the timing plunger.

8. Because the metering plunger is being driven downward hydraulically, rapid pressure rise begins in the metering chamber.

9. Ducting connects the metering chamber with the pressure chamber of the hydraulic, multi-orifice injector nozzle located at the base of the CELECT injector. When the pressure in the metering chamber (and therefore in the nozzle pressure chamber) reaches the NOP value (approximately 5,000 psi [340 atm; 34.442 MPa]) the nozzle valve opens and injection begins. Fuel is forced through the nozzle orifices and atomized directly into the engine cylinder. The minute sizing of the nozzle orifices means that they are unable to relieve the pressure as fast as it is created, and peak pressure is capable of rising well above the NOP value, depending on the length of the effective stroke.

10. Injection continues until the metering plunger passes the spill passage. This action causes a collapse of metering chamber pressure and permits abrupt nozzle valve closure. At this moment, the pressure relief valve will relieve, minimizing the effect of the high-pressure spike that occurs at metering spill. The relief valve passage connects to the fuel drain line.

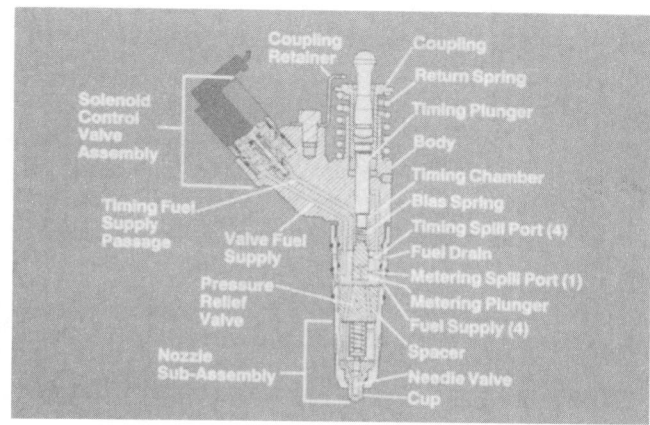

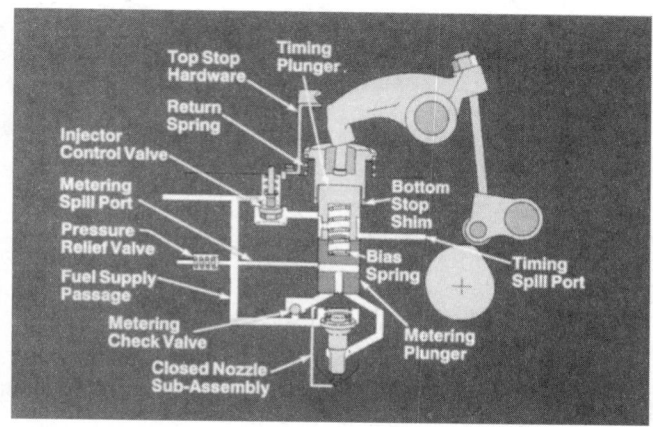

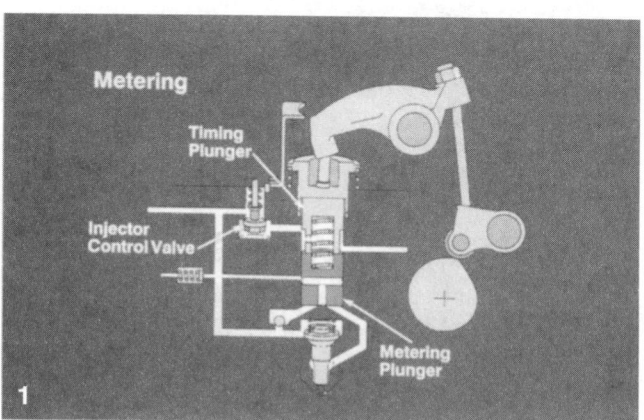

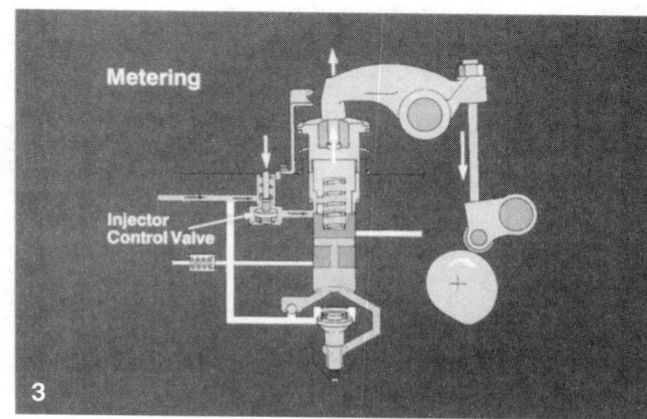

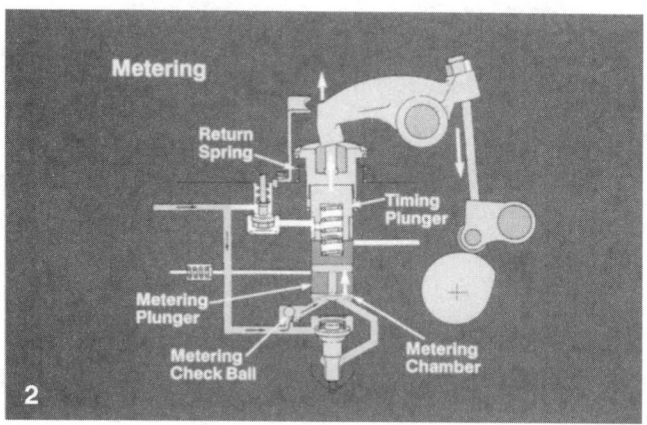

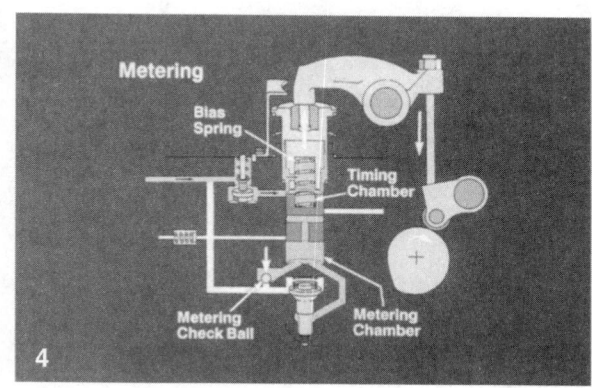

Figure 11-24 CELECT-PLUS injector operation. (Courtesy of Cummins Engine Company)

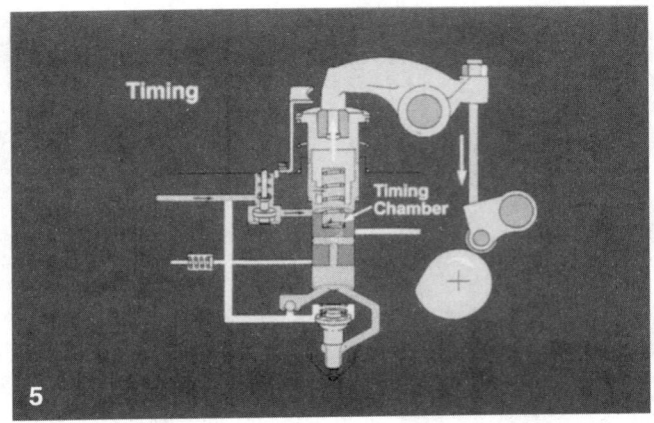

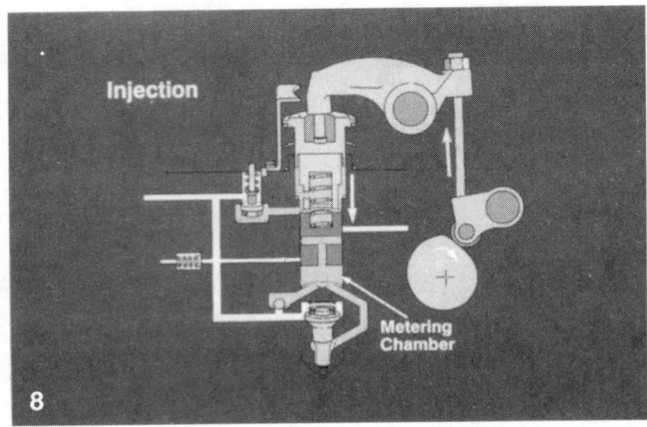

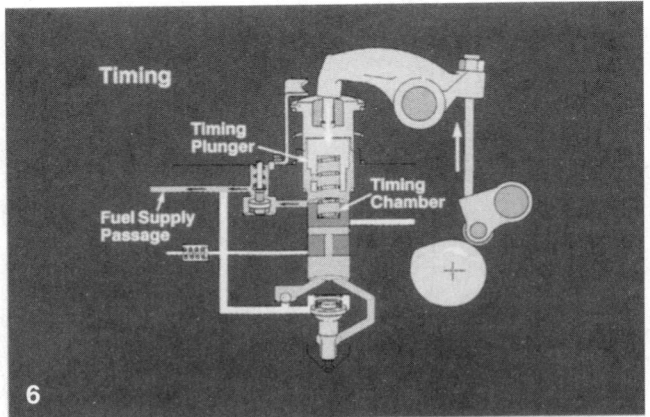

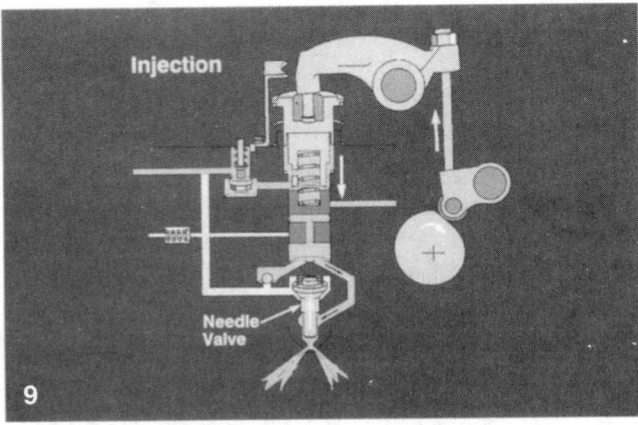

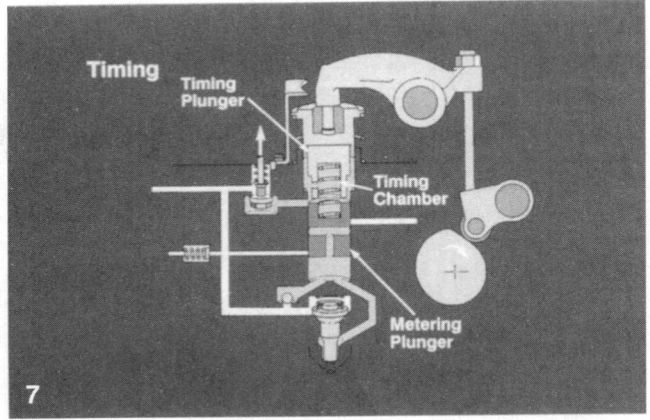

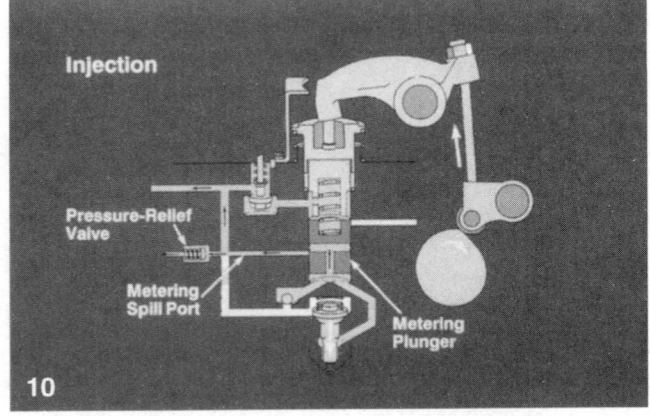

Figure 11-24 Continued.

11. Immediately after the metering spill port is exposed by the downward travel of the metering plunger, its upper edge exposes the timing spill port.

12. This action permits the fuel in the timing chamber that was used as the hydraulic medium to spill to the fuel drain. This completes the cycle.

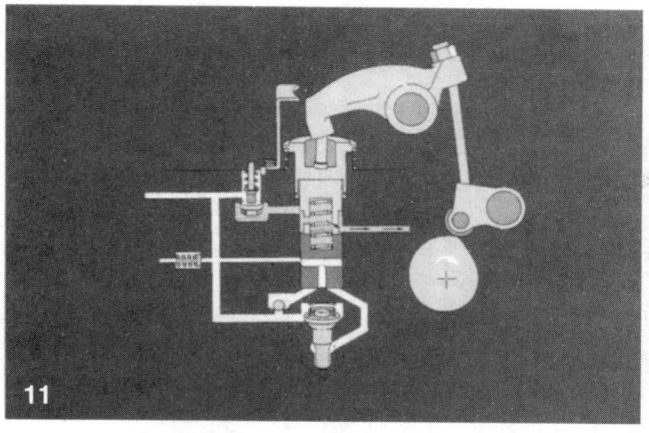

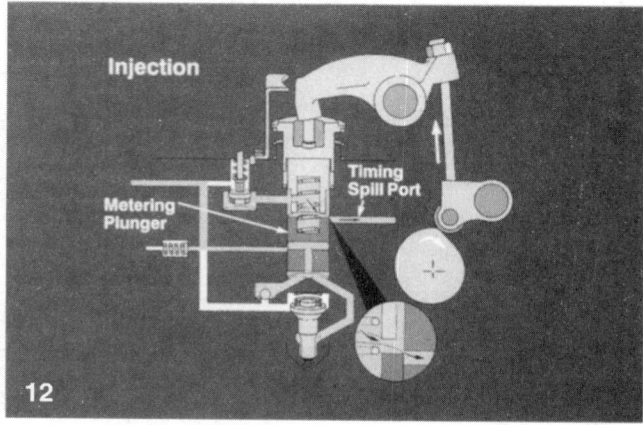

Figure 11-24 Continued.

The ECM controls the CELECT injector by varying the time the injector control valve remains open and closed. CELECT and CELECT Plus injectors, while sharing operating principles, are *not* interchangeable, the latter incorporating increased orifice sizing, modified spray angles, and other performance enhancements.

CELECT Plus Engine Protection System

The CELECT Plus monitors critical engine sensors and will log fault codes when an out-of-normal condition is detected. If an out-of-range condition is detected in a critical sensor an engine derate action may be initiated. Flashing of the appropriate dash light will alert the operator. The engine protection system monitors the following:

1. Coolant temperature
2. Oil temperature
3. Intake manifold temperature
4. Oil pressure
5. Coolant level
6. Engine overspeed

At engine derate, the operator is alerted by the flashing light, and engine power and speed deration will occur incrementally depending on the severity of the condition. The engine will not actually shut down unless the ECM is programmed with the engine shutdown option. The engine protection shutdown option will automatically shut down the engine when a fluid pressure or temperature out-of-normal range problem occurs that could lead to catastrophic damage. Should this occur, the dash engine protection lamp would flash for 30 seconds prior to the shutdown. Key off is required to restart the engine but if the shutdown trigger persists, the engine will be shut down again after 30 seconds.

Bosch EUP on DDC Series 55 and Mack Trucks V-MAC III-E-Tech Engines

Shop Manual Chapter 11, page 403.

Electronic unit pump (EUP) technology (Figure 11-25) has been used for a number of years in heavy equipment applications. Bosch EUP systems divide the functions of a standard EUI assembly by locating the control solenoid and cam-actuated (jerk) pump in the EUP and mounting the injector nozzle in the cylinder head, connecting them with a high-pressure pipe. The DDC series 55 engines controlled by DDEC III (introduced in 1996) and Mack Trucks V-MAC III on E-7 engines use this system. DDC made the series 55 engine available as an exclusive to Freightliner

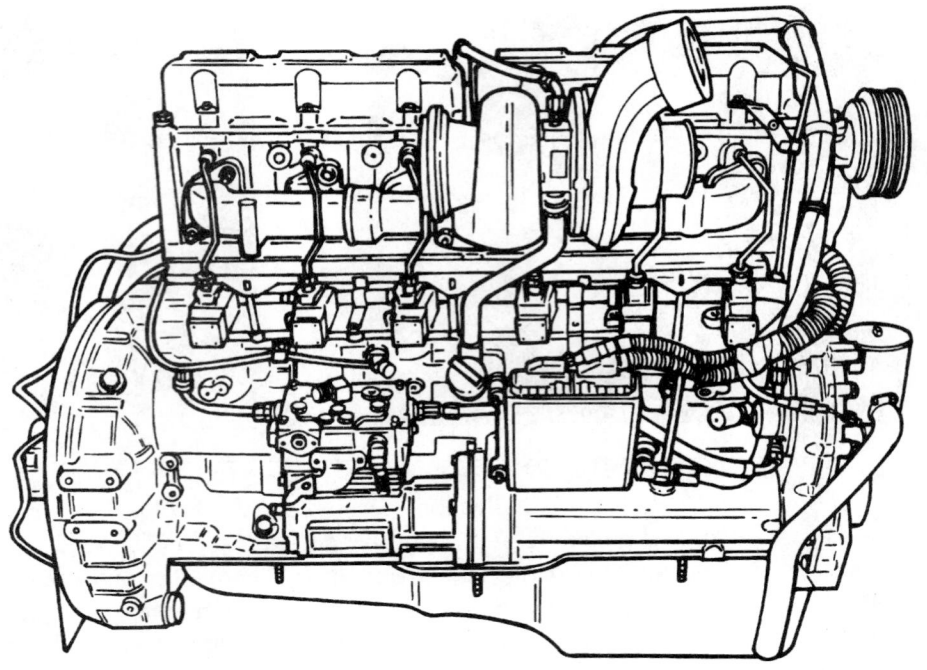

Figure 11-25 Side view of an E-Tech engine showing location of ECM and EUPs. (Courtesy of Mack Trucks, Inc.)

Bosch EUP systems combine the actions of a EUI with the old jerk pumps used on larger displacement marine engines.

Corporation on its introduction. This engine is based on a Mercedes Benz European engine platform. DDC ceased production of the engine after a little over a year. Industry talk has it that the engine will be repackaged and made available in the near future to be marketed possibly under the Mercedes Benz name with management electronics other than DDEC. Mack Trucks replaced their E-7 line of engines, equipped with V-MAC II management electronics, with a moderately re-engineered E-7 engine to accommodate the EUP fuel system managed with V-MAC III electronics. This engine was marketed as the Mack E-Tech.

A BIT OF HISTORY

Early diesel engines were large, low-speed engines used in marine and stationary applications. Stuart used an air blast from a pressurized tank to provide fuel injection. The lack of compactness and problems in flexibility and control associated with this type of fuel injection delayed the application of the compression-ignition engine to the automotive and truck field. Many early engines engaged solid fuel injection using a jerk pump. The engine manufacturers developed these units purely for developmental reasons.

System Overview

Any diesel engine EUI can be divided into three subsystems:

    1. Solenoid control

    2. Cam-actuated pumping

    3. Hydraulic injector nozzle

Bosch EUP technology incorporates the two EUI functions in a single unit. It is flange mounted directly over the camshaft that actuates the pumping stroke, and the electronic unit pump manages fueling for one engine cylinder. The EUP is connected by means of a high-pressure pipe to a hydraulic injector nozzle, much the same as would be found on any diesel engine with a port-helix metering injection pump. The injector nozzle is of the multi-orifice type. The EUP on DDC series 55 engines uses DDEC III management adapted to EUP. Mack Trucks has elected to use EUP partly because the system allows them to make a minimum of design changes to their successful E-7 engine while meeting stringent EPA emissions standards. It should be noted that V-MAC III on E-7-EUP/E-Tech could not be retrofit to E-7 engines managed by V-MAC II because there are notable hardware differences.

Fuel Subsystem

The EUO fuel subsystem that supplies fuel is fairly standard and is divided into suction and charge circuits (Figure 11-26). In the Mack Trucks E-Tech engine, a gear-type pump (driven off the accessory drive that the Econovance was coupled to in the E-7 V-MAC II engine) replaces the plunger-type transfer pump. The gear-type pump is required to produce the charging pressures of 100 psi (6.8 atm; 690 kPa) and up to 100 gallons per hour (gpm). Fuel is drawn from the fuel tank by the transfer pump and pulled through a primary filter and the ECU cooling plate, within which it acts as the cooling medium for the heat generated by the microprocessor and switching units. Fuel exits the gear-type pump at the charging pressure and is next pumped through a

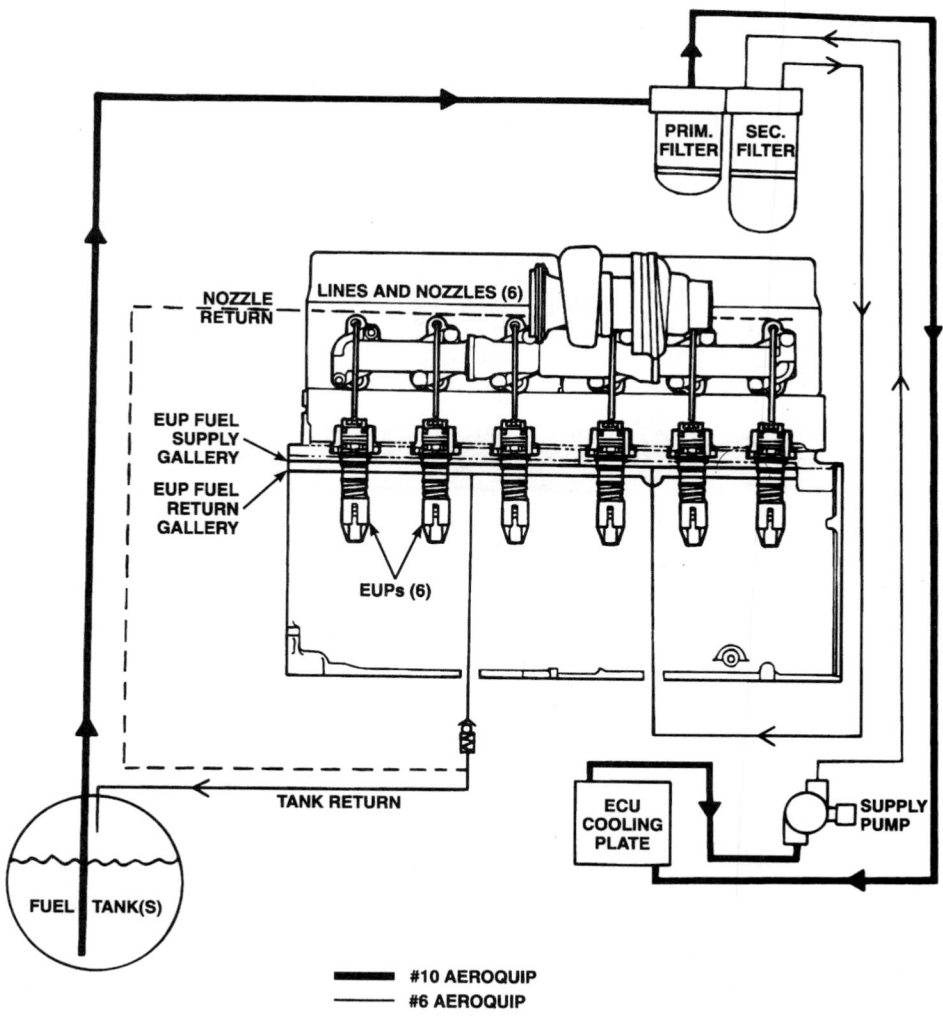

Figure 11-26 Schematic of EUP fuel system. (Courtesy of Mack Trucks, Inc.)

secondary filter. Mack Trucks uses different filters on their E-Tech engines, and they filter at higher efficiencies and have metric mounting threads. A pair of circumferential black bands identifies the E-Tech filters. Fuel exiting the secondary filter is then fed to the fuel supply gallery in the cylinder block, which runs the length of the cylinder block. This enables charging pressure fuel to be made available to the EUPs, which are located in cylindrical bores in the cylinder block. The EUPs are manufactured with exterior annular inlet and output recesses, separated, and sealed with O-rings. Charge fuel is pumped into the EUPs and circulated within (where some, of course, is used for fueling the engine) and then returned to the fuel tank by means of the cylinder block return gallery, which runs parallel with the supply gallery (Figure 11-27).

Input Circuit

Both DDC series 55 and Mack Trucks E-7 EUPs are full authority, electronic management systems with comprehensive engine/chassis monitoring and command circuits. The DDEC III electronic input circuit to the series 55 ECM is essentially the same as other DDEC systems. EUP uses the same two critical sensors for inputting engine position and rotational speed data in DDEC. They are the synchronous reference sensor (SRS) and the timing reference sensor (TRS). However, because V-MAC II was a partial authority system, there are some differences in the input circuit of V-MAC II compared with V-MAC III. In V-MAC, many of the critical system sensors were located in the PE8500 injection pump and integral rack actuator housing. Because this component is eliminated from V-MAC III, an engine position sensor is located in the front timing gear cover where the passage of holes in the front face of the camshaft gear is monitored. The function of the engine position sensor is similar to that of the timing event marker (TEM) on the E-7 RE30 rack actuator housing. The engine position sensor must be precisely installed using shims to ensure the correct gap. The V-MAC III engine speed sensor (RPM-TDC) is identical to the engine position sensor. It is located in the flywheel housing. The E-7-EUP flywheel has one more tooth than the E-7, two of which have half their width machined off. Again, correct installation of this sensor is critical, and the gap must be calculated and then set by shims. Additional sensors used with V-MAC III are a boost air temperature sensor and a fuel temperature sensor on chassis fitted with the optional Co-Pilot data display.

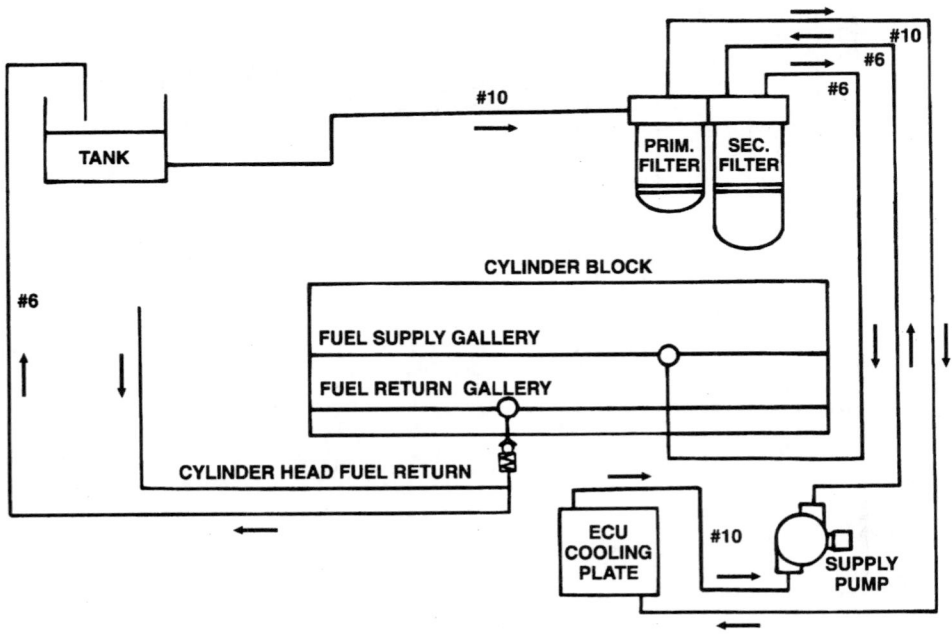

Figure 11-27 Fuel routing in the E-Tech fuel circuit. (Courtesy of Mack Trucks, Inc.)

Engine Controllers and Management Electronics

DDC series 55 engine is managed by DDEC III electronics, and the EUP drivers (switching units) are integral with the ECM housing. DDCs use the acronym ECM to describe their engine controller. The DDEC III ECM is located on the left side of the series 55 engine. It is cooled by the fuel subsystem that is routed through a heat exchanger on which the ECM in mounted.

Mack Trucks V-MAC III on E-7-EUP engines returns to a two-module system that separates vehicle control functions from engine control functions. Mack Trucks chooses to use the acronym ECU to describe their modules. The engine control module is located on the right side of the engine, which is mounted on a fuel-cooled heat exchanger. As the engine control module is located close to the engine exhaust manifold, a heat shield helps protect the device from exhaust heat. The engine controller is multiplexed to the vehicle control module, which is mounted under the dash in the cab. It should be stated that the engine controller module is perhaps better described as a fuel control module, because most of the engine control logic is computed in V-MAC and simply effected by the engine control module.

Both DDC series 55 and Mack Trucks E-7 EUPs may be multiplexed to other on-board, electronically managed systems. Multiplexing means the connecting of two or more electronic systems on board a vehicle to optimize performance and cut down on unnecessary hardware. For instance, both the engine controller and the transmission controller require a throttle input signal, and when the controllers are multiplexed, the TPS signal is shared.

> Multiplexing is the connecting of two or more electronic systems on a common wire using digital electronics.

Electronic Unit Pump (EUP)

The electronic unit pumps (Figure 11-28A) are flange mounted over a roller tappet that rides the center lobe of a set of cylinder cams (Figure 11-28B). There is a separate EUP for each engine

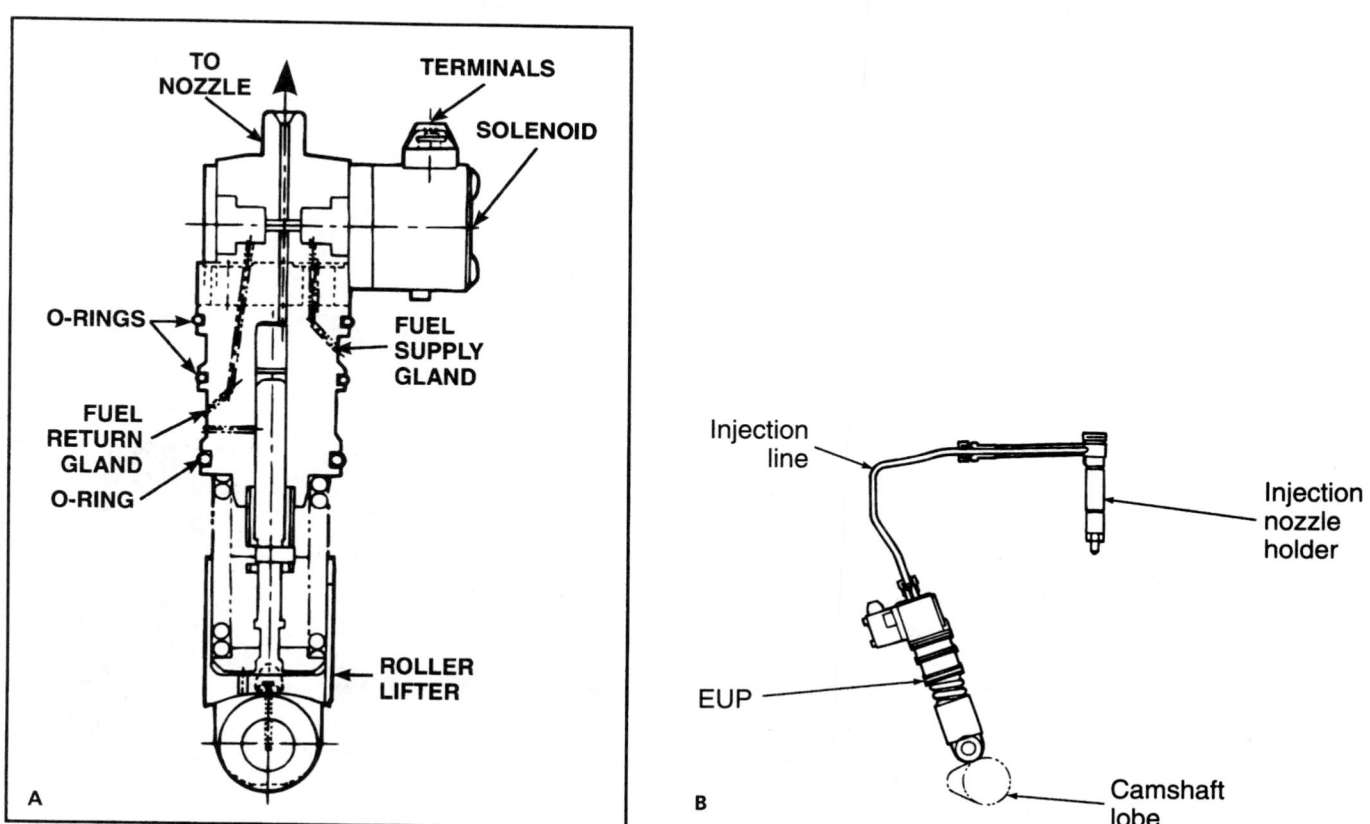

Figure 11-28 (A) Sectional view of EUP; (B) The actuation of the EUP with relationship to the engine camshaft and injector nozzle. (Courtesy of Mack Trucks, Inc.)

cylinder. Both the DDC series 55 and the Mack Trucks E-7-EUP are six-cylinder engines, so there are six separate EUPs located on the engine, each connected electrically to the ECM and hydraulically to an injector nozzle. Pumping action in the EUP is dictated by the actuating cam geometry. As the engine-driven cam rotates off its base circle ramping toward the nose, the roller lifter drives the EUP plunger into the pump chamber. Fuel is routed at charging pressure through the pump chamber at any time the engine is running. It enters through an upper exterior annulus, circulates through the EUP internal ducting, and exits through a lower exterior annulus. Control of the injection pulse is the responsibility of the EUP solenoid, which is switched by the engine-mounted V-MAC III module. When the EUP plunger is driven into the pump chamber by the actuating cam profile, fuel is merely displaced. However, the moment the EUP solenoid is energized, fuel is trapped in the pump chamber and pressure rise occurs. Connected to the EUP pump chamber by a high-pressure pipe is a multi-orifice, hydraulic injector nozzle. Actual injection to the engine cylinder begins when the preset NOP value is achieved.

The EUP plunger has a 10 mm diameter and potential for an 18 mm stroke. Actual plunger stroke is determined by the cam profile. The E-7 EUP operates at 1,770 atm (26,000 psi; 180 MPa) (Figure 11-29) and is supplied with low-pressure (charging pressure) fuel by a fuel supply gallery machined into the engine block. High-pressure fuel is routed from the EUP to the hydraulic injector nozzles in the cylinder head by a relatively short (compared to previous E-7 engines) high-pressure pipe. Each high-pressure pipe is of equal length and has an identical part number regardless of engine cylinder. Controlling effective stroke is the responsibility of the DDEC III ECM or the V-MAC III engine ECU. Operation of the EUP almost identically parallels that of any typical diesel engine EUI system with one exception.

The EUP is connected to the hydraulic injector nozzle (Figure 11-30) by means of a high-pressure pipe. The lag time between the beginning of energizing and beginning of injection is obviously extended in the EUP-managed engine when compared to the typical EUI system. The nozzles are set to a specified NOP value (around 340 atm; 5,000 psi; 34 MPa), and they are serviced as units in the same manner as any other hydraulic injector nozzles (as shown in Shop Manual Chapter 8).

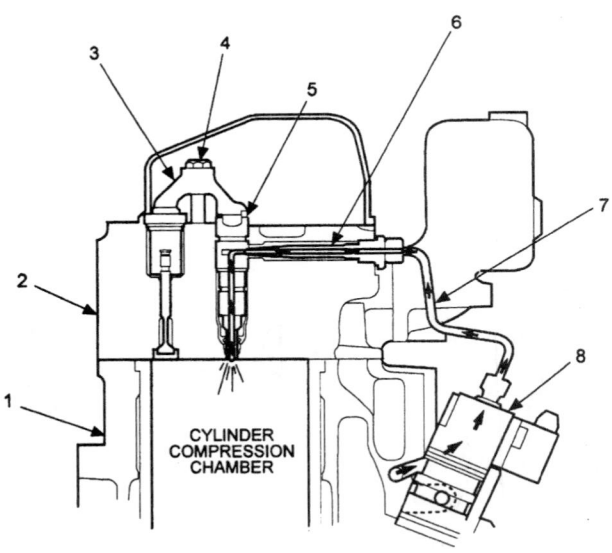

1. Cylinder Block
2. Cylinder Head
3. Nozzle Hold Down Crab
4. M10 Bolt
5. Injector Nozzle
6. Internal Injector Fuel Line
7. High Pressure Fuel Line
8. Electronic Unit Pump

Figure 11-29 Sectional view of DDC series 55 EUP showing the injector and EUP. (Courtesy of Mack Trucks, Inc.)

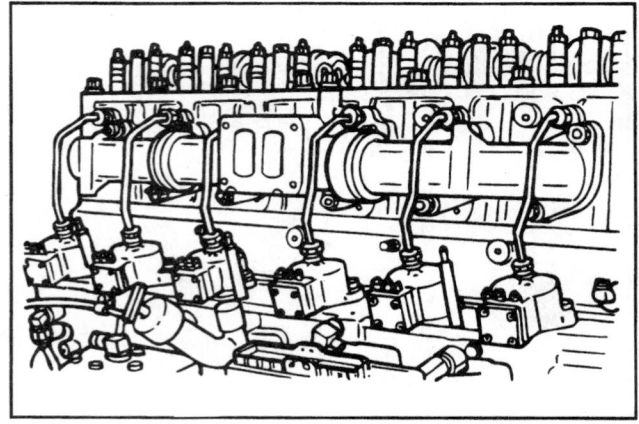

Figure 11-30 E-Tech engine showing the EUP line connection to the hydraulic injector. (Courtesy of Mack Trucks, Inc.)

Comparison Between E-7 V-MAC II and E-Tech V-MAC III Engines

The E-Tech engine will be Mack Trucks' primary truck engine for the next few years, so some of the design differences between the two engines are covered here. One of the reasons Mack Trucks adopted the EUP fuel system is that it required only minimal changes to their successful E-7 platform.

Cylinder Head Cover   The E-Tech cylinder head cover is more symmetrical and larger than previously, mainly to accommodate the larger sized Jacobs's compression brake used on the engine.

Cylinder Head Gasket   Both E-7 and E-Tech engines use the same fire ring but different cylinder head gaskets. The E-Tech head gasket has much larger protrusions at the pushtube cutouts.

Gear Train   The gear train on the E-Tech engine is entirely changed. All new gears are used, and an idler gear is added to the train. The auxiliary and camshaft gears rotate in opposite directions from the E-7 engine, as do the power steering pump and compressor.

Camshaft   The camshaft is still located on the cylinder block, but it is much larger on the E-Tech engine and is located upward and outward in the cylinder block compared to the E-7 engine. The placement of the camshaft further away from the crankshaft requires that larger timing gears are used, plus an idler gear is located between the two. The crankshaft drive gear to idler gear to camshaft gear is a hunting tooth relationship.

The E-Tech camshaft rotates in a direction opposite to the crankshaft. It is machined with cam profiles for actuating the cylinder valves and, in between, a cam profile responsible for actuating the EUP assemblies.

Oil Pump   The oil pump gear drive helix has been changed to keep the same oil pump direction of rotation. Because of some similarity of parts between the E-7 and E-Tech engines, special care must be taken to avoid mixing parts. If an improper drive gear is installed on the oil pump, it will be impossible to install it. However, if an oil pump and auxiliary shaft assembly were both replaced, two improper gears could be installed and the result would be engine failure. The drive ratios of the E-Tech oil pump provide for about a 6% increase in pump speed over the E-7 engine, thereby increasing the lubricating capacity of the engine.

Timing Gear Cover   The E-Tech timing gear cover eliminates the timing access cover required when an in-line pump is used. It adds bosses for an automatic belt tensioner and engine position sensor and provides room for the enlarged gears and addition of an idler gear.

Crankcase Breather   This is located on the timing gear cover and it functions to trap, collect, and return oil vapors to the crankcase that would otherwise be lost with blow-by gases. In operation, blow-by gases from the crankcase pass upward through the filter element, then down to exit through a central standpipe. When blow-by gases pass through the filter element, oil collects on the element surfaces and drains back down into the crankcase. The assembly can be removed to permit the cleaning of the element.

Conclusion

The DDC series 55 engine based on a Mercedes Benz design and adapted for DDEC III management was only available for a short period of time. However, the promise is that this engine will re-emerge in the marketplace with the EUP fuel system but with non-DDEC management electronics. Mack Trucks V-MAC III E-7 EUP is the third generation of management electronics applied to the E-7 engine. Mack Trucks claims that this engine will optimize performance, lower emissions, and simplify servicing and diagnostics. The E-7 EUP has a substantially beefed-up camshaft to accommodate the loading the EUPs will subject it to. The timing window within which V-MAC electronics may select an effective EUP stroke is increased beyond that obtainable in the PE8500 injection pump. In abandoning the partial authority, V-MAC II system with its electronically controlled in-line pumps, Mack Trucks has also simplified the pumping and metering hardware. EUPs have the simplicity of EUI systems and an extensive field service record in heavy duty applications of the technology.

HEUIs for Navistar 444E (Ford 7.3 L), DT466E, and 530E Engines, and the Caterpillar 3126 Engine

HEUI stands for hydraulic electronic unit injector (Figure 11-31). HEUI technology represents a new direction for the Caterpillar Engine Division; for the first time they have supplied fuel system engineering and components to competitor engine OEMs. The HEUI story began in 1987 when Caterpillar and Navistar signed a joint development agreement. This resulted in the introduction of HEUI on the Navistar 7.3 L engine (444E) in 1994. This engine is also widely marketed as a Ford 7.3 L engine. HEUI systems are more commonly found in Navistar applications, so the Navistar HEUI system will be used as a primary reference. The Caterpillar version will be referred to when there is a difference in terminology. Although the Navistar and Caterpillar HEUIs are almost identical, each OEM uses slightly different terminology and specifications in describing their system.

HEUI components are machined and assembled in a facility that uses extensive robotics, electrical discharge machining (EDM), and electrochemical machining (ECM) technology. EDM

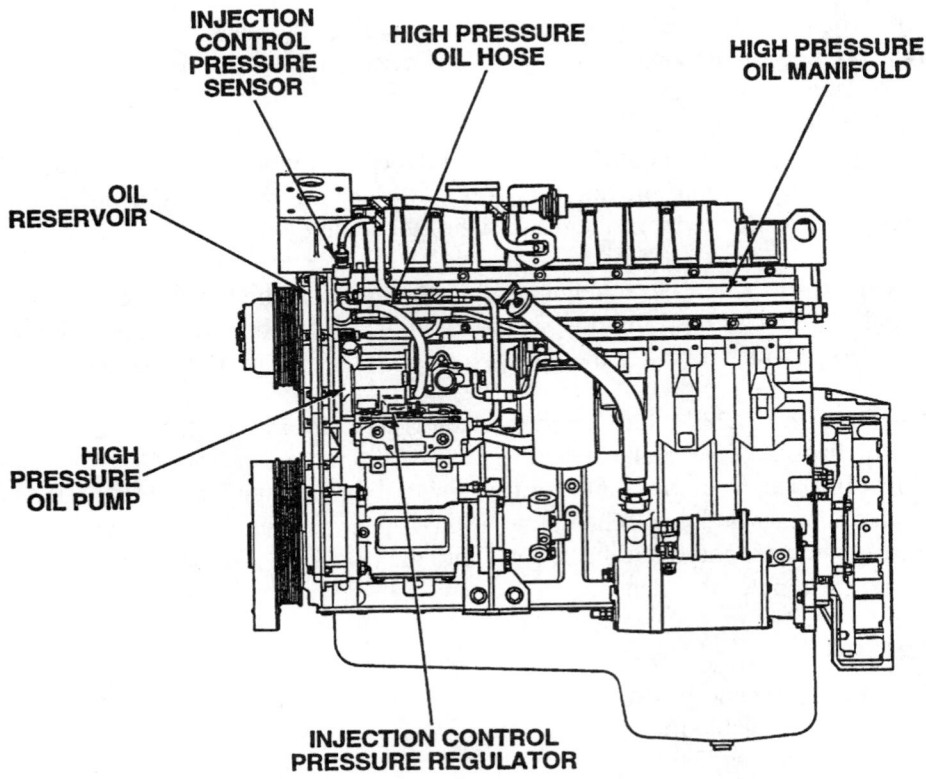

Figure 11-31 Component location of the Navistar DT466E with the HEUI common rail fuel system. (Courtesy of Navistar International Engine Group)

produces precision-sized orifices required in the HEUI hydraulic injector nozzles. This latest machining technology produces machine tolerances so precise that Caterpillar no longer matches injector bodies to the nozzle valves.

Principles of Operation

A disadvantage inherent in full authority EUI systems is that the fueling window is defined by the hard parameters of the injector train cam profile. The Caterpillar/Navistar HEUI system uses engine lube oil as hydraulic medium to actuate the fuel delivery pulse in their HEUI assemblies. The delivery stroke is therefore actuated hydraulically, switched by the engine management ECM(s), and thus is not confined to any hard limits. HEUI is therefore truly a full authority system in that the ECM can select the fuel pulse (pump effective stroke) to occur at anytime it computes that it is required. HEUI is a critical step toward eliminating the camshaft from a production diesel engine and achieving a fully regulated engine environment. HEUI technology is the first that permits injection rate shaping and represents a significant departure from all other current systems that are confined to the hard parameters of actuating cam geometry.

HEUI management contains four subsystems:

1. Fuel supply system
2. Hydromechanical injection system
3. HEUI assembly
4. Electronic management and switching

> HEUI systems are different from EUI systems because the delivery stroke is actuated hydraulically and switched by the ECM. It is not confined by cam profile hard limits as in EUI systems.

Fuel Supply System

The fuel supply system (Figure 11-32) serves to deliver fuel from the vehicle's tanks to the injector units. Fuel movement through the fuel supply system is the responsibility of a cam-driven plunger pump flange mounted to the engine block rear of the high-pressure oil pump. This

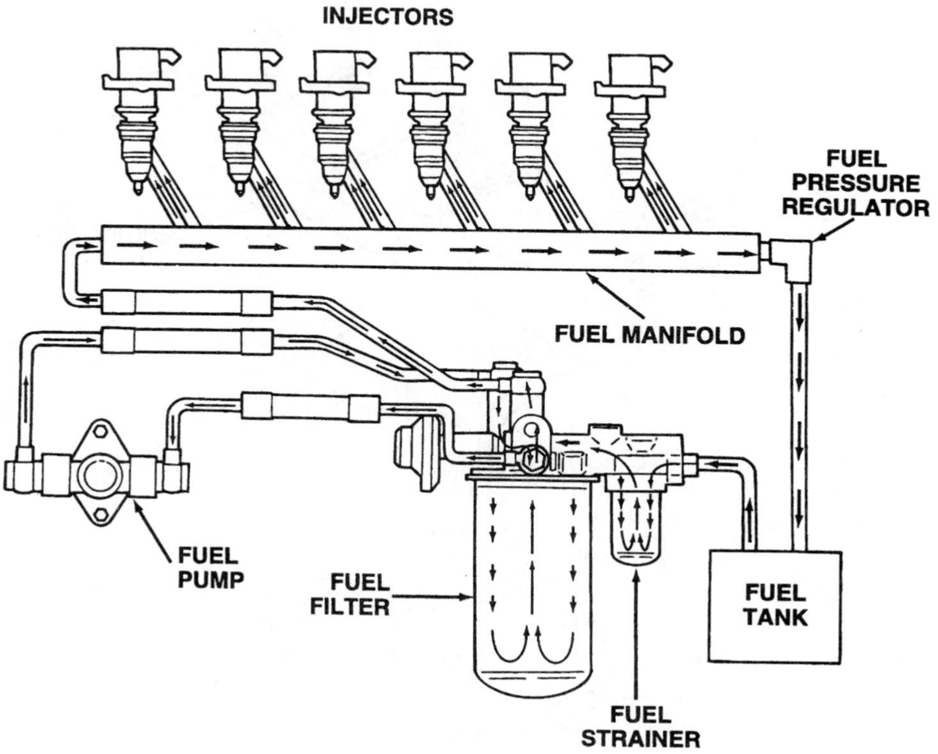

Figure 11-32 Navistar HEUI fuel supply system schematic. (Courtesy of Navistar International Engine Group)

transfer pump pulls fuel from the chassis fuel tank(s) through a fuel strainer. It then charges fuel through a disposable cartridge-type fuel filter and feeds it to the fuel gallery of fuel/oil supply manifold. A fuel pressure regulator at the fuel manifold outlet is responsible for maintaining a charging pressure of approximately 30–60 psi (2–4 atm; 206 kPa; 412 kPa). Fuel is cycled through the fuel supply system, and the HEUI's are mounted in parallel from the fuel manifold. A hand-priming pump is located on the filter pad in the event of loss of prime on the #I-6 engines (Caterpillar 3126 and Navistar 466E and 530E); there is no hand primer on the V-8 444E Navistar engine.

Injection Actuation System

The HEUI system uses hydraulically actuated, electronically controlled injector assemblies to deliver fuel to the engine's cylinders. The hydraulic medium used to actuate the pumping action required of the injector is engine oil (Figure 11-33). The engine lubrication circuit provides a continuous supply of engine lube to the HEUI high-pressure pump, a gear-driven, swash-plate hydraulic pump used to boost the lube oil pressure up to values exceeding 3,000 psi (200 atm; 20 MPa). The pump is actually capable of producing pressures of up to 4,000 psi (272 atm; 27.5 MPa), so future versions of HEUI may exploit this higher potential for peak pressures.

The **injection pressure regulator (IPR)** (Figure 11-34) manages actual high-pressure oil values, which are controlled electronically and actuated electrically. Swash-plate pumps use opposing cylinders and double-acting pistons driven by a swash plate: they are similar in principle of operation to common automotive air conditioner compressors. The injection pressure regulator manages the high-pressure oil pressure values between a low of 450 psi (3,103 kPa; 31 atm;

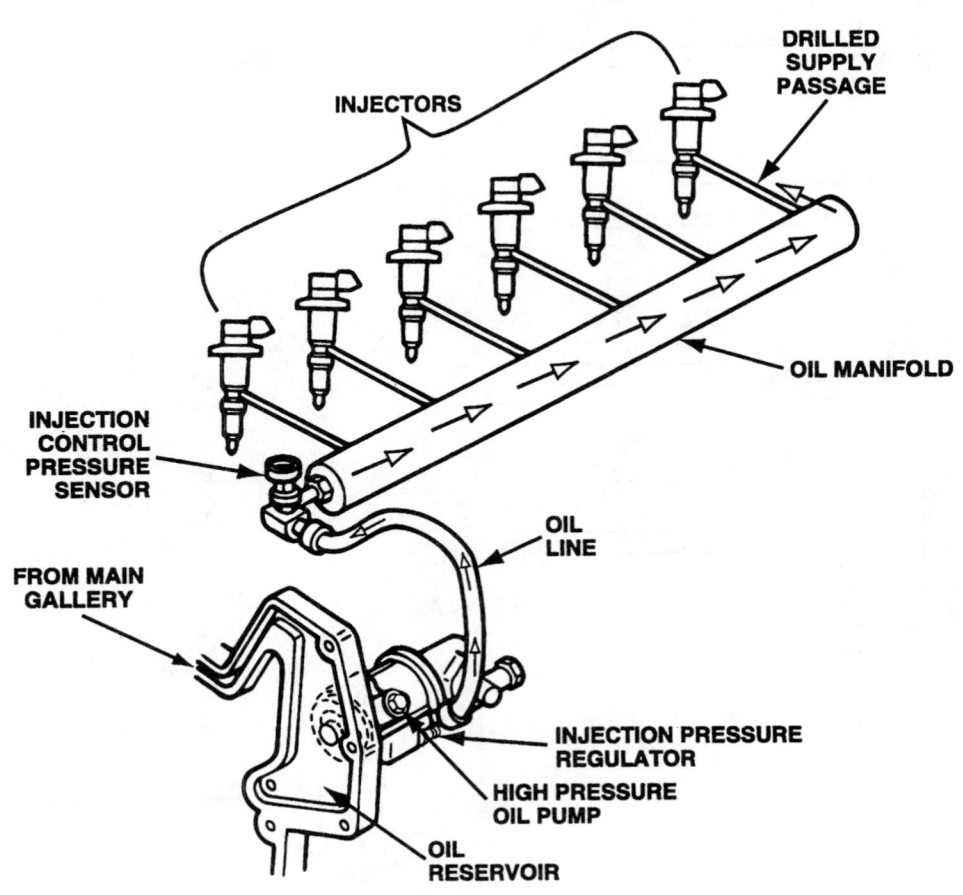

Figure 11-33 Navistar injection pressure oil flow schematic. (Courtesy of Navistar International Engine Group)

3.1 MPa) to highs exceeding 3,500 psi (240 atm; 24 MPa). The pump accomplishes this regulation by receiving all the lube pressurized by the high-pressure pump and spilling the excess to the oil reservoir.

The high-pressure oil is then piped to the oil gallery ducting within the fuel/oil supply manifold. From there, the oil is delivered to an exterior annulus in the upper portion of each HEUI. The HEUIs are mounted in parallel and fed by the high-pressure oil manifold. When the HEUI solenoid is energized, a poppet valve is opened by an electric solenoid within the HEUI. This action permits the high-pressure lube to flow into a chamber and act on the amplifier piston (Navistar) or intensifier piston (Caterpillar), actuating the pumping stroke required to convert the HEUI charging pressure to injection pressure values. At the completion of the duty cycle or pulse width, the HEUI is de-energized and the poppet valve retracts, spilling the oil acting on the amplifier piston to the rocker housing.

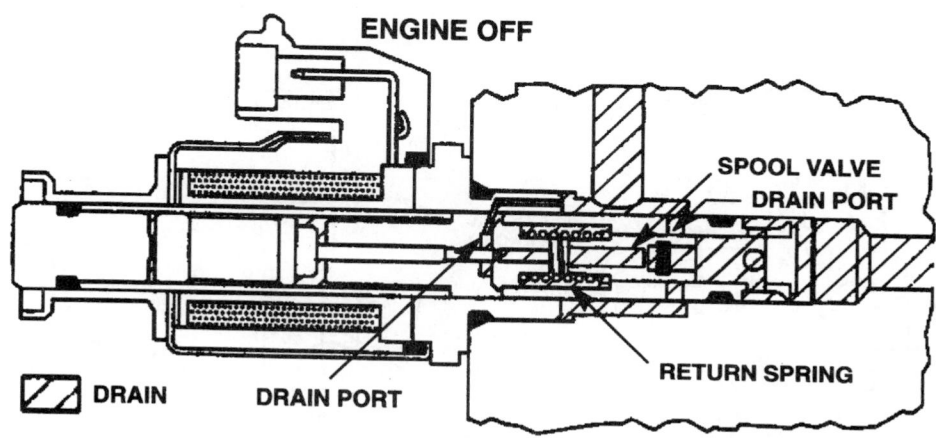

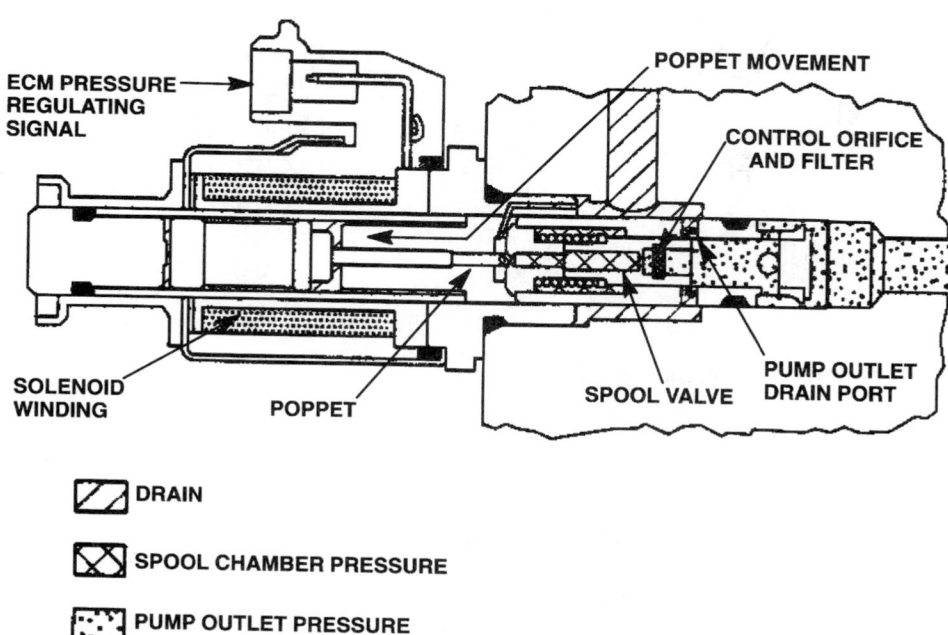

Figure 11-34 Navistar IPR valve operation. (Courtesy of Navistar International Engine Group)

Injection Pressure Regulator The IPR (Figure 11-34) is a spool valve, positional controlled by a pulse width modulated signal to achieve a range of **injection control pressure (ICP)** pressures. It operates at 400 Hertz. The ECM pressure-regulating signal will determine the magnetic field strength in the solenoid coil and, therefore, armature position. Integral with the armature is a poppet valve, which is designed to control the extent to which high pressure pump oil is spilled to the drain port and, therefore, the pressure of the oil that is used as the hydraulic medium for actuating the HEUI.

HEUI Assembly

The HEUI (Figure 11-35) is an integral pumping, metering, and atomizing unit controlled by ECM switching apparatus. The unit is essentially an EUI that is actuated hydraulically rather than by cam profile. At the base of the HEUI is a hydraulically actuated, multi-orifice nozzle. When the HEUI pumping element achieves the required NOP value acting on the sectional area of the nozzle valve exposed to the pressurizing annulus, the valve retracts, permitting fuel to pass around

Shop Manual Chapter 11, page 406.

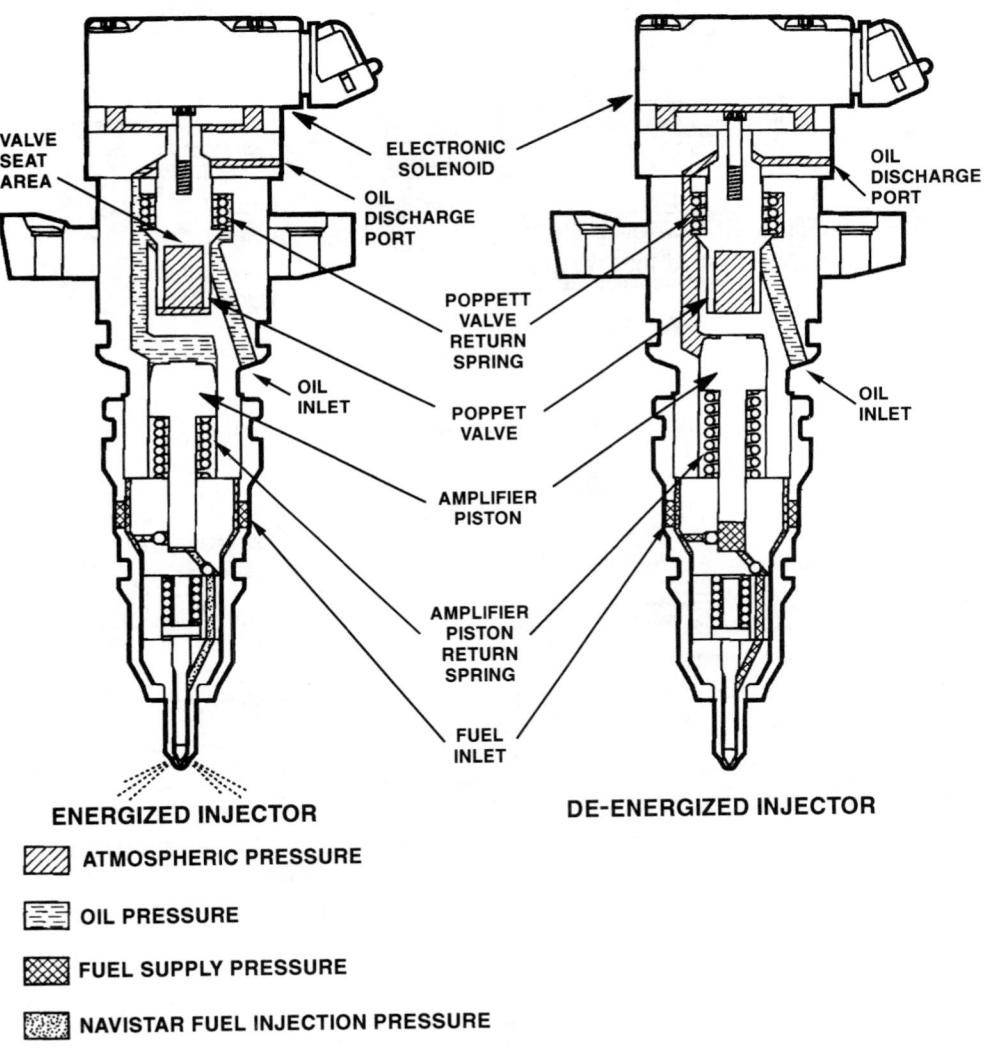

Figure 11-35 Navistar HEUI internal component identification and operating phases. (Courtesy of Navistar International Engine Group)

the nozzle seat and exit the nozzle orifices directly to the engine cylinder. A valve closes orifice (VCO) or sacless nozzle design is used.

The amplifier, or **intensifier piston,** is responsible for creating injection pressure values. This component is termed an amplifier piston in Navistar technical literature and an intensifier piston in Caterpillar versions. "Amplifier piston" will be the term generally used in this text. When the HEUI is energized, high-pressure oil supplied by a stepper pump acts on the amplifier piston and drives its integral plunger downward into the fuel in the pumping chamber.

A duct connects the pump chamber with the pressure chamber of the injector nozzle valve. The moment the HEUI is de-energized, the oil pressure acting on the amplifier piston collapses, and the amplifier piston return spring, plus the high-pressure fuel in the pump chamber, retracts the amplifier piston/plunger, causing the almost immediate collapse of the pressure holding the nozzle valve open. This results in rapid cessation of the injection pulse. In fact, the real time period between the moment the HEUI solenoid is de-energized and the point that droplets cease to exit the injector nozzle orifice is claimed to be less with HEUI than with equivalent EUI systems.

HEUIs typically have NOPs of 5,000 psi (340 atm; 35 MPa) with a potential for peak pressures of up to 24,000 psi (1,630 atm; 165 MPa), attainable depending on application. The oil pressure acting on the HEUI amplifier piston is amplified by six to seven times (depending on the application) in the fuel pump chamber. The droplet sizing decreases as the length of the injection pulse increases (as the real-time dimension available for combustion decreases). Rapid pressure collapse enabled by HEUIs avoids the injection of larger sized droplets toward the end of injection, which would be difficult to completely oxidize in the afterburn phase of combustion. At the completion of the HEUI duty cycle or pulse width, the pressurized oil that actuated the pumping action is spilled to the rocker housing. Emitted droplet sizing is directly factored by the injection rate, which is itself a factor of the amplifier piston descent rate (that is, the actual speed the plunger is driven downward into the pump chamber). Plunger descent rate is fully managed by the ECM, which switches and controls the actuating high-pressure oil value through the injection pressure regulator.

HEUI injectors are capable of being driven or switched at high rates, and the latest versions have plunger and barrel geometry that provides automatic pilot injection. Pilot injection is a term used to describe an injection pulse that is broken into two separate phases. In a pilot injection fueling pulse, the initial phase injects a short duration pulse of fuel into the engine cylinder, ceases until the moment of ignition, and at that point resumes injection, pumping the remainder of the fuel pulse into the engine cylinder. Pilot injection has been used as cold-start and warm-up strategy in EUI systems to avoid an excess of fuel in the engine cylinder at the point of ignition and minimize the tendency to cold-start detonation. HEUI systems with the pilot injection feature are designed to produce a pilot pulse for each injection. Figure 11-36 shows how cylinder head passages route the high-pressure oil and the fuel to the HEUI.

HEUI Subcomponents   The HEUI injector assembly can be subdivided as follows:

Solenoid   The ECM using a 115 V coil-induced voltage switches the solenoid. The HEUI electrical terminals connect the solenoid coil with the ECM injector drivers.

Poppet Valve   The HEUI poppet valve is integral with the solenoid armature. It is machined with an upper and lower seat. For most of the cycle, the poppet valve seat loads the lower seat into a closed position, preventing high-pressure engine oil from entering the HEUI. The upper seat is open, venting the oil actuation ducting. When the HEUI solenoid is energized, the poppet valve is drawn into the solenoid, opening the lower seat and admitting high-pressure oil from the IPR. When the poppet valve is fully open, the upper seat seals, preventing the oil from exiting the HEUI.

Intensifier or Amplifier Piston   The intensifier/amplifier piston is designed to actuate the injection plunger that is located below it. When the ECM to admit high-pressure oil into the HEUI

> The amplifier or intensifier piston multiplies the force creating the HEUI high fuel pressures six to seven times (depending on the application) in the fuel pump chamber.

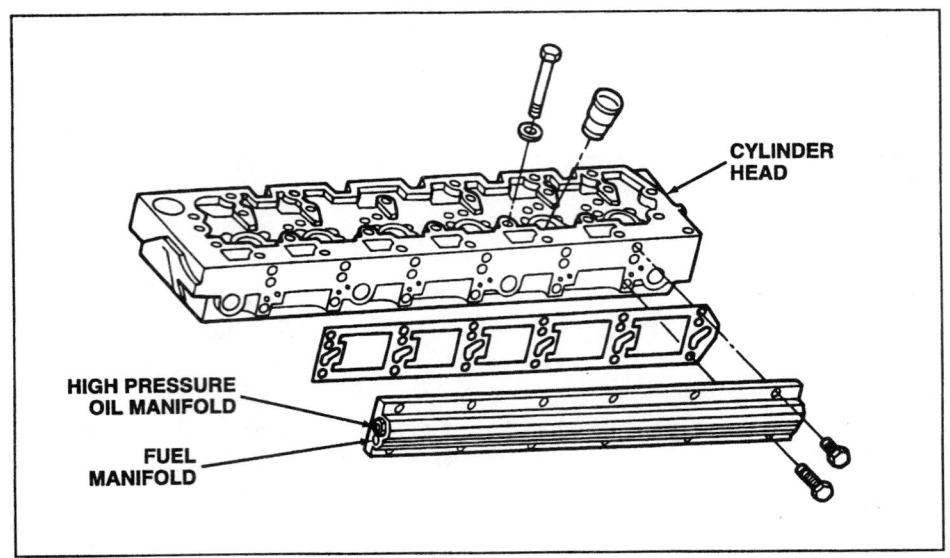

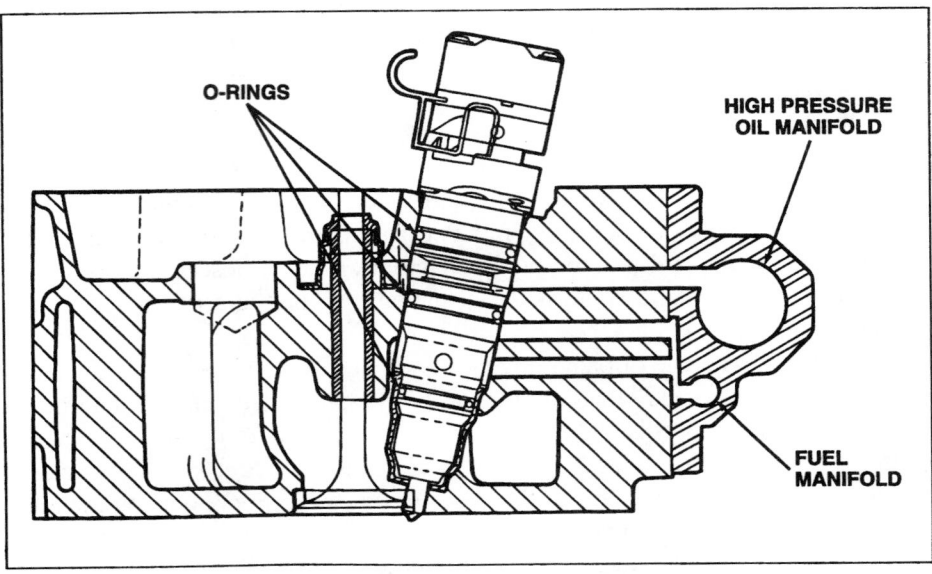

Cylinder Head and Cylinder Head Cross Section

Figure 11-36 Navistar HEUI DT466E in-line 6-cylinder oil/fuel manifolds and HEUI cross-section in cylinder head. (Courtesy of Navistar International Engine Group)

switches the poppet-control valve, the oil pressure acts on the sectional area of the amplifier piston. The actual oil pressure (managed by the ECM) determines the rate at which the plunger located below the amplifier piston is driven into the injection pump chamber. The sectional area of the amplifier piston determines how much the actuating oil pressure is multiplied in the injection pump chamber. This value is specified as six times by Caterpillar HEUI and seven times by Navistar HEUI. An actuating oil pressure of 3,000 psi (204 atm; 21 MPa) in a Caterpillar EUI would produce an injection pressure potential of 18,000 psi (1225 psi; 124 MPa). The amplifier piston and injection plunger are loaded into their retracted position by a spring.

Plunger and Barrel The plunger and barrel form the HEUI pump element. The first versions of the HEUI injectors did not offer the pilot injection feature. This description will use the recently introduced HEUI with the PRIME feature. The acronym **PRIME** is produced from the words **pre-**

injection metering. As the injection plunger is driven into the pump chamber, fuel is pressurized for a short portion of the stroke, actuating the injector nozzle. The pressure rise is of short duration because when the PRIME recess in the plunger registers with the PRIME spill port in the barrel, the pressure in the pump chamber collapses as fuel spills through the PRIME spill port. This closes the injector nozzle, and injection ceases. However, the moment the PRIME recess in the plunger passes beyond the spill port, fuel is once again trapped in the pump chamber and pressure rise resumes. This results in the injector nozzle being opened for the delivery of the main portion of the fuel pulse. The fueling pulse will continue until the ECM ends the effective stroke by de-energizing the HEUI solenoid. At this point, the poppet control valve is driven onto its lower seat, opening the upper seat and permitting the actuating oil to be vented. With no force acting on the amplifier piston, the plunger is driven upward by the combined force of the high-pressure fuel in the pump chamber and the plunger return spring. This causes an immediate collapse of pump chamber pressure and an almost immediate closure of the nozzle. A feature of the HEUI is its ability to almost instantly effect nozzle closure and end injection at the end of the plunger effective stroke.

Injector Nozzle   The HEUI injector nozzle is a multi-orifice injector nozzle of the VCO that is little different from any other injector nozzle used in a MUI or EUI assembly. A duct connects the nozzle pressure chamber with the HEUI pump chamber. A spring loads the injector nozzle valve onto its seat. The spring tension defines the NOP value. When the hydraulic pressure acting on the sectional area of the nozzle valve is sufficient to overcome the spring pressure, the nozzle valve unseats, permitting fuel to pass around the nozzle seat and through the nozzle orifices. The nozzle valve functions as a simple hydraulic switch. Because of the nozzle differential ratio, the nozzle closure pressure is always lower than the NOP. For instance, a Caterpillar version of the HEUI with a NOP identified at 4,500 psi (306 atm; 31 MPa) will not close until the pressure drops to 4,000 psi (272 atm; 28 MPa).

Stages of Injection   When the newer PRIME HEUIs are used, the injection pulse can be divided into five distinct stages, as described below.

Preinjection   The HEUI internal components (Figure 11-37) are all located in their retracted positions. In fact, they are in the preinjection position for most of the cycle. The poppet valve seat is spring loaded into the lower seat, preventing the high-pressure actuating oil from entering the HEUI; the amplifier piston and plunger are both in their raised position. Fuel enters the HEUI (Figure 11-38) to charge the pump chamber at the charging pressure value.

Pilot Injection   This phase begins when the plunger is first moved into the HEUI pump chamber. The pressure rise created opens the injector nozzle to deliver a short pulse of fuel. The pilot injection phase ends when the PRIME recess in the HEUI plunger is driven downward enough to register with the PRIME spill port, causing the pump chamber pressure to collapse and the nozzle valve to close.

Delay   The delay phase occurs between the ending of the pilot injection phase and the resumption of the fuel pulse. The objective is to cease fueling the engine cylinder while the PRIME pulse of fuel is vaporized and heated to its ignition point. It is important to note that the plunger is still being driven through its stroke during this phase because the poppet valve is in the open position and oil pressure continues to drive the amplifier piston downwards.

Main Injection   When the PRIME recess in the plunger passes beyond the PRIME spill port, fuel is once again trapped in the HEUI pump chamber because it can no longer exit through the spill port. The resulting pressure rise opens the injector nozzle a second time to deliver the main volume of fuel to be delivered in the cycle (Figure 11-39).

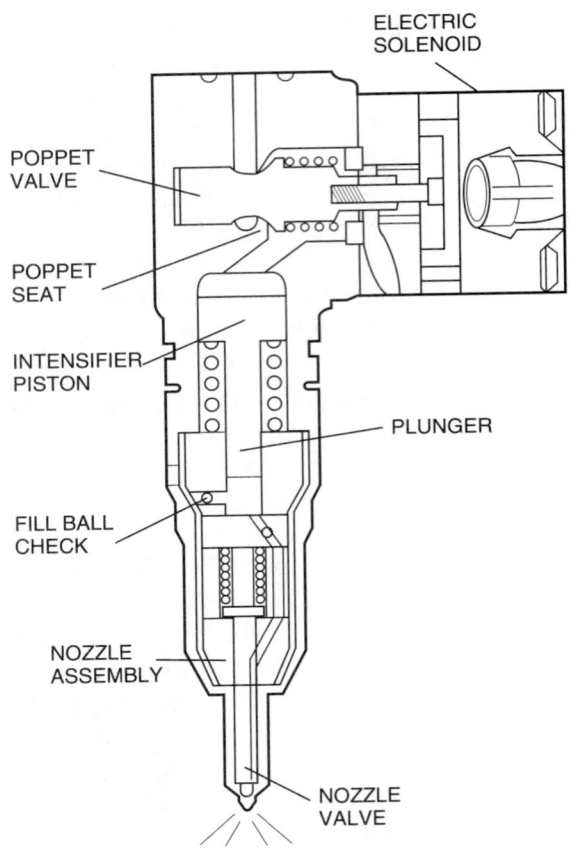

ELECTRIC
SOLENOID

POPPET
VALVE

POPPET
SEAT

INTENSIFIER
PISTON

PLUNGER

FILL BALL
CHECK

NOZZLE
ASSEMBLY

NOZZLE
VALVE

Figure 11-37 Caterpillar HEUI with side-mounted solenoid sub-components identification. (Courtesy of Caterpillar Inc.)

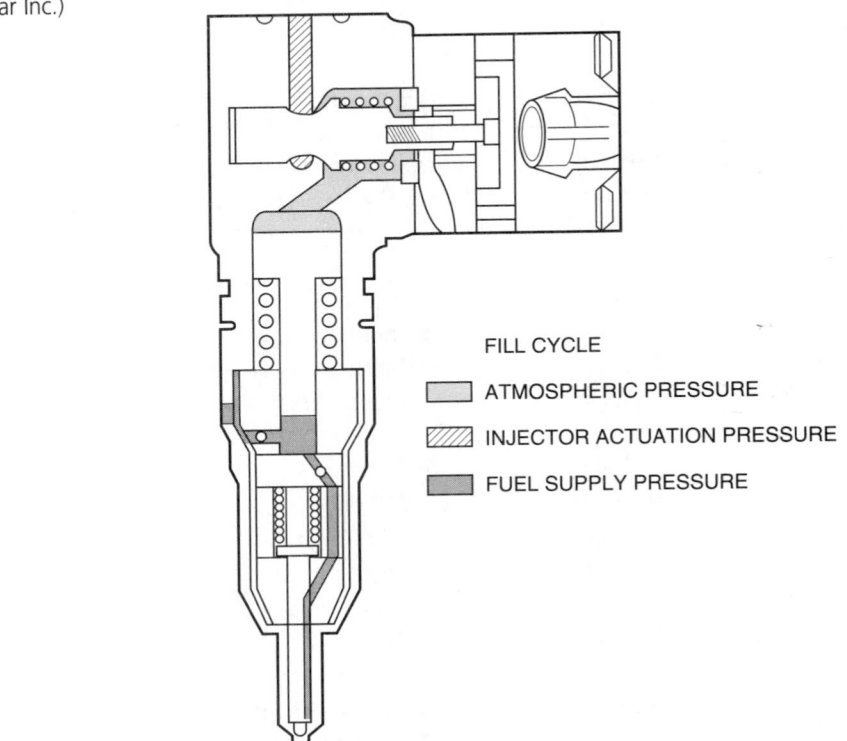

FILL CYCLE

ATMOSPHERIC PRESSURE

INJECTOR ACTUATION PRESSURE

FUEL SUPPLY PRESSURE

Figure 11-38 Caterpillar HEUI fill cycle. (Courtesy of Caterpillar Inc.)

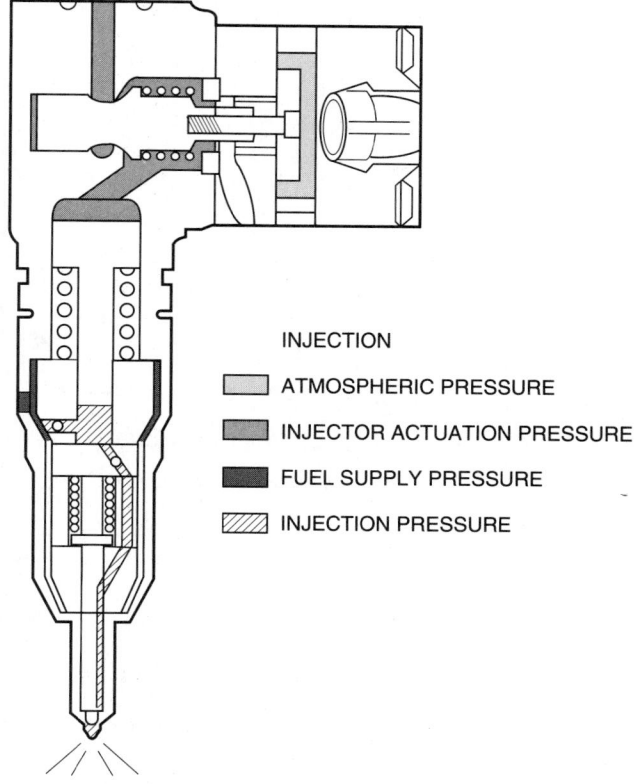

INJECTION

☐ ATMOSPHERIC PRESSURE

▨ INJECTOR ACTUATION PRESSURE

▨ FUEL SUPPLY PRESSURE

▨ INJECTION PRESSURE

Figure 11-39 Caterpillar HEUI injection cycle. (Courtesy of Caterpillar Inc.)

End of Injection   The end of injection (Figure 11-40) begins with the de-energizing of the HEUI solenoid. The solenoid coil releases the armature, and a spring drives the poppet valve downward to seat on its lower seat. The instant the poppet valve starts to move downward, the upper seat is exposed, permitting the actuating oil inside the HEUI to spill. When the actuating oil pressure acting on the amplifier piston is relieved, the fuel pressure in the HEUI pump chamber, combined with the plunger return spring, collapses the fuel pressure almost instantly. Injection ends when there is insufficient pressure to hold the nozzle valve in its open position, and the three moving assemblies (poppet valve, amplifier/plunger, and nozzle valve) in the HEUI are all in their return positions, as outlined in the preinjection phase.

HEUI Electronic Management and Switching

HEUI systems can be categorized as full authority electronic management systems with comprehensive monitoring, vehicle management, and self-diagnostic capabilities. In Navistar applications, logging of fault codes is the responsibility of nonvolatile RAM (NV-RAM), which is described as **keep alive memory (KAM)**. The circuit that provides continuous power to maintain KAM data is known as **KAMPWR**.

In addition, the HEUI management ECM, vehicle personality module, and switching apparatus are housed separately on Navistar engines manufactured up to 1997. In the Caterpillar 3126 engine, the HEUI switching apparatus is located in a single ECM housing with no fuel cooling. In 1997, Navistar introduced their consolidated engine controller that consolidates the microprocessing and switching functions of the ECM, the personality module, and the **injector drive module (IDM)** into a single engine-mounted unit. The housing will be referred to simply as the ECM.

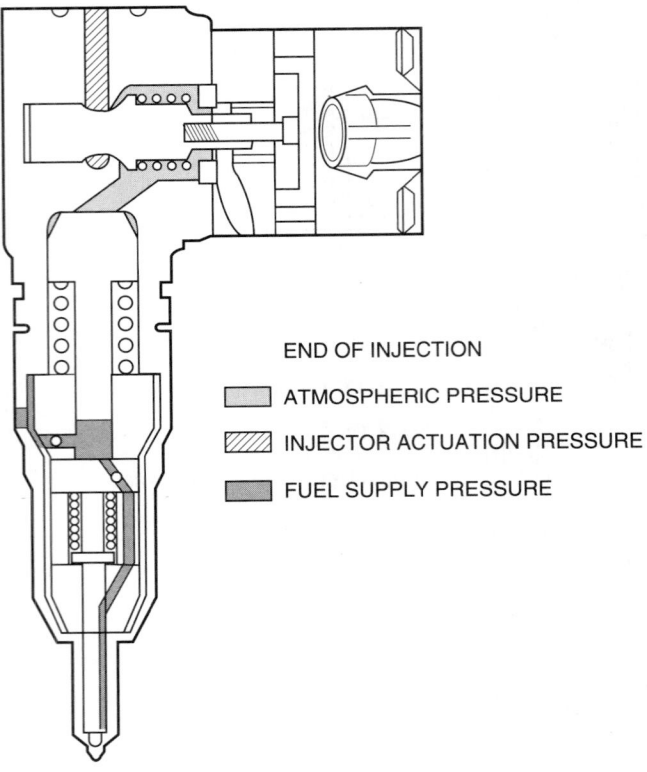

END OF INJECTION

ATMOSPHERIC PRESSURE

INJECTOR ACTUATION PRESSURE

FUEL SUPPLY PRESSURE

Figure 11-40 Caterpillar HEUI end of injection. (Courtesy of Caterpillar Inc.)

**Shop
Manual
Chapter 11,
page 408.**

ECM Functions (Figure 11-41)

1. Reference voltage regulator

2. Input conditioning

3. Microcomputer

4. Outputs

Reference Voltage　Reference voltage is delivered to system sensors that divide this input and return a percentage of it as a signal to the ECM. Thermistors (temperature sensors) and potentiometers (TPS) are examples of sensors requiring reference voltage. Reference voltage values used are at 5 V pressure and a current limiting resistor used to safeguard against a dead short to ground limits the flow. Reference voltage is also used to power up the circuitry in Hall-effect sensors used in the system, such as the **camshaft position sensor (CMP)**.

Input Conditioning　Signal conditioning consists of converting analog signals to digital signals, squaring up sine wave signals, and amplifying low-intensity signals for processing.

Microcomputer　The Navistar HEUI microprocessor functions similarly to other vehicle system management computers. It stores operating instructions called *control strategies* and tables of values called *calibration parameters*. It compares sensor monitoring and command inputs with the logged control strategies and calibration parameters and then computes the appropriate operating strategy for any given set of conditions.

　　ECM computations occur at two different speeds, referred to by Navistar as foreground and background calculations. Foreground calculations are more critical functions, and these are computed at a higher frequency than background calculations. Engine speed control and throttle position signals require foreground calculations (an immediate response to a changing condition or

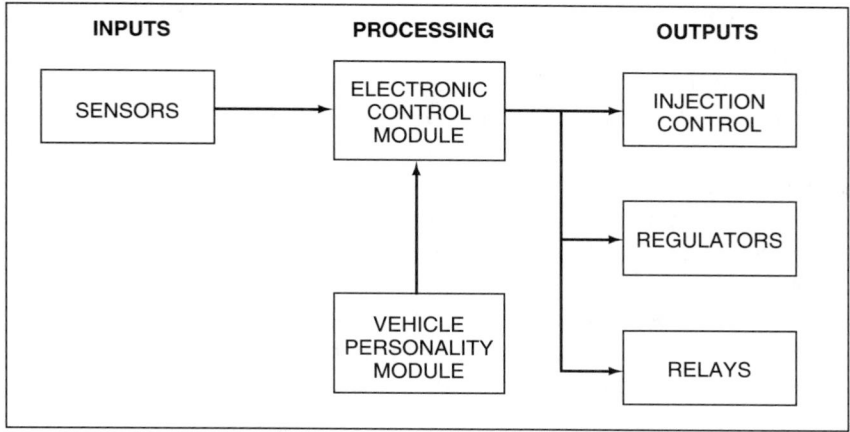

ELECTRONIC CONTROL SYSTEM OPERATION

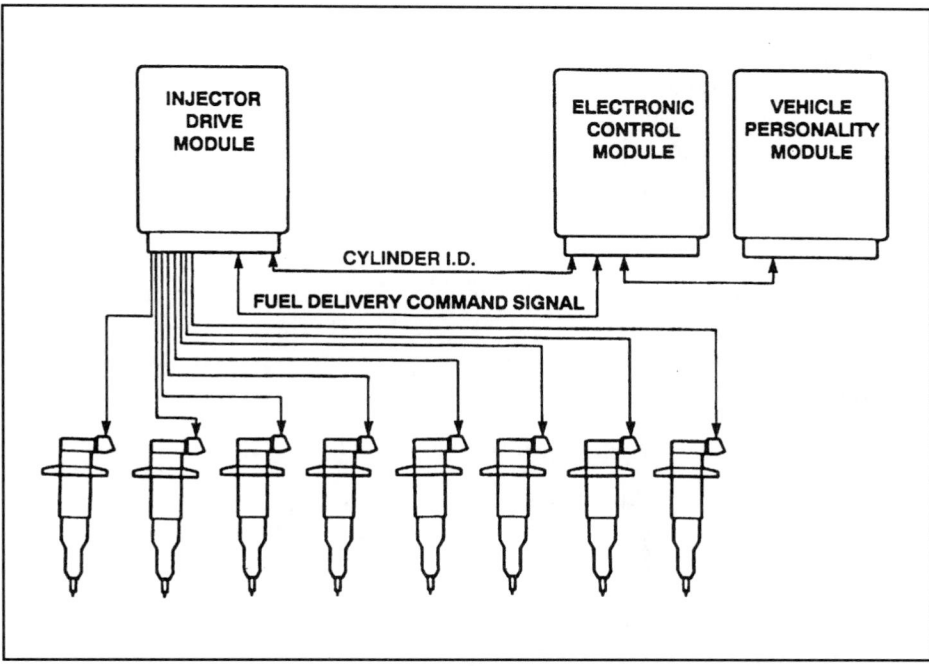

IDM Electronic Distributor Operation

Figure 11-41 HEUI ECM processing cycle; Navistar IDM operation: the FDCS emphasized. (Courtesy of Navistar International Engine Group)

command). Background calculations are processed at a lower frequency and include input signals, such as ambient temperature and engine temperature. The difference in foreground and background computations is simply the speed at which the microprocessor is required to react to a change in operating characteristics. A change in throttle position requires an immediate adjustment in engine fueling; therefore, this command input requires almost instant response by the ECM. However, while an increase in coolant temperature could have serious consequences if ignored, engine overheat conditions occur gradually, so an instant reaction by the ECM is not required.

Diagnostic strategies include monitoring input data on a continuous basis and flagging codes when an abnormal operating parameter is detected. Calibration tables and operating strategies are retained in ROM. Neither opening the ignition circuit nor disconnecting the vehicle batteries causes loss of these data, as they are magnetically retained. RAM data are electronically retained and thus are retained only while a circuit is energized.

Navistar ECM RAM stores information sourced from electronic monitoring and data processing/manipulation, which is volatile and, as such, is dumped each time the ignition circuit is opened. Navistar KAM is nonvolatile RAM and used to log fault codes. Out-of-normal parameters may also result in adaptive strategies being written into KAM. Subsystem failure or component wear would be examples. KAM data are retained when the ignition circuit is opened but dumped when the vehicle batteries are disconnected. Navistar PROM is described as the vehicle personality module (VPM) and is both customer and Navistar data programmable. The function of the VPM is to trim engine management to specific chassis application and customer requirements. The **engine family rating code** (EFRC) is located in the VPM calibration list and can be read with an EST; this identifies the engine power and emission calibration of the engine.

Outputs The switching apparatus within the ECM can be referred to as actuator control. The ECM controls the system actuators by delivering a signal to the base of the transistor output drivers. These drivers, when switched, ground the various actuator circuits. The actuators may be controlled through a duty cycle (that is, percent time on/off), controlled by modulating pulse width or simply switched on and off, depending on the actuator type.

Injection Driver Module (Siemens) The IDM (Figure 11-42) has two primary functions:
1. Electronic distributor for the HEUIs
2. Powers the HEUIs

Electronic Distributor for HEUIs The ECM determines engine position from the camshaft position sensor located at the engine front cover (Figure 11-43). The ECM uses this signal to determine cylinder firing sequence and then delivers this command data to the IDM as a **fuel demand command signal (FDCS)** signal. The FDCS contains injection timing and fuel quantity data.

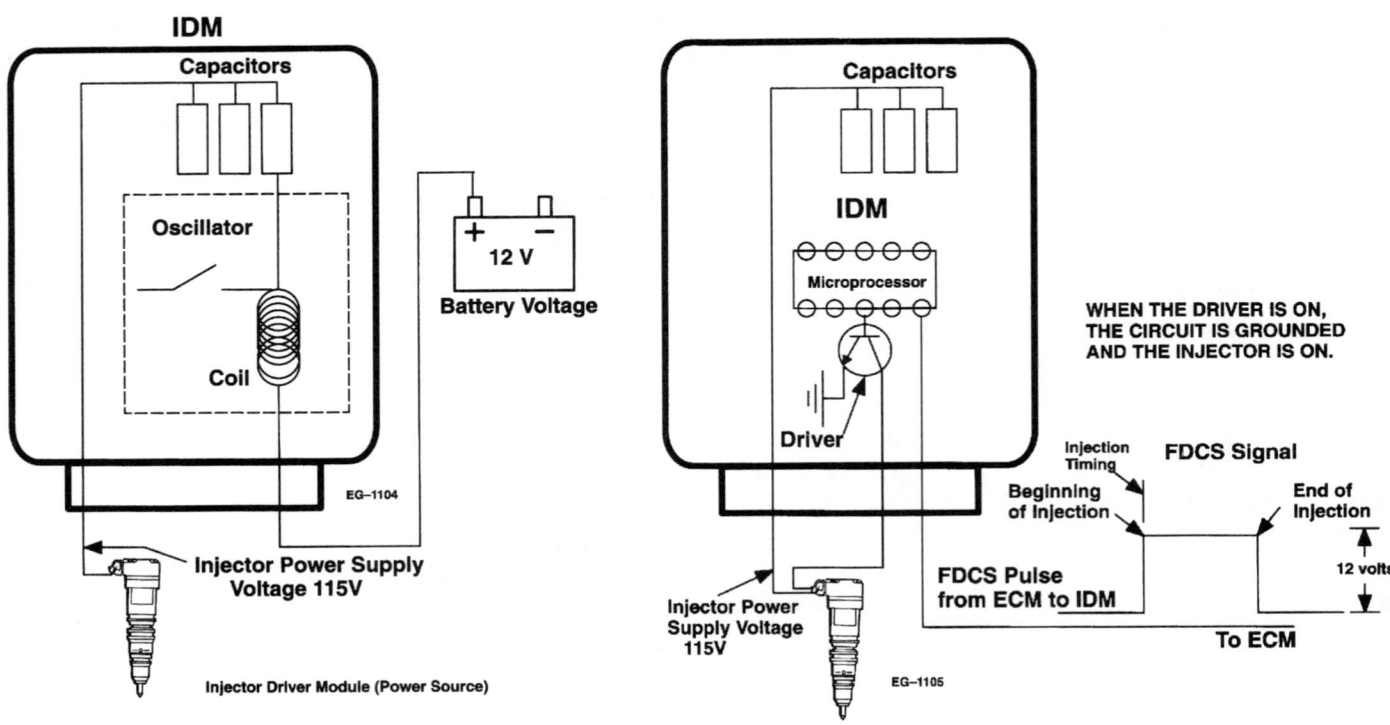

Figure 11-42 Navistar IDM: power source; Navistar driver module operation. (Courtesy of Navistar International Engine Group)

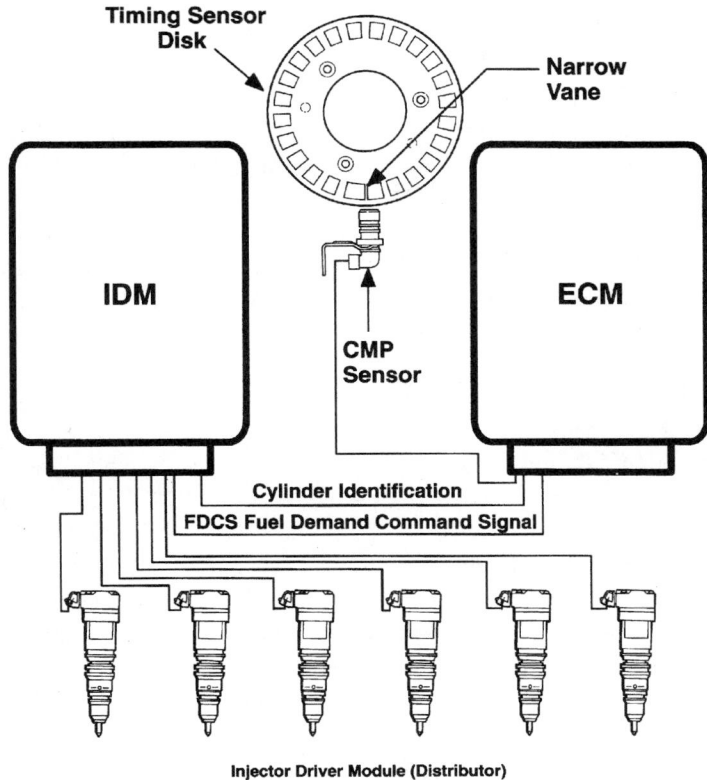

Figure 11-43 Navistar IDM and ECM communications signals and relationship with CMP input. (Courtesy of Navistar International Engine Group)

Powers HEUIs The IDM supplies a constant 115+ V direct current supply to each HEUI. This 115 V DC supply is created in the IDM by making and breaking a 12V source across a coil using the same principles employed by the ignition coil in a spark-ignition engine. The resultant 115 V induced is stored in capacitors until discharged to the HEUIs. All the functions of the IDM are incorporated into a single ECM housing in the consolidated engine controller, which was introduced in 1997.

Cummins HPI and Interact Series Engine Management

Cummins introduced several new engine series toward the end of 1997. Some of these engines are reengineered versions of previous engine series, such as the B and C series, which have entirely new fuel systems and electronics. The 600 Signature series is a new engine that promises to make a big splash in the high horsepower sector. In the past, higher brake power engines have been associated with disproportionately high fuel costs, shorter engine life, and more than their share of day-to-day problems. The new generation of high-power engines comes with the promise of all of the performance features expected in an electronically managed diesel engine. The new Interact System C series (ISC) engine comes with a variation of the common rail (injectors supplied from a common accumulator chamber or manifold) fuel system that is a patented full authority fuel management system known as the Cummins accumulator pump system (CAPS). The new Interact System B series (ISB) engine displaces 5.9 L and is capable of producing 275 bhp and 660 lb.-ft. of torque.

Signature 600 Series Engine

Cummins introduced the Signature series engine in 1997. The engine was introduced as the Signature 600 because of its nominal rated power of 600 bhp. The engine also boasts a peak torque output of 2,050 lb.-ft. The Signature series engine is both lighter and more compact than the N14 package. It is notably free of external plumbing and wiring, and this gives the engine an uncluttered appearance. The engine boasts a number of innovative features, perhaps the most notable of which is a return to a common rail, open-nozzle (plunger seals cup orifices) injection system distantly related to the PT system. The 19 L K-series engine (not used on highway applications) uses a electronically managed common rail fuel system even more closely related to the hydromechanical PT system, known as HPI-PT.

Signature Series Engine Hardware

**Shop
Manual
Chapter 11,
page 412.**

Cummins developed the Signature series from a clean slate, and it shares nothing other than manufacturer name with other Cummins engines. The engine is built on an in-line, six-cylinder platform with a bore of 137 mm, and strokes of 169 mm give it a 15-liter displacement. The cylinder head is a single slab, cast iron alloy construction that supports dual overhead camshafts. The first camshaft is responsible for actuating the unit injectors and imparting drive to the gear pump. The unit injectors are expected to produce injection pressure values exceeding 28,000 psi, so the cam lobes are of an extra-wide construction to deliver the high mechanical forces required. The second camshaft is responsible for actuating the cylinder head valves and the engine compression brake. The engine compression brake, known as Interbrake, was developed in conjunction with Jacobs Manufacturing to be completely integral with the cylinder head assembly; as a consequence, it adds a mere 20 lb. to the engine weight. The engine brake is ECM controlled and has six settings, providing up to 600 bhp of braking. Both camshafts are drilled through the center to reduce weight and distribute oil to the cylinder head components.

Cummins has used a unitized design that eliminates many external components by simply incorporating them into the cylinder head and cylinder block design. The engine uses articulating piston assemblies with steel crowns. The wet liners used are said to have high resistance to variations in coolant chemistry. Cummins uses a variable output turbocharger that is wastegate controlled to perform like a small turbo at low engine speeds and load with little response lag. At higher engine speeds and loads, the turbocharger performs like a large turbocharger with low flow restriction. The engine management electronics use comprehensive system monitoring of both main engine system performance and subsystems, such as coolant pump and compressor operation. In common with most full authority engine management systems, it contains a programmable engine protection system.

HPI-TP Electronics

Signature series is a full authority engine management system that uses SAE J1939 protocols to multiplex other chassis electronic systems. Two microprocessor-based modules are used. The ECM is mounted to the cylinder head and is a Motorola unit (CM750) that uses polybend technology. Polyblend technology uses printed circuit boards assembled flat then folded to increase the data processing and data retention capability. Cummins claims that ten times the computing power can be packaged in a more compact unit than their current CELECT ECM.

The ECM housing is air cooled and does not use a fuel-cooled heat exchanger like most other current ECMs. The housing is sealed with a Gore-Tex seal designed to equalize the pressure inside and out of the housing. Most competing ECMs are vacuum sealed. Vacuum sealed components have a way of attracting moisture, which is a primary cause of ECM failure. The second module is the **integrated fuel system module (IFSM)**, which manages the common rail fuel system. The IFSM is bolted to the side of the cylinder head, adjacent to the injection camshaft. The IFSM incorporates the gear pump, actuators, shutdown solenoids, and priming pump. The ECM manages IFSM functions. The input circuit to the ECM includes comprehensive command and chassis monitoring circuits.

The output circuit primarily involves the managing of the fuel system, but "smart cruise" and brake management are built into the system. The ECM and the IFSM are positioned close to each other and are easily accessible. Cummins also offers INFORM with the Signature electronics package. INFORM (Figure 11-44) is information management software that can take raw data from the on-board ECM by means of a serial link adapter and download it into a PC. INFORM is Windows based and can analyze numerous ECM parameters to produce reports in a number of formats. The following reports may be produced:

1. Fuel report
2. Driver report
3. Service report
4. Safety report
5. Executive summary report
6. Exception report
7. Comparison reports

HPI-TP Fuel System

With the introduction of the Signature series, Cummins reverted to a common rail, open-nozzle fuel system that has more in common with their previous PT system than with CELECT and other full authority, electronically managed fuel systems. It is known as the HPI-TP (high-pressure injection–time pressure) system. Cummins introduced their hydromechanical PT system by describing the hydraulic equation around which the metering of fuel is constructed. In the PT system, the factors of the hydraulic equation are the critical *flow areas* located in the PT injectors (held constant), the *time* of flow (of fuel) dependent on cam geometry and engine speed, and *pressure*. Pressure is the variable in the equation used to control actual fuel quantity delivered to the engine cylinders. Because pressure is the variable in the hydraulic equation that was used to

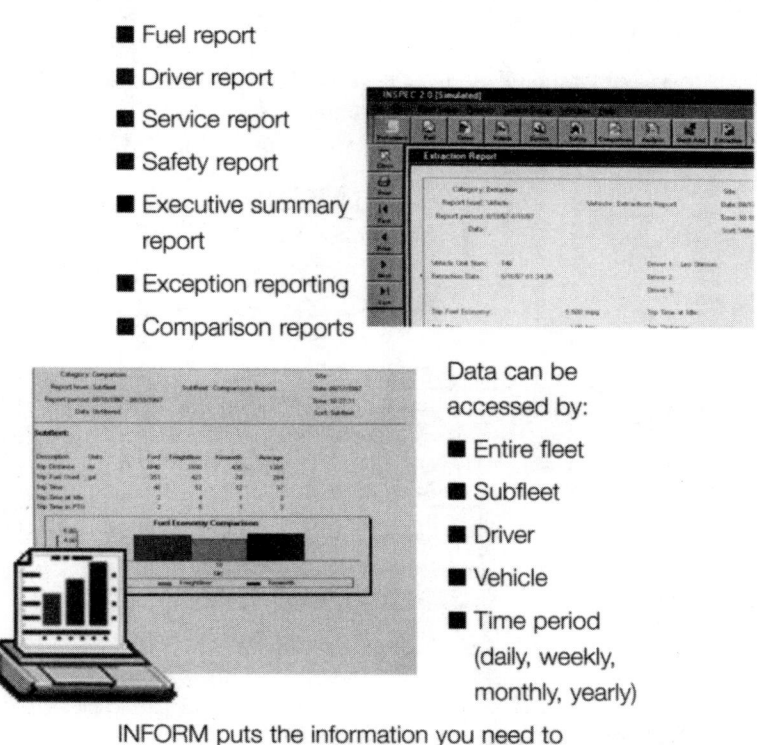

- Fuel report
- Driver report
- Service report
- Safety report
- Executive summary report
- Exception reporting
- Comparison reports

Data can be accessed by:

- Entire fleet
- Subfleet
- Driver
- Vehicle
- Time period (daily, weekly, monthly, yearly)

INFORM puts the information you need to manage for increased productivity and profits right at your fingertips.

Figure 11-44 Cummins INFORM software. (Courtesy of Cummins Engine Company)

control fueling, it appears first in the PT acronym. The Cummins hydraulic equation is constructed to determine engine fueling in the HPI-TP system. As in the PT systems, the critical flow areas are established in the injector unit (and are held constant). However, the pressure value is now held constant (rail pressure is regulated at a constant 250 psi) so the control variable is time. Specifically, time correlates to the amount of time the fueling actuator is held open to feed the injector-metering chamber with fuel. In the acronym used to describe the system, the letter T appears before the letter P to indicate it is the control variable. In terms of its ability to manage fueling and injection timing, the Signature series HPI-TP system is classified as a full authority, engine and chassis management system.

HPI-TP Operating Principles (Figure 11-45)   Fuel is drawn from the fuel pump through a suction side filter, which will entrap particulate sized 10 μ and larger. The function of the electric primary pump shown in the schematic is to purge air from the fuel subsystem after a filter change or in the event the pump prime is lost. The priming pump may not appear in released versions of the engine. A gear pump located in the IFSM is responsible for all movement of fuel through the fuel subsystem and to the point at which fuel is delivered to the TP injectors. The gear pump charges a fuel manifold within the IFSM, which supplies two pairs of actuators. Fueling and timing parameters are plotted by the ECM, which controls the actuators by means of a pulse width modulation signal. Two fueling actuators are responsible for delivering fuel to the pumping or fueling chambers of the unit injectors.

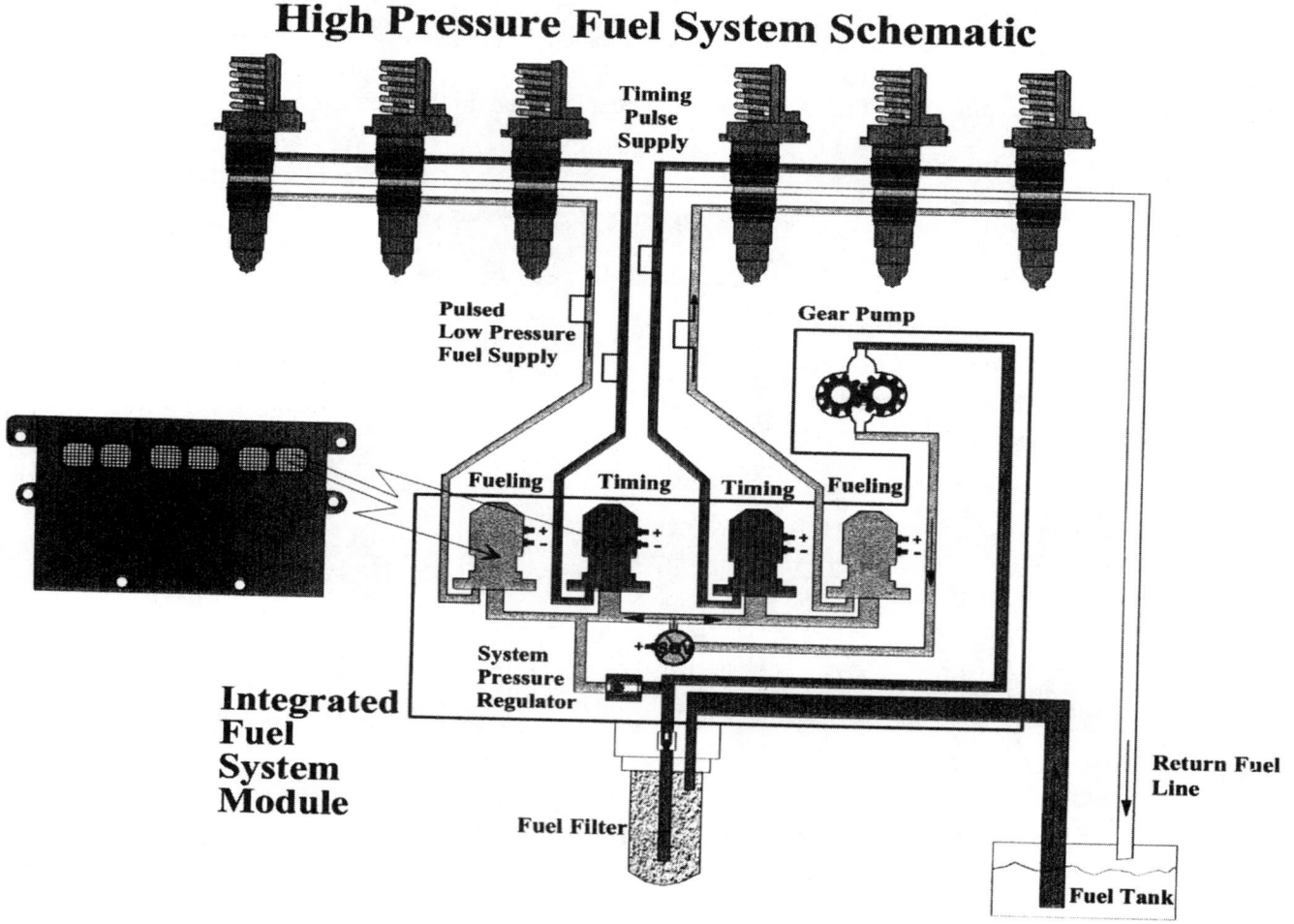

Figure 11-45 HPI-TP fuel system schematic. (Courtesy of Cummins Engine Company)

Each fueling actuator is an ECM-switched solenoid that controls fuel flow from the supply rail to the rail that feeds the injectors. Actual fuel quantities delivered to the pumping chambers of the unit injectors can be precisely controlled by ECM, which uses the fueling actuators as metering devices. Fuel is delivered from the actuators to the unit injectors by means of drillings in the cylinder head. The unit injectors are cylindrical and receive fuel from exterior annuli sealed by O-rings. A pair of ECM-controlled timing actuators is used to charge hydraulic tappets within the unit injectors and thereby control the beginning of the plunger effective stroke.

The unit injectors have a pump chamber that consists of a plunger and cup. The cup is drilled with orifices, which are sealed when the plunger is loaded into the cup. The unit injector cam uses cam geometry that gives it a profile that is mostly OBC. This means that the plunger is loaded into the cup with some mechanical crush for the larger portion of the cycle. When the actuating cam profile ramps off OBC toward its IBC, a spring lifts the plunger, permitting the fueling actuator to charge the cup.

A major advantage of the open nozzle injector principle used by these injectors is the elimination of the pressure collapse phase that occurs between the end of pump effective stroke and actual nozzle closure in pump-line-nozzle and unit injector systems using hydraulic injector nozzles. The pressure collapse phase results in larger injected droplets that cannot be completely combusted in the time available at higher rpm. In an open nozzle system, the injection pulse ends abruptly when the plunger tip seats in the cup sealing the cup orifice. This gives the system an edge when it comes to emissions control.

The timing actuator charges the hydraulic tappet within the unit injector. This fuel acts as hydraulic medium and the quantity of fuel in the tappet will define the point at which the plunger acts on the fuel metered to the cup. When the actuating cam ramps off its IBC toward OBC, the unit injector plunger is driven through it's pumping stroke, generating injection pressures that can exceed 30,000 psi. The fuel that charges the timing chambers is returned to the tank after the fueling pulse.

Operating Sequence Figure 11-46 highlights the HPI-TP unit injector operation, as explained below:

1. The HPI-TP system is split into two banks: forward and rear banks. The forward bank of the in-line, six-cylinder configuration engine comprises cylinders 1-2-3. The rear bank comprises cylinders 4-5-6.

2. Fuel is drawn from the tank through a suction side filter by the supply pump. In common with most Cummins fuel systems, the entire fuel subsystem is under suction. The supply pump is of the external gear type and therefore uses a positive displacement pumping principle. A pulsation damper smooths out the pressure surges that are a characteristic of gear type pumps. As the pump into the rail discharges each slug of fuel, the pressure tends to surge. The pulsation damper operates in the same manner as the device by the same name on a PT hydromechanical fuel system. Fuel exiting the supply pump has its pressure regulated at 250 psi (17 atm; 1.7 MPa). This charging pressure of 250 psi is maintained throughout engine operation.

3. Fuel exiting the gear pump passes through a rapid restart fuel shutoff solenoid. This solenoid is ECM controlled. The ECM can cease engine fueling in the event of any programmed malfunction strategy (engine overspeed would be an example). The engine is also shut down by means of this fuel shut-off solenoid.

4. The fuel is next forced through a charge side filter. This filter protects the ECM-controlled timing and fuel rail actuators from debris in the fuel.

5. Fuel is next routed into a passage linking the front and rear banks. This passage feeds both the front bank timing and fueling actuators and the rear bank timing and fueling actuators.

6. Each timing and fueling/rail actuator is ECM controlled. The timing and fueling actuators are switched to open for a specific amount of time measured in milliseconds (ms) by the ECM. In their open position, they simply permit fuel at the rail pressure

HPI-TP FUEL SYSTEM SCHEMATIC

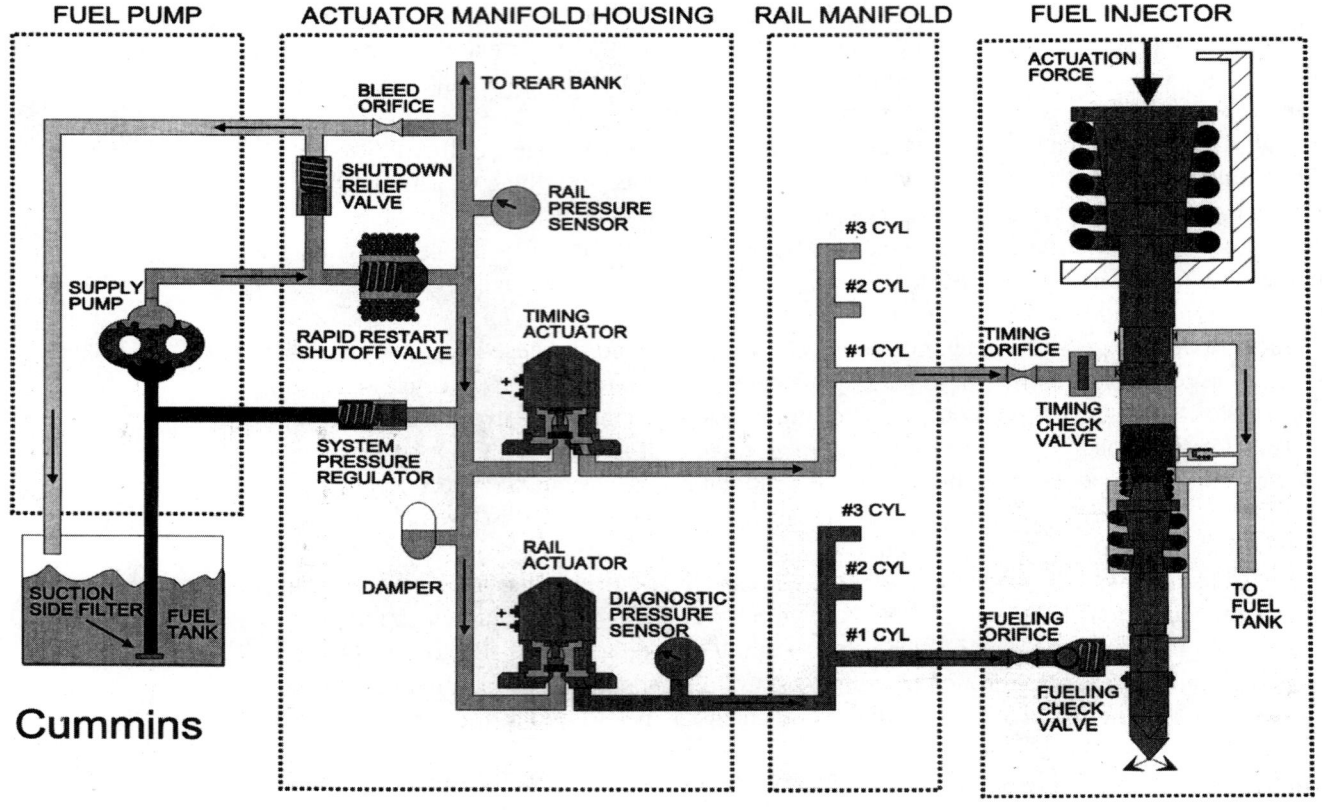

Figure 11-46 The HPI-TP fuel system highlighting circuit flow and fuel injector operation. (Courtesy of Cummins Engine Company)

to pass through. They are on/off devices and therefore are either open or closed. Both actuators are commanded to open for a specific amount of time, which must be correlated to engine position and rotational speed.

7. The timing actuator is connected to a passage that feeds the timing chamber within the upper portion of the TP injector. Fuel entering the timing chamber of the unit injector first passes through a sized timing orifice and a timing check valve. The check valve prevents fuel from being forced back out of the timing chamber (in which case it would back pressurize the rail) of the TP injector when it is mechanically actuated.

8. The fueling or rail actuator is connected to the metering chamber in the lower section of the TP injector. To enter the metering chamber of the TP injector, fuel must first pass through a fueling orifice and fueling check valve. The fueling orifice is carefully sized to define the flow area into the metering chamber. The fueling check valve is designed to isolate the TP pump chamber under the pumping plunger at any time the pressure exceeds metering pressure. This would be a factor during preinjection plunger stroke. Like the timing check valve, the fueling check valve protects the rail from being subject to back pressure generated in the TP injector.

9. When the ECM switches each actuator, fuel passes through the actuator circuit, through a passage to charge the relevant section of the TP injector.

10. Due to the firing order of the engine typical for a 1-6 engine (1-5-3-6-2-4), only one injector unit in each bank can be charged with timing and metering fuel at a given moment.

11. When metering is complete, the actuating cam mechanically drives the plunger downward through its stroke. The fuel that has been metered forms a solid hydraulic lock into the timing chamber, and this acts on the pumping plunger. The fuel that is to be injected into the engine cylinder has been metered into the fueling chamber located under the pumping plunger. As the pumping plunger is driven through its stroke, when it contacts the fuel, it is forced through the orifices in the TP injector directly into the engine cylinder. The moment the injection pulse begins is known as start of injection (SOI).

12. All of the fuel metered into the fueling chamber is pumped into the engine. At the end of the effective pumping stroke, the plunger seats in the cup of the TP injector. Because the plunger will usually seat in the injector cup before the actuating cam has attained peak lift, the fuel in the timing chamber is trapped between the seated injector plunger and the cam profile, which continues to rise. Once the pressure in the timing chamber achieves a predetermined value, a check ball is unseated and the fuel in the timing chamber is spilled to a return passage, which routes it back to the fuel tank. At the end of the injection cycle, the timing tappet collapses and the fueling plunger is loaded into the injector cup with some mechanical crush.

13. The timing of the injection stroke depends on the fuel quantity metered into the timing chamber (this will determine the "length" of the hydraulic column that will act on the pumping plunger) and the amount of fuel metered into the pumping chamber under the plunger. The hard parameters are dictated by cam profile. As the TP injector is actuated mechanically, cam profile geometry will define the window within which the ECM can manage an effective stroke.

14. None of the fuel used to charge the timing chamber is used to fuel the engine. This fuel is routed back to the fuel tank after a fueling cycle.

15. All of the fuel metered to the fueling chamber is injected to the engine cylinder. A major advantage of the open nozzle system when comparing it to hydraulic injector nozzles is the fact that injection ends when the plunger seats in the cup. In other words, there is no lag between the ending of pump stroke and the point at which the pressure has collapsed sufficiently to permit nozzle closure. Because orifice size and pressure define emitted droplet size, the droplet size will increase during the pressure drop-off phase in hydraulic injector nozzles.

Interact System

The Interact System (IS) is used by Cummins as a prefix to describe the electronic management systems on some of their reworked engine series. The term describes the command, monitoring, and data retention electronics designed to connect ECM data with depot-based information and systems management and analysis capability. Cummins has introduced full authority, electronic management systems on smaller bore engines ahead of their competitors. They account for this by pointing out that most electronic management systems for diesel engines share a high degree of input and processing commonality. For instance, most of the system monitoring and command inputs, diagnostic tracking, customer programmable data options, fault tracking, and analysis and failure strategies are not going to differ significantly from one engine to another. The exception is the fuel system management software and output drivers, and this accounts for one-fourth to one-third of the total programmed data.

CAPS and ISC

The acronym CAPS represents Cummins accumulator pump system introduced to the North American truck market in 1998. The fuel system is a unique approach to common rail fuel management and was the subject of a major patent. Initially, this fuel system is to be used as the management system on a reengineered C series engine. A brief description of the new version of this engine, the ISC, and its fuel system is provided here.

Interact System "C" Engine (ISC)

Engine Features   The cylinder block is manufactured from gray cast iron and is fitted with mid-stop wet liners. The cylinder head is a single-slab, cast iron design. As with the Signature series, fluid lines have been integrated into the cylinder head and block to give the engine a clean, un-cluttered appearance. The pistons use a symmetrical bowl design, and the high-horsepower versions use plasma-coated, top compression rings.

Fuel System   The ISC-managed engine combines certain features found in a common rail system with other features more common in a pump-line-nozzle system. A primary advantage of the CAPS fuel system is its ability to control engine fueling rates independent of engine rpm and hard limits such as cam profile geometry. Rate shaping refers to the speed at which a fuel pulse is delivered to the engine cylinder. In most cases, fuel injected to a diesel engine cylinder must be forced through orifices in an injector nozzle or cup. The pressure of the fuel defines the sizing of the droplets emitted from the nozzle. The higher the fuel pressure, the smaller the droplet. In most fuel systems, the pump used to generate the injection pressures is actuated mechanically by a cam profile. This means that the cam profile and the speed at which it is rotated define injection pressures and therefore rate shape. When a system is capable of rate shaping, it means that an ECM can control the injection pressure independently of rpm and hard limits such as cam geometry. CAPS also uses pilot injection. Pilot injection refers to the ability of a fuel system to break up the fuel pulse into a pilot pulse and main injection. The pilot pulse is of short duration. When the pilot pulse is delivered to the engine cylinder, fueling is cut off temporarily while the fuel in the cylinder vaporizes and is heated to ignition temperature. At ignition of the pilot fuel, main injection begins. Pilot injection was initially used as a cold start strategy to reduce diesel knock, but some current systems manage pilot injection through all fueling. Pilot injection and rate shaping combine not only to reduce emissions and provide more complete combustion, but also to greatly improve cold startability, overall engine noise, and transient response to operating variables such as a sudden increase or decrease in engine load.

The CAP fuel system is a common rail fuel system with a difference. The system operation is as follows:

1. A gear pump delivers fuel to a pair of pumping plungers driven by dedicated lobes.
2. These deliver fuel to an accumulator positioned at the top of the CAPS.
3. An ECM-controlled fuel valve then sends fuel to the distributor module.
4. The distributor module directs this fuel in firing order sequence to hydraulic, multi-orifice injectors located in the cylinder head by means of high-pressure pipes.
5. The injector nozzles are centered over the engine cylinder. CAPS peak pressures are stated to be in the region of 18,000 to 20,000 psi.

Fuel System Components

Gear Pump   The gear pump draws fuel from the fuel tank, pulls it through a filter, and delivers it to the high-pressure pumping elements. It outputs a regulated pressure of 180 psi. In common with most Cummins fuel management systems, the entire fuel subsystem is under suction.

High Pressure Pump   A pair of cam lobes with three lift profiles per lobe is engaged to two ceramic pumping plungers by means of roller tappets that ride the cam profile. The plungers pressurize the fuel and deliver it to the accumulator. A one-way check valve maintains accumulator pressure when the plunger retracts. Fuel is only pumped to high pressure when required.

Accumulator   The accumulator is managed to contain the precise quantity of fuel required to prevent fuel pressure from dropping during fuel injection pulses. Two pumping control valves located in the accumulator and controlled by the ECM regulate the pressure in the accumulator. An accumulator pressure sensor signals constant feedback to the ECM for regulation.

Injection Control Valve   The injection control valve is a two-position, three-way valve that connects the accumulator to the distribution head when energized. When de-energized, the valve connects the injectors to drain. Control of fuel quantity and injection timing is the responsibility of this ECM-controlled valve.

Distribution Head   The distribution head hydraulically connects the fuel from the injection control valve to each hydraulic injector nozzle in engine cylinder firing sequence.

Electronics   CAPS is a full authority engine management system and is equipped with SAE J1939 compatible hardware and data protocols. The ECM monitors pressure values in the accumulator and controls the operation of the single fuel control valve.

The newest version of the successful medium bore engine M-11 gets a new name, the interact system M series (ISM). It receives some reworking to improve fuel efficiency by a claimed 2 to 4%, and oil change intervals up to twice that of its predecessor with an option to use the CENTINEL oil management system.

The top power rating of the engine is 425 bhp with a peak torque capability of 1,550 lb.-ft., an impressive output for an 11 L engine. The engine is equipped with a variable geometry, wastegated turbocharger to increase the effective rpm and load range of the engine over a wider band. The ECM may be programmed with a number of smart options that can increase peak torque in the top two gear ratios and boost brake power on an as-required basis. The M-11 targets a fleet customer base and some of the programmable features are designed to reward fuel-efficient driving by increasing the maximum road speed value, and "punish" lower fuel efficiency performance by reducing the maximum road speed value.

Interact Series B Engine (ISB)   The Cummins B Series engine is classified as a small-bore highway diesel and managed by in-line, port-helix, and rotary fuel injection pumps. Cummins recently introduced an electronically managed version of this engine available in standard ratings of 230 bhp and 250 bhp, with a special applications rating of 275 bhp for RV and fire equipment markets. These horsepower ratings will surely promote the use of this engine into a wide range of commercial vehicle applications.

Engine operating speeds range up to 2,500 rpm and peak torque values have improved 35% to 660 lb.-ft. at 1, 600 rpm. The engine dimensional specifications remain largely unchanged, and total displacement remains at 5.9 L but a single-casting, 24-valve cylinder head is used on the new ISB engine. Multi-orifice hydraulic injectors, specifically Bosch pencil-type nozzle assemblies, are used for direct injection of fuel to the engine cylinders. The injector location and cylinder head design were developed so it could be considered for adapting to a common rail system at a later date. The ECM is manufactured by Motorola and supports a comprehensive input, diagnostic, and customer programmable options package not usually found in the smaller bore, highway diesel engines.

The ECM is located on the right side of the cylinder block. Its primary output is the Bosch electronic controller for the fuel pump. The fuel pump is a Bosch VP44 rotary pump, which is a radial piston rotary distributor pump that supplies Bosch pencil-type injectors at peak pressure values of up to 20,000 psi. The VP 44-injection pump is controlled by solenoid valve that permits precise metering of fuel and a wide fuel delivery timing window. This pump is also capable of rate shaping and pilot injection.

Summary

❏ DDEC I, II, III, and IV are all full authority electronic diesel engine management systems that have most of their primary components in common. The ECM houses a microprocessor. The EDU, is usually engine mounted and, despite being physically smaller than the DDEC II ECM, processes data at eight times the speed and is capable of seven times the memory retention.

Terms to Know

BOE—beginning of energizing

BOI—beginning of injection

Camshaft position sensor (CMP)

❑ The electronic distributor unit is the DDEC injector driver unit responsible for switching the EUIs, and electronic governing offers both limiting speed (LS) and variable speed (VS) governing. DDEC may be programmed with customer and DDC data electronically to EEPROM. The removable PROM chip plugged into to the ECM motherboard has been replaced by the current system's EEPROM capability.

❑ DDEC systems may be read at different levels using flash codes, the DDR, and personal computers. They may be programmed to protect the engine when threshold values are exceeded at three different levels: driver alert, ramp down to default, and shutdown. DDEC III supports ProDriver, a driver data display and ProManager a program that allows vehicle data to be downloaded from the ECM for analysis.

❑ The Caterpillar ADEM ECM incorporates two microprocessors and all of the necessary switching apparatus. Caterpillar ADEMs locate the injector drivers within the ECM housing and use induction coils to spike the EUI actuation voltage to around 100 V. Current version Caterpillar ADEMS ECMs no longer use a replaceable personality module but retain PROM type data in flash memory or EEPROM. This logs all proprietary data and many customer programmed options. Caterpillar ECMs have a nonvolatile electronic memory component (NV-RAM) and an integral battery to sustain it should battery power be interrupted for any reason.

❑ Caterpillar electronic management systems support smart programming options, which replace formerly hard limits, such as vehicle speed limit and programmed cruise settings, with soft limits to optimize fuel economy and driver state of mind. Standard in current Caterpillar electronic systems is a data log, which records critical engine and vehicle running parameters in three formats: lifetime totals, trip totals, and instantaneous readout.

❑ CELECT Plus management may be divided into four subsystems: hydromechanical fuel, input circuit, ECM, and output circuit. Charging pressures to CELECT injectors are the responsibility of a gear-type transfer pump. Charging pressures are converted to injection pressures by the CELECT injectors, which are actuated by cam profile. The ECM controls the effective stroke by switching a solenoid on the CELECT injector using a coil spiked actuation signal of 72 V. CELECT systems have self-diagnostic capability and are programmable with customer and Cummins data.

❑ The Bosch EUP system may be classified as a full authority, electronic management system. Two EUP engine series are currently available, the DDC series 55 and the Mack Trucks E-7-EUP. The DDC series 55 engine is managed by DDEC III electronics and the Mack Trucks E-7-EUP is managed by V-MAC III, a two-module vehicle and engine controller system.

❑ The DDEC III ECM integrates all the computing and switching functions required to manage the series 55 engine in a single housing mounted on a fuel-cooled heat exchanger. The V-MAC engine controller (ECU) is mounted on a fuel-cooled heat exchanger located on the right side of the engine. Both DDEC III and V-MAC III on EUP systems function similarly to DDEC and V-MAC on non-EUP systems.

❑ The EUP systems are supplied with fuel from the fuel subsystem at charging pressure. Fuel is constantly cycled through the system circuitry and the EUPs, following which it is returned to the tank. Actual plunger stroke is dictated by cam geometry. The DDEC ECM/V-MAC ECU controls plunger effective stroke by switching the EUP solenoids.

❑ The EUP plunger is a simple cylindrical piston with no milled recesses and drillings, with a high-pressure pipe connecting each EUP unit to a multi-orifice hydraulic injector nozzle located over each cylinder for direct injection.

❑ The HEUI engine management system is the first to use electronically controlled, hydraulically actuated injection pump units. Until the introduction of HEUI technology, pumping to diesel injection pressure values was always achieved mechanically. Effective injection pumping stroke in HEUI units is not limited by the hard limits of cam geometry as in other full authority, electronic management systems.

❑ Engine lubricating oil is used as the hydraulic actuating medium for the HEUIs. A swash-plate hydraulic pump boosts pressure values up to 3,000 psi (200 atm). The ECM switches the injection pressure regulator (IPR) to achieve oil pressure values of between 33 and 200 atms to control the HEUI actuating oil pressure values. The actuating oil pressure value determines the descent speed of the HEUI pumping plunger, and thereby manages injection rate and the emitted atomized droplet sizing. The HEUI solenoid controls a poppet valve that, when energized, traps rather than spills the high-pressure oil so that it acts on the amplifier piston. When the HEUI solenoid is de-energized, the oil is once again spilled to the rocker housing.

❑ The Cummins Signature series engine uses a hydraulic equation that is constructed to determine engine fueling in the HPI-TP system. The critical flow areas are established in the injector unit (and are held constant). However, the pressure value is now held constant (common rail pressure regulated at 250 psi) so the control variable is time. Specifically, time correlates to the amount of time the fueling actuator is held open to feed the injector-metering chamber with fuel. In the acronym used to describe the system, the letter T appears before the letter P to indicate it is the control variable. The Signature series HPI-TP system is classified as a full authority, common rail engine and chassis management system.

Terms to Know, Continued

Throttle position sensor (TPS)

Timing reference sensor (TRS)

Vehicle speed sensor (VSS)

Review Questions

Short Answer Essays

1. Outline the components and explain the operation of the Detroit Diesel Corporation (DDC) Detroit Diesel electronic controls (DDEC) fuel subsystem and the principles of operation of a DDEC ECM and its switching apparatus.

2. Define the components and explain the layout and operation of the Caterpillar full authority, advanced electronic engine management system (ADEMS).

3. Identify the major components and explain the principles of operation of the Cummins CELECT system and the differences between CELECT and CELECT Plus.

4. Explain the differences between full authority EUP and EUI diesel engine management.

5. Identify the four primary subsystems that manage an HEUI engine's output.

6. Identify the components and explain the operation of the Cummins HPI-TP electronic common rail fuel system on a Signature 600 engine.

7. Explain how the HEUI system actually controls the oil pressure values.

8. Differentiate between EUP and EUI.

9. Explain the operation of the HEUI injection pressure regulator.

10. Explain how regulating oil pressure meters fuel in the HEUI system.

Fill-In-the-Blanks

1. The electronic distributor unit (EDU) is the DDEC _____ _____ unit responsible for switching the EUIs and electronic governing offers both limiting speed (LS) and variable speed (VS) governing.

2. DDEC systems may be read at different levels using flash codes, the DDR and a _____ _____.

3. Caterpillar ADEMS locates the _____ _____ within the ECM housing and uses induction coils to spike the EUI actuation voltage to values around 100 V.

4. Caterpillar ECMs have a ___ _____ electronic memory component (NV-RAM) and an integral battery to sustain it should battery power be interrupted for any reason.

5. CELECT Plus management may be divided into four subsystems: _____, _____, _____, and _____.

6. Charging pressures are converted to injection pressures by the CELECT injectors that are actuated by _____ _____. Effective stroke is ECM controlled by switching a _____ on the CELECT injector.

7. The Bosch EUP system may be classified as a _____ _____ _____ _____ system.

8. The EUP plunger is a simple _____ _____ with no milled recesses and drillings.

9. The HEUI engine management system is the first to use _____ _____ _____ _____ injection pump units. Until the introduction of HEUI technology, pumping to diesel injection pressure values was always achieved mechanically.

10. The Cummins Signature series engine uses a hydraulic equation that is constructed to determine engine fueling in the HPI-TP system. The critical flow areas are established in the injector unit (and are held constant). However, the _____ _____ is now held constant, so the control variable is _____.

ASE-Style Review Questions

1. The Caterpillar proprietary EST is known as:
   A. Compulink
   B. ECAP
   C. ProLink 9000
   D. DDR

2. What is the primary function of the SRS?
   A. Signal engine position data to the ECM
   B. Signal engine rpm data to the ECM
   C. Signal vehicle road speed data to the ECM
   D. Signal ambient temperature data to the ECM

3. A Caterpillar ADEMS on a 3406 engine is being discussed. *Technician A* says that the latest Caterpillar electronics permit the programming of the cruise control speed limit at a higher value than maximum vehicle speed limit. *Technician B* says that VSL is a hard programmed parameter that cannot be exceeded. Who is correct?
   A. A only
   B. B only
   C. Both A and B
   D. Neither A nor B

4. Fuel is delivered to the EUP by the transfer pump at what pressure?
   A. 5,000 psi
   B. Charging pressure
   C. 26,000 psi
   D. Peak pressure

5. Which HEUI circuit provides NV-RAM with continuous power?
   A. KAMPWR
   B. IDM
   C. SIG GRD
   D. VEPS

6. A Cummins CELECT Plus electronic fuel system is being discussed. *Technician A* says the coolant and oil temperature sensors are both of the thermistor type. *Technician B* says both sensors must be supplied with a reference voltage value. Who is correct?
   A. A only
   B. B only
   C. Both A and B
   D. Neither A nor B

7. *Technician A* says the Caterpillar intensifier piston and the Navistar amplifier piston both refer to the same HEUI component. *Technician B* says an amplifier piston will produce fuel pressures seven times the value of the actuating oil pressure. Who is correct?
   A. A only
   B. B only
   C. Both A and B
   D. Neither A nor B

8. Which dash diagnostic light is used to flash active DDEC fault codes?
   A. Oil pressure warning
   B. Service engine soon (SEL)
   C. Electronic malfunction
   D. Check engine light (CEL)

9. What is the V-Ref value in DDEC highway engine management systems?
   A. 5 volts
   B. 12 volts
   C. 24 volts
   D. 36 volts

10. The hydraulic injector nozzle used in the CELECT electronic unit injector is set to have an normal opening pressure of:
    A. 10 atm
    B. 340 atm
    C. 1700 atm
    D. 2000 atm

Diagnostic Strategy

Upon completion and review of this chapter, you should be able to:

- ❑ Properly define failure analysis and diagnosis or troubleshooting.

- ❑ Using a diagnostic thought process (scientific process of elimination) and fault statements, reduce the possible causes of a fault, verify the complaint, and road/dyno test the vehicle; review past maintenance documents (if available); determine further diagnosis.

- ❑ Differentiate between the diagnostic fault code terms DTC, FMI, MID, PID, and SID.

- ❑ Verify the complaint, and either road-test the vehicle or run it on a dynamometer; review past maintenance documents (if available); determine further diagnosis.

Introduction

The terms **troubleshooting** and diagnosis generally mean same thing. They are used to describe noninvasive (as little of the engine is disassembled as possible) methods of determining the cause of an engine or fuel system problem. These methods can vary from educated guesswork to highly structured procedures, such as GM's strategy based diagnosis. Most manufacturers make diagnosis or troubleshooting guides available for purposes of diagnosing common engine problems. These guides provide a means of helping the technician work methodically and not overlook an obvious cause of a problem. Diagnosing electronically managed engines is necessarily a much more structured procedure requiring the use of scan tools (DDR), personal computers, and **sequential troubleshooting charts**. These require the technician to test circuits and components in a step-by-step procedure, in which the results of a test in one step determine the routing to the next step in the sequence. Skipping or inaccurately performing just one of the steps in a sequential troubleshooting chart can render the whole procedure meaningless. When troubleshooting hydromechanical engines and fuel systems, there is no substitute for a thorough knowledge of the specific engine system.

The term **failure analysis** normally refers to component analysis methods used to determine the cause of a failure. In most cases, failure analysis takes place *after* the engine or other component has been disassembled with the objective of avoiding recurrence. Failure analysis often relies on learned knowledge of the specific engine. What follows is a general look at typical engine component problems and their causes. The following diagnostic approaches provide a general guideline and avoid addressing manufacturer-specific product problems. Most original equipment manufacturers (OEMs) do this in their own technical literature.

> Diagnosis is researching a system breakdown using technical information to determine diagnostic procedures. By following strategy-based diagnostic routines the technician generates solutions to problems and determines needed repairs while recognizing when assistance is needed.

Strategy of Diagnosis

Six-Step Diagnostic Process

The technician's job would be easy if he or she could walk up to a truck and know instinctively what the problem and how to fix it. Sometimes that happens, especially if there is a history a particular problem occurring on a particular model of truck.

Usually, though, the problems you see only occur occasionally. There is no series of identical vehicles that all have the same problem. To save time in diagnosis and to avoid replacing functional parts, you need to follow a good troubleshooting plan. The six-step troubleshooting plan is a standard approach, recognized by a number of truck original equipment manufacturers. The six steps are:

1. Verify the complaint.
2. Determine the related symptoms.
3. Analyze the symptoms.
4. Isolate the trouble.
5. Fix the problem.
6. Check for proper operation.

1. Verify the Complaint The first step is to ask the customer or driver what the problem is, then check the vehicle (Figure 12-1). Determine if you can duplicate the problem the customer has reported. The average customer or truck driver is typically not technically oriented and is likely to describe the problem in nontechnical terms. The technician must make sure he or she understands what the customer has reported.

It's also very important to know how each system works. If you don't know how a system is supposed to operate, you won't know when it is broken. The **owner's manual** is a good place to start. The owner's manual is a booklet specific to the truck and model year, provided to the owner. It gives the owner information about how to operate the various systems on the vehicle. Also, there is usually a description of the system in the OEM **service manual**. The service manual is specific to the truck and model year, and is provided to the OEM dealership. It gives technicians information about how the systems work, diagnosis, and repair. You need to also check the OEM diesel engine service or technical manual, because it will contain much more detailed information about the engine and the specific fuel system.

Sometimes a customer will report a "problem" that is not really a problem—the system is working the way it is supposed to. When you verify a problem, you also have to find out whether the problem is **continuous** (it happens all the time) or **intermittent**. If the problem is continuous, you'll notice it right away. If the problem is intermittent, you'll need to find out under what conditions the problem occurs.

Figure 12-1 Verify the concern with the driver or truck owner.

2. Determine the Related Symptoms   Once you know there is a problem, you need to check the symptoms thoroughly. When you check related symptoms, you are checking how the system operates. At this point, the thing that will be most useful is the wiring diagram. The goal is to:

1. Find out how much of the circuit is affected.
2. Find clues as to where the problem is located.

Figuring out how a circuit is related to other circuits will save a lot of time. The technician operates the problem circuit to see exactly what is and is not working. Automotive circuits are related to other circuits because they have parallel connections:

1. Most electrical circuits have two or more loads connected in parallel.
2. Common power sources or grounds relate different circuits to each other.
3. Different circuits may share a sensor or a switch.

By checking the operation of related circuits, you can eliminate parts of a circuit or components as possible causes. Since you will have fewer items to check, you will spend less time isolating the problem.

Based on how the circuit operates, you can make some assumptions: If the entire circuit system is "dead," this indicates:

1. There is a possible problem with power (a fuse) or ground.
2. The load (component) is bad.

If any part of the circuit still works:

1. Power to the circuit and the main ground point are probably OK.
2. You need to find out which loads are working and which ones are not. That way you can see what wiring and connections the bad parts of the circuit have in common.

3. Analyze the Symptoms   At this point you need to figure out:

1. Which components or circuits are affected
2. What kind of problem to look for
3. When the problem occurs

If you have a copy of the wiring diagram, you can **line out** the circuit; that is, you can highlight the parts of the circuit that are working. These are the parts you do not need to check. The parts that are not highlighted are the places where a problem could exist. Lining out the circuit will save a lot of time.

4. Isolate the Trouble   To isolate the problem (Figure 12-2), take the wiring diagram you have lined out:

1. Find the possible problem areas.
2. Decide where to begin checking the circuit.
3. Make the necessary inspections.

Shop Manual Chapter 12, page 424.

Shop Manual Chapter 12, page 424.

Most original equipment manufacturers offer PC-based software programs for diagnosis.

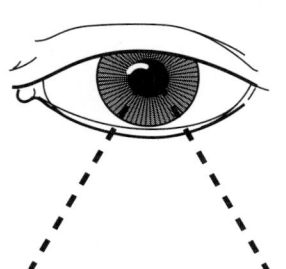
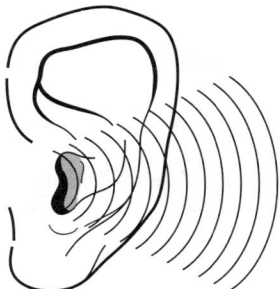

Figure 12-2 Look and listen to determine the cause.

Figure 12-3 Verify that the repair has resolved the original concern.

Anywhere you did not trace current flow is a possible problem area. Circle the places on the circuit where the problem could be located. Since any one of these places could be where the problem is, you will have to decide where to start. The best order to go in is based on:

1. How easy it is to get to a component
2. Whether you can inspect the component visually
3. If you know that many cars of this model have failed in this same way
4. If several components or circuits are inoperative, start with parts of the circuit that are common to all

5. Fix the Problem Fixing the problem is straightforward. Repairing an electrical problem involves:

1. Repairing or replacing a component
2. Wiring repair, or
3. Servicing a connection (connector, terminal, or ground)

The repair must be permanent. Temporary fixes will make the problem go away only for a short time.

6. Check for Proper Operation (Figure 12-3) After making the repair, always verify that the problem was actually fixed. Operate the circuit as thoroughly as you did when you first diagnosed the problem to make sure all of the functions of the circuit work normally. There may be several reasons why a circuit malfunctions. Rechecking the circuit will ensure that the driver will be satisfied with the repair.

Strategy-Based Diagnostics

Strategy-based diagnostics takes into account the truck engine's self-diagnostic capabilities and the service information provided by the manufacturer in addition to the wiring diagrams. This approach is particularly useful for technicians working in dealerships, where diagnostic tools and service information are readily available (Figure 12-4).

Detroit Diesel Corporation offers PC-based troubleshooting software called "Detroit Diesel Diagnostic Link Software that is a case-based reasoning program.

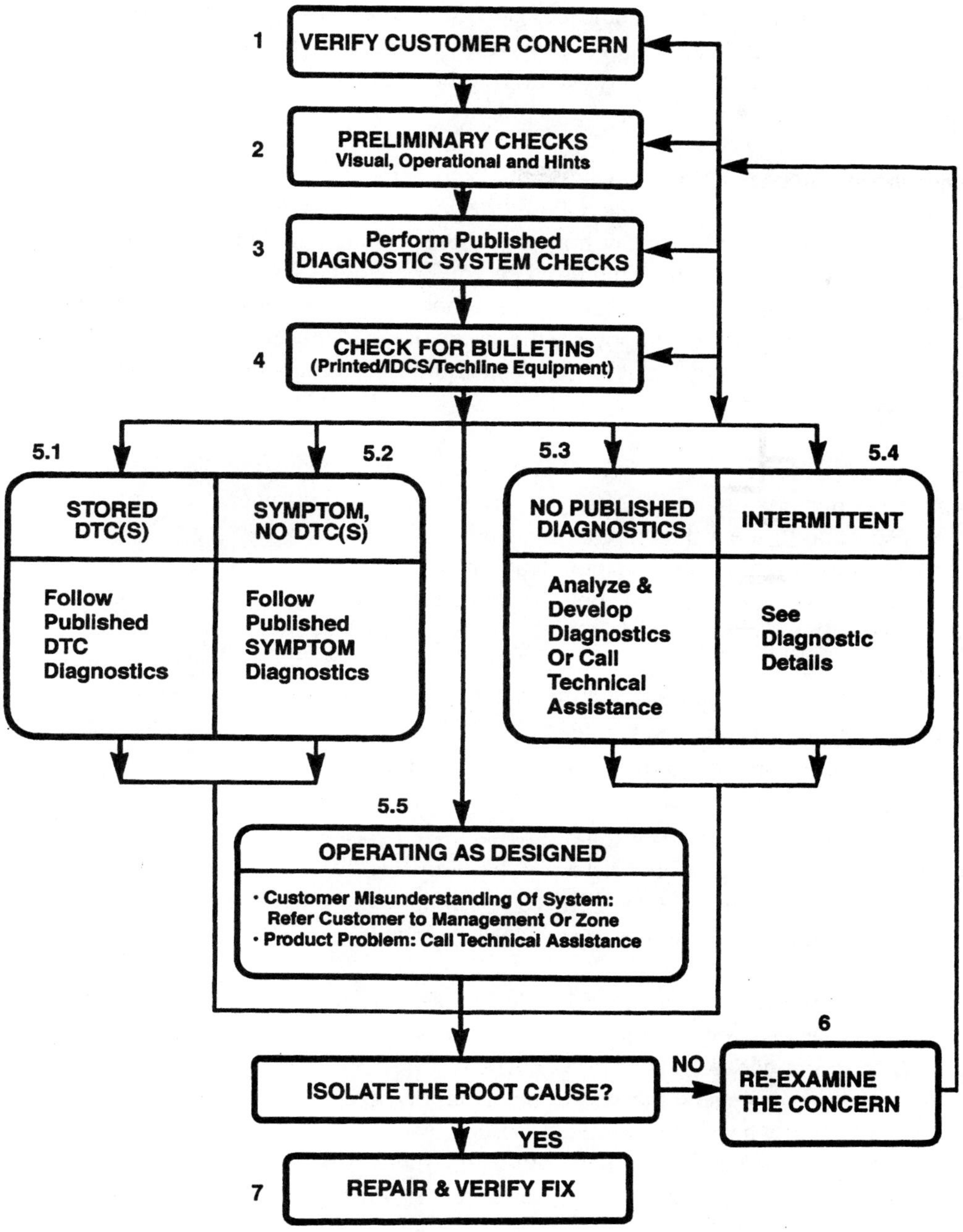

```
   ┌─────────────────────────────┐
 1 │ VERIFY CUSTOMER CONCERN     │◀──────────┐
   └─────────────────────────────┘           │
              │                               │
              ▼                               │
   ┌─────────────────────────────┐           │
 2 │ PRELIMINARY CHECKS          │◀──────────┤
   │ Visual, Operational and Hints│          │
   └─────────────────────────────┘           │
              │                               │
              ▼                               │
   ┌─────────────────────────────┐           │
 3 │ Perform Published           │◀──────────┤
   │ DIAGNOSTIC SYSTEM CHECKS    │           │
   └─────────────────────────────┘           │
              │                               │
              ▼                               │
   ┌─────────────────────────────┐           │
 4 │ CHECK FOR BULLETINS         │◀──────────┤
   │ (Printed/IDCS/Techline Equipment)│      │
   └─────────────────────────────┘           │
```

5.1 STORED DTC(S)	5.2 SYMPTOM, NO DTC(S)	5.3 NO PUBLISHED DIAGNOSTICS	5.4 INTERMITTENT
Follow Published DTC Diagnostics	Follow Published SYMPTOM Diagnostics	Analyze & Develop Diagnostics Or Call Technical Assistance	See Diagnostic Details

5.5

OPERATING AS DESIGNED

· Customer Misunderstanding Of System:
  Refer Customer to Management Or Zone
· Product Problem: Call Technical Assistance

```
   ┌──────────────────────────┐   NO   ┌──────────────┐  6
   │ ISOLATE THE ROOT CAUSE?  │──────▶ │ RE-EXAMINE   │
   └──────────────────────────┘        │ THE CONCERN  │
              │  YES                    └──────────────┘
              ▼
   ┌──────────────────────────┐
 7 │ REPAIR & VERIFY FIX      │
   └──────────────────────────┘
```

Figure 12-4 Strategy-based diagnosis chart.

A BIT OF HISTORY

General Motors instituted strategy-based diagnosis (SBD) in 1990. Truck technicians who work in independent repair shops also need to use diagnostic tools and service information when troubleshooting. Trucks are so complex that no technician can diagnose and repair every problem. Particularly with electrical and electronic systems, the technician is much more likely to be successful if he or she uses all available resources.

The Cummins Engine Company offers PC-based software for diagnosis called INSITE.

Shop Manual Chapter 12, page 424.

The goal of strategy-based diagnostics is to guide you when you make a plan for each specific diagnostic situation. This scientific process of elimination also mirrors the six-step process listed in Table 12-1. In strategy-based diagnostics the customer concern history and the vehicle's self-diagnostic record are used to fix the problem.

1. Verify the Customer's Concern   The first step is the same whether you are following the six-step toubleshooting plan or strategy-based diagnostics: verify the customer's concern (Figure 12-4). To do this, you need to know how the system operates normally, so you know whether the customer's concern really is a failure. When you operate the system, you might have to check the owner's manual or the service manual to see how the system is supposed to operate.

As an alternative, you could operate the system in an identical vehicle to see if it works differently. In any case, it is important to get as much information about the problem from the customer as possible.

2. Preliminary Checks   About 10% of successful vehicle repairs are diagnosed with this step during the preliminary inspection. After you have verified the customer's concern, you should do a visual inspection to see if there are any obvious clues. You may have to take some covers and trim off, but do not remove any system components.

In addition to inspecting, you should also notice any unusual smells, noises, or movements. If the service and maintenance history of the truck is available, check what work has been done before.

Depending on how complicated the system is, the quick checks should include:

1. Operate the suspect system.
2. Inspect the wiring harness, power, and ground.
3. Check for blown fuses.
4. Look for separated connectors.
5. Inspect the connectors (including the terminals) for damage and tightness.
6. Notice any unusual noises, smells or vibrations or other movements.
7. Check the service history of the truck.

3. Follow Diagnostic System Checks   Most diesel trucks have self-diagnostic capabilities. If you follow the OEM's procedure for checking that system, one of the steps will be to check for **diagnostic trouble codes (DTCs)** or SAE/ATA J1587/J1939 **PID** fault codes. If a DTC has been set, this will make diagnosing the system much easier, because it will tell you which part of the system is faulty and what sort of problem exists.

Even for systems that do not have self-diagnostic capabilities, in many cases the manufacturer provides a step-by-step procedure for troubleshooting the system. If you follow this procedure, it will save a great deal of time and effort.

4. Check for Service Bulletins   About 30% of successful vehicle repairs are diagnosed by checking service bulletins. Truck manufacturers and importers publish service bulletins about problems that have appeared on certain models. These bulletins may be published on paper, mi-

Table 12-1 Six-step process of diagnosis.

1. Verify the complaint.
2. Determine the related symptoms.
3. Analyze the symptoms.
4. Isolate the trouble.
5. Fix the problem.
6. Check for proper operation.

crofiche, CD-ROM, or on-line. Some manufacturers are beginning to publish service bulletins via the Internet. If you work in a dealership, you can search for a bulletin about the type of problem you are working on.

Often, a bulletin will exactly describes the problem you are working on and tell you how to fix it. Using service bulletins and other types of updated information for diagnosis and repair will save time and help you avoid replacing components unnecessarily. Once you have checked for service bulletins, you will need to select a diagnostic path. There are four paths to choose from:

5.1 Diagnose stored DTCs

5.2 Diagnose symptoms

5.3 No published diagnostics

5.4 Diagnose intermittents

Diagnose Symptoms About 40 percent of successful vehicle repairs are diagnosed with these steps. If the system you are working on has self-diagnostic capabilities, and you find one or more Diagnostic Trouble Codes, they will help you troubleshoot the system. DTCs tell you which part of the system is faulty and what sort of fault is occurring.

Truck manufacturers and importers publish step-by-step procedures in their service manuals, for use in diagnosing systems. As you follow the procedures do each step in order. Don't skip steps or do them out of order. If you do, you may diagnose the problem incorrectly and replace the wrong part. If special tools are required, be sure to use them.

Diagnostic Trouble Codes (DTC) **SAE/ATA** (Society of Automotive Engineers/American Trucking Association) developed a diagnostic trouble code system that has been adopted by all the North American truck engine/electronics OEMs. SAE J1587 covers common software protocols in electronic systems, SAE J1708 covers common hardware protocols, and SAE J1939 covers both hardware and software protocols. The acceptance and widespread usage of these protocols enables the interfacing of electronic systems manufactured by different OEMs on truck and bus chassis, plus allows any manufacturer's software to read the electronic systems of other manufacturers. The truck technician who works with multiple OEM systems may find it easier to work using these codes rather than use the OEM proprietary codes.

Shop Manual Chapter 12, page 425.

The SAE/ATA DTC is composed of the following: a MID, PID, SID, and a FMI. **MID** stands for **message identifier** and is used to describe a major vehicle electronic system, usually with independent processing capability and it appears as a separate identifier in the scan tool (DDR) or PC. It is used to identify the source of the data transmission. For example, MID 128 identifies an engine computer, whereas MID 130 identifies a transmission computer. The acronym **PID** stands for parameter identifier and is used to code components within an electronic subsystem. It is a two- or three-digit code that is assigned to each component to identify data via a data link to the engine computer (ECM/PCM). The PID becomes the main number in an SAE diagnostic trouble code (DTC); i.e., *P111-3* is a coolant level sensor with high-input voltage. The "P" in an SAE DTC identifies it as a PID, 111 is the identifier and the "3" is a **failure mode indicator (FMI)**, which will be explained later in this section.

The acronym **SID** stands for **subsystem identifier** and is a one-, two-, or three-digit code assigned to identify a section of a control system without a related parameter identifier (PID). The SID identifies the major subsystems of an electronic circuit. The SID is a single byte character used to identify field-repairable or replaceable subsystems for which failures can be detected or isolated. SIDs are used in conjunction with SAE standard diagnostic codes identified in J1587 within PID 194, which is "Transmitter system diagnostic code and occurrence count." For example, S250-12 identifies a J1587 data link fault, and S22-4 indicates a timing sensor problem with voltage below normal or a shorted sensor. The "3" is a failure mode indicator.

The failure mode identifier describes the type of failure detected in the subsystem and identified by either a PID or SID. The FMI and either a PID or SID combine to form a given DTC as defined in J1587 with PID 194. FMIs are indicated whenever an active or historic code is read using ProLink or a PC.

OBD II Diagnostic Trouble Codes (DTC) The on-board diagnostic second generation (OBD II) code process comes from the Clean Air Act Amendment that required OEMs to meet more stringent standards for diesel applications by 1997. OBD II requires specific emission-related systems to be monitored for malfunction by the engine computer. Diesel engine systems must monitor exhaust gas recirculation (EGR), misfire, and glow plug malfunction. They must further contain specific data organization, diagnostic trouble codes, and communication with the scan tool. OBD II requires a scan tool like the one in Figure 9-4 with specific features, such as class 2 communication in the kilo/mega Baud range (10.4 kilobaud is 10,400 bits per second), freeze frame, and stored failure records. A class 2 data line transfers information by toggling the line voltage from 0 to 7 V.

Figure 12-4 shows the structure of an OBD II diagnostic trouble code. OBD II federally mandates a standardized DTC for light duty as adopted by the American Trucking Association (ATA) in SAE J1587. These OBD II codes contain a letter and four digits. The letter indicates the function of the device at fault. The following is a list of these letters:

P Powertrain

C Chassis

B Body

U Network or data link code

The first digit indicates if the DTC is generic or OEM specific (0 = generic and 1 = specific). The second number indicates the specific vehicle system that has the fault. The following chart details the second number:

1. Fuel and air metering
2. Fuel and air metering
3. Ignition system or misfire
4. Auxiliary emission system
5. Speed and Idle control
6. Computer output circuit
7. Transmission
8. Transmission

The last two digits indicate the component or section of the system at fault. Finally, there are four types of DTCs indicated by the letters A, B, C, and D. A and B codes are emission related and will light the malfunction indicator light (MIL), also known as a check engine light. Types C and D are nonemissions related and will not light the MIL. However, all four types will store an inactive or historical DTC.

When no active or inactive diagnostic trouble codes are stored in the ECM/PCM, the technician will have to develop his or her own diagnostics based on operational knowledge of the system.

No Published Diagnostics When the technician encounters a problem for which there are no DTCs and no symptoms in the manufacturer's service information match the symptoms the truck operator reports, the technician will have to develop diagnostics.

The wiring diagrams in the service manual will help. They show the system power, ground, input, and output circuits. They also show the splices and other areas where different circuits come together. They indicate where components are located, so the technician can figure out whether components, connectors, or harnesses are exposed to extreme temperature, moisture, road salt, or other corrosive materials.

The following are steps in developing diagnostics:

1. Look at the wiring diagram to locate power, ground, input, and output circuits.
2. Make measurements on the circuit and compare the measurements to truck in working order.

3. Check the wiring, connectors, and terminals.
4. Check the controls (switches and relays).
5. Check the components. Unplug them one at a time to see which ones are working.
6. Using a scan tool or a diagnostic computer, look at the data for the system and operate the system.
7. Check other systems that share the same power, ground, or inputs.
8. Call the manufacturer's technical assistance hotline.

Diagnose Intermittents   An intermittent problem usually occurs under certain conditions, but the technician may not know what all the conditions are. These problems are generally caused by:

1. Faulty electrical connections and wiring
2. Malfunctioning components
3. **Electromagnetic interference (EMI)** (This is static in the circuit, which can cause it to behave erratically)
4. Aftermarket equipment

Particularly with intermittent problems, getting the details from the customer is very important. If you can duplicate the problem, you may be able to use the symptom-based diagnostics provided by the manufacturer. For example, if the customer says the problem occurs when the truck is going over a rough road, shake the connectors or harnesses in the portion of the circuit where you think the problem might exist. Tap on parts or sensors with your finger.

If the problem occurs when the component is hot from weather or operation, use a heat gun to apply moderate heat to the part you think might be failing. If the problem occurs when it is raining or the humidity is high, spray water on the part of the circuit where you think the problem might be occurring. If the problem occurs when there are other electrical loads (such as the blower, lights, or rear window defogger), turn on these loads and see what happens to the suspect circuit or part.

6. Operates as Designed   Sometimes a car is working the way it's supposed to, but the customer thinks there is a problem. In this case, you should make sure this is the case.

You can do this by comparing the truck with an identical one. You can also check the owner's manual, the service manual, and the service bulletins. The OEM's technical assistance hotline can also help. The hotline is a telephone service, staffed by technicians who answer questions from technicians in repair facilities.

Re-examine the Customer's Complaint   If you do not find the problem after you have performed a diagnostic procedure, you should look at the customer's complaint again. Now you need to backtrack. What information have you already gathered? Repeat any procedures that need special attention. Make sure any test equipment or special tools are working properly. Make sure you did not skip any procedures. Select a different diagnostic path (5.1, 5.2, 5.3, or 5.4). If you used 5.3, "no published diagnostics," go back to the service manual and look for the symptom that most closely describes the symptom of the vehicle you are working on.

Once you have tried everything else, you may have to call the OEM manufacturer's technical assistance hotline for help.

7. Repair and Verify Fix   Once you have found the problem, repair it following the procedures in the service manual. Then make sure you have fixed the problem by checking the system under the same conditions listed in the customer's complaint. You may also have to do some preventive work to make sure the same problem does not happen again. If you called the manufacturer's technical assistance hotline, you may need to call them again to let them know how you fixed the problem.

Summary

❑ The terms troubleshooting and "diagnosis" generally mean the same thing. They are used to describe noninvasive methods of determining the cause of an engine/fuel system problem. Most manufacturers make guides available for purposes of diagnosing common engine problems. These guides provide a means of helping the technician to work methodically and not overlook an obvious cause of a problem. Diagnosing electronically managed engines is a structured procedure requiring the use of scan tools (DDR), personal computers, and sequential troubleshooting charts.

❑ The term failure analysis normally refers to component analysis methods used to determine the cause of a failure.

❑ The first step in the six step diagnostic process is to verify the complaint by asking the customer or driver what the problem is, then checking the vehicle.

❑ The owner's manual is a booklet specific to the truck and model year. It gives the owner information about how to operate the various systems on the vehicle. Also, there is usually a description of the system in the OEM service manual. It gives technicians information about how the systems work and about diagnosis and repair. The OEM diesel engine service or technical manual will contain much more detailed information about the engine and specific fuel system.

❑ Once you've shown that there is a problem, you need to check the symptoms thoroughly. When you check related symptoms, you are checking how the system operates. The third step in the diagnostic process is to analyze the symptoms. At this point you need to stop and figure out which components/circuits are affected and what kind of problem to look for and when it occurs.

❑ The fourth step is to isolate the trouble. To isolate the problem, take the wiring diagram and find the areas that are possible problem areas. Decide where to begin checking the circuit. The fifth step is to fix the problem. Fixing the problem is straightforward. Step 6 in the diagnostic process is to verify the repair or check for proper operation.

❑ The goal of strategy-based diagnostics is to guide you when you make a plan for each specific diagnostic situation. This scientific process of elimination also mirrors the six-step process. The difference is that this diagnostic strategy approach examines the customer concern history and utilizes the vehicle's self-diagnostic ability and then symptom diagnosis begins in step 5 of the flow chart.

Review Questions

Short-Answer Essays

1. Properly define the terms failure analysis and diagnosis or troubleshooting.

2. Explain the step 1 of the six-step diagnostic process.

3. Explain the step 2 of the six-step diagnostic process.

4. Explain the step 3 of the six-step diagnostic process.

5. Explain the step 4 of the six-step diagnostic process.

6. Explain the step 5 of the six-step diagnostic process.

7. Explain the step 6 of the six-step diagnostic process.

8. Explain GM's strategy-based diagnosis process.

9. Differentiate between the six-step process and the GM's strategy-based diagnosis process.

10. Differentiate between the diagnostic fault code terms DTC, FMI, MID, PID, and SID.

Fill-In-the-Blanks

1. The terms _____ and _____ are generally the same thing. They are used to describe noninvasive methods of determining the cause of an engine/fuel system problem.

2. Diagnosing electronically managed engines is necessarily a structured procedure requiring the use of scan tools (DDR), personal computers, and _____ _____ _____.

3. The term _____ _____ normally refers to component analysis methods used to determine the cause of a failure.

4. The first step in the six-step diagnostic process is to _____ ___ _____ by asking the customer or driver what the problem is, then checking the vehicle.

5. The _____ _____ is a booklet specific to the truck and model year provided to the owner. It gives the owner information about how to operate the various systems on the vehicle. There is usually a description of the system in the OEM _____ _____. It gives technicians information about how the systems work and about diagnosis and repair.

6. Once the technician has shown that there is a problem, he or she needs to check the _____ thoroughly.

7. The fourth step in the six-step diagnostic process is to _____ the trouble.

8. _____ the problem is straightforward.

9. Step 6 in the diagnostic process is to _____ the repair or check for proper _____.

10. The goal of strategy-based diagnostics is to guide you when you make a _____ for each specific diagnostic situation. This scientific process of _____ also mirrors the six-step process. The difference is that this diagnostic strategy approach examines the _____ _____ _____.

ASE-Style Review Questions

1. The first step in the diagnostic process is which of these items:
   A. Determine the related symptoms.
   B. Verify the complaint.
   C. Analyze the symptoms.
   D. Isolate the trouble.

2. The process of strategy-based diagnosis is being discussed. *Technician A* says that preliminary checks means making a quick inspection for obvious problems. *Technician B* says preliminary checks means gathering enough information so you can effectively search for a related service bulletin. Who is correct?
   A. A only.   C. Both A and B.
   B. B only.   D. Neither A nor B

3. *Technician A* says good technicians usually use steps 5.1 and 5.2. *Technician B* says steps 5.1 and 5.2 use directed diagnostics from the service manual. Who is correct?
   A. A only.
   B. B only.
   C. Both A and B.
   D. Neither A nor B.

4. A customer complains that the cruise control will not disengage. You take the truck with the DDEC III fuel system for a road test to see if the cruise control stays on when the brake pedal is depressed. The cruise control disengages when the brake pedal is depressed. Which of these is the *most* likely next step:
   A. Replace the stop lamp fuse.
   B. Replace all the bulbs in the stop lamps.
   C. Replace the stop lamp switch.
   D. Determine the related symptoms.

5. The customer reports that the brake lights on a 1998 GMC C70 medium truck (base model) do not work. What is first step in the six-step diagnostic process?
   A. Determine the related symptoms.
   B. Verify the complaint.
   C. Analyze the symptoms.
   D. Isolate the trouble.

6. A technician verifies the customer condition or concern by taking the truck in question number 5 for a road test and steps on the brakes and finds that neither the right nor left tail/stop lamp lights up. Which of these is the next step in the six step process:

   A. Isolate the trouble.

   B. Fix the problem.

   C. Determine the related symptoms.

   D. Analyze the symptoms.

7. All of the following are acronyms associated with retrieving faults from a vehicle *except*:

   A. PID (parameter identifier)

   B. DTC (diagnostic trouble code)

   C. FMI (failure mode identifier)

   D. SBDS (single bit diagnostic subroutine)

8. When trying to erase the PID error codes without the diagnostic test set, the technician can do which of the following to accomplish this?

   A. Turn off the key switch

   B. Disconnect the battery cable

   C. Unplug engine sensors

   D. Remove the electronic control module

9. *Technician A* says most generic scan tools can access any OBD II vehicle MIL DTCs. *Technician B* says that vehicle manufacturers have to provide MIL DTC commands to be OBD II certified. Who is correct?

   A. A only.

   B. B only.

   C. Both A and B.

   D. Neither A nor B.

10. Which of the following failure code identifiers codes and identifies components in the diagnostic trouble code?

   A. PID

   B. FMI

   C. SID

   D. DTC

GLOSSARY/GLOSARIO

> *Note: Terms are highlighted in bold followed by Spanish translation in bold italic.*

ADEMS advanced diesel engine management system. Caterpillar acronym used to describe their management electronics. The current version is third generation that uses a 32 bit processor and has dual 70 pin connectors.
ADEMS sistema de gestión avanzada del motor diesel. Acrónimo de Caterpillar que se utiliza para describir su electrónica de gestión. La versión actual es de tercera generación y utiliza un procesador de 32 bits con conectores duales de 70 contactos.

afterburn term that can be used to describe the normal combustion of fuel in a diesel engine cylinder after injector nozzle closure or random ignitions of fuel after primary flame quench in an engine cylinder.
postcombustión término que se puede utilizar para describir la combustión normal de combustible en un cilindro de motor diesel tras el cierre de la boquilla de inyección o encendidos aleatorios de combustible tras la extinción de la llama primaria en un cilindro de motor.

air box the term used to describe the chamber charged by a Roots blower in a DDC two-stroke cycle engine; the air box supplies the engine cylinders with the scavenging air charge.
caja de aire término utilizado para describir la cámara que se carga con un compresor Roots en un motor de dos tiempos DDC; la caja de aire proporciona a los cilindros del motor la carga de aire de evacuación de gases.

air/fuel ratio the mass ratio of an air-to-fuel mixture; also AFR.
relación aire/combustible relación de masa de una mezcla de aire y combustible; también AFR.

air standard dual cycle corresponds to the cycle used in today's diesel engines. Heat energy is added at constant volume and constant pressure. The constant-volume burning process is carried out until a predetermined pressure is reached, after which any additional burring is carried out at constant pressure.
ciclo dual de aire estándar corresponde al ciclo utilizado en los motores diesel actuales. La energía térmica se agrega con un volumen y una presión constantes. El proceso de quemado de volumen constante se lleva acabo hasta que se alcanza una presión predeterminada, después de la cual cualquier quemado adicional se lleva a cabo con una presión constante.

Air turbulence is the extreme disturbance of the compressed air in the combustion chamber. It causes the air molecules to move in all directions. They collide with each other and cause friction and heat. A twisting force like small tornadoes changes the smooth laminar airflow into turbulent flow. This increases the transfer of heat between the cool liquid-fuel droplets and this hotter air.
turbulencia de aire es la perturbación extrema del aire comprimido en la cámara de combustión. Hace que las moléculas de aire se muevan en todas las direcciones. Chocan unas con otras y provocan fricción y calor. Una fuerza que gira como pequeños tornados convierte el flujo de aire laminado suave en un flujo turbulento. Esto aumenta la transferencia de calor entre las gotitas frías de combustible líquido y este aire más caliente.

all speed governor another term for variable speed governor.
regulador global de velocidad otro término para el regulador de velocidad variable.

amplifier piston hydraulically actuated piston that pumps fuel to injection pressure values in a Cat/Navistar HEUI; also known as *intensifier piston*.
pistón multiplicador pistón de acción hidráulica que bombea combustible a las válvulas de presión de inyección en un HEUI Cat/Navistar; conocido también como *émbolo multiplicador*.

aneroid a device used to sense light pressure conditions. The term is used to describe manifold boost sensors that limit fueling until there is sufficient boost air to combust it and usually consists of a diaphragm, spring, and fuel-limiting mechanism.
aneroide dispositivo utilizado para detectar condiciones de presión ligera. El término se utiliza para describir los sensores de sobrealimentación del colector que limitan el combustible hasta que haya suficiente aire para quemarlo y suele consistir en un diafragma, un muelle y un mecanismo de limitación de combustible.

antifreeze a liquid solution added to water to blend the engine coolant solution that raises the boil point and lowers the freeze point. Ethylene glycol (EG), propylene glycol (PG), and extended life coolants (ELC) are currently used.
anticongelante solución líquida que se agrega al agua para mezclarla con la solución refrigerante del motor que sube el punto de ebullición y baja el de congelación. Se suelen utilizar etilenglicol (EG), propilenglicol (PG) y refrigerantes de larga duración (ELC).

anti-thrust face used to describe the minor thrust face of a piston: the outboard side of the piston as its throw rotates off the crankshaft centerline through the power stroke.
cara de empuje contraria se utiliza para describir la cara de empuje menor de un pistón: parte externa del pistón cuando su codo gira para salir de la línea central de cigüeñal durante el tiempo de combustión.

articulating piston a two-piece piston with separate crown and skirt assemblies, linked by the piston wrist pin and afforded a degree of independent movement. The wrist pin is usually full floating or bolted directly to the conn rod, in which case it is known as a *crosshead piston*.
pistón articulado pistón de dos piezas con conjuntos de corona y de falda, unidos por el pasador de articulación y dotado de un grado de movimiento independiente. El pasador de articulación suele ser totalmente flotante o estar empernado directamente a la biela, en cuyo caso se le conoce como *pie de biela*.

ash
1. the powdery/particulate residues of a combustion reaction.
2. solid residues found in crude oils. Present in trace quantities in engine lubricating oils and diesel fuels.
ceniza
1. residuos de polvo o partículas de una reacción de combustión.
2. residuos sólidos que se encuentran en el petróleo crudo. Presentes en trazas en aceites lubricantes para motores y combustibles diesel.

ATAAC air-to-air charge air cooling.
RCAA refrigeración de carga aire-aire.

ATDC after top dead center.
TPMS tras punto muerto superior.

atm a unit of pressure commonly used in fuel injection comparator equipment equivalent to one unit of atmospheric pressure at sea level or 14.7 psi or 101.3 kPa; close but not exactly equivalent to 1 *bar*.
atmósfera unidad de presión comúnmente utilizado en un equipamiento comparador de inyección de combustible equivalente a una unidad de presión atmosférica a nivel del mar o a 14,7 psi o a 101,3 kPa; cercano, pero no exactamente equivalente a 1 *bar*.

atom the smallest part of a chemical element that can take part in a chemical reaction; composed of electrons, protons, and neutrons.
átomo la parte más pequeña de un elemento químico que puede tomar parte en una reacción química; se compone de electrones, protones y neutrones.

atomization the process of breaking liquid fuel into small droplets by pumping it at a high pressure through a minute flow area.

atomización proceso de dividir el combustible líquido en gotitas bombeándolo a una presión elevada a través de una zona de flujo muy pequeña.

atomized droplets the liquid droplets emitted from an injector nozzle.

gotitas atomizadas gotitas líquidas emitidas desde una boquilla de inyección.

balance orifice the inlet orifice of a Cummins PT injector. Collectively, the balance orifices (one in each PT injector) define the flow area the rail unloads to.

orificio de equilibrado orificio de entrada de un inyector Cummins PT. De forma colectiva, los orificios equilibrado (uno en cada inyector PT) definen el área de flujo a la que descarga el distribuidor.

BARO barometric pressure sensor.

BARO sensor de presión barométrica.

barometric capsule a barometer device used on some hydromechanical injection pumps to limit high altitude fueling.

cápsula barométrica dispositivo barométrico que se utiliza en algunas bombas de inyección hidromecánica para limitar la alimentación en altitudes elevadas.

barometric pressure sensor an electronic barometric pressure sensing device.

sensor de presión barométrica dispositivo electrónico de medición de la presión barométrica.

base circle the smallest radial dimension of an eccentric. Used to describe cam geometry, the train that the cam is responsible for actuating would be unloaded on the cam base circle; also known as *inner base circle* or IBC.

perfil de la leva la dimensión radial más pequeña de una excéntrica. Se utiliza para describir la geometría de la leva, el tren de cuya actuación es responsable la leva se descargará en el perfil de la leva; conocido también como *perfil interior de la leva* o IBC.

bearing shell a half segment of a friction bearing such as would be used as a crankshaft main bearing.

casquillo de cojinetes medio segmento de un cojinete de fricción como el que se utilizaría como cojinete principal de cigüeñal.

big end the crankshaft throw end of a connecting rod.

cabeza extremo del codo de cigüeñal de una biela.

black smoke smoke that appears black to the observer is caused by particulate (solids) emission in the exhaust gas stream: light is blocked by the particulate, making it appear black.

humo negro el humo que se ve negro es debido a emisión de partículas (sólidas) en la corriente de gases de escape: la partícula bloquea la luz, lo que hace que se vea negro.

blue smoke usually associated with engine oil combusted in the engine cylinder; caused by the mixture of condensing droplets and particulate emitted when oil is burned in an engine.

humo azul normalmente asociado con aceite de motor quemado en el cilindro del motor; se produce por la mezcla de gotitas en condensación y partículas emitidas cuando se quema aceite en un motor.

blink codes fault codes blinked out using diagnostic lights; also known as *flash codes.*

códigos intermitentes códigos de errores con luces intermitentes mediante luces de diagnóstico; conocido también como *códigos de destello.*

BOE beginning of energizing.

BOE principio de la revolución.

BOI beginning of injection.

BOI principio de la inyección.

boosted engine any turbocharged engine; turbo-boosted.

motor sobrealimentado cualquier motor turboalimentado; turbosobrealimentado.

boundary lubrication thin film lubrication characteristics of an oil.

lubricación de límites características de lubricación de película fina de un aceite.

brake horsepower power developed by an engine measured at the flywheel measured by a dynamometer or *brake*. Factored by *torque* and rpm.

potencia al freno potencia desarrollada por un motor medida en el volante con un dinamómetro o *freno*. Sus factores son el *par* y las rpm (revoluciones por minuto).

Brake Saver a Caterpillar engine mounted, hydraulic retarder.

Brake Saver retardador Caterpillar hidráulico montado en el motor.

BSFC brake specific fuel consumption. A measure of the fuel required to perform a unit of work; used in graphs of engine data designed to show fuel efficiency at specific engine loads and rpm.

BSFC consumo de combustible específico de frenos. Medida del combustible necesaria para realizar una unidad de trabajo; se utiliza en gráficos de datos de motor diseñados para mostrar el rendimiento del combustible en cargas de motor y rpm.

BTDC before top dead center.

APMS antes de punto muerto superior.

BTM brushless torque motor. Caterpillar rotary proportional solenoid used for PEEC timing and rack position control.

BTM motor de par sin escobilla. Solenoide proporcional rotatorio Carterpillar utilizado para la sincronización PEEC y el control de posición de cremallera.

bubble collapse the condition caused by wet liner combustion pressure impulses acting on the coolant resulting in vapor bubbles that collapse and cause *cavitation.*

colapso de burbujas problema causado por impulsos de presión de combustión de camisa de agua que actúa en el refrigerante, lo que provoca burbujas que colapsan y causan *cavitación.*

buffers memory locations used to store processed data before it is sent to output devices.

memorias intermedias ubicaciones de memoria utilizadas para almacenar los datos procesados antes de enviarlos a los dispositivos de salida.

buffer screw a DDC governor adjustment that lightly loads the governor differential lever to reduce engine rpm hunting.

tornillo amortiguador ajuste del regulador DDC que presiona ligeramente la palanca del diferencial del regulador para reducir la irregularidad de rpm del motor.

bundle multiple arrangement of cooling tubes that form the core of a heat exchanger.

haz tubular disposición múltiple de conductos refrigerantes que forman el núcleo de un intercambiador térmico.

bus an electronic connection; transit lines that connect the CPU, memory, and input/output devices; increasingly used to mean "connected".

bus Conexión electrónica; líneas de tránsito que conectan la CPU, la memoria y los dispositivos de entrada y salida; que se utiliza cada vez más para querer decir "conectado".

bushing any of a number of types of friction bearings designed to support shafts.

casquillo cualquiera de una serie de tipos de cojinetes de fricción diseñados para soportar ejes.

Button this is trade jargon for the Cummins PTG pump Idle Spring Plunger.

Button Es el término de la jerga comercial para el émbolo de resorte inactivo de la bomba PTG Cummins.

bypass filter a filter assembly plumbed in parallel with the lubrication circuit, usually capable of high filtering efficiencies.

filtro en derivación conjunto de filtros conectados en paralelo con el circuito de lubricación, que normalmente es capaz de alcanzar un alto rendimiento de filtrado.

bypass valve a diverter valve fitted to full flow filter (series) mounting pads, designed to reroute lubricant around a plugged filter element to prevent a major engine failure.

válvula de sobrecarga válvula de desvío ajustada para filtrar el flujo total (series) de cojines de montaje, diseñados para redistribuir el lubricante alrededor de un elemento de filtro conectado para evitar un error grave del motor.

calorific value the heating value of a fuel measured in Btu, calories, or joules.

potencia calorífica valor térmico de un combustible medido en Btu, calorías o julios.

calibrating orifice see *balance orifice.*

orificio de calibración véase *orificio de equilibrado.*

cam an eccentric. An eccentric portion of a shaft, often used to convert rotary motion into reciprocating motion.

leva excéntrica. Parte excéntrica de un eje, utilizado a menudo para convertir el movimiento rotatorio en movimiento de vaivén.

cambox the lower portion of a port helix metering injection pump in which the actuating camshaft is mounted and the lubricating oil sump is located.

caja de levas parte inferior de una bomba de inyección de dosificación helicoidal de abertura en la que está montado el árbol de levas de mando y en que está ubicado el colector de aceite lubricante.

cam follower housing timing the method used by Cummins on their PT N series engines to synchronize the actuation of the PT pumping stroke by the cam profile with engine position.

sincronización de alojamiento del empujador de leva el método utilizado por Cummins en sus motores de la serie PT N para sincronizar la actuación del tiempo de bombeo por medio del perfil de la leva con la posición del motor.

cam geometry the shaping of a cam profile and the effect it produces on the train it actuates.

geometría de la leva la forma de un perfil de la leva y el efecto que produce en el tren que impulsa.

cam ground trunk-type pistons that are machined slightly eccentrically. Because of the greater mass of material required at the wrist pin boss, this area will expand proportionally more when heated. Cam ground pistons are designed to assume a true circular shape at operating temperatures.

elípticos pistones de tipo de tronco que se disponen de forma ligeramente excéntrica. Debido a la gran cantidad de masa de material necesaria en la muñonera del pie de biela, esta área se expandirá más proporcionalmente cuando se caliente. Los pistones elípticos están diseñados para asumir una forma realmente circular en temperaturas de funcionamiento.

cam heel the point on a cam profile that is exactly opposite the *toe* or center point of the highest point on the cam.

talón de leva el punto de un perfil de la leva exactamente opuesto a la *punta* o punto central del punto más alto de la leva.

cam nose the portion of the cam profile with the largest radial dimension; its center point would be the *cam toe.* That portion of the cam profile that is *OBC.*

pico de leva la parte del perfil de la leva con la dimensión radial mayor; su punto central sería la *punta de leva.* La parte del perfil de la leva que es *OBC.*

cam profile the cam geometry; simply, the shape of the cam.

perfil de la leva la geometría de la leva; simplemente, la forma de la leva.

camshaft a crankshaft-driven shaft, machined with eccentrics (cams) designed to actuate trains positioned to ride the cam profiles; the engine feedback assembly actuator responsible for timing/actuating cylinder valves and fuel injection apparatus. Driven at half engine speed on four-stroke cycle engines and at engine speed on two-stroke cycles.

árbol de levas eje de cigüeñal, dispuesto con excéntricas (levas) diseñado para impulsar los trenes colocados para hacer funcionar los perfiles de la leva; el regulador de conjunto de retroalimentación del motor responsable de sincronizar/impulsar las válvulas de los cilindros y el aparato de inyección de combustible. Impulsado a mitad de velocidad de motor en motores de cuatro tiempos y a velocidad de motor en dos tiempos.

CAPS Cummins Accumulator Pump System. A Cummins innovative variation of common rail fueling that uses full authority electronic management and will be used initially on the ISC engine.

CAPS sistema de bombas acumuladoras Cummins. Una variación innovadora de Cummins de la alimentación de distribuidor común que utiliza una gestión electrónica de autoridad total y que se utilizará inicialmente en el motor ISC.

carbon (C) an element found in various forms including diamonds, charcoal, and coal. It is the primary constituent element in hydrocarbon fuels. Atomic #6.

carbono (C) un elemento que se encuentra en varias formas, incluidos los diamantes, el carbón vegetal y el carbón mineral. Es el elemento constituyente principal de los combustibles hidrocarburos. Número atómico 6.

carbon dioxide (CO_2) the product of combusting carbon in the oxidation reaction of a HC fuel. An odorless, tasteless gas that is nontoxic and not classified as a noxious engine emission, but which causes the greenhouse condition that concerns environmentalists.

dióxido de carbono (CO_2) el producto de la combustión en la reacción de oxidación de un combustible HC. Gas inodoro e insípido que no es tóxico y no es clasificado como emisión nociva de motores, pero que provoca el efecto invernadero que preocupa a los ecologistas.

carbon monoxide (CO) a colorless, odorless, and poisonous gas that is produced when carbon is not completely oxidized in combustion.

monóxido de carbono (CO) gas incoloro, inodoro y venenoso que se produce cuando el carbono no se oxida completamente en la combustión.

catalyst a substance that stimulates, accelerates, or enables a chemical reaction without itself undergoing any change.

catalizador sustancia que estimula, acelera o habilita una reacción química sin sufrir ningún cambio.

catalytic converter an exhaust system device that enables oxidation and reduction reactions; in lean burn truck diesel engines, only oxidation catalytic converters are used at this moment in time.

convertidor catalítico dispositivo de sistema de escape que habilita las reacciones de oxidación y reducción; en los motores diesel de camiones de combustión pobre, sólo se utilizan los convertidores catalíticos en este momento.

CAT-ID Caterpillar information display; the Caterpillar digital dash display that provides the driver with ECM feedback data such as fuel economy and engine parameters.

CAT-ID dispositivo de información de Caterpillar; dispositivo del tablero digital de Caterpillar que proporciona datos de respuesta ECM como ahorro de combustible y los parámetros del motor.

cavitation describes metal erosion caused by the formation and subsequent collapse of vapor pockets (bubbles) produced by physical pulsing into a liquid such as that of a wet liner against the wall of coolant that surrounds it. Bubble collapse causes high unit pressures and can quickly erode wet liners when the protective properties of the coolant diminish.

cavitación describe la erosión del metal causada por la formación y subsiguiente colapso de bolsas de vapor (burbujas) producidas por la pulsión física en un líquido como el de la camisa de agua contra el muro de lubricante que lo envuelve. El colapso de burbujas provoca altas presiones de unidades y puede erosionar rápidamente los camisas de agua cuando las propiedades protectoras del refrigerante disminuyen.

CCW counterclockwise or left hand rotation.

CCW rotación en la dirección opuesta a las agujas del reloj o hacia la izquierda.

CD-ROM an optically encoded data disk that is read by a laser in the same way an audio CD is read and is designed for read-only data.

CD-ROM disco de datos codificados ópticamente que se lee con un láser de la misma forma que un CD de sonido y está diseñado para datos de sólo lectura.

Cetane is a measure of the ignition quality of a Diesel fuel. A high cetane number indicates good ignition quality (Short Delay Period) and a low cetane number indicates poor ignition quality (Long Delay Period). Cetane is a colorless liquid hydrocarbon that has excellent ignition qualities. Cetane is rated at 100.

cetano es una medida de la calidad de encendido de un gasóleo. Un número alto de cetano indica una buena calidad de encendido (período de demora corto) y un número bajo indica una calidad de encendido baja (período de demora largo). El cetano es un hidrocarburo líquido incoloro que tiene calidades de encendido excelentes. El cetano tiene un valor de 100.

centrifugal filter a filter that uses a centrifuge consisting of a rotating cylinder charged with pressurized fluid and canted jets to drive it; centrifugal filters often have high efficiencies and are often of the *bypass* type.

filtro centrífugo filtro que utiliza un centrifugado que consiste en un cilindro rotatorio cargado con fluido presurizado y chorros sesgados para impulsarlo; el filtro centrífugo suele tener altos rendimientos y suele ser del tipo de *derivación*.

centrifugal force the force acting outward on a rotating body.

fuerza centrífuga fuerza que actúa hacia fuera en un cuerpo que gira.

CEL check engine light.

CEL luz de comprobación del motor.

CELECT Cummins electronic engine control.

CELECT control electrónico Cummins del motor.

cetane improvers see *ignition accelerators*.

mejoradores de cetano véase *aceleradores de encendido*.

cetane number (CN) the standard rating of a diesel fuel's ignition quality. It is a comparative rating method that measures the ignition quality of a diesel fuel versus that of a mixture of cetane (good ignition characteristics) and heptamethylnonane (poor ignition characteristics). A mixture of 45% cetane and 55% would have a CN of 45. Diesel fuels refined for use on North American highways are classified by the ASTM as #1D and #2D and must have a minimum CN of 40.

número de cetano (CN) medición estándar de la calidad de encendido de un gasóleo. Es un método de medición comparativa que mide la calidad de encendido de un gasóleo en comparación con esa mezcla de cetano (buenas características de encendido) y heptametilnonano (malas características de encendido). Una mezcla con el 45% de cetano y el 55% de heptametilnonano daría como resultado un CN de 45. Los gasóleos refinados para su uso en carreteras norteamericanas por el ASTM como D n° 1 y D n° 2 y deben tener un CN mínimo de 40.

charge air cooling the cooling of turbo boost air by means of ram air or coolant medium heat exchangers.

carga de refrigeración por aire refrigeración del aire de sobrealimentación turbo por medio de aire a presión o intercambiadores térmicos medios de refrigerante.

charging circuit the portion of the fuel subsystem that begins with the charging or transfer pump and is responsible for delivering fuel to the injection pumping/metering apparatus. In a port helix metering pump, this extends through the charging gallery of the injection pump.

circuito de carga parte del subsistema de combustible que empieza con la carga o con la bomba de transferencia y es responsable de proporcionar combustible al aparato de bombeo/dosificación de inyección. En una bomba de dosificación helicoidal de abertura, se extiende a la galería de carga de la bomba de inyección.

charging pressure a term used to describe the pressure on the charge side of the transfer pump in a fuel subsystem. Charging pressure parameters are defined by the cycle speed of the charging pump, the flow area it unloads to, and regulating valve.

presión de carga término que se utiliza para describir la presión en el lado de la carga de la bomba de transferencia de un subsistema de combustible. Los parámetros de presión de carga se definen con la velocidad del tiempo de la bomba de carga, el área de flujo en que se descarga y la válvula reguladora.

charging pump the pump responsible for moving fuel through the fuel subsystem. Plunger, gear, and less commonly, vane-type pumps are used.

bomba de carga bomba responsable de mover el combustible por el subsistema de combustible. Se utilizan bombas de émbolo, de engranajes y, más raramente, de excéntricas.

chassis dynamometer a test bed that measures brake power delivered to the vehicle wheels by having them drive roller(s) to which torque resistance is applied and accurately measured.

dinamómetro de chasis banco de pruebas que mide la potencia de freno enviada a las ruedas del vehículo haciendo que rueden en rodillos a los que se aplica resistencia de par y que se miden de forma precisa.

CEL is an acronym for check engine light that flashes blink codes and indicates a fault in the system.

CEL es un acrónimo de *check engine light* (luz de comprobación del motor) que emite códigos intermitentes e indica algún error en el sistema.

CI compression ignition; an engine in which the fuel/air mixture is ignited by the heat of compression.

CI encendido por compresión; un motor en el que se enciende la mezcla de combustible y aire con el calor de la compresión.

CKP is an acronym for the crankshaft position sensor.

CKP es un acrónimo de *crankshaft position sensor* (sensor de posición del árbol de levas).

cloud point the temperature at which wax crystals present in all diesel fuels become large enough to make the fuel appear hazy. It is also the point at which plugging of fuel filters becomes a possibility. The cloud point is usually 5° F (3° C) above the fuel's pour point.

temperatura de cristalización temperatura a la que los cristales de cera, presentes en todos los gasóleos crecen lo suficiente para hacer que el combustible se vea turbio. También es el punto en el que es posible conectar filtros de combustible. La temperatura de cristalización suele ser de 3° C (5° F) por encima del punto de fluidez del combustible.

CMAC chassis-mounted charge air cooling.

CMAC carga de refrigeración por aire montada en el chasis.

CMP sensor camshaft position sensor.

sensor CMP sensor de posición de árbol de levas.

coalesce to combine to form a single whole.

coalescer combinar para formar un todo.

coefficient of friction a means of rating the aggressiveness of friction materials; alters with temperature and the presence of any kind of lubricant.

coeficiente de fricción forma de medir la agresividad de los materiales de fricción; se altera con la temperatura y la presencia de cualquier tipo de lubricante.

coefficient of thermal expansion the manner in which a material behaves as it is heated and cooled. For instance, aluminum has a higher coefficient of thermal expansion than steel meaning that when a similar mass of each material is subjected to an identical amount of heat, the aluminum will expand more.

coeficiente de dilatación térmica comportamiento de un material al calentarse y enfriarse. Por ejemplo, el aluminio tiene un coeficiente de dilatación térmica más alto que el acero, lo que significa que cuando una masa similar de cada uno de los materiales está sujeta a una cantidad de calor idéntica, el aluminio se dilatará más.

cold start strategy a programmed start-up sequence in an electronic management system in which the timing, fuel quantity, and engine-operating parameters are managed on the basis of ambient and engine fluid temperatures. During this process, other inputs such as throttle position may be ignored by the ECM.

sistema de arranque en frío secuencia programada de arranque en un sistema de gestión electrónica en el que la sincronización, la cantidad de combustible y los parámetros de funcionamiento del motor se gestionan basándose en la temperatura ambiente y la del motor. Durante este proceso, la EMC puede omitir otras entradas como la entrada del acelerador.

combustion the act of burning a substance. An oxidation reaction.
combustión acto de quemar una sustancia. Reacción de oxidación.

command circuit used to describe input sensors such as the TPS (throttle position sensor) that commands (requests) an output from the ECM.
circuito de control se utiliza para describir sensores de entrada como el TPS (sensor de posición del acelerador) que pide (solicita) una salida al ECM.

common rail a fuel system in which injectors are arranged in parallel and supplied with fuel from a pump at a common pressure value. The Cummins PT and HPI-TP Series are two examples of common rail systems.
distribuidor común sistema de combustible en el que se disponen los inyectores en paralelo y se los suministra combustible desde una bomba con un valor de presión común. Las series Cummins PT y HPI-TP son dos ejemplos de sistemas de distribuidor común.

companion cylinders term used to describe pistons paired by their respective crank throws to rotate together through the engine cycle such as #1 and #6 in an in-line, 6-cylinder engine.
cilindros gemelos término utilizado para describir pistones emparejados por sus respectivos codos de palanca para que giren juntos a través del ciclo del motor, como los números 1 y 6 en un motor en línea de 6 cilindros.

compressor housing the section of a turbocharger responsible for compressing the intake air and feeding it into the intake circuit; also known as *impeller housing*.
alojamiento del compresor sección de un turboalimentador responsable de la compresión del aire de admisión y cargarlo en el circuito de admisión; conocido también como *alojamiento del impulsor*.

Compression ratio is the volume of the cylinder at the beginning of the compression stroke divided by the volume of the cylinder at the end of the compression stroke.
relación de compresión es el volumen del cilindro al principio del tiempo de compresión dividido por el volumen del cilindro al final del tiempo de compresión.

compression ring the ring(s) designed to seal cylinder gas pressure located in the upper ring belt.
anillo de compresión anillo(s) diseñado(s) para sellar la presión del gas del cilindro en la zona de segmentación superior.

Compulink Cummins CELECT EST.
Compulink EST CELECT de Cummins.

conduction heat transmission through solid matter.
conducción transmisión del calor a través de materiales sólidos.

connecting rod the rigid mechanical link between the piston wrist pin and the crankshaft throw.
biela unión rígida mecánica entre el pasador de articulación y el codo del cigüeñal.

consolidated engine controller an ECM that houses the microcomputer and output switching such as injector drivers.
controlador de motor consolidado ECM que aloja el microordenador y la conmutación de salida como los controladores de inyección.

constant horsepower sometimes used to describe a *high torque rise* engine.
potencia constante se utiliza a veces para describir un motor de *aumento alto del par*.

continuity an unbroken circuit; used to describe a continuous electrical circuit. A continuity test would determine if a circuit or circuit component was capable of current flow.
continuidad circuito ininterrumpido; se utiliza para describir un circuito eléctrico continuo. Una prueba de continuidad determina si un circuito o un componente de circuito es capaz de tener flujo de corriente.

control rack the fuel control mechanism on an MUI or multi-cylinder port helix metering pump that when moved linearly, rotates the pumping plunger(s) in unison.
cremallera mecanismo de control de combustible de una bomba de dosificación helicoidal de abertura MUI o de múltiples cilindros que cuando se mueve linealmente, gira el émbolo o émbolos de bombeo al unísono.

control rod in a DDC two-stroke cycle MUI engine, it links the governor with the control tube, which when rotated moves the MUI racks.
biela motora en un motor MUI de dos tiempos DDC, une el regulador con el conducto de control que, cuando se gira, mueve las cremalleras MUI.

control sleeve the component that is tooth meshed to the control rack and connects to the plungers by means of slots, used to rotate the plungers in the barrels.
manguito de regulación componente con engranajes de dientes para la cremallera, que conecta con los émbolos por medio de mellas, que se utilizan para girar los émbolos en los cilindros.

convection heat transfer by currents of gas or liquids.
convección transferencia de calor por corrientes de gas o de líquidos.

conventional theory (of current flow) asserts that current flows from a positive source to a negative source. Despite the fact that it is fundamentally incorrect, it is nevertheless widely accepted and used. See *electron theory*.
teoría convencional (de flujo de corriente) afirma que la corriente fluye de un origen positivo a uno negativo. A pesar del hecho de que es fundamentalmente incorrecta, es ampliamente aceptada y utilizada. Véase *teoría de electrones*.

CoPilot Mack Trucks V-MAC digital monitor and driver display unit.
CoPilot monitor digital y unidad de dispositivos de controladores de Mack Trucks V-MAC.

corrosive alkaline or acidic substances that dissolve metals and skin tissue.
corrosivas sustancias alcalinas o ácidas que disuelven metales y el tejido de la piel.

coulomb one coulomb is equal to 6.28×10^{18} electrons.
culombio un culombio es igual a $6,28 \times 10^{18}$ electrones.

counterflow radiator a double pass radiator in which coolant is cycled through U column tubes from usually a bottom-located intake tank to a bottom located output tank; they have higher cooling efficiencies than other radiator designs.
radiador de contracorriente radiador de doble paso en el que se traslada el refrigerante a través de conductos de columnas en U normalmente de un depósito de admisión de la parte inferior a un depósito de salida de la parte superior; tienen rendimientos más altos que otros diseños de radiadores.

covalent bonding the atomic condition that occurs when electrons are shared by two atoms.
enlace covalente condición atómica que se produce cuando dos átomos comparten electrones.

crankcase the lower portion of the engine cylinder block in which the crankshaft is mounted and under which is the lubrication oil sump.
cárter parte inferior del bloque de los cilindros del motor en el que se monta el cigüeñal y debajo del cual está el colector de aceite lubricante.

crankshaft a shaft with offset throws designed to convert the reciprocating movement of pistons into torque.
cigüeñal eje con codos excéntricos diseñado para convertir el movimiento de vaivén de los pistones en par.

creep describes the independent movement of two components clamped by fasteners when they have different coefficients of thermal expansion or have different mass, which means their expansion and contraction rates do not concur.
fluencia describe el movimiento independiente de dos componentes aferrados con sujetadores cuando tienen diferentes coeficientes de expansión térmica o diferente masa, lo que significa que sus niveles de expansión y contracción no coinciden.

crossflow radiator a usually low profile design of radiator (used with aerodynamic hood/nose), in which the entry and output tanks are located at either end and coolant flow is horizontal.

radiador de flujo horizontal diseño de radiador normalmente bajo (utilizado con capó o morro aerodinámico), en el que los depósitos de entrada y salida están en los dos extremos y el flujo de refrigerante es horizontal.

crossflow valves a cylinder head valve configuration in which the intake and exhaust valves are located in series in the cylinder head, meaning that gas flow from the inboard valve differs from (and may interfere with) that of the outboard valve.

válvulas de flujo horizontal configuración de válvulas de culata en la que las válvulas de admisión y de escape están colocadas en serie en la culata, lo que significa que el flujo de los gases entrantes difiere de (o puede interferir con) el de la válvula de salida.

crown valve a now obsolete DDC MUI nozzle valve.

válvula de corona válvula de boquilla MUI DDC ya obsoleta.

crosshead piston an articulating piston with separate crown and skirt assemblies in which the connecting rod is bolted directly to the wrist pin.

pie de biela pistón articulado con conjuntos de corona y de falda separados en los que la biela está empernado directamente al pasador de articulación.

crown the leading edge face of a piston or in articulating pistons, the upper section of the piston assembly. Crown geometry (shape) plays a large role in defining the cylinder gas dynamic.

corona la cara de borde de entrada de un pistón o, en pistones articulados, sección superior del conjunto de émbolo. La geometría de corona (forma) desempeña un papel importante en la definición de la dinámica de gases del cilindro.

crown valve nozzle an obsolete DDC hydraulic injector nozzle integral with early version MUIs.

boquilla de válvula de corona boquilla de inyección hidráulica DDC obsoleta integral con los MUI anteriores.

CW clockwise.

CW en el sentido de las agujas del reloj.

cycle
1. a sequence of events that recurs such as those of the *diesel cycle*.
2. one complete reversal of an alternating current from positive to negative.

ciclo
1. secuencia de sucesos que se repiten, como los del *ciclo diesel*.
2. inversión completa de una corriente alterna de positiva a negativa.

cylinder block the main frame of any engine to which all the other components are attatched.

bloque de cilindros estructura principal de cualquier motor al que están unidos los demás componentes.

cylinder head the components clamped to a cylinder block containing the engine breathing and fueling control mechanisms.

culata componentes aferrados a un bloque de cilindros que contienen los mecanismos de control de ventilación y alimentación.

cylinder leakage tester device used to test cylinder leakage by applying regulated air to the cylinder at a controlled volume and pressure and producing a percentage of leakage specification.

verificador de fugas en cilindros dispositivo que se utiliza para verificar las fugas en los cilindros aplicando aire regulado al cilindro con un volumen y una masa controlados y produciendo una especificación de porcentaje de fugas.

data link the connection point or path for data transmission in networked devices.

vínculo de datos punto de conexión o ruta de acceso para la transmisión de datos en dispositivos en red.

DCA diesel coolant additives. A proprietary supplemental coolant additive.

DCA aditivos refrigerantes para diesel. Aditivo refrigerante suplementario patentado.

DDR diagnostic data reader.

DDR lector de datos de diagnóstico.

dead volume fuel fuel that is statically retained for a portion of the cycle; usually refers to the fuel retained at residual line pressure in a high pressure injection pipe that connects injection pump elements with hydraulic injectors in PLN system.

combustible de volumen fijo combustible retenido estáticamente durante una parte del ciclo; se suele referir al combustible retenido en la presión de línea residual en una tubería de inyección a alta presión que conecta los elementos de la bomba de inyección con inyectores hidráulicos en el sistema PLN.

Delay period the time between the *Start of Injection* and the *Start of Ignition*, which starts the pressure rise due to combustion. It may also be called Ignition Lag, or Ignition Delay (ID). It happens at the end of compression, after start of injection, and the fuel does not ignite immediately.

período de demora tiempo transcurrido entre el *inicio de la inyección* y el *inicio del encendido*, que inicia la subida de la presión debido a la combustión. También se puede llamar retraso del encendido o demora del encendido (ID). Tiene lugar al final de la compresión, tras el inicio de la inyección y el combustible no se enciende inmediatamente.

delivery valve a combination check and pressure management valve that is used on many hydromechanical diesel fuel injection systems.

válvula de salida combinación de válvula de gestión de presión y de verificación que se utiliza en muchos sistemas de inyección de gasóleo.

Deutsch connector a widely used, weather proof, proprietary electrical and electronic connector.

conector Deutsch conector eléctrico y electrónico patentado muy utilizado, resistente a diferentes condiciones atmosféricas.

DI direct injection.

DI inyección directa.

diagnostic pressure sensor Cummins HPI-TP sensor located downstream from the rail actuator in each bank.

sensor de presión de diagnóstico Sensor Cummins HPI-TP ubicado por debajo del impulsor del distribuidor en cada fila.

diesel fuel a simple hydrocarbon fuel obtained from crude petroleum by means of fractioning and usually containing both residual and distillate fractions.

gasóleo un combustible hidrocarburo obtenido de petróleo crudo por medio del fraccionamiento y que suele contener fracciones residuales y destiladas.

diesel knock a detonation condition caused by prolonged ignition lag.

golpeteo Diesel condición de detonación provocada por un retraso de encendido prolongado.

diffuser the device in a turbocharger compressor housing that converts air velocity into air pressure.

difusor dispositivo de un alojamiento de compresor de turboalimentación que convierte la velocidad del aire en presión de aire.

digital signals data interchange/retention signals limited to two discernable states; combinations of ones and zeros into which data, video, or human voice must be coded for transmission/storage and subsequently reconstructed.

señales digitales señales de intercambio/retención de datos limitadas a dos estados diferenciables; combinaciones de unos y ceros en los que datos, vídeo o voz humana se deben codificar para su transmisión/almacenamiento y reconstruir después.

DFF direct fuel feed. A Cummins PT injector type that stops fueling under motoring conditions.

DFF alimentación de combustible directa. Tipo de inyector PT de Cummins que detiene la alimentación cuando hay problemas en el motor.

direct injection (DI) describes any engine in which fuel is injected directly into the engine cylinder and not to any kind of external prechamber. Most current diesel engines are direct injected.

inyección directa (DI) describe cualquier motor en el que el combustible se inyecta directamente al cilindro del motor y no a ningún tipo de precámara externa. La mayoría de los motores diesel actuales son de inyección directa.

distillate any of a wide range of distilled fractions of crude petroleum, some of which would be constituents of a diesel fuel. Refers to the more volatile fractions in a fuel. Sometimes used to refer to diesel fuels.

destilado cualquiera de las muchas fracciones destiladas del petróleo crudo, algunas de las cuales formarían parte de un gasóleo. Se refiere a las fracciones más volátiles de un combustible. A veces se utiliza para referirse a los gasóleos.

Distributor-type injection pump distributor-type fuel injection pump pressurizes and distributes a metered amount of fuel to each cylinder nozzle at the proper time based on the calibrated needs of the engine. The use of a separate pumping element for each cylinder has been the general practice of the In-Line pump. The distributor-type uses one pump barrel and a set of plungers to supply all cylinders in rotation.

bomba de inyección de tipo distribuidor la bomba de inyección de combustible de tipo distribuidor presuriza y distribuye una cantidad medida de combustible a cada boquilla de cilindro en el momento adecuado basándose en las necesidades calibradas del motor. El uso de un elemento de bombeo distinto para cada cilindro es la práctica general de la bomba en línea. El tipo distribuidor utiliza un cilindro de bomba y un conjunto de émbolos para alimentar todos los cilindros en rotación.

DMM digital multimeter. A voltage, resistance (ohms), and current (amperes) reading instrument.

DMM multímetro digital. Instrumento de lectura de tensión, resistencia (ohmios) y corriente (amperios).

double helix a port helix plunger design, with both upper and lower helix characteristics that results in a variable beginning and ending of the pump effective stroke.

doble espiral diseño de émbolo de helicoidal de abertura, con características de espiral inferior y superior que dan como resultado un principio y un final variables del tiempo efectivo de la bomba.

double pass radiator a counterflow radiator in which the coolant is routed to make two passes, therefore, entering and exiting from separate tanks both located either at the top or the bottom of the radiator. A high efficiency radiator.

radiador de doble paso radiador de contracorriente en el que el refrigerante se distribuye para dar dos pases, es decir, entrando y saliendo de depósitos distintos localizados en la parte superior y en la inferior del radiador. Radiador de alto rendimiento.

downflow radiator a typical radiator in which hot coolant from the engine enters at the top tank, flows downward, and exits through a bottom tank.

radiador de flujo vertical radiador normal en el que el refrigerante caliente procedente del motor entra en el depósito superior, fluye hacia abajo y sale a través de un depósito inferior.

download data transfer from one computer system to another—often used to describe proprietary data transfer when reprogramming vehicle ECMs.

descarga transferencia de datos de un sistema informático a otro (utilizado a menudo para describir la transferencia de datos patentados al reprogramar ECM de vehículos).

droop an engine governor term denoting a transient speed variation that occurs when engine loading suddenly changes.

caída término de regulador de motor que indica una variación temporal en la velocidad que se produce cuando la carga del motor cambia repentinamente.

droop curve a required hydromechanical governor characteristic in which fueling drops off in an even curve as engine speed increases from the rated power value to high idle.

curva de caída característica necesaria del regulador hidromecánico en el que la alimentación desciende con una curva uniforme a medida que disminuye la velocidad desde una potencia en régimen a un ralentí alto.

dry liners liners that are fitted either with fractional looseness or fractional interference that dissipate cylinder heat to the cylinder block bore and have no direct contact to the water jacket.

camisas secas camisas ajustadas bien con soltura fraccionaria o con interferencia fraccionaria que disipa el calor del cilindro al calibre del bloque de cilindros y no tiene contacto directo con la camisa de agua.

dry sump an engine that uses a remotely located oil sump; not often seen on highway diesel applications but used in some bus engines to reduce the profile of the engine.

colector del lubricante fuera del cárter motor que utiliza un colector de aceite de ubicación remota; no se suele utilizar en aplicaciones diesel de carretera, pero sí en motores de autobuses para reducir el perfil del motor.

dual helices a plunger geometric design with identical helices machined diametrically opposite each other on the plunger. Commonly used, it helps prevent side loading of the plunger at high pressure spill.

espirales gemelas diseño geométrico de émbolos con espirales idénticas fresadas de forma diametralmente opuesta entre sí en el émbolo. Se utilizan de normalmente y ayudan a evitar la carga lateral del émbolo a altas presiones.

DTC the SAE J1930 Clean Air Act Amendment OBD II acronym for Diagnostic Trouble Code.

DTC acrónimo de *Diagnostic Trouble Code* (Código de problema de diagnóstico) de SAE J1930 Clean Air Act Amendment OBD II.

Dynatard Mack internal engine compression brake; not used in Mack engines after 1996.

Dynatard freno de motor interno de compresión de Mack; no se utiliza en motores Mack desde 1996.

ECAP electronic control analyzer programmer. Caterpillar PC-based reader/programmer instrument.

ECAP programador analizador de control electrónico. Instrumento de lectura/programación basado en PC de Caterpillar.

Echeck Cummins EST.

Echeck EST de Cummins.

ECM electronic/engine control module. A system controller consisting of a microcomputer and integral output switching apparatus.

ECM módulo de control electrónico/de motor. Controlador de sistema que consiste en un microordenador y un aparato de conmutación de salida integral.

ECU electronic/engine control unit. Synonymous with ECM.

ECU unidad de control electrónico/de motor. Sinónimo de ECM.

Econovance Mack Trucks mechanical or electronic variable timing device for port helix metering injection pumps.

Econovance dispositivo de sincronización variable mecánico o electrónico de Mack Trucks para bombas de inyección de dosificación helicoidal de abertura.

EDU electronic distributor unit (DDEC). DDC term for injector drivers.

EDU unidad de distribuidor electrónica (DDEC). Término DDC para controladores de inyección.

effective stroke describes that portion of a constant travel plunger or piston stroke used to actually pump fluid.

carrera útil describe la parte de un tiempo del recorrido constante del émbolo o pistón que se utiliza para bombear fluido.

EG ethylene glycol. An antifreeze of higher toxicity that the EPA hopes to phase out.

EG etilenglicol. Anticongelante de alta toxicidad que la EPA (Agencia para la protección del medio ambiente) espera ir eliminando del mercado.

EFPA electronic foot pedal assembly.

EFPA conjunto de pedal electrónico.

EFRC engine family rating code.

EFRC código de medición de familias de motores.

ELAB Bosch fuel shutoff solenoid used on early V-MAC I PE7100 injection pumps.

ELAB solenoide Bosch de cierre de combustible de las primeras bombas de inyección V-MAC I PE7100.

ELC extended life coolant. Coolant premix that claims a service life of up to 6 years with almost no maintenance.

ELC refrigerante de larga duración. Premezcla de refrigerante que ofrece una duración de servicio de hasta 6 años sin necesitar casi ningún mantenimiento.

electricity a form of energy that results from charged particles, specifically electrons and protons, either statically (accumulated charge) or dynamically such as current flow in a circuit.

electricidad forma de energía que proviene de partículas cargadas, específicamente de electrones y protones, ya sea estática (carga acumulada) o dinámicamente, como el flujo de corriente de un circuito.

electron theory the theory that asserts that current flow through a circuit is by electron movement from a negatively charged point to a positively charged one. See *conventional theory*.

teoría de electrones teoría que afirma que el flujo de corriente de un circuito es por el movimiento de electrones desde un punto cargado negativamente a uno cargado positivamente. Véase *teoría convencional*.

electronic engine management computerized engine control.

gestión electónico del motor control computerizado del motor.

electronic governor any kind of governing using computer controls.

regulador electrónico cualquier tipo de regulador que utilice controles informáticos.

electronics branch of electricity concerned with the study of the movement of electrons through hard wire, semiconductor, gas, and vacuum circuits.

electrónica rama de la electricidad que estudia el movimiento de electrones a través de un cable, de un semiconductor, de gas y de circuitos al vacío.

element
1. any of more than 100 substances (most naturally occurring, some man-made) that cannot be chemically resolved into simpler substances.
2. a component part of something such as a pump *element*.

elemento
1. cualquiera de las más de 100 sustancias (algunas naturales, otras artificiales) que no se pueden dividir químicamente en sustancias más simples.
2. un componente de algo, como un *elemento* de una bomba.

emf electromotive force or voltage.

fem fuerza electromotriz o tensión.

emulsify the dispersion of one liquid to another or the suspension of a fine particulate in a solution.

emulsificar dispersión de un líquido en otro o suspensión de una partícula fina en una solución.

emulsion the dispersion of one liquid into another such as water in the form of fine droplets into diesel fuel.

emulsión dispersión de un líquido en otro, como agua en forma de gotas finas en gasóleo.

end gas the gas that results from combusting fuel in engine cylinders; usually means the gases present at flame quench, that is, before any exhaust gas treatment so a mixture of CO_2, H_2O, and whatever noxious gases are present.

gas de escape gas que procede de quemar combustible en los cilindros del motor; suele referirse a los gases presentes al apagarse una llama, es decir, antes de ningún tratamiento de gases de escape, o antes de que esté presente una mezcla de CO_2, H_2O y de gases nocivos.

engine a machine that converts one form of energy to another.

motor máquina que convierte una forma de energía en otra.

engine brake any type of engine retarder. The term usually describes an internal engine compression brake but may also refer to an exhaust compression brake or an engine-mounted hydraulic retarder.

retardador de compresión cualquier tipo de retardador de motor. Este término suele describir un freno interno de compresión del motor, pero puede que se refiera también a un freno de compresión de escape o a un retardador hidráulico montado en el motor.

engine dynamometer a dynamometer used for testing the engine on a test bed outside of the chassis.

dinamómetro de motor dinamómetro utilizado para verificar el motor en un banco de pruebas fuera del chasis.

engine hours a means of comparing engine service hours to highway mileage. Most engine OEMs equate 1 engine hour to 50 highway linehaul miles (80 km), so a service interval of 10,000 miles (16,000 km) would equal 200 engine hours. The term *service hours* is also used.

horas de motor medio de comparar las horas de servicio del motor con el kilometraje de carretera. La mayoría de los OEM equiparan 1 hora de motor con 80 kilómetros (50 millas) en carretera, por lo que un intervalo de 16.000 km (10.000 millas) equivaldría a 200 horas de motor. También se utiliza el término *horas de servicio*.

engine silencer a *muffler* that uses sound absorption and resonation principles to change the frequency of engine noise.

silenciador de motor *silencioso* que utiliza los principios de absorción y de resonancia para cambiar la frecuencia del ruido del motor.

EOE ending of energizing. Denotes the end of the switched duty cycle of an EUI.

EOE final de la revolución. Indica el final de un ciclo de servicio conectado de un EUI.

EOI ending of injection.

EOI final de la inyección.

EOL end of line—usually in reference to a programming procedure.

final de línea normalmente en referencia a un procedimiento de programación.

E-9 Mack Trucks V8 engine.

E-9 motor V8 de Mack Trucks.

EPA Environmental Protection Agency. Federal regulating body that sets and monitors noxious emission standards amongst other functions.

EPA agencia para la protección del medio ambiente. Corporación federal estadounidense reguladora que define y supervisa los estándares de emisiones nocivas, entre otras funciones.

EPS engine position sensor.

EPS sensor de posición del motor.

E 7 Mack Trucks in-line, 6-cylinder, 12-liter engine.

E 7 motor de Mack Trucks en línea, de 6 cilindros y de 12 litros.

E7-EUP Mack Trucks V-MAC III managed, in-line, 6-cylinder 12-liter engine.

E7-EUP motor de Mack Trucks gestionado por V-MAC III en línea, de 6 cilindros y de 12 litros.

ESP electronic smart power (Cummins).

ESP potencia electrónica inteligente (Cummins).

EST electronic service tool. Covers a range of instruments including DMMs, diagnostic lights, generic and proprietary reader-programmers, and PCs.

EST herramienta electrónica de servicio. Se refiere a diferentes instrumentos, que incluyen DMM, luces de diagnóstico, lectores y progamadores genéricos y patentados y ordenadores.

ET electronic technician. Caterpillar PC-based software that enables the technician to diagnose system problems, reprogram ECMs, and access system data for analysis to produce fuel mileage figures and driver performance profiles.

ET técnico en electrónica. Software basado en PC de Caterpillar que permite que el técnico diagnostique problemas de sistemas, que reprograme ECM y que tenga acceso a datos del sistema para su análisis para obtener cifras de kilometraje de combustible y perfiles de rendimiento del conductor.

etching bearing or other component failure caused by chemical action.

corrosión error en cojinetes u otros componentes provocado por una reacción química.

E-Tech Mack Trucks V-MAC III, electronic unit pump (EUP) fueled, E-7 engine.

E-Tech motor E-7, V-MAC III de Mack Trucks, alimentada por una bomba de unidad electrónica (EUP).

ETR energized to run.

ETR revolucionado para funcionar.

EUI electronic unit injector. The cam actuated, electronically controlled pumping mechanism used to fuel most full authority, electronically controlled truck diesel engines.

EUI inyector de unidad electrónica. Mecanismo de bombeo de acción de levas y controlado electrónicamente que se utiliza para alimentar la mayoría de los motores diesel de camiones de autoridad total y controlados electrónicamente.

EUP electronic unit pump.

EUP bomba de unidad electrónica.

execute effect an operation or procedure.

ejcutar efectuar una operación o procedimiento.

executive the resident portion of a computer or program operating system.

ejecutivo parte residente del ordenador del sistema operativo de programas.

exhaust blowdown the first part of the cylinder exhaust process that occurs at the moment the exhaust valves open.

evacuación del escape primera parte del proceso de escape del cilindro que se produce cuando se abren las válvulas de escape.

exhaust brake an external engine compression brake that operates by choking down the exhaust gas flow area; sometimes used in conjunction with an internal engine compression brake, meaning that the piston is contributing to retarding effort on both its upward strokes.

freno por compresión de aire freno externo por compresión del motor que funciona estrangulando el área del flujo de gas de escape; se utiliza a veces junto a un freno interno por compresión del motor, lo que significa que el pistón contribuye a retardar el esfuerzo en sus dos tiempos hacia arriba.

exhaust manifold the cast iron or steel component bolted to the cylinder exhaust tracts responsible for delivering the end gases to the turbocharger and the exhaust system.

colector de escape componente de hierro fundido o acero empernado a las zonas de escape del cilindro responsable de enviar los gases de escape al turboalimentador y al sistema de escape.

expansion board a circuit board added to a computer system to increase its capability.

tablero de expansión tablero de circuitos que se agrega a un sistema informático para aumentar su capacidad.

explosion an oxidation reaction that takes place rapidly; high speed combustion.

explosión reacción de oxidación que tiene lugar rápidamente; combustión de alta velocidad.

external compression brake refers to an engine exhaust brake.

freno por compresión externo se refiere a un freno por compresión de aire del motor.

fanstat a combination temperature sensor and switch (usually pneumatic) used to control the engine fan cycle.

fanstat combinación de sensor de temperatura y conmutador (normalmente neumático) utilizada para controlar el ciclo del ventilador del motor.

feedback assembly the engine's mechanical, self management components consisting of a gear train, camshaft, valve trains, MUI and EUI actuating trains, fuel injection pumping apparatus and valves.

conjunto de retroalimentación componentes mecánicos de autogestión del motor que consisten en un tren de engranajes, el árbol de levas, trenes de válvulas, trenes de mando MUI y EUI, el aparato de bombeo de inyección de combustible y válvulas.

FDCS fuel demand command signal (Navistar).

FDCS señal de comando de demanda de combustible (Navistar).

FIC module Mack Trucks V-MAC I injection pump controller.

módulo FIC controlador de bomba de inyección Mack Trucks V-MAC I.

fire point the temperature at which a combustible produces enough flammable vapor for a continuous burn; always a higher temperature than *flash point*.

punto de inflamación temperatura a la que un combustible produce suficiente vapor inflamable para un quemado continuo; es una temperatura siempre más alta que el *punto de inflamabilidad*.

fire ring normally used to refer to the fixed ring that may be integral with the cylinder head gasket responsible for sealing the cylinder. Sometimes used to refer to the top compression ring but this usage is uncommon.

segmento de fuego se utiliza normalmente para refererirse al segmento fijo que puede formar una parte integral de la junta de culata responsable de sellar el cilindro. Se utiliza a veces para referirse al anillo de comprensión, pero este uso no es común.

flame front during flame propagation, the leading edge of the flame in an engine cylinder.

frente de la llama durante la propagación de la llama, el borde de entrada de la llama en un cilindro de motor.

flame quench the moment that the flame ceases to propagate or extinguishes in an internal combustion engine.

extinción de la llama momento en el que la llama deja de propagarse o se apaga en un motor de combustión interna.

flame propagation the flame pattern from ignition to quench during a power stroke in an engine cylinder.

propagación de la llama patrón de la llama desde la ignición a la extinción durante un tiempo de potencia en un cilindro de motor.

flammable any substance that can be combusted.

inflamable cualquier sustancia que se pueda quemar.

flashback a highly dangerous condition that can occur in operating oxy-acetylene equipment in which the flame may travel behind the mixing chamber in the torch and explode the acetylene tank using the system oxygen. Most current oxy-acetylene torches are equipped with flashback arresters.

retroceso de la llama problema muy peligroso que se puede producir en un equipamiento de oxiacetileno operativo en el que la llama se puede trasladar a la cámara de mezcla del soplete y hacer explotar el depósito de acetileno utilizando el oxígeno del sistema. La mayoría de los sopletes oxiacetilénicos están equipados con supresores de retroceso de llama.

flash codes the ECM-generated fault codes that are usually displayed by means of diagnostic lights and alert the driver or technician as to the nature of an electronically monitored malfunction; also known as *fault codes, blink codes.*

códigos de destello códigos de error generados por el ECM que se suelen mostrar mediante luces de diagnóstico y alertan al conductor o al técnico de la naturaleza de un mal funcionamiento supervisado electrónicamente; conocidos también como *códigos de error* y *códigos intermitentes.*

flash point the temperature at which a combustible produces enough flammable vapor for momentary ignition.

punto de inflamabilidad temperatura a la que un combustible produce suficiente vapor inflamable para un encendido momentáneo.

flow area the most restricted portion of a fluid circuit; for instance, a water tap sets a flow area and as the tap is opened, the flow area increases, thereby increasing the volume flow of water.

área de flujo parte más restringida de un circuito de fluido; por ejemplo, un grifo de agua define un área de flujo y, cuando se abre el grifo, el área de flujo aumenta, con lo que aumenta el volumen de flujo de agua.

flow control refers to any device that can proportionally control flow through a circuit. A thumb over the end of a hose is a flow control device.

regulación de flujo se refiere a cualquier dispositivo que puede controlar proporcionalmente el flujo a través de un circuito. Una mariposa al final de una manguera es un dispositivo de regulación de flujo.

fluidity a substance in state that permits it to conform to the shape of the vessel in which it is contained. Both liquids and gases possess *fluidity.*

fluidez sustancia en un estado que le permite adaptarse a la forma del recipiente en el que está contenido. Tanto los líquidos como los gases poseen *fluidez.*

flutes protruding lands with grooves in between.

acanaladuras partes planas salientes con ranuras entre ellas.

flywheel an energy and momentum storage device usually bolted directly to the crankshaft.

volante de inercia dispositivo de almacenamiento de energía y de impulso que suele estar empernado directamente al eje de cigüeñal.

flywheel housing concentricity a critical specification that ensures that the relationship between the crankshaft and the flywheel is concentric.

concentricidad de alojamiento del volante de inercia especificación crítica que asegura que la relación entre el árbol de levas y el volante de inercia sea concéntrica.

FMI fault mode indicator (SAE). Defines a component or circuit failure numerically to an EST by ascribing to it one of fourteen possible failure modes.

FMI indicador de modo de errores (SAE). Define un error de componente o circuito numéricamente en una EST atribuyéndolo a uno de los catorce posibles modos de error.

follower used to describe a variety of devices that ride a cam profile and transmit the effects of the cam geometry to the train to be actuated; also known as *tappet.*

rodillo se utiliza para describir diferentes dispositivos que hacen funcionar un perfil de leva y transmiten los efectos de la geometría de la leva al tren que se debe accionar; conocido también como *empujador.*

force the action of one body attempting to change the state of motion of another. The application of force does not necessarily result in any work accomplished.

fuerza acción de un cuerpo al intentar cambiar el estado de movimiento de otro. La aplicación de fuerza no tiene necesariamente como resultado la consecución de un trabajo.

foreground computations computer-operating responses that are prioritized, such as the response to a critical command input such as the TPS (throttle position sensor) whose signal must be acted on immediately to generate the appropriate outcome.

cálculos en primer plano respuestas informáticas a las que se da prioridad, como la respuesta a la entrada de un comando crítico como el TPS (sensor de posición del acelerador) tras cuya señal se debe actuar inmediatamente para generar el resultado adecuado.

forward leakage an injector bench fixture test that tests the nozzle seat sealing integrity.

pérdidas delanteras banco de artefactos de pruebas de inyección que verifica la integridad del sellado del alojamiento de la boquilla.

Four-Stroke Cycle completes one cycle of operation with four strokes as the intake, compression, power, and exhaust strokes of the piston and two revolutions of the crankshaft.

ciclo de cuatro tiempos completa un ciclo de funcionamiento con cuatro tiempos: admisión, compresión, potencia y escape del pistón y dos revoluciones del árbol de levas.

FPS fuel pressure sensor. A pressure sensing mechanism usually of the variable capacitance type that measures the charging pressure in the fuel subsystem and signals its value to the ECM.

FPS sensor de presión de combustible. Mecanismo de medición de la presión que suele ser del tipo de capacitación variable que mide la presión de carga en el subsistema de combustible e indica su valor en el ECM.

fractions refers to separate compounds of crude petroleums separated by distillation and other fractioning methods such as catalytic and hydrocracking and classified by their volatility.

fracciones se refiere a compuestos diferentes de petróleos crudos separados por destilación y otros métodos de fraccionamiento como el catalítico o la hidrodesintegración y se clasifican por su volatilidad.

FRC fuel ratio control. Caterpillar aneroid mechanism for limiting fueling in low boost conditions.

FRC control de la relación de combustible. Mecanismo aneroide de Caterpillar para limitar la alimentación en condiciones de baja alimentación.

friction the resistance an object or fluid encounters in moving over or through another.

fricción resistencia que un objeto o fluido encuentra al moverse por encima o a través de otro.

friction bearing a shaft supporting bearing in which the rotating member can directly contact the bearing face or race.

cojinete de rozamiento eje que soporta el cojinete en el que el miembro que gira puede entrar en contacto directo con la cara o guía de cojinete.

fuel conditioner usually unknown quantities of cetane improver and pour point depressants suspended in an alcohol base.

acondicionador de combustible cantidades normalmente desconocidas de mejorador de cetano y agentes tensoactivos de punto de fluidez suspendidos en una base de alcohol.

fuel control actuator any of a number of electronically controlled devices used as fuel control mechanisms.

regulador de combustible cualquier número de dispositivos controlados que se utilizan como mecanismos de control de combustible.

fuel heater a heat exchanger device used in extreme cold to prevent diesel fuel from waxing in the fuel subsystem.

calentador de combustible dispositivo intercambiador térmico que se utiliza en condiciones de frío extremo para evitar que el gasóleo se parafine en el subsistema de combustible.

fuel map a diagram or graph used to indicate fueling through the entire performance range of an engine.

circuito de combustible diagrama o gráfico que se utiliza para indicar la alimentación por todo el margen de rendimiento de un motor.

fuel rate actual rate of fuel pumped through an injector to an engine cylinder; factored by cam geometry and engine rpm.

caudal de combustible caudal real de combustible que se bombea a través de un inyector a un cilindro de motor; sus factores son la geometría de la leva y las rpm del motor.

fuel subsystem the fuel circuit used to pump fuel from the vehicle fuel tank and deliver it to the fuel metering/injection apparatus. The fuel subsystem typically comprises a fuel tank, water separator, primary filter, transfer or charge pump, secondary filter, and the interconnecting plumbing.

subsistema de combustible circuito de combustible que se utiliza para bombear combustible del depósito del vehículo y llevarlo al aparato de dosificación/inyección. El subsistema de combustible suele constar de un depósito, un separador de agua, un filtro primario, una bomba de transferencia o carga, un filtro secundario y las tuberías de interconexión.

fuel tank the fuel storage reservoir on a vehicle.

depósito recipiente de almacenamiento de combustible de un vehículo.

fueling actuator Cummins HPI-TP metering control solenoid; also known as the rail actuator.

impulsor de alimentación solenoide de control de dosificación Cummins HPI-TP; conocido también como impulsor del distribuidor.

fueling chamber the lower chamber in the Cummins TP injector that forms the injector pumping chamber.

cámara de alimentación cámara inferior del inyector Cummins TP que compone la cámara de bombeo de inyección.

full flow filter a filter plumbed in series on the charge side of the pump that feeds a circuit.

filtro de paso único filtro conectado en serie en el lado de carga de la bomba que alimenta el circuito.

gas dynamics the manner in which gases behave during the compression and combustion strokes and the processes of engine breathing.

dinámica de gases forma en que los gases se comportan durante los tiempos de compresión y combustión y los procesos de ventilación.

gasoline a hydrocarbon fuel composed of the volatile petroleum fractions from the aromatic and paraffin ranges.

gasolina combustible hidrocarburo compuesto de las fracciones volátiles del petróleo de los rangos aromáticos y parafinosos.

gear pump a positive displacement pump consisting of intermeshing gears that uses the spaces between the teeth to move fluid through a circuit.

bomba de engranajes bomba de desplazamiento positivo que consiste en engranajes de toma constante que utiliza los espacios entre los dientes para mover el fluido por un circuito.

gerotor a type of gear pump that uses an internal crescent gear pumping principle.

gerotor tipo de bomba de engranajes que utiliza un principio interno de bombeo de engranajes creciente.

governor a component that manages engine fueling on the basis of fuel demand (accelerator) and engine rpm; may be hydromechanical or electronic.

regulador componente que gestiona la alimentación del motor basándose en la demanda de combustible (acelerador) y las rpm del motor; puede ser hidromecánica o electrónica.

governor barrel the stationary governor component in a PT pump, within which the governor plunger rotates and into which the fuel passages are machined.

cilindro de bomba componente del regulador fijo en una bomba PT, dentro de la cual gira el émbolo del regulador y en el que se disponen los pasos de combustible.

governor button the component in a Cummins PT pump that helps define pressure within the governor barrel and therefore greatly influences fuel flow to the rail.

botón regulador componente de una bomba Cummins PT que ayuda a definir la presión que hay dentro del cilindro de bomba y, por tanto, influye en el flujo de combustible al distribuidor.

governor differential lever a double-bell crank-type, lever device that pivots on a fulcrum.

palanca diferencial del regulador tipo de codo de palanca de campana doble, dispositivo de palanca que pivota en un punto de apoyo.

governor plunger the rotating member of the Cummins PT governor, driven by the governor weight carrier.

pistón del regulador miembro giratorio del regulador Cummins PT, impulsado por el soportador de masa del regulador.

governor spring the force, usually variable, that opposes centrifugal force in mechanical governors; often amplified by accelerator pedal travel.

resorte del regulador fuerza, normalmente variable, que se opone a la fuerza centrífuga en reguladores mecánicos; a menudo amplificada por el recorrido del pedal del acelerador.

governor weight forks a means of jamming governor centrifugal weights in their outermost position for purposes of tuning an engine.

horquillas de masa centrífuga del regulador forma de bloquear las masas centrífugas del regulador en su posición más externa para poner a punto un motor.

gph gallons per hour. The means of rating liquid flow in a hydraulic circuit.

gph galones por hora. Forma de medir el flujo de líquido en un circuito hidráulico.

ground describes the point or region of lowest voltage potential in a circuit; the portion of a vehicle electrical circuit serving multiple system loads by providing a return path for the current drawn by the load. Used in vehicle systems using 48 V or less and ideal for the commonly used 12 V vehicle systems.

tierra describe el punto o zona de menor potencial de tensión de un circuito; parte del circuito eléctrico de un vehículo que sirve múltiples cargas de sistema proporcionando una ruta de vuelta para la corriente arrastrada por la carga. Se utiliza en sistemas de vehículos de 48 V o menos y es ideal para los sistemas comúnmente utilizados de vehículo de 12 V.

ground strap a conductive strap, usually braided wire, that extends a common ground electrical system.

cable a tierra cable conductor, normalmente trenzado, que se extiende por un sistema común eléctrico de tierra.

gumming a term used to describe unburned fuel and lubrication oil residues when they sludge in piston ring grooves and other areas of the engine.

engomado término que se utiliza para describir los residuos del combustible sin quemar y de aceite lubricante cuando se sedimentan en las ranuras del anillo del pistón y otras áreas del motor.

Hall effect a method of accurately sensing rotational speed and digitally signaling it. A rotating metallic shutter alternately blocks and opens a magnetic field from a semiconductor sensor.

efecto Hall método de medición precisa de la velocidad de rotación que sirve también para señalarla digitalmente. Un obturador metálico giratorio bloquea y abre alternativamente un campo magnético de un sensor semiconductor.

headers the manifold deck to which the coolant tubes are attached in a heat exchanger bundle or the term used to describe low gas restriction, individual cylinder exhaust pipes that converge at a point calculated to maximize pulse effect.

cabezales caja del colector a la que están unidos los conductos refrigerantes en un haz tubular intercambiador térmico o término que se utiliza para describir el gas de paso rápido, tubos individuales de escape del cilindro que convergen en un punto calculado para maximizar el efecto de los impulsos.

headland the area above the uppermost compression ring and below the leading edge of the piston.

promontorio área que está por encima del anillo de compresión y por debajo del borde de entrada del pistón.

headland volume the headland gas volume in a cylinder.

volumen del promontorio volumen de gas del promontorio en un cilindro.

heat engine a mechanism that converts thermal energy into mechanical work.

motor térmico mecanismo que convierte energía térmica en trabajo mecánico.

heat exchanger any of a number of devices used to transfer heat from one fluid to another where there is a temperature difference using the principles of conduction and radiation.

intercambiador térmico cualquiera de los varios dispositivos utilizados para transferir calor de un fluido a otro cuando hay diferencia de temperatura mediante los principios de conducción y radiación.

heating value the potential heat energy of a fuel; also known as *calorific value.*

poder calorífico energía de calor potencial de un combustible; conocido también como *potencia calorífica.*

helical gear a gear with spiral cut teeth.

engranaje helicoidal engranaje con los dientes cortados en espiral.

helix a spiral groove or scroll. The helical cut recesses in some injection pumping plungers that are used to meter fuel delivery. Plural: *helices.*

espiral ranura o voluta helicoidal. salidas helicoidales talladas en algunos émbolos de bombeo de inyección que se utilizan para dosificar el suministro de combustible. Plural: *espirales.*

HEUI hydraulically actuated, electronic unit injector (Caterpillar and Navistar).

HEUI inyector de unidad electrónica de acción hidráulica (Caterpillar y Navistar).

Hg manometer a mercury (Hg) filled manometer.

manómetro de mercurio manómetro relleno de mercurio (Hg).

high idle speed the highest no load speed of an engine.

velocidad máxima de ralentí la velocidad sin carga más alta de un motor.

high pressure pipes the pipes or lines that deliver fuel from an injection pump element to the injector nozzle.

conductos de alta presión conductos o líneas que llevan combustible de un elemento de bombeo de inyección a la boquilla del inyector.

high pressure washer a high pressure water pump used to clean equipment and components before repair and inspection that has generally replaced steam cleaners.

bomba de alta presión bomba de agua de alta presión que se utiliza para limpiar el equipamiento y los componentes antes de la reparación y la inspección que reemplaza en general a los limpiadores de vapor.

high spring injector a type of hydraulic injector nozzle that locates the injector spring high in the injector/nozzle holder body. NOP is usually adjusted by an adjusting screw that acts directly on the spring.

inyector de resorte alto tipo de boquilla de inyección hidráulica que coloca el resorte del inyector en la parte superior del cuerpo del contenedor del inyector o de la boquilla. El NOP se suele ajustar con un tornillo de ajuste que actúa directamente en el resorte.

Highway Master Caterpillar remote communications technology that enables remote programming of vehicles. Used in conjunction with Caterpillar Fleet Information software.

Highway Master tecnología de comunicaciones remotas de Caterpillar que permite la programación remota de vehículos. Se utiliza junto al software Caterpillar Fleet Information.

historic codes fault codes that are no longer active but are retained in ECM memory (and displayed) for purposes of diagnosis until they are erased; also knwon as *inactive* codes.

códigos históricos códigos de error que ya no están activos pero que se conservan en la memoria del ECM (y se muestran) como diagnóstico hasta que se borran; conocidos también como códigos *no activos.*

horsepower the standard unit of power measurement used in North America defined as a work rate of 33,000 lb.-ft. per minute; equal to 0.746 kW.

caballo de vapor unidad estándar de medición de potencia que se utiliza en Norteamérica y que se define como un trabajo de 33.000 lb.-ft. por minuto; igual a 0,746 kW.

host computer a main computer that is networked to other computers or nodes.

ordenador central ordenador principal conectado en red con otros ordenadores o nodos.

HPI high pressure injection. Cummins acronym for their electronically controlled, common rail fuel systems.

HPI inyección a alta presión. Acrónimo de Cummins para sus sistemas controlados electrónicamente de sistemas de combustible de distribuidor común.

HPI-PT electronically managed, common rail, open nozzle fuel system used to fuel the Cummins K-19 engine in which *pressure* is the control variable.

HPI-PT sistema de combustible de boquilla abierta de distribuidor común gestionado electrónicamente que se utiliza para alimentar el motor Cummins K-19 en el que la *presión* es la variable de control.

HPI-TP electronically managed, common rail, open nozzle fuel system used to fuel the Cummins Signature Series engine, in which *time* is the control variable.

HPI-TP sistema de combustible de boquilla abierta de distribuidor común gestionado electrónicamente que se utiliza para alimentar el motor de Cummins de la series Signature en el que el *tiempo* es la variable de control.

hunting rhythmic fluctuation of engine rpm usually caused by unbalanced cylinder fueling.

penduleo fluctuación rítmica de las rpm de un motor provocada normalmente por una alimentación desequilibrada del cilindro.

hunting gears an intermeshing gear relationship in which after timing, the gears may have to be turned through a large number of rotations before the timing indices realign.

engranajes de penduleo relación de engranajes de toma constante en la que, tras la sincronización, es posible que haya que girar los engranajes mediante un gran número de rotaciones antes que se vuelvan a alinear los índices de sincronización.

hydraulic governor, non-servo a hydraulic governor that uses fuel pressure unloaded into a defined flow area by a positive displacement pump, as the basis for determining rpm; not used on any current truck engines.

regulador hidráulico regulador hidráulico que utiliza presión de combustible sin carga en un área de flujo definida por una bomba de desplazamiento positivo, como base para determinar las rpm; no se utiliza en ningún motor de camión actual.

hydraulic governor, servo type a hydraulic governor that uses centrifugal weights to sense rpm but the force responsible for actually moving the fuel control mechanism is hydraulic either engine oil or fuel pressure.

servorregulador hidráulico regulador hidráulico que utiliza masas centrífugas para medir las rpm, pero la fuerza responsable para mover realmente el mecanismo de control de combustible es presión hidráulica de combustible o de aceite de motor.

hydraulics the science and practice of confining and pressurizing liquids in circuits to provide motive power.

hidráulica ciencia o práctica de confinar y presurizar líquidos en circuitos para proporcionar fuerza motriz.

hydrodynamic suspension the principle used to float a rotating shaft on a bed of constantly changing, pressurized lubricant.

suspensión hidrodinámica principio utilizado para hacer flotar un eje giratorio en un lecho de lubricante presurizado que cambia constantemente.

hydromechanical engine management all engines managed without computers.

gestión hidrodinámica del motor todos los motores gestionados sin ordenadores.

IBC inner base circle. Describes the smallest radial dimension of a cam profile.

IBC perfil interior de la leva. Describe la dimensión radial más pequeña de un perfil de leva.

I/C integrated circuit. An electronic circuit constructed on a semiconductor chip, such as silicon, that can replace many separate electrical components and circuits.

C/I circuito integrado. Circuito electrónico montado en un *chip* de un material semiconductor, como el silicio, que puede reemplazar a muchos componentes y circuitos eléctricos diferentes.

ICP injection control pressure (Navistar).

ICP presión de control de inyección (Navistar).

IDI see *indirect injection*.

IDI véase *inyección indirecta*.

idle speed the lowest speed that an engine is run at usually requiring no input to the governor.

velocidad de ralentí velocidad más baja a la que funciona un motor sin requerir normalmente entrada alguna al regulador.

IDM injector driver module (Navistar). Separate injector driver unit used in versions of HEUI up to 1997. IDM functions are integrated in a single engine controller ECM in current applications.

IDM módulo controlador de inyección (Navistar). Unidad de controlador de inyección independiente que se utiliza en las versiones de HEUI hasta 1997. Las funciones del IDM se integran en un solo ECM de controlador de motor en las aplicaciones actuales.

IFSM integrated fuel system module. The Cummins HPI-TP processing and switching module.

IFSM módulo de sistema de alimentación integrado. El módulo de conmutación de procesamiento Cummins HPI-TP.

ignition accelerators volatile fuel fractions that are added to a fuel to decrease ignition delay. They increase CN.

aceleradores de encendido fracciones volátiles de combustible que se agregan al combustible para disminuir la demora de encendido. Aumentan el CN.

ignition delay or lag the time period between the entry of the first droplets of fuel to the engine cylinder and the moment of ignition based on the fuel chemistry and the actual temperatures of the engine components and the air charge.

demora o retraso de encendido período de tiempo transcurrido entre la entrada de las primeras gotitas de combustible en el cilindro de motor y el momento del encendido basándose en la química del combustible y las temperaturas reales de los componentes del motor y la carga de aire.

impeller
1. the driven member of a turbocharger, responsible for compressing the air charge.
2. the power input member of a pump such as on a torque converter or hydraulic retarder.

impulsor
1. miembro impulsado de un turboalimentador, responsable de comprimir la carga de aire.
2. miembro de entrada de potencia de una bomba, como en un convertidor de par o retardador hidráulico.

inactive codes fault codes that are no longer active but are retained in ECM memory (and displayed) for purposes of diagnosis until they are erased; also known as historic codes.

códigos no activos códigos de error que ya no están activos pero que se conservan en la memoria del ECM (y se muestran) como diagnóstico hasta que se borran; conocidos también como códigos históricos.

indicated horsepower gross power produced in the engine cylinders often arrived at by calculation and always greater than *brake power* because it does not factor in pumping and friction losses.

potencia indicada fuerza bruta producida en los cilindros del motor a la que se suele llegar por cálculo y que es siempre menor que la *potencia de freno* porque no es un factor en las pérdidas por bombeo o fricción.

indirect injection (IDI) describes any of a number of methods of injecting fuel to an engine outside of the cylinder. This may be to an intake tract in the intake manifold or to a cell adjacent to the cylinder such as a pre-combustion chamber.

inyección indirecta (IDI) describe cualquiera de los muchos métodos de inyección de combustible a un motor fuera del cilindro. Puede ser una zona de entrada del colector de admisión o a una célula adyacente al cilindro, como una cámara de precombustión.

induction circuit refers to the engine air intake circuit but more appropriately describes air intake on naturally aspirated engines than on boosted engines.

circuito de admisión se refiere al circuito de admisión de aire del motor pero describe más correctamente la admisión de aire en motores de aspiración natural de aire que los motores alimentados.

INFORM Cummins PC-based data management system.

INFORM sistema de gestión de datos de Cumminis basado en PC.

infrared the wavelength just greater than the red end of the visible light spectrum, but below the radio wave frequency.

infrarrojo longitud de onda inmediatamente superior al rojo visible en el espectro de luz visible, pero por debajo de la frecuencia de ondas de radio.

infrared thermometer accurate heat measuring instrument that can be used for checking cylinder fueling balance.

termómetro de infrarrojos instrumento de medición precisa de calor que se puede utilizar para comprobar el equilibrio de la alimentación de los cilindros.

injection lag a diesel fuel injection term describing the time lag between port closure in a pumping element and the actual opening of the injector.

retraso de la inyección término de inyección de gasóleo que describe el retraso temporal entre el cierre de la abertura de un elemento y la apertura real del inyector.

injection rate a diesel fuel injection term that is defined as the fuel quantity pumped into an engine cylinder per crank angle degree. In all systems except the HEUI, injection rate is determined by the pump actuating cam profile geometry.

caudal de inyección término de inyección de gasóleo que se define como la cantidad de combustible que se bombea a un cilindro de motor por grado del ángulo del codo de palanca. En todos los sistemas, excepto el HEUI, el caudal de inyección lo determina la geometría del perfil de leva de acción de bombeo.

injector a term broadly used to describe the holder of a hydraulic nozzle assembly. May also be used to describe PT, MUI, EUI, and HEUI assemblies.

inyector término ampliamente utilizado para describir el contenedor de un conjunto de boquillas hidráulicas. También se puede utilizar para describir los conjuntos PT, MUI, EUI y HEUI.

injector drivers the ECM-controlled components that electrically switch EUI and HEUI assemblies. Injector drivers may be integral with the main ECM housing or contained in a separate module or housing.
controladores de inyección componentes controlados por el ECM que conmutan electrónicamente entre los conjuntos EUI y HEUI. Los controladores de inyección pueden estar integrados con el alojamiento principal ECM o contenidos en un módulo o alojamiento diferente.

inlet restriction a measure of the pressure value below atmospheric, developed on the suction side of a pumping mechanism. Air inlet restriction and fuel inlet restriction are common specifications used by the diesel technician.
restricción de admisión medida del valor de presión inferior a la atmosférica, desarrollado en el lado de succión de un mecanismo de bombeo. La restricción de entrada de aire y la de combustible son especificaciones comunes utilizados por los técnicos en diesel.

inlet restriction gauge instrument that measures (usually air) inlet restriction often on-chassis.
manómetro de restricción de entrada instrumento que mide la restricción de entrada (normalmente de aire) que suele estar incluido en el chasis.

inner base circle in cam geometry, the portion of the cam profile with the smallest radial dimension; also known as *base circle*/IBC. When the train riding the cam profile is on IBC, it is unloaded.
perfil interior de la leva en la geometría de la leva, la parte del perfil de la leva con la menor dimensión radial; conocida también como *perfil de la leva*/IBC. Cuando el tren que hace funcionar el perfil de la leva está en IBC, está descargado.

input devices the hardware, such as a keyboard on a PC, or *sensors* on a vehicle system responsible for signaling/switching data to a computer system.
dispositivos de entrada hardware, como el teclado de un PC, o *sensores* en un sistema de vehículo responsable de indicar/conmutar datos en un sistema informático.

INSITE Cummins PC software.
INSITE software de PC de Cummins.

INSPEC Cummins Windows driven PC vehicle ECM diagnostics and programming software.
INSPEC software de Cummins diagnóstico y programación de ECM de vehículos para PC con Windows.

insulators materials that either prevent or inhibit the flow of electrons; usually nonmetallic substances that contain more than four electrons in their outer shell.
aislantes materiales que evitan o inhiben el flujo de electrones; normalmente sustancias no metálicas que contienen más de cuatro electrones en su capa externa.

intake manifold the piping that is clamped to the intake tract flange faces responsible for directing intake air into the engine cylinders.
colector de admisión conductos aferrados a las caras de la brida de unión de la zona de admisión responsables de dirigir el aire de admisión a los cilindros del motor.

integrated circuit see *IC*.
circuito integrado véase *C/I*.

intensifier piston Caterpillar HEUI hydraulically actuated piston that pumps fuel to injection pressure values; also known as *amplifier piston*.
émbolo multiplicador pistón de acción hidráulica HEUI de Caterpillar que bombea combustible a las válvulas de presión de inyección; conocido también como *pistón amplificador*.

Interact System Cummins term used to describe their integrated electronic engine management systems with the ability to connect with fleet management and analysis software. Acronym IS used ahead of the engine series letter.
Interact System término de Cummins que describe sus sistemas de gestión electrónica del motor con capacidad para conectar con software de gestión de flotilla y de análisis. Se utiliza el acrónimo IS antes de la letra de serie del motor.

Interbrake the internal engine compression brake manufactured by Jacobs and used on the Cummins Signature 600 Series engine.
Interbrake freno interno por comprensión del motor fabricado por Jacobs y utilizado en el motor Cummins Signature 600 Series.

interference fit the fitting of two components so that the od of the inner component fractionally exceeds the id of the outer component. Liners are sometimes interference fit to cylinder bores. Interference fitting requires the use of a press, chilling, heating, or other forceful means.
ajuste duro ajuste de dos componentes para que el diámetro exterior (od) del componente interno exceda fraccionalmente el diámetro interior (id) del componente externo. Las camisas a veces son ajustes duros de los calibres de cilindros. El ajuste duro requiere el uso de una prensa, enfriamiento, calentamiento u otros métodos de fuerza.

internal compression brake any of a number of engine brakes that use the principle of making the piston perform its usual work through the compression stroke and then negate the power stroke by releasing the compression air to the exhaust system at TDC on the completion of the compression stroke.
freno de compresión interna cualquiera de los muchos frenos de motor que utilizan el principio de hacer que el pistón realice su trabajo habitual a través del tiempo de comprensión y negar luego el tiempo de potencia liberando el aire de compresión al sistema de escape en el TDC al finalizar el tiempo de compresión.

ion an atom with either an excess or deficiency of electrons, that is, an unbalanced atom.
ión átomo con exceso o defecto de electrones, es decir, un átomo desequilibrado.

IPR injection pressure regulator (Navistar). ECM-controlled device that manages HEUI actuating oil pressure.
IPR regulador de presión de inyección (Navistar). Dispositivo controlados por el ECM que gestiona la presión de aceite de acción HEUI.

IRT injector response time.
IRT tiempo de respuesta del inyector.

ISB Interact System B series, in-line, 6-cylinder, 6.9-liter engine (Cummins).
ISB Motor Interact System B en serie, en línea, de 6 cilindros y de 6,9 litros (Cummins).

ISC Interact System C series, in-line, 6-cylinder, 8.3-liter engine (Cummins).
ISC Motor Interact System C en serie, en línea, de 6 cilindros y de 8,3 litros (Cummins).

ISM Interact System M series, in-line, 6-cylinder, 11-liter engine (Cummins).
ISM Motor Interact System serie M, en línea, de 6 cilindros y de 11 litros (Cummins).

isochronous governor a zero droop governor or one that accommodates no change in rpm on the engine it manages as engine load varies. In electronically managed truck engines, the term is sometimes used to describe engine operation in PTO mode.
regulador isocrono regulador de caída cero o uno que facilita que no haya cambios en las rpm en el motor que gestiona cuando varía la carga del motor. En motores de camión gestionados electrónicamente, este término se utiliza a veces para describir el funcionamiento del motor en modo PTO.

Jacobs retarders Jacobs are known mainly for their internal engine compression brakes but also manufacture driveline retarders.
frenos Jacobs Jacobs es conocida principalmente por sus frenos internos por comprensión del motor, pero también fabrican frenos del sistema de transmisión.

Jake brake see *Jacobs retarders*.
freno Jake véase *frenos Jacobs*.

jumper pipes a term used to describe the pipes that connect the charge and return galleries with DDC MUIs or with each other in multi-cylinder heads.
conductos de puente término que se utiliza para describir los conductos que conectan la carga y las galerías de vuelta con los DDC MUI o entre sí en cabezas de múltiples cilindros.

JWAC is an acronym for Jacket Water Air Cooler, which is a heat exchanger that uses engine coolant to cool the incoming air from the turbocharger.
JWAC es un acrónimo de *Jacket Water Air Cooler* (camisa de refrigeración por agua), que es un intercambiador térmico que utiliza refrigerante de motor para enfriar el aire entrante del turboalimentador.

KAM keep alive memory. Nonvolatile RAM.
KAM memoria estable. RAM no volátil.

KAMPWR electrical circuit that powers KAM (Navistar).
KAMPWR circuito eléctrico que alimenta la KAM (Navistar).

Kirchhoff's 1ˢᵗ law states that the current flowing into a point or component in an electrical circuit must equal the current flowing out of it.
primera ley de Kirchhoff afirma que el flujo de corriente que llega a un punto o componente de un circuito eléctrico debe ser igual al flujo de corriente que sale de él.

Kirchhoff's 2ⁿᵈ law states that the voltage will drop in exact proportion to the resistance in a circuit component and that the sum of the voltage drops must equal the voltage applied to the circuit; also known as Kirchhoff's law of voltage drops.
segunda ley de Kirchhoff afirma que la tensión caerá en exactamente la misma proporción que la resistencia de un componente del circuito y que la suma de las caídas de tensión debe ser igual a la tensión aplicada al circuito; conocida también como ley de Kirchhoff de caídas de tensión.

kerosene a petroleum derived fuel with a lower volatility than gasoline and fewer residual oils than diesel fuel.
queroseno un combustible derivado del petróleo con una volatilidad más baja que la gasolina y con menos aceites lubricantes que el gasóleo.

keystone the trapezoidal shape that gets its name from the trapezoidal stones used in a classic Roman arch bridge.
trapezoide forma trapezoidal que recibe su nombre de las piedras trapezoidales de un puente romano clásico de arcos.

keystone ring a trapezoidally shaped piston ring commonly used in diesel engine compression ring design.
anillo trapezoidal anillo del pistón de forma trapezoidal utilizado normalmente en el diseño del anillo de compresión de motores diesel.

keystone rod a connecting rod with a trapezoidal eye (small end) to increase the loaded sectional area.
biela trapezoidal biela de conexión con un ojo trapezoidal (pequeño extremo) para aumentar el área de secciones cargada.

kilowatt a unit of power measurement equivalent to 1,000 watts. Equal to 1.34 BHP.
kilovatio unidad de medida de potencia equivalente a 1.000 vatios. Equal to 1.34 BHP.

kinetic energy the energy of motion.
energía cinética energía del movimiento.

kinetic molecular theory states that all matter consists of molecules that are constantly in motion and that the extent of motion will increase at higher temperatures.
tería molecular cinética afirma que toda materia consiste en moléculas en constante movimiento y que el grado de movimiento aumentará con temperaturas más altas.

lamina a thin layer, plate, or film.
lámina capa, placa o película fina.

lands the raised areas between grooves especially on the ring belt of a piston.
partes planas áreas elevadas entre ranuras, especialmente en la zona de segmentación de un pistón.

latching solenoid a solenoid that locks to a position when actuated and usually remains in that position until the system is shut down.
solenoide cerrado solenoide que se bloquea en una posición cuando se acciona y suele permanecer en esa posición hasta que se apaga el sistema.

LED light-emitting diode.
LED diodo emisor de luz.

lever a rigid bar that pivots on a fulcrum and can be used to provide a mechanical advantage.
palanca barra rígida que pivota en un punto de apoyo y que se puede utilizar para proporcionar una ventaja mecánica.

limiting speed governor a standard automotive governor that defines the idle and high idle fuel quantities and leaves the intermediate fueling to be managed by the operator within the limitations of the fuel system.
limitador de velocidad regulador estándar de automoción que define las cantidades (de combustión) de ralentí y ralentí máximo y deja que la alimentación intermedia la gestione el operador dentro de los límites del sistema de combustible.

linear magnet a proportional solenoid. This term is used by Mack Trucks to describe their rack actuator mechanism.
magneto lineal solenoide proporcional. Los Mack Trucks utilizan este término para describir su mecanismo impulsor de cremallera.

liners the normally replaceable inserts into the cylinder block bores of most diesel engines that permit easy engine overhaul service and greatly extended cylinder block longevity.
camisas inserciones normalmente reemplazables de los calibres del bloque del cilindro de la mayoría de motores diesel que permiten un sencillo servicio de revisión general y una longevidad mayor del bloque del cilindro.

link and lever assembly Caterpillar intermediary between the governor and MUIs.
conjunto de uniones y palancas zona intermedia de Caterpillar entre el regulador y los MUI.

liquid petroleum gas (LPG) propane, butane, or mixtures of both used as an alternative fuel in SI or pilot-ignited engines.
gas licuado de petróleo (GLP) propano, butano o mezclas de ambos que se utilizan como combustible alternativo en motores SI o de encendido piloto.

load ratio of power developed versus rated peak power at the same rpm.
carga nivel de potencia desarrollado en comparación con la potencia máxima en régimen con las mismas rpm.

locking tang a tab on a component, such as a bearing, that may help position and lock it.
cola de bloqueo pestaña de un componente, como de un cojinete, que puede ayudar a colocarlo y bloquearlo.

lower helix the standard helix milled into most port helix plungers used in truck diesel applications. These produce a constant beginning, variable delivery characteristic when no external variable timing mechanism is used.
espiral inferior espiral estándar tallada en la mayoría de los émbolos helicoidales de abertura utilizada en aplicaciones diesel de camiones. Producen un principio constante, característica de suministro variable cuando no se utiliza ningún mecanismo de sincronización variable externo.

low spring injector an injector design that locates the spring directly over the nozzle valve, thereby reducing the mass of moving components compared to a high spring model. Injector spring tension is usually defined by shims.

inyector de resorte bajo diseño de inyector que coloca el resorte directamente encima de la válvula de la boquilla, con lo que reduce la complicación de mover componentes en comparación con un modelo de resorte alto. La tensión del resorte de inyección se suele definir con espaciadores.

lubricity literally, the *oiliness* of a substance.
lubricidad literalmente, lo *oleoso* de una sustancia.

lugging term used to describe an engine that is run at speeds lower than the base of the torque rise profile (peak torque) under high loads, that is, high cylinder pressures.

arrastre término que se utiliza para describir un motor que funciona a velocidades más bajas que la base del perfil de aumento del par (par máximo) en condiciones de cargas altas, es decir de presiones de cilindro.

M (Mann) Type combustion chamber the Mann type piston crown is named after the German Company that first designed them. It is usually used with trunk type pistons and consists of a spherical recess or bowl located under the injector, not necessarily in the center of the piston crown. Depending on the depth of the recess, the Mann type combustion chamber generally produces high turbulence but is more vulnerable to localized burn out in the bowl

cámara de combustión tipo Mann la corona de pistón tipo Mann recibe el nombre por la empresa alemana que las diseñó por primera vez. Se suele utilizar con pistones de tipo de tronco y consisten en una salida esférica o cuenca ubicada bajo el inyector, no necesariamente en el centro de la corona del pistón. Según la profundidad de la salida, la cámara de combustión tipo Mann suele producir una alta turbulencia pero es más vulnerable al quemado localizado en la cuenca.

magnetic flux test magnetic flux crack detection used to identify defects in crankshafts, connecting rods, cylinder heads, etc. An electric current is flowed through the component being tested and iron particles suspended in liquid are then sprayed over the surface. The particles will concentrate where the magnetic flux lines are broken up by cracks.

prueba del flujo magnético detección de fisuras en el flujo magnético que se utiliza para identificar defectos en cigüeñales, bielas de conexión, culatas, etc. Una corriente eléctrica fluye por el componente que se está probando y las partículas de hierro suspendidas se pulverizan después por la superficie. Las partículas se concentrarán donde las fisuras dividen las líneas de flujo magnético.

magnetomotive force the magnetizing force created by flowing current through a coil.

fuerza magnetomotriz fuerza magnetizadora que se crea haciendo fluir la corriente a través de una bobina.

Main Bearings the crankshaft is supported in a cradle of main bearings, and referred to as the crankcase.

cojinetes principales el cigüeñal se apoya en un armazón de cojinetes principales y se denomina cárter.

major thrust face the inboard side of the piston as its throw rotates off the crankshaft centerline through the power stroke.

cara de mayor empuje parte interna del pistón cuando su codo gira para salir de la línea central de cigüeñal durante el tiempo de combustión.

malleable possessing the ability to be deformed without breaking or cracking.

maleable que tiene la capacidad de deformarse sin romperse ni quebrarse.

mass the quantity of matter a body contains; weight.
masa cantidad de materia que contiene un cuerpo; peso.

mechanical advantage the ratio of applied force to the resultant work in any machine or arrangement of levers.

ventaja mecánica relación de fuerza aplicada al trabajo resultante en cualquier máquina o disposición de palancas.

mechanical efficiency a measure of how effectively *indicated power* is converted into *brake power*: factors in pumping and friction losses.

rendimiento mecánico medida de la eficacia con que la *potencia indicada* se convierte en *potencia de freno*: es un factor en las pérdidas por bombeo o fricción.

mechanical governor a governor in which the centrifugal force developed by the rotating flyweights used to sense rpm is the force used to move the fuel control mechanism.

regulador mecánico regulador en el que la fuerza centrífuga desarrollada por contrapesos giratorios utilizados para medir las rpm es la fuerza que se utiliza para mover el mecanismo de control de combustible.

metering the process of precisely controlling fuel quantity.
dosificación proceso de medición precisa de la cantidad de combustible.

metering chamber the lower chamber in a Cummins TP injector located under the pump plunger.

cámara de dosificación cámara inferior del inyector Cummins PT ubicada bajo el émbolo de la bomba.

metering orifice the component within a Cummins PT injector that defines the flow area to the cup pump chamber.

orificio de dosificación componente de un inyector Cummins PT que define el área de flujo en la cámara de bomba aclopada.

metering recesses the milled recesses in MUI and port helix plungers that are used to vary the fuel quantity and timing during pumping.

salidas de dosificación salidas talladas en émbolos MUI y helicoidales de abertura que se utilizan para variar la cantidad de combustible y la sincronización durante el bombeo.

meter resolution a measure of the power and accuracy of a DMM.
resolución del dosificador medida de la potencia y precisión de un DMM.

Mexican hat piston a piston design in which the center of the crown peaks in the fashion of a sombrero. Commonly used in DI diesel engines.

pistón en forma de sombrero diseño de pistón en el que el centro de la corona se eleva en forma de sombrero mejicano. Se suele utilizar en motores diesel DI.

microorganism growth a condition that may result from water contamination in fuel storage tanks.

proliferación de microorganismos problema que se puede producir por contaminación del agua en los depósitos de combustible.

MID message identifier (SAE). Identifies an on-vehicle electronic circuit by numeric code when read by an EST.

MID identificador de mensajes (SAE). Identifica un circuito electrónico del vehículo por códigos numéricos cuando se lee con un EST.

MIL is an acronym for malfunction indicator light, which is a J1930 term that flashes blink codes and indicates a fault in the system

MIL es un acrónimo de *malfunction indicator light* (luz de comprobación del motor) que es un término de J1930 que emite códigos intermitentes e indica algún error en el sistema.

minor thrust face the outboard side of the piston as its throw rotates away from the crankshaft centerline on the powerstroke. See *thrust faces*.

cara de menor empuje parte externa del pistón cuando su codo gira para salir de la línea central de cigüeñal durante el tiempo de combustión. Véase *caras de empuje*.

mixture the random distribution of one substance with another without any chemical reaction or bonding taking place. Air is a mixture of nitrogen and oxygen.

mezcla distribución aleatoria de una sustancia con otra sin que se produzcan reacción ni enlace químicos. El aire es una mezcla de nitrógeno y oxígeno.

Molecule is believed to be the smallest amount of a particular compound that retains the distinctive physical properties of the compound.

molécula se cree que es la menor cantidad de un compuesto en particular que mantiene las propiedades físicas distintivas del compuesto.

monitor the common output screen display used by a computer system; a CRT.

monitor pantalla de salida normal utilizada por un sistema informático; un CRT.

monatomic a molecule consisting of a single atom.

monoatómico molécula que consiste en un sólo átomo.

motoring running an engine at 0 throttle, with chassis momentum driving engine.

rodado sin encedido funcionamiento de un motor con acelerador 0, cuando la inercia del chasis impulsa el motor.

muffler an *engine silencer* that uses sound absorption and resonation principles to alter the frequency of engine noise.

silencioso silenciador de motor que utiliza los principios de absorción y de resonancia para alterar la frecuencia del ruido del motor.

MUI mechanical unit injector. Cam-actuated, governor- controlled unit injectors used by DDC and Caterpillar.

MUI inyector de unitario mecánico. Inyectores unitarios de acción de levas controlados por regulador utilizados por DDC y Caterpillar.

multi-orifii nozzle a typical hydraulic injector nozzle whose function it is to switch and atomize the fuel injected to an engine cylinder. Consists of a nozzle body machined with the orifii, a nozzle valve, and a spring. Used in most DI diesel engines using port helix injection pumps, MUIs, EUIs, and HEUIs.

boquilla de múltiples orificios boquilla de inyección hidráulica habitual cuya función es conmutar y atomizar el combustible inyectado a un cilindro de motor. Consiste en un cuerpo de boquilla fresado con los orificios, una válvula de boquilla y un resorte. Se utiliza en la mayoría de los motores diesel DI con bomba de inyección helicoidal de abertura, MUI, EUI y HEUI.

multiple splice an electrical connection that joins a number of wires at a single junction.

empalme múltiple conexión eléctrica que une varios cables en una sola junta.

multiplex used to describe the connecting of two (or more) electronic system controllers to synergize system operation and reduce the number of common components.

múltiplex se utiliza para describir la conexión de dos (o más) controladores de sistemas electrónicos para sinergizar el funcionamiento del sistema y reducir el número de componentes comunes.

Nalcool a brand of antifreeze solution.

Nalcool marca de solución anticongelante.

naturally aspirated (NA) describes any engine in which intake air is induced into the cylinder by the lower than atmospheric pressure created on the downstroke of the piston and receives no assist from boost devices such as turbochargers.

de aspiración natural (NA) describe cualquier motor en el que el aire de admisión se induce en el cilindro por la presión más baja que la atmosférica que se crea en la carrera descendente del pistón y no recibe ayuda de dispositivos de alimentación, como turboalimentadores.

needle valve nozzle another way of describing a multi-orifii, hydraulic injector nozzle (DDC).

boquilla de válvula de aguja otra forma de describir una boquilla de inyección hidráulica (DDC) de múltiples orificios.

new scroll pump a Caterpillar port helix metering injection pump.

nueva bomba de espiral bomba de dosificación helicoidal de abertura de Caterpillar.

newton unit of mechanical force defined as the force required to accelerate a mass of 1 kilogram through 1 meter in 1 second.

newton unidad de fuerza mecánica definida como la fuerza necesaria para acelerar una masa de 1 kilogramo en un metro en un segundo.

Ni-Resist insert a high strength, nickel alloy piston ring support insert in an aluminum trunk-type piston with a similar coefficient of heat expansion as aluminum.

pieza protectora de níquel pieza de soporte del anillo del pistón muy fuerte, de aleación de níquel en un pistón de tipo de tronco de aluminio con un coeficiente similar de dilatación térmica al del aluminio.

nitrogen (N) a colorless, tasteless, and odorless gas found elementally in air at a proportion of 76% by mass and 79% by volume. Atomic # 7.

nitrógeno (N) gas incoloro, insípido e inodoro que se encuentra como elemento del aire con una proporción del 76% de masa y 79% de volumen. Número atómico 7.

nitrogen dioxide one of the oxides of nitrogen produced in vehicle engines and a significant contributor in the formation of photochemical smog.

dióxido de nitrógeno uno de los óxidos de nitrógeno producidos en motores de vehículos y contribuye de forma significativa a la formación de la niebla de humo (*smog*) fotomecánica.

no-air set screw a PT pump adjustment that defines the maximum flow area to the rail until the AFC circuit is activated by manifold boost.

tornillo de cierre de aire ajuste de bomba PT que define el área de flujo máximo al distribuidor hasta que se activa el circuito AFC con la alimentación del colector.

normal rated power the highest power specified for continuous operation of an engine.

potencia en régimen normal potencia máxima especificada para un funcionamiento continuo de un motor.

NO₂ nitrogen dioxide. The primary constituent of NOx that results when the nitrogen in air is oxidized in the internal engine combustion process.

NO₂ dióxido de nitrógeno. Constituyente primario de NOx que se produce cuando el nitrógeno del aire se oxida en el proceso de combustión interna del motor.

NOx any of a number of oxides of nitrogen that may result from the combustion process: they are referred to collectively as NOx. When combined with HC and sunlight, reacts to form photoelectric smog.

NOx cualquiera de una serie de óxidos de nitrógeno que se pueden producir por un proceso de combustión: se denominan, de forma colectiva, NOx. Cuando se combinan con los hidrocarburos y la luz del sol, reacciona para formar la niebla de humo (*smog*) fotoeléctrica.

nozzle the component of most hydraulic and electronic injector assemblies responsible for switching and atomizing fuel.

boquilla componente de la mayoría de conjuntos de inyección hidráulica y electrónica responsable de conmutar y atomizar combustible.

nozzle seat the seat in an injector nozzle body sealed by the nozzle valve in its closed position.

portaboquilla alojamiento de un cuerpo de boquilla de inyección sellado con la válvula de boquilla en su posición cerrada.

OBC outer base circle. The portion of a cam profile with the largest radial diameter.

OBC perfil exterior de la leva. Parte de un perfil de leva con el mayor diámetro radial.

OCR optical character recognition. Scanners that read type by shape and convert it to a corresponding computer code.

OCR reconocimiento óptico de caracteres. Escáneres que leen las fuentes por la forma que las convierten en el código informático correspondiente.

OCV open circuit voltage.

OCV tensión en circuito abierto.

od outside diameter.

od diámetro exterior.

octane rating denotes the ignition and combustion behavior/rate of a fuel, usually gasoline. As the octane number increases, the fuel's anti-knock characteristics increase and the burn rate slows.

octanaje indica el comportamiento/nivel de encendido y combustión de un combustible, normalmente gasolina. A medida que aumenta el número de octanos, aumentan las características antidetonantes y disminuye el nivel de quemado.

OEM original equipment manufacturer.

OEM fabricante de equipo original.

offset camshaft key timing a Cummins method of timing engine position with cam-actuated, injector pump mechanisms.

sincronización de chaveta excéntrica del árbol de levas método de Cummins de sincronización de la posición del motor con mecanismos de bombeo de inyección de acción de levas.

ohm a unit for quantifying electrical resistance in a circuit.

ohmio unidad de medida de la resistencia eléctrica de un circuito.

Ohm's law the formula used to calculate electrical circuit performance. It asserts that it requires 1 V of potential to pump 1 A of current through a circuit resistance of 1. Named for George Ohm (1787–1854).

ley de Ohm fórmula utilizada para calcular el rendimiento de un circuito. Afirma que hace falta 1 V de potencial para bombear 1 A de corriente por una resistencia de circuito de 1. Tiene ese nombre en honor de George Ohm (1787–1854).

Oil Control Ring Oil control rings are responsible for lubricating the cylinder walls and also provide a path to dissipate piston heat to the cylinder walls.

segmento de lubricación los segmentos de lubricación son responsables de lubricar las paredes del cilindro y también de proporcionar una ruta para disipar el calor del pistón en las paredes del cilindro.

oil cooler a heat exchanger designed to cool oil usually using engine coolant as its medium.

refrigerador del aceite intercambiador térmico diseñado para enfriar el aceite, normalmente mediante un refrigerante.

oil pan the oil sump normally flange mounted directly under the engine cylinder block.

cárter de aceite colector de aceite montado normalmente bajo el bloque del cilindro de motor.

oil window the portion of the upper strata of the earth's crust in which crude petroleum is formed.

estrato petrolífero parte del estrato superior de la corteza terrestre en la que se forma el petróleo crudo.

opacimeter see *opacity meter.*

opacímetro véase *medidor de opacidad.*

open circuit any electrical circuit through which no current is flowing whether intentional or not.

circuito abierto cualquier circuito eléctrico a través del cual no fluye corriente, ya sea de forma intencionada o no.

Open combustion chamber the Direct Injection (DI) Diesel has no prechamber, the combustion chamber is open. Fuel is injected directly into the space between the cylinder head and the top of the piston. This is also referred to as an open chamber design. The piston often contains a bowl or has a specially shaped crown to aid in the mixing process for good combustion. One such design, the Toroidal Piston was a development of the plain menispherical piston cavity introduced by *Saurer Company* of Switzerland in 1934.

cámara de combustión abierta el diesel de inyección directa (DI) no tiene precámara: la cámara de combustión está abierta. El combustible se inyecta directamente en el espacio que hay entre la culata y la parte superior del pistón. También se denomina diseño de cámara abierta. El pistón contiene a menudo una cuenca o tiene una corona de forma especial que ayuda a que se produzca el proceso de mezcla para una combustión correcta. Un diseño así, el pistón Toroidal, fue una evolución de la cavidad de pistón menisférico plano presentado por la empresa *Saurer* de Suiza en 1934.

open nozzle refers to an injector that is sealed by its pump plunger such as Cummins PT and HPI-TP used in the Signature series.

inyector abierto se refiere a un inyector que está sellado por su émbolo de bomba como Cummins PT y HPI-TP en las series Signature.

opens an electrical term referring to open circuits/no continuity in a circuit, portion of the circuit, or a component.

abierto término eléctrico que se refiere a circuitos abiertos/falta de continuidad en un circuito, parte de un circuito o componente.

orifice a hole or aperture.

orificio agujero o abertura.

orifice nozzle a hydraulic injector nozzle that uses a single orifice (unusual) or a number of orifii through which high pressure fuel is pumped and atomized during injection.

inyector con agujeros boquilla de inyector hidráulico que utiliza un solo orificio (poco común) o varios orificios a través de los cuales se bombea el combustible a alta presión y se atomiza durante la inyección.

oscilloscope an instrument designed to graphically display electrical waveforms on a CRT or other display medium.

osciloscopio instrumento diseñado para mostrar gráficamente formas de onda eléctrica en un CRT u otro tipo de pantalla.

Outer base circle (OBC) the largest radial dimension from the camshaft centerline is known as outer base circle (OBC).

perfil exterior de la leva (OBC) la dimensión radial mayor desde la línea central del árbol de levas se conoce como perfil exterior de la leva (OBC).

output the result of any processing operation.

salida resultado de cualquier operación de procesamiento.

overspeed a governor condition in which the engine speed, for whatever reasons, exceeds the set high idle speed or top engine limit.

sobrevelocidad problema del regulador en el que la velocidad del motor, por las razones que sean, supera la velocidad de ralentí definida o el límite superior del motor.

oxygen colorless, tasteless, odorless gas the most abundant of elements on the earth occurring elementally in air and in many compounds including water.

oxígeno gas incoloro, insípido e inodoro, el más abundante de los elementos de la tierra, que se encuentra en el aire y en muchos compuestos, incluida el agua.

oxidation the act of oxidizing a material; can mean combusting or burning a substance.

oxidación acto de oxidar un material; puede suponer la combustión o el quemado de una sustancia.

oxidation catalyst a catalyst that enables an oxidation reaction. In the oxidation stage of a catalytic converter, the catalysts, platinum and palladium, are used.

catalizador de oxidación catalizador que permite una reacción de oxidación. En la fase de oxidación de un convertidor catalítico, se utilizan catalizadores, platino y paladio.

oxidation stability describes the resistance of a substance to be oxidized. It is a desirable characteristic of an engine lubrication oil to resist oxidation so one of its specifications would be its oxidation stability.

estabilidad a la oxidación describe la resistencia de una sustancia a oxidarse. Es una característica deseable para un aceite de lubricación de motor que se resista a la oxidación para que una de sus especificaciones sea la estabilidad a la oxidación.

oxides of nitrogen see *NOx.*

óxidos de nitrógeno véase *NOx.*

PACE/PACER Cummins partial authority, electronic PT fuel system management.

PACE/PACER Gestión electrónica de sistema de combustible PT de autoridad parcial de Cummins.

parallel circuits electrical circuits that permit more than a single path for current flow.

circuitos paralelos circuitos eléctricos que permiten más de una sola ruta para el flujo de corriente.

parallel port valve configuration engine cylinder valve arrangement that locates multiple valves parallel to crank centerline permitting equal gas flow through each (assuming identical lift).

disposición paralela de las válvulas de lumbrera configuración de válvulas de cilindro de motor que coloca múltiples válvulas en paralelo a la línea central, lo que permite que haya un flujo de gas igual por cada una de ellas (suponiendo que haya una elevación idéntica).

Parent bore cylinder block most automobile and small bore engines use integral or **parent bore** cylinder blocks. Typically, large bore dimension Diesel engines do not use a parent bore. When the integral cylinder bore is used in a medium bore application they are sometimes referred to as "throwaway blocks".

bloque de cilindro del calibre principal la mayoría de motores de automóvil y de calibre pequeño utilizan bloques de cilindro integrales o de **calibre principal**. Normalmente, los motores diesel de calibre grande no utilizan un calibre principal. Cuando el calibre de cilindro integral se utiliza en una aplicación de calibre medio, suelen recibir el nombre de "bloques desechables".

particulate matter (PM) solid matter. Often refers to minute solids formed by incompletely combusted fuel and emitted in the exhaust gas.

materia en partículas (PM) materia sólida. Suele referirse a partículas sólidas diminutas formadas por combustible sin quemar del todo y emitidas en el gas de escape.

peak pressure the highest pressure attained in a hydraulic system.

presión máxima presión más alta obtenida en un sistema hidráulico.

peak torque maximum torque. In an internal combustion engine, peak torque always occurs at peak cylinder pressure and in most cases this will be realized at a lower speed than rated power rpm.

par máximo par más alto. En un motor de combustión interna, el par máximo se produce siempre con la presión máxima del cilindro y en la mayoría de los casos tendrá lugar a una velocidad menor que las rpm de la potencia en régimen.

PEEC programmable electronic engine control (Caterpillar). Describes an electronically managed 3406 B engine.

PEEC control electrónico programable del motor (Caterpillar). Describe un motor 3406 B gestionado electrónicamente.

Period of Controlled Combustion is the period of time from maximum pressure to the point when combustion is measurably complete. Once the first fuel is injected, it vaporizes and mixes with the hot compressed air. It is burned and injection continues, a rich core of fuel remains along with zones of air. The remaining fuel is burned as it mixes with the air.

período de combustión controlada es el período de tiempo transcurrido desde la presión máxima hasta el punto en el que la combustión se haya completado de forma mensurable. Una vez que se inyecta el primer combustible, se vaporiza y se mezcla con el aire comprimido caliente. Se quema y continúa la combustión y permanece un rico núcleo de combustible junto con zonas de aire. El resto de combustible se quema cuando se mezcla con el aire.

Period of Rapid Combustion injection occurs at or just before top dead center (TDC) to take advantage of the highest compression temperature. The ID is always long enough (approximately .001 to .003 seconds that, when ignition happens there is an appreciable amount of evaporated and finely divided fuel well mixed with air. Once ignited, this fuel tends to burn very rapidly by reason of the multiplicity of ignition points and the high temperature already existing in the combustion chamber. After about 0.001 second, any zones that are hot enough and have the correct fuel-air mixture ratio will auto-ignite. Flame does not propagate through the combustion chamber as in a gasoline engine. This is because the core of fuel produced by the injector is too rich to burn and only air exists in other parts of the chamber. Thus combustion occurs only at the interface where fuel and air come together (commonly called a diffusion flame).

período de combustión rápida la inyección se produce en o antes del punto muerto superior (TDC) para aprovechar la temperatura de compresión más alta. La demora del encendido (ID) siempre dura lo suficiente, aproximadamente de 0,001 a 0,003 segundos en que, cuando se produce el encendido, hay una cantidad apreciable de combustible evaporado y finamente dividido bien mezclado con el aire. Una vez encendido, este combustible tiende a quemarse muy rápidamente por el gran número de puntos de encendido y la alta temperatura ya existente en la cámara de combustión. Después de aproximadamente 0,001 segundos, las zonas que no estén lo bastante calientes y tengan la relación correcta de mezcla de aire y combustible tendrán un encendido automático. La llama no se propaga por la cámara de combustión como en un motor de gasolina. Esto es porque el núcleo de combustible producido por el inyector es demasiado rico para quemarse y sólo hay aire en otras partes de la cámara. Así, la combustión sólo se produce en la interfaz en la que el combustible y el aire se unen (normalmente llamada llama de difusión).

periphery in cam geometry, the entire outer boundary of the cam; cam profile.

periferia en geometría de leva, todo el límite exterior de la leva; perfil de la leva.

personality module Caterpillar and Navistar PROM/ EEPROM component.

personality module componente PROM/ EEPROM de Caterpillar y Navistar.

personality ratings term used by Caterpillar and Navistar to describe PROM and EEPROM functions.

personality ratings término utilizado por Caterpillar y Navistar para describir las funciones PROM y EEPROM.

PG propylene glycol. A less toxic glycol base antifreeze solution than EG. PG mixture strength must be tested with a refractometer with a PG scale and not mixed with EG.

PG propilenglicol. Una solución anticongelante de base de glicol menos tóxica que EG. Debe comprobarse la fuerza de la mezcla PG con un refractómetro con una escala PG y no mezclarse con EG.

pH used to evaluate the acidity or alkalinity of a substance. From a logarithm of the reciprocal of the hydrogen ion concentration in a solution in moles per liter: p = power, H = hydrogen.

pH se utiliza para evaluar la acidez o alcalinidad de una sustancia. Desde un logaritmo del recíproco de la concentración de iones de hidrógeno en una solución de moles por litro: p = potencia, H = hidrógeno.

phasing the precise sequencing of events; often used in the context of *phasing* the pumping activity of individual elements in a multi-cylinder injection pump.

secuenciación secuencias precisas de sucesos; utilizado a menudo en el contexto de *secuenciación* de la actividad de bombeo de elementos individuales en una bomba de inyección de múltiples cilindros.

photochemical reaction a chemical reaction caused by radiant light energy acting on a substance.
reacción fotoquímica reacción química provocada por energía de luz radiante que actúa en una sustancia.

pick-up tube a suction tube or pipe in a fuel tank or oil sump.
tubo de recogida tubo o conducto de succión en un depósito de combustible o un colector de aceite.

PID the acronym for Parameter Identifier portion of the SAE/ATA fault codes.
PID acrónimo de la parte del identificador de parámetros (*Parameter Identifier*) de los códigos de error SAE/ATA.

pilot ignition a means of igniting a fuel charge that might normally require a spark, by injecting a short pulse of diesel fuel into a cylinder to ignite a premixed charge of gaseous fuel and air.
encendido piloto forma de encender una carga de combustible que normalmente podría necesitar una chispa, inyectando un corto impulso de gasóleo en un cilindro para encender una carga premezclada de combustible y aire gaseosos.

pilot injection the injection of a short duration pulse of diesel fuel, followed by a pause to await ignition, followed by the resumption of the fuel pulse. Used as a cold start strategy in some systems to prevent *diesel knock* and throughout the fueling profile by others, notably, later versions of HEUI. Can also be used as the ignition means in applications using an alternative fuel that does not readily compression ignite. In such instances, a short pulse of diesel fuel is injected to act as the ignition means for the primary fuel.
inyección piloto inyección de un impulso de gasoil de corta duración, seguido por una pausa para esperar al encendido, seguida de la reanudación del pulso de combustible. Se utiliza como estrategia de arranque en frío en algunos sistemas para evitar el *golpeteo Diesel* y durante el perfil de alimentación de otras versiones posteriores de HEUI. También se puede utilizar como forma de encendido en aplicaciones que utilizan un combustible alternativo que no se encienden bien por compresión. En tales casos, se inyecta un corto pulso de gasóleo para que actúe como forma de encendido para el combustible primario.

pin boss the wrist pin support bore in a piston assembly.
resalte del pasador calibre de soporte del pasador de articulación de un conjunto de émbolo.

pintle nozzle a type of hydraulic injector nozzle used in some IDI automobile, small bore diesel engines until recently.
boquilla de gancho tipo de boquilla de inyección hidráulica que se utilizaba hasta hace poco en motores diesel de calibre pequeño de automóviles IDI.

piston the reciprocating plug in an engine cylinder bore that seals and transmits the effects cylinder gas pressure to the crankshaft.
pistón obturador de vaivén de un calibre de cilindro de motor que sella y transmite los efectos de presión de gas del cilindro al cigüeñal.

piston pin a wrist pin that links the piston assembly to the connecting rod eye.
pasador del pistón pasador de articulación que une el conjunto de émbolo con el ojo de biela de conexión.

piston speed the distance traveled by one piston in an engine per unit of time.
velocidad del pistón distancia recorrida por un pistón en un motor por unidad de tiempo.

pitting a wear pattern that results in small pock marks or holes.
picadura desgaste que produce pequeños hoyuelos o agujeros.

PLN pump-line-nozzle. Acronym used to describe the hydromechanical or electronically managed, injection pump to line to nozzle fuel injection principle used in a majority of diesel fuel systems until the introduction of EUI engines. The term can, but is not usually, applied to some of the more recent systems such as the Mack E-Tech and DDC Series 55 EUP systems.
PLN bomba-conducto-inyector. Acrónimo utilizado para describir la bomba de inyección gestionada hidromecánica o electrónicamente que alinea la inyección de combustible con el inyector y que se utilizaba en la mayoría de sistemas de gasóleo hasta la aparición de los motores EUI. Este término se puede aplicar, aunque no es habitual, a algunos de los sistemas más recientes como los sistemas Mack E-Tech y DDC Series 55 EUP.

plunger the reciprocating member of a plunger pump element.
émbolo buzo miembro de vaivén de un elemento de bomba de émbolo.

plunger and cup the PT injector components that form the high pressure pump element.
émbolo y copa componentes del inyector PT que conforman el elemento de bomba de alta presión.

plunger geometry term used to describe the shape of the metering recesses/helices in a pumping plunger and therefore the pump timing and delivery characteristics.
geometría de émbolo término utilizado para describir la forma de los cortes o espirales de dosificación en un émbolo de bombeo y por tanto de las características de sincronización de bombeo y suministro.

plunger leading edge the point on a pumping plunger that is closest to the pump chamber.
borde de contacto del émbolo el punto de un émbolo de bombeo más cercano a la cámara de bombeo.

plunger link on a Cummins PT injector, the rod located in the plunger flange that prevents side loading of the plunger by the rocker arm.
articulación de pistón en un inyector Cummins PT, biela ubicada en la unión del émbolo que evita la carga lateral del émbolo por el balancín.

plunger pump any pump that uses a reciprocating piston or plunger and, in most cases, is hydraulically classified a *positive displacement*.
bomba de émbolo sumergido cualquier bomba que utilice un pistón o émbolo de vaivén y, en la mayoría de los casos, se clasifica hidráulicamente como *desplazamiento positivo*.

pneumatics the science of the mechanical properties of gases, especially in confined circuits designed to provide motive power.
neumática ciencia de las propiedades mecánicas de los gases, especialmente en circuitos confinados diseñados para proporcionar fuerza motriz.

port closure the beginning of effective stroke in a plunger and barrel pumping element, occurring when the plunger leading edge closes off the spill/fill port(s).
cierre de válvulas principio del tiempo efectivo de un elemento de bombeo de émbolo y cilindro, que se produce cuando el borde de contacto del émbolo cierra la(s) abertura(s) de vaciado/llenado.

port opening the ending of effective stroke in a plunger and barrel pumping element, occurring when the fill/spill ports are exposed to the chamber.
apertura de válvulas final del tiempo efectivo de un elemento de bombeo de émbolo y cilindro, que se produce cuando se exponen a la cámara las aberturas de vaciado/llenado.

positive displacement describes a pumping principle in which the quantity of fuel pumped (displaced) per cycle does not vary so the volume pumped depends on the rate of cycles per minute. When a positive displacement pump unloads to a defined flow area, pressure rise will increase in proportion to rpm or cycles per minute.
desplazamiento positivo describe un principio de bombeo en el que la cantidad de combustible bombeado (desplazado) por ciclo no varía, con lo que el volumen bombeado depende del nivel de ciclos por minuto. Cuando una bomba de desplazamiento positivo descarga en un área de flujo definida, el aumento de presión subirá en proporción a las rpm o ciclos por minuto.

positive filtration a filter in which all of the fluid (gas or liquid) to be filtered is forced through the filtering medium. Most air, fuel, coolant, and oil filters used today employ a positive filtration principle.

filtración positiva filtro en el que todo el fluido (gas o líquido) que se debe filtrar pasa por el medio de filtrado. La mayoría de los filtros de aire, combustible y aceite utilizados actualmente utilizan un principio de filtración positiva.

potential difference electrical *charge differential* measured in *voltage*.

diferencia de potencial *diferencial de carga* eléctrica medido en *tensión*.

potentiometer a three-terminal variable resistor or voltage divider used to vary the voltage potential of a circuit. Commonly used as a throttle position sensor.

potenciómetro resistor variable de tres terminales o divisor de tensión que se utiliza para variar el potencial de tensión de un circuito. Se utiliza normalmente como sensor de posición del acelerador.

pour point a means of evaluating a fuel or lubricants low temperature flow characteristics. The pour point of a fuel is slightly higher in temperature than its gel point.

punto de fluidez modo de evaluación de las características de flujo de combustible o de lubricantes a bajas temperaturas. El punto de fluidez de un combustible es ligeramente superior en temperatura a su punto de gelificación.

power the rate of accomplishing *work*; it is necessarily factored by time.

potencia nivel de consecución de *trabajo*; el factor es un factor necesario.

prelubricator a pump used to charge the lubrication circuit on a rebuilt engine before start-up.

prelubricador bomba utilizada para cargar el circuito de lubricación en un motor reconstruido antes de arrancarlo.

pressurizing the process of raising the pressure in a circuit.

presurización proceso de aumento de la presión de un circuito.

primary filter usually describes a filter on the suction side of a fuel subsystem where the term secondary filter describes the filter on the charge side of the transfer pump.

filtro primario suele describir un filtro del lado de la succión de un subsistema de combustible en el que el término filtro secundario describe el filtro del lado de la carga de la bomba de transferencia.

PRIME pre-injection metering. A Caterpillar term for the pilot injection concept used in their HEUI injectors.

PRIME dosificación preinyección. Término de Caterpillar para el concepto de inyección piloto utilizado en sus inyectores HEUI.

processing the procedure required to compute information in a computer system. Input data is processed according to program instructions and outputs are plotted.

procesamiento procedimiento necesario para calcular la información en un sistema informático. La entrada de datos se procesa según las instrucciones del programa y se imprimen los resultados.

ProDriver DDC-DDEC driver digital display.

ProDriver pantalla digital de transmisión DDC-DDEC.

program set of detailed instructions that organize processing activity.

programa conjunto de instrucciones detalladas que organizan la actividad de procesamiento.

ProLink a microprocessor-based, generic EST that has become an industry standard. Used with the appropriate proprietary software cartridge, it can be used to diagnose and reprogram vehicle ECMs. Manufactured by MPSI/ Kent Moore.

ProLink EST basada en microprocesador genérica que se ha convertido en un estándar del mercado. Se utiliza con el correspondiente cartucho de software patentado y se puede utilizar para diagnosticar y reprogramar ECM de vehículos. La fabrica MPSI/ Kent Moore.

ProManager DDC-DDEC program that permits ECM data to be downloaded to a PC for analysis.

ProManager programa DDC-DDEC que permite descargar datos ECM a un PC para su análisis.

propagate to breed, transmit, or multiply. The word is often used to describe the combustion process in an engine cylinder such as in *flame propagation*.

propagar generar, transmitir o multiplicar. Esta palabra se suele utilizar para describir el proceso de combustión de un cilindro de motor, como en *propagación de la llama*.

proportional solenoid a solenoid whose armature will be positioned according to how much current is flowed through its coil. Often an ECM-actuated output. Proportional solenoids may be linear such as the V-MAC rack actuator or rotary such as the Caterpillar BTM.

solenoide proporcional solenoide cuya armadura se colocará según la cantidad de corriente que fluya a través de su bobina. Es a menudo una salida de acción ECM. Los solenoides proporcionales pueden ser lineales, como el impulsor de cremallera, o giratorios, como el Caterpillar BTM.

PT pressure-time. Cummins common rail hydromechanical fuel system used in highway applications until 1994. An electronic version known as HPI-PT is used on K-19 engines.

PT presión-tiempo. Sistema de combustible hidromecánico de distribuidor común que se utilizaba en carretera hasta 1994. En los motores K-19 se utiliza una versión electrónica conocida como HPI-PT.

PTCM PT (pump) control module. The ECM used by Cummins in their PACE/PACER partial authority, electronic management system.

PTCM módulo de control (de bomba) PT. El ECM utilizado por Cummins en su sistema de gestión electrónica de autoridad parcial PACE/PACER.

PTO power takeoff. Refers to an auxiliary drive output on an engine or transmission or the coupling component from which drive is transmitted to a drive train.

PTO toma de fuerza. Se refiere a una salida de transmisión auxiliar de un motor, o a la transmisión, o al componente de acoplamiento desde el que se transmite la transmisión a un mecanismo de transmisión.

pulsation damper on a Cummins gear type supply pumps, pulsation dampers are used to smooth the pressure waves caused by a gear type pump as it loads fuel into its outlet.

amortiguador de impulsos en una bomba de suministro de tipo de engranajes de Cummins, los amortiguadores de impulsos se utilizan para suavizar las ondas de presión causadas por una bomba de tipo de engranajes al cargar combustible en su orificio de salida.

pulse exhaust a tuned exhaust system used to optimize the gas dynamic of exhaust gas delivered to the turbocharger.

escape por impulsos sistema de escape puesto a punto que se utiliza para optimizar la dinámica de gas del gas de escape que llega al turboalimentador.

pulse wheel the rotating disc used to produce rpm or rotational position data to an ECM. The term is most often applied to the rotating member of a Hall effect sensor, but at least one manufacturer uses the term to describe an AC reluctor wheel.

rueda de impulsos disco giratorio que se utiliza para producir datos de rpm o de posición rotacional para un ECM. El término se aplica casi siempre al miembro giratorio de un sensor del efecto Hall, pero al menos un fabricante utiliza el término para describir una rueda reluctora CA.

pulse width modulation the shaping of pulses and waveforms for purposes of digital signaling. Acronym PWM is often used.

modulación de impulsos en altura forma de impulsos y ondas para señalización digital. Se utiliza a menudo el acrónimo, PWM.

pump drive gear the gear responsible for imparting drive force to a pump.
engranaje de transmisión a la bomba engranaje responsable de impartir la fuerza de transmisión a una bomba.

push rods cylindrical solid rods located between a follower and a rocker assembly that transmit the effects of cam profile to action at the rocker arm.
bielas de empuje bielas sólidas cilíndricas ubicadas entre un rodillo y un conjunto de balancín que transmiten los efectos del perfil de la leva al balancín.

push tubes hollow, cylindrical tubes located between a follower and a rocker assembly that transmit the effects of cam profile to action at the rocker arm.
tubos de empuje tubos cilíndricos huecos ubicados entre un rodillo y un conjunto de balancín que transmiten los efectos del perfil de la leva al balancín.

PW pulse width. Usually refers to EUI duty cycle measured in milliseconds.
PW anchura entre impulsos. Se suele referir al ciclo de servicio EUI medido en milisegundos.

PWM pulse width modulation. Refers to on/off percentage time in certain electronic sensors and actuators.
PWM modulación de impulsos en altura. Se refiere al tiempo de porcentaje de conexión/desconexión de algunos sensores e impulsores electrónicos.

pyrometer a thermocouple type, high temperature sensing device used to signal exhaust temperature. Consists of two dissimilar wires (pure iron and constantan) joined at the hot end with a millivoltmeter at the read end. Increase in temperature will cause a small current to flow, which is read at the voltmeter as a temperature value.
pirómetro dispositivo de medición de altas temperaturas de tipo termopar que se utiliza para indicar la temperatura del escape. Se compone de dos cables diferentes (hierro puro y constantano) unidos en el extremo caliente por un milivoltímetro en el extremo de lectura. Un aumento en la temperatura hará que fluya una pequeña corriente, que se lee en el voltímetro como valor de temperatura.

quiescent a term used to describe any low turbulence engine cylinder dynamic. Its root is from the word *quiet*.
en reposo término utilizado para describir cualquier dinámica de cilindro de motor de baja turbulencia. La raíz es la palabra *quieto*.

rack actuator a proportional solenoid (Bosch) or hydraulic servo (Cat) responsible for moving the rack in a computer-controlled, port helix metering injection pump. An ECM output.
impulsor de cremallera solenoide proporcional (Bosch) o servohidráulico (Cat) responsable de mover la cremallera de una bomba de inyección de dosificación helicoidal de abertura controlada por ordenador. Salida ECM.

rack actuator housing the housing at the rear of a computer controlled port helix metering injection pump that contains sensors and the rack actuating mechanism. It is located in place of the governor on a hydromechanical engine.
alojamiento del impulsor de cremallera alojamiento en la parte posterior de una bomba de inyección de dosificación helicoidal de abertura controlada por ordenador que contiene sensores y el mecanismo impulsor de cremallera. Está colocado en lugar del regulador de un motor hidromecánico.

rack lever the actuating mechanism on a DDC MUI system that converts the rotary movement of the control tube into linear movement of the rack.
palanca de cremallera mecanismo impulsor de un sistema DDC MUI que convierte el movimiento giratorio del tubo de control en el movimiento lineal de la cremallera.

rack position sensor an electromagnetic sensor used to signal rack position data to the ECM on electronically controlled, port helix metering injection pumps.
sensor de posición de la cremallera sensor electromagnético que se utiliza para indicar los datos de posición de la cremallera en el EMC en bombas de inyección de dosificación helicoidal de abertura controladas electrónicamente.

radial vector the radial angle off a reference point say TDC in a crankshaft that indicates the mechanical advantage of a throw in its relationship with the crankshaft centerline.
vector radial ángulo radial desde un punto de referencia, como el TDC, de un cigüeñal que indica la ventaja mecánica de un codo en relación con la línea central del cigüeñal.

radiation the transfer of heat or energy by rays not requiring matter such as a liquid or a gas.
radiación transferencia de calor o energía por rayos sin necesidad de materias como el líquido o el gas.

radiator a heat exchanger used in liquid-cooled engines designed to dissipate some of the engine's rejected heat to atmosphere.
radiador intercambiador térmico utilizado en motores refrigerados por líquido diseñado para disipar algo del calor rechazado del motor a la atmósfera.

radioactive any substance or set of physical conditions capable of emitting radioactivity. Exposure to high level radioactivity can be life threatening while low level radioactivity (such as electrical or radar waves) represents debatable hazards.
radioactivo cualquier sustancia o conjunto de condiciones físicas capaces de emitir radioactividad. Una exposición a un alto nivel de radioactividad puede amenazar la vida, mientras que un nivel bajo de radioactividad (como ondas eléctricas o de radar) representan riesgos debatibles.

rail actuator see *fueling actuator*.
impulsor del distribuidor véase *impulsor de alimentación*.

rail pressure sensor Cummins HPI-TP rail fuel pressure sensor.
sensor de presión en el distribuidor sensor Cummins HPI-TP de presión de combustible en el distribuidor.

RAM random access memory. Electronically retained "main memory".
RAM memoria de acceso aleatorio. "Memoria principal" mantenida electrónicamente.

ram air air fed into engine cooling and intake circuits by the velocity of a moving vehicle; increases proportionally with vehicle speed.
aire dinámico aire proporcionado a los circuitos de refrigeración y admisión del motor por la velocidad de movimiento de un vehículo; aumenta proporcionalmente con la velocidad del vehículo.

ramps in cam geometry, the shaping of the cam profile between the IBC and the OBC. The ramp geometry will define the actuation/unload characteristics of the train that rides its profile.
rampas en geometría de la leva, forma del perfil de la leva entre el IBC y el OBC. La geometría de rampas definirá las características de impulso/descarga del mecanismo que hace funcionar su perfil.

rapid start shutoff valve Cummins common rail fuel system electric (solenoid) shutoff valve that traps fuel in the rail on shutdown to enable an almost instant restart.
válvula de retención de arranque rápido válvula de retención eléctrica (solenoide) del sistema de combustible de distribuidor común Cummins que atrapa combustible en el distribuidor para permitir que se pueda volver a arrancar de forma casi inmediata.

rate shaping a fuel injection term that describes the ability of a fuel system to control fuel delivery to the cylinder independent of the hard limitations of cam geometry and engine rpm. Because HEUI injectors are actuated hydraulically and the hydraulic actuation pressure can be controlled by the ECM, this system is capable of rate shaping. The Cummins CAPS is also capable of rate shaping.
conformación de la intensidad término de inyección de combustible que describe la capacidad de un sistema de combustible de controlar el suministro de combustible al cilindro independientemente de las limitaciones físicas de la geometría de la leva y de las rpm del motor. Como los inyectores HEUI son de acción hidráulica y la presión de impulso hidráulico se puede controlar con el ECM, este sistema es capaz de conformar la intensidad. El Cummins CAPS también es capaz de conformar la intensidad.

rated power the highest power specified for continuous operation.

potencia en régimen máxima potencia especificada para un funcionamiento continuo.

rated speed the rpm at which an engine produces peak power.

velocidad nominal rpm a las que un motor produce la potencia máxima.

RCRA Resource Conservation and Recovery Act. U.S. federal legislation that regulates the disposal of hazardous materials.

RCRA *Resource Conservation and Recovery Act* (Ley de conservación y recuperación de recursos). Legislación federal de EE.UU. que regula el desecho de materiales de riesgo.

RDI remote data interface. DDEC communications link between the vehicle electronics and a fleet's PC or PC network.

RDI interfaz de datos remota. Enlace de comunicaciones DDEC entre la electrónica del vehículo y un PC o red de PC de la flotilla.

reaction turbine an aeolipile, the first heat engine.

turbina de reacción eolipila, el primer motor térmico.

ream the machining process of accurately enlarging an orifice using a steel boring bit with straight or spiral fluted cutting edges.

mandrinar proceso de tallado que consiste en agrandar de forma precisa un orificio utilizando un calibrador de acero con bordes de cortes acanalados rectos o espirales.

reference coil Mack Trucks rack position sensor magnetic field temperature reference—validates input from the rack position sensor.

bobina de referencia referencia de temperatura de campo magnético del sensor de posición de la cremallera de Mack Trucks (valida la entrada desde el sensor de posición de la cremallera).

refraction the bending of a light ray when it enters a glass lens or drop of liquid.

refracción desvío de un rayo de luz al pasar por una lente de cristal o una gota de líquido.

register alignment or track point of two components.

ajuste punto de alineación de dos componentes.

rejected heat that portion of the potential heat energy of a fuel not converted into useful kinetic energy.

calor rechazado parte de la energía térmica potencial de un combustible que no se convierte en energía cinética útil.

relief valve a commonly used valve in hydraulic circuits (such as fuel subsystem and lubrication circuits) that defines maximum circuit pressure. The simplest type would consist of a ball check, loaded by a spring to seal a return line. When circuit pressure was sufficient to unseat the ball check, circuit fluid would be diverted from the main circuit to the return.

válvula de sobrepresión válvula utilizada habitualmente en circuitos hidráulicos (como subsistemas de combustible y circuitos de lubricación) que define la presión máxima de un circuito. El tipo más sencillo consistiría en un retén de bolilla, cargado por un resorte para sellar una línea de vuelta. Cuando la presión del circuito fuese insuficiente para desalojar el retén de bolilla, se desviaría el fluido del circuito del circuito principal al de vuelta.

reluctance resistance to the movement of magnetic lines of force.

reluctancia resistencia al movimiento de líneas magnéticas de fuerza.

reluctor a term used to describe a number of devices that use magnetism and motion to produce an AC voltage.

reluctor término utilizado para describir diferentes dispositivos que utilizan el magnetismo y el movimiento para producir una tensión CA.

residual line pressure the pressure that *dead volume fuel* is retained at in a high pressure pipe in a PLN fuel system that uses delivery valves at the injection pump.

presión residual del conducto presión con la que se retiene el *combustible de volumen fijo* en un conducto de alta presión en un sistema de combustible PLN que utiliza válvulas de suministro en la bomba de inyección.

resistance opposition to electrical current flow in a circuit.

resistencia oposición al flujo de corriente eléctica en un circuito.

retarder generally refers to braking action, that is, the retarding of vehicle movement.

retardador se refiere por lo general a la acción de frenar, es decir, de retardar el movimiento de un vehículo.

retraction collar/piston the component on a delivery valve core that is designed to seal before it seats and therefore helps define the *residual line pressure* value.

anillo/pistón retráctil componente de una válvula central de suministro diseñado para cerrarse antes de asentarse, con lo que ayuda a definir el valor de *presión residual del conducto*.

retraction spring any spring in any component that causes an assembly to mechanically withdraw or retract.

resorte retráctil resorte de cualquier componente que hace que un conjunto se retire o retracte mecánicamente.

ring belt the area of the piston in which the piston ring grooves are machined.

zona de segmentación área del pistón en la que están talladas las ranuras del anillo del pistón.

RMS root mean square.

RMS valor medio cuadrático.

road speed sensor a sensor usually of the pulse generator type located at the transmission tailshaft or a wheel assembly that signals the ECM road speed data.

sensor de velocidad en carretera sensor, normalmente del tipo de generador de impulsos, ubicado en eje de cola o en un conjunto de ruedas que indica al ECM los datos de velocidad de carretera.

rocker arm see *rockers*.

balancín véase *balancines*.

rocker assemblies the entire rocker assembly consisting of rockers, rocker shaft, and pedestals.

conjunto de eje y balancines todo el conjunto de balancines, que consiste en balancines, eje oscilante y pedestales.

rockers shaft mounted, pivoting levers that transmit the effects of cam profile to valves and injection pumping apparatus.

balancines palancas montadas en el eje y pivotantes que transmiten los efectos del perfil de la leva a válvulas y al aparato de bombeo de inyección.

rocker claw/lever PT or mechanical unit injector hold-down or short-out tool.

uña/palanca de balancín herramienta PT o de inyección de unidad mecánica de sostenimiento o corte.

rocker pallet the end of a rocker that contacts the injection pumping tappet, the valve stem, or the valve bridge.

trinquete de balancín extremo de un balancín que está en contacto con el empujaválvula de bombeo de inyección, el vástago de válvula o el puente de válvula.

rod eye the upper portion of a connecting rod that connects to the piston wrist pin; also known as *small end*.

ojo del pie de biela parte superior de una biela que se conecta con el pasador de articulación del pistón; conocido también como *pequeño extremo*.

ROM read-only memory. Data that is retained either magnetically or by optical coding and designed to be both permanent and read-only.

ROM memoria de sólo lectura. Datos que se mantienen magnéticamente o por cifrado óptico y que están diseñados para ser permanentes y de sólo lectura.

Roots blower a positive displacement air pump consisting of two gear-driven, intermeshing spiral fluted rotors in a housing; used to scavenge DDC two-stroke cycle engines.

compresor Roots bomba de aire de desplazamiento positivo que consta de dos rotores acanalados en espiral, de toma constante, dirigidos por engranajes y situados en un alojamiento; se usa para recuperar motores de dos tiempos DDC.

RSG road speed governing.
RSG limitación de velocidad en carretera.

RSL road speed limit.
RSL velocidad límite en carretera.

run-in usually describes the engine break-in procedure following a rebuild outlined by the OEM.
rodaje normalmente, describe el procedimiento de comienzo de uso de un motor después de una reparación realizada por el OEM.

sac a spherical cavity. Refers to the chamber in some multi-orifice injector nozzles beyond the seat and from which the orifii extend.
saco cavidad esférica. Se refiere a la cámara en algunas boquillas de inyectores de múltiples orificios junto al asiento y desde el cual se extienden los orificios.

SAE Society of Automotive Engineers. Organization responsible for setting many of the manufacturing standards and protocols of the motive power industries and dedicated to educating and informing its members.
SAE Sociedad de ingenieros de automoción. Organización responsable de establecer muchas de las normas y protocolos de fabricación de las industrias de fuerza motriz, dedicada a educar e informar a sus miembros.

SAE horsepower a structured formula used to calculate brake power data that can be used for comparison purposes.
potencia SAE fórmula estructurada que se usa para calcular datos de potencia de freno que se pueden usar para comparación.

SAE viscosity grades the industry standard for grading lubricating oil viscosity.
SAE grados de viscosidad que utiliza la industria como estándar para comparar la viscosidad del aceite lubricante.

saturation condition of an electromagnet in which a current increase results in no increase in the magnetic flux field.
saturación condición de un electroimán en el que un aumento de corriente no produce ningún aumento del campo de flujo magnético.

SCA supplemental cooling (system) additive.
SCA aditivo (del sistema) de refrigeración suplementario.

scavenge term used generally to describe the process used to expel end gases from an engine cylinder and specifically to describe:
1. the final stage of the exhaust process in a four-stroke cycle engine that occurs at valve overlap.
2. cylinder breathing on a two-stroke cycle diesel engine.
recuperación término usado en general para describir el proceso usado para expulsar los gases de escape de un cilindro del motor y en particular para describir:
1. la etapa final del proceso de escape en motores de cuatro tiempos que se produce en el solapamiento de válvulas.
2. ventilación del cilindro en motores diesel de dos tiempos.

scuffing a superficial scraping of metal against metal damage mode.
arañazo rozadura superficial que produce un daño de metal contra metal.

secondary filter usually refers to a filter downstream from the transfer or charge pump in a typical fuel subsystem. It is in most cases under pressure and capable of much finer filtration than a primary filter, which is usually under suction.
filtro secundario hace referencia totalmente a un filtro debajo de la transferencia o bomba de carga en un subsistema normal de combustible. En la mayoría de los casos está bajo presión y puede realizar una filtración mucho más fina que un filtro primario, que está normalmente bajo succión.

section modulus relates the shape of a beam, cylinder, or sphere to section and stiffness; the greater the section modulus, the higher the rigidity and resistance to deflection. A factor of RBM or resist bending moment.
coeficiente de forma relaciona la forma de una barra, cilindro o esfera con la sección y rigidez; cuanto mayor es el coeficiente de forma, mayor es la rigidez y la resistencia a la desviación. Es un factor de momento de resistencia de torcedura (RBM).

SEL stop engine light.
SEL luz de paro del motor.

SEO stop engine override.
SEO anulación de paro del motor.

semiconductor materials that neither conduct well nor insulate; they have four electrons in their outermost shell.
semiconductor materiales que no conducen bien ni aíslan; tienen cuatro electrones en su capa más externa.

sending unit a variable resistor and float assembly that signals a gauge and/or ECM the liquid level in a tank.
unidad de envío resistor de variable y ensamblaje de flotación que señala un indicador o ECM el nivel de líquido de un tanque.

sensor a term that covers a wide range of command and monitoring input (ECM) signal devices.
sensor término utilizado para numerosos dispositivos de señales de comando y de entrada de seguimiento (ECM).

series circuit a circuit with a single path for electrical current flow.
circuito en serie circuito con una única ruta para el flujo de la corriente eléctrica.

service hours a means of comparing engine service hours to highway mileage. Most engine OEMs equate 1 engine hour to 50 highway linehaul miles (80 km), so a service interval of 10,000 miles (16,000 km) would equal 200 engine hours. The term *engine hours* is also used.
horas de servicio medio de comparar las horas de servicio del motor con el kilometraje de carretera. La mayoría de los OEM equiparan 1 hora de motor con 80 kilómetros (50 millas) en carretera, por lo que un intervalo de 16.000 km (10.000 millas) equivaldría a 200 horas de motor. También se utiliza el término *horas de motor*.

shutterstat a temperature sensing, pneumatic switch used to manage air shutter operation.
obturador conmutador neumático sensible a la temperatura que se usa para gestionar la operación de obturación de aire.

SID the acronym for Sub-system Identifier portion of the SAE/ATA fault codes.
SID acrónimo (en inglés) de la parte Identificador de subsistema de los códigos de error de SAE/ATA.

signals codes, signs, or symbols used to convey information.
señales códigos, signos o símbolos usados para transmitir información.

single pass radiator any radiator through which flow is unidirectional.
radiador de paso único cualquier radiador en el que el flujo es unidireccional.

sinter a means of alloying metals in which the constituent materials are mixed in powdered form and then coalesced by subjecting them to heat and pressure. Produces more uniform metallurgical characteristics than alloying.
sinter medio de alear metales en el que los materiales constituyentes se mezclan en forma de polvo y luego se fusionan exponiéndolos a calor y presión. Produce características metalúrgicas más uniformes que la aleación.

sleeves see *liners*.
manguitos véase *camisas*.

small end the connecting rod eye.
pie de biela el ojo de la biela de conexión.

smart
1. used to describe computed outcomes that use a broad range of input and memory factors to produce "soft" outcomes rather than adhere to hard values. *Smart cruise control* will learn road terrain patterns and permit some latitude around the programmed road speed value to improve fuel economy.
2. may describe a computer peripheral that possesses some processing capability.
inteligente
1. se usa para describir resultados computados que utilizan un amplio rango de entradas y factores de memoria para producir resultados "procesados" en lugar de adherirse a valores sin procesar. El *control de marcha inteligente* aprenderá modelos de carreteras y permitirá una cierta flexibilidad en el valor de velocidad en carretera programado para mejorar el ahorro de combustible.
2. puede describir un periférico de ordenador que posee una cierta capacidad de procesamiento.

smog a word formed by combining the words fog and smoke. A haze produced by suspended airborne particulates. Two major types exist, sulfurous smog produced combusting sulfur laden fuels such as coal and heavy oils and photochemical smog, a primary cause of which, are vehicle emissions.

smog palabra formada por la combinación de las palabras *fog* (niebla) y *smoke* (humo). Nube formada por partículas suspendidas en el aire. Existen dos tipos principales: el *smog* sulfuroso, producido al quemar combustibles cargados de azufre como el carbón y los aceites pesados, y el *smog* fotoquímico, cuya causa principal son las emisiones de los vehículos.

snap rail test Cummins rail pressure spike test for PT systems.

prueba de raíl de sujeción prueba de pico de presión de raíl de Cummins para sistemas PT.

soft cruise a cruise control mode programmed into some vehicle/engine management electronics in which the road speed is managed within a window extending both above and below the set speed. Soft cruise can increase fuel economy and is often used in conjunction with vehicle maximum speed programming *below* maximum cruise speed.

marcha suave modo de control de marcha programado en la electrónica de gestión de vehículos y motores en el que la velocidad en carretera se gestiona dentro de un intervalo que se extiende por encima y por debajo de la velocidad establecida. La marcha suave puede aumentar el ahorro de combustible y se utiliza a menudo junto con la programación de velocidad máxima del vehículo *por debajo* de la velocidad de marcha máxima.

soft parameter a value that varies and depends on input and processing variables (see *fuzzy logic*). The term is often used to describe current cruise control systems that permit a cushion both above and below the set value. See *soft cruise*.

parámetro de suavidad valor que varía y depende de las variables de procesamiento y entrada (véase *lógica de aproximación*). El término se utiliza a menudo para describir sistemas de control de marcha actual que permiten superar por encima y por debajo el valor establecido. Véase *marcha suave*.

SOI start of injection.

SOI inicio de la inyección.

solenoid an electromagnet with a movable armature.

solenoide electroimán con armadura móvil.

solid state components that use the electronic properties of solids such as semiconductors to replace the electrical functions of valves.

estado sólido componentes que utilizan las propiedades electrónicas de sólidos como semiconductores para sustituir las funciones eléctricas de las válvulas.

solid state storage volatile storage of data in RAM chips.

almacenamiento de estado sólido almacenamiento volátil de datos en chips de RAM.

spark ignited (SI) any gasoline-fueled, spark-ignited engine usually using an Otto cycle principle.

encendido por bujía (SI) cualquier motor de gasolina encendido por bujía, que utiliza generalmente un principio de ciclo de Otto.

specific fuel consumption (SFC) fuel consumed per unit of work performed.

consumo específico de combustible (SFC) combustible consumido por unidad de trabajo realizado.

specific gravity the weight of a liquid or solid versus that of the same volume of water.

gravedad específica el peso de un líquido en relación con el mismo volumen de agua.

spectrographic analysis a low level radiation test that can accurately identify trace quantities of matter in a fluid; used to analyze engine oils.

análisis espectrográfico prueba de radiación de bajo nivel que puede identificar de forma exacta cantidades traza de materia en un fluido; se usa para analizar aceites de motor.

Speed is really angular velocity (rate of change of position at a particular angle) and defined in crankshaft revolutions per minute (RPM) of time and shown in formulas as the letter "N."

Velocidad es realmente la velocidad angular (relación de cambio de posición en un determinado ángulo); se define en revoluciones del cigüeñal por minuto (RPM) de tiempo y aparece en las fórmulas con la letra "N".

spike an electrical (voltage) or hydraulic pressure surge.

pico subida eléctrica (tensión) o de la presión hidráulica.

spill deflector a cylindrical sleeve that fits outside the bushing on a DDC MUI, and protects the MUI body from high pressure spill at port opening.

deflector de descarga manguito cilíndrico que se ajusta por fuera del casquillo de un DDC MUI y que protege al cuerpo del MUI de la descarga de alta presión al abrir la puerta.

spindle an intermediary, responsible for transmitting force. In a hydraulic injector, it relays spring force to the nozzle valve.

aguja de inyector intermediario responsable de transmitir fuerza. En un inyector hidráulico, transmite la fuerza del resorte a la válvula de boquilla.

splice to join.

empalmar unir.

spur gear a gear with radial teeth.

engranaje recto engranaje con dientes radiales.

SRS synchronous reference sensor. DDEC engine location sensor consisting of a single dowel located on the cam gear on Series 60/50 engines that inputs an analog voltage signal to the ECM.

SRS sensor de referencia síncrona. Sensor de localización de motores DDEC que consta de una única espiga ubicada en el engranaje de la leva en motores de Serie 60/50 que introducen una señal de tensión analógica en ECM.

SSC single speed control—Mack Trucks isochronous PTO governing.

SSC control de velocidad única, regulador PTO isocrono de Mack Trucks.

star network network set up to operate from a central hub computer.

red en estrella red configurada para operar desde un ordenador de servidor central.

starting aid screw a device used on DDC two-stroke cycle to limit fuel during cranking on turbocharged engines when the turbocharger acts as a restriction.

dispositivo auxiliar de arranque dispositivo usado en motores de dos tiempos DDC para limitar el combustible durante el arranque en motores turboalimentados cuando el turboalimentador actúa como restricción.

STC step timing control. Cummins PT injector, hydraulic timing advance tappet, integral with a PT injector.

STC control de sincronización de fases. Inyector Cummins PT, empujador de avance de sincronización hidráulico, integral con inyectores PT.

STC control valve the Cummins STC component that controls the supply of engine oil to the STC tappets and determines whether the system operates in advanced or normal modes.

válvula de control STC el componente STV de Cummins que controla el suministro de aceite de motor a empujadores STC y determina si el sistema opera en modos avanzado o normal.

stoichiometric ratio the exact ratio of reactants participating in a reaction required to complete the reaction. Most often used in the context of explaining the mass of air required to completely combust a fuel.

relación estequiométrica la relación exacta de los reactivos necesarios para completar la reacción en la que participan. Se usa principalmente al explicar la masa de aire necesaria para la combustión completa de un combustible.

stoichiometry the science of determining the ratio of reactants required to complete a chemical or physical reaction.
estequiometría la ciencia que determina la relación de reactivos necesaria para completar una reacción química o física.

STOP stop engine light.
STOP luz de paro del motor.

substrate
   1. the supporting material on which an electric/electronic circuit is constructed/infused.
   2. thermally stable, inert material on which active catalysts are embedded on a vehicle catalytic converter.
sustrato
   1. el material de soporte sobre el que se construye un circuito eléctrico o electrónico.
   2. material inerte y térmicamente estable en el que están incrustados los catalizadores activos del convertidor catalítico de un vehículo.

suction circuit the portion of a lubrication or fuel subsystem that is on the suction side of the transfer pump.
circuito de admisión la parte de un subsistema de lubricación o combustible que está en la parte de admisión de la bomba de transferencia.

sulfur an element present in most crude petroleums, but refined out of most current highway fuels. During combustion, it is oxidized to sulfur dioxide, and classified as a noxious emission.
azufre elemento presente en la mayoría de los petróleos crudos y que desaparece en la refinación realizada para obtener los combustibles de vehículos más actuales. Durante la combustión, se oxida a dióxido de azufre, lo que se clasifica como emisión nociva.

sulfur dioxide the compound that is formed when sulfur is oxidized that is the primary contributor to sulfurous type smog. Vehicles contribute little to sulfurous smog problems due to the use of low sulfur fuels.
dióxido de azufre compuesto formado al oxidar el azufre y que es el principal causante del *smog* de tipo sulfuroso. Los vehículos contribuyen en poco grado a los problemas del *smog* sulfuroso, debido al uso de combustibles de bajo contenido en azufre.

sump the lubricating oil storage device on an engine more commonly referred to as an *oil pan*.
cárter inferior el dispositivo de almacenamiento de aceite lubricante de un motor; se suele denominar *cárter de aceite*.

supercharger technically any device capable of providing manifold boost, but in practice used to refer to gear-driven blowers such as the Roots blower.
sobrealimentador técnicamente, cualquier dispositivo de proporcionar alimentación del colector, pero en la práctica se usa para hacer referencia a compresores dirigidos por engranajes, como el compresor Roots.

synchronizing position the Caterpillar MUI fuel balancing procedure.
posición de sincronización el procedimiento de equilibrar el combustible de MUI de Caterpillar.

synthetic oil petroleum-based and other elemental oils that have been chemically compounded by polymerization and other laboratory processes.
aceite sintético aceites extraídos del petróleo y otros elementos y formados químicamente mediante polimerización y otros procesos de laboratorio.

system pressure regulator a usually hydromechanical device responsible for maintaining a consistent line pressure; located downstream from a pump.
regulador de presión del sistema dispositivo, normalmente hidromecánico, responsable del mantenimiento de una presión de línea consistente; ubicado debajo de una bomba.

system unit the main computer housing and its internal components.
unidad del sistema el alojamiento principal del ordenador y sus componentes internos.

tappets used to describe a variety of devices that ride a cam profile and transmit the effects of the cam geometry to the train to be actuated; also known as followers.
empujadores se utiliza para describir diferentes dispositivos que hacen funcionar un perfil de leva y transmiten los efectos de la geometría de la leva al tren que se debe accionar; conocido también como *rodillo*.

TBS turbo boost sensor.
TBS sensor de sobrealimentación del turbocompresor.

TDC top dead center.
TDC punto muerto superior.

TDS total dissolved solids. Measured in a coolant by testing the conductivity with a current probe (TDS tester). High TDS counts can damage moving components in the cooling system such as water pumps.
TDS sólidos en suspensión totales. Se mide en los refrigerantes comprobando la conductividad midiendo la corriente (medidor de TDS). Niveles altos de TDS pueden dañar el desplazamiento de componentes en sistemas de refrigeración como las bombas de agua.

TEM timing event marker—Mack Trucks Bosch engine position sensor located in the rack actuator housing.
TEM marcador de sincronización. Sensor de posición del motor Mack Trucks Bosch ubicado en el alojamiento del impulsor de cremallera.

template torquing a torquing procedure that usually involves torquing to a specified value with a torque wrench followed by turning the fastener through an arc of a specified number of degrees measured by a protractor or template. Produces more consistent clamping pressures than torque-only methods.
torsión medida procedimiento de torsión que conlleva normalmente la torsión en un valor especificado con una llave dinamométrica seguido del giro del cierre en un arco de un número determinado de grados medido con un transportador de ángulos o plantilla. Produce una presión de sujeción más consistente que métodos de solo torsión.

terminal
   1. a computer station or network node.
   2. an electrical connection point.
terminal
   1. estación de ordenador o nodo de red.
   2. punto de conexión eléctrica.

thermatic fan a fan with an integral temperature sensing mechanism that controls its effective cycle.
ventilador térmico ventilador con un mecanismo sensor de temperatura integral que controla su ciclo efectivo.

thermistor a commonly used temperature sensor that is supplied with a reference voltage and by using a temperature sensitive variable resistor, signals back to the ECM a portion of it.
termistor sensor de temperatura muy utilizado que se proporciona con una tensión de referencia y que, mediante el uso de un resistor variable sensible a la temperatura, señala a ECM una parte de él.

Thermal Efficiency is the percentage of the fuel supplied heat energy that appears at the crankshaft as useful work. If two engines produce the same horsepower at the crankshaft, and one burns less fuel than the other, the thermal efficiency is greater for the more economical engine. The heat energy supplied by the fuel is not all converted into power for useful work.
rendimiento térmico es el porcentaje de la energía térmica suministrada por combustible que aparece en el cigüeñal como trabajo útil. Si dos motores producen la misma potencia en el cigüeñal y uno consume menos combustible que el otro, el rendimiento térmico es superior en el motor más económico. La energía térmica suministrada por el combustible no se convierte totalmente en potencia para trabajo útil.

thermocouple a device made of two dissimilar metals, joined at the "hot" end and capable of producing a small voltage when heated.

temopar dispositivo formado por dos metales distintos, unidos en el extremo donde se mide la temperatura y capaz de producir una pequeña tensión cuando se calienta.

thermostat a self-contained, temperature sensing/coolant flow modulating device used to manage coolant flow within the engine cooling system.

termostato dispositivo de modulación de flujo de refrigerante, autocontenido y sensible a la temperatura, que se usa para gestionar el flujo de refrigerante en sistemas de refrigeración de motores.

thick film lubrication lubrication of components where clearance factors tend to be large and unit pressures low.

lubricación de capa gruesa lubricación de componentes en los que los factores de espacio tienden a ser altos y las presiones de unidades, bajas.

throttle air flow to the intake manifold control mechanism used in SI gasoline and diesel engines with pneumatic governors. The term is commonly used to describe the speed control/accelerator/fuel control mechanism in a diesel engine.

acelerador flujo de aire del mecanismo de control de colector de admisión que se usa en motores de gasolina encendidos por bujía y diesel con reguladores neumáticos. El término se utiliza normalmente para describir el mecanismo de control de velocidad, aceleración o control de combustible en motores diesel.

throttle delay a mechanical device used to create a lag between accelerator demand and fuel delivered, usually to cut down on smoke emission.

retraso de acelerador dispositivo mecánico utilizado para crear una demora entre la demanda del acelerador y el combustible suministrado, normalmente para reducir la emisión de humos.

thrust bearing a bearing that defines the longitudinal or end play of a shaft.

cojinete de empuje cojinete que define el movimiento longitudinal o extremo de un eje.

thrust collar in a mechanical governor, the intermediary between the centrifugal force exacted by the flyweights and the spring forces that oppose it. The thrust collar in a governor is usually connected to the fuel control mechanism.

anillo de empuje en un regulador mecánico, el intermediario entre la fuerza centrífuga ejercida por los contrapesos y la fuerza del resorte que se opone a ella. El anillo de empuje de un regulador está conectado normalmente al mecanismo de control de combustible.

thrust faces a term used to describe loading of surface area generally but most often of pistons. When the piston is subject to cylinder gas pressure there is a tendency for it to cock (pivot off a vertical centerline) and load the contact faces off its axis on the pin.

superficie de empuje término usado para describir, en general, la carga del área de superficie, pero con más frecuencia de los pistones. Cuando el pistón está sujeto a la presión del gas del cilindro, tiende a ladearse (hacia fuera de una línea central vertical) y presionar las superficies de contacto fuera de su eje en el pasador.

thrust washer see *thrust bearing*.

arandela de empuje véase *cojinete de empuje*.

thyristor three terminal solid state switches.

tiristor conmutadores de estado sólido de tres terminales.

timing actuator term used by several manufacturers to describe different devices.
  1. Mack Trucks variable timing, proportional solenoid within the Econovance assembly.
  2. Cummins HPI-TP timing solenoid that controls the quantity of fuel charged to the injector timing chamber.

dispositivo de sincronización término usado por varios fabricantes para describir dispositivos diferentes.
  1. Sincronización variable de Mack Trucks, solenoide proporcional dentro del ensamblaje Econovance.
  2. Solenoide de sincronización Cummins HPI-TP que controla la cantidad de combustible cargado en la cámara de sincronización del inyector.

timing chamber term used by Cummins to describe the upper chamber in their two-stage injector units such as STC, CELECT, and TP injectors.

cámara de sincronización término usado por Cummins para describir la cámara superior en sus unidades de inyectores de dos etapas como los inyectores STC, CELECT y TP.

timing check valve device found in Cummins two-stage injector units that holds fuel in the timing chamber during downstroke.

válvula de retención de sincronización dispositivo encontrado en unidades de inyector Cummins de dos etapas que aloja el combustible en la cámara de sincronización durante la carrera descendente.

timing advance unit a hydromechanical or electronically controlled timing advance mechanism used with port helix metering pumps.

unidad de avance sincronizado mecanismo de avance sincronizado controlado hidromecánica o electrónicamente que se utiliza con bombas de dosificación helicoidal de abertura.

tip turbine a charge air cooler fan device driven by turbo-boosted air designed to blow filtered air through the heat exchanger and increase boost air cooling efficiencies.

turbina de filtro dispositivo de ventilador de carga de refrigerador por aire dirigida por aire turbosobrealimentado diseñado para introducir aire filtrado a través del intercambiador térmico y aumentar las eficiencias de refrigeración por aire de sobrealimentación.

tone wheel the rotating disc used to produce rpm and rotational position data to an ECM. The term is most often applied to the rotating member of a Hall effect sensor, but at least one manufacturer uses the term to describe an AC reluctor wheel.

rueda de señal el disco giratorio que se utiliza para llevar los datos de rpm y de posición de giro a un ECM. El término se aplica principalmente al miembro giratorio de un sensor de efectos Hall, pero al menos un fabricante utiliza el término para describir una rueda de reluctor de CA.

top stop injector a Cummins PTD injector in which plunger stroke is set internally and not as part of the injector train adjustment.

inyector de paro superior inyector Cummins PTD en el que el tiempo del émbolo se establece internamente y no como parte del ajuste del tren del inyector.

Torsional Damper (harmonic balancer) mounted on the free end of the crankshaft, usually at the front of the engine. Both terms describe the same component. Its function is to reduce the level of vibration and add to the flywheel's mass in establishing rotary inertia.

Amortiguador torsional (equilibrador armónico) montado en el extremo libre del cigüeñal, normalmente en la parte delantera del motor. Ambos términos describen el mismo componente. Su función es reducir el nivel de vibraciones y sumarse a la masa del volante de inercia para establecer la inercia de giro.

torque twisting effort or force. Torque does not necessarily result in accomplishing *work*.

par esfuerzo o fuerza de torsión. El par no viene acompañado necesariamente de un *trabajo*.

torque rise the increase in torque potential designed to occur in a diesel engine as it is lugged down from the rated power rpm to the peak torque rpm, during which the power curve remains relatively flat. It expresses in a percentage value the increase in engine torque as the engine speed is reduced from its maximum no-load RPM or rated speed. High torque rise engines are sometimes described as *constant horsepower* engines.

aumento del par el aumento en el potencial del par diseñado para que se produzca en un motor diesel cuando pasa de las rpm de potencia en régimen a las rpm de par máximo, durante el cual la curva de potencia permanece relativamente plana. Expresa en un valor porcentual el incremento del par del motor cuando la velocidad del motor se reduce de las RMP sin carga máximas o velocidad nominal. Los motores de aumento del par alto se describen a veces como motores *de potencia constante*.

torque rise profile diagrammatic representation of torque rise on a graph or fuel map.

perfil de aumento del par representación en diagrama del aumento del par en un gráfico o mapa de combustible.

torsion twisting force.

torsión fuerza de giro.

torsional stress twisting stresses. A crankshaft is subject to torsional stress because a throw through its compression stroke will travel at a speed fractionally lower than mean crank speed, whereas a throw through its power stroke will accelerate to a speed fractionally higher than mean crank speed. These occur at high frequencies.

esfuerzo torsional esfuerzos torsionales. Un cigüeñal está sometido a esfuerzo torsional, pues un codo en su tiempo de compresión se despalazará a una velocidad ligeramente inferior que la media, mientras que un codo en el tiempo de potencia se acelerará a una velocidad ligeramente superior que la media. Esto se produce a frecuencias altas.

tower computer a PC housed in an upright case, usually capable of greater system expansion than the horizontal desktop style.

ordenador de torre PC alojado en una carcasa vertical, normalmente con una mayor capacidad de expansión del sistema que el estilo horizontal de escritorio.

toxic materials that may cause death or illness if consumed, inhaled, or absorbed through the skin.

tóxico materiales que pueden causar la muerte o enfermedades si se consumen, se inhalan o se absorben por la piel.

TPS throttle position sensor.

TPS sensor de posición del acelerador.

transducer module (Caterpillar) component responsible for transducing all engine pressure values (such as oil/turbo boost, etc.) to electrical signals for ECM input.

módulo transductor (Caterpillar) componente responsable de transducir todos los valores de presión del motor (como aceite, sobrealimentación del turbocompresor, etc.) a señales eléctricas para la entrada de ECM.

transfer pump describes the fuel subsystem pump used to pull fuel from the fuel tank and deliver it to the injection pumping/metering apparatus.

bomba de trasiego describe la bomba del subsistema de combustible utilizada para extraer éste del tanque y llevarlo a la bomba de inyección o al aparato de bombeo/dosificación de inyección.

transformer an electrical device consisting of electromagnetic coils used to increase/decrease voltage/current values or isolate subcircuits.

transformador dispositivo eléctrico que consta de bobinas electromagnéticas usadas para aumentar o reducir los valores de tensión o intensidad o para aislar subcircuitos.

transient short lived, temporary; often refers to an electrical spike or pressure surge.

transitorio de corta vida, temporal; a menudo hace referencia a un pico eléctrico o a una subida de presión.

transistors any of a large group of semiconductor devices capable of amplifying or switching circuits.

transistores cualquiera de un gran grupo de dispositivos semiconductores capaces de amplificar o conmutar circuitos.

transponder a ground-based satellite uplink—can be either mobile or stationary.

transpondedor un enlace ascendente de satélite basado en tierra, que puede ser móvil o estacionario.

trapezoid a quadrilateral with one pair of parallel sides.

trapezoide un cuadrilátero con un par de lados paralelos.

trapezoidal ring see *keystone ring*.

segmento trapezoidal véase *anillo trapezoidal*.

trapezoidal rod a connecting rod with a trapezoidal small end or rod eye designed to maximize the sectional area of the rod subject to compressive pressures; more commonly referred to as a *keystone rod*.

biela trapezoidal biela de conexión con un pequeño extremo trapezoidal u ojo de biela diseñado para maximizar el área de sección de la biela sujeto a presiones de compresión; se denomina de forma más habitual *biela trapezoidal*.

tribology the study of friction, wear, and lubrication.

tribología el estudio de la fricción, el desgaste y lubricación.

TRS timing reference sensor. The acronym is used by most of the diesel engine electronics OEMs.

TRS sensor de referencia de sincronización. Este acrónimo lo usa la mayoría de los fabricantes de equipos electrónicos de motores diesel OEM.

trunk-type piston a single piece piston assembly usually machined from aluminum alloys in diesel engine applications.

pistón tubular ensamblaje de pistón de una pieza fabricado normalmente con aleaciones de aluminio en aplicaciones de motores diesel.

TT tailored torque.

TT par adaptado.

turbine a rotary motor driven by fluid flow such as water, oil, or gas.

turbina motor giratorio impulsado por un flujo de fluido como agua, aceite o gas.

turbocharger an exhaust gas driven, centrifugal air pump used on most truck diesel engines to provide manifold boost. Consists of a turbine housing within which a turbine is driven by exhaust gas, and a compressor housing within which an impeller charges the air supply to the intake manifold.

turbocompresor bomba de aire centrífuga impulsada por gas de escape utilizada en la mayoría de los motores diesel de camiones para proporcionar alimentación al colector. Consta de un alojamiento de turbina dentro del cual una turbina está impulsada por el gas de escape y un alojamiento de compresor dentro del cual un impulsor carga el suministro de aire al colector de admisión.

two stage filtering any filtering process that takes place in separate stages.

filtrado en dos fases cualquier proceso de filtrado que tiene lugar en fases separadas.

Two-stroke cycle completes two strokes or actions in one turn of the engine; i.e., compression and power. The four-stroke cycle requires two turns of the engine to complete the cycle.

ciclo de dos tiempos completa dos tiempos o acciones en una vuelta del motor; por ejemplo, compresión y potencia. El ciclo de cuatro tiempos precisa dos vueltas del motor para completar el ciclo.

uplink signal transmission from a stationary or mobile ground station to a telecommunications satellite.

enlace ascendente transmisión de señal desde una estación de tierra estacionaria o móvil a un satélite de telecomunicaciones.

upper helix a helix milled into the upper portion of a pumping plunger and giving the characteristic of a variable beginning of pump effective stroke.

espiral superior una espiral tallada en la parte superior de un émbolo de bombeo que le proporciona la característica de principio variable del tiempo efectivo de la bomba.

unit injector a combined pumping, metering, and atomizing device.

inyector unitario dispositivo combinado de bombeo, dosificación y atomización.

vacuum restriction used to describe a restriction on the suction side of a fluid circuit; for instance, a plugged filter.

restricción de vacío se utiliza para describir una restricción en la parte de succión de un circuito de fluido; por ejemplo, un filtro conectado.

valves any device that controls fluid flow through a circuit.
válvulas dispositivos que controlan el flujo de fluidos en un circuito.

valve bridges a means of actuating a pair of valves with a single rocker; also known as valve yoke.
puentes de válvula medio de actuar un par de válvulas con un único balancín; también se conoce como marco del distribuidor.

valve float a condition that is caused by running an engine at higher than specified rpm where valve spring tension becomes insufficient, causing asynchronous (out of time) valve closing.
flotación de válvula condición producida por hacer funcionar un motor a unas rpm superiores a las especificadas, con lo que la tensión de resorte de válvula se hace insuficiente, produciendo un cierre de válvula asíncrono (fuera de tiempo).

valve pockets recesses machined into the crown of a piston designed to accommodate cylinder valve protrusion when the piston is at TDC.
bolsas de válvulas salidas dispuestas en la corona de un pistón y diseñadas para alojar la parte superior de la válvula del cilindro cuando el pistón está en el punto muerto superior, TDC.

Valve Rotators Valve rotators use a ratchet principal or a ball and coaxial spring to fractionally rotate the valve each time it is actuated to minimize carbon build up on the seat and promote even wear.
rotadores Los rotadores usan un direccionador de trinquete o una bola y un resorte coaxial para girar de forma fraccionada la válvula cada vez que actúa, para reducir la acumulación de carbón en el asiento y hacer que sea uniforme.

Valve Seat Insert most current truck and bus diesel engines use **valve seat inserts** rather than integral valve seats, machined into the cylinder head.
anillo de asiento de válvula la mayoría de los motores diesel actuales de camiones y autobuses utilizan **anillos de asiento de válvula** en lugar de asientos integrales de válvula, diseñados en la cabeza del cilindro.

valve train all the component between the cam and the valve; typically would include followers/tappets, push tubes/rods, rocker assemblies and valve bridges/yokes.
mecanismo de válvula todos los componentes entre la leva y la válvula, que incluyen normalmente rodillos o empujadores, tubos de empuje o bielas, conjunto de eje y balancines y puentes de válvula o marcos del distribuidor.

valve yoke see *valve bridge*.
marco del distribuidor véase *puente de válvula*.

vaporization changing the state of a liquid to a gas.
vaporización cambio del estado de líquido a gas.

variable speed governor a governor in which the speed control/throttle mechanism inputs an engine speed value and the governor attempts to maintain that speed as the engine load changes.
regulador de velocidad variable regulador en el que el mecanismo de acelerador o control de velocidad introduce un valor de velocidad del motor y el regulador trata de mantener dicha velocidad mientras cambia la carga del motor.

VCO valve closes orifice (nozzle). A sacless hydraulic injector nozzle.
VCO válvula cerrada (boquilla). Boquilla de inyector hidráulica sin saco.

VCP valve closing pressure.
VCP presión de cierre de la válvula.

VDT video display terminal. A monitor or CRT.
VDT terminal de vídeo. Monitor o CRT.

vector a straight line between two points in space; a line that extends from the axis of a circle to a point in its periphery.
vector línea recta entre dos puntos del espacio; línea que se extiende desde el eje de un círculo hasta un punto de su alrededor.

venting the act of breathing an enclosed vessel or circuit to atmosphere to moderate or equalize pressure.
ventilación el acto de introducir aire en un recipiente o circuito cerrado para moderar o igualar la presión.

VEPS vehicle electronics programming system (Caterpillar/Navistar).
VEPS sistema de programación electrónica del vehículo (Caterpillar/Navistar).

viscosity often used to describe the fluidity of lubricant but correctly defined, it is a fluid's resistance to sheer.
viscosidad se suele utilizar para describir la fluidez del lubricante, pero, definido correctamente, es la resistencia de un fluido a la desviación.

V-MAC vehicle management and control—Mack Trucks engine and chassis electronic management systems; current version is V-MAC III.
V-MAC gestión y control del vehículo. Sistemas de gestión de la electrónica del chasis y de motores Mack Trucks; la versión actual es V-MAC III.

volatile organic compounds the boiled-off, more volatile fractions of hydrocarbon fuels. The evaporation to atmosphere occurs during production, pumping, and refueling procedures; also known as VOCs.
compuestos orgánicos volátiles las fracciones más volátiles de los combustibles de hidrocarburos. La evaporación a la atmósfera ocurre durante los procedimientos de producción, bombeo y reabastecimiento de combustible; también se conocen como VOC.

volatile memory RAM data that is only retained when a circuit is switched on.
memoria volátil datos de RAM que solamente se mantienen cuando está encendido un circuito.

volatility the ability of a liquid to evaporate. Gasoline has greater volatility than diesel fuel.
volatilidad la capacidad de un líquido para evaporarse. La gasolina tiene una mayor volatilidad que el combustible diesel.

volt a unit of electrical potential; named after Alessandro Volta (1745–1827).
voltio unidad de potencial eléctrico; recibe su nombre en honor a Alessandro Volta (1745–1827).

voltage drop voltage drops in exact proportion to the resistance in a component or circuit. The voltage drop calculation is made to analyze component and circuit conditions.
caída de tensión la tensión cae en la misma proporción que la resistencia de un componente o circuito. El cálculo de la caída de tensión se realiza para analizar las condiciones de componentes y circuitos.

volute a snail-shaped diminishing sectional area such as used in turbocharger geometry.
voluta área de sección reducida con forma de caracol como la usada en la geometría del turboalimentador.

VPM vehicle personality module (Caterpillar/Navistar).
VPM módulo de comportamiento del vehículo (Caterpillar/Navistar).

V-Ref reference voltage. The ECM controlled output to on-board sensors.
V-Ref tensión de referencia. La salida controlada por ECM a sensores internos.

VSC variable speed control—Mack Trucks electronic setting of rpm using cruise switches.
VSC control de velocidad variable. Valor electrónico de Mack Trucks de rpm que utiliza conmutadores de marcha.

VSG variable speed governor.
VSG regulador de velocidad variable.

VSL variable speed limit—rpm limiting on a moving vehicle for PTO operation.
VSL límite de velocidad variable. Limitación de rpm en un vehículo que se desplace para la operación de PTO.

VSS vehicle speed sensor.

VSS sensor de velocidad del vehículo.

water separator a canister located in a fuel subsystem used to separate water from fuel and prevent it from being pumped through the injection circuit.

separador de agua bote situado en un subsistema de combustible que se usa para separar el agua del combustible y evitar que se bombee al circuito de inyección.

watt a unit of power commonly used to measure mechanical and electrical power. Named after James Watt (1736–1819).

vatio unidad de potencia usada habitualmente para medir la potencia mecánica y eléctrica. Recibe su nombre en honor a James Watt (1736–1819).

wet liners cylinder block liners that have direct contact with the water jacket and therefore must support cylinder combustion pressures and seal the coolant to which they are exposed.

camisas de agua camisas de bloque cilíndrico que tienen contacto directo con la camisa de agua y por tanto deben soportar presiones de combustión de cilindro y sellar el refrigerante al que están expuestos.

white smoke caused by liquid condensing into droplets in the exhaust gas stream. Light reflects or refracts from the droplets making them appear white to the observer.

humo blanco producido por el líquido que se condensa en gotas en la corriente de gases de escape. La luz de las gotas se refleja o refracta haciendo que aparezcan blancas para el observador.

work when force produces a measurable result, work is accomplished. Work is defined as the product of a force and the distance through which the force acts.

trabajo cuando la fuerza produce un resultado mensurable, se obtiene un trabajo. El trabajo se define como el producto de una fuerza por la distancia en la que actúa la fuerza.

WOT wide open throttle. A term usually used in the context of SI gasoline-fueled engines to mean full fuel request. Used by one diesel OEM to describe *high idle speed*.

WOT acelerador completamente abierto. Término usado normalmente en el contexto de motores de gasolina encendidos por bujía para indicar la solicitud total de combustible. Un fabricante de diesel la utiliza para describir la *velocidad máxima de ralentí*.

wrist pin the pin that links the connecting rod eye to the piston pin boss; also known as piston pin.

pasador del pistón el pasador que enlaza el ojo de la biela de conexión con el resalte del pasador del pistón.

zener diode a diode that will block reverse bias current until a specific breakdown voltage is achieved.

diodo zener diodo que bloquea la corriente inversa de voltaje de polarización hasta que se consigue un determinada tensión de fallo.

INDEX

Pour point, 120, 205
Power control, 59–62
Powertrain control module (PCM), 302
Pre-injection metering, 368–69
Prechambers, 23, 45–46
Precleaners (air), 163
Preheaters
    air intake system, 180, 181
    fuel system, 222–23
Pressure lubrication system, 116–35. *See* Lubrication system
Pressure-volume (pV) curves, 41–42
Primary (fuel) filter, 210, 218
PRIME, 368–69
Proportional solenoid, 315, 318–19
Propylene glycol-based coolants, 140
Pulse width (PW), 326
Pump-line-nozzle (PLN), 240
    electronically managed PLN systems, 290–322. *See* Electronically managed PLN systems
    in-line type fuel injection (port-helix) systems, 239–50
Pump mounted driver (PMD), 294–95, 298
Pumps
    fuel charging (transfer) pumps, 219, 223–27
    fuel injection. *See* Fuel injection
    lubricating oil, 27, 124–25
    water, 27, 145
Push rods and tubes, 27, 106
pV curves, 41–42
Pyrometer, 177

Q
Quench (Squish) Area, 44, 72

R
Rack actuator, 307, 315–16
Rack position sensor, 308
Rack travel sensor, 316
Radiator caps, 144
Radiators, 27, 141–43
Rapid combustion period, 57–58
Rectangular piston rings, 85
Reference coil, 316–17
Rejected heat, 136, 160
Retarders, driveline, 185–86. *See* Braking systems
Ricardo Comet Head, 46–47
Ricardo Turoidal pistons, 77–78
Right-to-know laws, 1, 15–16
Rings, 25
    piston, 80–87. *See* Piston rings
Rocker arms, 27
Rocker assemblies, 106
Rods
    connecting, 26, 86–89. *See* Connecting rods
    push rods, 27, 106
Roller-type cam followers, 106
Roots blowers, 28, 165–66
Rudolf Diesel, 24

S
Safety, 1–20
    alcohol and drug use, 5
    antifreeze and solvent handling, 17, 18
    electrical equipment, 6
    eyes, 12–13

fire, 9–12
first-aid kits, 13
gasoline and diesel fuel, 6–7
general truck shop safety, 8–9
hand tools, 6
hazardous wastes, 14–18
housekeeping, 7–8
jewelry, 69
laws, 1, 2
lifting and carrying, 5–6
personal, 2–4
smoking, 5
Safety glasses, 12, 13
Scavenge, 165
Scavenge pumps/pick-ups, 125
    roots blowers, 28, 165–66
Secondary (fuel) filter, 210, 218–19
Sequential troubleshooting charts, 387
Service manual, 388
Shaft horsepower, 35–36
Shop hazards, 2
Shop safety, 1–20. *See* Safety
Shutters, 150
Side bypass thermostats, 148
Silencers, 178
Sleeves, cylinder, 25, 66, 68–71
Smoking in shop, 5
Solvent handling, 17, 18
Speed, 31
    control, 59–62
    range, 21
    timing advance device, 60–61
Springs, valve, 109
Square engine, 29
Squish, 43, 44, 72
Stanadyne DS pump, 290–304
Starting
    cold starting aids, 180–83
    starter, 28
Stoichometric, 22
Stroke, 21, 23, 28–30, 71
    effective stroke, 234, 243
    four-stroke cycle, 48–51, 52
    two-stroke cycle, 51–52
Substrates, 171, 179
Subsystem identifiers (SIDs), 393
Sump oil pan, 27, 123
Supplemental cooling additives (SCAs), 138–39
Swirl, 43, 44, 72
Synchronous reference sensor (SRS), 326
Synthetic oils, 121–22

T
Taper faced piston rings, 85
Tappets, cam, 27, 104
Thermal efficiency, 22, 25, 37, 38
Thermatic viscous drive fan hubs, 152
Thermistors, 146, 302
Thermodynamics, 22
Thermostats, 146–49
Thick film lubrication, 118
Throttle position sensor, 327
Throwaway blocks, 68
Timing
    fuel delivery, 60–62, 227
    gears, 27, 112–13
    valve and port, 52–54

Timing reference sensor (TRS), 326
Tip turbine heat exchangers, 174
Top bypass thermostats, 147
Top dead center (TDC), 29
Toroidal, 43, 44
Torque, 32–33
Torque rise, 33–34
Torsional (vibration) damper, 28, 92–93, 94
Transducer module, 312
Trapezoidal piston rings, 85
Troubleshooting, 387–98
Truck shop safety, 1–20. *See* Safety
Trunk-type pistons, 72–73
Tubes, push, 106
Turbines, 166
Turbo boost sensor, 328
Turbochargers, 28, 166–72
    construction, 169
    function of, 159, 160
    gas flow, 169, 170, 171
    geometry and performance, 171, 172
    heat exchangers, 160, 172–75
    lubrication, 169, 170
    principles of operation, 166–68

U
Undersquare engine, 30
Unvented thermostats, 148–49
Upright engines, 42

V
V-MAC systems, 312–20
Vacuum, 39–40
Valve pockets, 72
Valve rotators, 109
Valves
    cylinder. *See* Cylinder heads and valves
    intake manifold design, 175–76
    oil filter bypass, 125–27
    oil pressure regulating, 125–27
Variable capacitance (oil pressure) sensor, 131, 132
VCO (valve closes orifice) nozzles, 236
Vehicle speed sensor (VSS), 330
Vented thermostats, 148–49
Vibration damper, 28, 92–93, 94
Viscosity, 119, 205
Viscosity index (VI), 119–20
Viscous impingement oil air filters, 164
Volatility, 206
Volumetric efficiency, 38–40

W
Wastegating, 171
Water manifolds, 152, 154
Water pumps, 27, 144, 145
Water separators, 220–22
Wet/dry combination sleeve, 71
Wet liner (sleeve), 69–70
Williams exhaust brakes, 188
Winter fronts, 150
Work, 31–32
Workplace Hazardous Materials Information Systems (WHMIS), 2
Wrist pins, 74, 86, 87

Y
Yield point, 98